Self Assessment and Review of
BIOCHEMISTRY

Self Assessment and Review of BIOCHEMISTRY

Fifth Edition

Rebecca James Perumcheril MD (Biochemistry)
Assistant Professor
Medical College Kozhikode
Kozhikode, Kerala, India

Edited by
Jomy P Thomas

Foreword
Fathima Beevi O

JAYPEE BROTHERS MEDICAL PUBLISHERS
The Health Sciences Publisher
New Delhi | London | Panama

Jaypee Brothers Medical Publishers (P) Ltd

Headquarters
Jaypee Brothers Medical Publishers (P) Ltd
4838/24, Ansari Road, Daryaganj
New Delhi 110 002, India
Phone: +91-11-43574357
Fax: +91-11-43574314
Email: jaypee@jaypeebrothers.com

Overseas Offices

J.P. Medical Ltd
83, Victoria Street, London
SW1H 0HW (UK)
Phone: +44 20 3170 8910
Fax: +44 (0)20 3008 6180
E-mail: info@jpmedpub.com

Jaypee-Highlights Medical Publishers Inc
City of Knowledge, Bld. 235, 2nd Floor, Clayton
Panama City, Panama
Phone: +1 507-301-0496
Fax: +1 507-301-0499
E-mail: cservice@jphmedical.com

Jaypee Brothers Medical Publishers (P) Ltd
Bhotahity, Kathmandu, Nepal
Phone: +977-9741283608
E-mail: kathmandu@jaypeebrothers.com

Website: www.jaypeebrothers.com
Website: www.jaypeedigital.com

© 2019, Jaypee Brothers Medical Publishers

The views and opinions expressed in this book are solely those of the original contributor(s)/author(s) and do not necessarily represent those of editor(s) of the book.

All rights reserved. No part of this publication may be reproduced, stored or transmitted in any form or by any means, electronic, mechanical, photocopying, recording or otherwise, without the prior permission in writing of the publishers.

All brand names and product names used in this book are trade names, service marks, trademarks or registered trademarks of their respective owners. The publisher is not associated with any product or vendor mentioned in this book.

Medical knowledge and practice change constantly. This book is designed to provide accurate, authoritative information about the subject matter in question. However, readers are advised to check the most current information available on procedures included and check information from the manufacturer of each product to be administered, to verify the recommended dose, formula, method and duration of administration, adverse effects and contraindications. It is the responsibility of the practitioner to take all appropriate safety precautions. Neither the publisher nor the author(s)/editor(s) assume any liability for any injury and/or damage to persons or property arising from or related to use of material in this book.

This book is sold on the understanding that the publisher is not engaged in providing professional medical services. If such advice or services are required, the services of a competent medical professional should be sought.

Every effort has been made where necessary to contact holders of copyright to obtain permission to reproduce copyright material. If any have been inadvertently overlooked, the publisher will be pleased to make the necessary arrangements at the first opportunity.
The **CD/DVD-ROM** (if any) provided in the sealed envelope with this book is complimentary and free of cost. **Not meant for sale.**

Inquiries for bulk sales may be solicited at: jaypee@jaypeebrothers.com

Self Assessment and Review of Biochemistry

First Edition: 2015
Second Edition: 2016
Third Edition: 2017
Fourth Edition: 2018
Fifth Edition: 2019
ISBN: 978-93-5270-917-5

Printed at Rajkamal Electric Press, Plot No. 2, Phase-IV, Kundli, Haryana.

Dedication

This book is dedicated to my children, my husband, my parents and in Laws, my brothers and to all the students of my classes, online CME by daily rounds/Marrow subscribers, Facebook group, Telegram group, and the students who could not attend my live or video classes who inspired me to write a book.

Foreword to the Fifth Edition

It is with great pleasure that I am introducing the 5th edition of the book by Dr Rebecca James Perumcheril, **Self Assessment and Review of Biochemistry.** It has been an excellent course material for various postgraduate medical entrance examinations conducted by Central Institutes, National Board of Examinations (NEET PG). Dr Rebecca James has been an astute and dedicated teacher whose ingenious and creative learning techniques has won acclaim from her students and colleagues alike. The lucid presentation and inclusion of latest and original references make this book authentic. It is indeed a treasure for medical postgraduate aspirants. I wholeheartedly recommend this book to them. My hearty congratulations and sincere prayers for the success of her excellent work.

Dr Fathima Beevi O
Professor and Head
Department of Biochemistry
Medical College, Calicut, Kerala
Director
Priyadarshini Institute of Paramedical Sciences (PIPMS)
Medical College, Thiruvananthapuram, Kerala, India

Forewords to the Third and Fourth Edition

I feel privileged to write the foreword to the 3rd edition of Dr Rebecca James' excellent book.

It is common knowledge that books play a major contributing role in any educational process. While several books are available on biochemistry, not all of them promote learning; some indeed hinder it. To be useful and worthwhile, a book has to be so designed as to present an appropriate body of knowledge in a style that suits students in different stages of learning. Accordingly, a book on biochemistry should contain 'must know' and 'nice to know' levels of factual and conceptual information. Designing such a book is a challenging task, especially if it is to be concise and comprehensive in scope. Commendably, Dr Rebecca James has accomplished this in her book, *Self Assessment and Review of Biochemistry*. A fairly large number of charts, diagrams and illustrations in the text amply demonstrate this.

With pleasure, I compliment Dr Rebecca James for such a fine piece of work.

Dr Gobind Rai Garg
Director, Ayush Institute of Medical Sciences
Author of Review of Pharmacology
Author of Review of Pathology and Genetics
Author of Ophthalmology (SARP series)

Self Assessment and Review of Biochemistry by Dr Rebecca James is the best compilation of authentic biochemistry content. This book is the greatest innovation by a great author in being very easy to recollect during stressful exam environment. The book no doubt is the best seller amongst all the books and is famous by the following lines. If it is here it may be found somewhere but if something is not here it surely will be nowhere 'In short it's complete and beautiful'.

Dr Apurv Mehra
MBBS MS (Orthopedics)
Author of Orthopedics Quick Review
Author of Review of Pediatrics
CEO, Med Miracle app

Dr Rebecca James is one of the most hardworking, dedicated and student friendly Biochemistry faculty who teaches every student with passion. She is famous for molding complicated biochemistry facts into simplified concepts by her excellent teaching and book.

Her iconic style of teaching is superbly reflected through her book, *Self Assessment and Review of Biochemistry*. This book is much simplified, updated and playing instrumental role in helping to achieve top ranks in various exams.

Dr Devesh Mishra
MBBS MD (Pathology)
Author of Concepts in Pathology

Forewords to the First and Second Edition

TMCAA (Thrissur Medical College Alumni Association) is a CME Program started in 2000 for Junior Doctors guiding and coaching for various PG entrance exams. Our principle is professionalism and structured orientation. Beyond guiding thousands of students to their dream PG and being one of the top institutes in the field we have also generated and promoted many deserving faculties and authors since the inception. Dr Rebecca had started her career as a PG guidance faculty from this institute nine years back. Her teaching methodology and updated notes and techniques are well appreciated by students and her notes are of much demand as a "must for exam points". We welcome and appreciate her effort in compiling her notes into students friendly entrance guide. Wishing all success to her and students reading *"Self Assessment and Review of Biochemistry"*.

Dr Shibu TS
Chief Coordinator, TMCAA
Dr Ravindran Chirukandath coordinator and Secretary, TMCAA
Dr Tajan Jose President and Coordinator, TMCAA

It gives me immense pleasure to introduce this book *Self Assessment and Review of Biochemistry* written by Dr Rebecca James. She is one of the best faculties in her subject in Calicut Medical College alumni CME entrance guidance program. The book comprises a vast collection of questions from the recent exams and thoroughly described. I am sure that now this book will become a necessity among the PG entrance Aspirant's books. Wishing all the success for the book.

Dr NC Cherian
Additional Professor of Pediatrics &
Co-ordinator Calicut Medical College alumni Medical updates program

The book *Self Assessment and Review of Biochemistry* has been written as a course material for the various national level (NBE pattern), central institutes, state level, and private medical entrance exams. Dr Rebecca James has been a very keen and devoted teacher in her career. Her new and innovative visual and easy learning techniques are much useful for her students. The references given are up-to-date and authentic and the presentation and charts are much student friendly. I do recommend this book for PG entrance aspirants. I heartily congratulate her and wish her all the success for this excellent piece of work.

Dr MG Jose Raj MD Biochemistry
Professor and Head, Department of Biochemistry
Medical College, Kozhikode, Kerala

Preface to the Fifth Edition

"To improve is to change, to be perfect is to change often....." **Winston Churchill**.

After the overwhelming reception of the previous editions of the book across India, I am more than excited for the release of Fifth edition. The responses from toppers of various PGMEE speak for the success of the previous editions. With each new edition, I try to give more knowledge and input that will be useful to the students and change the face of Biochemistry. Waiting for your reviews, comments and criticisms.

Dr Rebecca James
MD Biochemistry
drrebeccajomy@gmail.com

Preface to the First Edition

Writing a book for Postgraduate Medical Entrance in Biochemistry was not a dream for me when I was doing my PG career. But when I started postgraduate medical entrance coaching; gradually I recognized how hard it was for students to learn the subject. There is no subject expert written Book in Biochemistry for the major exams like *PGI, AIIMS, AIPGME, JIPMER*, or even state entrance exams. None of the students had learned biochemistry properly or were not taught properly in their MBBS course and now for entrance Students were forced for learning from books not by a subject expert which are just pick some points from standard books without explanation or internet reference. Biochemistry comprises up to 8% in NBE EXAMS and up to 30% in PGI exams and lots of researches and updates are coming up in this field. My students often would come asking with doubts quoting some of these guides which many answers I found wrong. I aimed to write a book which will create a difference, a new approach to see biochemistry mere than a subject. As nine years passed in Postgraduate medical entrance coaching I made my own notes based on standard textbooks with clinical scenarios and emphasized on teaching to think and correlate rather than mugging up biochemistry. Gradually students from other institutes started asking for my notes and requested me to write a book so that they could also read it. This book is meant for it. I approached Jaypee publishers and they accepted my proposal, the remaining is in front of you.

This is PGMEE book and don't compare it to any standard books, but you will get up to 96% if you learn this properly. Biochemistry is a very vast subject with future of research, genetics, which will be overtaking the future of the medical field in time, this book is just a small flower in a big garden. I know that learning in my class would be better than reading this book but due to practical difficulties, it is not possible. I have done much justice in including my own made pictures and mnemonics to make it as understandable as possible. I wish that you thoroughly go through this book and gain maximum from it as it has all the points for scoring top ranks in major exams. Wishing you best of luck for your dream PG in the coming exams. Genuine doubts are always welcome.

Dr Rebecca James MD
Biochemistry

Acknowledgements

Writing this book has never been easy to me. "No tears in the writer, no tears in the reader. No surprise in the writer, no surprise in the reader." —Robert Frost.

First and foremost I would like to thank God for enabling me in completing this work. In the process of putting this book together I realized how true the gift of writing is in me. You have given me the power to believe in my passion and pursue my dreams. This would not have been possible without the faith I have in you, the Almighty.

To my husband Dr Jomy. You have been the greatest inspiration and pillar of strength throughout the work. Without your relentless support this book would have never been materialized.

To my children, John and Joel: You're the best thing that ever happened to me in my life! This book would never been complete without you both letting me do. Thank you so much.

To my beloved Father Mr James P George, I am speechless! I can barely find the words to express all the wisdom, love and support you've given me. I am forever grateful for being born to and raised by such an amazing person. If I am blessed to live long enough, I hope I will be a good parent as you are and always been to me. I love you too, Amma. I also extend my gratitude to my brothers, Mr George James and Mr Philip James.

To my in-laws Mrs & Mr Thomas P John, thank you for all the support as my parents and for encouraging me.

I acknowledge Mr Jitendar P Vij (Group Chairman) and Mr Ankit Vij (Managing Director), Mrs Chetna Malhotra Vohra (Associate Director–Content Strategy) to Jaypee Brothers Medical Publishers, New Delhi, Mr Venugopal, Director-Sales South, Mr Sujeesh, Branch Manager, Jaypee, Kochi and Ms Payal Bharati (Senior Manager–Professional Publishing), in making this dream a reality

Deep gratitude to all the Jaypee team all over India for their tremendous efforts in marketing.

Special thanks to my colleague, Dr Shibu TS, Assistant Professor, Department of Biochemistry, Medical College Thrissur for his guidance and mental support.

I thank the colleagues of my department (Dr Geetha PA, Dr Asha E, Dr Muhammed Ashraf, Dr Shaji Sreedhar KP, Dr Lavanya Madhavan, Dr Sandeep Appunni, Dr Elza Boby, Dr Harish S, Dr Harish Kumar), for their expert opinions and valuable suggestions.

I thank all Senior residents (Dr Dhanya AK) and Junior Residents (Dr Hajfa Eranhikkal, Dr Neethi R Krishnan, Dr Maekha Ju Nath, Dr Diana Mariam, Dr Rosmi Johnachan, Dr Sibiya Odayapurath, Dr Raseema, Dr Veena G, Dr Allen John) in Department of Biochemistry, Government Medical College, Calicut for their helping hands in the tedious job of proof reading.

Also thanking

Dr Geetha PA, Additional Professor, Government Medical College, Kozhikode.

Dr Deepu Sebin, Dr Shujad Ahmed and all staff of Daily Rounds/Marrow team

Dr Tajan Jose, Co-ordinator and President TMCAA, Thrissur

Dr Ravindran Chirukandath, Co-ordinator and Secretary, TMCAA, Thrissur

Dr NC Cherian, Calicut Medical College, Weekly Medical Updates, Coordinator

Dr Arun Kumar, ADR Plexus-Digital strides Coordinator

Mr Daljit Singh and Mr Dapinder Singh DSL team

All my colleagues in PGMEE training field (Dr Sakshi Arora Hans, Dr Manisha Sinha Budhiraja, Dr Thameem Saif, Dr Devesh Mishra, Dr Apurv Mehra, Dr Vivek Jain, Dr Gobind Rai Garg, Dr Rajesh K Kaushal, Dr Apurba Sankar Sastry, Dr Ranjan Kumar Patel, Dr Vandana Puri Tiwari, Dr Soumen Manna, Dr Krishna Kumar Solaipan, Dr Utsav Bansal, and the legendary team of Marrow video app).

Staff and Residents of Department of Biochemistry, Government Medical College, Kozhikode.

Mr Praveen TMCAA staff

Mr Srinivas & Mrs Maniprabha Hyderabad ADR Plexus—Digital strides team

Mr Maulik Patel and Team of CCIAMS

Mr Manikandan, Chennai ADR Team

Mrs Jagruthi, ODD team

Team of Med miracle app

Mr Moin Bangalore Bright Medicos team

Mr Anil Dhiman, ARP team

To all my students especially to the students of 61st batch MBBS Mohammed Sahal K, Shana Fathima, Neelofar, Abhinav Prem, Ananya Arun, Sachin K, Laya Francis, Aiswarya K, Smerah Chammayil, Avani Unni, Anjitha Vinayakumar, Hadi Ameen, Manu Narayan for helping me whole heartedly for book updation.

All my students who supported and contributed questions of various PGMEE. Their immense support for making the Fifth edition of the book is impressive. I thank all those who had been on my side on this journey.

How to Use this Book

Dear Student,

I consider you as one of my students by using this book and I believe it's my responsibility to guide you through in your postgraduate medical exams and I wish you all the best to come out with flying colors. I know that many of you feel that biochemistry is one of the most boring subjects you have learned in MBBS. This is primarily due to the methodology you have been taught and you tend to think of it as a subject of cycles and test tubes. But the fact is that if it was taught with all clinical correlations and the practical applications the subject becomes interesting and beautiful. I hope you will have a renewed perception about biochemistry after reading this book.

This is a concept-oriented book where you learn biochemistry and its applications. This is made in a review book style to enable you to answer questions with ease. Another aspect of it is the clinical approach which makes it easier to learn, understand and recollect. Previous questions from all the major exams (PGI, AIIMS, Central Institute exams, Recent Exam) are included for discussion.

Tips to learn biochemistry for PGMEE from this book:

1. Always start from an easy topic for you to gain confidence in the beginning itself.
2. The chapter gradation based on its importance is given in the table of contents, give extra importance to chapters with maximum stars.
3. Big chapters in the previous editions are divided to subchapters for easy learning.
4. High Yield boxes, tables and bold letters are must learn.
5. Extra Edge topics are optional.
6. First do questions of a subchapter find the most important topics then learn the theory of that subchapter again do the review questions.
7. If possible make self-written notes based on chapter review.
6. At the end of every chapter, quick review points are added which help you to quickly revise for exams.
8. Check list for revision at the end of chapter help you how to learn each chapter and not to miss any High Yield Topics
9. Be a part of my Facebook group, Telegram group where I put all latest trend new questions, facts, etc.
10. Attending a live lecture always boost your memory and clear your concepts.
11. If not possible to attend live session subscribe my online CME in Daily rounds/Marrow app. My book is consistent with the online CME lectures, so using book along with lectures will be a holistic way of learning.
12. Most importantly revise at least three or four times before each exams.
13. Most importantly what to learn in biochemistry depends on which exams you are targeting.
    #For NEET PG exam, metabolism and vitamins carry maximum marks.
    #For AIIMS, metabolism, vitamins and genetics are equally important.
    #30 of 200 PGI exam questions are from biochemistry, so cover all topics including Extra Edge portions. That is the reason why I never come up with book in a capsule form. I want this book to be complete book for all PGMEE exams.

This book is different from the usual guide for postgraduate medical entrance exams since every part of it are referred from standard books.

Learning is a continuous process. Active participation in my Facebook group, Telegram group is very essential to get online author support and also to get updates. Practice questions just prior to PGMEE exams.

You may feel free to ask your questions, you can e-mail me or post in my Facebook discussion forum, Telegram group, messenger. Always your feedback is important to me. If you believe you have identified an error in the book, please send an email to *drrebeccajomy@gmail.com*. If you have general comments or suggestions please drop me a line directly to my e-mail/messenger. We are continually striving to meet the needs of all individuals preparing for the entrance exams. I request you to join my Facebook group or Telegram for further updates, and revision.

Unique features of this edition:
- Concept connect diagrams
- More flowcharts
- Subchapters are divided separately with their MCQs for students quick review and practise
- Quick revision after each chapter
- Check list for revision
- Chapter grading for students convenience.

Wishing you all the best.

Dr Rebecca James
Mob 9447440193
drrebeccajomy@gmail.com

For author online support, follow
'Dr Rebecca biochemistry discussion group' in Facebook
NEET PG Biochemistry discussion-Dr Rebecca in Telegram

Contents

Image-Based Questions

Chapter 1: Amino acids — 1

1.1 Chemistry of Amino Acids***** 3
1.2 General Amino Acid Metabolism***** 17
1.3 Aromatic amino Acids, Simple Amino Acids and Serine***** 31
1.4 Sulphur Containing Amino Acids***** 49
1.5 Branched Chain Amino Acids, Acidic Amino Acids, Basic Amino Acids and Amides*** 54

Chapter 2: Proteins — 67

2.1 Chemistry of Proteins*** 69
2.2 Structural Organization of Proteins and its Study*** 72
2.3 Biochemical Techniques*** 79
2.4 Fibrous Proteins***** 86
2.5 Protein Folding and Degradation 92
2.6 Plasma Proteins and Glycoproteins 97
2.7 Protein Sorting 105

Chapter 3: Enzymes — 109

3.1 General Enzymology***** 111
3.2 Clinical Enzymology**** 126

Chapter 4: Carbohydrates — 133

4.1 Chemistry of Carbohydrates*** 135
4.2 Major Metabolic Pathways of Carbohydrates***** 156
4.3 Minor Metabolic Pathways of Carbohydrates**** 186

Chapter 5: Lipids — 197

5.1 Chemistry of Lipids**** 199
5.2 Phospholipids and Glycolipids*** 207
5.3 Metabolism of Lipids***** 216
5.4 Lipoprotein Metabolism***** 238

Chapter 6: Bioenergetics — 259

6.1 TCA Cycle***** 261
6.2 Electron Transport Chain**** 267
6.3 Integration of Metabolism***** 274
6.4 Shuttle Systems 277

Chapter 7: Heme and Hemoglobin — 281

- 7.1 Heme Synthesis and Porphyrias***** 283
- 7.2 Heme Catabolism and Hyperbilirubinemia**** 293
- 7.3 Hemoglobin*** 298

Chapter 8: Nutrition — 305

- 8.1 Fat Soluble Vitamins***** 307
- 8.2 Water Soluble Vitamins***** 318
- 8.3 Minerals**** 333
- 8.4 Basics of Nutrition*** 343

Chapter 9: Special Topics — 345

- 9.1 Metabolism of Alcohol 347
- 9.2 Free Radicals*** 349
- 9.3 Xenobiotics 353
- 9.4 Biomembranes and Cell Organelle*** 355

Chapter 10: Molecular Genetics — 361

- 10.1 Chemistry of Nucleic Acid 363
- 10.2 Metabolism of Nucleotides*** 367
- 10.3 Organization and Structure of DNA**** 375
- 10.4 DNA Replication and Repair**** 385
- 10.5 Transcription***** 395
- 10.6 Different Classes of RNA***** 406
- 10.7 Translation*** 413
- 10.8 Regulation of Gene Expression**** 422
- 10.9 Mutations**** 432
- 10.10 Mitochondrial DNA***** 439
- 10.11 Patterns of Inheritance***** 441
- 10.12 DNA Polymorphism*** 444

Chapter 11: Molecular Biology Techniques — 447

- 11.1 Recombinant DNA Technology**** 449
- 11.2 Amplification and Hybridization Techniques**** 457
- 11.3 Cytogenetic Techniques***** 464
- 11.4 DNA Sequencing Techniques, Transgenic Technique and Hybridoma*** 469
- 11.5 Other Molecular Biology Techniques and Recent Advances*** 474

Most Important*** (Learn Completely)
Very Important** (Learn Completely)
Important* (Selective Reading)
No Star Topics – Selective Reading Based on Check List Given after Every Subchapter

Image-based Questions

Image-Based Questions

1. A man on hunger strike brought to the casualty, a test done on his urine sample is given here. Identify this test:

   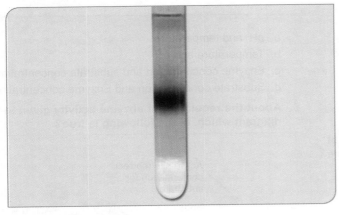

   a. Benedict's test
   b. Rothera's test
   c. Aldehyde test
   d. Molisch's test

2. What is the principle of the given test for Urine reducing substance?

   a. Cu^{2+} ions are reduced to Cu^+ ions by enediols
   b. Cu^+ ions are oxidised to Cu^{2+} ions by enediols
   c. Fe^{2+} ions are reduced to Fe^+ ions by enediols
   d. Fe^+ ions re oxidised to Fe^{2+} ions by enediols

3. A 3-year-old male child presented to paediatric OPD with generalised edema started as puffiness of face few weeks back. He is also irritable, and having decreased appetite. Urine analysis done by the test given in the picture showed positive results. What is the given test?

   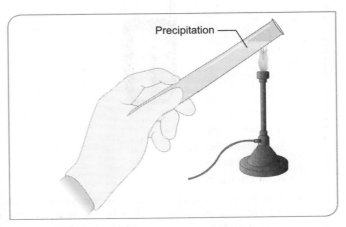

   a. Bradford test
   b. Heat and acetic acid test
   c. Isoelectric test
   d. Barfoed test

4. A 70-year-old man presented with itching, jaundice and mass palpable per abdomen and clay coloured stools. Two tests done in his urine sample is shown here. It gave positive results. What is the type of Jaundice he is suffering from?

   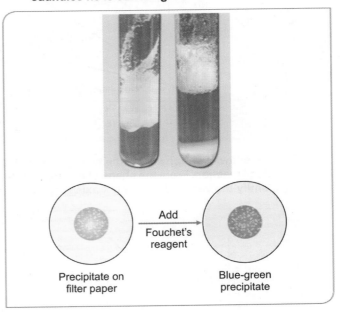

   a. Obstructive jaundice
   b. Hemolytic jaundice
   c. Prehepatic jaundice
   d. Hepatic jaundice

5. Identify the given test for proteins:

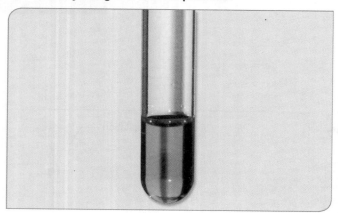

a. Barfoed test
b. Pettenkofer test
c. Biuret test
d. Molisch test

6. The following graph is the line weaver burk plot of an enzyme inhibition. What is the type of enzyme inhibition that is depicted in the graph?
a. Allosteric inhibition
b. Competitive inhibition
c. Noncompetitive inhibition
d. Uncompetitive inhibition

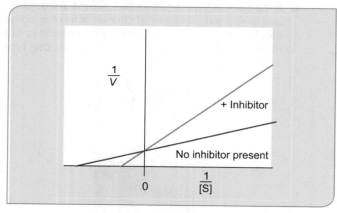

7. The following graph shows the two factors that affect rate of reaction. From the shape of the graph obtained identify the variable given on X axis respectively:

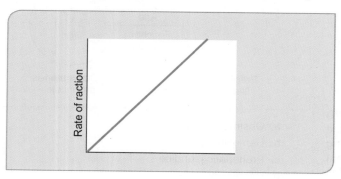

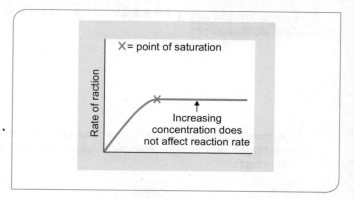

a. pH and temperature
b. Temperature and pH
c. Enzyme concentration and substrate concentration
d. Substrate concentration and Enzyme concentration

8. About the regulation of enzyme activity given in the diagram which of the following is true?

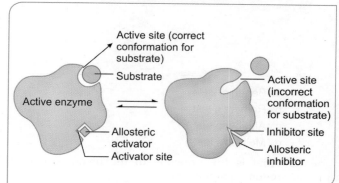

a. Substrate and modifier binds to same site
b. Substrate and modifiers are structural analogues
c. They exhibit cooperative binding
d. Regulate the enzyme quantity

9. Identify the given amino acid:

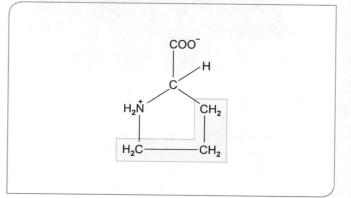

a. Tryptophan
b. Tyrosine
c. Proline
d. Histidine

10. Which of the following property most suit the given amino acid?

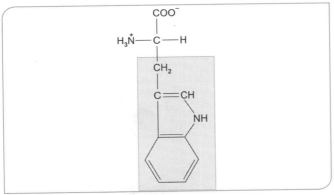

   a. Uncharged polar aliphatic
   b. Charged polar aromatic
   c. Nonpolar aromatic amino acid
   d. Uncharged polar aliphatic

11. A 2-year-old intellectutually disabled child is having blue eyes, blonde hair and fair skin. He also have a peculiar body odour. What is the diagnosis?

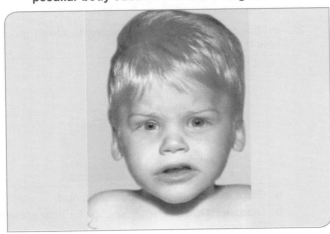

   a. MSUD
   b. Isovaleric aciduria
   c. Phenyl ketonuria
   d. Canavan's disease

12. A 40-year-old man presented with back ache. His sclera of eyes and dorsum of hands have black pigmentations. The urine collected is also shown in the picture. What is the diagnosis?

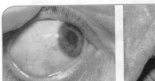 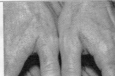

   a. Melaiinuria
   b. Malignant melanoma
   c. Ochronosis
   d. Homogentisic acidosis

13. The picture given is a case of Albinism. What is the enzyme that is defective?

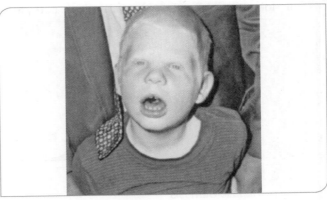

   a. Tyrosine hydroxylase
   b. Tryptophan hydroxylase
   c. Tyrosinase
   d. Tyrosine transaminase

14. Identify the metabolic disorder from the pictures given?

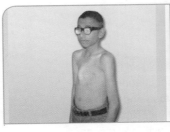 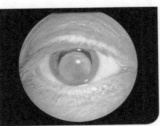

   a. Pheochromocytoma   b. Homocystinuria
   c. Cystinuria   d. Rickets

15. A 2-week-old neonate with complete hypotonia, convulsions, failure to thrive and metabolic acidosis. The baby has small of burnt sugar in urine. The test called DNPH test is positive. What is the enzyme deficiency in this metabolic disorder?

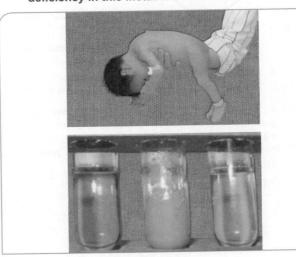

   a. Isovaleryl CoA dehydrogenase
   b. Dihydrolipoamide dehydrogenase
   c. Branched chain keto acid dehydrogenase
   d. Transacylase

16. Identify the abnormal electrophoresis pattern:

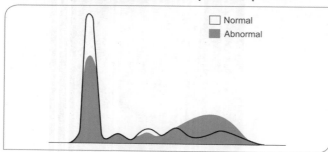

a. Nephrotic Syndrome
b. Hepatic Cirrhosis
c. Acute Inflammation
d. Hypogammaglobulinemia

17. All the following are true about the disaccharide present in the milk except:

a. It is a reducing disaccharide
b. Made up of Galactose and Glucose
c. Contraindicated in galactosemia
d. Aldose sugar joins with a ketosugar

18. Identify the heteropolysaccharide from the disaccharide repeat unit given in the picture

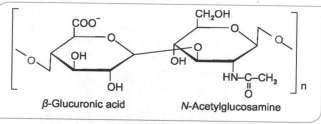

β-Glucuronic acid   N-Acetylglucosamine

a. Heparan Sulphate
b. Heparin
c. Hyaluronic acid
d. Dermatan Sulphate

19. A 3-year-old child presented with swelling in the groins and defective vision. O/E Intellectual disability, coarse facial features, hepatosplenomegaly, prominent forehead, short stature is present. Urine spot test for GAG is positive

What is the most probable diagnosis?

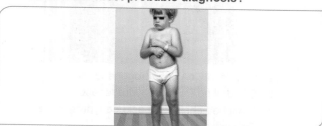

CASE-5a

a. MPS-IH
b. MPS-I S
c. MPS-II
d. MPS-III

20. A 12-year-old intellectually disabled boy having short stature, protuberant abdomen with umbilical hernia, prominent forehead. His vision is normal. His parents are normal. What is metabolic defect in this disorder?

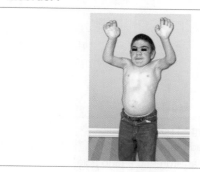

CASE-5b

a. L –Iduronidase
b. Iduronate Sulfhatase
c. Aryl Sulfatase B
d. Beta Glucoronidase

21. In the test given bellow to estimate blood glucose, which of the following is true:

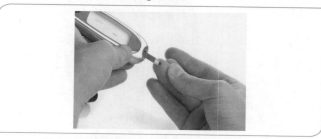

a. Glucose is converted to glucuronic acid
b. Glucose oxidase is highly specific for beta anomer of Glucose
c. The terminal Carbon is oxidised.
d. This is an example of Reducometric method

22. The osazones given in the diagram are respectively:

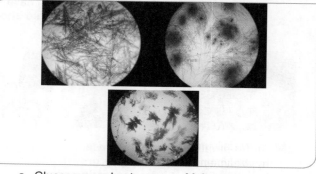

a. Glucosazone, Lactosazone, Maltosazone
b. Lactosazone, Maltosazone, Glucosazone
c. Maltosazone, Lactosazone, Glucosazone
d. Glucosazone, Maltosazone, Lactosazone

23. A 2 year old boy presented with fasting hypoglycemia, hepatomegaly, doll like facies and thin extremities. Kidneys are enlarged but no splenomegaly. What is the most probable GSD?

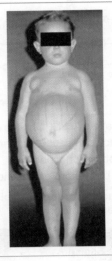

a. Type I GSD
b. Type II GSD
c. Type III GSD
d. Type IV GSD

24. A 2-year-old child presented with hepatosplenomegaly, failure to thrive and progressive cirrhosis with portal hypertension. What is the diagnosis?

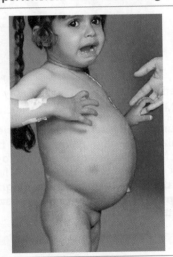

a. Type I GSD
b. Type II GSD
c. Type III GSD
d. Type IV GSD

25. An 8-month-old child presented with fasting hypoglycemia, hepatomegaly, growth retardation and hyperlipidemia. Liver transaminases are elevated. Once this child reached puberty, he no longer has hypoglycemia, muscle weakness, hepatomegaly. He reached normal adult height also. What is the most probable GSD?

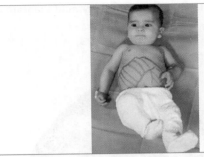

a. Type I GSD
b. Type II GSD
c. Type III GSD
d. Type IV GSD

26. A 2 month old infant presented with generalized muscle weakness with a "floppy infant" appearance. He also has other features like macroglossia, feeding difficulty, hepatomegaly and hypertrophic cardiomyopathy. What is the diagnosis?

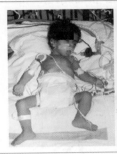 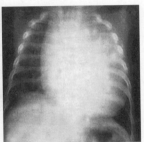

a. Type I GSD
b. Type II GSD
c. Type III GSD
d. Type IV GSD

27. A 3-week-old neonate who began vomiting 2 days after birth, usually within 30 minutes after breastfeeding. He also has abdominal distension with enlargement of liver, with jaundice. The consulting doctor did two urine dipstick test, one specific for glucose was negative, second test specific for reducing sugar was positive. What is the diagnosis?

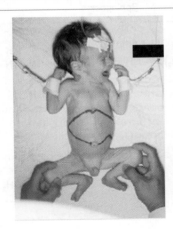

a. Hereditary fructose intolerance
b. Classic galactosemia
c. Essential fructosuria
d. Essntial pentosuria

28. All the following about the fatty acid diagram are true except:

   a. Used in the treatment of type I hyperlipoproteinemia
   b. Decrease triacyl glycerol level
   c. Decrease HDL cholesterol
   d. Lower the inflammatory risk

29. The essential fatty acid present abundantly in the given oil is:

Safflower oil

   a. Arachidonic acid
   b. α Linolenic acid
   c. g Linolenic acid
   d. Linoleic acid

30. Identify the structure given in the diagram:

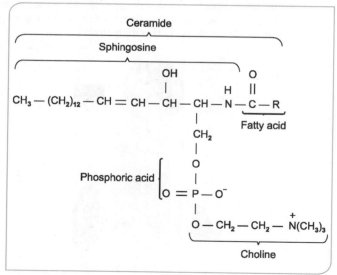

   a. Lecithin
   b. Sphingosine
   c. Cephalin
   d. Sphingomyelin

31. Which of the following is NOT true regarding the structure given below?

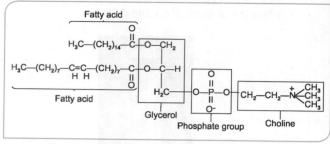

   a. Phosphatidic Acid + Choline
   b. Deficiency can cause respiratory distress in newborn
   c. It is a sphingophospholipid
   d. Largest body store of Choline

32. Identify the disorder associated with both these clinical presentation:

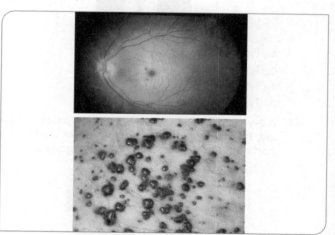

   a. Farber's Disease
   b. Fabry's Disease
   c. GM1 Gangliosidoses
   d. Gaucher Disease

33. A young boy presented to paediatric OPD with severe agonizing burning pain in both hands associated with fever. O/E Angiokeratoma on the shoulders, lower abdomen and buttocks. A lysosomal storage disorder, Fabry's disease was diagnosed. What is the accumulating substance in lysosomes of vascular endothelium?

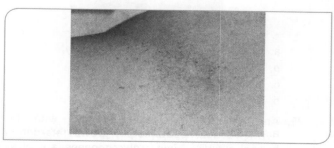

   a. Ceramide
   b. Ganglioside
   c. Globotriaosyl Ceramide
   d. Galactocerebroside

34. Identify the lysosomal storage disorder presenting like rheumatoid arthritis:

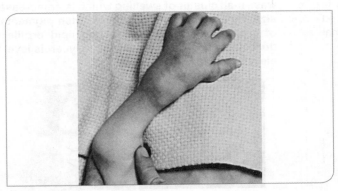

   a. Tay Sach's Disease
   b. Farber's Disease
   c. Fabry's Disease
   d. Gaucher's Disease

35. The given feature is pathognomonic of which storage disorder:

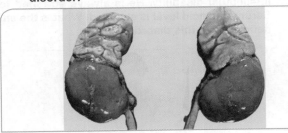

   a. Mucolipidoses
   b. Sphingolipidosis
   c. Wolman's Disease
   d. Niemann Pick Disease

36. An eight-month-old female infant presented with recurrent episodes of hypoglycemia, especially if time interval of feeding is increased. Dicarboxylic acid is present in the urine. Urine ketone bodies is negative. The child responded well to IV Glucose, less fat and more carbohydrate diet, frequent feeding. The child was diagnosed to be MCAD deficiency. What is the reason for hypoglycemia?

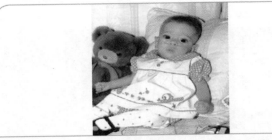

   a. Increased dicarboxylic acid inhibit glycogenolysis
   b. Lack of ATP to support gluconeogenesis
   c. Lack of acetyl-CoA to favour glycogenolysis
   d. Glycogen stores are inadequate in infants

37. A child few hours after ingestion of a fruit, vomiting started later developed convulsion. In the hospital his blood glucose was very low, but no ketone bodies. The fruit he ingested was identified as a fruit found in Africa. What is the diagnosis?

   a. MCAD deficiency
   b. Zellweger syndrome
   c. Carnitine deficiency
   d. Jamaican vomiting sickness

38. What is the biochemical defect in the Zellweger syndrome?

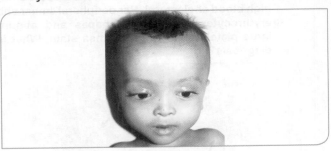

   a. Peroxisomal biogenesis disorder
   b. Lysosomal targeting disorder
   c. Defect in glycosylation of proteins
   d. Trisomy 21

39. A 2-year-old girl presented with recurrent abdominal pain. She was admitted in pediatrics for detailed work up. When her blood was drawn for investigation it was milky white. Yellowish white papules noted on the dorsum of hands. On fundoscopic examination opalescent retinal vessels were seen. Fasting triglycerides >1000 mg/dL, but cholesterol level was normal. What is the diagnosis?

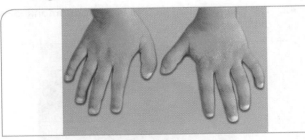

   a. Familial chylomicronemia syndrome
   b. Familial defective Apo B
   c. Sitostrelomia
   d. Familial dysbetalipoproteinemia

40. A 10-year-old boy is presented to ophthalmology OPD for white ring around the black of the eye. His father died of coronary heart disease. O/E Corneal arcus is found, xanthoma on Achilles tendon. His fasting blood cholesterol >300 mg/dl, Triglycerides within normal limit. What is the diagnosis?

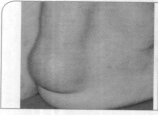
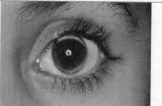

a. Type I Hyperlipoproteinemia
b. Type II A Hyperlipoproteinemia
c. Type II B Hyperlipoproteinemia
d. Type III Hyperlipoproteinemia

41. A child that presented with tendinous xanthoma, high blood cholesterol. Her peripheral smear showed erythrocytes of different shapes and abnormally large platelets in Wright Giemsa stain. What is the diagnosis?

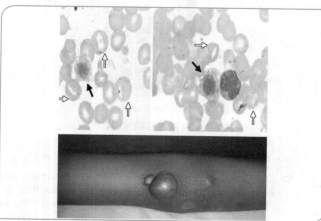

a. Familial chylomicronemia syndrome
b. Familial defective apo B
c. Sitostrelomia
d. Familial dysbetalipoproteinemia

42. The given clinical picture is pathognomonic of which dyslipidemia?

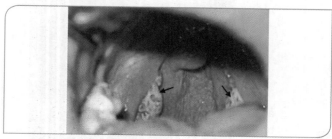

a. Abetalipoproteinemia
b. Tangier's disease
c. Fish eye disease   d. Sitosterolemia

43. A 30-year-old male patient came to Medicine OPD for swellings which looks like grapes, he said initially it was small cluster of swelling which later increased in size. He also complained of yellowish pigmentation of creases of palms. His fasting lipid profile was done, both cholesterol and triacylglycerols level was elevated. What is the diagnosis?

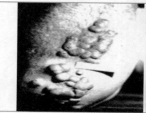

a. Familial chylomicronemia syndrome
b. Familial defective apo B
c. Sitostrelomia
d. Familial dysbetalipoproteinemia

44. A 7-year-old boy with compulsive self mutilation, intellectual disability. He is always strapped to bed Serum uric acid level is elevated. What is the enzyme deficiency in this disorder?

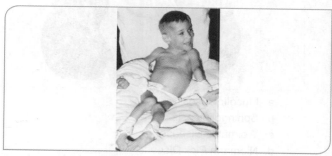

a. PRPP Synthetase
b. Xanthine oxidase
c. Hypoxanthine Guanine Phosphoribosyl transferase
d. Glucose 6 Phosphatase

45. A boy presented to ortho OPD with a painful swelling on the dorsum of foot. His synovial fluid examination revealed the crystals given in the picture. What is the diagnosis?

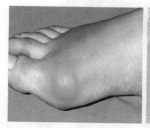

a. Xanthinuria
b. Hypercalcemia
c. Gout
d. Von Gierke's Disease

46. A 13 month old child presented with developmental delay. Her peripheral smear show macrocytic anaemia. No B_{12} or folic acid deficiency. The compound given in the picture is excreted in the urine. Her peripheral smear is also given. What is the diagnosis?

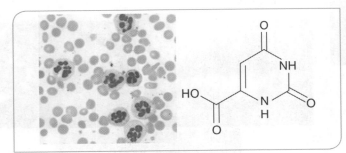

   a. Gout
   b. SCID
   c. Orotic aciduria
   d. Copper deficiency

47. What is the contribution of this scientist to molecular genetics?

   a. Chemical synthesis of ribonucleotide
   b. Sequencing of amino acid
   c. Base pairing rule
   d. Structure of DNA

48. What is the contribution of these three Nobel laureates to genetics?

49. Which is the cytogenetic technique used? What is the interpretation of Image A (Red colour is for Chromosome 21 and Green for chromosome 13).

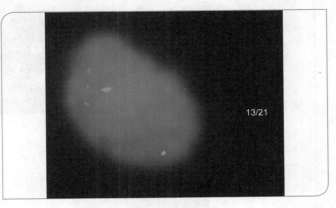

50. Identify the molecular biology technique used in A, B and C:

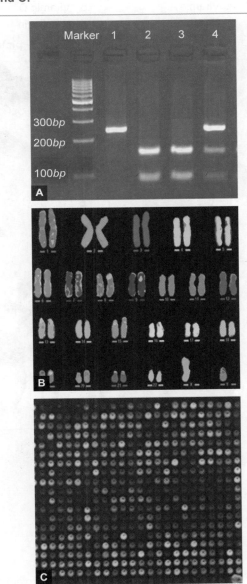

51. **Identify the Vitamin deficiency from the pictures given**

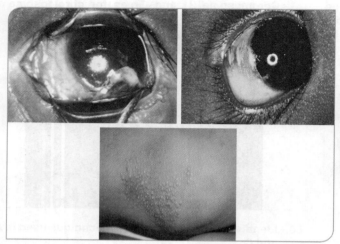

a. Vitamin A
b. Vitamin D
c. Thiamine
d. Riboflavin

52. **Identify the disorder from the clinical pictures given:**

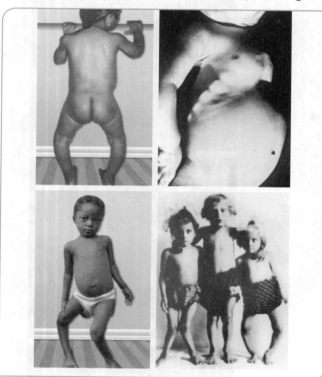

a. Scurvy
b. Rickets
c. Osteomalacia
d. Beri beri

53. **Identify the vitamin deficiency from the given clinical pictures:**

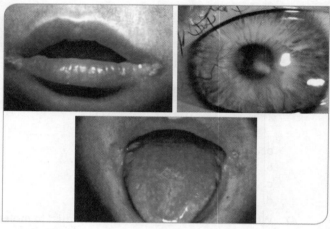

a. Thiamine
b. Pyridoxine
c. Riboflavin
d. Biotin

54. **Which is the vitamin deficiency disorder, in people who take the given cereals as their staple diet?**

a. Scurvy
b. Beriberi
c. Pellagra
d. Pyridoxine deficiency

Explanatory Answers

1. **Ans. b. Rothera's test.**
   The patient is on hunger strike hence starvation ketosis should be ruled out. Hence the test given is Rothera's test for ketone bodies

2. **Ans. a. Cu^{2+} ions are reduced to Cu^+ ions by enediols**
   - This test is Benedict's test. The principle of the test is
   - Cu^{2+} ions present in Copper sulphate is reduced to Cuprous hydroxide and later to Cuprous oxide by enediols formed from reducing sugar.

3. **Ans. b. Heat and acetic acid test**
   - Heat and acetic acid test is the test for Urine protein. The most probable diagnosis for this child is Nephrotic syndrome. Massive proteinuria is a feature of this disease. Hence will give Heat and acetic acid test positive.

4. **Ans. a. Obstructive jaundice**
   Two tests given here are:
   - Hays test –Tests for bile salts
   - Fouchet's test –Test for bile bigment.
   - So the best answer is obstructive jaundice

   Urine tests for jaundice

Urine test	Hemolytic Jaundice/ Prehepatic Jaundice	Hepatic Jaundice	Obstructive/ Post hepatic Jaundice
Hay's test-Bile salts	Negative	Positive/ Negative	Positive
Ehrlich's test- Urobilinogen	Positive	Positive/ Negative	Negative
Fouchet's test- Bile pigment	Negative	Positive/ Negative	Positive

5. **Ans. c. Biuret test**
   - Biuret test is the general test for all proteins.
   - Ninhydrin test is the test for alpha amino acids

6. **Ans. b. Competitive inhibition**
   - In Competitive inhibition X intercept -1/Km decreases and Y intercept 1/Vmax remains a constant.

7. **Ans. c. Enzyme concentration and substrate concentration**
   - Enzyme concentration is directly proportional to rate of reaction.
   - Increasing substrate concentration increases the rate of reaction initially but later as active sites of the enzyme are fully saturated with substrate rate of reaction plateaus.

8. **Ans. c. They exhibit cooperative binding.**
   - In allosteric enzyme, modifier binds to a distinct site that of active site. Modifier and substrates are not structural analogues. This is an enzyme regulation of quality or intrinsic catalytic activity.

9. **c. Proline**
   Identifying feature alpha NH_2 group is incorporated in the pyrrolidine ring.

10. **c. Nonpolar aromatic amino acid**
    The given amino acid is tryptophan. So its nonpolar and Aromatic.

11. **c. Phenyl ketonuria**
    Identifying features:
    - Blue eyes, blonde hair and fair skin is a classical desctription of a case of PKU. The child is intellectually disabled and they have mousy body odour.

12. **c. Ochronosis**
    Identifying features are:
    - Age of presentation
    - No intellectual disability
    - Back ache because of deposition of alkaptone bodies in IV disc
    - Blackish discoloration in eyes, pinna and other cartilaginous tissues
    - Blackish discoloration of urine.

13. **c. Tyrosinase**
    The case given in picture is Albinism. The enzyme defect is Tyrosinase.

14. **b. Homocystinuria**
    The given picture shows pectus carinatum. Elongated limbs and dislocated lens. All these features fit in to diagnosis of Homocystinuria.

15. **c. Branched chain keto acid dehydrogenase**
    Identifying features are early age of ptesentation with failure to thrive, convulsion, hypotonia.
    Smell of burnt sugar or caramel odour.
    Positive DNPH test.

16. **b. Hepatic Cirrhosis**
    Identifying features of hepatic cirrhosis
    - Decreased albumin (due to decreased synthesis of albumin)
    - Balanced by polyclonal increase gamma globulin (to maintain oncotic pressure)
    - Beta–Gamma bridging (due to increased IgA which comes in slow beta region).

17. **d. Aldose sugar joins with a ketosugar**
    The disaccharide is Lactose (Milk sugar)
    Lactose is
    - Reducing disaccharide
    - Made up of Galactose and glucose

- Contraindicated in Classic galactosemia
- Both monosaccharide units are aldose sugar.

18. c. Hyaluronic acid

GAG	Disaccharide Repeat Unit
Hyaluronic Acid (Hyaluronan)^Q	N Acetyl Glucosamine + Glucuronic Acid^Q
Chondroitin Sulphate	N Acetyl Galactosamine + Glucuronic Acid
Keratan Sulphate I & II	N Acetyl Glucosamine, Galactose
Heparin	Glucosamine, Iduronic Acid
Heparan sulphate (HS)	Glucosamine, Glucuronic Acid
Dermatan sulphate (DS)	N Acetyl Galactosamine, Iduronic Acid/ Glucuronic Acid

19. a. MPS IH, Hurler's Disease
- Identifying features are
- Swelling in groins-inguinal hernia
- Defective vision-Corneal clouding
- Intellectual disability, coarse facial features, short stature, prominent forehead.

20. b. Iduronate Sulfatase
- Hunter's Disease (MPS-II)
- Distinguishing features are vision normal as there is no corneal clouding. Males are affected.

21. b. Glucose oxidase is highly specific for beta anomer of Glucose

The reaction in glucometer is

$$\text{Glucose} + H_2O + O_2 \xrightarrow{GOD} \text{Gluconic acid} + H_2O_2$$
$$2H_2O_2 + 4\text{ aminoantipyrine} + PHBS \xrightarrow{POD} \text{Quinoneimine dye} + H_2O$$

- This reaction is specific for beta anomer of glucose.
- Here the first carbon atom is oxidised
- This is an example of enzymatic method (Glucose Oxidase Peroxidase method).

22. a. Glucosazone, Lactosazone, Maltosazone

Shape	Sugar
Needle shaped/Broomstick/Sheaves of Corn	Glucose, Fructose, Mannose
Pincushion with pins/Hedgehog/ Flower of Touch me not	Lactose
Sunflower Petal shaped	Maltose

23. a. Type I GSD (Von Gierke's Disease)
Identifying features
- Doll like facies with fat cheeks.
- Thin extremities
- Short stature
- Protuberant abdomen
- Hepatomegaly
- Renomegaly
- No splenomegaly

Biochemical hallmarks are:
1. Hypoglycemia
2. Lactic acidosis
3. Hyperlipidemia
4. Hyperuricemia

24. d. Type IV GSD (Anderson Disease)
- This disorder is clinically variable.
- The most common and classic form is characterized by progressive cirrhosis of the liver and is manifested in the 1st 18 mo of life as hepatosplenomegaly and failure to thrive.
- The cirrhosis progresses to portal hypertension, ascites, esophageal varices, and liver failure that usually leads to death by 5 yr of age.

25. c. Type III GSD

Type III GSD (Limit Dextrinosis) (Cori's Disease)

Identifying features
- Hypoglycemia
- Hepatomegaly
- Hyperlipidemia
- Short stature
- Variable skeletal myopathy
- Elevated transaminases
- Liver failure and cirrhosis
- But remarkably hepatomegaly and hepatic symptoms in most patients with type III GSD improve with age and usually resolve after puberty.

26. b. Type II GSD (Pompe's Disease)
Identifying features
- Affected infants present in the 1st few months of life with hypotonia, a generalized muscle weakness with a "floppy infant" appearance,
- Neuropathic bulbar weakness, feeding difficulties, macroglossia,
- Hepatomegaly
- A hypertrophic cardiomyopathy followed by death from cardiorespiratory failure or respiratory infection usually by 1 yr of age.

27. b. Classic Galactosemia
Identifying features
- Classic galactosemia is a serious disease with onset of symptoms typically by **the 2nd half of the 1st wk of life.**
- With jaundice, vomiting, seizures, lethargy, irritability, feeding difficulties, poor weight gain or failure to regain birth weight.
- Hepatomegaly **Oil drop cataracts**, Hepatic failure.
- Mental retardation.
- Demonstrating a reducing substance in several urine specimens collected while the patient is receiving human milk, cow's milk, or any other formula containing lactose.
- **Benedict's Test Positive^Q**

- Clinistix urine test results are usually negative because the test materials rely on the action of glucose oxidase, which is specific for glucose and is nonreactive with galactose (**Glucose Oxidase Test is negative**).

28. **c. Decreases HDL Cholesterol**
    - Fish oils are rich in omega 3 fatty acids.
    - Omega 3 fatty acids reduces the Triacyl Glycerol level, hence used in hyperlipoproteinemia with hypertriglyceridemia, they also reduces the inflammatory risks.

29. **d. Linoleic acid (Safflower oil contain 75% linoleic acid)**

30. **d. Sphingomyelin.**
    - Sphingomyelin contains Sphingosine + Fatty acid + PO_4 + Choline

31. **c. It is a sphingophospholipid**
    The given diagram is Lecithin
    - Belongs to Glycerophospholipid.
    - It is Phosphatidic acid + Choline
    - It is the major lung surfactant, hence deficiency can cause RDS.
    - It is the largest body store of Choline.

32. **c. Cherry red spot and Angiokeratoma is seen in GM1 gangliosidoses**
    - Cherry red spot is not seen in Fabry's Disease and Gaucher disease
    - Angiokeratoma is not seen in Farber's disease and Gaucher disease

33. **c. Globotriaosyl Ceramide**
    - The case given is Fabry's Disease
    - Accumulating substance is Globotriaosyl Ceramide.

34. **b. Farber's Disease**
    - Sphingolipidosis that has features similar to Rheumatoid arthritis.

35. **c. Wolman's Disease**
    - Calcification of adrenals is pathognomonic of Wolman's disease.

36. **b. Lack of ATP to support gluconeogenesis**
    - This is a case of MCAD deficiency.
    - Identifying features are:
    - Recurrent episodes of hypoglycaemia is time interval of feeding increased
    - Dicarboxylic acids in urine
    - Absence of ketone bodies in urine
    - Reasons for hypoglycemia are:
        - Due to MCAD deficiency, beta oxidation isaffected. This is the source of ATP for gluconeogenesis, when glycogen stores are depleted. So lack of ATP is one reason.
        - Due to lack of **ace**tyl-CoA, which is released by beta oxidation. Acetyl-CoA is the activator of pyruvate carboxylase, one of the key enzymes of gluconeogenesis.

37. **d. Jamaican vomiting sickness**
    - The fruit given in picture is Ackee apple fruit seen in Africa.
    - The fruit contains a toxin hypoglycin that inhibit acyl-CoA dehydrogenase.
    - Lack of beta oxidation results in lack of ATP and acetyl-CoA, which decreases gluconeogenesis.
    - This result in hypoglycaemia.
    - No ketone bodies due to lack of beta oxidation.

38. **a. Peroxisomal biogenesis disorder.**
    Biochemical defect in Zellweger Syndrome
    There is gene defects, involving mainly the import of proteins that contain the PTS1 targeting signal
    - Zellweger syndrome (most severe)
    Clinical picture
    - Typical facial appearance (high forehead, unslanting palpebral fissures, hypoplastic supraorbital ridges, and epicanthal folds)
    - Severe weakness and hypotonia, neonatal seizures
    - Eye abnormalities (cataracts, glaucoma, corneal clouding, Brushfield spots, pigmentary retinopathy, and nerve dysplasia)
    - Because of the hypotonia and "mongoloid" appearance, Down syndrome may be suspected.

39. **a. Familial chylomicronemia syndrome**
    Identifying features are
    - Recurrent abdominal pain due to pancreatitis
    - Milky white plasma
    - Eruptive xanthoma
    - Lipemia retinalis
    - Very high triglycerides
    - Normal cholesterol

40. **b. Type IIA Hyperlipoproteinemia**
    Identifying features
    - Corneal arcus
    - History of coronary artery disease
    - Tendinous xanthoma
    - Very high cholesterol level
    - Triglycerides within normal limit.

41. **c. Sitosterolemia**
    Identifying features
    - Usual presentation of familial hypercholesterolemia like tendon xanthoma, premature atherosclerotic cardiovascular disease.
    - Anisocytosis, poikilocytosis of erythrocytes, megathrombocytes due to incorporation of plant sterols in to cell membrane.
    - Episodes of hemolysis and splenomegaly are distinctive features.
    - Severe hypercholesterolemia not responding to statins but dramatic response to dietary therapy or ezetemibe.

42. **b. Tangier's disease**

    The given picture is yellow or orange tonsils seen in Tangier's disease.

    Explained in the chapter.

43. **d. Familial dysbetalipoproteinemia**

    The picture given is tuberoeruptive xanthoma and palmar xanthoma of Type III hyperlipoproteinemia.

    Clinical presentation of familial dysbetalipoproteinemia (Type III hyperlipoproteinemia)
    - Patients with FDBL usually present in adulthood with incidental hyperlipidemia.
    - **Premature coronary disease or peripheral vascular disease.**
    - Lipoprotein elevated is chylomicron and VLDL remnant.
    - Lipid elevated is cholesterol and triacylglycerol.

    Xanthomas in FDBL
    - Two distinctive types of xanthomas, tuberoeruptive and palmar, are seen in FDBL patients.
    - **Tuberoeruptive xanthomas** begin as clusters of small papules on the elbows, knees, or buttocks and can grow to the size of small grapes.
    - **Palmar xanthomas** (alternatively called xanthomata striata palmaris) are orange-yellow discolorations of the creases in the palms and wrists.
    - Both these xanthomas are virtually pathognomonic of FDBL.

44. **c. Hypoxanthine Guanine Phosphoribosyl transferase**

    Identifying features
    - Seen only in males
    - Compulsive self mutilation
    - Intellectual siability
    - Hyperuricemia

45. **c. Gout**

    Identifying features
    - Acute inflammatory arthritis typically affect Ist Metatarsophalangeal joint.
    - Synovial fluid shows negatively birefringent needle shaped Monosodium UrateCrystals.

46. **c. Orotic aciduria**

    Identifying features
    - Developmental delay
    - Megaloblastic anaemia, so peripheral smear show hypersegmented neutrophils
    - Absence of B_{12} and Folic acid deficiency
    - The compound given in the picture is Orotic acid, which is excreted in urine.

47. **a. Chemical synthesis of ribonucleotide**
    - The scientist given in picture is Har Gobind Khorana
    - Developed chemical method to synthesize polyribonucleotide.
    - This lead to the cracking of genetic code.

48. **Answer is Elucidation of high resolution structure of ribosome.**

    The scientists are Venkatraman Ramakrishnan, Thomas A Steitz and Ada E Yonath

    They got nobel prize in the year 2009 for chemistry.

49. **Answer is**

    The cytogenetic technique is Interphase FISH with chromosome specific probes.

    Image A shows aneuploidy, i.e., Trisomy 21 and Normal 13 chromosome.

50. **Answer is**

    Image A is Restriction fragment Length Polymorphism

    Image B is Metaphase FISH

    Image C is Microarray technique

51. **a. Vitamin A**

    The pictures show the clinical manifestations in Vitamin A deficiency.
    1. Dryness of conjunctiva and Cornea
    2. Bitot's spot in the conjunctiva.
    3. Phrynoderma

52. **b. Rickets**

    Pictures given are
    1. Anterior bowing of tibia and femur
    2. Rachitic rosary
    3. Windswept deformity
    4. Different skeletal malformations like coxa vara, coxa valgum, etc.

53. **c. Riboflavin**

    Pictures given are:
    - Angular stomatitis, Cheilosis
    - Corneal Vascularisation
    - Glossitis, Magenta coloured tongue

54. **c. Pellagra**

    Picture given are of maize and Jowar
    - Maize-Niacin present in unavailable form Niacytin
    - Sorghum vulgare (Jowar)-High Leucine content inhibit QPRTase, rate limiting enzyme in Niacin synthesis.

CHAPTER 1

Amino Acids

Chapter Outline

1.1 Chemistry of Amino Acids
1.2 General Amino Acid Metabolism
1.3 Aromatic Amino Acids, Simple Amino Acids and Serine
1.4 Sulphur Containing Amino Acids
1.5 Branched Chain Amino Acids, Acidic Amino Acids, Basic Amino Acids and Amides

CHAPTER 1

Amino Acids

1.1 CHEMISTRY OF AMINO ACIDS

- Structure of Amino Acids
- Beta Alanine
- Classification of Amino Acids.
- Decarboxylation of Amino Acids
- Abbreviations of Amino Acids
- Colour Reactions of Amino Acids
- Derived Amino Acids
- Buffering Action of Amino Acids
- Properties of Amino Acids
- Titration Curve

GENERAL STRUCTURE OF ALPHA AMINO ACID

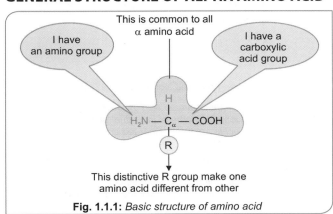

Fig. 1.1.1: Basic structure of amino acid

Alpha Amino Acid

Amino group and carboxyl group attached to the alpha carbon atom.

Most of the amino acids are **alpha amino acid**.

Non-Alpha Amino Acid

Unlike alpha amino acids either carboxyl group or Amino group is not attached to the alpha carbonatom.

Non-alpha amino acids present in tissues in free form are:
- β Alanine
- β Amino Isobutyrate
- γ-Amino Butyrate

Imino Acid—Proline

The α amino nitrogen form a rigid five membered **pyrrolidine ring**. Then this amino group is called a **secondary amino group**. So proline is referred to as an **imino acid**. Still it can form a peptide bond.

- It favors the unique triple helix in collagen.
- But it interrupts the alpha helix found in most globular proteins.

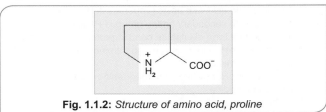

Fig. 1.1.2: Structure of amino acid, proline

CLASSIFICATION OF AMINO ACIDS[Q]

- Based on the variable side chain (R group)
- Based on side chain characteristics (polarity)
- Based on nutritional requirement.
- Based on metabolic fate.

Based on Variable Side Chain[Q]

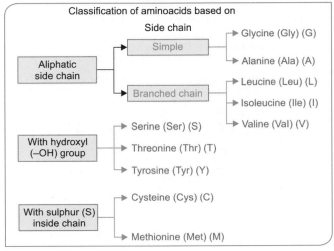

Contd...

Contd...

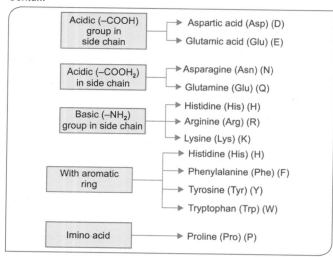

Concept box

Why amino acids are classified based on its side chain?

Alpha amino acids
At physiological pH, alpha carboxyl group is deprotonated and negatively charged COO⁻ and amino group is protonated and positively charged (NH_3^-) They are reactive group.

Alpha amino acids forming peptide bond
But in proteins the alpha carboxyl group and alpha amino group are combined through peptide bond. So ultimately the side chain R group determines the role of amino acids in proteins. Hence amino acids are classified based on side chain.

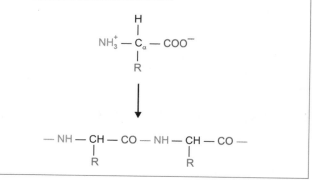

High Yield Points—Amino acids

- Aromatic amino acid with hydroxyl group is Tyrosine
- Aromatic amino acid with basic properties is Histidine
- Amino acid with secondary amino group is Proline
- Amino acid that form disulphide bond in proteins is Cysteine
- At physiologic pH, negatively charged amino acids are Aspartic acid and Glutamic acid
- At physiologic pH, positively charged amino acids are Arginine, Lysine and Histidine
- Most basic amino acid is Arginine
- Amino acid with maximum number of amino group is Arginine

Based on Side Chain Characteristic (Polarity)^Q

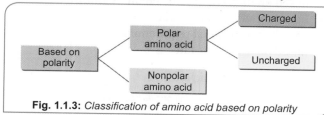

Fig. 1.1.3: Classification of amino acid based on polarity

Tips to memorise–Polar and Nonpolar amino acid

- Learn polar and nonpolar amino acid by classifying the amino acid rather than learning name of individual amino acid.
- Charged amino acids are Polar. Charged amino acids are Acidic and Basic amino acids.
- Mnemonic ABC: Acidic and Basic amino acids are Charged amino acids
- All branched chain amino acids are nonpolar.
- All aromatic amino acids except Histidine are nonpolar.

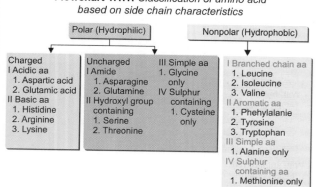

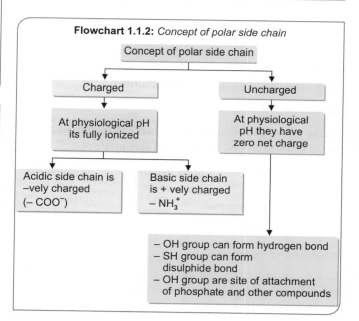

Flowchart 1.1.2: Concept of polar side chain

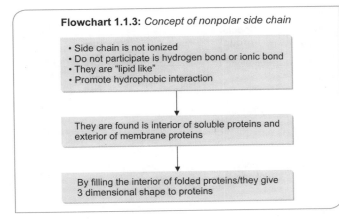

Flowchart 1.1.3: Concept of nonpolar side chain

- Side chain is not ionized
- Do not participate is hydrogen bond or ionic bond
- They are "lipid like"
- Promote hydrophobic interaction

↓

They are found is interior of soluble proteins and exterior of membrane proteins

↓

By filling the interior of folded proteins/they give 3 dimensional shape to proteins

Classification of Amino	Amino Acid
Purely Ketogenic	Leucine[Q] **Lysine**
Both Ketogenic and Glycogenic	**P**henylalanine, **I**soleucine, **T**yrosine **T**ryptophan
Glycogenic	Any amino acid that do not belong to the above groups

> **Tips to memorise—Classification based on metabolic fate**
>
> *First learn the amino acids which are Ketogenic, and then learn amino acids which are both Glucogenic and Ketogenic.*
> *Mnemonic to learn both Ketogenic and Glycogenic—PITT (Lysine, Phenylalanine, Isoleucine, Tyrosine, Tryptophan).*
> *All the rest amino acids are purely glycogenic*

Practical tips to approach controversial questions.

1. **Glycine polar or nonpolar?**
   Yes, Glycine is polar; but it is least polar among the polar amino acid.
   It exhibits some nonpolar nature also.
2. **Tyrosine polar or nonpolar?**
   Tyrosine is significantly polar amino acid among the nonpolar amino acid.

How to approach such questions?

1. Which of the following amino acid is polar?
   a. Glycine
   b. Arginine
   c. Leucine
   d. Isoleucine

Choose arginine as answer NOT Glycine.

2. Which of the following amino acid is nonpolar?
   a. Tyrosine
   b. Leucine
   c. Arginine
   d. Lysine

Choose Leucine as the answer

Among the nonpolar amino acids, Tyrosine and Leucine choose leucine as the best answer as Tyrosine has some polar nature due to the hydroxyl group.

Based on Metabolic Fate[Q]

Classified into:
1. **Ketogenic:** Amino acids that are converted to Acetyl CoA and thereby to ketogenic pathway
2. **Glycogenic:** Amino acids that enter into glucogenic pathway
3. **Both glycogenic and ketogenic:** That can enter into both ketogenic and glucogenic pathway.

Practical tips to approach controversial questions

Example: 1 (multiple response PGI type)
1. **Which of the following amino acids are ketogenic?**
   a. Lysine
   b. Leucine
   c. Alanine
   d. Tyrosine
   e. Isoleucine

Answers are a, b, d, e, i.e. Lysine, Leucine, Tyrosine, Isoleucine. Here you have to consider all ketogenic amino acids. Among these options, exclude only Alanine as it is purely glucogenic.

Based on Nutritional Requirement[Q]

- **Essential:** Those amino acids which cannot be synthesized in the body[Q]. Hence these amino acids are to be supplied in the diet.
- **Semiessential:** Growing children require them in the food, but not essential in adults.
- **Nonessential:** Amino acids which can be synthesized in the body[Q], hence not required in the diet.

Nutritionally Essential	Nutritionally Nonessential
Methionine	All the other amino acids
Threonine	
Tryptophan	
Valine	
Isoleucine	
Leucine	
Phenylalanine	
Lysine	
Histidine	
Arginine***	

***Arginine is nutritionally semiessential. Because it is inadequately synthesized in growing children.

Self Assessment and Review of Biochemistry

Tips to memorise—Essential amino acids

Mnemonic to learn essential amino acids—MeTT VIL PHLY (read as Met will fly).
   Methionine, Threonine, Tryptophan, Valine, Isoleucine, Leucine, Phenylalanine, Lysine.

Practical tips to approach controversial questions

Is histidine essential or semiessential?

Although histidine is considered essential, unlike the other essential it does not fulfill the criteria of inducing negative nitrogen balance promptly upon removal from the diet.

How to approach such question?

Example 1:
- Which of the following amino acid is semiessential?
    a. Lysine   b. Tyrosine
    c. Arginine   d. Histidine

    For single response type of question choose Arginine as the answer NOT Histidine.

Example: 2:
- Which of the following amino acid is semiessential?
    a. Histidine   b. Glycine
    c. Tyrosine   d. Glutamate

    From these options, Histidine is the single best answer.

Concept Box

What makes certain amino acids nutritionally essential?
The lengthy pathway to synthesize certain amino acids make certain amino acids essential. All nonessential amino acids need 1 or 2 enzymes for its synthesis. But essential amino acids need more than 5 enzymes.

Special Groups Present in Amino Acids

Amino acid	Special group	Structure
Arginine	Guanidinium^Q	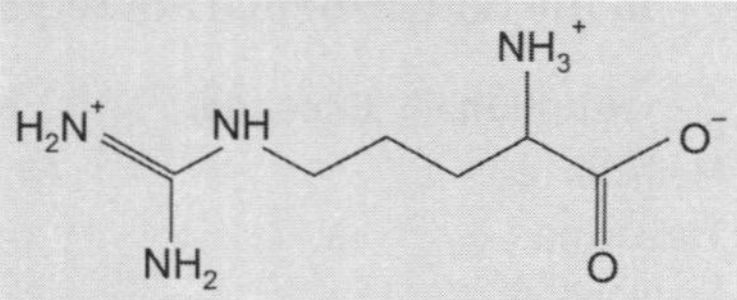 (structure of arginine)
Phenyl-alanine	Benzene	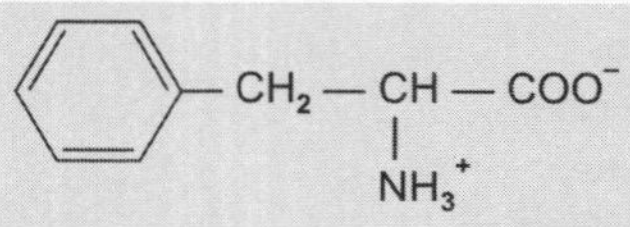 (structure of phenylalanine)
Tyrosine	Phenol	(structure of tyrosine)

Contd...

Contd...

Amino acid	Special group	Structure
Histidine	Imidazole^Q	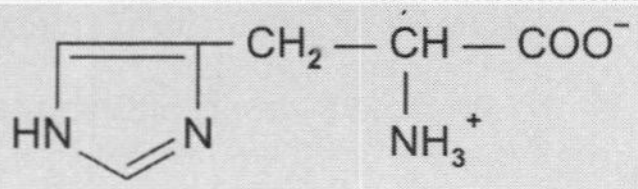 (structure of histidine)
Proline	Pyrrolidine	(structure of proline)
Methionine	Thioether Linkage	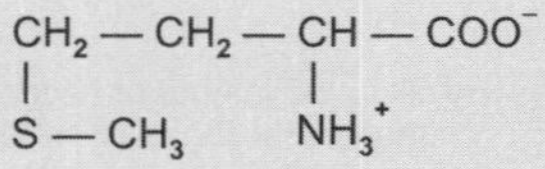 (structure of methionine)
Tryptophan	Indole	(structure of tryptophan)
Cysteine	Thioalcohol (SH) or Sulfhydryl group or Thiol	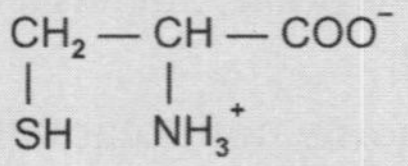 (structure of cysteine)

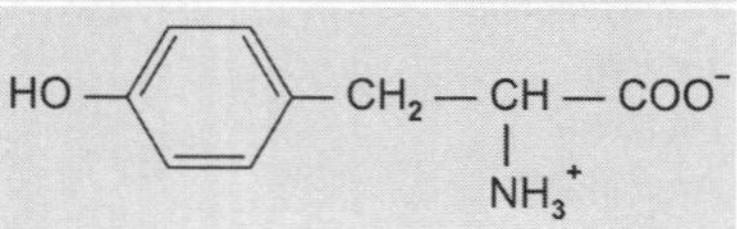

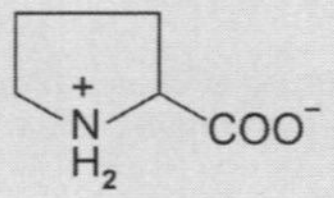

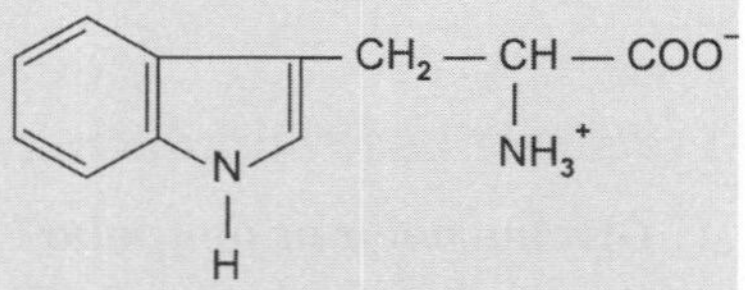

Conservative (Homologous) Substitution

One amino acid replaced by another amino acid of similar characteristics.

   Examples of homologous substitution is shown in the diagram given below.

Conservative Mutation

Hydrophilic, Acid	Asp	Glu			
Hydrophilic, Basic	His	Arg	Lys		
Polar, Uncharged	Ser	Thr	Gln	Asn	
Hydrophobic	Ala	Phe	Leu	Ile	Val	Pro

Nonconservative (Nonhomologous) Substitution

One amino acid replaced by another amino acid of different characteristics.

21st and 22nd Amino Acids^Q

Selenocysteine

- 21st protein forming Amino Acid^Q
- Precursor amino acid for selenocysteine is Serine^Q
- **Serine** is modified to cysteine. Selenium replaces sulphur of cysteine cotranslationally
- In humans approximately 2 dozen selenoproteins are there, that includes Peroxidase^Q and Reductases^Q.

Seen in the active site of following enzymes and proteins:^Q
- Thioredoxin reductase
- Glutathione peroxidase
- Iodothyronine deiodinase
- Selenoprotein P

Recoding
- Selenocysteine is coded by a stop codon, UGA
- This process of converting stop codon to a coding codon is called Recoding
- SECIS element in the mRNA help in this process

Pyrrolysine
- 22nd protein forming Amino Acid
- By recoding **UAG** stop codon, helped by PYLIS element in the mRNA.

DERIVED AMINO ACIDS

Classified into:
- Derived amino acids seen in proteins
- Derived amino acids not seen in proteins

Derived Amino Acid seen in Protein^Q

4-Hydroxy Proline	• Found in Collagen
5-Hydroxy Lysine	• Vitamin C is needed for hydroxylation.
Methyl lysine	• Found in Myosin
Gamma carboxy glutamate	• Found in clotting factors, like Prothrombin that bind Ca^{2+} • Vitamin K is needed for Gamma carboxylation
Cystine	• Found in proteins with disulphide bond.^Q • Two cysteine molecules join to form cystine • For example, Insulin, Immunoglobulin
Desmosine	• Found in Elastin^Q

Derived Amino Acid not seen in Protein^Q

Ornithine	Intermediates of Urea Cycle
Arginosuccinate	
Citrulline	
Homocysteine	Derived from Methionine^Q
Homoserine	Product of Cysteine biosynthesis
Glutamate-γ Semialdehyde	Serine catabolite

PROPERTIES OF AMINO ACID

More than 300 naturally occurring amino acids exist in nature out of which 20 amino acids constitute monomer units of proteins.

- Derived amino acids do not have a genetic code.^Q
- Amino acids coded by stop codon are: Selenocysteine, Pyrrolysine

I. Amino Acids Exist in Different Charged State

Depends on the two factors:
- Isoelectric pH of the amino acid.
- pH of the surrounding medium.

Isoelectric pH of Amino Acids

1. **At pH = Isoelectric pH(pI)**
   - The amino acid carry equal number of positive and negative charge, i.e. **NO NET CHARGE.**
   - Amino acid exists as **ZWITTER ION (AMPHOLYTE)**

Zwitter Ions or Ampholytes

Molecules which carry equal number of ionizable groups of opposite charge and therefore bear no net charge are called **Zwitter ions or ampholytes**^Q. Zwitter is a german word which means hermaphrodite.

 HIGH YIELD POINTS

Properties of Amino acid at Isoelectric pH(pI)
- No mobility in electric field.^Q
- Minimum solubility.
- Maximum precipitability.^Q
- Minimum Buffering capacity.

2. **At pH less than isoelectric pH (pI)**
   Amino acid exists as protonated or positively charged.
3. **At pH greater than isoelectric pH (pI)**
   Amino acid exists as deprotonated or negatively charged.

The charge of carboxyl group and amino group at physiological pH (pH = 7.4)
- Carboxyl group is negatively charged
- Amino group is positively charged.

> **HIGH YIELD POINTS**
>
> **How to calculate isoelectric pH of amino acid?**
> **Fact-1**
> Isoelectric pH (pI) is average of pKa of ionisable groups.
> First we calculate the pI of Alanine
> pK_1 of αCOO^- group is 2.35
> pK_2 of αNH_3^+ is 9.69
>
> So pI = $\dfrac{pK_1 + pK_2}{2}$
>
> pI of Alanine = $\dfrac{9.69 + 2.35}{2} = 6.02$
>
> **Fact-2**
> If the amino acid has ionisable group other than alpha carboxylic and alpha amino group, then isoelectric pH is the average of pKa of isoionic group.
> **Calculate pI of Aspartic acid**
> Aspartic acid has an extracarboxylic group.
> pKa of αCOO^- group = 2.09
> pKa of αNH_3^+ group is = 9.9
> pKa of COO^- group in the side chain (R) = 3.96
> pI = average pKa of isoionic species, which means we have to find the average of pKa of two COO^- group
>
> pI = $\dfrac{2.09 + 3.96}{2}$
>
> So pI of Aspartic acid is 3.02

II. Amino Acids Exhibit Isomerism

Amino acids have asymmetric (chiral) alpha carbonatom. The mirror images produced with reference to alphacarbon atom, are called D and L forms or enantiomers.

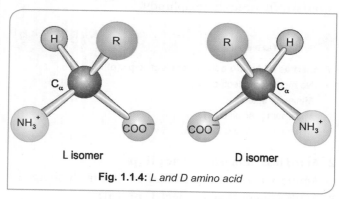

Fig. 1.1.4: *L and D amino acid*

- Almost all naturally occurring amino acids are **L-Isomers**
- Some naturally occurring amino acids are D **Amino acids.**

Naturally Occurring D Amino Acid

- Free D Aspartate and Free D Serine in brain tissue
- D-Alanine and D Glutamate in cell walls of gram positive bacteria
- *Bacillus subtilis* excretes D-methionine, D-tyrosine, D-leucine, and D-tryptophan to trigger biofilm disassembly
- *Vibrio cholerae* incorporates D-leucine and D-methionine into the peptide component of their peptidoglycan layer.

> **HIGH YIELD POINTS**
>
> - Amino Acid with No Chiral/No Asymmetric/No Optically Active Carbon **Glycine**[Q]
> - Source of D-Amino acids in humans is exogenous[Q].
> - The enzyme that interconvert D and L isomers is Racemase.

III. Amino Acids Absorb UV Light

Amino acids which absorb **250–290 nm** (Maximum at **280 nm**) UV light are **tryptophan, phenylalanine, tyrosine.** Maximum absorption of UV light by **tryptophan.**[Q]

> **HIGH YIELD POINTS**
>
> - Aromatic amino acids absorb UV light.
> - Amino acids are color less because they do not absorb visible light.

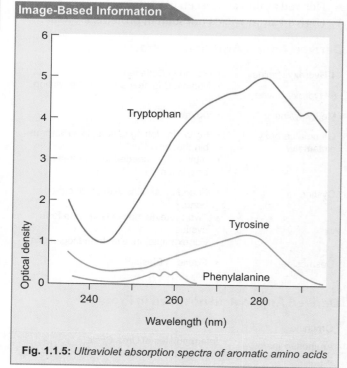

Fig. 1.1.5: *Ultraviolet absorption spectra of aromatic amino acids*

BETA-ALANINE

Formed from **Cytosine and Uracil.**[Q]

Other sources of Beta Alanine is hydrolysis of Beta alanyl dipeptides.

Beta Alanine is seen in^Q
- Pantothenic Acid
- Coenzyme A
- Acyl Carrier Protein
- Beta Alanyl Dipeptides.

Beta Alanyl Dipeptides are
- Carnosine [Histidine + Beta alanine]
- Anserine [N methyl Carnosine]
  Both present in **Skeletal muscle**

Uses of Carnosine
- Activate Myosin ATPase.
- Chelate Copper
- Enhance Copper uptake
- Buffers the pH of anaerobically contracting muscle.

HIGH YIELD POINTS
- Homocarnosine is GABA + Histidine
- There is no beta alanine in homocarnosine

DECARBOXYLATION OF AMINO ACID
- The amino acid undergo alpha decarboxylation to form corresponding Amines
- **PLP^Q is the coenzyme for this reaction.**

Examples of Amino Acid Decarboxylation

Amino acid	Biologic Amines
Histidine	Histamine
Tyrosine	Tyramine
Tryptophan	Tryptamine
Lysine	Cadaverine
Glutamic acid^Q	Gamma Amino Butyric Acid (GABA)
Serine	Ethanolamine
Cysteine	Betamercapto Ethanolamine

HIGH YIELD POINTS
- Amino acids undergo decarboxylation to form corresponding amines.
- **PLP** is the coenzyme for Amino Acid Decarboxylation
- Amino acids undergo deamination to form corresponding Ketoacids.
- Most common amino acid that undergo Oxidative deamination is **Glutamic Acid (Glutamate)**
- Glutamic acid undergo decarboxylation to form **GABA**
- Glutamic Acid undergo deamination to form **Alpha Keto Glutarate**

COLOUR REACTIONS OF AMINO ACIDS
Biuret Test
- General test for Proteins
- Cupric ions in alkaline medium forms violet colour with peptide bond nitrogen.

Dipeptides and individual amino acid do not answer biuret test because this test needs a minimum of two peptide bonds.

Ninhydrin Test

General test for all alpha Amino Acid + 2 mols of Ninhydrin ⟶ Aldehyde with 1 carbon atom less + CO_2 + Purple Complex (Ruhemann's Purple).

- Amino acid which do not give purple colour are:
  - Proline and Hydroxy proline (Yellow colour)
  - Glutamine and Asparagine (Brown colour)

Colour Reactions	Test answered by
Xanthoproteic Test (**Conc HNO$_3$ is a reagent^Q**)	Aromatic Amino Acid^Q DNB (Phenylalanine, Tyrosine, Tryptophan)
Millon's test	Tyrosine (Phenol)
Aldehyde test can be done in two methods: Acree Rosenheim Test (Formaldehyde and Mercuric Sulphate is used) Hopkin's Cole Test^Q (Glyoxylic Acid is used)	Tryptophan (Indole group)
Saka **G**uchi's test	Arginine (**G**uanidinium group) Mnemonic–G is common to all
Sulphur test	Cysteine
Cyanide Nitroprusside Test	Homocysteine
Pauly's Test	Histidine (Imidazole) Tyrosine (Phenol)

Methionine does not answer Sulphur test because sulphur in methionine is in the thioether linkage which is difficult to break.

BUFFERING ACTION OF AMINO ACIDS

- Buffers are solutions which can resist changes when acid or alkali is added.

Henderson Hasselbalch Equation

pH = pKa + log [Base]/[Acid]
When [Base] = [Acid] pH = pKa

Maximum buffering capacity is at **pH = pKa**. So amino acid which has pKa range near physiologic pH can act as an effective buffer.

pKa of Imidazole group of histidine is 6.5–7.4.

Hence at physiologic pH, **Imidazole** group of Histidine has the maximum buffering capacity.^{Q}

TITRATION CURVE

Titration is done to find out the amount of acid in a given solution. To find out that a measured volume of acid is titrated against a strong alkali. The endpoint of titration is the point at which the pH of solution is 7. A plot called titration curve is obtained.

Definition of titration curve

A plot of (OH–) added (represented in equivalents) against pH is called Titrarion curve.

- Let us see the important landmarks in Titration curve of weak acids and certain amino acids in the Image-based Information boxes.

Image-Based Information
Titration curve of a compound with single ionisable group (Acetic acid)

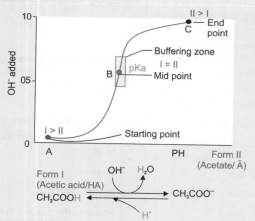

- At point A-Most of the Acetic acid is **unionized**
- At point B-the pH = pKa and Acetic acid is **partially ionized** ($CH_3COOH = CH_3COO^-$).
- Buffering region of Acetic acid is at pH = pKa +/-1
- The **horizontal segment of titration curve shown in blue colour is the buffering region.**
- At point C-Acetic acid completely ionized (CH_3COO^-)

Image-Based Information
- Titration Curve of Amino acid, Glycine
- Titration curve of a compound with 2 ionizable group

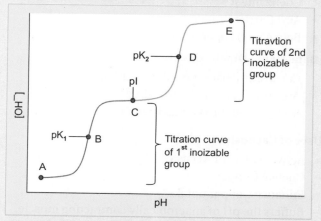

pK_1 = Ionization constant of Ist ionizable group
pK_2 = Ionization constant of IInd ionizable group

$$pI = \frac{pK_1 + pK_2}{2} = \text{isoelectric pH}$$

- Titration curve of compound with two ionizable groups
- Ionization constant of each ionisable group is the midpoint of each curves.
- Buffering region is at pH ± 1 of pKa of ionisable group.
- Buffering region is shown in blue colour.
- Point A is the isoelectric pH, pI

Image-Based Information
Titration Curve of Amino acid, Histidine
Titration curve of a compound with three ionisable group
- TC-1 Titration Curve of first ionisable group
- TC-2 Titration curve of second ionisable group
- TC-3 Titration curve of third ionisable group
- pK_1-Ionisation constant of first ionisable group
- pK_2-Ionisation constant of second ionisable group
- pK_3-Ionisation constant of third ionisable group

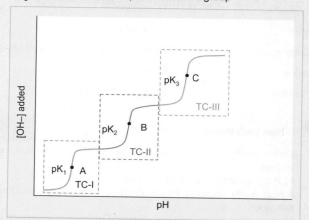

- As histidine has three ionisable group three curves TC-I, TC-II and TC-III are present.

TC = Titration curve; pK = ionization constant

Amino Acids

> **HIGH YIELD POINTS**
>
> **Amino Acids**
> - Simplest Amino acid: Glycine
> - Most hydrophobic (nonpolar) Amino acid: Isoleucine
> - Second most nonpolar amino acid is Valine
> - Most polar amino acid is Arginine
> - Most abundant amino acid in the proteins present in the body is Alanine.
> - Most abundant amino acid in the plasma: Glutamine
> - Least polar amino acid is Glycine
> - Least nonpolar is Proline

Amino Acids and its derivatives as Neurotransmitters
- Glycine: Major inhibitory neurotransmitter in brainstem and spinal cord
- Glutamate: Major excitatory neurotransmitter.

Amino Acid derivative as Neurotransmitter
- Dopamine
- Epinephrine
- Norepinephrine
- Serotonin
- Gamma Amino Butyric Acid (GABA).

DIGESTION OF PROTEINS

Native proteins are resistant to digestion because few peptide bonds are accessible to the proteolytic enzymes without prior denaturation of dietary proteins (by heat in cooking and by the action of gastric acid).

Enzymes Catalyze the Digestion of Proteins

There are two main class of proteolytic digestive enzymes (proteases).

I. **Endopeptidases** hydrolyze peptide bonds between specific amino acids throughout the molecule.
   - **Pepsin** in the gastric juice catalyzes hydrolysis of peptide bonds adjacent to amino acids with bulky side-chains (aromatic and branched-chain amino acids and methionine).
   - **Trypsin, chymotrypsin, and elastase** are secreted into the small intestine by the pancreas.
   - Trypsin catalyzes hydrolysis of **lysine and arginine esters**.
   - Chymotrypsin catalyzes hydrolysis esters of **aromatic amino acids**.
   - Elastase catalyzes hydrolysis esters of small **neutral aliphatic amino acids.**

II. **Exopeptidases** *catalyze the hydrolysis of peptide bonds, one at a time, from the ends of peptides.*
   - **Carboxypeptidases**, secreted in the pancreatic juice, release amino acids from the free carboxyl terminal.
   - **Aminopeptidases**, secreted by the intestinal mucosal cells, release amino acids from the amino terminal.
   - **Dipeptidases and tripeptidases** in the brush border of intestinal mucosal cells catalyze the hydrolysis of di- and tripeptides, which are not substrates for amino- and **carboxypeptidases.**

The proteases are secreted as inactive **zymogens**; the active site of the enzyme is masked by a small region of the peptide chain that is removed by hydrolysis of a specific peptide bond.

Pepsinogen is activated to pepsin by gastric acid and **by activated pepsin.**

In the small intestine, trypsinogen, the precursor of trypsin, is activated by **enteropeptidase**[Q], which is secreted by the duodenal epithelial cells; trypsin can then activate chymotrypsinogen to chymotrypsin, proelastase to elastase, procarboxypeptidase to carboxypeptidase, and proaminopeptidase to aminopeptidase.

Absorption of Amino Acid

Free amino acids are absorbed across the intestinal mucosa by **sodium-dependent active transport**. There are several different amino acid transporters, with specificity for the nature of the amino acid side-chain.

Transporters of Amino Acids
- For Neutral Amino acids
- For Basic Amino acids and Cysteine.
- For Imino Acids and Glycine
- For Acidic Amino acids
- For Beta Amino Acids (Beta Alanine).

- Most of the amino acids are alpha amino acids.
  - Imino acid—Proline has Pyrrolidine ring
  - Two amino acids are that are coded by stop codons are
    1. Selenocysteine-by UGA
    2. Pyrolysine –by UAG
  - Amino acids have maximum buffering capacity at pH = pKa.

Contd...

Self Assessment and Review of Biochemistry

Contd...

- Imidazole group of histidine has maximum buffering action at physiological pH.
- Aromatic amino acids (Trp, Phe, Tyr) absorb UV light at 250–290 nm.
- Pantothenic acid contain beta alanine.

This table is a workbook model table for quick review before exams. Two are done for you. Try the rest! Based on the classification of amino acids we have learnt in the chapter.

Amino acid	Based on side chain	Based on side chain characteristic	Nutritional classification	Metabolic fate
Glycine	Simple amino acid	Polar	Nonessential	Glycogenic
Alanine				
Cysteine				
Methionine				
Serine				
Threonine				
Aspartate				
Glutamate				
Asparagine				
Glutamine				
Arginine				
Histidine	Heterocyclic Aromatic, Basic	Polar	Essential (Can be semiessential)	Glycogenic
Lysine				
Phenylalanine				
Tyrosine				
Tryptophan				
Proline				
Leucine				
Isoleucine				
Valine				

Check List for Revision

- This chapter is high yield topic.
- Classification of amino acids, Selenocysteine are the must learn topic
- Isoelectric pH, Derived amino acids, Beta alanine-text with bold letters is most important.
- Titration curve is an IBQ hence learn in that aspect.

REVIEW QUESTIONS MCQ

Amino Acid Classification

1. **Amino acids with hydroxyl group:** *(PGI Nov 2016)*
   a. Threonine
   b. Tyrosine
   c. Serine
   d. Tryptophan
   e. Valine

2. **Fibropeptidase A & B are highly negative due to presence of which amino acids?**
   (Recent Question Nov 2017)
   a. Glutamate and Aspartate
   b. Serine and Threonine
   c. Lysine and Arginine
   d. Valine and Lysine

Amino Acids

3. Which of the following special amino acid is not formed by post-translational modification?
   (AIIMS Nov 2017)
   a. Triodothyronine
   b. Hydroxyproline
   c. Hydroxylysine
   d. Selenocysteine

4. Which of the following have a positive charge in physiological pH? *(AIIMS Nov 2016)*
   a. Arginine
   b. Aspartic acid
   c. Isoleucine
   d. Valine

5. What is the pH of the solution if the Hydrogen ion concentration is 5 millimoles/L? *(AIIMS Nov 2016)*
   a. 2.3
   b. 3.7
   c. 6.6
   d. 3.5

6. Selenocysteine is coded by: *(AIIMS Nov 2015)*
   a. UAG
   b. UGA
   c. UAA
   d. GUA

7. All of the following are essential amino acids except:
   (AIIMS May 2006)
   a. Methionine
   b. Lysine
   c. Alanine
   d. Leucine

8. Polar amino acids is/are: *(PGI May 2012)*
   a. Serine
   b. Tryptophan
   c. Tyrosine
   d. Valine
   e. Lysine

9. Nonpolar amino acid are: *(PGI Nov 2010)*
   a. Alanine
   b. Tryptophan
   c. Isoleucine
   d. Lysine
   e. Tyrosine

10. Hydrophobic amino acids are: *(PGI May 2010)*
    a. Methionine
    b. Isoleucine
    c. Tyrosine
    d. Alanine
    e. Asparagine

11. Basic amino acids is/are: *(PGI Dec 2013)*
    a. Leucine
    b. Arginine
    c. Lysine
    d. Histidine

12. Guanidinium group is associated with:
    (PGI June 2009)
    a. Tyrosine
    b. Arginine
    c. Histidine
    d. Lysine
    e. Tryptophan

13. Sulphur containing amino acid is:
    a. Cysteine
    b. Leucine
    c. Arginine
    d. Threonine

14. Which of the following is a non-aromatic amino acid with a hydroxyl R-group?
    a. Phenylalanine
    b. Lysine
    c. Threonine
    d. Methionine

15. Which is not an essential amino acid?
    a. Tryptophan
    b. Threonine
    c. Histidine
    d. Cysteine

16. Which of the following is not an aromatic amino acid?
    a. Phenylalanine
    b. Tyrosine
    c. Tryptophan
    d. Valine

17. Which of the following group contains only non-essential amino acid? *(Recent Question 2012)*
    a. Acidic amino acid
    b. Basic amino acid
    c. Aromatic amino acid
    d. Branched chain amino acid

18. Amide group containing amino acid is:
    (Recent Question)
    a. Glutamate
    b. Glutamic acid
    c. Glutamine
    d. Aspartate

19. Which of the following is semiessential amino acid?
    (Recent Question)
    a. Arginine
    b. Histidine
    c. Glycine
    d. Phenylalanine

20. Aminoacyl t-RNA is required for all except: *(AI 2000)*
    a. Hydroxyproline
    b. Methionine
    c. Cysteine
    d. Lysine

Properties of Amino Acids

21. pKA = pH when: *(Recent Question Nov 2017)*
    a. Solute is completely ionised
    b. When the concentration of ionised and unionized form is same
    c. Solute is completely unionized
    d. All of the above

22. HCO_3^-/H_2CO_3 is considered most effective buffer at physiological pH because: *(AIIMS Nov 2016)*
    a. It has pKa close to physiological pH
    b. It is formed from a weak acid and base
    c. Its components can be increased or decreased by the body
    d. It can donate and accept H+

14 Self Assessment and Review of Biochemistry

23. The graph shown below is the titration curve of a biochemical compound. Which of the following statement is true? *(AIIMS May 2016)*

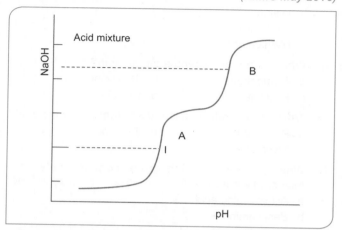

   a. The maximum buffering capacity of the compound is represented by points A and B
   b. The points A and B represent the range of maximum ionisation of the amine and carboxyl group
   c. The compound has three ionisable side chains
   d. The compound has one ionisable group

24. Which of the following proteins cannot be phosphorylated using Protein kinase in prokaryotic organisms? *(AI 2012)*
   a. Leucine   b. Proline
   c. Arginine   d. Tryptophan

25. Carboxylation of clotting factors by vitamin K is required to be biologically active. Which of the following amino acid is carboxylated?
   (AIIMS Nov 2008)
   a. Histidine   b. Histamine
   c. Glutamate   d. Aspartate

26. Which of the following is/are not optically inactive amino acids? *(PGI May 2014)*
   a. Histidine   b. Histamine
   c. Glutamate   d. Aspartate

27. Property of photochromisity is seen amongst the following amino acids: *(AI 1997)*
   a. Threonine   b. Tyrosine
   c. Valine   d. Glycine
   e. Serine

28. The property of proteins to absorb ultraviolet rays of light is due to: *(AIIMS June 99)*
   a. Unsaturated amino acid
   b. Aromatic amino acid
   c. Monocarboxylic acid
   d. Dicarboxylic acid

29. All biologically active amino acids are: *(AIIMS Nov 93)*
   a. Peptide bond   b. Imino group
   c. Disulphide bond   d. Aromatic amino acid

30. Replacing alanine by which amino acid will increase UV absorbance of protein at 280 nm wavelength?
   (AIIMS Nov 2008)
   a. L-forms   b. D-forms
   c. Mostly D-forms   d. D and L forms

31. Flexibility of protein depends on: *(AI 1994)*
   a. Glycine   b. Tryptophan
   c. Phenylalanine   d. Histidine

32. Which amino acid can protonate and deprotonate at neutral pH? *(AIIMS May 95)*
   a. Histidine   b. Leucine
   c. Glycine   d. Arginine

33. Phosphorylation of amino acid by: *(PGI June 98)*
   a. Serine   b. Tyrosine
   c. Leucine   d. Tryptophan

34. Which of the following amino acid is purely Glucogenic? *(Recent Question)*
   a. Valine   b. Lysine
   c. Alanine   d. Glycine

Answers to Review Questions

Amino Acid Classification

1. **a, b, c. Thr, Tyr, Ser** *(Ref: Harper 30/e page 18 table 3.2)*
   - Tryptophan has indole group
   - Valine is a branched chain amino acid.

2. **a. Glu. Asp** *(Ref: Harper 30/e page 18 Table 3.2)*
   - Negatively charged amino acids are Aspartate and Glutamate.

3. **d. Selenocysteine** *(Ref: Harper 30/epage)*
   - Selenocysteine is formed by cotranslational modification. All others are derived amino acids. Derived amino acids are formed by post-translational modification.

4. **a. Arginine** *(Ref: Harper 30/e page 18 Table 3-2)*
   At physiological pH positive charge is for Histidine, Arginine and Lysine.
   At physiological pH negative charge is for Aspartic acid and Glutamic acid.

Charge of an amino acid depends on its isoelectric pH
- If pH of medium >pI, the amino acid is negatively charged
- If pH of medium <pI the amino acid is positively charged

5. **a. 2.3**

   pH = –log [H+] = log 1/[H+]
   [H+] = 5 millimoles/L = 5 × 10–3 moles/L
   pH = log 1/5 × 10–3 moles/L = log 0.2 + log 1/10–3 = –0.6987 + 3 = 3 – 0.6987 = 2.3

6. **b. UGA**
   - Stop codon UGA codes Selenocysteine
   - Stop codon UAG codes Pyrrolysine

7. **c. Alanine** *(Ref: Harper 30/e page 282 Table 27–1)*

Essential (MettVilPhly Read As Met Will Fly)	All the other amino acid	Nonessential
Methionine	Arginine	All the other amino acid
Threonine		
Tryptophan		
Valine		
Isoleucine		
Leucine		
Phenylalanine		
Lysine		

8. **a, e. Serine, Lysine** *(Ref: Harper 30/e page 18 Table 3–2)*

   Classification of amino acids based on side chain characteristics (Polarity)
   Polar amino acids (Hydrophilic):
   Uncharged amino acids are serine, threonine, glutamine, asparagine, cysteine, glycine
   Charged amino acids are aspartic acid, glutamic acid, histidine, arginine, lysine.
   Nonpolar Amino Acid (Hydrophobic)
   alanine, leucine, isoleucine, valine, phenylalanine, tyrosine, tryptophan, proline, methionine.

9. **a, b, c, e. Alanine, Tryptophan, Isoleucine, Tyrosine**
   (Ref: Harper 30/e page 183-2)

10. **a, b, c, d. Methionine, Isoleucine, Tyrosine, Alanine**
    (Ref: Harper 30/e page 183-2)

11. **b, c, d. Arginine, Lysine, Histidine**
    (Ref: Harper 30/e page 17, Table 3-1)
    - Basic amino acids are Histidine, Arginine and Lysine
    - Acidic amino acids are Aspartic Acid (Aspartate), Glutamic Acid (Glutamate)

12. **b. Arginine** *(Ref: Harper 30/e page 18 Table 3–2)*

Special Groups Present in Amino Acids

Amino Acid	Special Group
Arginine	Guanidinium^Q
Phenylalanine	Benzene
Tyrosine	Phenol
Histidine	Imidazole^Q
Proline	Pyrrolidine
Methionine	Thioether Linkage
Tryptophan	Indole
Cysteine	Thioalcohol (SH)

13. **a. Cysteine**
    - Sulphur containing amino acids are Cysteine and Methionine.
    - The Sulphur of Cysteine is provided by Methionine.
    - Special group in Cysteine is Sulfhydryl group (Thioalcohol (-SH)
    - Special group in Methionine is Thioether (C-S-C)

14. **c. Threonine** *(Ref: Harper 30/e page 17 Table 3.1)*
    - Aromatic amino acid with hydroxyl group—Tyrosine Nonaromatic amino acid with hydroxyl group are Serine and Threonine.

15. **d. Cysteine** *(Ref: Harper 30/e page 282)*

    Essential Amino acids are Methionine, Threonine, Tryptophan, Valine, Isoleucine, Leucine, Phenyl Alanine, Lysine (Mnemonic MeTT VIL PhLy) and Histidine Semiessential Amino acids is Arginine.

16. **d. Valine** *(Ref: Harper 30/e Page 17 Table 3-1)*

    Aromatic amino acids are
    - Histidine (with Imidazole ring)
    - Phenylalanine (Benzene ring)
    - Tyrosine (Phenol ring)
    - Tryptophan (Indole ring)

17. **a. Acidic amino acid** *(Ref: Harper 30/e page 282)*
    - Group of amino acid that contain only essential amino acid is Branched chain amino acids (Leucine, Isoleucine, Valine)
    - Group of amino acid that contain only nonessential amino acid is Acidic Amino acids, Amide group containing amino acids, Iminoacid, Simpleamino acids

18. **c. Glutamine**
    - Glutamine and Asparagine are Amide group containing amino acids.
    - Aspartate and Glutamate are Acidic amino acid.

19. **a. Arginine** *(Ref: Harper 30/e page 282 Table 27–1)*

16 Self Assessment and Review of Biochemistry

20. a. Hydroxyproline
- Derived amino acids do not require Aminoacyl tRNA.
- Among the above options Hydroxyproline is a derived amino acid

4-Hydroxy Proline	Found in Collagen
5-Hydroxy Lysine	Vitamin C is needed for hydroxylation

Properties of Amino Acids

21. b. When the concentration of *(Ref: Harper 30/e page)*

This question is an application of Henderson Hasselbalch equation.
It is
$pH = pKa + \log [A-]/[HA]$
[A-] is concentration of ionised form
[HA] is concentration of unionised form
When [A-] = [HA], then
$pH = pKa + \log [A-]/[HA]$
$pH = pKa + \log 1$ i.e. $pH = pKa + 0$

22. c. Its components can be increased or decreased by the body.

According tp Henderson Hasselbalch equation $pH = pKa + \log [HCO_3^-]/[H_2CO_3]$
The physiological pH is 7.4
In case of bicarbonate buffer
$pH = 6.1 + \log 24/1.2 = 6.1 + \log 20 = 6.1 + 1.3 = 7.4$, which is exactly equal to physiological pH.
Hence bicarbonate buffer is most effective buffer due to two reasons
1. The concentration of [HCO⁻] is very high, i.e. 24 mmol/l hence the ratio is maintained at 20, hence pH can be maintained.
2. The components is under physiological control, i.e.
   - [HCO⁻] can be regulated by kidneys
   - CO_2, hence Carbonic acid [H_2CO_3] is regulated by lungs. So any imbalance in [HCO⁻]/[H CO] can be regulated by renal or respiratory method and ratio 20 can be maintained

23. a. The maximum buffering capacity is represented by the points A and B

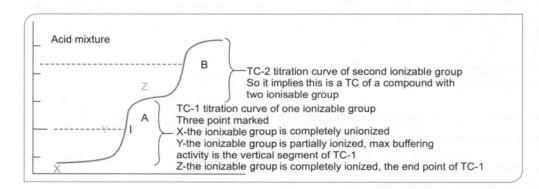

24. d. Tryptophan *(Ref: Harper 30/e page 21,22)*

Amino Acid Absorbs UV Light
Amino Acids which absorb 250–290 nm (Maximum at 280 nm) UV light are tryptophan, phenylalanine, tyrosine. Maximum absorption of UV light by tryptophan.
Remember—aromatic amino acids absorb UV light

25. d. Asparagine *(Ref: Harper 30/e Chapter 9 page 93)*

Protein kinases phosphorylate proteins by catalyzing transfer of the terminal phosphoryl group of ATP to the hydroxyl groups of seryl, threonyl, or tyrosyl residues, forming O-phosphoseryl, O-phosphothreonyl, or O-phosphotyrosyl residues, respectively
- Commonest site of phosphorylation is Serine and Threonine followed by Tyrosine.

26. c. Glutamate *(Ref: Harper 30/e page 717)*
- The Vitamin that act as coenzyme for carboxylation is Biotin
- The vitamin that act as coenzyme for gamma carboxylation is Vitamin K

The Proteins that are gamma carboxylated by Vitamin K are
- Factor II (Prothrombin),
- Factor VII (Proconvertin or Serum Prothrombin conversion Accelerator, SPCA)
- Factor IX (Antihemophilic factor or Christmas factor)
- Factor X (Stuart Prower factor)
- Protein C, Protein S,
- Osteocalcin, Nephrocalcin
- Product of gene gas6

27. a, b, c, e. Threo…, Tyr, Val, Ser… *(Ref: Harper 30/e page 19)*
- Glycine is the only optically inactive amino acid

28. b. Aromatic amino acid *(Ref: Harper 30/e page 21,22)*
- Amino Acid Absorb UV Light
- Amino Acids which absorb 250–290 nm (Maximum at 280 nm) UV light are tryptophan, phenylalanine, tyrosine.
- Maximum absorption of UV light by tryptophan.
Remember—aromatic amino acids absorb UV light

29. **d. Aromatic Amino acid** (Ref: Harper 30/e page 21, 22)

30. **a. L-forms** (Ref: Harper 30/e page 18)
    - Amino acids mostly exist in L-forms
    - Carbohydrates exist in D-forms

31. **a. Glycine** (Ref: Harper 30/e page 39)
    - Glycine having the smallest R group fit into small spaces and induces bends in the alpha helix.
    - Glycine is usually present in beta turns.
    - Kinks in protein structure is due to proline.

32. **a. Histidine**
    - Amino acid which can protonate and deprotonate means those which can act as buffer.
    - Amino acid whose pKa = pH of the medium has maximum buffering capacity.
    - pKa of imidazole group of histidine is 6.5 -7.4.
    - At pH = 7, Imidazole group of histidine can act as buffer.

33. **a, b. Ser, Tyr** (Ref: Harper30/e page 93)

34. **c. Alanine>Valine/Glycine**
    - Lysine is both ketogenic and gluco(glyco)genic
    - Glycine, Alanine and Valine are purely Glucogenic.
    - But Alanine is the principal Glucogenic amino acid.
    - Remember—Glucose Alanine cycle in starvation for provision of substrate for gluconeogenesis.

1.2 GENERAL AMINO ACID METABOLISM

- Biosynthesis of Urea—Stages
- Disposal of Ammonia
- Transamination
- Urea Cycle
- Oxidative Deamination
- Urea Cycle Disorders
- Transport of Ammonia

CONCEPT BOX

Why ammonia is toxic to brain?
Ammonia is trapped by Glutamate and get converted to Glutamine. So in hyperammonemia glutamate is depleted. Glutamate is replenished by Alpha Ketoglutarate. So Alpha ketoglutarate, an intermediate of TCA cycle is depleted. Hence TCA cycle is affected. Brain depends on oxidative pathway like TCA for ATP. Hence it's toxic to brain. Other reasons are:
1. Decreased Glutamate, Hence decreased GABA, the inhibitory neurotransmitter.

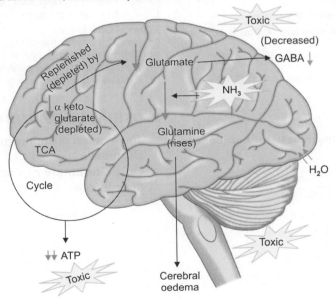

2. Glutamine accumulation, lead to cerebral oedema as it absorb more water.

BIOSYNTHESIS OF UREA

Urea biosynthesis occurs in four stages:
1. Transamination
2. Oxidative deamination of Glutamate
3. Ammonia Transport
4. Disposal of Ammonia

I. TRANSAMINATION

Definition

- Transfer of alpha amino group from one amino acid to a ketoacid to form another pair of amino acid and ketoacid.
- Amino group from amino acids are concentrated in the form of **Glutamate**.
- Because only **Glutamate can undergo oxidative deamination** to significant amount thereby releasing ammonia that enter into urea cycle.

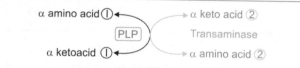

Fig. 1.2.1: General reaction of Transamination

 HIGH YIELD POINTS

Important Points—Transamination
- Transaminase (Amino transferase) is the enzyme
- **Pyridoxal Phosphate (PLP)** a derivative of Vitamin B_6 is the Coenzyme
- Occur in all the tissues
- This reaction occurs via **double displacement (Ping Pong) mechanism**
- No free ammonia is liberated
- Freely **reversible**
- Play an important role in **biosynthesis of nutritionally nonessential amino acid**.

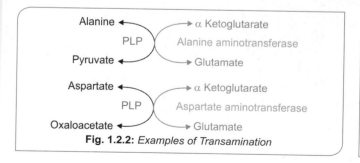

Fig. 1.2.2: Examples of Transamination

CONCEPT BOX

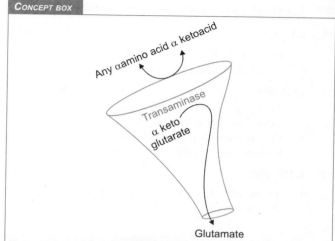

Any α amino acid, under going transamination reaction, transfer the α amino group to α ketoglutarate to form glutamate. Hence glutamate is the concentrated form of toxic α amino group of any amino acid

 HIGH YIELD POINTS

Amino Acid that do not undergo Transamination
- Proline
- Hydroxyproline
- Threonine
- Lysine

Clinical Box

Delta Ornithine Aminotransferase
- Apart from α amino group of amino acid, δ amino group of Ornithine can undergo transamination

Clinical Correlation
- Ornithine δ Aminotransferase deficiency can lead to Gyrate
- Atrophy of Retina and Choroid
- Treatment involve restriction of dietary Arginine
- Pyridoxine is given as a treatment

II. OXIDATIVE DEAMINATION

- The removal of amino group from amino acid is called Deamination.
- Only **Glutamate** can undergo significant Oxidative deamination.

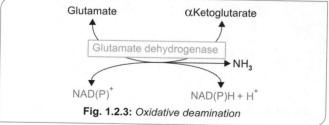

Fig. 1.2.3: Oxidative deamination

HIGH YIELD POINTS

Oxidative Deamination-Important Points
- **Glutamate dehydrogenase** (GDH) is the enzyme
- **Takes place in liver and kidney**
- Organelle-Mitochondria
- **This is a unique enzymes as either NAD^+ or $NADP^+$ can act as coenzyme.**
- **Releases Nitrogen as Ammonia which enter into Urea Cycle.**
- **Reversible.**
- Liver GDH is allosterically inhibited by **ATP, GTP, and NADH**[Q]
- Liver GDH is activated by **ADP**.

L-Amino Acid Oxidase

Minor pathway of deamination of amino acids.
- Takes place in the **liver and kidney**.
- **FMN** is the coenzyme of this reaction.
- H_2O_2 is formed.

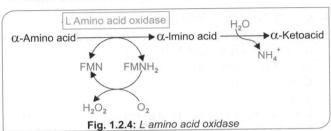

Fig. 1.2.4: L amino acid oxidase

EXTRA EDGE

SOME EXAMPLES OF NONOXIDATIVE DEAMINATION[Q]
1. **Amino acid Dehydrases** for amino acids with hydroxyl group (Serine, Threonine)
2. **Histidase** for histidine
3. **Amino acid Desulfydrases** for amino acids with sulphydryl group, Cysteine and Homocysteine

Transdeamination

Conversion of α Amino nitrogen to ammonia is by concerted action of amino transferase and Glutamate Dehydrogenase is often termed as Transdeamination.

Transamination + Oxidative Deamination = Transdeamination

CONCEPT BOX

Concept of Transdeamination
Transamination can be considered coupled with oxidative deamination. Because Transamination concentrate toxic amino group of all amino acids as Glutamate. This takes place in all the cells. This concentrate of amino group finally reach liver (as Glutamine or Alanine). In the liver toxic amino group collected by Glutamate is released freely by oxidative deamination. As liver has the detoxifying machinery, the urea cycle, toxic amino group is detoxified as Urea. So toxic amino group is handled carefully. Hence Transamination and oxidative deamination is even though anatomically separated but physiologically coupled together.

Contd...

Contd...

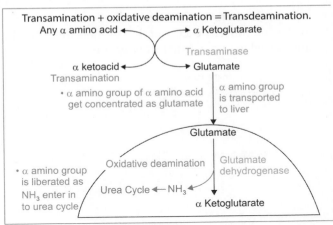

III. TRANSPORT OF AMMONIA

Free ammonia is toxic to cells especially to brain. Excess ammonia generated has to be converted to nontoxic form then transported to liver to enter into urea cycle.

Transport of Ammonia from most of the tissues including the brain.
- As **Glutamine**[Q] with the help of the enzyme Glutamine synthetase.

Glutamine Synthetase
- Ammonia formed in most tissues including the brain is trapped by Glutamate to form **Glutamine**.
- This is called first line trapping of ammonia.
- **ATP** is required for this reaction.

Glutaminase
- In the liver, Glutaminase removes the ammonia from Glutamine.
- Ammonia enter into urea cycle in the liver.

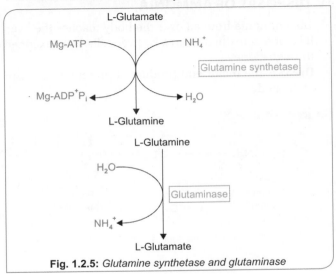

Fig. 1.2.5: Glutamine synthetase and glutaminase

Transport of Ammonia from Skeletal Muscle

- From skeletal muscle-as **Alanine.**Q
- In skeletal muscle, excess amino groups are generally transferred to pyruvate to form **alanine.**

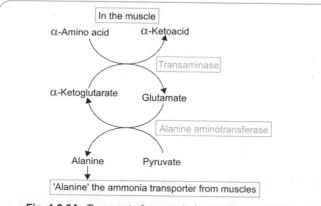

Fig. 1.2.6A: *Transport of ammonia from different organs*

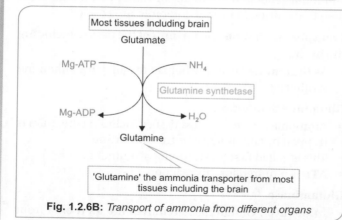

Fig. 1.2.6B: *Transport of ammonia from different organs*

IV. DISPOSAL OF AMMONIA

- The ammonia from all over the body reaches the liver. It is then detoxified to urea by liver cells, then excreted through kidney.
- **Urea**Q is the major end product of protein catabolism in the body.

Sources of UreaQ

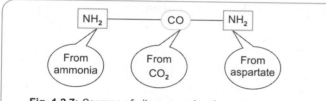

Fig. 1.2.7: *Sources of nitrogen and carbon atoms of urea*

UREA CYCLE

The urea cycle is the first metabolic pathway to be elucidated by Sir Hans Krebs and a medical student associate, Kurt Henseleit hence as **Krebs Henseleit Cycle**.

- Ornithine consumed in the reaction 2 is regenerated in the reaction 5 (see Fig 1.13). Hence called **Ornithine Cycle**.

Site of Urea Cycle
- Organ: Takes place in liver.
- Organelle: Partly **mitochondrial and partly cytoplasmic**.

Reactions of Urea Cycle

The first two reactions take place in the mitochondria. The rest of the reactions take place in the cytoplasm.

I. Carbamoyl Phosphate Synthetase –I(CPS-I)

- Carbamoyl Phosphate is formed from the condensation of CO_2, Ammonia and ATP.
- Takes place in the **mitochondria**.
- **CPS-I** is the rate limiting (pacemaker) enzyme in this pathway.
- Cytosolic CPS-II is involved in pyrimidine synthesis.
- CPS-I is active only in the presence of **N-Acetyl Glutamate, an allosteric activator**.Q
- This step require **2 mols of ATPs**.
- This enzyme is a **Ligase**.

All Synthetases are Ligases. They require ATP.

II. Ornithine Transcarbamoylase (OTC)

- Transfer carbamoyl group of carbamoyl phosphate to Ornithine forming **Citrulline**.
- Takes place in the **mitochondria**
- Subsequent steps take place in the **cytoplasm.**

III. Argininosuccinate Synthetase

- Links Amino nitrogen of Aspartate to Citrulline
- **Aspartate** provides second nitrogen of Urea
- This enzyme is a **Ligase**
- This reaction requires **ATP**
- **2 inorganic phosphates** are utilized.

Transporters of Urea Cycle

- **Ornithine/citrulline Transporter:** For entry of Ornithine and exodus of Citrulline
- **Citrin Transporter:** Aspartate-Glutamate transporter. For exodus of Aspartate from mitochondria.

IV. Argininosuccinate Lyase

Cleavage of Arginosuccinate to Arginine and Fumarate. This enzyme is a **Lyase.**

V. Arginase^Q

Hydrolytic cleavage of Arginine, releases Urea and reforms Ornithine which reenter into mitochondria.

Arginase is a Hydrolase^Q

All the enzymes in the Cytoplasm start with the letter "A"

N-Acetyl Glutamate Synthase

- Enzyme which catalyses the formation of N Acetyl Glutamate (NAG).
- Generally considered as the **sixth enzyme** of Urea Cycle.
- Because **CPS-I is active only in the presence of NAG**^Q.

Energetics of Urea Cycle

- Urea cycle requires **4 high energy phosphates or 4 ATP equivalents.**
- Urea cycle requires **3 ATPS**^Q directly.

Urea Bicycle

Urea cycle is linked to TCA cycle through **Fumarate and Aspartate**. Hence this cycle is called **urea bicycle** or Krebs bicycle.

> **HIGH YIELD POINTS**
>
> - CO_2, **ATP**, NH_4 and **Aspartate** consumed.
> - **Ornithine, Citrulline, Argininosuccinate, Arginine** not consumed
> - The compound that enter into urea cycle and regenerated is **Ornithine**.

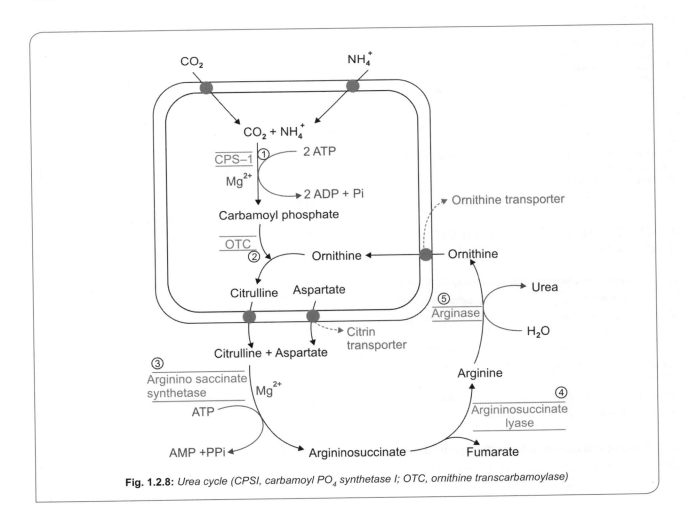

Fig. 1.2.8: *Urea cycle (CPSI, carbamoyl PO_4 synthetase I; OTC, ornithine transcarbamoylase)*

CLINICAL CORRELATIONS—UREA CYCLE DISORDERS

Key Points of Urea Cycle Disorders

Characterized by
- Hyperammonemia
- Encephalopathy
- Respiratory alkalosis.

> **Concept box**
> **Why respiratory alkalosis in Hyperammonemia?**
> Due to hyperammonemia, glutamate is depleted as it is used for synthesis of Glutamine (explained earlier). Hence decreased GABA, as it is synthesized by decarboxylation of Glutamate. GABA is an inhibitory neurotransmitter. So inhibition on respiratory centres is lost. Hence tachypnoea. This leads to CO_2 wash out and hence respiratory alkalosis.

> **Symptom box**
> **Clinical Symptoms Common to all Urea Cycle Disorders in the neonatal period,**
> - Refusal to eat
> - Vomiting
> - **Tachypnea**
> - Lethargy
> - Convulsions are common
> - Can quickly progress to a deep coma
>
> **In infants and older children**
> - Vomiting
> - Neurological abnormalities (ataxia, mental confusion, agitation, irritability, and combativeness)

> **Concept box**
> **Ammonia Intoxication is most severe in the deficiency of first two enzymes**
> Because once citrulline synthesized some ammonia is already been covalently linked to an organic metabolite.

BIOCHEMICAL DEFECT IN UREA CYCLE DISORDERS

Urea Cycle Disorders due to Enzyme Deficiency	
Disorder	**Enzyme defective**
Hyperammonemia Type I	Carbamoyl Phosphate Synthetase I (CPS-I)
Hyperammonemia Type II	Ornithine Transcarbamoylase (OTC)
Citrullinemia Type I (Classic Citrullinemia)	Argininosuccinate synthetase
Argininosuccinic aciduria	Argininosuccinate lyase
Hyperargininemia	Arginase

Urea Cycle Disorders due to Transporter Defect	
Citrullinemia Type II	Citrin (Transport Aspartate and Glutamate) defect

Contd...

Contd...

Hyperammonemia Hyperornithinemia Homocitrllinuria (HHH) syndrome	Ornithine Transporter defect

UREA CYCLE DISORDERS—AT A GLANCE

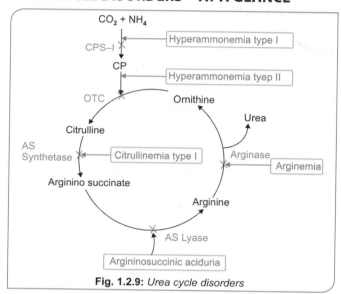

Fig. 1.2.9: Urea cycle disorders

Hyperammonemia Type II (OTC Deficiency)

- **Most common** Urea Cycle disorder[Q]
- Disorder with **X-linked partially dominant inheritance** (All other urea cycle disorders are **Autosomal Recessive**)
- Urea cycle disorder with **Orotic Aciduria**
- Marked elevations of plasma concentrations of glutamine and alanine with low levels of citrulline and arginine
- Orotate may precipitate in urine as a pink colored gravel or stones.

Argininosuccinic Aciduria

- **Trichorrhexis nodosa** (dry and brittle hair) is a common finding.

Hyperargininemia (Argininemia), A Distinct Urea Cycle Disorder

- Hyperargininemia is the urea cycle disorder with least Hyperammonemia. Because by the time Arginine is formed the two nitrogen are already incorporated.
- The clinical manifestations of this condition are quite different from those of other urea cycle enzyme defects.

- A progressive spastic diplegia with scissoring of the lower extremities, choreoathetotic movements, and loss of developmental milestones in a previously normal infant.

The compounds excreted in urine in hyperargininemia:
- **Cystine, Ornithine, Lysine, Arginine [COLA]**
- **Alpha ketoguanidino valeric acid**

Remember COLA (Cystine, Ornithine, Lysine and Arginine) also excreted in Cystinuria.

Hyperammonemia-Hyperornithinemia-Homocitrullinemia (HHH) Syndrome

- Autosomal recessively inherited disorder
- **Biochemical Defect** is mutation in the **ORNT 1 gene** that encodes mitochondrial membrane **Ornithine Permease**
- This results in defect in the transport system of ornithine from the cytosol into the mitochondria
- This leads to accumulation of ornithine in the cytosol causes **hyperornithinemia**
- Deficiency of ornithine in the mitochondria. Results in disruption of the urea cycle and **hyperammonemia**
- **Homocitrulline** is presumably formed from the reaction of mitochondrial carbamoyl phosphate with lysine.

N-Acetyl Glutamate Synthase Deficiency

- The sixth enzyme deficiency which lead to a urea cycle disorder.
- The condition is almost **similar to Hyperammonemia Type I.**
- **Arginine**^{Q} an allosteric activator of NAG Synthase improves CPS I defect as N-Acetyl Glutamate activates CPS-I.
- But Arginine does not improve N-Acetyl Glutamate deficiency, as the enzyme itself is defective.

Citrullinemia Type II

- The adult form (type II) is caused by the deficiency of a mitochondrial transport protein named **citrin**.
- **Citrin (aspartate-glutamate carrier protein)** is a mitochondrial transporter encoded by a gene *SLC25A13*
- One this protein's function is to transport aspartate from mitochondria into cytoplasm;
- Aspartate is required for converting citrulline to argininosuccinic acid
- So Citrulline accumulates.

Biochemical Investigation in a Case with Hyperammonemia

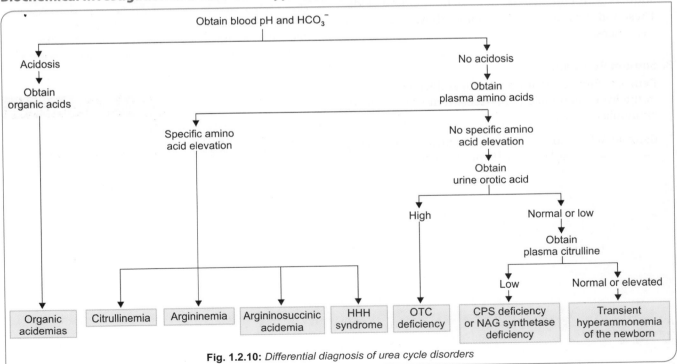

Fig. 1.2.10: *Differential diagnosis of urea cycle disorders*

Normal Blood Ammonia level: 20–40 µg/dL

Methods of estimation of Blood ammonia:
- Chemical Method: Berthelot Method
- Enzymatic method: Glutamate Dehydrogenase Method
- Using **Ammonia selective Electrodes**.

Methods of estimation of Urea:
- Chemical Method-Diacetyl Monoxime-Thiosemicarbazide Method
- Enzymatic Method-Using Urease

Tandem Mass Spectrometry is the most sensitive tool to detect metabolic Disorders.

Biochemical Basis of Treatment of Urea Cycle Disorder

1. **Arginine:**
   - Essential Amino Acid provide Ornithine
   - Arginine is an activator of N Acetyl glutamate Synthase but contraindicated in Arginase Defect.

2. **Acylation therapy:**

   The main organic acids used for this purpose are sodium salts of benzoic acid and **phenylacetic acid.**

   Principle

   Exogenously administered organic acids form acyl adducts with endogenous nonessential amino acids. These adducts are nontoxic compounds with high renal clearances.

3. **Sodium Benzoate**

   Benzoate forms hippuric acid with endogenous glycine in the liver. Each mole of benzoate removes 1 mole of ammonia as glycine.

   Benzoic Acid + CoA ⟶ Benzoyl CoA + Glycine ⟶ Benzoyl Glycine [Hippuric Acid]

4. **Phenylacetate**

   Given as a prodrug, **Phenylbutyrate.** This is rapidly converted to Phenylacetate.

   Phenylacetate conjugates with glutamine to form **phenylacetyl glutamine,** which is readily excreted in the urine. One mole of phenylacetate removes 2 moles of ammonia as glutamine from the body.

   Phenylacetic Acid + CoA ⟶ Phenylacetyl CoA + Glutamine ⟶ Phenylacetyl Glutamine

Quick Review

- Transamination concentrates the alpha amino group as Glutamate.
- Glutamate can undergo significant oxidative deamination.
- PLP is the coenzyme for transamination.
- Ammonia is transported as glutamine from most tissues.
- Ammonia is transported as Alanine from muscle.
- Sources of nitrogens present in urea are ammonia and Aspartate.
- N Acetyl Glutamate is the allosteric activator if CPS-I.
- The pacemaker enzyme for urea cycle is CPS-I.
- The most common urea cycle disorder is Hyperammonemia Type II
- Pyrimidines are excreted in urine in Hyperammonemia Type II.
- Orotic aciduria present in Hyperammonemia Type II
- Only X linked urea cycle disorder is Hyperammonemia Type II
- Plasma Glutamine level rises in urea cycle disorders.
- The earlier the enzyme deficiency in the urea cycle pathway the highest is the glutamine level in and Ammonia level in the blood.

Check List for Revision

- This is a very important chapter for all exams.
- So conceptually learn this chapter.
- Transamination, transport of ammonia and urea cycle and its disorders are high yield topics.

Review Questions MCQ

Digestion and Absorption of Proteins, Transamination and Transport of Amino Acids

1. **Ammonia from brain is detoxified as:** *(AIIMS Nov 2016)*
   a. Glutamate
   b. Glutamine
   c. Alanine
   d. Urea

2. **True about Glutamate Dehydrogenase:** *(Recent Question)*
   a. Can use NADH or NADPH
   b. PLP is the coenzyme
   c. Enzyme of transamination
   d. Ammonium ion is not released in the free form

3. **Increased alanine during prolonged fasting represents:** *(AIIMS Nov 2011)*
   a. Increased breakdown of muscle proteins
   b. Impaired renal function
   c. Decreased utilization of amino acid from Gluconeogenesis
   d. Leakage of amino acids from cells due to plasma membrane damage

4. **Transfer of an amino group from an amino acid to an alpha ketoacid is done by:** *(AI 2011)*
   a. Transaminases
   b. Aminases
   c. Transketolases
   d. Deaminases

5. **The amino acid which serves as a carrier of ammonia from skeletal muscle to liver is:** *(AI 2006)*
   a. Alanine
   b. Methionine
   c. Arginine
   d. Glutamine

6. **Glutamine in blood acts as:** *(PGI Dec 98)*
   a. NH_3 transporter
   b. Toxic element
   c. Stored energy
   d. Abnormal metabolite

7. **Amino acid absorption is by:** *(Recent Question)*
   a. Facilitated transport
   b. Passive transport
   c. Active transport
   d. Pinocytosis

8. **The transporter gene defective in Hartnup's disease:** *(JIPMER Dec 2016)*
   a. SLC 6A 19
   b. SLC 6A 18
   c. SLC 36 A2
   d. SLC 7A7

9. **Nontoxic form of storage and transportation of ammonia:** *(Recent Question)*
   a. Aspartic acid
   b. Glutamic acid
   c. Glutamine
   d. Glutamate

Urea Cycle

10. **Which of the following defect is associated with increase in glutamine in blood, urine and CSF?** *(Recent Question Jan 2019)*
    a. CPS-I
    b. OTC
    c. Argininosuccinate synthetase
    d. Arginase

11. **Substrate linking Kreb's cycle and urea cycle is:** *(Recent Question Jan 2019)*
    a. Fumarate
    b. Aspartate
    c. Alanine
    d. Arginine

12. **Ammonia is toxic to the brain because it leads to depletion of which substrate?** *(Recent Question Jan 2019)*
    a. Succinate
    b. Alpha ketoglutarate
    c. Isocitrate
    d. Fumarate

13. **CPS-I used in which pathway?** *(Recent Question)*
    a. Pyrimidine synthesis
    b. Purine synthesis
    c. Urea cycle
    d. TCA cycle

14. **Urea cycle enzymes are:** *(PGI May 2010)*
    a. Glutaminase
    b. Asparaginase
    c. Argininosuccinate synthetase
    d. Ornithine transcarbamylase
    e. Glutamate dehydrogenase

15. **Which enzymes are part of urea cycle?** *(PGI 2012)*
    a. Ornithine transcarbamylase
    b. Asparaginase
    c. Glutamate synthase
    d. Argininosuccinase

16. **Urea cycle occurs in:** *(AI 2011)*
    a. Liver
    b. GIT
    c. Spleen
    d. Kidney

17. **In which of the following condition there is increased level of ammonia in blood?**
    a. Ornithine transcarbamoylase deficiency
    b. Galactosemia
    c. Histidinemia
    d. Phenyl ketonuria

18. **Urea cycle occurs in:**
    a. Cytoplasm
    b. Mitochondria
    c. Both
    d. Endoplasmic reticulum

19. **Which of the following enzymes(s) is/are not involved in Urea Cycle?** *(PGI May 2012)*
    a. Glutamate dehydrogenase
    b. Argininosuccinate dynthetase
    c. α Ketoglutarate dehydrogenase
    d. Isocitrate dehydrogenase
    e. Fumarase

20. **Glutamate dehydrogenase in mitochondria is activated by:** *(Recent Question)*
    a. ATP
    b. GTP
    c. NADH
    d. ADP

21. **Nitrogen atoms of Urea contributed by:** *(Recent Question)*
    a. Ammonium and aspartate
    b. Ammonium and glutamate
    c. Ammonium and glycine
    d. Ammonium and asparagine

22. **Phenyl butyrate is used in urea cycle disorder because:** *(AIIMS May 2017)*
    a. Scavenges nitrogen
    b. Increases enzyme activity
    c. Maintain energy level
    d. Increases renal output of ammonia

23. **A 6-month-old boy admitted with failure to thrive with high glutamine and Uracilin urine. Hypoglycemia, high blood ammonia. Treatment given for 2 months. At 8 months again admitted for failure to gain weight. Gastric tube feeding was not tolerated. Child became comatose. Parenteral Dextrose given. Child recovered from coma within 24 hours. What is the enzyme defect?** *(AIIMS May 2015)*
    a. CPS1
    b. Ornithine transcarbamoylase
    c. Arginase
    d. Argininosuccinate synthetase

24. **Which of the following is true in relation of urea cycle:** *(PGI Dec 05)*
    a. First 2 steps in cytoplasm
    b. First 2 steps in mitochondria
    c. Defect of enzyme of any step can cause deficiency disease
    d. Urea is formed by NH, glutamic acid and CO_2
    e. Citrulline is formed by combination of carbamoyl phosphate and L. ornithine

25. **A baby presents with refusal to feed, skin lesions, seizures, ketosis, organic acids in urine with normal ammonia; likely diagnosis:** *(AI 2001)*
    a. Proprionicaciduria
    b. Multiple carboxylase deficiency
    c. Maple syrup urine disease

26. **True about urea cycle:** *(PGI May 2015)*
    a. Nitrogen of the urea comes from alanine & ammonia
    b. Uses ATP during conversion of argininosuccinate to arginine
    c. On consumption of high amount of protein, excess urea is formed.
    d. Occur mainly in cytoplasm
    e. Synthesis of argininosuccinate consumes energy

27. **All are true regarding urea cycle except:** *(PGI Nov 2014)*
    a. Urea is formed from ammonia
    b. Rate limiting enzyme is ornithine transcarbamylase
    c. Require energy expenditure
    d. Malate is a byproduct of urea cycle
    e. One nitrogen of urea comes from aspartate

28. **Enzyme involved in nonoxidative deamination is:** *(Recent Question)*
    a. L-amino acid oxidase
    b. Glutamate dehydrogenase
    c. Glutaminase
    d. Amino acid dehydratase

29. **Which of these is a conservative mutation?** *(AIIMS Dec 98)*
    a. Glutamic acid-glutamine
    b. Histidine-glycine
    c. Alanine-leucine
    d. Arginine-aspartic acid

Answers to Review Questions

Digestion and Absorption of Proteins, Transamination and Transport of Amino Acids

1. b. Glutamine
- Ammonia from the brain and most other tissue is detoxified as Glutamine. The enzyme is Glutamine
- Synthestase. This is called as first line trapping of Ammonia.
- Ammonia is transported as Alanine from muscle. Urea cycle does not operate in the brain

2. a. Can use NADH or NADPH
- Other options PLP is not the coenzyme of Oxidative deamination by GDH.
- GDH is the enzyme of Oxidative deamination.
- Ammonium ion is released in free form.

3. a. Increased breakdown of muscle protein

During prolonged fasting, there is increased gluconeogenesis. Alanine provided by muscle is one of the substrates for gluconeogenesis.
This is called Glucose Alanine Cycle or Cahill Cycle.
So plasma level of Alanine rises in prolonged starvation.

Remember:
- In prolonged fasting plasma level of Alanine rises.
- In hyperammonemia plasma level of Glutamine rises.

4. a. Transaminases *(Ref: Harper 30/e page 290)*

Key points

Transamination
- Interconvert pair of α amino acids and α Ketoacid.
    - Ketoacid formed by transamination from Alanine is Pyruvate
    - Ketoacid formed by transamination from Aspartate is Oxaloacetate.
    - Ketoacid formed by transamination from Glutamate is α Keto Glutarate
- Freely reversible.
- Transamination concentrate α amino group of nitrogen as L-Glutamate.
- L-Glutamate is the only enzyme that undergo significant amount of oxidative deamination in mammals.
- Takes place via ping pong mechanism.
- Takes an important role in biosynthesis of nutritionally nonessential amino acids.
- Specific for one pair of substrate but nonspecific for other pair of substrates.
- Pyridoxal Phosphate is the coenzyme

5. a. Alanine *(Ref: Lippincott 6/e page 253)*

- Transport form of Ammonia from most tissues including brain is Glutamine
- Transport form of Ammonia from skeletal muscle is Alanine.

6. a. Ammonia transporter

Transport form of ammonia from brain and most other tissues.

7. c. Active Transport *(Ref: Harper 30/e page 539)*

Free amino acids are absorbed across the intestinal mucosa by sodium-dependent active transport. There are several different amino acid transporters, with specificity for the nature of the amino acid side-chain.

Transporters of Amino Acids
- For Neutral Amino acids
- For Basic Amino acids and Cysteine.
- For Imino Acids and Glycine
- For Acidic Amino acids
- For Beta Amino Acids (Beta-Alanine)

Meister's Cycle
- For absorption of Neutral Amino acids from Intestines, Kidney tubules and brain.
- The main role is played by Glutathione (GSH).
- For transport of 1 amino acid and regeneration of GSH 3 ATPs are required.

Disorders associated with Meister's Cycle Oxoprolinuria
- 5 Oxoprolinase deficiency leads to Oxoprolinuria

Disorders associated with absorption of amino acids

Hartnup's Disease	Malabsorption of neutral amino acids, including the essential amino acid tryptophan **SLC6A19**, which is the major luminal sodium-dependent neutral amino acid transporter of small intestine and renal tubules, has been identified as the defective protein
Blue Diaper Syndrome or Drummond Syndrome Indicanuria	Tryptophan is specifically malabsorbed and the defect is expressed only in the intestine and not in the kidney. Intestinal bacteria convert the unabsorbed tryptophan to indican, which is responsible for the bluish discoloration of the urine after its hydrolysis and oxidation
Cystinuria	Dibasic amino acids, including **cystine, ornithine, lysine, and arginine** are taken up by the Na-independent SLC3A1/ SLC7A9, in the apical membrane which is defective incystinuri(a) Most common disorder associated with Amino acid malabsorption.

Contd...

Contd...

Lysinuric Protein Intolerance	(SLC7A7) carrier at the basolateral membrane of the intestinal and renal epithelium is affected, with failure to deliver cytosolic dibasic cationic amino acids into the paracellular space in exchange for Na+ and neutral amino acids.
Oasthouse Urine Disease (**Smith Strang** Disease)	A methionine-preferring transporter in the small intestine was suggested to be affected. Cabbage-like odor, containing 2-hydroxybutyric acid, valine, and leucine.
Iminoglycinuria	Malabsorption of proline, hydroxyproline, and glycine due to the proton amino acid transporter SLC36A2 defect
Dicarboxylic Aciduria	Excitatory amino acid carrier SLC1A1 is affected. Associated with neurologic symptoms such as POLIP (polyneuropathy, ophthalmoplegia, leukoencephalopathy, intestinal pseudo-obstruction

8. a. SLC 6A19

Hartnup's Disease	Malabsorption of neutral amino acids, including the essential amino acid tryptophan **SLC6A19**, which is the major luminal sodium-dependent neutral amino acid transporter of small intestine and renal tubules, has been identified as the defective protein

9. c. Glutamine

- Transport form of Ammonia from most tissues is Glutamine.
- The enzyme responsible is called Glutamine Synthetase.
- Belong to Ligase class.
- Require ATP.

Urea Cycle

10. a. CPS-1

If there is ammonia intoxication, excess glutamate is converted to Glutamine. The higher the ammonia intoxication higher will be glutamine content. The earlier the enzyme defiency the ammonia intoxication is high. Ammonia intoxication is higher with first two enzymes of Urea cycle CPS-I and OTC. But the earliest enzyme being CPS I, it is the best answer.

11. a. Fumarate　　　　　　　*(Ref: Harper 31/e page 275)*

The compounds that link Kreb's cycle and Urea Cycle is Aspartate and Fumarate.

12. b. Alpha glutarate　　　*(Ref: Harper 31/e page 274)*

Ammonia reacts with Glutamate to form Glutamine. The Glutamate is formed from alpha keto glutarate an intermediate of TCA Cycle. So alpha ketoglutarate is depleted.

13. c. Urea Cycle

14. c, d. Arginino succinate Synthetase, Ornithine transcarbamoylase
　　　　　　　　　　　　　　(Ref: Harper 30/e page 293)

Reactions of Urea Cycle

The first two reaction takes place in the mitochondria. The rest of the reactions takes place in the cytoplasm.
1. **Carbamoyl Phosphate Synthetase –I (CPS-I)**
2. **Ornithine Transcarbamoylase (OTC)**
3. **Arginino Succinate Synthetase**
4. **Arginino Succinate Lyase**
5. **Arginase**^{Q}

Remember: *All the enzymes in the Cytoplasm starts with the letter "A".*

15. a, d. Ornithine Transcarbamoylase, Argininosuccinase
　　　　　　　　　　　　　　(Ref: Harper 30/e page 293)

Enzymes of urea Cycle and its classes

Enzymes name	Class of enzyme it belongs
Carbamoyl-phosphate synthase I	Class 6 (Ligase)
Ornithine carbamoyl transferase	Class 2 (Transferase)
Argininosuccinate synthase	Class 6 (Ligase)
Argininosuccinate lyase (Argininosuccinase)	Class 4 (Lyase)
Arginase	Class 3 (Hydrolase)

16. a. Liver

- Site of urea synthesis in liver mitochondria and cytosol.
- Derived amino acids which has almost exclusive role in urea cycle are Ornithine, Citrulline, Argininosuccinate.
- Four amino acids which has no net loss or gain in urea cycle is ornithine, citrulline, argininosuccinate, arginine.

17. a. Ornithine Transcarbamoylase

Increased ammonia in blood is suggestive of a urea cycle disorder. So answer is an enzyme of urea cycle.

Hyperammonemia Type II (OTC Deficiency)

- Most common Urea Cycle disorder^Q
- Disorder with X-linked partially dominant inheritance (All other Urea Cycle Disorders are Autosomal Recessive)
- Urea cycle disorder with OroticAciduria
- Marked elevations of plasma concentrations of glutamine and alanine with low levels of citrulline and arginine
- Orotate may precipitate in urine as a pink colored gravel orstones.

18. c. Both

The pathways that take place in two compartments are:
- Heme Synthesis
- Urea Cycle
- Gluconeogenesis

19. **a, c, d, e. Glu..., α keto..., Iso..., Fumarase**
(Ref: Harper 30/e page 276, 277)

- Glutamate Dehydrogenase: Oxidative deamination
- Argininosuccinate Synthetase: Urea Cycle
- Alpha Ketoglutarate Dehydrogenase & Isocitrate Dehydrogenase: TCA Cycle
- Fumarase: TCA Cycle

20. **d. ADP** (Ref: Harper 30/e page 291)

Glutamate Dehydrogenase (GDH)

- Liver Glutamate Dehydrogenase (GDH) is allosterically inhibited by ATP, GTP, NADH.
- Liver Glutamate Dehydrogenase (GDH) is allosterically activated by ADP
- Reversible reaction but strongly favour Glutamate formation
- Can use either NAD+ or NADP+.

21. **a. Ammonium and aspartate** (Ref: Harper 30/e page 293)

- First nitrogen by Ammonium ion—by the reaction CPS-I
- Second nitrogen by Aspartate—by the reaction Argininosuccinate Synthetase

22. **a. Scavenges nitrogen**

- Phenyl Butyrate is a nitrogen scavenger by combining with Glutamine

23. **b. Ornithine transcarbamoylase**
(Nelson 20/e Defects in metabolism of amino acids)

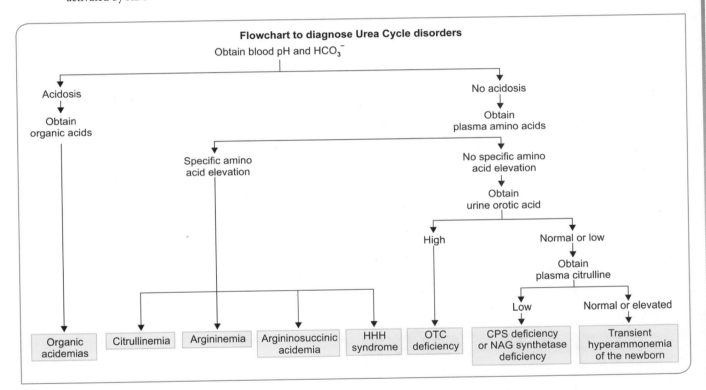

In the given case clue to diagnosis are:

- **High Glutamine:** Usually seen in hyperammonemia. Because glutamine is the transport form of ammonia from brain and most other tissues. So in hyperammonemia Glutamine level is elevated.
- **Increased uracil** in urine can be seen in Ornithine Transcarbamoylase defect becauseas OTC defective, carbamoyl phosphate in mitochondria spills to cytoplasm. Then it enter into Pyrimidine synthesis. Pyrimidine intermediates and pyrimidines can accumulate. Hence Uracil inurine.

24. **b, c, e. First 2 steps in mitochondria, Defect of enzyme of any step can cause deficiency disease, Citrulline is formed by combination of carbamoyl phosphate and L. ornithine**

Urea Cycle

- First two steps in mitochondria, rest three steps in the cytoplasm.
- Ornthine condenses with Carbamoyl Phosphate to form citrulline by the action of the enzyme OTC.
- Disorder is associated with all the steps of urea cycle disorders

Urea Cycle Disorders Due to Enzyme Deficiency	
Disorder	**Enzyme Defective**
Hyperammonemia Type I	Carbamoyl Phosphate Synthetase I (CPS-I)
Hyperammonemia type -II	Ornithine Transcarbamoylase (OTC)

Urea Cycle Disorders Due to Enzyme Deficiency	
Citrullinemia Type I (Classic Citrullinemia)	Argino succinate synthetase
Argininosuccinic aciduria	Argininosuccinate lyase

Contd...

Contd...

Urea Cycle Disorders Due to Enzyme Deficiency	
Hyperargininemia	Arginase
Urea Cycle Disorders due to Transporter Defect	
Citrullinemia Type II	Citrin (Transport Aspartate & Glutamate) Defect
Hyperammonemia Hyperornithinemia Homocitrullinuria (HHH) Syndrome	Ornithine Transporter Defect

25. **b. Multiple Carboxylase deficiency** *(Nelson 20/e Defects in Amino acid metabolism)*

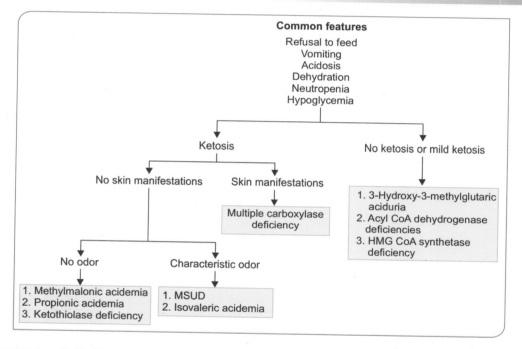

Common features
Refusal to feed
Vomiting
Acidosis
Dehydration
Neutropenia
Hypoglycemia

Ketosis branch:
- No skin manifestations
  - No odor: 1. Methylmalonic acidemia 2. Propionic acidemia 3. Ketothiolase deficiency
  - Characteristic odor: 1. MSUD 2. Isovaleric acidemia
- Skin manifestations: Multiple carboxylase deficiency

No ketosis or mild ketosis:
1. 3-Hydroxy-3-methylglutaric aciduria
2. Acyl CoA dehydrogenase deficiencies
3. HMG CoA synthetase deficiency

26. **c, d, e. On consumption of high amount of protein, excess urea is formed, Occur mainly in cytoplasm, Synthesis of arginin osuccinate consumes energy**

- Nitrogen of urea comes from Ammonia and Aspartate
- ATP is required for CPS-I and Arginino Succinate Synthetase
- Out of the 5 reactions, 3 reactions occur in cytoplasm. So occur mainly in the cytoplasm.
- On consumption of high protein, urea synthesis is increased.

27. **b, d. Rate limiting enzyme is Ornithine transcarbamoylase, Malate is a by product of urea cycle**

- Nitrogen of Urea are contributed by Ammonia and Aspartate.
- 3ATPs are directly required for urea cycle
- Rate limiting step is Carbamoyl Phosphate Synthase-I
- Fumarate is a by product of urea cycle.

28. **d. Amino acid dehydrases**

Some examples of Nonoxidative Deamination^Q NBE pattern:
1. Amino acid Dehydrases for amino acids with hydroxyl group (Serine, Threonine)
2. **Histidase** for histidine
3. Amino acid Desulfhydrases for amino acids with sulfhydryl group, Cysteine & Homocysteine

29. **c. Alanine-leucine**

Conservative mutation means an amino acid replaced by another amino acid of same characteristics.

Glutamic acid—negatively Charged Polar	Glutamine—Uncharged polar
Histidine—Positively charged polar	Glycine—Uncharged, nonpolar
Alanine—Uncharged nonpolar	Leucine –Uncharged nonpolar
Arginine—Positively charged polar	Aspartic acid—Negatively charged polar

1.3 AROMATIC AMINO ACIDS, SIMPLE AMINO ACIDS AND SERINE

- Phenylalanine and Tyrosine
- Synthesis of Tyrosine
- Catabolism of Tyrosine
- Specialised Products from Tyrosine
- Metabolic Disorders Associated with Tyrosine
- Tryptophan
- Catabolism of Tryptophan
- Specialised Products from Tryptophan
- Metabolic Disorders Associated with Tryptophan
- Simple Amino Acids
- Synthesis of Glycine
- Catabolic Pathways of Glycine
- Metabolic Disorders Associated with Glycine
- Alanine
- Serine

PHENYLALANINE AND TYROSINE

Key Points of Chemistry of Phenylalanine
- Aromatic amino acid
- Essential amino acid
- Hydrophobic amino acid
- Partly glycogenic partly ketogenic.

Structure of Phenylalanine

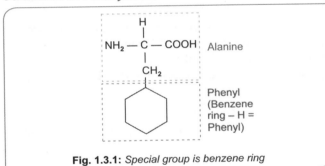

Fig. 1.3.1: *Special group is benzene ring*

Key Points of Chemistry of Tyrosine
- Aromatic amino acid
- Synthesized from Phenylalanine
- Nonessential
- Partly glycogenic and partly ketogenic.

Structure of Tyrosine

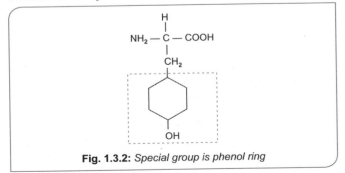

Fig. 1.3.2: *Special group is phenol ring*

Overview of Phenylalanine and Tyrosine Metabolism

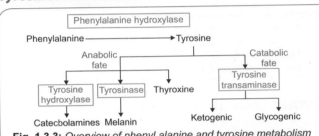

Fig. 1.3.3: *Overview of phenyl alanine and tyrosine metabolism*

CONCEPT BOX

Concept of Phenylalanine to Tyrosine Conversion

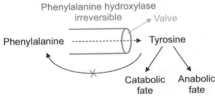

Phenylalanine hydroxylase is like a valve, so tyrosine cannot be converted to phenylalanine; So phenylalanine is essential; Phenylalanine can do its catabolic and anabolic function only by getting converted to tyrosine

SYNTHESIS OF TYROSINE FROM PHENYLALANINE

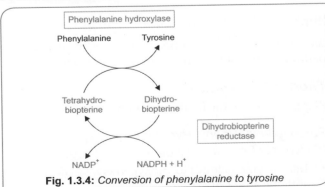

Fig. 1.3.4: *Conversion of phenylalanine to tyrosine*

Phenylalanine Hydroxylase

- Enzyme belongs to mixed function oxidase (Mono-oxygenase)
- Require coenzymes Tetrahydrobiopterin, NADPH
- One mol of oxygen is incorporated
- Irreversible reaction.

Tetrahydrobiopterin

- Resemble folic acid but is not a vitamin.
- Precursor of Tetrahydrobiopterin is **Guanosine Triphosphate (GTP)**.
- Rate limiting enzyme in the pathway is **GTP Cyclohydrolase**.

> **HIGH YIELD POINTS**
>
> **Enzymes with Tetrahydrobiopterin as Coenzyme**
> - Phenylalanine Hydroxylase
> - Tyrosine Hydroxylase
> - Tryptophan Hydroxylase
> - Nitric Oxide Synthase

CATABOLISM OF TYROSINE

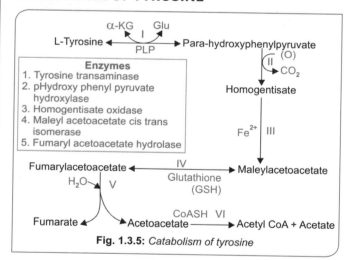

Fig. 1.3.5: *Catabolism of tyrosine*

Important Points in the Catabolism of Tyrosine

As phenylalanine is converted to tyrosine, the degradative pathway is the same for both Phenylalanine and Tyrosine.

Tyrosine Transaminase

PLP is the coenzyme for this reaction.

Para Hydroxy Phenyl Pyruvate Hydroxylase (4 Hydroxy Phenyl Pyruvate Dioxygenase)

- This enzyme belongs to **Dioxygenase**, i.e incorporates both the atoms of oxygen.
- Cofactor for this enzyme is **Copper**.
- **Ascorbic Acid** is also needed for this reaction.

Homogentisate Oxidase

- Belongs to **Dioxygenase**
- Contain **Iron** at the active site.

Maleyl Acetoacetate Cis Trans Isomerase

- Belongs to Isomerase
- Need **Glutathione (GSH)** as cofactor

> **CONCEPT BOX**
>
> Catabolic pathway of almost all amino acids starts with a transamination (except those amino acid which do not undergo transamination). This is because amino group prevents oxidative break down of amino acids.

SPECIALISED PRODUCTS FROM TYROSINE

- Melanin
- Catecholamines
- Thyroxine.

Synthesis of Melanin

- Takes place in the **melanosome of melanocyte present in the deeper layers of epidermis.**
- Under the influence of MSH.
- **Melanin gives pigmentation to the skin and hair.**

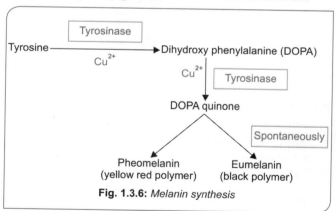

Fig. 1.3.6: *Melanin synthesis*

Tyrosinase

- Rate limiting step
- Mono-oxygenase
- **Copper** is the cofactor for this enzyme
- Single enzyme catalyse two reactions.

Synthesis of Catecholamines

Catecholamines are:
- Dopamine

- Epinephrine
- Norepinephrine.

Catecholamines are compound which contain Catechol nucleus.

Site of synthesis: Chromaffin cells of Adrenal Medulla and Sympathetic Ganglia.
- In adrenal medulla major product is **Epinephrine** (80%).
- In organs innervated by Sympathetic nerves major product is **Norepinephrine** (80%).

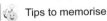

Tips to memorise
Epinephrine is also called Adrenaline, so adrenaline in adrenal medulla.

Conversion of Tyrosine to Epinephrine

Involves 4 sequential steps
- Ring hydroxylation
- Decarboxylation
- Side chain hydroxylation.
- N-Methylation.

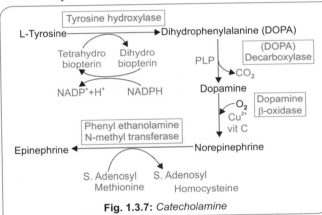

Fig. 1.3.7: *Catecholamine*

Important Points—Catecholamine Synthesis

Tyrosine Hydroxylase
- **Rate limiting Step** in Catecholamine Synthesis
- Similar to Phenylalanine Hydroxylase
- Mono-oxygenase
- Require **tetrahydrobiopterin**

HIGH YIELD POINTS

Tyrosinase vs Tyrosine Hydroxylase
- Both the enzymes convert Tyrosine to DOPA
- Tyrosinase is expressed only in **melanocyte** where DOPA is used to synthesize Melanin.
- Tyrosine Hydroxylase is expressed only in the sites where catecholamines are synthesized, where DOPA utilised for Catecholamine synthesis.
- Tyrosinase is a **mono-oxygenase containing Cu^{2+} in the active site.**
- Tyrosine hydroxylase is **a mono-oxygenase with Tetrahydrobiopterin** as the cofactor.

DOPA Decarboxylase
- Present in all the tissues.
- **PLP** is the coenzyme for this enzyme^Q.

Degradation of Catecholamines
- The half-life of catecholamines are very short, only 2–5 minutes.
- Epinephrine and norepinephrine is catabolized by **Catechol O Methyl Transferase (COMT)^Q then by Monoamine Oxidase (MAO)^Q.**
- The major end product of Epinephrine and norepinephrine is **Vanillyl Mandelic Acid (VMA)^Q.**
- Normal level of VMA excretion in urine is 2–6 mg/24 hour.
- **The major end product of Dopamine is** Homo Vanillic Acid (HVA).

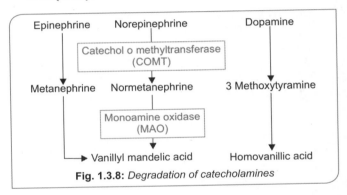

Fig. 1.3.8: *Degradation of catecholamines*

Synthesis of Thyroid Hormones

Thyroid hormones are synthesized on **thyroglobulin**, a large iodinated glycosylated protein.

It contains **115 tyrosine residues.**

Tyrosine residues are iodinated to form Mono-Iodo-Tyrosine (MIT) and Di-iodo Tyrosine (DIT).

Coupling of **MIT and DIT** on the thyroglobulin produce Thyroxine
- IT + DIT → Tri-iodothyronine (T3)
- DIT + DIT → Tetra-iodothyronine (T4) or Thyroxine.

CLINICAL CORRELATIONS (PHENYLALANINE AND TYROSINE METABOLISM)

Metabolic Disorders associated with catabolic pathway of Phenylalanine and Tyrosine
- Phenylketonuria
- Alkaptonuria
- Tyrosinemias
- Hawkinsinuria
- Segawa syndrome.

Disorders associated with melanin synthesis
- Albinism.

Disorders associated with excess Catecholamines
- Pheochromocytoma.

Metabolic Disorders Associated with Catabolic Pathway of Phenylalanine and Tyrosine

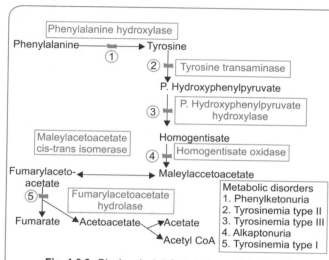

Fig. 1.3.9: *Biochemical defects in the metabolism of aromatic amino acids*

Phenyl Ketonuria

Classic Phenyl Ketonuria (Type I PKU)
- Most Common Metabolic disorder concerned with Amino Acid.

Biochemical defect
- **Phenylalanine hydroxylase** deficiency.
- Phenylalanine could not be converted into tyrosine.
- Phenylalanine in the blood rises.
- Alternate metabolic pathways are opened.

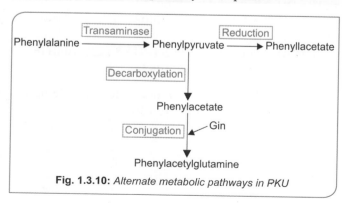

Fig. 1.3.10: *Alternate metabolic pathways in PKU*

The reason for the name Phenyl ketonuria
Excess phenylalanine is metabolized to phenylketones (phenylpyruvate and phenylacetate) that are excreted in the urine, giving rise to the term *phenyl ketonuria* (**PKU**).

Symptom box

Clinical Presentation of Phenyl Ketonuria
- The affected infant is normal at birth.
- Profound intellectual disability
- Vomiting, sometimes severe enough to be misdiagnosed as pyloric stenosis
- Hyperactive with autistic behaviours
- Lighter complexion (due to decreased melanin synthesis) unpleasant mousey or musty odor (due to **phenylacetic acid**.^{Q})

The brain is the main organ affected by hyperphenylalaninemia.

The high blood levels of phenylalanine in PKU saturate the transport system across the blood-brain barrier causing inhibition of the cerebral uptake of other large neutral amino acids such as tyrosine and tryptophan.

Clinical Features of PKU

Lab Diagnosis of PKU^Q

Guthries Test (Bacterial Inhibition Assay of Guthrie)
- Rapid screening Test in the blood sample.
- First method used for this purpose.
- Certain strains of **Bacillus Subtilis** need Phenylalanine as an essential growth factor.
- Bacterial growth is proportional to blood phenylalanine.

Ferric Chloride Test
- **Screening** test in urine sample.
- Identifies phenyl ketones in urine sample.
- Nowadays it has no place in any screening program especially in developed countries.
- These tests have been replaced by more precise and quantitative methods (**fluorometric and tandem mass spectrometry**).

Tandem Mass Spectrometry
- The method of choice is tandem mass spectrometry, which identifies all forms of hyperphenylalaninemia.

Other methods
- Molecular Biology Techniques like **Phenylalanine Hydroxylase specific probes**.
- Quantitative measurement of Blood Phenylalanine. (Blood level >20 mg/dL in PKU)
- Enzyme assay in dry blood spot also done.

Treatment of Classical PKU
- A low-phenylalanine diet
- Administration of large neutral amino acids (LNAAs) is another approach to diet therapy.
- **Sapropterin dihydrochloride (Kuvan),** a synthetic form of BH4 is approved by the FDA.

Rationale for using Large Neutral Amino Acids (LNAAs) as Treatment for PKU

The rationale for use of LNAA is that these molecules compete with phenylalanine for transport across the blood-brain barrier; therefore, large concentrations of other LNAAs in the intestinal lumen and in the blood reduce the uptake of phenylalanine into bloodstream and the brain.

- Preliminary trials with **recombinant phenylalanine ammonia lyase**.

Nonclassical Phenyl Ketonuria

Hyperphenylalaninemia due to Tetrahydrobiopterin defect

Type II and Type III PKU
Due to **Dihydrobiopterin Reductase^Q** defect.

Type IV and Type V PKU
Due to defect in the **enzymes that synthesize Tetrahydrobiopterin**
a. 6-pyruvoyltetrahydropterin synthase (Most Common)
b. Guanosine Triphosphate (GTP) Cyclohydrolase.

Lab Diagnosis of Nonclassical PKU
- Meausurement of **Neopterin and Biopterin**
- Tetrahydrobiopterin (BH4) loading
- Enzyme Assay in dry bloodspots
- Genetic mutation analysis help to confirm the diagnosis.

Segawa Syndrome (Hereditary Progressive Dystonia)

- Tetrahydrobiopterin deficiency due to defect in the enzyme **GTP Cyclohydrolase.**
- But interestingly **no Hyperphenylalaninemia**.
- **Autosomal Dominant** Inheritance.
- Dystonia with diurnal variation.
- Females are affected more than males.

Alkaptonuria

- Autosomal recessive disorder
- 1st inborn error detected
- Belongs to **Garrod's Tetrad (Alkaptonuria, Albinis, Pentosuria, Cystinuria).**

Biochemical Defect
- Homogentisate Oxidase deficiency
- Accumulation of Homogentisic Acid (Homogentisate) which polymerises to form Alkaptone bodies.

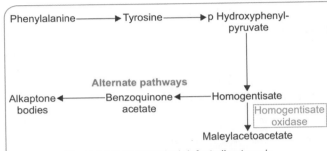

Fig. 1.3.11: *Biochemical defect alkaptonuria*

Symptom box

Clinical Presentation of Alkaptonuria
- Normal Life till 3rd or 4th decade.
- **Urine Darkens on standing** is the only manifestation in children.
- In adults **Ochronosis,** i.e. Alkaptone bodies deposited
- Organ damage is believed to result from accumulation in intervertebral disc, cartilage of nose, pinna, etc leading to pigmentation.
- Arthritis.
- **No intellectual defect^Q**.

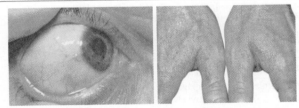

Fig. 1.3.12: *Blackish spots in eyes and dorsum of hands*

Figs. 1.3.13A and B: *Blackish discoloration of urine on standing in alkaptonuria*

Laboratory Diagnosis^Q

- Alkalanisation increase darkening of urine.
- Benedicts test positive in urine because homogentisic acid is reducing agent.
- Ferric chloride test positive.
- Silver Nitrate Test positive.

Treatment

- New Drug is Nitisinone [NTBC] which inhibit para Hydroxyl Phenyl Pyruvate hydroxylase which leads to the decreased accumulation of homogentisic acid
- Symptomatic Treatment.

> **Tips to memorise**
>
> **F**irst Tyrosinemia defective enzyme's name also start with letter **F**, Fumaryl Acetoacetate Hydrolase deficiency.
> **T**ype **T**wo tyrosinemia defective enzyme also start with **T & T**, i.e. Tyrosine Transaminase.

Different Types of Tyrosinemias

Tyrosinemia	Metabolic defect	Important features
Tyrosinemia Type I (Hepatorenal Tyrosinemia, Hereditary Tyrosinemia)	Fumaryl Acetoacetate Hydrolase	Most common Tyrosinemia^Q Resemble Porphyria Cabbage like odour due to metabolite succinyl acetone New treatment is Nitisinone (NTBC)
Tyrosinemia Type II (Oculocutaneous Tyrosinemia, Richner Hanhart Syndrome)	Tyrosine transaminase	Palmoplantar hyperkeratosis
Tyrosinemia Type III (Neonatal Tyrosinemia)	Para Hydroxy Phenyl Pyruvate Hydroxylase (Dioxygenase)	

Hawkinsinuria

- **Defect in Parahydroxyl Phenyl Pyruvate Hydroxylase** (4 Para Hydroxy Phenyl Pyruvate Dioxygenase)
- The mutant enzyme forms an intermediate that reacts with cysteine to form the unusual organic acid **hawkinsin**
- An unusual **swimming pool odor**.

Disorders Associated with excess catecholamines

- Pheochromocytoma
- Paraganglioma
- Pheochromocytoma-associated Syndromes
- Neurofibromatosis type 1(NF1)
- Multiple endocrine neoplasia type 2A and type 2B (MEN 2A, MEN2B).

Pheochromocytoma

Symptomatic catecholamine-producing tumors, in **adrenal and extra-adrenal** retroperitoneal, pelvic, and thoracic sites.

Clinical Presentation

The classic triad of Pheochromocytoma:
- Episodes of palpitations
- Headaches
- Profuse sweating are typical and constitute a classic triad.
- All three symptoms are associated with hypertension.

Biochemical testing of Pheochromocytoma and paraganglioma

Elevated plasma and urinary levels of:
- Catecholamines
- Metanephrines
- Vanillyl Mandelic acid

Biochemical Methods Used for Pheochromocytoma and Paraganglioma Diagnosis		
Diagnostic Method	Sensitivity	Specificity
24 hour Urinary Testing		
1. Vanillylmandelic acid	++	++++
2. Catecholamines	+++	+++
3. Fractionated metanephrines^Q	++++	++
4. Total metanephrines	+++	++++
Plasma Tests		
1. Catecholamines	+++	++
2. Free metanephrines	++++	+++

Albinism

Deficiency of **Tyrosinase**.

Classification of Albinism

I. Generalised Albinism or Oculocutaneous Albinism (OCA)
   - **OCA-1:** *Tyrosinase deficient*
   - **OCA-2:** *Tyrosinase positive (Most common Albinism)*
   - **OCA-3:** *(Rufous, red OCA)*

II. Ocular Albinism
   - Ocular albinism (Nettleship falls type)

III. Localised Albinism.

TRYPTOPHAN

- Aromatic Amino Acid
- Essential Amino Acid

- Special group present is Indole group
- Glycogenic and Ketogenic

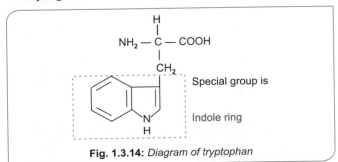

Fig. 1.3.14: *Diagram of tryptophan*

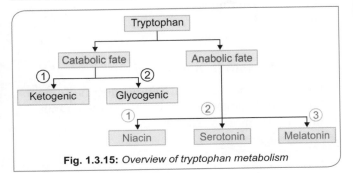

Fig. 1.3.15: *Overview of tryptophan metabolism*

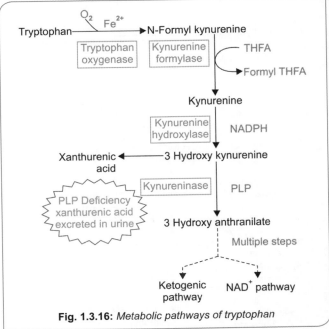

Fig. 1.3.16: *Metabolic pathways of tryptophan*

CATABOLIC PATHWAY OF TRYPTOPHAN (KYNURENINE-ANTHRANILATE PATHWAY) (SEE FIG. 1.3.16)

- Major metabolic fate of Tryptophan is to be oxidized by Tryptophan Pyrrolase

Tryptophan Pyrrolase (Tryptophan Oxygenase)
- Dioxygenase
- **Iron Porphyrin** Metalloprotein (i.e. it is a heme containing Protein)

Kynureninase
- Coenzyme is **PLP**.

Clinical Correlation-Kynurenine Anthranilate Pathway

Pellagra like symptoms in PLP deficiency
- Decreased Kynureninase leads to decreased NAD^+ pathway.
- Hence Niacin deficiency, which lead to Pellagra like symptoms.

Xanthurenate is excreted in urine in PLP deficiency
- This is because PLP deficieny leads to decreased Kynureninase activity.
- Hence Kynurenine accumulate which is converted to Xanthurenate.

SPECIALISED PRODUCTS FROM TRYPTOPHAN

- Niacin (Nicotinic Acid)
- Serotonin
- Melatonin.

Nicotinic Acid Pathway of Tryptophan
- 3% of Trptophan enter this pathway
- 60 mg Tryptophan is converted to 1 mg of Niacin.[Q]
- **Quinolinate Phosphoribosyl Transferase is the rate limiting step** in this pathway.

Serotonin (5 Hydroxy Tryptamine)

Synthesized in the **Argentaffin cells** in the intestine, mast cells, platelets and in the brain.

Functions of Serotonin
- Neurotransmitter in the Brain
- Mood Elevation
- GI Motility
- Temp Regulation
- Potent Vasoconstrictor.

Melatonin

Synthesized in the Pineal gland.

Functions of Melatonin
- Diurnal Variation
- Biological Rhythm
- Sleep Wake Cycle

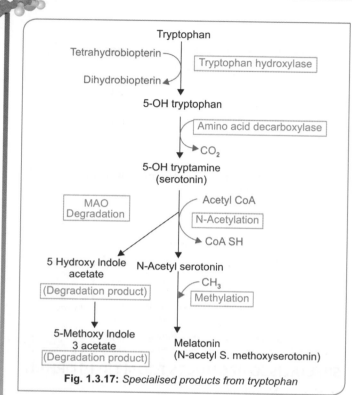

Fig. 1.3.17: *Specialised products from tryptophan*

Important Enzymes in the Synthesis of Serotonin

Tryptophan Hydroxylase
- **Rate limiting step** in the Serotonin and Melatonin Synthesis.
- By this enzyme Tryptophan is converted to 5 Hydroxy Tryptophan.
- **Tetrahydrobiopterin** is the coenzyme for this enzyme
- Monoxygenase.

> **HIGH YIELD POINTS**
>
> **Hydroxylases Dependent on Tetrahydrobiopterin**
> - Phenylalanine Hydroxylase
> - Tyrosine Hydroxylase
> - Tryptophan Hydroxylase
>
> Remember all are Aromatic Amino Acids

Amino Acid Decarboxylase
- 5 OH Trytophan is decarboxylated to 5 OH Tryptamine, or Serotonin.

Catabolism of Serotonin
- **Monoamine Oxidase** is the enzyme.
- **5 Hydroxy Indole Acetic Acid (5HIAA)** is the degradatory product of Serotonin.
- Normal urinary Excretion of 5 HIAA is <5 mg/day.

Synthesis of Melatonin
- N-Acetylation of serotonin followed by N-methylation in the pineal body forms Melatonin.
- Methyl donor is S-Adenosyl Methionine (SAM).

Excretory Product of Tryptophan
Normal excretory product of Tryptophan in Urine is **5 Hydroxy Indole Acetate and Indole 3 Acetate**.

METABOLIC DISORDERS ASSOCIATED WITH TRYPTOPHAN METABOLISM
- Carcinoid syndrome
- Hartnup's disease
- Blue Diaper syndrome.

Carcinoid Tumour (Argentaffinoma)
- Belongs to Gastrointestinal Neuroendocrine Tumours
- Tumour of Argentaffin Cells that secrete Serotonin.
- Increased Synthesis of Serotonin.

> **Symptom box**
>
> **Clinical Symptoms**
> - Most common symtoms are Intermittent Diarrhoea (32–84%) and Flushing (63–75%)
> - Sweating
> - Fluctuating Hypertension
> - Pellagra like Symptoms

Diagnosis
- Serum Serotonin increased
- Urinary 5 HIAA increased
- Neuroendocrine markers used for diagnosis are:
  – Chromogranin A
  – Neuron Specific Enolase
  – Synaptophysin

Comparison of Typical and Atypical Carcinoid	
Typical carcinoid	**Atypical carcinoid**
Caused by **Midgut** Carcinoid	Caused by **foregut** carcinoid
No enzyme defect.	Aromatic amino acid decarboxylase enzyme defect, hence 5 OH Tryptophan cannot be converted to Serotonin
Serotonin level **increased** in blood and platelet.	Serotonin level is **normal in blood**. But in Kidneys 5 OH Tryptophan is converted to Serotonin, hence increased
Urine 5HIAA is **increased**	Urinary 5 HIAA **is only slightly elevated**

Hartnup Disorder

Autosomal recessive condition.
Named after first family in which the disorder identified.

Biochemical Defect
- Defective absorption of **Tryptophan and other Neutral Amino Acid from intestine and renal Tubules**
- The transporter protein for these amino acids (B0AT1) is encoded by the *SLC6A19* gene.

Clinical Features
- Asymptomatic
- **Cutaneous Photosensitivity** is the most common presenting complaint
- Intermittent Ataxia manifested as unsteady wide based gait
- Pellagra likes symptoms

Pellagra Like Symptoms in Hartnup's Disorder

Decreased absorption of Tryptophan from intestine. Decreased availability of Trp for NAD^+ pathway leading to Niacin deficiency.

Laboratory Diagnosis of Hartnup Disease
- **Obermeyer Test** *(Test for indole compounds in the urine) positive.*

Treatment
- Lipid-soluble esters of amino acids and tryptophan ethyl ester.
- Treatment with **nicotinic acid or nicotinamide** (50–300 mg/24 hr) and a high-protein diet.

Blue Diaper Syndrome (Drummond Syndrome)
- Tryptophan is specifically malabsorbed.
- The defect is expressed **only in the intestine** (unlike Hartnup Disease) and not in the kidney.

Blue Staining of the Diaper in Drummond Syndrome

This is due to bacterial breakdown of unabsorbed Tryptophan to Indican and Indigo Blue.

SIMPLE AMINO ACIDS

Glycine
- Simplest Amino Acid
- Nonessential
- Glucogenic
- Optically inactive Amino acid.[Q]

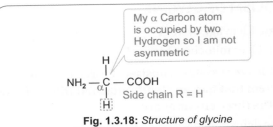

Fig. 1.3.18: *Structure of glycine*

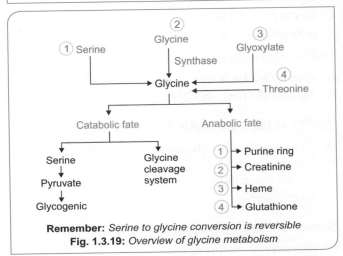

Remember: *Serine to glycine conversion is reversible*
Fig. 1.3.19: *Overview of glycine metabolism*

BIOSYNTHESIS OF GLYCINE[Q]

- Glycine Amino Transferase catalyse the synthesis of Glycine from **Glyoxylate, Glutamate and Alanine**[Q]
- From **Serine** by **Serine Hydroxy Methyl Transferase**. This is a reversible reaction.

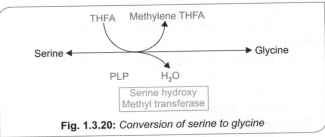

Fig. 1.3.20: *Conversion of serine to glycine*

- By **Glycine Synthase System** in Invertebrates
- From **Threonine** by Threonine Aldolase

 HIGH YIELD POINTS

Serine Hydroxy Methyl Transferase
- Belongs to class II enzyme.
- Freely reversible
- Vitamins involved in conversion of Serine to Glycine are **PLP and Folic Acid.**[Q]
- When Serine is converted to Glycine, **the β carbon** atom of Serine is donated to THFA.

CATABOLISM OF GLYCINE

By Glycine Cleavage system^Q

Present in Liver mitochondria.

Glycine Cleavage system consists of three enzymes and an H Protein that has covalently attached Dihyro lipoyl moiety. The three enzymes are:

1. Glycine dehydrogenase
2. Amino methyl Transferase
3. Dihydrolipomide Dehydrogenase

The overall reaction is

Glycine + THFA + NAD+ ⟶ CO_2 + NH_3 + N_5, N_{10} *Methenyl THFA + NADH + H^+*

SPECIALISED PRODUCTS FROM GLYCINE

- Creatine, Creatine Phosphate and Creatinine
- Heme
- Purine nucleotide
- Glutathione.

Other Functions of Glycine

Glycine as conjugating agent

- Conjugation of Bile acid (Glycocholic Acid, Glycochenodeoxy Cholic Acid)
- Conjugation of Benzoic Acid

Glycine + Benzoyl CoA ⟶ *Benzoyl Glycine*

Glycine as Neurotransmitter

Both excitatory and inhibitory Neurotransmitter

Glycine is the recurring amino acid present in the collagen

- Every **third amino acid** in Collagen is **Glycine**.

Creatinine

- Synthesized from 3 Amino Acids **(Glycine, Arginine & Methionine^Q)**

Steps of Synthesis of creatinine

Step I Glycine Arginine Amide Transferase

- First step in the Kidney.
- **Guanidino** group of Arginine is transferred to Glycine to form Guanidino Acetic Acid.

Step II Guanidino Acetate Methyl Transferase

- Second step in the Liver
- Creatine is formed
- S Adenosyl Methionine is the methyl donor.

Step III Creatine Kinase

- Third step in the Muscle
- Creatine Phosphate is formed.

Step IV

- Occur spontaneously
- Creatinine is formed.

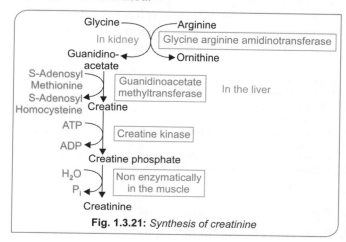

Fig. 1.3.21: *Synthesis of creatinine*

Heme

- Succinyl CoA + Glycine ⟶ Heme
- In the liver and erythroid precursor cells.

Formation of Purine Ring

- C4, C5, N7 of Purine ring is contributed by Glycine.

Remember Glycine do not contribute to Pyrimidine Ring.^{Q}

Glutathione^Q

- It is otherwise called **Gamma Glutamyl Cysteinyl Glycine**
- **Tripeptide**^{Q} from three Amino Acids-Glutamic Acid, Cysteine and Glycine.
- Pseudopeptide.^{Q}
- Abbrevated as **GSH**.
- Business part of Glutathione is **sulfhydryl group of Cysteine.**

Functions of Glutathione^Q

1. **Meister's Cycle or Gamma Glutamyl Cycle**
   - Absorption of neutral Amino acids in the Intestine, Kidney tubules and brain.
   - 3 mols of ATP utilised for the transport of Amino acid.
2. **Free Radical Scavenging**
   - Especially in the RBC, hence responsible for RBC membrane integrity.^{Q}

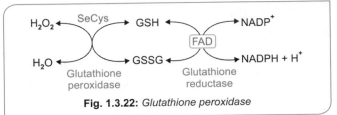

Fig. 1.3.22: *Glutathione peroxidase*

3. **Reduction of Methemoglobin**[Q]
   – Keep the iron in the Heme in the Ferrous state by Reduced Glutathione.
4. **Conjugation reactions in Phase II Xenobiotic reactions**
   – Glutathione S Transferase is the enzyme.
   – Acts as coenzyme for some reactions.

> **HIGH YIELD POINTS**
> - **Sarcosine** is N- Methyl Glycine
> - **Betaine** is Trimethyl Glycine

METABOLIC DISORDERS ASSOCIATED WITH GLYCINE

- Primary Hyperoxaluria Type I
- Primary Hyperoxaluria Type II
- Non-Ketotic Hyperglycinemia

Primary Hyperoxaluria Type I

- The most common form of primary hyperoxaluria.
- It is due to a deficiency of the peroxisomal enzyme [**alanine-glyoxylate aminotransferase** expressed only in the liver peroxisomes and requires **pyridoxine (vitamin B$_6$)** as its cofactor]
- Protein targeting defect.

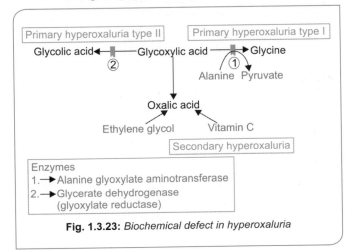

Fig. 1.3.23: *Biochemical defect in hyperoxaluria*

Primary Hyperoxaluria Type II (Glyceric Aciduria)

- Due to a deficiency of **D-glyceratedehydrogenase** (glyoxylate reductase enzyme complex).

Secondary Hyperoxaluria

- **Pyridoxine deficiency** (cofactor for alanine-glyoxylate aminotransferase)
- After ingestion of ethylene glycol
- High doses of vitamin C
- After administration of the anesthetic agent methoxyflurane (which oxidizes directly to oxalic acid)
- In patients with inflammatory bowel disease or extensive resection of the bowel (*enteric hyperoxaluria*).

Nonketotic Hyperglycinemia

- Due to a defect in the **Glycine cleavage system**.

ALANINE

Key Points of Chemistry

- Simple Amino Acid
- Nonessential Amino Acid
- Principal Glycogenic Amino Acid
- Transports amino group from Skeletal Muscle.[Q]
- Participate in Glucose-Alanine Cycle (Cahill Cycle).

Biosynthesis of Alanine

- From Pyruvate by Transamination

SERINE

Key Points of Chemistry

- Hydroxyl Group containing Amino Acid
- Glucogenic Amino acid
- Nonessential Amino Acid
- Polar Amino Acid.

Biosynthesis of Serine

- From Glycine by Serine Hydroxymethyl Transferase. PLP is a coenzyme in this reaction.
- From Glycolytic intermediate 3 Phosphoglycerate.

Metabolic Functions of Serine

1. **Primary donor** of one carbon group.

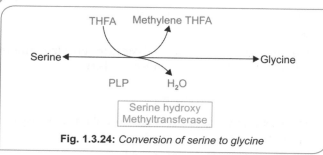

Fig. 1.3.24: *Conversion of serine to glycine*

2. Serine is used for formation of **Cysteine**

   Serine + Homocysteine ----------> Cysteine + Homoserine

3. For **Phospholipid synthesis Phosphatidyl Serine**
4. Serine analogs as drugs Cycloserine-Antituberculous drug Azaserine-Anticancer drug
5. Serine is used for:
   - **Ethanolamine** Synthesis
   - **Choline (Trimethyl Ethanolamine)** synthesis
   - **Betaine (Trimethyl Glycine) synthesis**

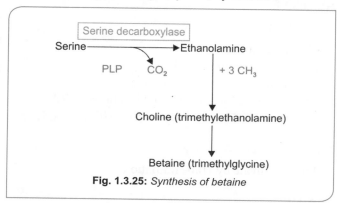

Fig. 1.3.25: *Synthesis of betaine*

6. Serine is the precursor of **Selenocysteine**^{Q}
7. Serine used for Glycoprotein synthesis

   O Glycosylation takes place at **Serine and Threonine** residues.
8. Most common sites for phosphorylation are **Serine and Threonine**.
9. Serine and Palmitoyl CoA are the starting material for the synthesis of **Sphingosine**, thereby Ceramide.

 HIGH YIELD POINTS

All Sphingolipids are formed from Ceramide

Vitamins required for conversion of serine to Glycine are Folic Acid^Q and Pyridoxine.^{Q}

Quick Revision

- Tyrosine can be synthesised from **Phenylalanine**
- Compounds derived from Tyrosine are **Melanin, Catecholamines, and Thyroxine.**
- Tetrahydrobiopterin is derived from **GTP**.
- Classic PKU is due to deficiency of **Phenylalanine Hydroxylase.**
- Alkaptonuria is due to deficiency of **Homogentisate oxidase.**
- The major end product of Epinephrine and norepinephrine is **Vanillyl Mandeilic acid**
- Most common Tyrosinosis is **Type I** due to deficiency of **Fumaryl Acetoacetate Hydrolase.**
- Compounds derived from Tryptophan are **Niacin, Serotonin, Melatonin.**
- Only amino acid that can be converted to a vitamin is **Tryptophan.**
- **Tryptophan Pyrrolase** is a heme containing enzyme.
- Rate limiting enzyme of niacin synthesis is **quinolinate phosphoribosyl transferase.**
- Serotonin is **5 hydroxytryptamine**.
- **Nitisinone** is used in the treatment of Alkaptonuria and Hartnups disease.
- **5HIAA** is derived from Serotonin
- Compounds derived from Glycine are **Glutathione, creatinine, heme and purine**
- Cysteine is derived from **Methionine and Serine**

Check List for Revision

Phenylalanine and Tyrosine
- Structure and key points of chemistry
- Enzymes and disorders associated with each step of catabolism of Phenylalanine and tyrosine.
- Compounds derived from Tyrosine.
- Phenylketonuria and Alkaptonuria learn all points.
- Pheochromocytoma.

Tryptophan
- Structure and key points of its chemistry.
- Catabolism of Tryptophan learn the clinical correlates.
- Name of the compounds derived from Tryptophan.
- Name of the metabolic disorders associated with Tryptophan with its metabolic defect.
- Learn all the details of Hartnup's disease and Carcinoid Syndrome.

Glycine
- Structure and key points of chemistry
- Glycine synthesis and catabolism only relevant points
- Compounds derived from Glycine
- Glutathione
- Hyperoxaluria.

Alanine
- Learn only key points.

Serine
- Glycine to Serine conversion
- Metabolic functions of serine all bold letter points.

Review Questions MCQ

Aromatic Amino Acids

1. **Norepinephrine to epinephrine conversion requires which amino acid?** *(Recent Question Jan 2019)*
   a. Phenylalanine
   b. Methionine
   c. Lysine
   d. Cysteine

2. **Serotonin is:** *(Recent Question Jan 2017)*
   a. 5 Hydroxy tryptophan
   b. 5 Hydroxy tryptamine
   c. 5 Carboxy tryptamine
   d. 5 Carboxy tryptophan

3. **Tyrosinosis most common cause is:** *(Recent Question Jan 2017)*
   a. Fumarylacetoacetate hydrolase
   b. Tyrosine transaminase
   c. Parahydroxy phenyl pyruvate hydroxylase
   d. Homogentisate oxidase

4. **5 HIAA in urine is due to:** *(Recent Question Nov 2017)*
   a. Pheochromocytoma
   b. Carcinoid syndrome
   c. Phenyl ketonuria
   d. Alkaptonuria

5. **VMA is excreted in urine in which condition?** *(Recent Question Jan 2017)*
   a. Pheochromocytoma
   b. Carcinoid syndrome
   c. Phenyl ketonuria
   d. Alkaptonuria

6. **PKU true statement is:** *(Recent Question Jan 2017)*
   a. Neurological symptoms are due to excess phenyl alanine in blood
   b. Blood phenylalanine more than 20 mg/dl is suggesting bad prognosis
   c. Urinary phenylalanine is used for screening
   d. No neurological symptoms

7. **Hyperphenylalaninemia occurs due to:** *(PGI Nov 2016)*
   a. Phenylalanine hydroxylase deficiency
   b. Phenylalanine hydroxylase overactivity
   c. Dihydrobiopterin reductase deficiency
   d. Tyrosine hydroxylase deficiency
   e. Defect in dihydrobiopterin biosynthesis

8. **Melanin derived from:** *(Recent Question)*
   a. Tryptophan
   b. Tyrosine
   c. Methionine
   d. Alanine

9. **Melatonin derived from:** *(Recent Question)*
   a. Tryptophan
   b. Tyrosine
   c. Methionine
   d. Alanine

10. **Treatment of tyrosinemia type 1 is:** *(Recent Question)*
    a. NTBC
    b. Vitamin B6
    c. Large neutral amino acids
    d. Tyrosine restricted diet

11. **Which is elevated in PLP deficiency?** *(Recent Question)*
    a. FIGLU
    b. Xanthurenic acid
    c. Methyl malonic acid
    d. VMA

12. **Dopamine is synthesized from:** *(Recent Question)*
    a. Tryptophan
    b. Threonine
    c. Tyrosine
    d. Lysine

13. **In Phenyl ketonuria the main aim of first line therapy is:** *(AIIMS Nov 2010)*
    a. Replacement of the defective enzyme
    b. Replacement of the deficient product
    c. Limiting the substrate for deficient enzyme
    d. Giving the missing amino acid by diet

14. **A 40-year-old woman presents with progressive palmoplantar pigmentation. X-ray spine shows calcification of IV disc. On adding benedicts reagent to urine, it gives greenish brown precipitate and blue-black supernatant fluid. What is the diagnosis?** *(AIIMS Nov 2008)*
    a. Phenyl ketonuria
    b. Alkaptonuria
    c. Tyrosinemia type 2
    d. Argininosuccinic aciduria

15. **Dopamine hydroxylase catalyse:**
    a. Dopamine → norepinephrine
    b. Dopa to dopamine
    c. Norepinephrine to epinephrine
    d. Tyrosine to Dopa

16. **Type I Tyrosinemia is caused by:** *(Recent Question)*
    a. Tyrosine transaminase
    b. Fumarylacetoacetate hydrolase
    c. 4 Hydroxy phenyl pyruvate hydroxylase
    d. Maleyl acetoacetate isomerase

17. **Terminal product of Phenylalanine metabolism is:** *(PGI May 2014)*
    a. Fumarate
    b. Acetyl CoA
    c. Oxaloacetate

18. **Enzyme deficiency in albinism is:** *(Recent Question)*
    a. Tyrosinase
    b. Tyrosine hydroxylase
    c. Phenylalanine hydroxylase
    d. Homogentisate oxidase

19. **Mousy body odour is due to:** *(JIPMER May 2015)*
    a. Phenylalanine
    b. Phenyl acetate
    c. Phenyl butazone
    d. Phenyl acetyl Glutamine

20. **The amino acid that can be converted into a vitamin:** *(Recent Question)*
    a. Glycine
    b. Tryptophan
    c. Phenylalanine
    d. Lysine

21. **Which of the following amino acids is involved in the synthesis of thyroxine?**
    a. Glycine
    b. Methionine
    c. Threonine
    d. Tyrosine

22. **Tyrosinemics are more susceptible to develop:** *(AIIMS Feb 97)*
    a. Adenocarcinoma colon
    b. Melanoma
    c. Retinoblastoma
    d. Hepatic carcinoma

23. **Metabolites of tryptophan can give rise to:** *(PGI June 02)*
    a. Diarrhoea
    b. Vasoconstriction
    c. Flushing
    d. Can predispose to albinism
    e. Phenylketonuria

24. **Correct combination of Urine odour in various metabolic disorder:** *(PGI Nov 2013)*
    a. Phenyl ketonuria: Mousy body odour
    b. Tyrosinemia: Rotten cabbage
    c. Hawkinsinuria: Potato smell
    d. Maple syrup disease: Rotten tomato
    e. Alkaptonuria: Rotten egg

25. **Which of the following is true regarding Phenyl Ketonuria?** *(PGI Nov 2014)*
    a. Dietary phenylalanine restriction is used as a treatment
    b. Occur due to deficiency of phenylalanine hydroxylase
    c. Occur due to increase activity of phenylalanine hydroxylase
    d. Diet should contain high phenylalanine containing food items
    e. Tyrosine should be supplied in the diet simple amino acids

26. **Hyperoxaluria associated with which amino acid?** *(Recent Question)*
    a. Glycine
    b. Serine
    c. Threonine
    d. Lysine

27. **Which of the following is true about glycine?**
    a. Glycine is an essential amino acid
    b. Sulphur containing at 4th position
    c. Has a guanidine group
    d. Optically inactive

28. **Which of the following would not act as source of glycine by transamination?** *(Recent Question)*
    a. Alanine
    b. Aspartate
    c. Glutamate
    d. Glyoxylate

29. **Glycine cleavage system in liver mitochondria is associated with which enzyme?** *(NBE Pattern Question)*
    a. Glycine dehydrogenase
    b. Glycine transaminase
    c. Glycine decarboxylase
    d. Glycine dehydratase

30. **Guanido acetic acid is formed in..... from......** *(JIPMER 2000, DNB 98)*
    a. Kidney; Arginine + glycine
    b. Liver; Methionine + glycine
    c. Liver; Cysteine + arginine
    d. Muscle; Citrulline + aspartate

31. **Conversion of glycine to serine requires:** *(PGI Dec 02)*
    a. Folic acid
    b. Thiamine
    c. Vit. C
    d. Fe^{2+}
    e. Pyridoxal phosphate

32. **N Methyl Glycine is known as:** *(Recent Question)*
    a. Ergothionine
    b. Sarcosine
    c. Carnosine
    d. Betaine

33. **What is the metabolic defect in Primary Oxaluria Type II?** *(Recent Question)*
    a. Glycine cleavage system
    b. Alanine glyoxalate amino transferase
    c. DGlycerate dehydrogenase
    d. Excess vitamin C

34. **All are true about glutathione except:** *(AIIMS Nov 2008)*
    a. It is a tripeptide
    b. It converts hemoglobin to methemoglobin.
    c. It conjugates xenobiotics
    d. It is co-factor of various enzymes

Answers to Review Questions

Aromatic Amino Acids

1. **b. Methionine** *(Ref: Harper 31/e Page 302)*

   In the conversion of Norepinephrine to epinephrine a methylation reaction occur. This methyl group is donated by S Adenosyl Methionine a derivative of Methionine.

2. **b. 5Hydroxytryptamine** *(Ref: Harper 30/e)*

   5 Hydroxy Tryptophan is decarboxylated by Aromatic amino acid decarboxylase to form 5 Hydroxytryptamine or Serotonin. Coenzyme is PLP.

3. **a. Fumaryl Acetoacetate Hydrolase** *(Ref: Nelson 20/e)*

   Tyrosinosis or Tyrosinemia most common type is Type I Tyrosinemia. The enzyme defect is Fumaryl Acetoacetate Hydrolase deficiency. Incidence is 1 in 1846 live birth. Type II and III Tyrosinemia is rare types.

4. **b. Carcinoid syndrome** *(Ref: Harper 30/e)*
   - In Pheochromocytoma VMA is excreted in excess in urine.
   - In Carcinoid syndrome 5 HIAA is excreted in excess in urine.
   - In PKU Phenyl Ketones like Phenyl acetate and Phenyl Pyruvate is excreted in urine.
   - In Alkaptonuria Homogentisate is excreted in urine.

5. **a. Pheochromocytoma** *(Ref: Harper 30/e)*
   - In Pheochromocytoma VMA is excreted in excess in urine.
   - In Carcinoid syndrome 5 HIAA is excreted in excess in urine.
   - In PKU Phenyl Ketones like Pheny Acetate and Phenyl Pyruvate is excreted inurine.
   - In Alkaptonuria Homogentisate is excreted inurine.

6. **b>>a. Blood phenylalanine more than 20 mg/dl....> Neurological symptoms are ...**
   - High plasma concentration > 20 mg/dl is an indicator of severity of enzyme deficiency.
   - Neurological symptoms are not directly due to excess phenyl alanine in blood but due to excess phenylalanine in brain reducing Tyrosine and Tryptophan

   Few lines given in Nelsons Textbook of Paediatrics is given here

   Severity of PKU

   The severity of hyperphenylalaninemia depends on the degree of enzyme deficiency and may vary from very high plasma concentrations (>20 mg/dL).

   Neurological deficits in PKU

   The **brain** is the main organ affected by hyperphenylalaninemia. The CNS damage in affected patients is caused by the elevated concentration of phenylalanine in brain tissue. The high blood levels of phenylalanine in PKU saturate the transport system across the blood-brain barrier causing inhibition of the cerebral uptake of other large neutral amino acids such as **tyrosine and tryptophan.**

   Screening for PKU

   A few drops **of blood**, which are placed on a filter paper and mailed to a central laboratory, are used for assay. The bacterial inhibition assay of Guthrie, which was the 1st method used for this purpose, has been replaced by more precise and quantitative methods (fluorometric and tandem mass spectrometry). The method of choice is tandem mass spectrometry (MS/MS), which identifies all forms of hyperphenyl alaninemia with a low false-positive rate, and excellent accuracy and precision.

7. **a, c, e. Phenyl alanine Hydro..., Dihydrobiopterin...., Defect in...** *(Ref: Nelson 20/e)* Causes of PKU

   Classic PKU is Phenylalanine Hdroxylase defect Nonclassic PKU are due to Cofactor BH4 defect
   - Causes are Dihydrobiopterin reductase deficiency
   - Defective synthesis of Dihydrobiopterin

8. **b. Tyrosine**

9. **a. Tryptophan**

10. **a. NTBC**

11. **b. Xanthurenic acid**
    - Urinary metabolite in Vitamin B6 deficiency: Xanthurenic acid
    - Urinary Metabolite in Folic Acid Deficiency: Formimino Glutamic acid, Homocysteine
    - Urinary metabolite in Vitamin B12 deficiency: Homocysteine, Methyl Malonic Acid

12. **c. Tyrosine**

    Metabolic products formed from Tyrosine are:
    - Melanin
    - Thyroxine
    - Catecholamines (Dopamine, Epinephrine, Norepinephrine)

    Albinism is due to Defect in Tyrosinase

13. **c. Limiting substrate for the deficient enzyme**
    (Ref: Nelson 20/e Defects in metabolism of Amino acids)
    - The primary goal of therapy is to reduce phenylalanine levels in the plasma and brain.

    Treatment of Classical PKU
    - A low-phenylalanine diet
    - Administration of large neutral amino acids (LNAAs) is another approach to diet therapy.

- Sapropterin dihydrochloride (Kuvan), a synthetic form of BH4, which acts as a cofactor in patients with residual PAH activity, is approved by the FDA to reduce phenylalanine levels in PKU.

Preliminary trials with recombinant phenylalanine ammonia lyase have been encouraging and demonstrated reduced blood levels of phenylalanine during treatment

14. b. Alkaptonuria
(Ref: Nelson 20/e Defects in metabolism of Amino acids)

Alkaptonuria
- Autosomal Recessive disorder is due to a deficiency of Homogentisic Acid Oxidase
- 1st inborn error detected
- Belongs to Garrod's Tetrad [Alkaptonuria, Albinism, Pentosuria, Cystinuria]

Biochemical Defect

Homogentisate Oxidase deficiency leads to accumulation of Homogentisic Acid (Homogentisate) which polymerises to form Alkaptone bodies.

Clinical Presentation
- Normal Life till 3rd or 4th decade.
- Urine Darkens on standing is the only manifestation in children.
- In adults Ochronosis-Alkaptone Bodies in Intervertebral Disc, cartilage of nose, pinna. etc.

Laboratory Diagnosis
- Alkalanisation increase darkening of urine.
- Benedicts test positive in urine because homogentisic acid is reducing agent.
- Ferric Chloride test positive
- Silver Nitrate Test positive.
- No Mental Retardation

Treatment
- New Drug is Nitisinone [NTBC] which inhibit para Hydroxyl Phenyl Pyruvate hydroxylase which prevent the accumulation of homogentisic acid.
- Symptomatic Treatment.

15. a. Dopamine → Norepinephrine
(Ref: Harper 30/e page 320)

Conversion of Tyrosine to Epinephrine involves 4 sequential steps
1. Ring Hydroxylation
2. Decarboxylation
3. Side chain hydroxylation.
4. N-Methylation

16. b. Fumarylaceto Acetate Hydrolase
(Ref: Nelson 20/e Defects in metabolism of Amino acids)

Amino acidurias and enzyme defect

Classic Phenyl Ketonuria	Phenylalanine Hydroxylase
Alkaptonuria	Homogentisate Oxidase
Tyrosinemia Type I (Most common Tyrosinemia)	Fumaryl Aceto Acetate Hydrolase
Tyrosinemia Type II	Tyrosine Transaminase
Tyrosinemia Type III	Para Hydroxy Phenyl Pyruvate hydroxylase/Para hydroxyl Phenyl Pyruvate Dioxygenase
Hawkinsinuria	Para Hydroxy Phenyl Pyruvate hydroxylase/Para hydroxyl Phenyl Pyruvate Dioxygenase is mutant, so that it catalyse only partial reaction.
Segawa Syndrome	GTP Cyclohydrolase
Albinism	Tyrosinase

17. a, b. Fum…, AcetylCoA
(Ref: Harper 30/e page 304)

- Terminal end products of Phenylalanine and Tyrosine metabolism is Fumarate, acetate and Acetyl CoA

Amino Acid	Terminal end products
Asparagine, Aspartate	Oxaloacetate
Glutamine, Glutamate	α Ketoglutarate
Proline	α Ketoglutarate
Arginine, Ornithine	α Ketoglutarate
Histidine	α Ketoglutarate
Glycine, Serine	CO_2, NH_3, N5N10 Methylene THFA or Pyruvate
Alanine	Pyruvate
Threonine	Glycine, Acetaldehyde
Methionine	Cysteine, Succinyl CoA
Cysteine	Pyruvate, 3 Mercaptolactate
Phenylalanine, Tyrosine	Fumarate, Acetyl CoA, Acetate
Tryptophan	Acetyl CoA
Leucine	Acetoacetate, Acetyl - CoA
Isoleucine	Acetyl CoA, Succinyl CoA
Valine	Succinyl CoA, β Amino isobutyrate

18. a. Tyrosinase
(Ref: Nelson 20/e Defects in metabolism of Amino acids)

Aminoaciduria	Enzyme deficiency
Albinism	Tyrosinase
Phenyl Ketonuria	Phenylalanine hydroxylase
Alkaptonuria	Homogentisate Oxidase
Homocystinuria	Cystathionine Beta Synthase
Maple syrup Urine Disease	Branched Chain ketoacid Dehydrogenase

19. **b. Phenyl acetate**
    (Ref: Nelson 20/e Defects in metabolism of Amino acids)

    Clinical Manifestations of PKU
    - The affected infant is normal at birth. Profound mental retardation develops gradually if the infant remains untreated. Cognitive delay may not be evident for the 1st few months.
    - Vomiting, sometimes severe enough to be misdiagnosed as pyloric stenosis, may be an early symptom.
    - The infants are lighter in their complexion than unaffected siblings.
    - Some may have a seborrheic or eczematoid rash, which is usually mild and disappears as the child grows older.
    - These children have an unpleasant odor of phenyl acetic acid, which has been described as musty or mousey.
    - Neurologic signs include seizures (≈25%), spasticity, hyperreflexia, and tremors; more than 50% have electroencephalographic abnormalities.
    - Microcephaly, prominent maxillae with widely spaced teeth, enamel hypoplasia, and growth retardation are other common findings in untreated children.

20. **b. Tryptophan**
    - Tryptophan can be converted to Niacin.
    - The rate limiting enzyme in Niacin synthesis is Quinolinate Phosphoribosyl Transferase (QPRTase)

    Specialised products of Tryptophan
    - Serotonin (5 Hydroxytryptamine)
    - Melatonin
    - Niacin

21. **d. Tyrosine**

    Specialised products of Tyrosine are
    Melanin, Thyroxine, Catecholamines (Dopamine, Epinephrine, Norepinephrine).

22. **d. Hepatic Carcinoma**
    (Ref: Nelson 20/e genetic disorders of metabolism)

    Tyrosinemia Type I (Tyrosinosis, Hereditary Tyrosinemia, Hepatorenal Tyrosinemia)

    Clinical Manifestations of Tyrosinemia Type I
    - Untreated, the affected infant appears normal at birth and typically presents between 2 and 6 months of age
    - An acute **hepatic crisis** commonly heralds the onset of the disease and is usually precipitated by an intercurrent illness that produces a catabolic state. Fever, irritability, vomiting, hemorrhage, hepatomegaly, jaundice, elevated levels of serum transaminases, and hypoglycemia are common. An odor resembling boiled cabbage may be present, due to increased methionine metabolites. Cirrhosis and **eventually hepatocellular carcinoma** occur with increasing age. Carcinoma is unusual before 2 years of age.
    - Episodes of acute **peripheral neuropathy** resembling acute porphyria occur in ≈ 40% of affected children.
    - **Renal involvement** is manifested as a Fanconi-like syndrome with normal anion gap metabolic acidosis.
    - Hypertrophic cardiomyopathy and hyperinsulinism are seen in some infants.

23. **a, b, c. Diarrhoea, Vasoconstriction, Flushing Actions of Serotonin (Metabolite of Tryptophan) are**
    - Neurotransmitter in the Brain
    - Mood Elevation
    - GI Motility
    - Temp Regulation
    - Vasoconstriction but in excess can cause Vasodilation in pathologic conditions like Carcinoid Syndrome

24. **a,b. Phenyl Ketonuria—Mousybodyodour, Tyrosinemia-Rotten cabbage**

    Peculiar odours in different Amino acidurias

Inborn Error of Metabolism	Urine Odor
Glutaric acidemia (type II)	Sweaty feet, acrid
Hawkinsinuria	Swimming Pool
Isovaleric Acidemia	Sweaty feet, Acrid
3-Hydroxy-3-methylglutaric aciduria	Cat urine
Maple syrup urine disease	Maple syrup
Hypermethioninemia	Boiled cabbage
Multiple carboxylase deficiency	Tomcat urine
Oasthouse urine disease	Hops-like
Phenyl ketonuria	Mousey or musty
Trimethylaminuria	Rotting fish
Tyrosinemia	**Boiled cabbage**, rancid butter

25. **a, b, e. Dietary...., Occur...., Tyrosine.**
    - Phenyl Ketonuria is due to deficiency of Phenylalanine Hydroxylase
    - Dietary restriction of Phenylalanine with supplementation of Tyrosine is needed as Tyrosine is a nonessential amino acid synthesized from Phenylalanine by the action of Phenylalanine Hydroxylase.

Simple Amino Acids

26. **b. Glycine**

27. **d. Optically inactive** *(Ref: Harper 30/e page 19)*
    - Glycine is the only optically inactive amino acid.
    - Sulphur containing amino acids are Cysteine and Methionine
    - Guanidinium group is present in Arginine.
    - Glycine is a nonessential amino acid

28. b. Aspartate (Ref: Harper 30/e page 283)

Biosynthesis of Glycine
- Glycine Amino Transferase catalyse the synthesis of Glycine from **Glyoxylate, Glutamate and Alanine**^{Q}
- From **Serine** by Serine Hydroxy Methyl Transferase. This is a reversible reaction

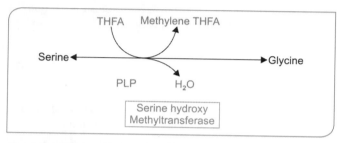

- By **Glycine Synthase System** in Invertebrates
- From **Threonine** by Threonine Aldolase

29. a. Glycine Dehydrogenase (Ref: Harper 30/e page 302)

Glycine Cleavage system consists of three enzymes and an H Protein that has covalently attached Dihyrolipoyl moiety. The three enzymes are:
1. Glycine Dehydrogenase
2. Amino methyl Transferase
3. Dihydrolipomide Dehydrogenase

30. a. Kidney; Arginine + Glycine

Steps of synthesis of Creatinine

Step I Glycine Arginine Amido Transferase
- First step in the Kidney.
- **Guanidino** group of Arginine is transferred to Glycine to form Guanidino Acetic Acid.

Step II Guanidino Acetate Methyl Transferase
- Second step in the Liver
- Creatine is formed
- S Adenosyl Methionine is the methyl donor

Step III Creatine Kinase
- Third step in the Muscle
- Creatine Phosphate is formed

Step IV
- Occur spontaneously
- Creatinine is formed.

31. a, e. Folic acid, Pyridoxal Phosphate
- Glycine is converted to Serine by Serine Hydroxy methyl Transferase
- Coenzymes required are Folic acid, Pyridoxal Phosphate

32. b. Sarcosine

Sarcosine	N Methyl Glycine
Betaine	Trimethyl Glycine
Choline	Trimethyl Ethanolamine
Ethanolamine	Serine on decarboxylation
Ergothionine	Derivative of Histidine
Beta mercaptoethanolamine	Cysteine on decarboxylation
Carnosine	Beta Alanyl Histidine
Anserine	Carnosine on Methylation
Homocarnosine	GABA + Histidine
Serotonin	5 Hydroxytryptamine

33. c. D Glycerate Dehydrogenase

Primary Hyperoxaluria Type I
- The most common form of Primary hyperoxaluria.
- It is due to a deficiency of the peroxisomal enzyme alanine-glyoxylate amino transferase (expressed only in the liver peroxisomes and requires pyridoxine (vitamin B6) as its cofactor)
- Protein targeting defect.

Primary Hyperoxaluria Type II (GlycericAciduria)
- Due to a deficiency of Dglycerate dehydrogenase (glyoxylate reductase enzyme complex)

Secondary Hyperoxaluria
- Pyridoxine deficiency (cofactor for alanine glyoxylate amino transferase)
- After ingestion of ethylene glycol
- High doses of vitamin C,
- After administration of the anesthetic agent methoxyflurane (which oxidizes directly to oxalic acid)
- In patients with inflammatory bowel disease or extensive resection of the bowel (enteric hyperoxaluria).

Nonketotic Hyper Glycinemia
a. Due to a defect in the Glycine Cleavage System

34. b. It converts hemoglobin to methemoglobin (Ref: Harper 30/e page 23)

Functions of glutathione
- Glutathione is a tripeptide
- Free radical scavenging
- Transport of Amino acid across cell membrane
- Keep iron in the ferrous state, so prevent methHb formation.
- Act as coenzyme for certain enzymes.
- Phase II Xenobiotic reaction in Conjugation

1.4 SULPHUR CONTAINING AMINO ACIDS

- Methionine and Cysteine
- Metabolism of Sulphur Containing Amino Acids
- Metabolic Disorders Associated with Sulphur Containing Amino Acids

METHIONINE
Key Points of Chemistry

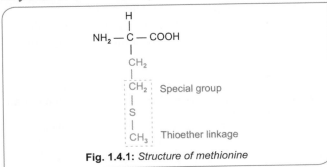

Fig. 1.4.1: *Structure of methionine*

- Sulphur containing amino acid
- Essential amino acid
- Glycogenic amino acid
- Thioether linkage present.

CYSTEINE
Key Points of Chemistry

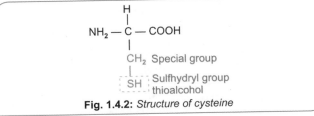

Fig. 1.4.2: *Structure of cysteine*

- Nonessential amino acid
- Glycogenic amino acid
- Thioalcohol/Thiol group is present.

METABOLISM OF SULPHUR CONTAINING AMINO ACIDS

Steps of Methionine metabolism

1. Conversion of Methionine to S-Adenosyl Methionine (SAM) and Transmethylation reactions
2. S Adenosyl Methionine to Homocysteine
3. Two fates of Homocysteine:
   a. Synthesis of Cysteine
   b. Resynthesis of Methionine
4. Degradation of Cysteine

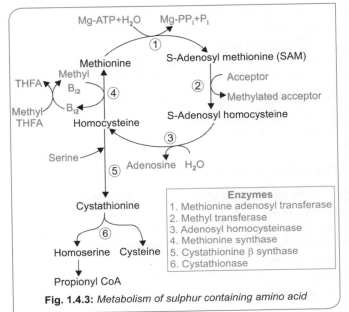

Fig. 1.4.3: *Metabolism of sulphur containing amino acid*

Step I: Methionine Adenosyl Transferase (MAT)

- Methionine converted to **S-adenosyl Methionine**, the principal methyl donor of the body.
- ATP donates the Adenosyl group to methionine.
- **3 Isoenzyme**^{Q} forms for MAT are MAT-I, MAT-II and MAT-III
- MAT-I and MAT-III in the liver. MAT-II in the extrahepatic tissue.

Methionine has to be activated to S-Adenosyl Methionine (SAM)

In methionine, the thioether linkage (C-S-C) is very stable. Adenosyl group is transferred to sulphur atom makes the methyl group labile. Hence methyl group can be easily transferred to acceptors.

Step II: Fates of Homocysteine

1. **Resynthesis of Methionine**
   - By transferring a methyl group to Homocysteine, Methionine is resynthesized.
   - **N5 Methyl THFA** and **Vitamin B12** is involved. Folate trap is discussed below.

2. Synthesis of Cysteine (Trans sulfuration reactions) Cystathionine Beta synthase
- Homocysteine condenses with Serine to form Cystathionine by removing H_2O by the enzyme Cystathionine Beta Synthase.
- **PLP** is the coenzyme.

Cystathionase
- Cystathionine to Cysteine and Homoserine by Cystathionase
- **PLP** is the coenzyme
- By further reactions Homoserine is converted to Propionyl CoA then to Succinyl CoA.

Functions of S-Adenosyl Methionine
- Transmethylation reactions
- DNA Methylation
- Polyamine Synthesis

Transmethylation Reactions

Acceptor of Methyl group	Methylated Compound
Guanidinoacetate	Creatine
Norepinephrine	Epinephrine
Epinephrine	Metanephrine
Ethanolamine	Choline
Carnosine	Anserine
Acetyl Serotonin	Melatonin

Polyamine Synthesis
Polyamines are organic compounds having multiple amino groups. They are:
- **Cadaverine** derived from decarboxylation of Lysine
- **Putrescine** derived from decarboxylation of Ornithine.
- **Spermidine** derived from Ornithine and Methionine.
- **Spermine** derived from Ornithine and Methionine.

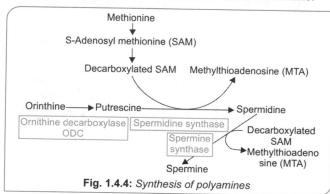

Fig. 1.4.4: Synthesis of polyamines

Steps of Polyamine Synthesis
- Ornithine is decarboxylated to form Putrescine, by the enzyme **ornithine Decarboxylase**.
- S adenosyl Methionine is decarboxylated to form Decarboxylated SAM.
- Decarboxylated SAM donates 3 carbon atom and 1 α amino group to Putrescine to form Spermidine.
- Decarboxylated SAM donates 3 carbon atom and 1 α amino group to Spermidine to form Spermine.

Significance of Polyamines
- They bear multiple positive charges, they associate readily with DNA and RNA.
- Function in cell proliferation and growth.
- Act as growth factors for cultured mammalian cells.
- Stabilize intact cells and membranes and orgeneles.
- Role in carcinogenesis.

> **CONCEPT BOX**
> **Methionine can be synthesized from Homocysteine but it is an essential Amino Acid.**
> This is because Homocysteine is derived from Methionine.

Vitamins in the Metabolism of Sulphur Containing Amino Acids
- Three vitamins needed are **Vitamin B_{12}, Folic Acid and Vitamin B_6**.
- Vitamin B_{12} and Folic Acid for Methionine Synthase reaction.
- **Vitamin B_6 for Cystathionine Beta Synthase (Transsulfuration reaction)** and Cystathionase.

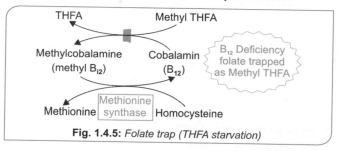

Fig. 1.4.5: Folate trap (THFA starvation)

Folate Trap
In Vitamin B_{12} deficiency the conversion of N5 Methyl THFA to THFA is blocked. This is the only reaction in which free THFA is released. Most of the body Folate is trapped as N5 Methyl THFA. This is called Folate Trap. This results in unavailability of free THFA for one carbon metabolism. Hence this is also called **THFA Starvation**.
- In Folate trap, the folic acid is trapped as its *Methyl derivative*.

In Vitamin B_{12} deficiency Homocysteine cannot be converted to Methionine. Hence Homocysteine accumulate. This is a risk factor for acute coronary Syndrome.

Specialized products derived from cysteine[Q]

- Cysteine on decarboxylation gives Betamercapto-ethanolamine.
- Coenzyme A
- Taurine
- Glutathione
- Cystine-Condensation product of two Cysteine.

> **HIGH YIELD POINTS**
>
> - Serine on decarboxylation gives Ethanolamine
> - Cysteine on decarboxylation gives Betamercaptoethanolamine
>
> **The amino acids that decreases Ageing**
> - Cysteine, hence ageing is otherwise called Cysteine deficiency syndrome.
> - Taurine.
>
> **The amino acid that accelerate aging**
> - Homocysteine.

METABOLIC DISORDERS ASSOCIATED WITH SULPHUR CONTAINING AMINO ACIDS

- **Homocystinuria:**
  - Classic Homocystinuria.
  - Nonclassic Homocystinuria.
- Cystathioninuria (Cystathioninemia)
- Hypermethioninemia

Defective reabsorption:
- Cystinuria
- Oasthouse syndrome

Lysosomal Storage Disorder:
- Cystinosis

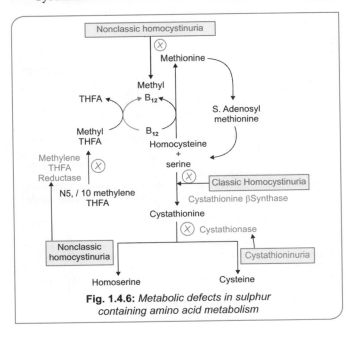

Fig. 1.4.6: *Metabolic defects in sulphur containing amino acid metabolism*

Classic Homocystinuria

Most common inborn error of methionine metabolism.

Biochemical defect

- Due to deficiency of **Cystathionine beta synthase**.
- Homocysteine is not converted to Cysteine, so there is cysteine deficiency
- More homocysteine is available for methionine synthesis, so there is hypermethioninemia.

> **Symptom box**
>
> **Clinical Features**
> - Normal at birth
> - Failure to thrive and developmental delay
> - The diagnosis is usually made after 3 years of age, when subsubuxation of the ocular lens (**ectopia lentis**) occurs. This causes severe myopia and iridodonesis (quivering of the iris).
> - Progressive **intellectual disability** is common
> - **Skeletal abnormalities** resembling those of **Marfan syndrome** tall and thin, with elongated limbs and arachnodactyly scoliosis, pectus excavatum or carinatum, genu valgum, pes cavus, high-arched palate, and crowding of the teeth
> - These children usually have fair complexions, blue eyes, and a peculiar malar flush
> - Thromboembolic episodes

Diagnosis

- **Elevations of both methionine and homocystine** (or homocysteine) in body fluids are the diagnostic
- Cystine level is low in the plasma
- Screening test for Homocystinuria-**Cyanide Nitroprusside Test**[Q] in freshly voided urine as homocystine is highly unstable.
- **Enzyme analysis** in liver biopsy specimen or cultured fibroblasts.
- **DNA mutation analysis** can be done.
- Prenatal diagnosis by Enzyme assay of cultured amniotic cells or chorionic villi or by DNA analysis.

Treatment

- **High doses of vitamin B_6** (200–1,000 mg/24 hr) causes dramatic improvement in most patients
- Restriction of methionine intake in conjunction with **cysteine supplementation** is recommended for patients who are unresponsive to vitamin B_6
- **Betaine (trimethylglycine)** lowers homocysteine levels in body fluids by remethylating homocysteine to methionine
- Administration of large doses of **vitamin C** (1 g/day) has improved endothelial function.

Nonclassic Homocystinuria

Biochemical defect

- Due to defects **in Methylcobalamin** formation
- Due to Deficiency of **Methylene tetra hydrofolate Reductase (MTHFR).**

I. Homocystinuria due to defect in Methylcobalamin formation

- Methylcobalamin is the cofactor for the enzyme methionine synthase, which catalyzes remethylation of homocysteine to **methionine.**
- Homocysteine cannot be remethylated to Methionine.
- Homocysteine accumulate.
- Methionine level decreases.

Laboratory findings

- **Megaloblastic anemia:** The presence of megaloblastic anemia differentiates Methyl Cobalamin formation defects from homocystinuria due to methylenetetrahydrofolate reductase deficiency.
- Homocystinuria.
- Hypomethioninemia.

> **HIGH YIELD POINTS**
>
> The absence of hypermethioninemia differentiates nonclassical homocystinuria from cystathionine β-synthase deficiency.

Treatment

Vitamin B_{12} in the form of hydroxycobalamin (1–2 mg/24 hr) is used to correct the clinical and biochemical findings

II. Homocystinuria caused by deficiency of Methylenetetrahydrofolate Reductase

- This enzyme reduces N5, N10-methylenetetrahydrofolate to form N5 Methyl THFA
- N5 Methyl THFA provides the methyl group needed for remethylation of homocysteine to methionine
- Hence hypomethionema
- Homocysteinemia and homocystinuria
- Absence of megaloblastic anaemia (Unlike Methyl Cobalamin formation defect).

Treatment

- Combination of folic acid, vitamin B_6, vitamin B_{12}.
- Methionine supplementation (Because Methionine is not resynthesized)
- Betaine (early treatment with betaines seems to have the most beneficial effect).

Cystathioninuria

Cystathionase Deficiency

Mental Retardation, Anaemia, Thrombocytopenia.

Cyanide Nitroprusside Test Negative.

Cystinuria

- Included in Garrod's Tetrad (Cystinuria, Albinism, Pentosuria, Alkaptonuria)
- Defect in **Dibasic Amino Acid Transporter.**
- Defective reabsorption of Cystine, Ornithine, Lysine and Arginine (Remember-COLA)
- **Cystine, Ornithine, Lysine and Arginine** (COLAQ in urine).
- Cystine Stones in urine.
- Cyanide Nitroprusside Test positive.
- Treated with ample hydration and alkalinisation of urine.

Oasthouse Syndrome

- Malabsorption of Methionine and other neutral amino Acid
- This is otherwise called **Smith Strang Disease.**

Primary Hypermethioninemia

- Due to deficiency of hepatic Methionine Adenosyl Transferase (MAT I and III)
- Peculiar smell of **Boiled Cabbage**.

Cystinosis

- Lysosomal storage disorder.
- Systemic disease caused by a defect in the metabolism of cystine.
- Caused by mutations in the **CTN Sgene**, which encodes a novel protein, **cystinosin.**
- Cystinosin is a H^+-driven lysosomal cystine transporter.
- Results in accumulation of cystine crystals in most of the major organs of the body:
  - Kidney
  - Liver
  - Eye
  - Brain.

Treatment

- Specific therapy is available with **cysteamine,** which binds to cystine and converts it to cysteine.

> **Quick Revision**
>
> - The principal methyl donor of the body is S Adenosyl methionine
> - Coenzyme of Cystathionine beta Synthase is PLP
> - Folate trap is due to deficiency of Vitamin B_{12}.
> - Homocysteine is derived from methionine.
> - Cysteine is derived from Methionine and Serine.
> - Coenzyme of transsulfuration reaction is B_6.
> - Vitamins needed for metabolism of homocysteine is B_{12}, Folic acid and B_6.
> - Polyamines are Cadaverine, Putrescine, Spermidine, Spermine
> - Putrescine is derived from Ornithine.
> - Spermine and Spermidine is derived from Ornithine and Methionine.
> - Rate limiting step of polyamine synthesis is Ornithine decarboxylase.
> - Enzyme defective in Classic Homocystinuria is Cystathionine beta Synthase.
> - Classic homocystinuria resemble Marfan's Syndrome.
> - Amino acids excreted in Cystinuria is Cystine, Ornithine, Lysine and arginine
> - Systemic disease caused by defect in metabolism of Cystine is Cystinosis
> - Treatment for cystinosis is Cysteamine.

Amino Acids

Check List for Revision

- Key points of chemistry of sulphur containing amino acids
- Metabolic pathway of methionine
- Homoscysteine
- Folate trap
- Polyamines
- Classic Homocystinuria
- Cystinuria
- Biochemical defects of other disorders.

Review Questions MCQ

1. Which amino acid is not excreted in Cystinuria?
   (Recent Question Nov 2017)
   a. Lysine
   b. Ornithine
   c. Cysteine
   d. Cystine

2. Tripeptide is: *(Recent Question)*
   a. Glutathione
   b. Anserine
   c. Carnosine
   d. Homocarnosine

3. In a case of classic homocystinuria what should be supplemented in the diet to prevent heart attacks?
   (JIPMER May 2016)
   a. Pyridoxine
   b. Methionine
   c. Methyl cobalamine
   d. Niacin

4. Sulphur of cysteine are not used/utilised in the body for the following process/product: *(PGI May 2015)*
   a. Help in the conversion of cyanide to thiocyanate
   b. Thiosulphate formation
   c. Introduction of sulphur in methionine
   d. Disulphide bond formation between two adjacent peptide

5. N Acetyl Cysteine replenishes: *(JIPMER 2012)*
   a. Glutathione
   b. Glycine
   c. Glutamate
   d. GABA

6. Which of the following is true about Glutathione?
   (PGI 2000)
   a. Contain sulfhydral group
   b. Forms met Hb from Hb
   c. It does not detoxify superoxide radicals
   d. Transport amino acid across cell membrane
   e. Part of enzymes

7. In glutathione which amino acid is reducing agent?
   (AIIMS June 1997)
   a. Glutamic acid
   b. Glycine
   c. Cysteine
   d. Alanine

Answers to Review Questions

1. **c. Cysteine** *(Ref: Harper 30/e page 23)*

   Amino acids excreted in Cystinuria is Cystine, Ornithine, Lysine and Arginine

2. **a. Glutathione**

3. **a. Pyridoxine**

   Coenzyme for Cystathionine beta synthase is PLP.

4. **c. Introduction of sulphur in methionine**

   Methionine is an essential amino acid, so it cannot be synthesised from Cysteine.
   - But Sulphur of cysteine is donated by sulphur of methionine.
   - This is called transsulfuration reaction.
   - PLP is the coenzyme of transulfuration.
   - The reaction is catalysed by Cystathionine beta Synthase and Cystathionase enzyme.

5. **a. Glutathione**
   - The active part of both Glutathione and N Acetyl Cysteine is Sulfhydryl group of Cysteine. So NAcetyl Cysteine replenishes Glutathione.

6. **a, d, e. Contain sulfhydral group, Transport aminoacid across cell membrane, Part of enzymes**

   Functions of glutathione:
   - Free radical scavenging
   - Transport of Amino acid across cell membrane
   - Keep iron in the ferrous state, so prevent methHb formation.
   - Act as coenzyme for certain enzymes.

7. **c. Cysteine** *(Ref: Harper 30/e page23)*
   - Glutathione is a tripeptide (Gamma Glutamic acid + Cysteine + Glycine)
   - Gamma glutamylcysteinyl Glycine
   - Atypical peptide bond is present between Gamma Glutamic acid and cysteine.
   - SH (Sulfhydryl) group of cysteine is the active part of glutathione.

1.5 BRANCHED CHAIN AMINO ACIDS, ACIDIC AMINO ACIDS, BASIC AMINO ACIDS AND AMIDES

- Branched Chain Amino Acids
- Acidic Amino Acids and their Amides
- Basic Amino Acids
- Entry of Amino Acids to TCA Cycle

BRANCHED CHAIN AMINO ACIDS (BCAA)

Key Points of Chemistry of BCAA
BCAA are valine, leucine and isoleucine

Structure of BCAA

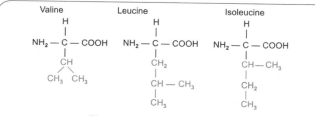

Fig. 1.5.1: *Structure of BCAA*

- All BCAA are nonpolar.
- All BCAA are essential.

Metabolic fates of BCAA

Branched chain amino acid	Metabolic fate
Valine	Glucogenic
Leucine	Ketogenic
Isoleucine	Both Ketogenic and Glycogenic

> **Tips to memorise—Classification based on metabolic fate**
> - Just like branches of a tree branched amino acid go into different metabolic fate.
> - All branched amino acids are essential.^{Q}

METABOLISM OF BRANCHED CHAIN AMINO ACIDS

There are three common steps in the metabolism of all branched chain amino acids. After these common steps each branched chain amino acid enter into different metabolic fate.

Reaction	Enzyme	Coenzyme
1. Transamination	Branched Chain Amino Acid Transaminase	PLP
2. Oxidative Decarboxylation	Branched Chain Ketoacid Dehydrogenase	Thiamine Pyrophosphate, FAD, NAD$^+$, Lipomide and CoA
3. Dehydrogenation	Acyl CoA Dehydrogenase	FAD

After the First Three Common Steps

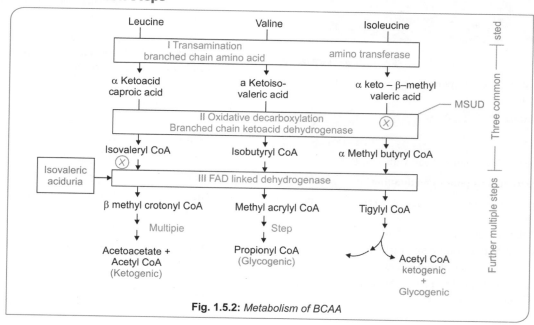

Fig. 1.5.2: *Metabolism of BCAA*

MAIN METABOLIC DISORDERS ASSOCIATED WITH BRANCHED CHAIN AMINO ACID

- Maple Syrup Urine Disease
- Isovaleric Aciduria

Maple Syrup Urine Disease

Biochemical Defect

- Deficiency of the enzyme **Branched Chain Ketoacid Dehydrogenase**.
- Defective reaction is defective oxidative decarboxylation.

Components of Branched Chain Ketoacid Dehydrogenase Complex and Defective Components in MSUD^Q

Gene	Component	MSUD Types
E1α	Branched Chain α Ketoacid decarboxylase (contains TPP)	Type IA MSUD
E1β	Branched Chain α Ketoacid decarboxylase	Type IB MSUD
E2	Dihydrolipoyl Transacylase (contains Lipomide)	Type II MSUD
E3	Dihydrolipomide Dehydrogenase (Contains FAD)	Type III MSUD

Symptom box

Clinical Features
- Affected infants who are normal at birth develop poor feeding and vomiting in the 1st week of life.
- Lethargy and coma, convulsions may ensue within a few days
- Metabolic Acidosis.
- Periods of hypertonicity may alternate with bouts of flaccidity manifested as repetitive movements of the extremities (boxing and bicycling).
- The peculiar odor of **maple syrup (burnt sugar)** found in urine, sweat, and cerumen.
- Intellectual disability.

Lab Diagnosis

- Plasma shows marked elevation of leucine, isoleucine, valine, and alloisoleucine (a stereoisomer of isoleucine not normally found in blood)
- Urine contains high levels of **leucine, isoleucine, and valine and their respective ketoacids**.
- Ketoacids are detected **by Di Nitro Phenyl Hydrazine (DNPH)** Test
- Rothera's Test
- Enzyme Analysis in leukocytes and cultured fibroblast
- Tandem Mass Spectrometry.

Treatment

Restrict branched chain amino acid give high doses **thiamine**.

Isovaleric Aciduria

Biochemical Defect

- Defective Leucine metabolism.
- Defective Enzyme is **Isovaleryl CoA Dehydrogenase**
- Characteristic **odor of Sweaty Feet^Q** is present.

Intermittent Branched Chain Ketonuria

- Retains some activity of Branched Chain α Ketoacid decarboxylase.

BASIC AMINO ACID

They are Lysine, Arginine and Histidine.

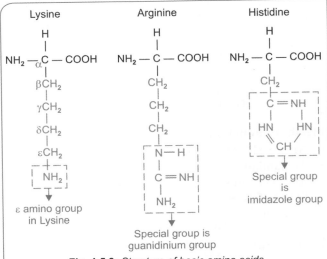

Fig. 1.5.3: *Structure of basic amino acids*

Key Points of Chemistry of Lysine

Structure of Lysine

- Represented by the letter **K**
- Essential amino acid
- **Saccharopine** is an intermediate in the Lysine catabolic pathway
- Amino Acid deficient in **Cereals**
- Purely ketogenic.

Metabolic functions of Lysine

- Hydroxy Lysine is important in **covalent cross links in Collagen** and **Desmosine crosslinks in Elastin**.
- ε Amino group of Lysine forms Schiff's bases.
- Lysine along with Methionine are the precursors of **Carnitine**.
- Bacterial putrefaction (decarboxylation) of Lysine forms **Cadaverine**.
- **Histone** proteins are Lysine rich.

Key Points of Chemistry of Arginine

- Purely glycogenic
- Semiessential amino acid
- Most polar amino acid
- Most basic amino acid
- Histones are rich in Arginine.

Catabolic Pathway of Arginine

- Arginine is catabolised to L Glutamate Semialdehyde.
- L Glutamate Semialdehyde to αKetoglutarate to Glucogenic pathway.

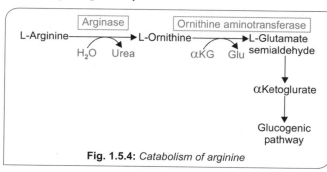

Fig. 1.5.4: Catabolism of arginine

Metabolic Functions of Arginine

- Nitric Oxide Synthesis
- Agmantine
- Arginine splits to ornithine and urea (Terminal step in Urea Cycle)
- Creatine.

Nitric Oxide

- Uncharged molecule having an unpaired electron, so it is **highly reactive, free radical**
- Very short half life (0.1 seconds)
- Formerly called **Endothelium Derived Relaxing Factor**
- Gaseous molecule
- Second messenger is **cGMP**.

Functions of Nitric Oxide

- Potent vasodilator.
- Involved in penile erction.
- Neurotransmitter in brain and Peripheral Nervous System.
- Low level of NO involved in Pylorospasm in Congenital Hypertrophic Pyloric Stenosis.
- Inhibit adhesion, activation and aggregation of platelets.

Therapeutic Uses of Nitric Oxide

- Inhalation of Nitric Oxide in the treatment of Pulmonary Hypertension.
- Treatment of Impotence (Sildenafil inhibit cGMP Phosphodiesterase)
- Glyceryl Nitrite which is converted to Nitric Oxide is used in Angina Pectoris.

Synthesis of Nitric Oxide

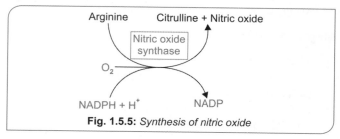

Fig. 1.5.5: Synthesis of nitric oxide

Nitric Oxide synthase

- Cytosolic enzyme
- Mono-oxygenase

Five Redox Cofactors are

1. NADPH
2. FAD
3. FMN
4. Heme
5. Tetrahydrobiopterin

Three major isoforms of Nitric Oxide Synthase

Subtype	Name	Characteristics	Deficiency leads to
1.	nNOS	First identified in the neurons. Activated by increase in Ca^{2+}	Pyloric stenosis Aggressive sexual behaviour
2.	iNOS	Prominent in **macrophages** Independent of elevated Ca^{2+}	More susceptible to certain types of infection
3.	eNOS	First identified in endothelial cells. Activated by Ca^{2+}	Elevated mean blood pressure.

Mechanism of Action of Nitric Oxide

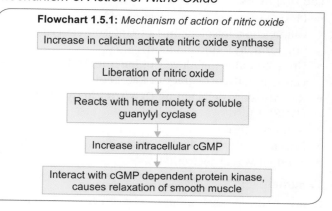

Flowchart 1.5.1: Mechanism of action of nitric oxide

Amino Acids

Agmatine
- Derived from Arginine by decarboxylation
- Properties of neurotransmitter.
- May have antihypertensive properties.

Key Points in Chemistry of Histidine

Structure of Histidine
- Semiessential amino acid
- Contains imidazole ring
- Maximum buffering capacity at physiological pH
- Basic amino acid
- Polar amino acid
- Purely glycogenic amino acid.

Metabolism of Histidine

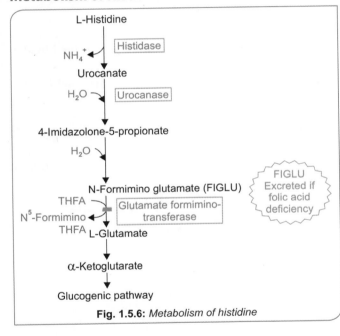

Fig. 1.5.6: *Metabolism of histidine*

Important points of the Histidine Metabolism Pathway
- Urocanate is a derivative of Histidine
- FIGLU is Formimino Glutamic Acid
- **FIGLU** is derived from Histidine
- In Folic Acid deficiency FIGLU is excreted in Urine.

Histidine Load Test
- To identify Folic Acid Deficiency.
- FIGLU excreted in urine is measured following a Histidine load.

Biologically important compounds derived from Histidine:
- **Histamine** from histidine by decarboxylation. PLP is a coenzyme.
- Carnosine (Beta Alanyl Histidine)
- **Anserine** (Methyl Carnosine)
- **Homocarnosine** (Gamma Amino Butyryl Histidine)
- Ergothioneine

Function of Histamine and Receptor Responsible for its Action

Type of receptor	Effect
H1	Smooth muscle contraction. Increased vascular permeability.
H2	Gastric HCl secretion.
H3	Synthesis and release of histamine in the brain.

ACIDIC AMINO ACIDS AND AMIDE GROUP CONTAINING AMINO ACIDS

Acidic amino acids are Glutamic acid and Aspartic acid. Amide group containing amino acids are Glutamine and Asparagine.

GLUTAMIC ACID (GLUTAMATE)
- Nonessential amino acid
- Glucogenic amino acid
- Central role in metabolism of amino acid.
- Amino group of all amino acids are concentrated as **Glutamate**^{Q} by transamination.

Biosynthesis of Glutamate

By reductive amidation of α Ketoglutarate catalysed by Glutamate dehydrogenase.

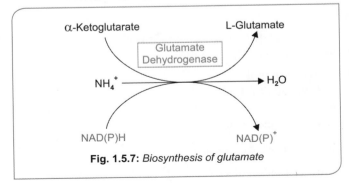

Fig. 1.5.7: *Biosynthesis of glutamate*

Metabolic Functions of Glutamic Acid

1. **Synthesis of N-Acetylglutamate**
   Positive regulator of Carbamoyl Phosphate synthetase-1 of urea cycle
   Glutamic Acid + Acetyl CoA ⟶ N-Acetyl Glutamate + CoASH
2. Synthesis of Glutathoine (Gamma Glutamyl Cysteinyl Glycine)
3. Synthesis of Gamma Amino Butyric Acid (GABA)
   – Glutamic Acid on decarboxylation gives GABA.
   – PLP is the coenzyme.

GLUTAMINE

Biosynthesis of Glutamine

Glutamine synthesized from Glutamic Acid by Glutamine synthetase.

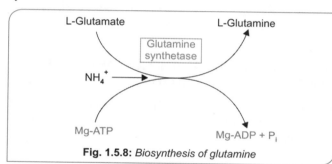

Fig. 1.5.8: Biosynthesis of glutamine

Metabolic functions of Glutamine

- Carry amino group from brain and most other tissues.
- N3 and N9 of Purine ring derived from Glutamine.
- N3 of Pyrimidine is derived from glutamine.
- Source of NH_2 group of Guanine and Cytosine.
- Glutamine is a conjugating agent.
- Source of Ammonia excretion for Kidney which has a role in renal regulation of acid base balance.

Key Points in Chemistry of Aspartic Acid (Aspartate)

- Nonessential
- Glycogenic amino acid
- Acidic
- Polar amino acid.

Synthesis of Aspartate

Transamination of oxaloacetate forms Aspartate.

Functions of Aspartate

- Contribute its alpha amino group for urea synthesis
- Contributes to purine synthesis
- Contributes to pyrimidine synthesis.

Canavan Disease

Biochemical defect

- Deficiency of **aspartoacylase,** leads to Canavan disease,
- Aspartoacylase, cleaves the N-acetyl group from acetyl aspartic acid.

🩺 Symptom box

- Leukodystrophy
- Persistent headlag
- Developmental delay
- Macrocephaly.

Diagnosis

- Aspartoacylase deficiency can be determined in skin fibroblasts
- Increased excretion of N-acetyl aspartic acid in the urine.

ASPARAGINE

Synthesis of Asparagine

Aspartate is converted to Asparagine by Asparagine Synthetase.

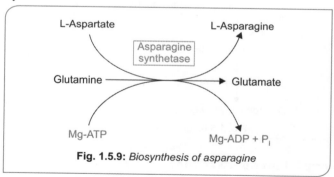

Fig. 1.5.9: Biosynthesis of asparagine

Asparagine Synthetase

- Asparagine synthetase is analogous to Glutamine synthetase
- In Asparagine synthetase, Glutamine rather than ammonium ions, provides nitrogen.

Catabolism of Glutamate, Glutamine, Aspartate and Asparagine

- Glutamine and Glutamate forms α-Ketoglutarate.^{Q}
- Asparagine and Aspartate forms Oxaloacetate.^{Q}

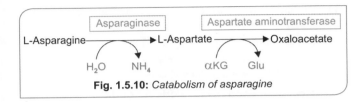

Fig. 1.5.10: *Catabolism of asparagine*

AMINO ACIDS ENTER INTO TCA CYCLE AT DIFFERENT LEVELS

1. **As Pyruvate to Oxaloacetate:**
   - Hydroxy Proline, Serine, Cysteine, Threonine, Glycine to Pyruvate
   - Lactate to Pyruvate

   HIGH YIELD POINTS
   - Pyruvate is converted to Oxaloacetate by **Pyruvate Carboxylase**.
   - **Acetyl CoA** is an allosteric activator of Pyruvate Carboxylase.
   - **This is the major filling up (anaplerotic) reaction.**

2. **As Alanine to Pyruvate to Oxaloacetate**
   - Tryptophan to Alanine to Pyruvate
3. **Directly to Oxaloacetate**
   - Aspartate
4. **As Aspartate to Oxaloacetate**
   - Asparagine
5. **As Glutamate to Alpha Keto Glutarate (5C)**
   - Histidine, Proline, Glutamine and Arginine
   - Glutamine and Glutamate are the major anaplerotic substrates of Alpha Ketoglutarate

6. **At the level of Succinyl CoA (4C)**
   - Valine, Isoleucine, Methionine and Threonine.
   - These are the compounds that form Propionyl CoA.
7. **At the level of Fumarate (4C)**
   - Tyrosine and Phenylalanine

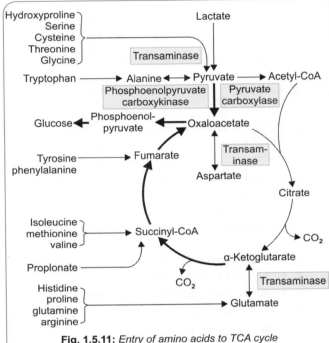

Fig. 1.5.11: *Entry of amino acids to TCA cycle*

Quick Revision

- Branched chain amino acids are Leucine, Isoleucine & Valine.
- MSUD is due to defect in Branched chain ketoacid dehydrogenase.
- The smell of urine in MSUD is burnt sugar or caramel like or Maple Syrup odour.
- DNPH test and Rothera's tests can detect MSUD.
- Nitric oxide is derived from Arginine.
- FIGLU is derived from Histidine.
- FIGLU is excreted in urine in Folic acid deficiency.
- Canavan disease is due to deficiency of Asparto Acylase.

Check List for Revision

- Metabolic fates of BCAA
- MSUD
- Functions of Lysine
- Nitric oxide
- Biologically important compounds from Histidine
- Metabolic functions of Aspartate
- Metabolic functions of Glutamate
- Catabolic fates of Glutamine, Glutamate, Aspartate and Asparagine
- Entry of amino acids to TCA cycle

Review Questions MCQ

Acidic and Basic Amino Acids

1. Creatinine, Urea and Nitric oxide are derived from which amino acid? *(Recent Q Jan 2019)*
   a. Glycine
   b. Aspartate
   c. Arginine
   d. Citrulline

2. Vasodilator produced by decarboxylation of: *(Recent Question)*
   a. Histidine
   b. Glutamic acid
   c. Aspartic acid
   d. Lysine

3. Nitric Oxide synthesised from: *(Recent Question)*
   a. Arginine
   b. Citrulline
   c. Alanine
   d. Cysteine

4. Histidine load test is used for: *(Recent Question)*
   a. Folate deficiency
   b. Histidine deficiency
   c. Histamine deficiency

5. True about Nitric Oxide are all except: *(Recent Question)*
   a. Produced from arginine
   b. Nitric oxide synthase has three isoforms
   c. Otherwise called endothelium derived relaxing factor
   d. Acts through cAMP

6. Creatinine is formed from: *(PGI June 06)*
   a. Arginine
   b. Lysine
   c. Leucine
   d. Histamine

7. Histidine is converted to Histamine by which reaction? *(NBE Pattern Question)*
   a. Carboxylation
   b. Oxidation
   c. Decarboxylation
   d. Amination

Branched Chain Amino Acid

8. In MSUD amino acid that is excreted: *(Recent Question)*
   a. Histidine
   b. Methionine
   c. Leucine
   d. Lysine

9. Branched chain ketoacid decarboxylation is defective in: *(AI 2010)*
   a. Maple syrup urine disease
   b. Hartnup disease
   c. Alkaptonuria
   d. GMI gangliosidosis

10. MSUD type I A is due to mutation of: *(Recent Question)*
    a. E I α
    b. E I β
    c. E 2
    d. E 3

11. Which is not formed from branched chain amino acid? *(Recent Question)*
    a. Xanthurenate
    b. Tiglyl CoA
    c. Acetoacetyl CoA and acetyl CoA
    d. Acetyl CoA and Succinyl CoA

12. Treatment used in Isovaleric Aciduria: *(Recent Question)*
    a. Arginine
    b. Lysine
    c. Glycine
    d. Methionine

13. Which of the following amino acid is excreted in urine in maple syrup urine disease? *(AI 1999)*
    a. Tryptophan
    b. Phenylalanine
    c. Leucine
    d. Arginine

14. Diseases of branched chain amino acid includes: *(PGI Nov 2013)*
    a. Phenyl ketonuria
    b. Maple syrup urine disease
    c. Tay-Sachs disease
    d. Isovaleric acidemia
    e. Niemann-Pick disease

Other Amino Acids & Entry of Amino Acid to TCA Cycle

15. Fish odour syndrome can be prevented by intake of: *(Recent Question Nov 2017)*
    a. Choline
    b. Niacin
    c. Pantothenic acid
    d. Riboflavin

16. Proline is formed from: *(Recent Question)*
    a. Alpha ketoglutarate
    b. Glutamate
    c. Pyruvate
    d. Alanine

17. The nitrogen atom of aspartate formed from asparagines using enzyme asparaginase is from: *(Recent Question)*
    a. Ammonium
    b. Glutamate
    c. Glutamine
    d. Alpha ketoglutarate

18. Oxaloacetate is formed from: *(Recent Question)*
    a. Proline
    b. Histidine and arginine
    c. Glutamate and glutamine
    d. Aspartate and asparagine

19. Amino acid responsible for Thioredoxin reductase activation: *(Recent Question)*
    a. Serine
    b. Selenocysteine
    c. Cysteine
    d. Alanine

20. Oxaloacetate is derived from which amino acids? *(Recent Question)*
    a. Glutamine and glutamate
    b. Asparagine and aspartate
    c. Histidine and arginine
    d. Glutamine and proline

21. **Smell of sweaty feet is seen in:** *(Recent Question)*
    a. MSUD
    b. Phenyl ketonuria
    c. Homocystinuria
    d. Glutaric acidemia

22. **During the formation of hydroxyl proline and hydroxyl lysine, the essential factors required is/are:**
    (PGI Dec 2003)
    a. Pyridoxal phosphate
    b. Ascorbic acid
    c. Thiamine pyrophosphate
    d. Methylcobalamine
    e. Biotin

23. **Succinyl CoA is formed by:** *(PGI June 1998)*
    a. Histidine
    b. Leucine
    c. Valine
    d. Lysine

24. **In one carbon metabolism Serine loses which carbon atom?** *(Recent Question)*
    a. Alpha
    b. Beta
    c. Gamma
    d. Delta

Answers to Review Questions

Acidic and Basic Amino Acids

1. c. Arginine *(Ref: Harper 31/e page 301, 302)*

The metabolic derivatives of Arginine are Urea, Creatine, Creatinine, Creatine Phosphate, Nitric oxide & Agmantine.

2. a. Histidine

Histidine is decarboxylated to Histamine, which is a vasodilator

3. a. Arginine *(Ref: Harper 30/e page 661)*

Synthesis of Nitric Oxide

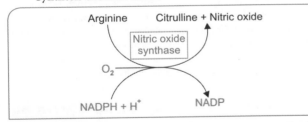

4. a. Folate deficiency *(Ref: Harper 30/e page 299)*

Important Points of the histidine metabolism pathway
- Urocanate is a derivative of Histidine.
- FIGLU is Formimino Glutamic Acid
- FIGLU is derived from Histidine.
- In Folic Acid deficiency FIGLU is excreted in Urine.

Histidine Load Test
- To identify Folic Acid Deficiency.
- FIGLU excreted in urine is measured following a Histidine load.

5. d. Acts through cAMP *(Ref: Harper 30/e page 661)*
- Nitric Oxide acts through cGMP
- Formed from Arginine
- iNOS, eNOS, nNOS are three isoforms of Nitric Oxide Synthase

Nitric Oxide

Uncharged molecule having an unpaired electron, so it is highly reactive, free radical.
- Very short half life (0.1 seconds)
- Formerly called Endothelium Derived Relaxing Factor.
- Gaseous molecule.
- Second messenger is cGMP.

Functions of Nitric Oxide
- Potent Vasodilator.
- Involved in Penile erection
- Neurotransmitter in brain and Peripheral Nervous System.
- Low level of NO involved in Pylorospasmin Congenital Hypertrophic Pyloric Stenosis.
- Inhibit adhesion, activation and aggregation of Platelets.

6. a. Arginine *(Ref: Harper 30/e page 320)*

Three amino acids from which Creatine and creatinine is synthesized are:
1. Glycine
2. Arginine
3. Methionine

7. c. Decarboxylation
- Amino acid is converted to ketoacid by Deamination or transamination.
- Amino acid converted to biological amines by decarboxylation.

Amino acid	Biologic Amines
Histidine	Histamine
Tyrosine	Tyramine
Tryptophan	Tryptamine
Lysine	Cadaverine
Glutamic Acid^Q	Gamma Amino Butyric Acid (GABA)
Serine	Ethanolamine
Cysteine	Betamercapto Ethanolamine

Branched Chain Amino Acid

8. a, c. Histidine, Leucine

9. a. Maple Syrup urine disease *(Ref: Harper 30/e page 276-278)*

Maple Syrup Urine Disease
Biochemical Defect
- Deficiency of the enzyme Branched Chain Ketoacid Dehydrogenase.
- Defective reaction is Defective Decarboxylation.

Clinical Features
- Mental Retardation
- Convulsion
- Acidosis, Coma
- Smell of Burnt Sugar [Maple Syrup]

Tests for MSUD
- DiNitro Phenyl Hydrazine Test (DNPH Test)
- Rothera's Test
- Enzyme Analysis

Treatment
- Restrict Branched Chain Amino Acid
- Give high doses Thiamine.

10. a. E1 α *(Ref: Nelson 20/e Defects in metabolism of Aminoacids, Ref: Harper 30/e page 311)*

- Types of MSUD^Q

Gene	Component	MSUD Types
E1α	Branched Chain α Ketoacid decarboxylase (contains TPP)	Type I A MSUD
E1β	Branched Chain α Ketoacid decarboxylase	Type I B MSUD
E2	Dihydrolipoyl Transacylase (contains Lipomide)	Type II MSUD
E3	Dihydrolipomide Dehydrogenase (Contains FAD)	Type III MSUD

11. a. Xanthurenate *(Ref: Harper 30/e page 309)*

- Xanthurenate is formed from Tryptophan if Kynureninase enzyme is defective.

12. c. Glycine
(Ref: Nelson's text book of Pediatrics 20/e chapter 79.6)

Treatment of Isovaleric Acidemia
Hydration
Reversal of the catabolic state (by providing adequate calories orally or intravenously), correction of metabolic acidosis (by infusing sodium bicarbonate)
Removal of the excess isovaleric acid.
By Administering Glycine
Because isovaleryl glycine has a high urinary clearance, administration of glycine (250 mg/kg/24 hr) is recommended to enhance formation of isovaleryl glycine. **By administering L-Carnitine**
L-carnitine (100 mg/kg/24 hrorally) also increases removal of isovaleric acid by forming is ovaleryl carnitine, which is excreted in the urine.

13. c. Leucine

Lab Diagnosis of MSUD
- Plasma shows marked elevation of leucine, isoleucine, valine, and alloisoleucine (astereoisomer of isoleucine not normally found in blood)
- Urine contains high levels of leucine, isoleucine, and valine and their respective ketoacids

14. b, d. MSUD, Isovaleric Acidemia

- Phenyl Ketonuria associated with Aromatic Amino acid
- Taysach's Disease and Niemann Pick disease are Sphingolipidoses

Other Amino Acids & Entry of Amino Acid to TCA Cycle

15. a. Choline *(Ref: Ref Nelsons 20/e)*

Trimethylaminuria (Fish odour Syndrome) Trimethylamine is normally produced in the intestine from the breakdown of dietary choline and trimethylamine oxide by bacteria. Eggyolk and liver are the main sources of choline, and fish is the major source of trimethylamine oxide. Trimethylamine is absorbed and oxidized in the liver by trimethylamine oxidase (flavin containing monooxygenases) to trimethylamine oxide, which is odorless and excreted in the urine. Deficiency of this enzyme results in massive excretion of trimethylamine in urine. There is a foul body odor that resembles that of a **rotten fish**, which may have significant social and psychosocial ramifications. Restriction of fish, eggs, liver, and other sources of **choline** (such as nuts and grains) in the diet significantly reduce the odor.

16. b. Glutamate *(Ref: Harper 30/e page 284)*

Proline
The initial reaction of proline biosynthesis converts glutamate to the mixed acid anhydride of glutamate phosphate). Subsequent reduction forms glutamate semialdehyde, which following spontaneous cyclization is reduced to L-proline.

17. c. Glutamine *(Ref: Harper 30/e page 267)*

Remember: The nitrogen atom of Glutamate formed from Glutamine is from Ammonium ion.

Asparagine Synthetase
- Asparagine Synthetase is analogous to Glutamine Synthetase.
- In Asparagine Synthetase, Glutamine rather than ammonium ions, provides nitrogen.
- Hence cannot fix ammonia like Glutamine Synthetase.
- Bacterial Asparagine Synthetase can however, also use ammonium ion

18. d. Aspartate and asparagine *(Ref: Harper 30/e page298)*

Amino Acids

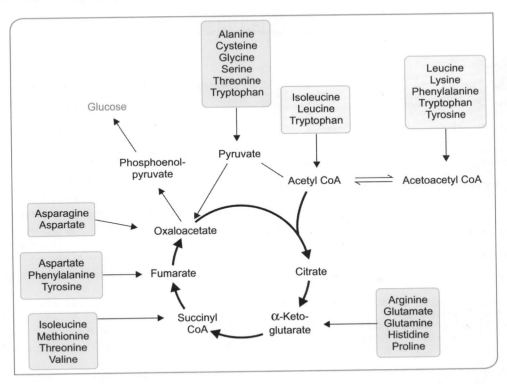

Entry of Amino Acids into TCA Cycle

19. **b. Selenocysteine** (Ref: Harper 30/e page 16)

    Selenocysteine is seen in the active site of following Enzymes and Proteins[Q]
    - Thioredoxin reductase
    - Glutathione Peroxidase
    - Iodothyronine Deiodinase
    - Selenoprotein P

20. **b. Asparagine and aspartate** (Ref: Harper 30/e page 299)
    - Asparagine and Aspartate forms Oxaloacetate
    - Glutamine and Glutamate forms alpha Keto Glutarate.

21. **d. Glutaric Acidemia**
    (Ref: Nelson's text book of Pediatrics 19/e Table 78.3)

    Peculiar odours in different Amino acidurias

Inborn Error of Metabolism	Urine Odor
Glutaric acidemia (type II)	Sweaty feet, acrid
Hawkinsinuria	Swimming Pool
Isovaleric Acidemia	Sweaty feet, Acrid
3-Hydroxy-3-methylglutaric acid uria	Cat urine
Maple syrup urine disease	Maple syrup
Hypermethioninemia	Boiled cabbage
Multiple carboxylase deficiency	Tomcat urine

    Contd...

    Contd...

Oasthouse urine disease	Hops-like
Phenylketonuria	Mousey or musty
Trimethylaminuria	Rotting fish
Tyrosinemia	**Boiled cabbage**, rancid butter

22. **b. Ascorbic acid**
    - Hdroxylation of Proline and Lysine
    - Enzyme: Prolyl and Lysyl Hydroxylase
    - Coenzyme is Vitamin C

23. **c. Valine**
    - Succinyl CoA is formed by Valine, Leucine, Methionine and Threonine

24. **b. Beta Carbonatom** (Ref: Harper 30/e page284)

    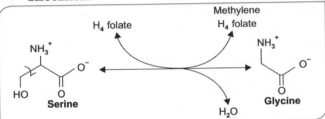

    Enzyme is Serine Hydroxy Methyl Transferase
    Serine loses the beta Carbon atom to form Glycine and Methylene THFA

ANNEXURE

Quick Revision

- Amino acid that absorbs UV light-Tryptophan, Phenylalanine, Tyrosine
- Amino acid with no asymmetric carbon atom-Glycine
- Beta Alanine is derived from Uracil and Cytosine
- Amino acid at isoelectric pH has no net charge
- Most common amino acid that undergo oxidative deamination is Glutamate
- Coenzyme for transamination reaction is Pyridoxal Phosphate (PLP)
- Amino acid that transport ammonia from most organs including the brain is Glutamine
- Amino acid that transport ammonia from skeletal muscle is Alanine.
- The nitrogen atoms of Urea are contributed by Ammonia and Aspartate.
- The rate limiting step of Urea Cycle is Carbamoyl Phosphate Synthetase I
- Most common urea cycle disorder is Hyperammonemia Type II (Ornithine Transcarbamoylase Defect)
- Polyamines are derived from Ornithine, Methionine and Lysine
- Amino acid involved in Cahill Cycle is Alanine
- Amino acid that play an important role during Starvation as a gluconeogenic AA is Alanine
- Transamination concentrate amino group as Glutamate.
- Precursor of carnitine is Lysine and Methionine
- Selenocysteine is derived from Serine
- Glutamic acid is decarboxylated to GABA
- Glutamic acid is deaminated to Alpha Ketoglutarate
- Folate trap traps the THFA as its Methyl Derivative.
- Amino acids that enter TCA Cycle as Succinyl Choline is Valine, Isoleucine and Methionine

Metabolic Disorders at a Glance

Metabolic disorder	Biochemical defect
Hyperammonemia Type I	Carbamoyl Phosphate Synthetase I
Hyperammonemia Type II	Ornithine Transcarbamoylase
Citrullinemia Type I	Argininosuccinate Synthetase
Citrullinemia Type II	Citrin (Aspartate- Glutamate) Transporter
Argininosuccinic Aciduria	Argininosuccinate Lyase
Argininemia	Argininase
HHH syndrome	ORNT-I defect (Ornithine Permease)
Classic Phenyl Ketonuria	Phenyl Alanine Hydroxylase
Alkaptonuria	Homogentisate Oxidase
Tyrosinemia Type I	Fumaryl Acetoacetate Hydrolase
Tyrosinemia Type II	Tyrosine Transaminase
Tyrosinemia Type III	Para Hydroxy Phenyl Pyruvate hydroxylase/Para hydroxyl Phenyl Pyruvate Dioxygenase

Contd...

Contd...

Metabolic disorder	Biochemical defect
Hawkinsinuria	Para Hydroxy Phenyl Pyruvate hydroxylase/Para hydroxyl Phenyl Pyruvate Dioxygenase is mutant, so that it catalyse only partial reaction
Segawa Syndrome	GTP Cyclohydrolase
Albinism	Tyrosinase
Pheochromocytoma	Excess production of Catecholamines
Carcinoid Syndrome	Excess production of Serotonin
Hartnup's Disease	Defective absorption of Tryptophan and other neutral amino acids from renal tubules and intestines
Primary Hyperoxaluria Type I	Alanine-Glyoxylate Amino Transferase
Primary Hyperoxaluria Type II	D-Glycerate Dehydrogenase/ Glyoxylate reductase Enzyme Complex
Non-Ketotic Hyperglycinemia	Glycine Cleavage System
Classic Homocystinuria	Cystathionine Beta Synthase
Nonclassic Homocystinuria	Methyl Cobalamin formation defect Methylene THFA Reductase
Cystathioninuria	Cystathionase
Cystinuria	Defective reabsorption of Cystine, Ornithine, Lysine and Arginine
Oasthouse Syndrome	Malabsorption of Methionine and other neutral amino acids
Type I A MSUD	$E_{1\alpha}$ gene that codes for Branched Chain Ketoacid Decarboxylase
Type I B MSUD	$E_{1\beta}$ gene that codes for Branched Chain Ketoacid Decarboxylase
Type II MSUD	E_2 gene that codes for Dihydrolipoyl Transacylase
Type III MSUD	E3 gene that codes Dihydrolipomide Dehydrogenase
Isovaleric Aciduria	Isovaleryl CoA Dehydrogenase
Canavan Disease	N Asparto Acylase

Specialised Products from Amino Acids -at a Glance

Amino acid	Metabolic products
Tyrosine	Melanin
	Catecholamines (Epinephrine, Norepinephrine, Dopamine)
	Thyroxine

Contd...

Contd...

Amino acid	Metabolic products
Tryptophan	Serotonin Melatonin Niacin
Cysteine	Cystine Taurine Glutathione Betamercaptoethanolamine
Glycine	Purine Heme Glutathione Creatinine
Arginine	Nitric oxide Arginine Arginine splits to Ornithine and Urea Creatine
Histidine	FIGLU Histamine
Glutamate	N acetyl Glutamate Glutathione Gamma Amino Butyric Acid
Glutamine	N3 & N9 of Purine N3 of Pyrimidine
Aspartate	Purine Pyrimidine Urea Synthesis

Peculiar Odours in Different Amino Acid Urias

Inborn Error of Metabolism	Urine Odor
Glutaricacidemia (type II)	Sweaty feet, acrid
Hawkinsinuria	Swimming Pool
Isovaleric Acidemia	Sweaty feet, Acrid
3-Hydroxy-3-methylglutaric aciduria	Cat urine
Maple syrup urine disease	Maple syrup
Hypermethioninemia	Boiled cabbage
Multiple carboxylase deficiency	Tomcat urine
Oasthouse urine disease	Hops-like
Phenylketonuria	Mousey or musty
Trimethylaminuria	Rotting fish
Tyrosinemia	Boiled cabbage, rancid butter

CHAPTER 2

Proteins

Chapter Outline

2.1 Chemistry of Proteins
2.2 Structural Organization of Proteins and its Study
2.3 Biochemical Techniques
2.4 Fibrous Proteins
2.5 Protein Folding and Degradation
2.6 Plasma Proteins and Glycoproteins
2.7 Protein Sorting

CHAPTER 2

Proteins

2.1 CHEMISTRY OF PROTEINS

☞ Peptide Bond ☞ Classification of Proteins

INTRODUCTION

Proteins are polymers of amino acid. Proteins contain Carbon, Hydrogen, Oxygen and Nitrogen as the major components. Nitrogen is characteristic of Protein. On an average, the **nitrogen content of ordinary proteins is 16% by weight.**

PEPTIDE BOND

All Proteins are linked by Peptide bond.

Alpha Carboxyl group of one amino acid reacts with alpha amino group of another amino acid to form a peptide bond or **CO-NH** bridge.

- N terminal end—The free NH_2 group of the terminal amino acid is called as N terminal end.
- C terminal end—The free COOH end is called C terminal end.
- The amino acids are sequenced from N-terminal end to C-terminal end.

Formation of Peptide Bond

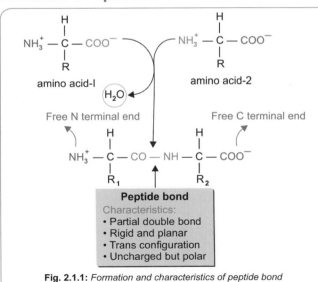

Fig. 2.1.1: Formation and characteristics of peptide bond

Atypical Peptide Bond (Pseudopeptide Bond) (Isopeptide Bond)

An amide bond formed between an amino group and a carboxyl group at least one of which is not an alpha group, seen in the side chains of proteins.

Characteristic Features of an Atypical Peptide Bond

- Occur **post-translationally**
- Can be formed **spontaneously or enzymatically**
- Can produce stably linked protein dimers, multimers or complexes
- Makes the **protein resistant** as proteases cannot hydrolyse isopeptide bond.

> 📍 **HIGH YIELD POINTS**
> **Examples of Pseudopeptides**
> - Glutathione
> - Thyrotropin releasing hormone
> - Ubiquitin attached to protein
> - Blood clots
> - Cyclic peptide antibiotic like Tyrocidin and Gramicidin
> - Heptapeptides like Dermorphin, Deltorphin.

Some Biologically Important Peptides

Peptide	Example
Tripeptide	Thyrotropin releasing hormone (TRH) Glutathione
Pentapeptide	Enkephalin
Octapeptide	Angiotensin II
Nonapeptide	Oxytocin vasopressin [ADH] bradykinin
Decapeptide	Angiotensin-I

CLASSIFICATION OF PROTEINS

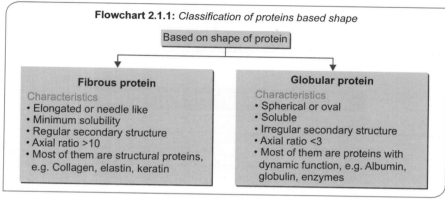

Flowchart 2.1.1: Classification of proteins based shape

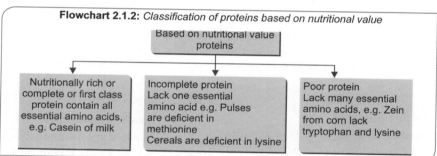

Flowchart 2.1.2: Classification of proteins based on nutritional value

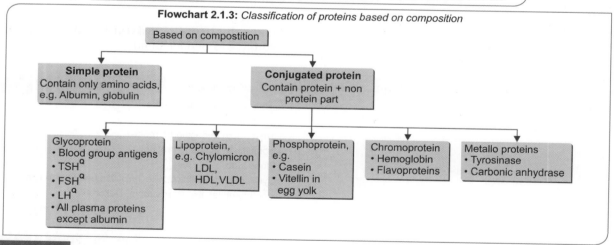

Flowchart 2.1.3: Classification of proteins based on composition

Quick Review

- Nitrogen content in proteins is 16% by weight.
- TRH and Glutathione are atypical peptide.
- Proteins with structural function are fibrous protein.
- Fibrous protein has regular secondary structure.
- Most enzymes are globular proteins.
- TSH, FSH & LH are glycoproteins.
- Albumin is a simple protein
- Casein of milk is a first class protein.
- Limiting amino acid in cereals is Lysine.
- Limiting amino acid in pulses is methionine.

Review Questions MCQ

1. **True about Isopeptide bond is:** *(PGI Nov 2011)*
   a. It makes protein resistant.
   b. Bond is formed between carboxy terminal of one protein and amino group of a lysine residue on another.
   c. Involves in post-transcriptional modification of protein
   d. Enzyme act as catalyst for bond formation

2. **Which of the following is/are storage protein?** *(PGI May 2011)*
   a. Myoglobin
   b. Ovalbumin
   c. Ricin
   d. Ferritin
   e. Glutelin

3. **Which one of the following can be homologous substitution for isoleucine in a protein insequence?** *(AI 2006)*
   a. Methionine
   b. Asparticacid
   c. Valine
   d. Arginine

4. **In a mutation if valine is replaced by which of the following would not result in any change in the function of protein?** *(AIIMS May 02)*
   a. Proline
   b. Leucine
   c. Glycine
   d. Asparticacid

5. **At isoelectric pH protein:** *(PGI June 03)*
   a. Have net charge '0'
   b. Do not migrate
   c. Are positively charged
   d. Are negatively charged

6. **Isoelectric point is when:**
   a. Net charge of protein is zero
   b. Mass of protein is zero
   c. Protein
   d. Denaturation of protein occurs

7. **Biuret test is used for detection of:**
   a. Protein
   b. Cholesterol
   c. Steroid
   d. Sugar

8. **What type of protein is Casein?** *(CMC Ludhiana 2014)*
   a. Lipoprotein
   b. Phosphoprotein
   c. Glycoprotein
   d. Flavoprotein

9. **Which is/are not transport Protein?** *(PGI May 2012)*
   a. Transferrin
   b. Collagen
   c. Ceruloplasmin
   d. Hemoglobin
   e. Albumin

10. **Which of the amino acid is responsible for peptide bond?** *(PGI Nov 2014)*
    a. Amino group
    b. Carboxyl group
    c. Side chain
    d. Amide group
    e. Aldehyde group

Answers to Review Questions

1. **a, b, d.** It makes protein resistant, Bond is formed between carboxy terminal of one protein and amino group of a lysine residue on another, Enzyme act as catalyst for bond formation

   Atypical Peptide Bond (Pseudopeptide Bond) (Isopeptide Bond)
   A bond formed between an amino group and a carboxyl group at least one of which is not an alpha group. Seen in the side chains of proteins.

   Characteristic features
   - Occur post-translationally
   - Can be formed spontaneously or enzymatically.
   - Can produce stably linked protein dimers, multimers or complexes
   - Makes the protein resistant as proteases cannot hydrolyse isopeptide bond.

   Examples
   - Glutathione
   - Thyrotropin releasing hormone
   - Ubiquitin attached to protein
   - Blood clots

   Application of spontaneous Isopeptide Bond formation
   - Develop a new peptide tag called Isopeptag. Used in:
     ▫ In vivo Protein targeting
     ▫ Fluorescence Microscopy Imaging

2. **b, d, e.** Ovalbumin, Ferritin, Glutelin
   (Ref: Chatterjee & Shinde 8/e page 82-85)

Storage protein

Proteins that act as store house of amino acids and metal ions that can be easily mobilised.
- Casein of milk
- Vitellin of egg yolk
- Ovalbumin of egg white
- Glutelin of wheat.
- Oryzenin of Rice
- Gliadin of Wheat
- Ferritin-stores iron

Other options

Myoglobin:
- It is a transport protein of Oxygen to Skeletal Muscle

Ricin:
- Inhibitor of mammalian Protein Synthesis.
- From cast or be an in activates eukaryotic 28S ribosomal RNA

3. **c. Valine**

Conservative [Homologous] Substitution

- One amino acid replaced by another amino acid of similar characteristics
- Examples of homologous substitution is shown in the diagram given below.

Hydrophilic, Acidic	Asp	Glu			
Hydrophilic, Basic	His	Arg	Lys		
Polar, Uncharged	Ser	Thr	Gln	Asn	
Hydrophobic	Ala	Phe	Leu	Ile	Val	Pro

Nonconservative (Nonhomologous) Substitution
- One amino acid replaced by another amino acid of different characteristics

4. **b. Leucine**

5. **a, b. Have net charge 0, Do not migrtate**
   (Ref: Harper 31/e page 20)

   Properties of proteins at isoelectric pH:
   - They have no net charge.
   - They do not have electrophoretic mobility
   - Maximum precipitability
   - Minimum solubility
   - Minimum buffering action

6. **a. Net charge of a protein is Zero**
   (Ref: Harper 31/e page 19)

7. **a. Protein**

   General test for Proteins
   Cupric ions in alkaline medium forms violet colour with peptide bond nitrogen.

   Ninhydrin Test
   General test for all alpha Amino Acid

   Amino acid + 2 mols of Ninhydrin ⟶ Aldehyde with 1 carbon atom less + CO_2 + Purple Complex (Ruhemann's Purple)

8. **b. Phosphoprotein**   *(Ref: Chatterjea and Shinde 8/e page 85)*

   Two important Phosphoproteins are
   - Casein found in milk
   - Ovovitellin found in egg yolk

9. **b. Collagen**   *(Ref: Harper 31/e p 630 Table 52-3)*

10. **a, b. Amino group, Carboxyl group**
    - Peptide bond is formed between amino group of one amino acid with carboxyl group of the next amino acid.

2.2 STRUCTURAL ORGANIZATION OF PROTEINS AND ITS STUDY

- Primary Structure
- Secondary Structure
- Tertiary Structure
- Quaternary Structure
- Study of Structure of Proteins

Proteins have different levels of structural organization; primary, secondary; tertiary and quaternary.

PRIMARY STRUCTURE

It is the linear sequence of amino acid held together by peptide bonds in its peptide chains.

Bond involved in primary structure is **Peptide Bond**, a type of covalent bond.

Amino acid sequence determines the 3D structure of the protein.[Q]

SECONDARY STRUCTURE

- The folding of short (3- to 30-residue), contiguous segments of polypeptide into geometrically ordered units.

HIGH YIELD POINTS

- Bonds involved in the secondary structure are primarily **noncovalent** bonds like
  - Hydrogen bond (Most important Bond)
  - Hydrophobic bond
  - Electrostatic bond (Ionic bond, salt bridges)
  - Van der Waals forces.

Secondary Structures of Proteins

- Alpha helix
- Beta pleated sheet
- Loops
- Bends
- Turns

Alpha Helix[Q]

- Alpha helix is the **most common and stable** secondary structure.
- **Right-handed spiral structure**
- Structure stabilized primarily by **intra chain hydrogen bond** between carbonyl oxygen of 1st and amide nitrogen of 4th amino acid.
- Each turn formed by **3.6 Amino acyl** residues.
- Distance of 1turn of alpha helix (called Pitch) is 0.54 nm.
- **Proline** can only be stably accommodated within the first turn of an α-helix
- Examples of proteins whose major secondary structure is α-helix
  - Hemoglobin
  - Myoglobin.

Beta-Pleated Sheet[Q]

- The **second most common** (hence "beta") recognizable regular secondary structure in proteins.
- Polypeptide chain is almost fully extended.
- They have a **zig-zag or pleated pattern**.
- In contrast to intrachain hydrogen bond in alpha helix, here it is **interchain hydrogen bond** between carbonyl oxygen and amide nitrogen of two adjacent chains.
- Adjacent strands in a sheet can run in the same direction (parallel) or opposite direction (antiparallel).

Example of proteins whose major secondary structure is
- Parallel β-sheet—Flavodoxin.
- Antiparallel β-sheet—Silk Fibroin.
- Both parallel and antiparallel β-sheet—Carbonic Anhydrase.

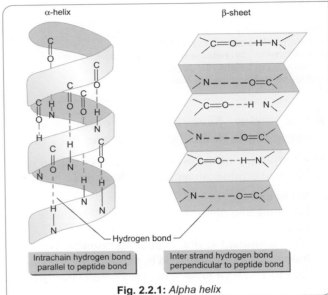

Fig. 2.2.1: Alpha helix

Turns and Bends

Short segments of amino acid that join two units of secondary structures.

Example: Beta turn

Loops

- Unlike bends and turns; loops are long segments of amino acid that joins two secondary structures.

Image-Based Information

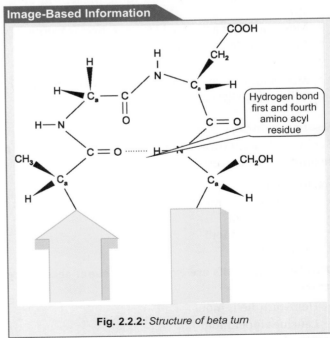

Fig. 2.2.2: Structure of beta turn

Contd...

Contd...

Beta Turn
- Involves four amino acyl residues
- First amino acyl residue is hydrogen bonded to the fourth resulting in a tight 180° turn.
- Proline and Glycine are often present in β turn

Amino Acids have different propensity for forming Alpha Helix, Beta Sheets and Beta Turns

The frequency of occurrence of certain amino acid residues determine the secondary structure formed.

Alpha helix
- Residues like Alanine, Glutamate, Methionine and Leucine tend to be present in the alpha helix.
- Most abundant amino acid is **Methionine** (relative frequency is 1.47) followed by Glutamate (1.44)
- Amino acid least present in alpha helix is **Proline**.

Amino acids that do not favor alpha helix are[Q]
- Amino acids with branches at β-carbon atom **valine, threonine and isoleucine** disrupt the stability.
- Other amino acids that disrupt the stability are **serine, aspartate and asparagine**.
- **Proline** also disrupt the stability of alpha helix.
- **Glycine** induces bends in alpha helix.
- **Amino acids with bulky R group like Tryptophan**

Beta sheet
- Valine and isoleucine tend to be present in beta strands
- Most abundant amino acid in beta sheet is **Valine**
- Amino acid least present in beta sheet is **Proline**.

Turns
- Most abundant amino acid in turns is **Proline** (1.91) followed by Glycine (1.64).

Super Secondary Structure (Motifs)
Secondary Structural elements join to form **Super secondary structures**:

Examples are
- Beta-alpha-beta motif
- Greek key motif
- Beta meander motif
- Beta barrel:

DNA Binding motifs are examples of super secondary structure. They are:
- Helix-turn-helix motif
- Leucine zipper motif
- Zinc finger motif.

HIGH YIELD POINTS

Points to Ponder–DNA Binding Motifs
- The first motif described is the **Helix-turn-helix**.
- The second DNA binding motif is the **Zinc finger**.
- DNA binding motif which require Zinc for its activity Zinc Finger.
- DNA binding motif with leucine residues at every seventh position is **Leucine Zipper**.

Tertiary Structure
The entire three-dimensional conformation of a polypeptide is referred to as tertiary structure.

Domain
A domain is a section of protein structure sufficient to perform a particular chemical or physical task such as binding of a substrate.

Rossmann Fold
- It is a domain seen in the family of Oxidoreductases.
- They share a common N terminal **NAD(P)+ binding region called Rossmann fold**.

Oxidoreductase with Rossmann Fold[Q]
- Lactate dehydrogenase
- Alcohol dehydrogenase
- Glyceraldehyde-3-phosphate dehydrogenase
- Malate dehydrogenase
- 6-phosphogluconate dehydrogenase
- D-glycerate dehydrogenase

Quaternary Structure
If more than one polypeptides aggregate to form one functional protein, the spatial relationship between the polypeptide subunits is referred to as Quaternary structure.

HIGH YIELD POINTS

Bonds involved in Tertiary and Quaternary Structures are primarily non-covalent bonds.
- Hydrophobic Interaction
- Hydrogen Bond
- Electrostatic Bond
- Van der Waals Forces

Insulin has two polypeptide chains but it does not have Quaternary structure.

In Quaternary structure, the bond involved is primarily non-covalent bond. In insulin two polypeptide chains are connected by disulphide bond which is a Covalent Bond. Hence, even if Insulin has 2 polypeptide chain, it does not have quaternary structure.

Structure of Insulin

The first Hormone to be extracted in pure form
- This was done by **Banting and Best**
- **Banting** along with the director of the institute **John Macleod received Nobel Prize for the work.**
- The first protein in which complete sequencing was done
  - **Mr Frederick Sanger** was the man behind this work.
  - He used Sanger's Reagent for this.
  - He won Nobel Prize for his work
  - The first protein to be produced by Recombinant DNA Technology.

Primary Structure of Insulin
- Consist of two polypeptide chains
- Number of amino acids is 51
- A chain with 21 amino acids
- B chain-30 amino acids

Disulfide Bonds in Insulin

Two interchain Disulphide Bonds:
- 7th Amino Acid in A chain to 7th Amino Acid in B chain.
- 20th Amino Acid in A chain to 19th Amino Acid in B chain.

One Intrachain Disulphide Bond
- 6th amino acid in A chain to 11th amino acid in A chain itself.

Species Variation in Insulin
- Restricted to 8, 9, 10 in A chain and C terminal amino Acid of B chain.
- Porcine and human insulin vary only in the terminal amino acid of B chain

Denaturation of Proteins

Nonspecific alteration in **secondary, tertiary and quaternary structures** of protein molecule when treated with a denaturing agent.

Two types of denaturation
1. **Reversible denaturation:** Denatured proteins are sometimes renatured when physical agent is removed.
2. **Irreversible Denaturation:** Denatured proteins are not renatured when physical agent is removed.
   - For example, Albumin heated is irreversibly denatured called **Heat Coagulation**.

> **HIGH YIELD POINTS**
>
> **Characteristic features of denaturation are**^{Q}
> - Loss of biological Activity
> - Primary structure (i.e. the peptide bond) is not lost.
> - Loss of secondary and tertiary and quaternary structures.
> - Loss of Folding
> - They assume a Random Coil Structure.

> **CONCEPT BOX**
>
> - Everything lost in denaturation except the primary structure (i.e. the peptide bond).
> - Remember the peptide bond is a covalent bond which is the strongest bond.

STUDY OF PROTEIN STRUCTURE

Study of Primary Structure/Sequencing of Proteins

Methods of protein sequencing
1. End group analysis
2. Mass spectrometry
3. Molecular biology techniques

End Group Analysis

Identification of N-terminal and C-terminal amino acid in a polypeptide chain is called end group analysis.

> **HIGH YIELD POINTS**
>
> **Identification of N Terminal Amino Acid:**
> - **Sanger's Technique** using Sanger's reagent (1, Fluoro 2, 4 Dinitro Benzene, FDNB)^Q
> - **Edman's Degradation Technique** using **Edman's reagent (Phenyl Isothiocyanate)**
>
> **Identification of C Terminal Amino Acid**
> - Using **Carboxypeptidase A and B**

Sanger's Technique
- This was the first technique to determine the sequence of protein.
- Sanger's reagent is **1, Fluoro 2, 4 Dinitrobenzene**^{Q}
- Sanger's Reagent derivatizes the amino terminal residues.
- The first protein to be sequenced by the method is **Insulin by Fredrick Sanger.** He got Nobel Prize in **1958.**
- Only dipeptides or tripeptides can be sequenced.

Edman's Technique
- By using Edman's reagent (Phenyl Isothiocyanate).
- Phenyl isothiocyanate derivatizes the amino terminal of polypeptide.

- Edman's Technique can sequence many residues (5–30 residues) of a single polypeptide sample unlike Sanger's sequencing.

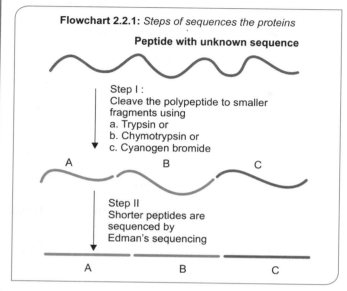

Flowchart 2.2.1: *Steps of sequences the proteins*

Site of Cleavage of Polypeptide

- **Trypsin:** Carboxyl side of basic amino acids like **Lysine** and **Arginine**.
- **Chymotrypsin:** Bulky nonpolar amino acids like **Phe, Tyr, Trp, Leu, Met**
- **Cyanogen bromide:** Carboxyl side of **methionine**.

Mass Spectrometers

Today mass spectrometers has emerged the **method of choice of protein identification.**

The principle used to identify protein based on mass (precisely saying on **mass/charge ratio**).

The molecular mass of each amino acid is unique, the sequence of the peptide can be reconstructed from the masses of its fragments.

The exceptions are molecular mass of (1) leucine and isoleucine and (2) glutamine and lysine.

Methods for Dispersion of Analyte into Vapor Phase

Analytes has to be converted to vapour phase by using various technique.

- Heating in a vacuum, but proteins and oligonucleotides are destroyed by heat.
- Electrospray Ionization
- Matrix assisted laser desorption and ionization (MALDI)^Q
- Fast atom bombardment (FAB).

Types of Mass Spectrometers

Mass Spectrometers come in various configurations

1. Simple quadrupole mass spectrometers
2. Time of flight mass Spectrometers
3. Tandem mass spectrometry

Quadrupole Mass Spectrometers

- Quadrupole mass spectrometers generally are used to determine the masses of molecules of **4000 Da or less**

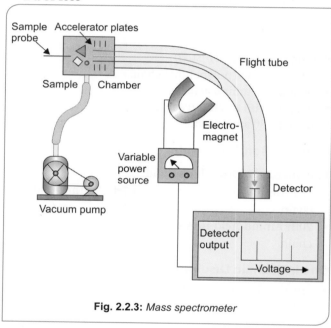

Fig. 2.2.3: *Mass spectrometer*

Time of Flight (TOF) Mass Spectrometer

Time-of-flight mass spectrometers are used to determine the large masses of complete proteins (**> 4 KDa**)

Tandem Mass Spectrometry

Two mass spectrometers linked in series. For this reason, such tandem instruments are often referred to as MS–MS.

Advantage of Mass Spectrometers

- Method of choice Protein determination.
- Superior Sensitivity, speed and versatility.
- Detect **Post-translational modification** unlike Edman's reaction and DNA derived Protein sequence.
- Can be used for other biomolecules like oligonucleotide, carbohydrates as mass and charge are common properties of all this.

STUDY OF SECONDARY AND TERTIARY STRUCTURES

Study of Secondary Structure

- Circular dichroism
- Optical Rotatory dispersion chromatography
- Study of tertiary structures
- X-ray crystallography
- NMR spectroscopy
- UV light spectroscopy (most rapid method)
- Fluorescence spectroscopy
- Molecular modelling

Quick Review

- Bond of primary structure is peptide bond.
- Bonds of secondary structure is primarily noncovalent bonds like Hydrogen bond, hydrophobic bond, ionic bond and Vanderwaal's forces.
- Proline inserts kinks in alpha helix hence it disrupt the stability of it.
- Glycine is also alpha helix breaker as it confer high flexibility.

Contd...

Contd...

- Most common secondary structure is alpha helix.
- Beta turn is an example of secondary structure.
- Beta barrel, Beta meander are super secondary structure.
- NAD(P) binding domain in certain oxidoreductase is called Rossmann fold.
- First protein to be sequenced completely is insulin by Frederick Sanger.
- Sanger's reagent is 1 Fluoro 2,4 DiNtrobenzene, (FDNB)
- Edman's reagent is phenyl isothiocyanate.
- Mass spectrometers identify proteins, oligonucleotides and carbohydrates
- Principle used to identify any compound by mass spectrometry is mass to charge ratio.
- Two mass spectrometers in series is called Tandem mass spectrometry.

Check List for Revision

- Alpha helix
- Examples of secondary, tertiary and quaternary structures.
- Characteristics of denaturation of proteins.
- Structure of Insulin
- Sanger's and Edman's Technique.
- Bonds that stabilises different levels of organisation.
- Names of techniques to determine the structure of proteins.
- Mass spectrometers

REVIEW QUESTIONS MCQ

1. **Amino acid sequence is not found by:**
   (NBE Pattern Question)
   a. Sanger's reagent
   b. Benedict's reagent
   c. Trypsin
   d. Cyanogen bromide

2. **Method used to study the structure of proteins include all except:**
   a. UV Spectroscopy
   b. NMR Spectroscopy
   c. X-ray crystallography
   d. Edman's technique

3. **Sanger's reagent is chemically:** *(NBE Pattern Question)*
   a. 2, 4 Dinitrobenzene
   b. 2, 4 Dinitrocresol
   c. 1, Fluoro 2, 4 DinitroBenzene
   d. 2, 4 FluorodinitroCresol

4. **Which one of the following about protein structure is correct?** *(PGI May 2015)*
   a. Protein consisting of one polypeptide can have quaternary structure
   b. The formation of disulphide bond in a protein requires that the two participating cysteine residues be adjacent to each other in the primary sequence of the protein
   c. The stability of quaternary structure in protein is mainly the result of covalent bonds among the subunits.
   d. The denaturation of proteins always leads to irreversible loss secondary and tertiary structure.
   e. The information required for the correct folding of a protein is contained in the specific sequence of amino acid along the polypeptide chain.

5. **An alpha helix of a protein is most likely to be disrupted if a missense mutation introduces the following amino acid within the alpha helical structure:** *(AIIMS Nov 2002)*
   a. Alanine
   b. Aspartic acid
   c. Tyrosine
   d. Glycine

6. **Proteins are linear polymers of amino acids. They fold into compact structures. Sometimes, these folded structures associate to form homo-or-heterodimers. Which one of the following refers to this associated form?** *(AI 2006)*
   a. Denatured state
   b. Molecular aggregation
   c. Precipitation
   d. Quaternary structure

7. Which among the following is the structure of myoglobin?
   a. Monomer
   b. Homodimer
   c. Heterodimer
   d. Tetramer

8. Denaturation is resisted by which of the following bond? *(NBE Pattern Question)*
   a. Peptide bond
   b. Hydrogen bond
   c. Disulphide bond
   d. Electrostatic bond

9. Polypeptide formation in amino acid is by: *(NBE Pattern Question)*
   a. Primary structure
   b. Secondary structure
   c. Tertiary structure
   d. Quaternary structure

10. Rossman fold associated NADH domain is found in which enzyme? *(NBE Pattern Question)*
    a. Pyruvate Dehydrogenase
    b. Lactate Dehydrogenase
    c. Alpha ketoglutarate Dehydrogenase
    d. Isocitrate Dehydrogenase

11. In forming 3D structure following factors help: *(PGI May 2014)*
    a. Peptide bond
    b. Amino acid sequence
    c. Interaction between polypeptide
    d. Chaperone
    e. Side chain

Answers to Review Questions

1. **b. Benedict's reagent** *(Ref: Harper 31/e page 27,32)*

   Methods of protein sequencing Sanger's technique
   - Sanger's reagent is 1, Fluoro 2, 4 Dinitrobenzene^Q
   - Sanger's reagent derivatizes the amino terminal residues.
   - The first Protein to be sequenced by the method is **Insulin by Fredrick Sanger. He got Nobel prize in 1958.**
   - Only dipeptides or tripeptides can be sequenced.

   Edman's Technique
   - By using Edman's Reagent (Phenyl Iso-thiocyanate).
   - Phenyl Isothiocyanate derivatizes the amino terminal of Polypeptide.

   Edman's Technique can sequence many residues (5–30 residues) of a single polypeptide sample unlike Sanger's

2. **d. Edman's Technique** *(Ref: Harper 29/e Table 1.1)*
   - Edman's reaction is used to detect sequence of amino acids in a polypeptide.

3. **c. 1, Fluoro 2, 4 DinitroBenzene**
   - Sanger's reagent is 1 Fluoro 2, 4Dinitrobenzene
   - Edman's reagent is Phenyl Isothiocyanate

4. **e. The information required for the correct folding of a protein is contained in the specific sequence of amino acid along the polypeptide chain**
   - Proteins with more than one polypeptide chain can only have quaternary structure.
   - The cysteine residues need not be adjacent for the formation of disulphide bond

   The stability of quaternary structure is by non covalent bonds.
   - Denaturation can be reversible also.

5. **d. Glycine** *(Ref: Harper 31/e page 35)*
   - Glycine induces bends in the alphahelix.
   - Proline also disrupts the conformation of alpha helix.
   - But proline is present in the first turn of alpha helix.

6. **d. Quaternary structure** *(Ref: Harper 31/e page 38)*

7. **a. Monomer**
   - Myoglobin is a monomer
   - Hemoglobin is a tetramer

8. **a. Peptide bond**

   Denaturation of proteins
   - Primary structure is not lost
   - Hence peptide bond is not broken.
   - Secondary and tertiary structure is lost.
   - There is loss of folding

9. **a. Primary structure** *(Ref: Harper 31/e page 34)*

   Primary structure
   It is the linear sequence of amino acid held together by peptide bonds in its peptide chains. Bond involved in primary structure is Peptide Bond, a type of covalent bond.

   Secondary Structure
   - Configurational relationship between residues which are about 3-4 amino Acids apart in linear sequence.
   - The folding of short (3- to 30-residue), contiguous segments of polypeptide into geometrically ordered units

   Tertiary structure
   The entire three-dimensional conformation of a polypeptide is referred to as tertiary structure.

   Quaternary structure
   If more than one polypeptides aggregate to form one functional protein, the spatial relationship between the polypeptide subunits is referred to as Quaternary structure.

10. **b. Lactate dehydrogenase**

    Oxidoreductase with Rossmann Fold
    - Lactate dehydrogenase
    - Alcohol dehydrogenase

- Glyceraldehyde-3-phosphate dehydrogenase
- Malate dehydrogenase
- Quinoneoxidoreductase
- 6-phosphogluconate dehydrogenase
- D-glycerate dehydrogenase

11. **b, c, d, e. Amino acid sequence, Interaction between polypeptide, Chaperone, Side chain**
 (Ref: Harper 31/e page 39,574; Stryer 7/e page 60)

- Amino acid sequence determines the tertiary structure of proteins
- Side chains help in the formation of bonds involved in tertiary structure of proteins
- Interaction of polypeptide also helps the three dimensional structure of proteins.
- Chaperones helps in protein folding hence the three dimensional structure.
- But Peptide bond helps in the primary structure.

2.3 BIOCHEMICAL TECHNIQUES

- Separatory Techniques of Proteins
- Precipitation of Proteins
- Electrophoresis
- Methods to Quantitate Proteins
- Chromatography

SEPARATORY TECHNIQUES OF PROTEINS

Different separatory techniques of Proteins
- Salt fractionation
- Ultra Centrifugation
- Electrophoresis
- Chromatography

Salt Fractionation (Salting Out)
Principle
The solubility of proteins is generally lowered at high salt concentrations, an effect called "**salting out**." The addition of a salt in the right amount can selectively precipitate some proteins, while others remain in solution.
Ammonium sulfate $((NH_4)_2SO_4)$ is often used for this purpose because of its high solubility in water.

Ultracentrifugation
Principle
Method to separate the protein based on **the mass, density**, to an extend to the shape also.

Electrophoresis
- **Migration of charged particle in an electric field is called Electrophoresis.**

Concept of Electrophoresis
- Sample is applied usually at the cathode end and the analyte moves towards the anode.
- So negatively charged particles moves faster.
- If a mixture of amino acids are separated then the negatively charged amino acid moves faster.

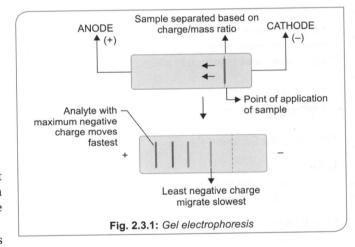

Fig. 2.3.1: *Gel electrophoresis*

Important Types of Electrophoresis

Type of electrophoresis	Supporting media used	Property used to separate the proteins
Agarose gel Electrophoresis	Agarose gel	Based on charge to mass ratio
Cellulose acetate Electrophoresis	Cellulose acetate membrane	
Poly acrylamide Gel Electrophoresis (PAGE)	Polymer of Acrylamide	Based on charge to mass ratio and molecular weight (Size)
Sodium Dodecyl Sulphate (SDS)-PAGE	Sodium dodecyl Sulphate and Polyacrylamide	Based on molecular weight (Size)
Capillary electrophoresis	Separation done in a capillary tube.	Based on the method of electrophoresis used
Isoelectric focusing	Supporting media with pH gradient	Based on isoelectric pH

Two-Dimensional Electrophoresis

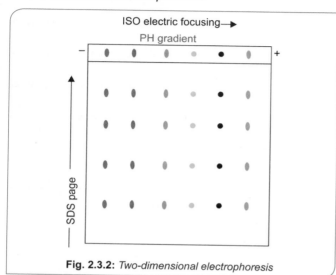

Fig. 2.3.2: *Two-dimensional electrophoresis*

In one direction SDS-PAGE and in the other direction isoelectric focusing. So separation based on both **molecular weight (Size) and isoelectric pH**.

Densitometry
- Method to quantitate the separated protein.

Compare the Different PAGES

PAGE
- Protein is separated based on molecular mass or molecular weight/size and charge.

SDS-PAGE
- SDS impart equal negative charge so that it masks the inherent charge of the Protein
- Now Proteins separate based on molecular weight (Size) only.

SDS-PAGE in conjunction with 2 Mercaptoethanol or dithiothreitol
- Oxidatively cleave disulphide bond.
- So separate the components of multimeric proteins.

CHROMATOGRAPHY

The process by which components of a mixture are separated by differential distribution between a mobile phase and stationary phase. This technique invented by Mikhail Tswett. Chroma = Color Graphein = to write. So Chromatography is color writing.

Classification of Chromatography

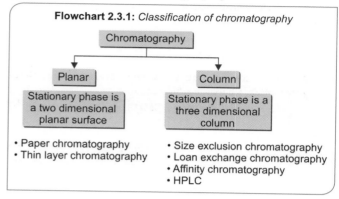

Flowchart 2.3.1: *Classification of chromatography*

PLANAR CHROMATOGRAPHY

Thin Layer/Paper Chromatography
- Stationary phase is water or polar solvent coated on to paper fibres in paper chromatography.
- Mobile phase is a mixture of nonpolar solvents.
- Sample is applied on stationary phase.
- Sample is distributed on stationary phase by the mobile phase.
- The sample is applied on the support media.
- As the non polar solvent ascend on the stationary phase, the non polar component of the analyte move along the mobile phase, but polar component will remain in the stationary phase.

Paper Chromatography

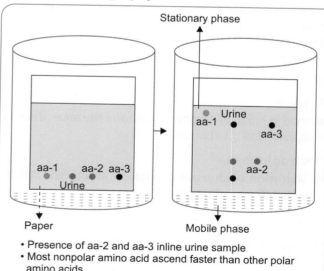

- Presence of aa-2 and aa-3 inline urine sample
- Most nonpolar amino acid ascend faster than other polar amino acids

Fig. 2.3.3: *Paper chromatography*

COLUMN CHROMATOGRAPHY

Size Exclusion Chromatography

- Stationary phase is a column of porous beads.
- Mobile phase is a mixture of analytes with different sizes.
- Smaller particles enter the porous beads and hence it take a labyrinthine path through the column.
- Larger particles cannot enter the pores hence will migrate faster in the column.
- So separation is based on molecular weight or size of the particle.

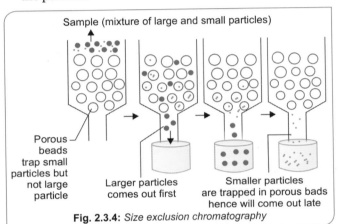

Fig. 2.3.4: *Size exclusion chromatography*

Affinity Chromatography

The beads in the column have a covalently attached chemical group. A protein with affinity for this particular chemical group will bind to the beads in the column, and its migration will be retarded as a result. So this is based on the biological activity.

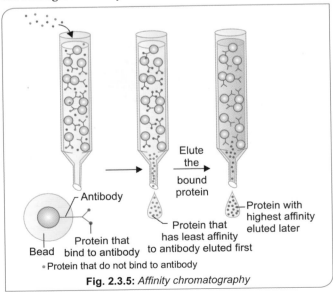

Fig. 2.3.5: *Affinity chromatography*

Ion Exchange Chromatography

This can be anion exchange or cation exchange chromatogrphy. In cation exchange chromatography, the solid matrix has negatively charged groups. In the mobile phase, proteins with a net positive charge migrate through the matrix more slowly than those with a net negative charge, because the migration of the former is retarded more by interaction with the stationary phase. In anion exchange chromatography vice versa. So separation is based on the **charge of the analyte.**

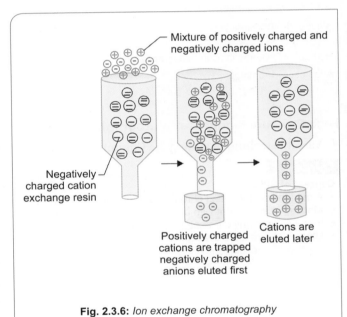

Fig. 2.3.6: *Ion exchange chromatography*

Cation exchange resin
Negatively charged cation exchange resin bind to positively charged cations (⊕)

Anion exchange resin
Positively charged anion exchange resin bind to negatively charged anions (⊖)

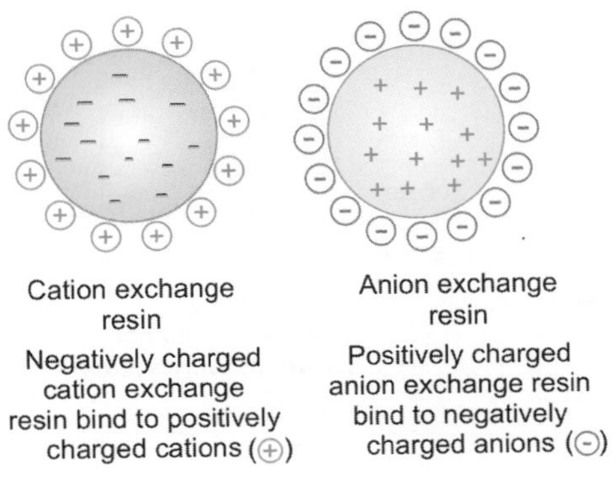

Fig. 2.3.7: *Ion exchange resins*

Important Chromatographic Techniques

Chromatography	Stationary phase used	Property used for separation
Paper chromatography	Water held on a solid support of filter paper (or Cellulose)	Based on the polarity Least polar moves faster
Thin layer chromatography	Silica gel (Kieselguhr) spread on a glass plate or a plastic sheet or aluminum sheet.	Based on polarity Least Polar moves faster
Ion exchange chromatography	Column of Ion exchange resins Anion exchange or Cation exchange resins	Based on charge-charge interaction
Size exclusion chromatography other names Molecular sieve chromatography gel filtration chromatography Gel Permeation Chromatography	Column of porous beads	Based on molecular weight (Size) Particles emerge in the descending order of Stokes Radius^Q
Affinity chromatography^Q	Column of resins bound to specific ligands used	Based on specific ligand binding behavior or biological activity
Hydrophobic interaction chromatography		Based on hydrophobic interaction
Adsorption chromatography		Adsorption and desorption of solute at a solid particle

High Pressure Liquid Chromatography (HPLC)

- Versatile technique among the Column methods of Chromatography

> **Concept Box**
>
> **Concept of HPLC**
> - A modern refinement in chromatographic methods is **HPLC**, or **high-performance liquid chromatography.**
> - Mobile phase solvent with pumped at a high pressure to the column.
> - In the column any chromatographic methods like affinity column method or ion exchange resins can be used
> - Using high pressure the transit time in the column is reduced hence less diffusional spreading of protein bands, So it greatly improve the resolution.

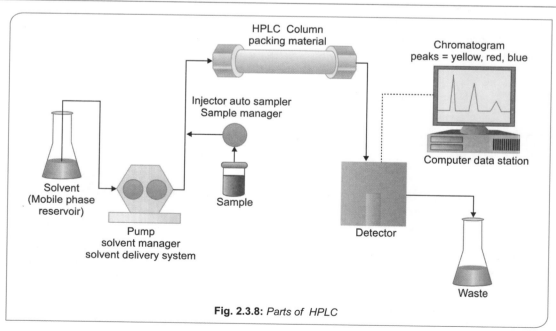

Fig. 2.3.8: *Parts of HPLC*

Advantages

- Separation can be based on different property based on the column (affinity chromatography, ion exchange columns, Size exclusion etc.) used.
- Better resolution and Less Transit Time
- Reproducible results
- Hence high performance (So also called high performance liquid chromatography)
- For separating hemoglobin fraction HPLC^Q is used.
- Method of separation of proteins based on the differential distribution of analyte between stationary and mobile phase.

PRECIPITATION REACTIONS OF PROTEINS

Polar groups of proteins tend to attract water molecules around them to produce a shell of hydration. This makes the protein soluble in water. Any factor that neutralize the charge of protein or remove the shell of hydration will cause precipitation of proteins.

Methods to precipitate the protein by neutralizing the charge are
- Precipitation by heavy metallic salt.
- Precipitation by acids.

Methods to precipitate protein by removing the shell of hydration:
- Precipitation by neutral salts
- Precipitation by organic solvents

- Precipitation by heavy metallic salt
  – Heavy metals like Hg, Zn, Pb etc. provide positively charged ions neutralize the negative charge of the protein.
- Precipitation by acids
  – Acids bring the pH of the medium to isoelectric pH, precipitability is maximum at isoelectric pH. This is because at isoelectric pH proteins carry no net charge, hence no shell of hydration.
  – The reagents used are **Phsophotungstic acid, Sulphosalicylic acid, Phosphomolybdic acid, Trichloroacetic acid.**
- Precipitation by neutral salts
  – Concentrated salt solution removes the shell of hydration.
  – Reagents used are **Ammonium sulphate**. This is called Salting out.
- Precipitation by organic solvents
  – Organic solvents reduce the dielectric constant of the water and decreases the water available for protein, hence it is precipitated.
  – The reagent used are ether, alcohol, acetone, etc.

Methods to Quantitate Proteins

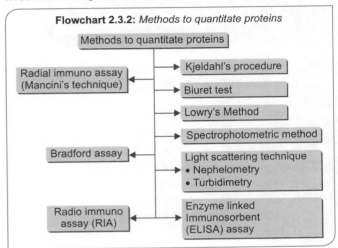

Flowchart 2.3.2: *Methods to quantitate proteins*

Quick Review

- Most specific Chromatographic technique—affinity chromatography
- Most rapid method of separation of proteins is—capillary Electrophoresis
- Method to quantitate the separated protein is densitometry.
- Hydrophobic amino acid moves fastest along the stationary phase of thin layer chromatography or paper chromatography.
- The amino acid which moves fastest in a Thin layer Chromatography/ Paper Chromatography—Isoleucine
- SDS-PAGE used in conjunction 2Mercaptoethanol or dithiothreitol break component polypeptides of multimeric proteins.
- In electrophoresis, negatively charged amino acids and proteins move faster.
- In paper and thin layer chromatography, nonpolar amino acids move faster.
- Methods of separation in various separatory techniques
  - Electrophoresis: Charge to mass ratio
  - Isoelectric focusing: Isoelectric point
  - PAGE: Charge and Molecular weight/size
  - SDS-PAGE: Molecular weight/size
  - 2D Electrophoresis: Isoelectric point and molecular weight/Size
  - Size exclusion chromatography-molecular weight/size
  - Ion exchange chromatography-Charge
  - Affinity Chromatography: Ligand binding behavior

Check List for Revision

- Concept of electrophoresis
- Concept of different chromatographic techniques.
- Methods of separation in all chromatographic techniques.
- Basis of precipitation techniques.
- Names of methods to quantitate proteins

Review Questions MCQ

1. **Best investigation for Metabolic disorders is:** *(AIIMS May 2018)*
   a. Western blot
   b. Tandem mass spectrometry
   c. ELISA
   d. Immunoturbidimetry

2. **Best investigation for HbA1C is:** *(AIIMS May 2018)*
   a. Affinity chromatography
   b. Ion exchange chromatography
   c. High performance liquid chromatography
   d. Electrophoresis

3. **Confirmatory test for proteins are:** *(PGI May 2014)*
   a. Western blot
   b. ELISA
   c. Chip assay
   d. Dot blot

4. **Precipitation of proteins occur in all except:** *(AIIMS Nov 2015)*
   a. Adding, alcohol and acetone
   b. pH changes is moved away from isoelectric pH
   c. With Trichloro acetic acid
   d. With heavy metals

5. **In HbS, Glutamic acid replaced by valine. What will be its electrophoretic mobility?** *(AIIMS Nov 2015)*
   a. Increased
   b. Decreased
   c. No change
   d. Depends on level of concentration of HbS

6. **All of the following are true about Sickle cell disease, Except:** *(AI 2008)*
   a. Single nucleotide change results in change of Glutamine to Valine
   b. RFLP result from a single base change
   c. 'Sticky patch' is generated as a result of replacement of a nonpolar residue with a polar residue
   d. HbS confers resistance against malaria in heterozygotes.

7. **Following SDS PAGE electrophoresis, protein is found to be 100 kDa. After treatment with mercaptoethanol, it shows 2 bands of 20 KDa and 30 KDa widely separated. True statement is:**
   a. Protein has undergone hydrolysis of S-S linkage
   b. It is a dimer of 2 subunits of 20 and 30KDa
   c. It is a tetramer of 220KDa and 230KDa
   d. Protein break down due to noncovalent linkage

8. **Protein is purified using ammonium sulphate by:** *(AIIMS Nov 2010)*
   a. Salting out
   b. Ion exchange chromatography
   c. Mass chromatography
   d. Molecular size exclusion

9. **All of the following can determine the protein structure, except:** *(AIIMS Nov 2008)*
   a. High performance liquid chromatography
   b. Mass spectrometry
   c. X-ray crystallography
   d. NMR spectrometry

10. **Protein separation based on mass/molecular weight (size) is/are done in all except:** *(PGI May 2012)*
    a. Ultrafiltration
    b. Native gel electrophoresis
    c. 2D gel electrophoresis
    d. Gel filtration chromatography
    e. Ultra centrifugation

11. **Method(s) to determine protein structure is/are:** *(PGI 2014)*
    a. X-ray crystallography
    b. NMR spectroscopy
    c. Electrophoresis
    d. Ultra sonography
    e. Infra red spectroscopy

12. **In SDS-PAGE, proteins are separated on basis of:** *(PGI June 2009)*
    a. Mass
    b. Charge
    c. Density
    d. Molecular weight
    e. Solubility

13. **Method of chromatography in which molecules that are negatively charged are selectively released from stationary phase into the positively charged molecules in mobile phase is termed as?** *(AI 2010)*
    a. Affinity chromatography
    b. Ion-exchange chromatography
    c. Adsorption chromatography
    d. Size-exclusion chromatography

14. **Movement of protein from nucleus to cytoplasm can be seen by:** *(AIIMS 2010)*
    a. FISH
    b. FRAP
    c. Confocal microscopy
    d. DNA microscopy

15. **Molecules up to size 4 KD is identified by:**
    a. Gene array chip
    b. Electron spray ionization
    c. Quadruple mass spectrometry
    d. Matrix assisted laser desorption ionisation

16. **Protein fragment separation is/are done by:**
    a. Western blot
    b. Chromatography
    c. Centrifugation
    d. Ultrafiltration
    e. Electrophoresis

Answers to Review Questions

1. **b. Tandem mass spectrometry** *(Ref: Harper 31/e page 278)*

   The most powerful and sensitive test to screen for any metabolic disorder is Tandem Mass Spectrometry)

2. **Ans c. High performance liquid chromatography**

3. **a, b, c, d. Western blot, ELISA, Chip assay, Dot blot**

   Western blot, ELISA, Chip assay and dot blot is based on Antigen antibody interaction. Hence they are confirmatory test for proteins. Chip is the other name for Microarray. Just like DNA Chip, where DNA –DA Hybridisation is done, there Protein Microarray or Protein Chip where Antigen antibody interaction is done.
   - Dot blot or slot blot is a blot technique in which the step blotting to nitrocellulose membrane is not done.
   - This can be used for proteins also.

4. **b. pH changes away from isoelectric pH**
   - Precipitability maximum at isoelectric pH, so pH of medium should be brought to isoelectric pH.

   Precipitation reactions of Proteins

   Methods to precipitate the protein by neutralising the charge are:
   - Precipitation by heavy metallic salt.
     - The reagents used are Mercuric nitrate, Zinc- Sulphate, Lead acetate, Ferric Cholride.
   - Precipitation by acids.
     - Acids bring the pH of the medium to isoelectric pH, precipitability is maximum at isoelectric pH. This is because at isoelectric pH proteins carry no net charge, hence no shell of hydration.
     - The reagents used are Phsophotungstic acid, Sulphosalicylic acid, Phosphomolybdic acids, Trichloroacetic acid.

   Methods to precipitate protein by removing the shell of hydration
   - Precipitation by neutral salts
     - Concentrated Salt solution removes the shell of hydration.
     - Reagents used are Ammonium sulphate. Thisis called Salting out.
   - Precipitation by organicsolvents
     - The reagent used are ether, alcohol, acetone etc.

5. **b. Decreased**

   In HbS, Glutamic acid is replaced by Valine. Hence the negative charge is decreased. So HbS moves slower than HbA1.

The relative mobility of various Hb fractions in Hb electrophoresis.

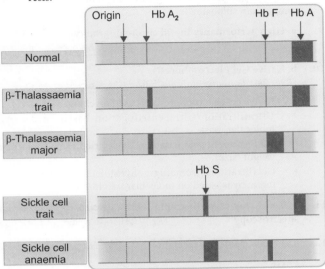

Relative mobility of common Hb fractions from origin or point of application is
1. HbA2,   2. HbS,   3. HbF,   4. HbA1

6. **c. 'Sticky patch' is generated as a result of replacement of a non polar residue with a polar residue**

   Molecular basis of Sickle Cell Disease

   Point mutation in the sixth codon of β-globin (GAG→GUG) that leads to the replacement of a Glutamate (Polar) residue with a Valine (Nonpolar) residue

   Investigations
   - Hb electrophoresis—HbS migrates lower in an electric field because less negative due to glu replaced by valine
   - Sickling Test
   - HPLC to fractionate Hb
   - Isoelectric Focussing
   - Allele-specific **oligonucleotide** probe detects hemoglobin (HbS)allele.
   - Direct diagnosis of sickle cell disease using RFLP

7. **c. It is a tetramer of 220KDa and 230KDa** *(Ref: Harper 31/e page 26)*

   SDS-PAGE in conjunction with 2 Mercaptoethanol or dithiotreitol
   - Oxidatively cleave disulphide bond.
   - So separate the components of multimeric proteins

   Option A: Protein is undergoing oxidative cleavage of S-S bond not hydrolysis

   Option B: The protein is 100KDa. Two subunits of 20KDa and 30KDa will not make a 100 KDa protein.

   Option D: Disulphide bond is a covalent bond not a noncovalent bond.

8. **a. Salting out** *(Ref: Harper 31/e page24 Table 1.1)*

   Salt Fractionation (salting out): Protein is purified using ammonium sulphate.

9. **a. High performance liquid chromatography**
   (Ref: Harper 31/e page24 Table 1.1)

10. **b. Native Gel Electrophoresis**
    (Ref: Tietz Fundamentals of Clinical Chemistry 5th page135)
    - Native gel electrophoresis is based on charge.
    - Ultrafiltration/Ultracentrifugation-based on density, molecular mass
    - 2D Electrophoresis based on isoelectricp H and molecular weight (size)
    - Gel filtration chromatography/Molecular sieve chromatography is based on molecular weight/size

11. **a, b, e. X-ray crystallography, NMR spectroscopy, Infra red spectroscopy** *(Ref: Harper 31/e page 39 Table 1.1)*

12. **a, d. Mass, Molecular weight** *(Ref: Harper 31/e page 26)*

    PAGE
    - Protein is separated based on molecular mass or molecular weight/size and charge

    SDS PAGE
    - SD Simp art equal negative charge so that it masks the inherent charge of the Protein
    - Now Proteins separate based on molecular weight (Size) only SDS-PAGE in conjunction with 2 Mercaptoethanol or dithiothreitol

- Oxidatively cleave disulphide bond.
So separate the components of multimeric proteins

13. **b. Ion-exchange chromatography**

14. **b. FRAP**

    FRA Pisa technique used to study Fluid mosaic model of cell membrane, movement of proteins etc.

    FRAP is fluorescence recovery after photobleaching

    The technique

    Flourescent dyes emit coloured light when it is illuminated, but if a very high intensity light is used then the sedyes are unable to fluorescence. Otherwise called photo bleached. Later it recover fluorescence

    This recovery after photobleaching is used to study movement of proteins lipids carbohydrates. etc.

15. **c. Quadruple mass spectrometry** *(Ref: Harper 31/e page 29)*

    Quadruple mass spectrometers generally are used to determine the masses of molecules of 4000 Da or less, whereas time-of-flight mass spectrometers are used to determine the large masses of complete proteins.

16. **b, c, d, e. Chromatography, Centrifugation, Ultrafiltration, Electrophoresis**
    - Chromatography, Centrifugation, Ultrafiltration, electrophoresis are separatory techniques.
    - Western blot is method to detect protein by Antigen antibody interaction

2.4 FIBROUS PROTEINS

- Collagen
- Fibrillin
- Elastin
- Laminin
- Keratin

COLLAGEN

The **major structural protein found in Extracellular matrix** (Connective tissue)
- Most abundant protein in the body.
- Present in all the tissues of the body.
- Highest concentration in the Skin (74%), followed by Cornea (64%).

Distribution of Collagen

Types of Collagen and their Tissue Distribution

Type	Distribution	Type	Distribution
I	Non-cartilaginous **connective tissues, including bone, tendon, skin**	XV	Associated with collagens close to basement membranes in many tissues including in eye, muscle, microvessels
II	**Cartilage, vitreous humor**	XVI	Many tissues

Contd...

Contd...

Type	Distribution	Type	Distribution
III	Extensible connective tissues, including skin, lung, **vascular system**	XVII	Epithelia, skin **hemidesmosomes**
IV	**Basement membranes**	XVIII	Associated with collagens close to basement membranes, close structural homologue of XV
V	Minor component in tissues containing collagen I	XIX	Rare, basement membranes, **rhabdomyosarcoma cells**
VI	Muscle and most connective tissues	XX	Many tissues, particularly corneal epithelium
VII	**Dermal-epidermal junction**	XXI	Many tissues
VIII	Endothelium and other tissues	XXII	Tissue junctions, including cartilage-synovial fluid, hair follicle-dermis
IX	Tissues containing collagen II	XXIII	Limited in tissues, mainly transmembrane and shed forms
X	Hypertrophic cartilage	XXIV	Developing cornea and bone
XI	Tissues containing collagen II	XXV	**Brain**
XII	Tissues containing collagen I	XXVI	**Testis, ovary**
XIII	Many tissues, including neuromuscular junction and skin	XXVII	Embryonic cartilage and other developing tissues, cartilage in adults
XIV	Tissues containing collagen I	XXVIII	Basement membrane around Schwann cells

HIGH YIELD POINTS- DISTRIBUTION OF COLLAGEN

Types of Collagen[Q]
- Major Collagen present in bone-Type I (90%)
- Collagen absent in bone is Type I Collagen
- Major Collagen present in Dermis, ligaments and tendons- Type I (80%)
- Major Collagen present in Cartilage-Type II (40–50%)
- Collagen Type in dermoepidermal junction is Type VII
- Major Collagen present in Hypertrophic Cartilage-Type X
- Major Collagen present in Aorta-Type I and Type III (20–40% each)
- Major Collagen present in Basement Membrane- Type IV
- Major Collagen present in Skin hemidesmosomes - Type XVII
- Major Collagen present in Rhabdomyosarcoma cells-Type XIX
- Most abundant Collagen-Type I

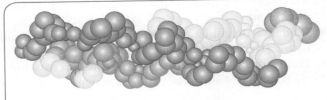

Fig. 2.4.1: *Triple helix–collagen*

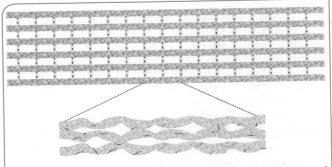

Fig. 2.4.2: *Covalent cross links and quarter staggered arrangement of collagen*

Structure of Collagen
- **Characteristic features glycine-X-Y repeats**
- Every third amino acid residue in collagen is a **glycine residue.**
- **Alpha Chain**-Polyproline helix of three residues per turn twisted in **left-handed direction.**
- Each polypeptide chain contains **1000 amino acids.**
- Triple helical structure
- Three of the alpha chains are then wounding to a **right-handed superhelix.**

"Quarter Staggered" Arrangement[Q]
- Lateral association of the triple helical units
- Each is displaced longitudinally from its neighbor by slightly less than one-quarter of its length
- Responsible for tensile strength of Collagen Fibres.

Synthesis of Collagen

Synthesis of collagen can be divided into
1. Intracellular events
   - Takes place inside Rough endoplasmic reticulum of fibroblast.
   - Procollagen is formed
2. Extracellular events
   - Takes place in the extra cellular matrix.
   - Tropocollagen is formed

Intracellular Events[Q]

- Cleavage of signal peptide
- **Hydroxylation** of prolyl residues and some lysyl residues
- **Glycosylation** of some hydroxylysyl residues
- Formation of intrachain and interchain S–S bonds in extension peptides
- Formation of **triple helix**

Extracellular Events[Q]

- Cleavage of amino and carboxyl terminal propeptides
- Assembly of collagen fibers in **quarter-staggered alignment**
- Oxidative deamination of amino groups of lysyl and hydroxylysyl residues to aldehydes.
- Formation of **intra and interchain covalent cross-links** via Schiff bases and aldol condensation products.

Unique Events in Collagen Formation

- Hydroxylation
  - Post-translational modification occurring intracellularly
  - Enzyme—Prolyl and Lysyl Hydroxylase
  - Coenzyme—**Vitamin C (Ascorbic Acid) and alpha Ketoglutarate.**
  - Essential for the three chains of the monomer to fold into a triple helix at body temperature.
- Glycosylation
  - Intracellular event
  - **Hydroxylysine** residues are glycosylated with galactose or glucose
  - **By type III O-glycosidic linkage.**
- Oxidative deamination
  - Extracellular event
  - Enzyme -**Lysyl Oxidase**
  - Cofactor-Copper
  - Reaction-**Oxidative deamination of Lysyl and Hydroxylysyl** residues to form Aldehydes

- Covalent Cross links
  - Covalent Cross links by aldol condensation of modified lysyl and hydroxylysyl residues by oxidative deamination.
  - Provide tensile strength to collagen.

Diseases Associated with Collagen[Q]

Type of collagen	Gene or enzyme	Disease
Type I	COL1A1 and COL1A2	Osteogenesis imperfecta Osteoporosis Ehlers-Danlos syndrome (Type VII EDS)
Type II	COL2A1	Chondrodysplasias Osteoarthritis
Type III	COL3A1	Ehlers-Danlos syndrome (Type IV EDS)
Type IV	COL4A3–COL4A6	Alport syndrome (including both autosomal and X-linked forms)
Type V and Type I	COL 5A1, COL5A2 COL1A1	Classical EDS
Type III	COL3A1 Tenascin XB (TNXB)	Hypermobile EDS (Type III EDS)
Type VII	COL7A1	Epidermolysis bullosa, dystrophic
Type X	COL10A1	Schmid metaphyseal chondrodysplasia
Lysyl hydroxylase	Lysyl hydroxylase	Ehlers-Danlos syndrome (Type VI EDS) (Kyphoscoliotic EDS) Scurvy
	Procollagen N-proteinase (also called as ADAM TS2)	Ehlers-Danlos syndrome (Type VII autosomal recessive) Dermatosparaxis type
Lysyl oxidase	Lysyl oxidase	Menkes disease

Clinical Correlation: Fibrous Proteins

Ehlers-Danlos Syndrome (EDS) or Cutis Hyper Elastica

Types of EDS

Type of EDS	Metabolic defect/Collagen affected	Villefranche subtype
Type I EDS	Type I and Type V Collagen	Classical EDS (Severe form)
Type II EDS	Type I and Type V Collagen	Classical EDS (Mild form)
Type III EDS	Type III Collagen/Tenascin X	Hypermobile EDS
Type IV EDS	Type III Collagen	Vascular EDS (Most serious EDS)
Type VI EDS	Lysyl Hydroxylase	Kyphoscoliotic
Type VII EDS	Procollagen N Proteinase/ ADAM TS2	Dermatosparaxis

High Yield Points

Points to Ponder–EDS
Most Common Collagen affected is **Type III Collagen**.
Most common inheritance is **Autosomal dominant**.
Most serious is Type IV EDS (Vascular) affecting Type III Collagen. Classical manifestation of EDS **are hyperelasticity of skin and hypermobile joints**

Alport Syndrome (AS)^Q (Hereditary Nephritis)

X-linked disorder:
- **Type IV Collagen** is affected
- Clinical features
  - Hematuria
  - Sensorineural deafness
  - Conical deformation of the anterior surface of the lens (lenticonus)

The pathognomonic of classic Alport Syndrome lenticonus together with hematuria.

Achondroplasia

- Best known cause of Chondrodysplasia
- Most common cause of short limb dwarfism.
- Caused by mutation in a gene that codes for the receptor for **fibroblast Growth factor-3 (FGFR-3)**.

ELASTIN

- A connective tissue protein that is responsible for properties of extensibility and elastic recoil in tissues.
- Present in lung, large arterial blood vessels, and some elastic ligaments.

Unique Features of Elastin

Oxidative Deamination and Desmosine^Q Cross-Links in Elastin

- Enzyme-**lysyl oxidase**
- Lysyl residues of tropoelastin are oxidatively deaminated to aldehydes.
- Condensation of three of these lysine-derived aldehydes with an unmodified lysine to form a tetra functional cross-link unique to elastin.

Hydroxylation of Proline

Differences between Collagen and Elastin

Collagen	Elastin
Different types of Collagen present	Only one type exists
Triple helix Structure	No triple helix instead random coil conformations
(Gly-XY)n repeating structure	No (Gly-X-Y) n repeating structure

Contd...

Collagen	Elastin
Presence of hydroxylysine	No hydroxylysine
Glycosylation present	No Glycosylation
Intramolecular aldol cross-links	Intramolecular desmosine cross-links
Presence of extension peptides during biosynthesis	No extension peptides present during biosynthesis

Clinical Correlation-Elastin

Mutations in the **Elastin Gene (Eln)** leads to
- Supravalvular aortic stenosis (**William-Beuren Syndrome**)
- Cutis laxa.

FIBRILLIN

- Glycoprotein
- Structural component of **microfibrils**
- Secreted into the extracellular matrix by **fibroblasts**
- Incorporated into the insoluble **microfibrils**
- Provide a **scaffold** for deposition of elastin
- **Fibrillin-1** is the major fibrillin present
- Fibrillin-2 is important in deposition of microfibrils early in the development

Clinical Correlation-Fibrillin

Marfan Syndrome

By mutations in the **gene (on chromosome 15) for fibrillin-1**.

Symptom box

Triad of Marfan's Syndrome:
- Skeletal changes
- Ectopia lentis
- Aortic aneurysms.

BIOCHEMICAL BASIS OF CHARACTERISTIC CLINICAL FEATURES IN MARFAN SYNDROME

Explained by location of Fibrillin 1
- Zonular fibers of the **lens**, hence ectopia lentis
- **Periosteum**, hence arachnodactyly
- Elastin fibers in the **aorta**, hence aortic aneurysms.

Ghent Criteria:

An international Criteria to classify the Marfan Syndrome. Fibrillin-1 gene mutation is also recently found to be associated with
- Acromicric dysplasia
- Geleophysic dysplasia

Congenital Contractural Arachnodactyly

- Mutation in **Fibrillin 2** located in Chromosome 5
- Fibrillin-2 may be important in deposition of microfibrils early in development
- Presence of **contractures.**

KERATIN

- **Alpha Helix coiled coil structure**, i.e. two alpha-helix are wind around one another to form a super helix.
- They belong to the family of **Intermediate filament (IF).**
- As it is an alpha helix coiled coil, Keratin is rich in **hydrophobic amino acids** like **Ala, Leu, Met, Val, Phe**.
- Cross links are formed by **disulphide Bond.**
- **Cysteine is involved in the disulphide bond.**
- The **more the disulphide bond,** the harder the Keratin
- Protein present in the **hair nails and outer layer of skin.**

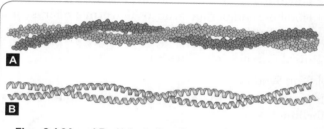

Figs. 2.4.3A and B: *Alpha helix coiled coil structure of Keratin*

Clinical Correlation-Keratin

Epidermolysis Bullosa Simplex

Mutations in the genes for the major keratins of basal epithelial cells **(keratins 5 and 14).**

LAMININ

- It is a major protein component of renal glomerular and other basal laminas.
- **Elongated cruciform shape.**

Basal Lamina

The primary components of the basal lamina are three proteins:
1. Laminin
2. Entactin
3. Type IV collagen

Glycosaminoglycans present in the Basal Lamina are:
1. Heparin
2. Heparan sulphate.

Quick Review

- Most abundant protein in humans is collagen
- Major collagen in bones is Type I
- Major collagen in cartilage is Type II
- Major collagen in basal lamina is Type IV.
- Type II Collagen is absent in the bones
- Most abundant amino acid in collagen is Glycine.
- Triple helix is present is collagen
- Quarter staggered arrangement of fibres is present in collagen.
- Prolyl and Lysyl hydroxylases need Vitamin C.
- Collagen is a Glycoprotein.
- Glycosylation takes place in a hydroxylysine residue.
- Classical EDS affect Type I and V collagen.
- Most EDS inherited autosomal dominant pattern.
- Most common collagen affected in EDS is Type III.
- Elastin has no hydroxylysine and no Gly-X-Y repeat.
- Fibrillin -1 mutation affect Marfan Syndrome.
- Alpha helix coiled coils are seen in Keratin

Check List for Revision

- Collagen learn all the points.
- For Elastin, Fibrillin, Keratin learn only bold letters and tables.

Review Questions (MCQ)

1. Type I Collagen is not present in: *(AIIMS May 2018)*
   a. Bone
   b. Hyaline Cartilage
   c. Ligament
   d. Aponeurosis

2. A 12-year-old boy have cut in forearms 4 days back. Now granulation tissue is seen there. The type of collagen found predominantly in wound healing granulation tissue after trauma is: *(AIIMS May 2018)*
   a. Type 1
   b. Type 2
   c. Type 3
   d. Type 4

3. Which enzyme defect is seen in Menke's disease?
   (Recent Question Jan 2019)
   a. Lysyl hydroxylase
   b. Lysyl oxidase
   c. Prolyl hydroxylase
   d. Prolyl oxidase

4. Type of collagen maximum in skin is:
   (Recent Question Jan 2019)
   a. Type I
   b. Type II
   c. Type III
   d. Type IV

5. **Copper containing enzyme is:**
   (Central Institute Nov 2018)
   a. Lysyl oxidase
   b. Lysyl Hydroxylase
   c. Prolyl hydroxylase
   d. Prolyl oxidase

6. **Collagen of which type is found in hyaline cartilage?**
   (AIIMS Nov 2007)
   a. Type I
   b. Type II
   c. Type III
   d. Type IV

7. **True about collagen:** *(PGI May 2011)*
   a. Triple helix
   b. β pleated structure
   c. Vitamin C is necessary for post-translational modification
   d. Glycine residue at every third position

8. **Keratin is present in both skin and nail. But nail is harder than skin. The reason is:** *(AI 2012)*
   a. Increased no of disulphide bonds
   b. Decreased no of water molecules
   c. Increased Na content
   d. Increased hydrogen bonds

9. **The structural proteins are involved in maintaining the shape of a cell or in the formation of matrices in the body. The shape of these protein is:** *(AI 2006)*
   a. Globular
   b. Fibrous
   c. Stretch of beads
   d. Planar

10. **Quarter staggered arrangement is seen in:**
    a. Immunoglobulin
    b. Hemoglobin
    c. Collagen
    d. Keratin

11. **All of the following are required for hydroxylation of proline in collagen synthesis except:** *(AI 1997)*
    a. O_2
    b. Vitamin C
    c. Monooxygenases
    d. Pyridoxal phosphate

12. **Major type of collagen in basement membrane:**
    (AIIMS May 2015)
    a. Type I
    b. Type II
    c. Type III
    d. Type IV

Answers to Review Questions

1. **b. Hyaline cartilage** *(Ref: Harper 31/e page 593)*

   Type I collagen is present in non cartilaginous tissues like bone, tendon, aponeurosis etc
   But Type II collagen is present in Cartilage & Vitreous humour

2. **c. Type III** *(Ref: Harper 31/e page 593)*

3. **b. Lysyl oxidase** *(Ref: Harper 31/e page 595)*

   Menke's disease affect collagen crosslinking of Collagen & Elastin
   Collagen disorders due to enzyme defect
   - Lysyl oxidase: Menke's disease
   - Lysyl hydroxylase: EDS Kyphoscoliotic, Scurvy

4. **a. Type I Collagen**

   In dermis 80% is Type I Collagen

5. **a. Lysyl oxidase** *(Ref: Harper 31/e page 595,596)*
   - Lysyl hydroxylase and Prolyl hydroxylase are alpha keto-glutarate containing iron linked hydroxylases that require Vit C also.

6. **b. Type II** *(Ref: Harper 31/e page 617)*

   Types of Collagen
   - Major Collagen present in Bone: Type I (90%)
   - Major Collagen present I Dermis, ligaments & tendons- Type I (80%)
   - Major Collagen present in Cartilage: Type II (40–50%)
   - Major Collagen present in Hypertrophic Cartilage: Type X
   - Major Collagen present in Aorta: Type I & Type III (20–40% each)
   - Major Collagen present in Basement Membrane: Type IV
   - Most abundant Collagen: Type I
   - Major Collagen in Keloid is Type 3>> Type 1

   Collagen in Wound Healing[Q]

   At first a provisional matrix containing fibrin, plasma fibronectin, and type III collagen is formed, but this is replaced by a matrix composed primarily of type I collagen

7. **a, c, d. Triple helix, Vitamin C is necessary for post-translational modification, Glycine residue at every third position**
   (Ref: Harper's 31/e page 593)

   Structure of Collagen
   Characteristic Features
   Glycine-X-Y repeats
   Every third amino acid residue in collagen is a glycine residue.

Alpha Chain
Polyproline helix of three residues per turn twisted in left-handed direction.
Each polypeptide chain contains 1000 Amino Acids
Triple helical structure
Three of these alpha chains are then wound into a right-handed super helix.
"Quarter Staggered" arrangement
Lateral association of the triple helical units
Each is displaced longitudinally from its neighbour by slightly less than one-quarter of its length
Responsible for tensile strength of Collagen Fibres

8. **a. Increase no disulphide bond**
   - The more the disulphide bond the harder the protein.
   - Keratin is rich in cysteine.

9. **b. Fibrous** *(Ref: Vasudevananr Sreekumari 7/e page 43)*

   Based on the shape of protein are classified into:
   - Fibrous Protein
     - Elongated or Needle shaped or long cylindrical or rodlike
     - Minimum Solubility in water
     - Regular Secondary Structure
     - Axial Ratio >10
     - They are Structural Proteins, e.g. Collagen, Elastin, Keratin
   - **Globular Proteins**
     - Spherical or oval or Spheroidal in shape
     - Easily Soluble
     - Axial Ratio < 3
     - They perform dynamic functions, e.g. Albumin, Globulin, most enzymes

10. **c. Collagen** *(Ref: Harper 31/e page 593)*
    - Triple Helix in Collagen
    - Quarter Staggered arrangement in Collagen
    - Covalent cross links in collagen
    - Desmosine cross links in elastin

11. **d. Pyridoxal phosphate** *(Ref: Harper 31/e page 593)*
    - Prolyl and Lysyl Hydroxylases are Monooxygenases, require molecular O_2, Vitamin C and, α Keto Glutarate.

12. **d. Type IV** *(Ref: Harper 31/e page 594)*

    Key points distribution of collagen
    - Type of collagen present in the noncartilaginous connective tissue, like bone, tendon: Type I
    - Type of collagen present in the Cartilage: Type II
    - Type of collagen present in the vitreous humour: Type II
    - Type of collagen present in the basement membrane: Mainly Type IV (rarely Type XIX)
    - Type of collagen present in the dermal epidermal junction: Type VII
    - Type of collagen present in the extensible connective tissue like skin, lung, vascular system: Type III
    - Type of collagen present in the hypertrophic cartilage: Type X
    - Type of collagen present in the skin hemidesmosomes: Type XVII
    - Type of collagen present in the rhabdomyosarcoma cell: Type XIX
    - Type of collagen present in the brain: Type XXV
    - Type of collagen present in the testis and ovary: Type XXVI

2.5 PROTEIN FOLDING AND DEGRADATION

- Chaperones
- Protein Degradation
- Protein Misfolding Diseases

CHAPERONES

Proteins are conformationally dynamic molecule that can fold into functionally competent conformation.
Auxilliary Proteins assist protein folding, they are called **Chaperones**.

Properties of Chaperone Proteins
Present in a wide range of species from bacteria to humans
- Many are so-called **heat shock proteins (Hsp)**
- Are **inducible by** conditions that cause unfolding of newly synthesized proteins (e.g. elevated temperature and various chemicals)
- They bind to **predominantly hydrophobic** regions of unfolded proteins and prevent their aggregation.
- They act in part as **a quality control or editing** mechanism for detecting misfolded or otherwise defective proteins.
- Most chaperones show associated **ATPase** activity.

Molecular Chaperones[Q]
- Hsp 70
- Hsp 90
- Hsp 40 (Cochaperone)
- BiP (Immunoglobulin heavy chain binding protein)
- Glucose regulated Protein (GRP-94)
- Calreticulin
- Calnexin

Enzymes assist folding are:
- Protein disulphide isomerase[Q]
- Peptidyl Prolyl Isomerase[Q]

Chaperonins

- The second major class of chaperones.
- Hsp 60 family of chaperones, sometimes called **Chaperonins**.
- They form complex **barrel-like structures** in which an unfolded protein is retained, giving it time and suitable conditions in which to fold properly.
- For example mt GroEL chaperonin.

PROTEIN DEGRADATION

- Intracellular proteases hydrolyze internal peptide bonds.
- The resulting peptides are then degraded to amino acids
  - By endopeptidases that cleave internal peptide bonds
  - By aminopeptidases and carboxypeptidases that remove amino acids sequentially from the amino- and carboxyltermini, respectively.
- *PEST sequences, regions rich in proline (P), glutamate (E), serine (S), and threonine (T), target some proteins for rapid degradation.*

Two Types of Proteins Degradation

1. ATP independent
2. ATP dependent

ATP Independent Degradation

Proteins that undergo ATP independent degradation are:
- Extracellular proteins
- Membrane-association proteins
- Long lived intracellular proteins

Site: Lysosomes

For example: Blood glycoprotein.

ATP Dependent Degradation

Proteins that undergo ATP dependent degradation:
- Regulatory proteins with short half-lives.
- Abnormal or misfolded proteins

Site: In the cytosol by proteasomal complex:
- ATP dependent mechanism
- Requires **Ubiquitin**.

This is also termed as ERAD [Endoplasmic reticulum Associated degradation] of Proteins

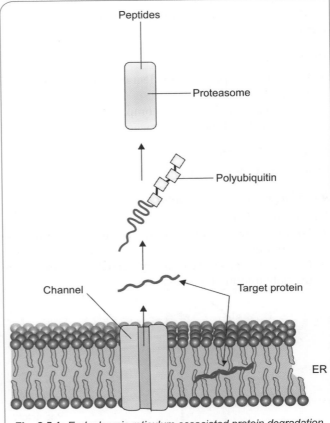

Fig. 2.5.1: *Endoplasmic reticulum associated protein degradation*

Endoplasmic Reticulum Associated Degradation

- Misfolded or incompletely folded proteins interact with chaperones, which retain them in the ER and prevent them from being exported to their final destinations.
- The misfolded proteins are usually disposed of by **endoplasmic reticulum associated degradation (ERAD)**.

Ubiquitin

- Key molecule in **Protein degradation**.
- Small protein with 76 amino acids.
- Highly conserved protein.
- Attachment of Ubiquitin to Protein to be degraded is called **Kiss of Death**.

- Ubiquitin bind to the epsilon amino group of Lysine of the target protein hence it is a **Pseudopeptide or Isopeptide bond or non α peptide bond.**
- Minimum of **four Ubiquitin** molecules must be attached to commit target molecule to degradation.

N end Rule of Ubiquitin Binding

Ubiquitin bind to proteins with PEST [Proline, Glutamic Acid, Serine and Threonine] sequence in the amino terminal.

Proteasome

- Ubiquitinated proteins are degraded in Proteasome.
- Located in the Cytosol.
- Large cylindrical structure composed of 50 subunits.
- This is an ATP dependent process.

Structure of Proteasome

Proteasomes are protein complexes

It is a large cylindrical structure

It is composed of:
- Four rings with a hollow **core** containing the protease active sites
- One or two **caps** or **regulatory particles** that recognize the polyubiquinated substrates

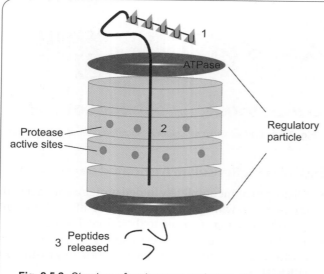

Fig. 2.5.2: Structure of proteasome and steps of proteasomal degradation

Clinical Correlation

Proteasome Inhibitor [Bortezomib]

Used in multiple myeloma for hepatocellular carcinoma.

PROTEIN MISFOLDING DISORDERS

- Prion diseases
- Prion related protein diseases

Human Prion Diseases

Disease	Host	Mechanism of pathogenesis
Kuru	Fore people	Infection through ritualistic cannibalism
Creutzfeldt-Jakob disease (CJD)		
• Iatrogenic CJD	Humans	Infection from prion- contaminated hGH, dura mater grafts, etc.
• Variant CJD	Humans	Infection from bovine prions
• Familial CJD	Humans	Germ-line mutations in PrP gene located in Chromosome 20
• Sporadic CJD	Humans	Somatic mutation or spontaneous conversion of cellular isoform of the prion protein (PrPC) into disease—causing isoform of the prion protein (PrPSc)
Gerstmann-Sträussler- Scheinker (GSS) disease	Humans	Germ-line mutations in PrP gene located in Chromosome 20
Fatal Familial Insomnia (FFI)	Humans	Germ-line mutation in PRNP
Sporadic Familial Insomnia	Humans	Somatic mutation or spontaneous conversion of PrPC into PrPSc

📍 HIGH YIELD POINTS

- **The most common prion disorder in humans Sporadic CJD (sCJD)**
- **The most common etiology of Prion Diseases—Sporadic (85%)**
- **The second most common etiology of Prion Diseases-Germline Mutation (10–15%)**
- **The Prion diseases with noninfectious etiology are sCJD**

Some Terms to Remember

PRNP	PrP gene located on human chromosome 20
PrP	Human prion-related protein
PrPC	Cellular isoform of the prion protein *Monomeric and rich in α helix*
PrPSc	Disease causing isoform of the prion protein. Is *rich in β-sheet*

Biochemical Basis of Prion Diseases

By a conformational Chain Reaction

One pathologic prion or prion-related protein can serve as template for the conformational transformation of many times its number of PrPc molecules.

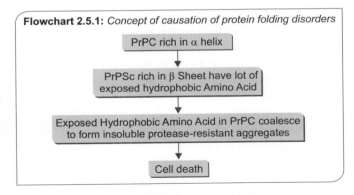

Flowchart 2.5.1: *Concept of causation of protein folding disorders*

PRION RELATED PROTEIN DISEASES[Q]

Prion like changes underlie many neurodegenerative disorders.

Basic mechanism is protein rich in α helix changes to protein rich in β sheet.

- Alzheimer's disease
- Parkinson's disease
- Huntington's disease
- Frontotemporal dementia
- Dementia with Lewy bodies
- Amyloidosis
- Beta thalassemia

Prion Related Protein Diseases and Abnormally Aggregated Proteins

Disease	Abnormally aggregated protein
Alzheimer's disease	Aβ42 Tau
Huntington's disease	Huntingtin
Frontotemporal dementia [FTD]	Tau inclusions pick bodies TDP-43 inclusions, FUS inclusions
Dementia with Lewy Bodies [DLB]	α-synuclein inclusions (Lewy bodies)

Beta-Thalassemias

- During the burst of hemoglobin synthesis that occurs during erythrocyte development, a specific chaperone called α-hemoglobin-stabilizing protein (AHSP) binds to free hemoglobin α-subunits awaiting incorporation into the hemoglobin multimer.
- In the absence of this chaperone, **free α-hemoglobin subunits aggregate,** and the resulting precipitate has cytotoxic effects on the developing erythrocyte.

Protein Misfolding results in Amyloidosis

The proteins that form amyloid fall into two general categories:

1. **Increased production of normal proteins** that have an inherent tendency to fold improperly, associate and form fibrils

   Examples:
   - SAA is synthesized by the liver cells under the influence of cytokines such as IL-6 and IL-1 that are increased during long standing inflammation leads to **AA Amyloid.**
   - **Immunoglobulin light chain** synthesized by the plasma cells increased in Monoclonal B lymphocyte proliferation results in **AL Amyloid**

2. **Mutant proteins** that are prone to misfolding are produced and subsequent aggregation. (NB: No increased production)

   Example:
   - Mutant TTR aggregation in Familial amyloidosis.

Quick Review

- Proteins that assist its folding are called Chaperones.
- Most chaperones are heat shock proteins.
- Enzymes that assist folding are Protein disulphide Isomerase and Peptidyl Prolyl Isomerase.
- Lysosomal degradation is ATP independent.
- Key molecule of protein degradation is ubiquitin.
- Ubiquitin linked proteins are degraded in proteasomes.
- ERAD is Endoplasmic Reticulum associated Degradation of proteins.
- Basic mechanism of prion related protein diseases is protein rich in alpha helix to a protein rich in beta sheet
- PRPSc is rich in beta sheet.

Check List for Revision

- Chaperones definition, names and properties.
- Types of protein degradation
- ERAD
- Protesomes
- Basic pathology of Prion related protein disorders
- Name of Prion diseases and Prion related protein disorders

Review Questions

1. Which is not a protein misfolding disease?
   a. Prion disease
   b. Alzheimer's disease
   c. Beta thalassemia
   d. Ehler's Danlos syndrome

2. Which of the following groups of proteins assist in the folding of other proteins? *(AI 2009)*
   a. Proteases
   b. Proteosomes
   c. Templates
   d. Chaperones

3. All are TRUE about chaperones, except:
   a. Many of them are known as heat shock proteins
   b. They use energy during the protein-chaperone interaction
   c. Ubiquitin is one of the most important chaperone
   d. They are present in wide range of species from bacteria to human

4. Amyloid protein in human beingis: *(Recent Question)*
   a. A naturally present protein in normal individuals
   b. Involves selectively blood vessels
   c. Is visible by naked eyes as whitish cheesy material
   d. A material which gets deposited in extracellular spaces

5. True about chaperones: *(PGI May 2012)*
   a. Belong to heat shock proteins
   b. Wide range of Expression
   c. Present from bacteria to human
   d. Ubiquitin is the most important chaperones
   e. Also known as Stress Proteins

6. These quence that target proteins to lysosome is: *(Recent Question)*
   a. Mannose 6 Phosphate
   b. PTS
   c. KDEL
   d. NLS

Answers to Review Questions

1. **d. Ehler's Danlos syndrome** *(Ref: Harper 31/e page 42)*

   The Prion Related Protein Diseases[Q]

   Prion like changes underlie many neurodegenerative diseases like protein rich in α helix changes to protein rich in β sheet.
   - Alzheimer's disease
   - Parkinson's disease
   - Huntington's disease
   - Frontotemporal dementia
   - Dementia with Lewy bodies
   - Amyloidosis
   - Beta thalassemia

2. **d. Chaperones** *(Ref: Harpers 31/e page574)*

   Molecular Chaperones are:
   - Hsp70
   - Hsp90
   - Hsp 40 [Co-chaperone]
   - BiP [Immunoglobulin heavy chain binding protein]
   - Glucose Regulated Protein[GRP-94]
   - Calreticulin
   - Calnexin

   Enzymes assist folding are:
   - Protein disulphide Isomerase
   - Peptidyl prolyl Isomerase

3. **c. Ubiquitin is the most importantchaperone** *(Ref: Harpers 30/e page 609)*

   - Congo red staining shows apple-green birefringence under polarized light.
   - By electron microscopy amyloid is seen to be made up largely of continuous, nonbranching fibrils with a diameter of approximately 7.5 to 10 nm.

   Chemical Nature of Amyloid
   - 95% of the amyloid material consists of fibril proteins,
   - 5% of the amyloid material consists of P component and other glycoproteins

4. **d. A material which gets deposited in extracellular spaces** *(Ref: Robbinsand Cotran Pathologicbasis of Disease 9/e Chapter 6)*

   Amyloid Fibrils
   Physical nature of amyloid
   X-ray crystallography and infra red spectroscopy demonstrate acharacteristic cross-β-pleated sheet conformation.

5. **a, b, c, e. Belong to heat shock Proteins, Wide range of Expression, Present from bacteria to human, Also known as Stress Proteins** *(Ref: Harper 31/e page 576 Table 49.2)*

   - Ubiquitin is the key molecule of protein degradation.

6. **a. Mannose 6 Phosphate** *(Ref: Harper 31/e page 574 Table49-1)*

   Protein folding

   Proteins are con formationally dynamic molecule that can fold into functionally competent conformation. Auxilliary proteins assist Protein Folding, they are called Chaperones.

   Properties of Chaperone proteins
   - Present in a wide range of species from bacteria to humans

- Many are so-called heat shock proteins (Hsp)
- Are inducible by conditions that cause unfolding of newly synthesized proteins (e.g., elevated temperature and various chemicals)
- They bind to predominantly hydrophobic regions of unfolded proteins and prevent their aggregation.
- They act in part as a quality control or editing mechanism for detecting misfolded or otherwise defective proteins.
- Most chaperones show associated ATPase activity.

Ubiquitin is a key molecule in protein degradation

Ubiquitin
- Key molecule in protein degradation.
- Small protein with 76 amino acids.
- Highly conserved protein.
- Attachment of ubiquitin to protein to be degraded is called Kiss of Death.
- Ubiquitin bind to the ε amino group of Lysine of the target protein hence it is a Pseudopeptide or Isopeptide bond.
- Minimum of four ubiquitin molecules must be attached to commit target molecule to degradation.

2.6 PLASMA PROTEINS AND GLYCOPROTEINS

- ☞ Structure of Immunoglobulin
- ☞ Acute Phase Proteins
- ☞ Classes of Immunoglobulin
- ☞ Plasma Proteins
- ☞ Clinical Correlation—Plasma Proteins
- ☞ Glycoproteins

IMMUNOGLOBULINS

Immunoglobulins, are synthesized mainly in plasma cells (specialized cells of B cell lineage) in response to exposure to a variety of antigens.

Structure of Immunoglobulin

- Consist of 2 heavy chains and 2 light chains
- **Light chain and Heavy chain** is divided into **Constant Region** towards the Carboxyl end and **Variable Region** towards the Amino terminal end.
  – In the Light chain VL and CL
  – In the Heavy chain VH and C_H^1, C_H^2, C_H^3
- Hinge region
  – The region between the C_H^1, and C_H^2, domains
  – The hinge region confers flexibility and allows both Fabarms to move independently, thus helping them to bind to antigenic sites.

Heavy Chain Types in Immunoglobulins

Based on the heavy chain types Immunoglobulins are divided into five classes.
- IgG γ-Chain
- IgA α-Chain
- IgM μ-Chain
- IgD δ-Chain
- IgE ε-Chain

Light Chain Types in Immunoglobulins

Two types:
- κ and λ light chain.
- In a given immunoglobulin either 2 κ or 2 λ and never a mixture of κ and λ
- Most abundant light chain in humans is κ.

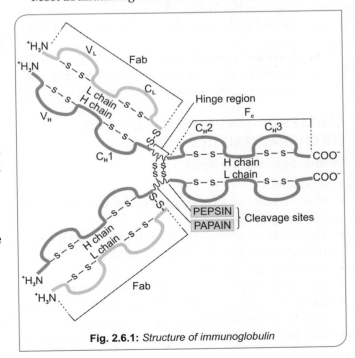

Fig. 2.6.1: Structure of immunoglobulin

Variable Regions of Immunoglobulin
- Consist of the VL and VH domains
- They are quite heterogeneous.
- **Variable regions are comprised of**
  1. Hypervariable regions
     - Hypervariable regions comprise the antigen-binding site (Fab)
     - Hypervariable regions are also termed complementarity-determining regions (CDRs).
  2. Relatively Invariable regions
     - The surrounding polypeptide regions between the hypervariable regions are termed as framework regions.

Proteolytic Cleavage of Immunoglobulin Papain Digestion
- Site of cleavage—beyond the disulphide bond in the Hinge region.
- Products of cleavage—2 separate Fab fragments and 1 Fc fragment.

Pepsin Digestion
- Site of Cleavage—before the disulphide bond in the hinge region.
- Products of Cleavage—Bivalent Fab fragments, $F(ab)_2$ and digested Fc fragments.

Fc and Fab Fractions of Immunoglobulin
- **Fab (Fraction Antibody)**
  - Fragment that bind with antigen.
  - Located in the variable region of Heavy and Light chain.
- **Fc (Fraction Crystallizable)**
  - Remaining part of the immunoglobulin molecule.
  - Concerned with activation of complement cascade.

CLASSES OF IMMUNOGLOBULIN AND ITS CHARACTERISTICS

Immunoglobulin G (IgG)
- Monomer
- Most versatile as IgG can perform all the functions of Immunoglobulin
- Major Immunoglobulin in the Serum (75–80%)
- Subclasses are IgG1, IgG2, IgG3 and IgG4
- Most abundant IgG Subclass is IgG1 (50%)
- Main antibody in secondary immune response
- IgG is the only class of Ig that crosses the placenta. Transfer is mediated by a receptor on placental cells
- Only IgG Subclass that do not cross the placenta is IgG2
- Complement activation is present in the order of
  - IgG3 > IgG1 > IgG2 > IgG4
- Good for opsonization as Fc receptor present on Phagocytic cells
  - IgG1 and IgG3 good for opsonization
  - IgG2 and IgG4 less affinity for Fc receptors

Immunoglobulin A (IgA)
- Two types:
  1. Serum IgA—Monomer
  2. Secretory IgA—Dimer joined by J Chain.
- The main effector of the mucosal immune system.
- The most abundant immunoglobulin of body secretions such as saliva, tears, colostrum and gastrointestinal secretions.
- Normally IgA does not fix complement, unless aggregated [? Alternate pathway]

Secretory Component (Piece) of Secretory Immunoglobulin
- It is made in mucosal epithelial cells and is added to the IgA as it passes into the secretions.
- Function of Secretory piece
  - Protects it from degradation in the secretions
  - Ensures, the appropriate tissue localization of SIgA by anchoring the antibody to mucus lining the epithelial surface.

Immunoglobulin M (IgM)
- Pentamer joined by J chain.
- Largest Immunoglobulin
- IgM is the first Immunoglobulin to be made by the fetus.
- Immunoglobulin involved in primary Immune response.
- Most effective activator of classical complement pathway.

Immunoglobulin D (IgD)
- Found in low levels in serum
- Role in serum uncertain.
- Primarily found on B cell surfaces where it functions as a receptor for antigen.
- Does not bind complement.

Immunoglobulin E (IgE)
- Cytophilic antibody
- Least common Immunoglobulin in the serum.
- Involved in allergic reactions as a consequence of its binding to basophils and mast cells.
- IgE also plays a role in parasitic helminth diseases.
- Does not fix complement.

Proteins

HIGH YIELD POINTS

Quick Revision Immunoglobulins
- Most Common Immunoglobulin present in the serum–IgG
- Least Common Immunoglobulin in the serum–IgE
- Largest Immunoglobulin–IgM
- Immunoglobulin which is a pentamer–IgM
- Immunoglobulin which is a dimer-Secretory—IgA
- **Immunoglobulin which fixes the complement–IgG and IgM.**
- ?IgA by Alternate Pathway
- Immunoglobulin which is present in Secretions–SecretoryIgA
- Shape of an Immunoglobulin monomer–Y Shape.
- Immunoglobulin which crosses placenta–IgG
- IgG which do not crosses the placenta–IgG2
- Immunoglobulin involved in Primary Immune response–IgM
- Immunoglobulin involved in Secondary Immune response–IgG
- Immunoglobulins with J Chain—IgM and IgA.
- Immunoglobulin with secretory piece—IgA

PLASMAPROTEINS

Fractions of Plasma Proteins

By using the method of Salting out Plasma Proteins can be divided into three fragments:

1. Fibrinogen
2. Albumin
3. Globulins

By using the Electrophoresis technique Serum Proteins can be divided into five bands. They are (from anode to cathode)

1. Albumin,
2. $\alpha 1$ Globulins
3. $\alpha 2$ Globulins
4. β Globulins
5. γ Globulins

Instead of Serum Electrophoresis if Plasma Electrophoresis done. What is the difference in electrophoresis pattern?

- To answer this question first know the difference between Serum and Plasma
- Blood collected without adding anticoagulants is Serum and with anticoagulants is Plasma
- Serum = Plasma–Clotting factors
- In Plasma Electrophoresis–Fibrinogen with other clotting factors form a prominent band in the Gamma region
- Clinical Significance: Confused with M Band in Multiple Myeloma.

Image-Based Information

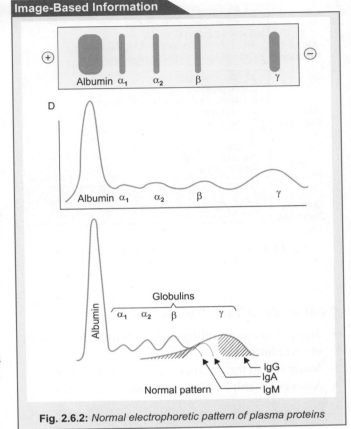

Fig. 2.6.2: *Normal electrophoretic pattern of plasma proteins*

Image-Based Information

Abnormal Electrophoretic Pattern- Hepatic Cirrhosis

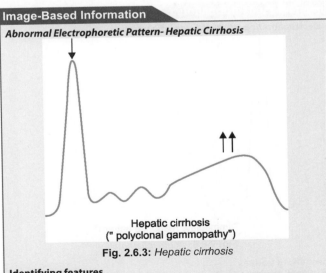

Fig. 2.6.3: *Hepatic cirrhosis*

Identifying features
- Decreased albumin (due to decreased synthesis of albumin)
- Balanced by polyclonal increase gamma globulin (to maintain oncotic pressure)
- Beta–Gamma bridging (due to increased IgA which comes in slow beta region)

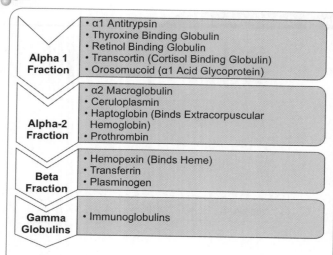

Fig. 2.6.4: *Plasma proteins present in each fractions of electrophoretogram*

Prealbumin or Transthyretin

- Slightly faster mobility than Albumin fraction in electrophoresis.
- Major role in transport of Thyroxine and retinol.
- Associated with Familial and Senile Amyloidosis.

Image-Based Information

Electrophoretic pattern of Acute inflammation

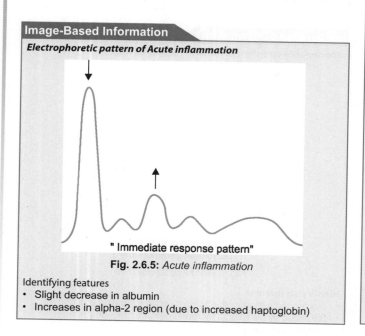

Fig. 2.6.5: *Acute inflammation*

Identifying features
- Slight decrease in albumin
- Increases in alpha-2 region (due to increased haptoglobin)

Image-Based Information

Electrophoretic pattern Nephrotic Syndrome

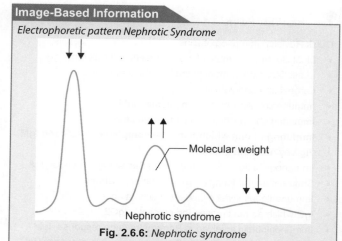

Fig. 2.6.6: *Nephrotic syndrome*

In nephrotic syndrome smaller proteins are lost more rapidly than large molecular weight protein like Alpha 2 Macroglobulin (AMG)
Identifying features
- Decreased albumin, alpha-1, beta and gamma fraction
- But slight increase in alpha-2 fraction (due to retension of AMG)

Image-Based Information

Electrophoretic pattern of Monoclonal gammopathy

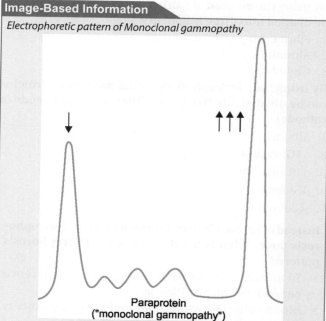

Fig. 2.6.7: *Monoclonal gammopathy*

- Monoclonal gammopathy (Multiple Myeloma) Identifying features
- A sharp monoclonal spike in the gamma region (called as M band)
- Slight decrease in albumin

Functions of Plasma Proteins

Function of the plasma protein	Plasma protein
Maintain the colloid osmotic pressure of the Plasma	Albumin
Nutritional Function	Albumin
Buffering Action in the Plasma	Albumin
Antiproteases	Antichymotrypsin α1-Antitrypsin (α1-antiproteinase) α2-Macroglobulin Antithrombin
Blood clotting	Various coagulation factors, fibrinogen
Hormones	Erythropoietin
Immune defense	Immunoglobulins, complement proteins, α2-microglobulin
Involvement in inflammatory Responses	Acute phase response proteins (e.g. C-reactive protein, α1-acid glycoprotein [orosomucoid])
Oncofetal	α1-Fetoprotein (AFP)

Plasma Proteins in Transport Function^Q

Transport protein	Compound it binds
Albumin	• Bilirubin • Free fatty acids • Ions [Ca^{2+}] • Metals [e.g. Cu^{2+}, Zn^{2+}], • Metheme • Steroids • Hormones
Ceruloplasmin	• Cu^{2+}
Corticosteroid-binding globulin (transcortin)	• Cortisol
Haptoglobin	• Extra corpuscular hemoglobin
Lipoproteins	• Plasma Lipids
Hemopexin	• Heme
Retinol-binding protein	• Retinol
Sex-hormone-binding globulin	• Testosterone • Estradiol
Thyroid-binding globulin	• T_4 • T_3
Transferrin	• Iron
Transthyretin (formerly prealbumin)	• T_4 and forms a complex with retinol-binding protein

HIGH YIELD POINTS

Quick Review of Plasma Proteins
Most abundant Plasma Protein—Albumin
Least abundant Plasma Protein—Alpha-1 Globulin
Plasma Protein with fastest electrophoretic mobility- Prealbumin (Transthyretin) followed by albumin.
Plasma Protein with least electrophoretic mobility- Gamma Globulin.
Most Plasma Proteins are synthesized in the liver except Immunoglobulins by Plasma cells of B Lymphocyte lineage.
Most Plasma Proteins are Glyco proteins except Albumin, which is a Simple protein.
Many Plasma Proteins exhibit Polymorphism. They are:
- α1-antitrypsin
- Haptoglobin
- Transferrin
- Ceruloplasmin
- Immunoglobulins

ACUTE PHASE REACTANTS

Group of plasma proteins whose concentration increases or decreases in response to inflammatory and neoplastic conditions.

Positive Acute Phase Reactants

Plasma proteins whose concentration increases in response to inflammatory and neoplastic conditions.
- C reactive Protein
- Ceruloplasmin
- Haptoglobulin
- Fibrinogen
- α1 Acid Glycoprotein
- α1 antiprotease.

Negative Acute Phase Reactants

Plasma proteins whose concentration decreases in response to inflammatory and neoplastic conditions.
- Albumin
- Transthyretin
- RBP
- Transferrin.

CLINICAL CORRELATION—PLASMA PROTEINS PLASMA CELL DISORDERS

Monoclonal Neoplasm arising from common progenitors in the B lymphocyte lineage. Synonyms are:
- Monoclonal Gammopathy
- Paraproteinemias
- Plasma cell Dyscrasias
- Dysproteinemias.

Biochemical Investigations in Plasma Cell Disorders

- Serum Protein Electrophoresis
  - Sharp spike (Church Spire Spike) in the globulin region is called an **M component (M for monoclonal)**
  - The minimum concentration of monoclonal antibody for M Component to be seen is 5 g/L (0.5g/dL)
  - This is a method for **quantitative**, assessment of the M component.
  - The amount of M component in the serum is a reliable measure of the tumor burden. This makes the M component an excellent **tumor marker.**
  - In approximately 1% of patients with myeloma, **biclonal or triclonal** gammopathy is observed.
- Immunoelectrophoresis
  - This is a method for qualitative assessment of Immunoglobulins.
  - The nature of the M component is variable in plasma cell disorders
  - Type of immunoglobulin is determined by immunoelectrophoresis.
  - IgG myelomas (53%) are more common than IgA and IgD myelomas.
- Urine Bence Jones Protein (BJP)
  - They are Monoclonal Immunoglobulin light chain excreted in urine
  - Seen in 20-30% of Myeloma.

Bradshaw's Test: Test to detect urine BJP
  - Urine layered over a few mL of conc HCl—White ring of Precipitate.

Special heating test for urine BJP
  - This test is based on special property of BJP
  - BJP precipitate when heated between 45°C and 60°C but redissolve when heated >80°C and <45°C
  - 50% False negative test.
- Serum Free Light Chain Assay
  - New Technique to quantitate Serum light chain.
- Serum Alkaline Phosphatase (ALP)
  - It is normal since there is no osteoblastic activity.
- Serum α2 Microglobulin

Single most powerful predictor of Survival Can substitute for staging:
  - Levels <0.004 g/L have a median survival of 43 months
  - Levels >0.004 g/L have a median survival of 12 months.

HIGH YIELD POINTS

Quick Review Points—Plasma Cell Disorders
- Patients secreting lambda light chains have a significantly shorter overall survival than those secreting kappa light chains.
- IgM Myeloma has the greatest tendency for Hyper viscosity.
- Among IgG myelomas, it is the IgG3 subclass that has the highest tendency to hyper viscosity and cold agglutination.
- High labeling index and high levels of lactate dehydrogenase are also associated with poor prognosis.

α1 Antitrypsin (α1 Antiproteinase) Deficiency

- α1-Antiproteinase was formerly called α1-antitrypsin.
- The major component (> 90%) of the α1 fraction of human plasma.
- It is synthesized by **hepatocytes and macrophages.**
- The **principal serine protease inhibitor (serpin, or Pi)** of human plasma.
- It inhibits trypsin, elastase, and certain other proteases, hence accurately called α1 Antiproteinase.
- The major genotype is **MM**, and its phenotypic product is PiM.

Two disorders associated with α1 Antitrypsin (α1 Antiproteinase) Deficiency:
1. Emphysema
2. Cirrhosis of Liver.

Emphysema

- 5% of **emphysema is associated with** α1 Antitrypsin deficiency.
- This occurs mainly in subjects with the ZZ genotype, who synthesize PiZ, and also in PiS Zheterozygotes.
- Both of whom secrete considerably less protein than PiMM individuals.
- Polymorphonuclear white blood cells increase in the lung (e.g. during pneumonia).
- The affected individual lacks a countercheck to proteolytic damage of the lung by proteases such as elastase.

Active Elastase + α1 AT → Inactive Elastase-α1 AT Complex → No Proteolysis of Lung

Active Elastase + Decreased α1 AT → Active Elastase → Proteolysis of Lung → Emphysema

Treatment

- Intravenous administration of α1-antitrypsin (augmentation therapy).
- Gene therapy

Smokers have more chance for developing emphysema

- Methionine residue at 358th position of α 1-antitrypsin is involved in its binding to **proteases.**
- **Smoking** oxidizes this methionine to methionine sulfoxide and thus inactivates it.
- As a result, affected molecules of α1-antitrypsin no longer neutralize proteases.
- This is particularly devastating in patients (e.g. PiZZ phenotype) who already have low levels of α1-antitrypsin.
- Hence smoking results in increased proteolytic destruction of lung tissue, accelerating the development of emphysema.

Cirrhosis of Liver

α1 Antitrypsin deficiency leads to liver disease:
- GAG to AAG mutation in α1 AT gene (Mechanism unknown)
- Glu to Lys substitution results in PiZZ Phenotype
- PiZZ accumulate in the cisternae of Endoplasmic Reticulum and aggregates of Liver.
- Results in hepatitis leading to Cirrhosis.

GLYCOPROTEINS

Proteins that contain oligosaccharide chain (glycans) covalently attached to polypeptides.

Carbohydrate content is less than 5%.

Three major classes of glycoproteins are
1. O-linked Glycoprotein
2. N-linked Glycoprotein
3. GPI-linked Glycoprotein

> **High Yield Points**
>
> Biologically important glycoproteins^Q
> - Plasma proteins (except albumin)
> - Blood group substances
> - Hormones (hCG, TSH, LH, FSH^Q)
> - Enzyme-Alkaline Phosphatase
> - Structural Protein—Collagen
> - Transport Proteins—Ceruloplasmin, Transferrin

Glycation vs Glycosylation

Glycation-Nonenzymic attachment of sugars to amino group of proteins.

Glycosylation-enzymic attachment of sugars to protein.

Comparison of N-linked and O-linked Glycoprotein

Linkage	O-linked Glycoprotein	N-linked Glycoprotein
Amino acid taking place in the linkage	Serine or Threonine i.e. Hydroxyl gp containing amino acids	Asparagine or Glutamine i.e. amide group containing amino acid
Site	Golgi apparatus	Endoplasmic reticulum
Manner in which synthesis takes place	Glycan group is added in stepwise manner	Glycan group is added en bloc
Involvement of Dolichol P-P oligosaccharide	Not involved	Dolichol P-P is involved
Time of synthesis	Post-translationally	Cotranslationally
Inhibition of oligosaccharide transfer by Tunicamycin	Not inhibited	Inhibited

GPI-anchored, or GPI-linked Glycoproteins

Third major class of Glycoproteins.
They link several proteins to Plasma membrane.

Some examples of GPI-linked Proteins^Q
- Acetylcholinesterase (red cell membrane)
- Alkaline phosphatase (intestinal, placental)
- Decay-accelerating factor (red cell membrane)
- 5'-Nucleotidase (T lymphocytes, other cells)

CLINICAL CORRELATION OF GLYCOPROTEINS

Paroxysmal Nocturnal Hemoglobinuria (PNH)

- Acquired mild anemia characterized by the presence of hemoglobin in urine due to hemolysis of red cells, particularly during sleep.
- Basic defect is somatic mutation in the **PIG-A** (Phosphatidyl Inositol Glycan class A) gene.
- Two protein linked by GPI, i.e. defective
- Hemolysis occur particularly during sleep in PNH.

- This is due to a slight drop in plasma pH during sleep, which increases susceptibility to lysis by the complement system.
- A test done to diagnose PNH is **Hams Test**
- A treatment option is Monoclonal Antibody to C5 (a terminal component of Complement System).

> **EXTRA EDGE**
>
> **Advanced Glycation End-Products (Ages)**
> - The end-products of glycation reactions are termed advanced glycation end-products (AGEs).
> - When glucose attaches to a protein, intermediate products formed include Schiff bases.
> - These can further re-arrange by the Amadori re-arrangement to ketoamines.
> - The overall series of reactions is known as the Maillard reaction.
>
> **Medical Importance of AGEs**
> - Aging
> - Atherogenesis
> - Microvascular and macrovascular damage in diabetes mellitus.
>
> **Aminoguanidine**
> - An inhibitor of the formation of AGEs
> - Reduce the complication in Diabetes Mellitus.

> **EXTRA EDGE**
>
> **"SUGAR CODE OF LIFE"**
> Certain oligosaccharide chains encode **biologic Information**. For example, mannose 6-phosphate residues target newly synthesized lysosomal enzymes to that organelle.

Check List for Revision

- Structure of Immunoglobulin
- Papain and Pepsin digestion
- High yield points Immunoglobulins
- Plasma proteins present in each fractions of electrophoretogram
- Overall idea of images of electrophoretogram
- Table of Transport functions of Plasma proteins
- Names of positive and negative acute phase reactants
- Glycoproteins an overview

REVIEW QUESTIONS MCQ

1. **Negative acute phase reactants:** *(PGI May 2017)*
   a. Transferrin
   b. Albumin
   c. Ceruloplasmin
   d. CRP
   e. Ferritin

2. **Immunoglobulins are:** *(Recent Question)*
   a. Proteins
   b. Glycoproteins
   c. Proteoglycan
   d. Glycoside

Glycoproteins

3. **Which of the following amino acid can have o-Glycoxylation linkage in oligosaccharide molecule:** *(AI 2012)*
   a. Asparagine
   b. Glutamine
   c. Serine
   d. Cysteine

ANSWERS TO REVIEW QUESTIONS

1. **a, b. Transferrin, Albumin** *(Ref: Harper 31/e page 630)*

   Negative acute phase proteins are
   - Transferrin
   - Transthyretin
   - Albumin
   - Retinol binding protein

2. **b. Glycoprotein.** *(Ref: Harper 31/e page 639)*

   Biologically Important Glycoproteins^Q
   - Plasma proteins (except albumin)
   - Blood group substances
   - Hormones (hCG, TSH, LH, FSH^Q)
   - Enzyme: Alkaline Phosphatase
   - Structural protein: Collagen
   - Transport proteins: Ceruloplasmin, Transferrin

Glycoproteins

3. **c. Serine** *(Ref: Harper 31/e page 547)*

   Major Classes of Glycoproteins O-linked Glycoprotein
   - O-glycosidic linkage between hydroxyl side chain of serine or threonine and N-acetyl galactosamine is present

   N-linked Glycoprotein
   - N-glycosidic linkage between amide nitrogen of asparagine and N–acetyl glucosamine is present.

   Glycosyl phosphatidylinositol-anchored (GPI-anchored, or GPI-linked) glycoproteins
   - Third major class of Glycoproteins.
   - They link several proteins to Plasma membrane.
   - Linked to the carboxyl terminal amino acid of a protein via a phosphoryl-ethanolamine moiety joined to an oligosaccharide (glycan), which in turn is linked via glucosamine to phosphatidylinositol(PI).

2.7 PROTEIN SORTING

☞ Sorting Decision ☞ Clinical Correlation—Protein Sorting

PROTEIN SORTING

Proteins must travel from polyribosomes, where they are synthesized, to many different sites in the cell to perform their particular functions. This process is called Intracellular Traffic of Proteins or simply Protein Sorting.

Golgi Apparatus is involved in the Glycosylation and Sorting of Proteins.

SORTING DECISION

- The major decision is at the site of synthesis.
- The second by the Signal sequence in the protein.

Site of Synthesis

Two Sorting branches based on the site of synthesis
- Cytosolic Branch
- Rough Endoplasmic Reticulum branch.

Cytosolic Branch (Free Polyribosome)

Proteins synthesized on free polyribosomes lack this particular signal peptide and are delivered to **organelle.**

Organs Targeted in Cytosolic branch

Signal Sequences direct proteins to specific organelles.
- Mitochondria
- Nuclei
- Peroxisomes
- Cytosol

Signal Sequences Direct Proteins to Specific Organelles

Target sequence	Organelle targeted
N terminal Signal Peptide Sequence	ER
Carboxy terminal KDEL Sequence (Lys-Asp-Glu-Leu) HDEL Sequence (His-Asp-Glu-Leu)	Lumen of ER
Diacidic Sequence (Asp-X-Glu)	Golgi Membranes
Amino terminal Sequence or Matrix Targeting Sequences	Mitochondrial Matrix
Nuclear Localization Sequence (NLS)	Nucleus
Peroxisomal Matrix targeting Seguence (PTS) (Ser-Lys-Leu)	Peroxisome
Mannose 6 Phosphate	Lysosome

Rough Endoplasmic Reticulum (RER) Branch

Proteins synthesized on membrane-bound polyribosomes contain a **signal peptide** that mediates their attachment to the membrane of the ER called **Signal Peptide Hypothesis** (explained later).

All the proteins synthesized in membrane bound Polyribosome (RER) destined for various membranes.

Membranes/Organelle Targeted in RER Branch

- Membrane of Endoplasmic Reticulum (ER)
- Membrane of Golgi apparatus (GA)
- Plasma membrane (PM)
- Lysosome

Usually proteins synthesized in RER branch is destined to various membranes. But an exception is proteins to lysosomes.

Signal Peptide Hypothesis

- The signal hypothesis was proposed by **Blobel and Sabatini.**
- To explain the distinction between free and membrane-bound polyribosomes.
- It is proposed that proteins synthesized on membrane-bound polyribosomes contained an N terminal peptide extension (signal peptide) at their amino terminal which causes them to be attached to the membranes of ER, and facilitates the protein transfer into the ER lumen.
- From ER lumen they are further sorted to various membranes and lysosomes.
- Certain ER membrane proteins are transferred directly into the membrane of ER, without reaching the lumen.

Properties of Signal Peptides[Q]

- Located at the amino terminal.
- Contain approximately 12–35 amino acids.
- Methionine is usually the amino terminal amino acid.
- Contain a central cluster of hydrophobic amino acids.
- Contain atleast one positively charged amino acid near their amino terminal.

Exocytic Pathway or Secretory Pathway of Rough Endoplasmic Reticulum Branch

- This was first delineated by George Palade and colleagues.
- The entire pathway of proteins traveling from **ER → GA → Plasma Membrane** is often called Exocytic or Secretory pathway.
- The protein is carried in **vesicles.**

Vesicle	Function and characteristics
Transport Vesicle	They carry proteins to Plasma membrane. The transport of protein occur continuously hence called Constitutive *Secretion.*
Secretory Vesicles	Transport of proteins to be secreted out of the cell, like Insulin from β cells of Pancreas. The secretion is regulated by external signals, hence called Regulated Secretion.
COP I Vesicle	They carry proteins from Golgi apparatus to the Endoplasmic Reticulum. This is called Retrograde Transport.
COP II Vesicle	They carry proteins from Endoplasmic Reticulum to Golgi Apparatus or Endoplasmic Reticulum – Golgi Intermediate Complex, ERGIC. This is called **Anterograde transport.**
Clathrin coated vesicle	Involved in Endocytosis of proteins from late endosome to Lysosomes. This is called *Receptor mediated Endocytosis.*

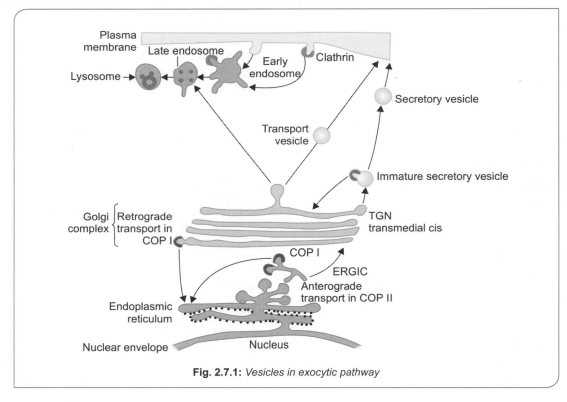

Fig. 2.7.1: *Vesicles in exocytic pathway*

CLINICAL CORRELATION—PROTEIN SORTING

Peroxisome Biogenesis Disorder

- Disorders associated with defect in import of 1 or more proteins to Peroxisome
- Due to mutations in **PEX genes** encoding certain proteins—so-called **peroxins**, involved in various steps of peroxisome biogenesis, such as the import of proteins to peroxisomes.

Peroxisomal Ghosts

Absence or reduction in the number of peroxisomes is pathognomonic for disorders of peroxisome biogenesis. In most disorders, there are membranous sacs that contain peroxisomal integral membrane proteins, which lack the normal complement of matrix proteins; these are peroxisome "ghosts."

High Yield Points

PEROXISOME TARGETING DISORDERS

Due to gene defects, which involve mainly the import of proteins that contain the PTS1 targeting signal
- Zellweger syndrome (most severe)
- Neonatal adrenoleukodystrophy
- Infantile Refsum disease (least severe)

Due to defect in PEX7, which involves the import of proteins that utilize PTS2
- Rhizomelic chondrodysplasia punctata

Zellweger Syndrome (Cerebrohepatorenal Disease)

- Typical facial appearance (high forehead, unslanting palpebral fissures, hypoplastic supraorbital ridges, and epicanthal folds)
- Severe weakness and hypotonia, neonatal seizures
- Eye abnormalities (cataracts, glaucoma, corneal clouding, **Brushfield spots,** pigmentary retinopathy, and nerve dysplasia)
- Because of the hypotonia and "mongoloid" appearance, Down syndrome may be suspected.

Abnormal laboratory findings common to disorders of Peroxisome Biogenesis
Peroxisomes absent to reduced in number
Peroxisomal enzyme Catalase in cytosol
Deficient synthesis and reduced tissue levels of plasmalogens
Defective oxidation and abnormal accumulation of **very long chain fatty acids**
Deficient oxidation and age-dependent accumulation of **phytanic acid**
Defects in certain steps of bile acid formation and accumulation of bile acid intermediates
Defects in oxidation and accumulation of l-pipecolic acid
Increased urinary excretion of dicarboxylic acids

Lysosomal Targeting Disorders

Due to defective targeting of proteins to Lysosomes I (Inclusion) Cell Disease

- Due to lack of targeting signal, **Mannose 6 phosphate.**
- Deficient in the activity of the cis-Golgi-located **GlcNAc phosphotransferase.**
- Lack of transfer of GlcNAc1P residue to specific mannose residue of certain enzymes destined for Lysosomes.
- These enzymes **lack Mannose 6 Phosphate.**
- These enzymes are secreted in to plasma rather than to Lysosomes.
- Lysosomal enzymes are deficient results in partially digested cellular material, as **inclusion bodies.**
- Clinical picture almost similar **to Mucopolysacchridoses.**

Quick Review

- Organelle involved in sorting is **Golgi apparatus**
- Glycosylation of proteins organelle involved is **Golgi apparatus > EPR**

Cytosolic branch synthesize protein in free ribosome. They are targeted to **specific organelle**
- Mitochondria, Nuclei, Peroxisome, Cytosol

Proteins synthesised **in RER branch** are targeted to **membranes**
- Membranes of EPR, GA, Plasma membrane
- Exception is one lysosomes which is not a membrane

Signal peptide is different from signal sequence
- **Signal sequence** target the protein in **cytosolic branch** to organelle.
- Whereas **signal peptide** target protein in **RER branch** to lumen of ER then sorted to various membranes and lysosomes. This is **signal peptide hypothesis** by Blobel and Sabatini.

Secretory Pathway or Exocytic Pathway
- Proteins inside lumen by Signal peptide hypothesis need vesicles
- For **constitutive** secretion – Transport vesicle
- For **regulated secretion** – Secretory vesicle
- From GA to EPR i.e. certain **misfolded proteins** travelling backwards to EPR for degradation i.e. ERAD called **retrograde transport need COP I vesicle**
- From EPR to GA **COP II vesicle**
- For **receptor mediated endocytosis** like uptake of IDL, LDL, Chylomicron remnant etc need **Clathrin coated vesicle**

Zellweger syndrome (Cerebrohepatorenal disease)
- This is peroxisomal biogenesis disorder or peroxisome targeting disorder.
- Defect is **Peroxisome targeting** sequence or PTS-I.

This is similar to **Downs Syndrome.**

Review Questions

1. **Proteins are sorted by:** (AI 2009)
   a. Golgi bodies
   b. Mitochondria
   c. Ribosomes
   d. Nuclear membrane

2. **Endoplasmic reticulum–signal transduction is through:** (Recent Question)
   a. Translocon
   b. Chaperones
   c. Ubiquitin
   d. Mannose 6 phosphate

3. **Not a function of endoplasmic reticulum:** (Recent Question)
   a. Protein synthesis
   b. Muscle contraction
   c. Protein sorting
   d. Glycoproteins

4. Targeting sequence that direct endoplasmic reticulum resident protein in retrograde flow to ER in COPI vesicles:
   a. KDEL
   b. KDAL
   c. DALK
   d. KDUL

5. Secretory proteins are synthesized in: *(AIIMS Nov 05)*
   a. Cytoplasm
   b. Endoplasmic reticulum
   c. First in cytoplasm and then in endoplasmic reticulum
   d. First in endoplasmic reticulum and then in cytoplasm

6. I cell disease is associated with
   (Central Institute Nov 2018)
   a. Lysosomes
   b. Peroxisomes
   c. Mitochondria
   d. Nucleus

7. Zellweger Syndrome is associated with
   (Recent Q Jan 2019)
   a. Lysosome
   b. Peroxisome
   c. Mitochondria
   d. Golgi apparatus

8. In cerebrohepatorenal syndrome, which of the following accumulate in brain? *(AIIMS Nov 2018)*
   a. Pyruvate
   b. Short chain Fatty acid
   c. Very long chain fatty acid
   d. Acetyl CoA

9. Enzyme deficient in I cell disease: *(Recent Question)*
   a. GlcNAc phosphotransferase
   b. Mannose Phosphotransferase
   c. Phosphodiesterase
   d. Mannose 6 phosphate transferase

Answers to Review Questions

1. **a. Golgi bodies** *(Ref: Harper 31/e Page 574)*

   Functions of Golgi bodies
   - O-Glycosylation of proteins
   - Protein sorting
   - Processing of oligosaccharide chains of Glycoproteins

2. **a. Translocon** *(Ref: Harper 31/e page 580)*

   The signal hypothesis proposed that the protein is inserted into the ER membrane at the same time as its mRNA is being translated on polyribosomes, so-called cotranslational insertion.

   Principal components involved in endoplasmic reticulum translocation

   N-terminal signal peptide
   - Polyribosomes
   - SRP, signal recognition particle
   - SR, signal recognition particle receptor
   - Sec 61, the translocon^Q
   - Signal peptidase
   - Associated proteins (e.g. TRAM and TRAP)
   - TRAM is translocating chain-associated membrane Protein. TRAM accelerates the translocation of certain proteins.
   - TRAP, Translocon-associated Protein Complex. The function of TRAP is not clear.

3. **b. Muscle Contraction** *(Ref: Harper 31/e page 574)*
   - Rough endoplasmic reticulum is involved in protein synthesis
   - N-linked glycoproteins are glycosylated in the EPR
   - RER along with GA is a part of exocytic pathway of protein sorting

4. **a. KDEL** *(Ref: Harper 31/e page 574 Table 49-1)*

   Retrograde Transport from the GA

   A number of proteins possess the amino acid sequence KDEL (Lys-Asp-Glu-Leu) at their carboxyl terminal. KDEL-containing proteins first travel to the GA in COPII transport vesicles and interact there with a specific KDEL receptor protein, which retains them transiently. They then return in COPI transport vesicles to the ER, where they dissociate from the receptor, and are thus retrieved.

   HDEL sequences

   (H = histidine) serve a similar purpose. The above processes result in net localization of certain soluble proteins to the ER lumen.

   SEC61 gene, which encodes a channel through which secretory proteins under construction pass into the endoplasmic reticulum lumen

5. **b. Endoplasmic reticulum** *(Ref: Harper 31/e page 574)*
   Secretory proteins are synthesized in the Rough endoplasmic reticulum.

6. **a. Lysosomes**

   I Cell disease (Inclusion cell disease) is a lysosomal targeting disorder.

7. **b. Peroxisome**

8. **c. Very Long Chain Fatty acid**

   Cerebroheptorenal syndrome is Zellweger Syndrome, Peroxisomal targeting disorder. VLCFA oxidation takes place in peroxisomes hence increased VLCFA in Zellweger syndrome.

9. **a. GlcNAc phosphotransferase**

   Inclusion (I) cell disease
   - Lysosomal enzymes targeting disorder
   - Recognition marker of Lysosomal enzymes is Mannose 6 Phosphate
   - I cell disease lack the Golgi located NAcetyl Glucosamine phosphotransferase.

CHAPTER 3

Enzymes

Chapter Outline

3.1 General Enzymology
3.2 Clinical Enzymology

CHAPTER 3

Enzymes

3.1 GENERAL ENZYMOLOGY

- Enzymes
- Coenzymes and Cofactors
- Enzymes Classes
- Mechanism of Enzymes Action
- Factors Affecting Enzyme Activity
- Michaelis Constant
- Enzyme Inhibition
- Regulation of Enzymes

ENZYMES

Enzymes are highly specialized proteins that can act as catalyst in biological system without itself undergoing any chemical change. The word enzymes was coined by Frederick Kuhne meaning "in Yeast".

- **Substrate**: The substance on which enzyme act.
- **Product**: The substance produced by the action of enzyme on the substrate.
- Enzymes are thermo labile and proteinaceous in nature.
- Exception to proteinaceous nature of enzymes are ribozymes

> **HIGH YIELD POINTS**
>
> Ribozyme[Q] is RNA with catalytic activity.
> e.g.: Sn RNA in Spliceosome
>   Ribonuclease P
>   Peptidyl Transferase
>   RNAse H
> *Abzyme are antibodies with catalytic activity*

Enzymes can be of two types

1. Simple Enzyme—Consists of only proteins.
2. A complex enzyme consists of:
   - Protein part: **Apo enzyme**
   - Non-protein part

Apoenzyme + Non-protein part (Prosthetic group/ Cofactor/Coenzyme) = Holoenzyme

Non-protein part can be
- Coenzyme
- Cofactor
- Prosthetic group

COENZYME AND COFACTORS

Coenzyme

- **Thermostable, low mol wt, non-protein organic substances** are called Coenzymes.
- A coenzyme can bind covalently or non-covalently to the enzymes.
- If covalently bound then it is called Prosthetic group.
- Involvement of coenzyme with substrate is so intimate that, coenzyme is called **cosubstrates.**[Q]
- Most coenzymes are water soluble vitamins

Examples of Coenzymes

Enzyme	Coenzyme
Kinases	ATP/GTP
Dehydrogenases	NAD+/FAD
Pyruvate dehydrogenase, Alpha keto dehydrogenase, Branched chain keto acid Dehydrogenase	TPP, Lipoic Acid, CoA, FAD, NAD+
Transketolase	Thiamine pyrophosphate
Transaminase	Pyridoxal phosphate
Carboxylases	Biotin

Cofactor and Prosthetic Group

Prosthetic groups are non-protein **part tightly integrated** to the enzyme structure by covalent forces.

- **Metals are the most common prosthetic group.**
- Enzyme which is tightly bound to metal is called **metalloenzyme**.
- Cofactors associate **reversibly** with enzymes or substrates.

- The most common cofactors are also metals.
- Enzymes which have metal as cofactors, i.e. loosely bound to them are termed **Metal-activated Enzymes**.

Metal as Cofactors and Prosthetic Group

Metal	Enzymes
Zinc[Q]	Carbonic anhydrase[Q], Carboxypeptidase, Alcohol Dehydrogenase[Q], Alkaline Phosphatase, ALA Dehydratase, Lactate Dehydrogenase
Magnesium	Phosphotransferase, (Hexokinase, Phosphofructokinase) Mutase, Enolase, Glucose 6 Phosphatase
Copper	Tyrosinase, Cyt c Oxidase, Lysyl Oxidase, Superoxide Dismutase, L. Amino Acid Oxidase
Molybdenum	Xanthine Oxidase, Sulfite Oxidase, Dinitrogenase
Manganese	Enolase, Arginase, Phosphotransferase (Hexokinase, Phosphofructokinase), Mitochondrial Super Oxide Dismutase, Ribonucleotide reductase
Iron	Succinate Dehydrogenase
Ni	Urease
Calcium	Lipase, Lecithinase

IUBMB CLASSIFICATION—ENZYME CLASSES

International Union of Biochemistry and Molecular Biology (IUBMB) assign a 4 digit unique number to all enzymes. The first digit is the Enzyme Class number or Enzyme Commission Number or Enzyme Code Number or EC Number

Enzymes Classes

IUBMB classify enzymes to 6 major classes. They are as follows:

Enzyme class number	Name of the classes
I	Oxidoreductases
II	Transferases
III	Hydrolases
IV	Lyases
V	Isomerases
VI	Ligases

Oxidoreductases

Enzymes that catalyse oxidation of one substrate with reduction of another substrate.

Subclasses of Oxidoreductases are:
1. Dehydrogenase
2. Oxygenase
   a. Monooxygenase
   b. Dioxygenase
3. Oxidase
4. Catalase
5. Peroxidase

Dehydrogenase

- Transfer of hydrogen from the substrate to reducing equivalents
- Usually reducing equivalents like NAD+ or FAD are the hydrogen acceptors.
- They cannot use oxygen as a hydrogen acceptor.

Examples are:
- Alcohol Dehydrogenase
- Lactate Dehydrogenase
- Succinate Dehydrogenase

Oxidases

- Removal of hydrogen from a substrate with **oxygen as the acceptor** of Hydrogen.
- $2AH_2 + O_2 \longrightarrow A + 2H_2O$

Examples are:
- Cytochrome c Oxidase
- Xanthine Oxidase

Oxygenases

- Catalyze the direct transfer and incorporation of oxygen into a substrate molecule.

Mono Oxygenases (Mixed Function Oxidases)

Incorporate **one atom** of molecular oxygen into the substrate:

$A-H + O_2 + H_2 \longrightarrow A-OH + H_2O$

Examples are:
- Phenyl alanine Hydroxylases
- Tyrosine Hydroxylase
- Tryptophan Hydroxylase
- 7 alpha Hydroxylases
- Cytochrome p450

Remember: Most hydroxylases are monooxygenases. All Cytochromes are monooxygenases

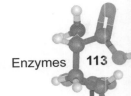

Dioxygenases
Incorporate **both atoms** of molecular Oxygen into the substrate. The basic reaction is shown below:

$$A + O_2 \rightarrow AO_2$$

Examples of Dioxygenase
- Homogentisate oxidase
- Tryptophan pyrrolase (Dioxygenase)

Transferases
Transfer functional group from one substrate to another.

Examples
- Kinases: Transfer phosphate group usually from ATP to a substrate.
- Transmethylases: Transfer methyl group
- Transaminases/Aminotransferases: Transfer of amino group from an amino acid to a ketoacid
- Phosphorylase: Transfer phosphate group from inorganic phosphate to substrate
- Trans ketolase: Transfer ketonic groups
- Transaldolase: Transfer aldehyde groups.

Hydrolases
Enzymes that catalyze hydrolytic cleavage of C—C, C—O, C—N and other covalent bonds.

So this class needs water to break the bond.

To learn this very easily remember the covalent bonds and the enzyme that break these bonds

Macromolecule	Covalent bond	Enzyme
Protein	Peptide bond	Proteases which include Trypsin, Pepsin, Chymotrypsin, Elastase, Carboxypeptidase
Carbohydrates	Glycosidic bond	Amylase
Nucleic acid	3'-5' phosphodiester bond	Endonucleases Exonucleases
Lipids	Ester bond	Esterases Lipases
Arginine		Arginase

Also remember Phosphatases belong to hydrolases

Lyases
Enzymes that catalyze cleavage of C—C, C—O, C—N and other covalent bonds by atom elimination, generating double bonds.
- Decarboxylases which cleave C-C bond and remove CO_2
- Formation of double bond by removal or addition of groups. Examples—Aconitase, Enolase, Fumarase, Aldolase
- All enzymes with lyase in its name i.e. HMG CoA Lyase, Argininosuccinate lyase
- Most synthases belong to lyases.

Isomerases
Enzymes that catalyze geometric or structural changes within a molecule.

Examples are:
- All enzymes with isomerase in its name.
- Racemase
- Mutase –Transfer functional group within the molecule.

Phosphoglucomutase

$$\text{Glucose 6 PO4} \longrightarrow \text{Glucose 1 PO4}$$

Ligase
Enzymes that catalyze the joining together (ligation) of two molecules in reactions coupled to the hydrolysis of ATP.

Examples

Synthetases are ligases. Example, Glutamine synthetase, Carbamoyl Phosphate synthetases

Carboxylases are ligases. Examples are:
- Pyruvate Carboxylases
- Acetyl-CoA Carboxylase
- Propionyl-CoA Carboxylase

All these carboxylases enzymes given above require **ATP and Biotin**

Recent advance
A new class is added **to Enzyme classification is EC No-7 i.e. Translocase** from August 2018. This class describes group of enzymes that transfer ions or molecules across the membrane. They were previously classified as ATPases under EC No 3, Hydrolases. The subclasses are designated based on types of ions or molecules translocated.

- Synthetases belongs to ligases
- Synthases belongs to lyase

MECHANISM OF ENZYME ACTION

Explained by Different Theories
- Lowering of activation energy
- Michaelis-Menten theory
- Fischer's Template theory
- Koshland's Induced Fit theory

Lowering of Activation Energy

Enzymes lower the activation energy. Activation energy is defined as the energy required to convert all molecules of a reacting substance from ground state to transition state.

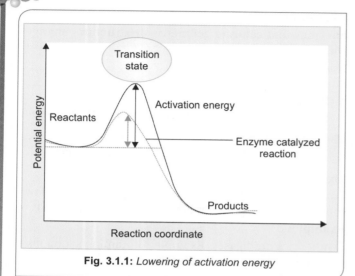

Fig. 3.1.1: *Lowering of activation energy*

Michaelis-Menten Theory (Enzyme-substrate Complex Theory)

Enzyme combines with a substrate to form a transient Enzyme-substrate Complex which immediately break in to Enzyme and products.

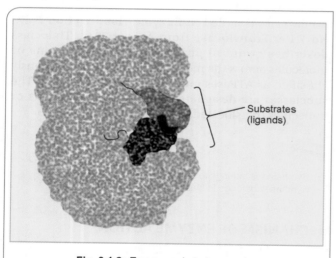

Fig. 3.1.2: *Enzyme substrate complex*

Fischer's Template Theory
- Lock and Key model
- **Three dimensional structure of the active site of unbound enzyme is complementary to the substrate.**
- Thus Enzyme and substrate fit each other.
- Failed to explain dynamic changes that accompany catalysis.

Koshland's Induced Fit Theory
- Binding of substrate to specific part of enzyme **induce conformational** changes in the active site of the enzyme.
- Enzyme changes shape during or after binding with the substrate.
- Can explain the dynamic changes that accompany catalysis.

FACTORS AFFECTING ENZYME ACTIVITY

Temperature
- Raising the temperature increases the rate of both uncatalyzed and enzyme-catalyzed reactions by increasing the kinetic energy and the collision frequency of the reacting molecules.
- **Bell-shaped curve** is obtained by plotting temperature against velocity of reaction.
- The optimum temperature for most human enzymes is between **35 and 40°C.**
- Human enzymes start to denature at temperatures above **40°C.**

Temperature Coefficient (Q^{10})

The temperature coefficient (Q^{10}) is the factor by which the rate of a biologic process increases for a 10°C increase in temperature. For the temperatures over which enzymes are stable, the rates of most biological processes typically double for a 10°C rise in temperature ($Q^{10} = 2$)

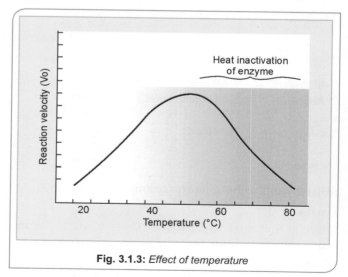

Fig. 3.1.3: *Effect of temperature*

Hydrogen Ion Concentration
- The rate of almost all enzyme-catalyzed reactions exhibits a significant dependence on hydrogen ion concentration.

- Most intracellular enzymes exhibit optimal activity at pH values between **5 and 9.**
- The relationship of enzyme activity to hydrogenion concentration gives a **bell-shaped curve**.

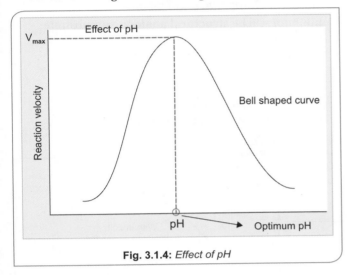

Fig. 3.1.4: *Effect of pH*

Enzyme Concentration

In the beginning velocity of enzyme reaction is directly proportional to the enzyme concentration.

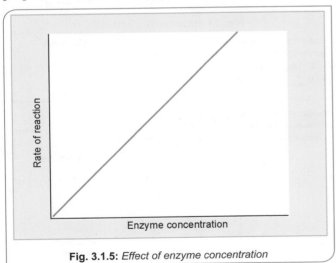

Fig. 3.1.5: *Effect of enzyme concentration*

Substrate Concentration

- For a fixed Enzyme concentration, rate of reaction is **directly proportional** to the Substrate concentration up to certain concentration of substrate, but later there is no further increase in velocity.
- Initially its **Ist order kinetics later it will be zero order kinetics.**

Most enzymes follow **Michaelis Menten kinetics,** a plot of initial reaction velocity against substrate concentration is **Hyperbolic**

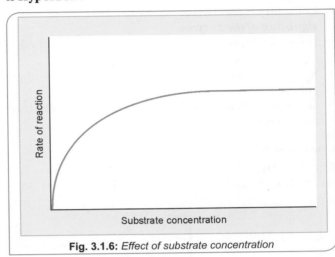

Fig. 3.1.6: *Effect of substrate concentration*

Michaelis-Menten Equation

$$Vi = \frac{V_{max} \times S}{K_m + S}$$

- Where Vi is the initial velocity
- V_{max} is the maximal velocity
- K_m is the Michaelis constant
- S is the substrate concentration

MICHAELIS CONSTANT (K_m)^Q

Michaelis Constant (K_m)

- Substrate concentration required to produce half-maximal velocity (½ V_{max})

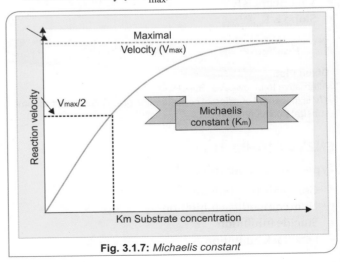

Fig. 3.1.7: *Michaelis constant*

Characteristics of Michaelis Constant

- **Independent of enzyme concentration**
- Unique for each enzyme—substrate pair
- Constant for an enzyme—Substrate pair, hence called *signature of the enzyme*.
- Denotes affinity of enzyme to substrate. Lower the K_m higher will be the affinity of the substrate
- K_m helps to understand the natural substrate of an enzyme.
- Substrate with lower K_m will be the natural substrate of the enzyme.

Lineweaver-Burk Plot

A graphical representation of $1/V$ on y axis and $1/S$ on x axis is called Lineweaver-Burk Plot or Double Reciprocal plot.

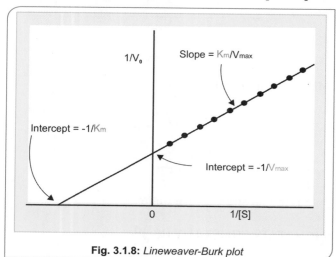

Fig. 3.1.8: *Lineweaver-Burk plot*

In Lineweaver-Burk Plot
- X intercept is $-1/K_m$
- Y intercept is $1/V_{max}$
- Slope is K_m/V_{max}

> ### HIGH YIELD POINTS
>
> **Dixon Plot**
> Alternative to Lineweaver-Burk Plot
> $1/V$ is measured at different concentration of inhibitor (I), but at same substrate concentration (S).

ENZYME INHIBITION

Types of Enzyme Inhibition

- Competitive inhibition
- Non-competitive inhibition
- Suicide inhibition
- Feedback inhibition

Competitive Inhibition

A type of inhibition in which the inhibitor compete directly with a normal substrate for an enzyme's substrate binding site.

Features of Competitive Inhibition

- Inhibitor will be **structural analogue** of substrate
- **Reversible.**
- Excess substrate abolishes inhibition.
- K_m **increases.**
- V_{max} remains the same.
- X intercept $-1/K_m$ decreases
- Y intercept $1/V_{max}$ remains the same

Image-Based Information

Competitive Inhibition

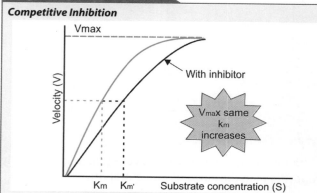

Fig. 3.1.9: *Competitive inhibition*

Substrate saturation curve with and without competitive inhibitor. With competitive inhibitor
- K_m increases
- V_{max} remains the same

Image-Based Information

Lineweaver-Burk Plot of Competitive Inhibition

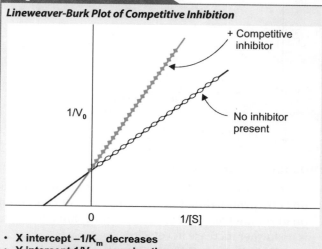

- X intercept $-1/K_m$ decreases
- Y intercept $1/V_{max}$ remains the same

Examples of Competitive Inhibition

Competitive inhibitors of enzymes are mostly drugs

Drug	Enzyme inhibited
Statins	HMG-CoA Reductase
Dicoumarol	Vitamin K Epoxide.
Methotrexate	Dihydrofolate Reductase
Succinyl Choline	Acetyl Choline Esterase

Some competitive inhibitors which are not drugs are:

Enzyme	Substrate	Competitive inhibitor
Lactate Dehydrogenase	Lactate	Oxamate
Aconitase	Cisaconitate	Transaconitate
Succinate Dehydrogenase	Succinate	Malonate
HMG-CoA Reductase	HMG-CoA	HMG

Non-Competitive Inhibition[Q]

A type of inhibition in which the inhibitor bind to a site distinct from the substrate binding site.

Two different types are:
1. Reversible non-competitive Inhibition (Only few Non-competitive inhibition are reversible)
2. Irreversible Non-competitive Inhibition (Most of Non-competitive are irreversible).

Features of Non-competitive Inhibition[Q]
- Inhibitor have **no structural resemblance** to substrate.
- **Mostly Irreversible** (Except a few reversible non-competitive inhibition)
- Excess substrate do not abolish the inhibition.
- K_m **remains the same.**
- V_{max} **decreases.**
- X intercept $-1/K_m$ remains the same.
- Y intercept intercept $1/V_{max}$ increases

Image-Based Information

Noncompetitive Inhibition

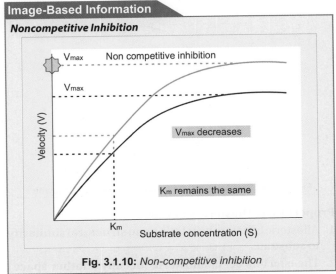

Fig. 3.1.10: Non-competitive inhibition

Contd…

Contd…

Substrate saturation kinetics with and without non-competitive inhibitor
With non-competitive inhibitor
- V_{max} decreases
- K_m remains the same

Image-Based Information

Lineweaver-Burk Plot of Non-competitive Inhibition

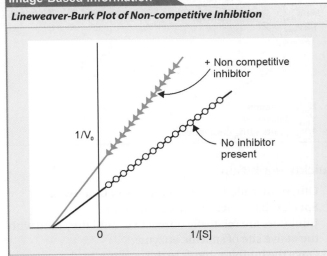

- X intercept $-1/K_m$ remains the same
- Y intercept $1/V_{max}$ increased

Examples of Irreversible Non-Competitive Inhibition
- Are mostly **poisonous agents**

Non-competitive inhibitor	Enzyme
Cyanide[Q]	Cytochrome C Oxidase
Iodoacetate	Glyceraldehyde 3 Phosphate
Flouride	Enolase
Disulfuram (Antabuse)	Aldehyde dehydrogenase
British Anti Lewisite (Dimercaprol)	SH group of several enzymes
Arsenite	Alpha Keto Glutarate Dehydrogenase
Flouroacetate	Aconitase
Di Isopropyl Fluorophosphate	Serine Proteases

Tips to memorise
Almost all inhibitors of Electron Transport Chain are examples of irreversible Non-competitive inhibition

Uncompetitive Inhibition
- Inhibitor need not resemble the substrate and it doesn't have any affinity for free enzyme

- Inhibitor binds to E-S complex.
- K_m and V_{max} are decreased

For example, Inhibition of placental alkaline phosphatase by phenylalanine

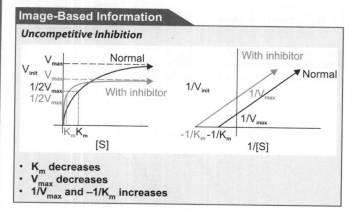

Uncompetitive Inhibition
- K_m decreases
- V_{max} decreases
- $1/V_{max}$ and $-1/K_m$ increases

Suicide Inhibition

- Otherwise called **Mechanism-based inactivation.**
- Special class of **irreversible** inhibition.
- Inhibitors are relatively unreactive, until they bind with the active site of specific enzyme.
- Once the inhibitor binds to the enzyme, by the action of the enzyme it is converted to a potent inhibitor.
- Irreversibly bind to the enzyme and inhibit the enzyme.

Examples of Suicide Inhibition

1. **Allopurinol inhibit Xanthine Oxidase**
   Allopurinol converted to all oxanthine which irreversibly inhibit the enzyme.
2. Treatment of Trypanosomiasis by Diflouromethyl ornithine (DFMO) inhibit Ornithine Decarboxylase
3. **Aspirin** acetylates the active site of Cyclooxygenase, there by inhibiting prostaglandin synthesis.

Feedback Inhibition
- Also called **end-product inhibition**.
- The activity of the enzyme is inhibited by the final product of the pathway.
  For example: AMP inhibits first step in purine synthesis

REGULATION OF ENZYMES

Regulation of Enzyme Quality (Intrinsic Catalytic Efficiency):
1. Allosteric regulation
2. Covalent modification

Regulation of Enzyme Quantity:
1. Control of Enzyme Synthesis (by Induction and Repression)
2. Control of Enzyme Degradation

Definition of Allosteric Enzyme
Allosteric enzymes are those enzymes in which catalysis at active site is modulated by the presence or absence of allosteric effectors at the allosteric site.

The allosteric regulation can be
1. **Allosteric activation**—The modifier/effector is a positive modifier.
2. **Allosteric inhibition**—The modifier/ effector is a negative modifier.

Allosteric Enzymes

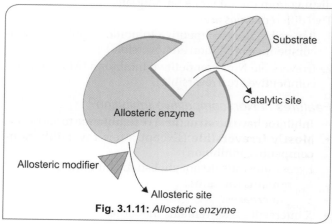

Fig. 3.1.11: *Allosteric enzyme*

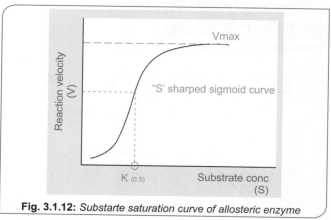

Fig. 3.1.12: *Substarte saturation curve of allosteric enzyme*

Properties of Allosteric Enzyme
1. Allosteric effectors bear little or **no structural similarity** to substrate.
2. The effectors act at **allosteric (occupy another space)** site not isosteric (same space of substrate).

3. Binding of an allosteric regulator, influences the catalysis by inducing **conformational changes in the active site.**
4. Kinetics of Allosteric inhibition is **competitive or noncompetitive or mixed.**
5. Allosteric effects binding is **reversible and noncovalent.**
6. Most of the allosteric enzymes have **2 or more subunits.**
7. Kinetics of allosteric enzyme diverge from hyperbolic Michaelis-Menten behavior, instead it gives **Sigmoid saturation curve or S-shaped curve**, due to **cooperative binding.**
8. K_m is better termed as $K_{0.5}$ or $S_{0.5}$ or **binding constant**
9. High activity state of enzyme is called **R (Relaxed) state** and low activity state is called **T (Taut) state.**
10. **A positive modulator decreases the $K_{0.5}/K_m$ and Negative modulator increases the $K_{0.5}/K_m$**

Allosteric Enzymes are Classified as

a. **K series—K_m is increased** without an effect on V_{max}. Hence its **Substrate saturation kinetics is similar to Competitive.** They weaken the bond between substrate and Substrate binding site. Hence increase K_m by **conformational change at catalytic site.**
b. **V series**—Allosteric inhibitor **lower V_{max} without altering K_m.** Hence its Substrate saturation kinetics is similar to **Noncompetitive.** They alter the **orientation or charge of catalytic residues.** Hence decrease V_{max}.
c. **Intermediate effects on K_m and V_{max} is also possible.**

Differences between noncompetitive and allosteric inhibition

Noncompetitive inhibition	Allosteric inhibition
Follow Michaelis-Menten Kinetics	Does not follow Michaelis-Menten Kinetics
Effect of Substrate concentration give a hyperbolic curve	Effect of Substrate concentration give a sigmoid curve

Image-Based Information

Effect of allosteric modfiers

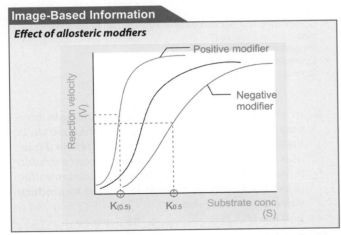

Contd...

A. Sigmoid curve of allosteric enzyme
B. Effect of a negative allosteric effector in an allosteric enzymes, i.e. curve shifts to right
C. Effect of a positive allosteric effector in an allosteric enzyme, i.e. curve shifts to left

Examples of Allosteric Enzymes[Q]

Enzyme	Activator	Inhibitor
ALA Synthase		Heme
Aspartate Transcarbamoylase	ATP	CTP
HMG-CoA Reductase		Cholesterol
Phosphofructokinase	Fructose 2,6 Bisphosphate	Citrate
Pyruvate Carboxylase	Acetyl-CoA	ADP
Acetyl-CoA Carboxylase	Citrate	Acyl-CoA
Carnitine Palmitoyl Transferase-1		Malonyl-CoA
Citrate Synthase		ATP
Carbamoyl Phosphate Synthetase I	N Acetyl Glutamate (NAG)	
Carbamoyl Phosphate Synthetase II	PRPP	UTP

Covalent Modification

- Method of **regulation of Enzyme activity.**
- Addition or removal of a group by making or breaking a covalent bond.
- By covalent modification enzyme activity is either increased or decreased.

Two Types of Covalent Modification

- **Irreversible**—Partial proteolysis/**zymogen activation**[Q]
- Reversible—Addition/removal of a particular group

Common Methods of Reversible Covalent Modification[Q]
- **Phosphorylation/dephosphorylation (most common covalent modification)**
- Methylation
- Adenylation
- ADP ribosylation
- Acetylation

Examples of Covalent modifications

Usually enzymes are in phosphorylated state when body is fasting under the influence of Glucagon.

High Yield Points

The enzymes that are active in phosphorylated state are:
- Glycogen Phosphorylase
- Key enzymes of Gluconeogenesis
- Citrate Lyase
- Phosphorylase b Kinase
- HMG-CoA Reductase Kinase

High Yield Points

The enzymes that are active in dephosphorylated state are:
- Glycogen synthase
- Key enzymes of Glycolysis
- Acetyl-CoA carboxylase
- Pyruvate dehydrogenase
- HMG-CoA reductase

High Yield Points

Protein Acetylation: A Ubiquitous Covalent Modification of Metabolic Enzymes
- Enzymes are subject to modification by covalent acetylation.
- Epsilon amino group of Lysine is acetylated.
- Depends on energy status of the cell.
- In a well-nourished cell, high level of Acetyl-CoA would promote lysine acetylation.
- When nutrients are lacking, acetyl-CoA levels drop favoring protein deacetylation.
- Lysine acetyl transferases catalyze the acetylation.
- Nonenzymatic acetylation in mitochondria.

Deacetylation
Two Classes of protein deacetylases have been identified:
- Histone deacetylases and sirtuins.
- Histone deacetylases catalyze the removal of acetyl group from proteins
- Sirtuins, remove acetyl group from NAD+.

Quick Review

- Enzymes are highly specialised proteins that can act as biological catalyst.
- Enzymes are thermolabile

Contd...

Contd...
- Coenzymes are thermostable
- Coenzymes are organic substances
- All kinases are transferases
- All dehydrogenases need NAD+ or FAD as hydrogen acceptors
- Most hydroxylases are monooxygenases
- All synthetases are ligases
- Most synthases are lyases
- Carboxylases are ligases
- Decarboxylases are lyases
- Mutase is an isomerase
- Phosphatases are hydrolases
- K_m is substrate concentration at half maximal velocity
- V_{max} remains constant and K_m increases in competitive inhibition
- V_{max} decreases and K_m remains constant in noncompetitive inhibition.
- Allosteric enzyme follow a sigmoid curve.
- Effects of substrate concentration on reaction velocity give a hyperbolic curve.
- Effect of temperature and pH on reaction velocity give a bell shaped curve
- Effect of enzyme concentration on reaction velocity give straight line as it is directly proportional to each other
- Most common covalent modification is phosphorylation and dephosphorylation
- Enzymes active under the influence of insulin is active in dephosphorylated state
- Enzymes active under the influence of glucagon is active in the phosphorylated state.

Check List for Revision

- Definition of enzyme and cofactor
- Ribozyme
- Enzymes with metal as cofactor
- 6 classes of enzymes with example
- Michaelis constant
- Effect of substrate concentration, temperature, enzyme concentration and pH
- Competitive and Noncompetitive inhibition features and examples.
- Properties of allosteric enzymes
- Examples of allosteric activators and inhibitors.
- Concept of covalent modification

Review Questions

Classification of Enzymes, Enzyme Kinetics

1. **Coenzymes of PDH complex is:**
   (Central Institute Nov 2018)
   a. PLP
   b. Biotin
   c. Folic acid
   d. Thiamin pyrophosphate

2. **In an enzyme mediated reaction, substrate concentration was 1000 times K_m. 1% of substrate is metabolised to form 12 µmol of substrate in 9 minutes. If in the same reaction enzyme concentration is reduced to 1/3rd and substrate concentration is doubled. How much time is needed to produce same amount of product?** *(AIIMS May 2018)*
   a. 9 min
   b. 13.5 min
   c. 18 min
   d. 27 min

3. The Susbstrate saturation curve given below characterizes an allosteric enzyme system:
   (AIIMS May 2016)

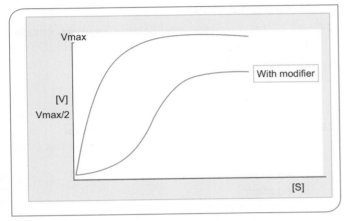

   a. Allosteric modifier binds in a concentration dependent manner
   b. Modifier can affect the catalytic site by binding to the allosteric site
   c. Adding more substrate to the enzyme can displace the allosteric modifier
   d. Allosteric modifiers changes the binding constant of the enzyme but not the velocity of reaction

4. Which enzyme is deficient in c/c alcoholics?
   (Recent Question 2016)
   a. Aconitase
   b. Citrate synthase
   c. Isocitrate dehydrogenase
   d. Alpha ketoglutarate dehydrogenase

5. Alcohol Dehydrogenase comes under which class of enzyme?
   (Recent Question 2016)
   a. Oxidoreductase
   b. Dehydrogenase
   c. Hydrolase
   d. Oxidase

6. Suicidal enzyme is: (AIIMS May 2013)
   a. Lipoxygenase
   b. Cycloxygenase
   c. Thromboxane synthase
   d. 5'Nucleotidase

7. Which of the following is a Lyase? (JIPMER 2014)
   a. Aldolase B
   b. Acetyl-CoA synthetase
   c. Fatty Acyl-CoA dehydrogenase
   d. Acetyl-CoA carboxylase

8. All are true about oxygenases, except:
   (AIIMS Nov 2011)
   a. Can incorporate 2 atoms of O_2 in a substance
   b. Can incorporate 1 atom of O_2 in a substance
   c. Important in hydroxylation of steroids
   d. Catalyse carboxylation of drugs

9. All of the following enzymes are involved in oxidation reduction, except: (AI 2009)
   a. Dehydrogenases
   b. Hydrolases
   c. Oxygenases
   d. Peroxidases

10. Enzyme which cleave CC bond:
    a. Lyase
    b. Oxidoreductase
    c. Ligase
    d. Isomerase

11. Velocity at K_m is:
    a. Half the substrate concentration
    b. Same as V_{max}
    c. Quarter the V_{max}
    d. Half the V_{max}

12. Coenzyme in decarboxylaton reaction:
    (Recent Question 2016)
    a. Niacin
    b. Biotin
    c. Pyridoxine
    d. Riboflavin

Enzyme Inhibition

13. In noncompetitive antagonism which of the following is correct? (AIIMS Nov 2018)
    a. V_{max} decreases
    b. K_m decreases
    c. No change in V_{max}
    d. Both K_m and V_{max} increases

14. The type of enzyme inhibition in which Succinate dehydrogenase reaction is inhibited by malonate is an example of: (AIIMS May 2006)
    a. Non-competitive
    b. Uncompetitive
    c. Competitive
    d. Allosteric

15. Features of competitive inhibition is/are:
    (PGI May 2014)
    a. V_{max} increases
    b. K_m increases
    c. V_{max} decreases
    d. K_m decreases
    e. V_{max} constant

16. **Which is true about enzyme kinetics for competitive inhibition?**
    (JIPMER 2014)
    a. Low K_m high affinity
    b. High K_m high affinity
    c. High K_m low affinity
    d. Low K_m low affinity

17. **Non-competitive enzyme inhibition leads to:**
    a. $V_{max}\uparrow$
    b. $V_{max}\downarrow$
    c. V_{max} unchanged
    d. $K_m\uparrow$
    e. $K_m\downarrow$

18. **True about competitive inhibition of enzyme:**
    (PGI May 2010)
    a. $\uparrow K_m$
    b. $\downarrow K_m$
    c. $\uparrow V_{max}$
    d. No change in K_m and V_{max}
    e. V_{max} remain same

19. **Non-competitive reversible inhibitors:**
    a. Raise K_m
    b. Lower K_m
    c. Lower V_{max}
    d. Raise both V_{max} and K_m
    e. Do not affect either V_{max} or K_m

20. **True about competitive antagonism:**
    (PGI June 2009)
    a. V_{max} increased
    b. Substrate analogue
    c. Reversible
    d. K_m increased
    e. V_{max} decreased

21. **K_m changes and V_{max} remains the same. What is the type of enzyme inhibition?**
    a. Competitive inhibition
    b. Non-competitive inhibition
    c. Uncompetitive inhibition
    d. Suicide inhibition

Enzyme Regulation

22. **Allosteric regulation true is?**
    (Recent Question Nov 2017)
    a. Binds to site other than active site
    b. Regulated by acting on catalytic site
    c. Follow Michaelis Menten Kinetics
    d. Substrate and modifier are structural analogues

23. **Which enzymes catalytic activity is by dephosphorylation?**
    (PGI Nov 2009)
    a. HMG-CoA reductase
    b. Glycogen Phosphorylase
    c. Citrate Lyase
    d. Glycogen Synthase

24. **All of the covalent modification regulate enzyme kinetics except:**
    a. Phosphorylation
    b. Acetylation
    c. ADP Ribosylation
    d. Glycosylation

25. **The following affect enzyme activity except:**
    (Recent Question 2016)
    a. Methylation
    b. Acetylation
    c. Induction
    d. Phosphorylation

26. **Chymotrypsinogen is a:**
    a. Zymogen
    b. Carboxypeptidase
    c. Transaminase
    d. Exopeptidase

Serine Proteases

27. **Chymotrypsin cleaves carbonyl terminal of:**
    (PGI May 2011)
    a. Phenylalanine
    b. Arginine
    c. Lysine
    d. Tryptophan
    e. Tyrosine

28. **Trypsin cleaves:**
    (PGI May 2010)
    a. Arginine
    b. Glutamate
    c. Lysine
    d. Proline

29. **A common feature of all serine proteases is:**
    (AI 2006)
    a. Autocatalytic activation of zymogen precursor
    b. Tight binding of pancreatic trypsin inhibitor
    c. Cleavage protein on the carboxyl site of serine residues
    d. Presence of Ser-His-Asp catalytic triad at the active site

30. **Trypsin is a:**
    a. Serine protease
    b. Lecithinase
    c. Phospholipase
    d. Elastase

Enzyme as Markers of Organelle and Membrane

31. **Markers of Plasma membrane is/are:** *(PGI June 2009)*
    a. Galactosyl transferase
    b. 5′ Nucleotidase
    c. Adenyl cyclase
    d. ATP synthetase

    e. Tyrosine

32. **Marker enzyme for Golgi apparatus:**
    a. Galactosyl Transferase
    b. Glucose 6 Phosphatase
    c. 5′ Nucleotidase
    d. Catalase

Answers to Review Questions

Classification of Enzymes, Enzyme Kinetics

1. **d. Thiamin Pyrophosphate**

   Coenzymes of multienzyme complex Pyruvate Dehydrogenase are
   - Thiamine pyrophosphate
   - Coenzyme A
   - Lipomide
   - NAD+
   - FAD

2. **d. 27 min**

   We can consider this question in two parts

   First part
   [S] = 1000 K_m 12 µmol of substrate in 9 minutes
   When S = K_m in a hyperbolic curve it follow Ist order kinetics i.e. [S] is directly proportion to velocity of reaction. But here its [S]=1000 K_m means its zero order kinetics, i.e. even if [S] increased velocity do not increase.

   Second part
   Enzyme concentration (E') is 1/3 of initial E concentration
   E is directly proportional to velocity of reaction
   [S] is = 2000 K_m
   So this is zero order kinetics hence velocity of reaction depends only E
   E' = 1/3E, hence New Velocity (V') is 1/3 V initial
   Hence time taken is three times more than initial time taken
   Time = 9x3 = 27 min.

3. **b. Modifier can affect the catalytic site by binding to the allosteric site**
   - Allosteric modifier is not binding in a concentration dependent manner.
   - Adding more substrate will not displace the allosteric modifier.
   - Allosteric modifier can change either binding constant or Velocity depending on K series or V series allosteric enzyme.

4. **d. Alpha ketoglutarate dehydrogenase**

   Because in alcoholics there is thiamine deficiency, so a thiamine dependent enzyme will be deficient.
   So answer is Alpha Ketoglutarate Dehydrogenase.

5. **a. Oxidoreductase**

   Here the question is asking about class of enzyme, hence answer oxidoreductase. Dehydrogenase is the subclass not class of enzyme.

6. **b. Cycloxygenase** *(Ref: Harper 31/e page 224)*

   Cyclooxygenase is a "Suicide Enzyme" "Switching off" of prostaglandin activity is partly achieved by are markable property of cyclooxygenase—that of self-catalyzed destruction; i.e., it is a "suicide enzyme." **Quickreview-Cyclooxygenase**
   - Cyclooxygenase (COX) (also called prostaglandin H synthase)
   - Involved in Prostanoid (Prostaglandin Thromboxane and Prostacyclin) synthesis
   - This enzyme has two activities, acyclooxygenase and peroxidase.
   - COX is present as two isoenzymes, COX-1 and COX-2.

   Drugs acting as inhibitors COX
   I. NSAID
      - Aspirin-inhibits COX-1 and COX-2.
      - Indomethacin and ibuprofen-inhibit cyclooxygenases by competing with arachidonate.
      - Coxibs—selectively inhibit COX-2
      - Some coxibs have been withdrawn or suspended from the market due to undesirable side effects and safety issues.
   II. Anti-inflammatory Corticosteroids
      - Transcription of COX-2—but not of COX-1—is completely inhibited by corticosteroids.

7. **a. Aldolase B**

   Some examples of Lyases are:
   - HMG-CoA Lyase
   - Arginino Succinate Lyase
   - ATP Citrate Lyase
   - Aldolase
   - Fumarase

8. d. Catalyse carboxylation of drugs
(Ref: Harper 31/e page 114)
- Oxygenases can be monooxygenase or dioxygenase
- Monooxygenase incorporate 1 atom of Oxygen molecule to the substrate.
- Dioxygenase incorporate both the atoms of Oxygen molecule to the substrate
- Phase 1 Xenobiotic reactions, **hydroxylation,** catalyzed mainly by members of a class of enzymes referred to as **monooxygenases** or **cytochrome** P450s.

9. b. Hydrolases
(Ref: Harper 31/e page 112)

Oxidoreductases
Can be:
- Dehydrogenase
- Oxygenase
- Monooxygenase
- Dioxygenase
- Oxidase
- Catalase
- Peroxidase

10. a. Lyase
(Ref: Harper 31/e page 157)

11. d. Half the V_{max}

12. c. Pyridoxine
(Ref: Harper 31/e page 528; Table 44-1)
- Coenzyme for carboxylation reaction—Biotin
- Coenzyme for decarboxylation—Pyridoxine

Coenzyme for decarboxylation reactions of Alpha Ketoglutate dehydrogenase and Branched chain Ketoacid dehydrogenase—Thiamine.

Enzyme Inhibition

13. a. V_{max} decreases
(Ref: Harper 31/e page 77)

Features of noncompetitive inhibition:
- K_m remains the same
- V_{max} decreases

14. c. Competitive
(Ref: Harper 31/e page 77)

Examples of Competitive Inhibition
Competitive inhibitors of enzymes are mostly drugs

Drug	Enzyme inhibited
Statins	HMG CoA Reductase
Dicoumarol	Vitamin K Epoxide
Methotrexate	Dihydrofolate Reductase
Succinyl Choline	Acetyl Choline Esterase

15. b, e. K_m increases, V_{max} constant

Features of competitive inhibition:
- V_{max} remains a constant
- K_m increases
- Reversible

16. c. High K_m and low affinity

Features of competitive inhibition:
- K_m increases, hence the affinity is lowered
- V_{max} remains the same

Features of noncompetitive inhibition:
- K_m remains the same
- V_{max} decreases

17. b. $V_{max}\downarrow$
(Ref: Harper 31/e page 77; Figure 8-11)

Features of noncompetitive inhibition
- Inhibitor have no structural resemblance to substrate
- Irreversible
- Excess substrate do not abolish the inhibition
- K_m remains the same
- V_{max} decreases

18. a, e. $\uparrow K_m$, V_{max} remain same
(Ref: Harpers 31/e page 77; Figure 8-10)

19. c. Lower V_{max}
(Ref: Harper 31/e page 77)

20. b, c, d. Substrate analogue, Reversible, K_m increased
(Ref: Harper 31/e page 77)

21. a. Competitive inhibition
(Ref: Harper 31/e page 77)
- Competitive inhibition—K_m increases and V_{max} remains the same.
- Noncompetitive inhibition—K_m remains the same and V_{max} decreases.
- Uncompetitive inhibition—Both K_m and V_{max} decreases.

Enzyme Regulation

22. a. Binds to site other than active site

23. a, d. HMG-CoA reductase, Glycogen synthase
(Ref: Harper 31/e page 89, Table 9.1)

Enzyme	Activity state	
	Low	High
Acetyl-CoA carboxylase	EP	E
Glycogen synthase	EP	E
Pyruvate dehydrogenase	EP	E
HMG-CoA reductase	EP	E
Glycogen phosphorylase	E	EP
Citrate lyase	E	EP
Phosphorylase b kinase	E	EP
HMG-CoA reductase kinase	E	EP

24. **d. Glycosylation** (Ref: Harper 31/e page 87)

   Reversible covalent modifications
   - Phosphorylation Dephosphorylation
   - ADP Ribosylation
   - Methylation
   - Acetylation

   Irreversible covalent modification
   - Zymogen activation/partial Proteolysis

25. **c. Induction** (Ref: Harper 31/e page 87)

   Induction is method of regulation of enzyme quantity.
   Regulation of enzymes
   Can be classified as:
   Regulation of enzyme quality (intrinsic catalytic efficiency)
   - Allosteric regulation
   - Covalent modification

   Covalent modification
   - Phosphorylation/dephosphorylation (most common covalent modification)
   - Methylation
   - Adenylation
   - ADP ribosylation
   - Acetylation

   Regulation of enzyme quantity
   1. Control of enzyme synthesis
   2. By induction and repression
   3. Control of enzyme degradation
   4. Enzymes are degraded by Ubiquitin–Proteasome pathway.

26. **a. Zymogen**

   Zymogen activation is an example of irreversible covalent modification.

Serine Proteases

27. **a, d, e. Phenylalanine, Tryptophan, Tyrosine** (Ref: Harper 31/e page 61)

   Serine proteases
   - Proteolytic enzymes with serine at their active site
   - Amino acid triad in the active site of serine proteases—Ser, His, Asp

   Examples of serine proteases
   - Chymotrypsin
   - Trypsin
   - Elastase
   - Thrombin
   - Plasmin
   - Complements
   - Factors X and XI

   Serine proteases differ in substrate specificity
   - Trypsin cleave basic amino acid like Arg, Lys
   - Chymotrypsin cleave hydrophobic bulky amino acid like Trp, Tyr, Phe
   - Elastase cleave small neutral amino acids like alanine, glycine.

28. **a, c. Arginine, Lysine** (Ref: Harper 31/e page 61)

29. **d. Presence of Ser-His-Asp catalytic triad at the active site** (Ref: Harper 31/e page 61)

30. **a. Serine protease** (Ref: Harper 31/e page 61)

Enzyme as Markers of Organelle and Membrane

31. **b, c. 5′ Nucleotidase, Adenyl cyclase** (Ref: Harper 31/e page 463; Table 40.2)

   Enzymes as markers of organelle and membranes

Enzymatic markers of different membranes	
Membrane	Enzyme
Plasma	5' Nucleotidase
	Adenylyl cyclase
	Na⁺-K⁺-ATPase
Endoplasmic reticulum	Glucose-6-phosphatase
Golgi apparatus	
Cis	Glc NAc transferase I
Medial	Golgi mannosidase II
Trans	Galactosyl transferase
Trans Golgi Network	Sialyl transferase
Inner mitochondrial membrane	ATP synthase

32. **a. Galactosyl Transferase**

3.2 CLINICAL ENZYMOLOGY

- Isoenzymes
- Cardiac Biomarkers
- Enzyme Profile for Liver Diseases
- Enzyme Profile in Prostate Cancer
- Enzyme Profile in Pancreatitis
- Novel Biomarkers of Acute Kidney Injury
- Serine Proteases
- Bi-Bi Reaction
- Enzymes as Markers of Organelle and Membranes
- Enzymes as Diagnostic Reagents
- Enzymes in Body Fluids

ISOENZYMES

- Physically distinct forms of the same enzyme, but catalyse the same reaction.
- Different molecular forms of the same enzyme synthesized from same/various tissues.

High Yield Points

Features of Isoenzymes
- They catalyse the same chemical reactions.
- They differ in heat stability.

For example: Heat stable ALP and heat labile ALP.
- They differ in electrophoretic mobility.

For example: CK-I moves faster than CK-3.
- They differ in the susceptibility to an inhibitor.

For example: Tartarate labile Acid Phosphatase and tartarate stable Acid Phosphatase.
- They differ in subunits they are made up of.

For example: LDH-1 (H4), LDH 5 (M4)
- They differ in tissue localization.

For example: LDH-1 located in the heart and LDH-5 located in the Muscle.
- They differ in K_m value.

For example: Glucokinase (an isoenzyme of Hexokinase) has high K_m and but Hexokinase has low K_m.

Functional Enzymes and Non-Functional Enzymes^Q

Functional Enzymes^Q
- Enzymes which have specific function in the plasma.

Examples of functional Enzymes
- Coagulation Factors
- Lipoprotein Lipase

Non-Functional enzymes
- No specific function in the plasma.
- Comes out from the tissue as a result of normal wear and tear.
- Their level is very low in the serum.
- But during tissue injury their level rises in the serum.
- Hence they help to diagnose the site of tissue injury.

Examples of non-functional enzymes
- LDH, Creatine Kinase, Alkaline Phosphatase.

Isoenzymes of Lactate Dehydrogenase (LDH)

There are five isoenzymes for LDH. It is a tetramer made up of two types of subunits, H&M.

Name of the isoenzyme	Subunit	Tissue localization
LDH-1	H4	Heart, Muscle, RBC, Kidney
LDH-2	H3M1	
LDH-3	H2M2	Spleen, Lungs, Lymph nodes, Leukocytes, Platelets
LDH-4	HM3	Liver and Skeletal Muscle
LDH-5	M4	

High Yield Points

- LDH-5 predominates in Liver
- LDHx or LDc—A sixth atypical LDH found in post pubertal testis
- LDH 6 is found in sera of severely ill patients.

Isoenzymes of Creatine Kinase (CK)

There are three isoenzymes for Creatine Kinase, made up of two types of subunits, M and B.

Name of the Isoenzyme	Subunit	Tissue localization
CK-1	BB	Brain
CK-2	MB	Heart
CK-3	MM	Skeletal Muscle

Two atypical Creatine Kinases are:

1. CK Macro (Macro–CK)

Formed by aggregation of CK-BB with immunoglobulins like IgG.

2. CK-Mi (Mitochondrial CK)

Found in the exterior surface of Inner mitochondrial membrane of muscle, liver and brain.

Isoenzymes of Alkaline Phosphatase (ALP)

Isoenzyme forms of ALP	Tissue of origin
Alpha-1 ALP	Synthesized by epithelial cells of Biliary Canaliculi
Alpha-2 Heat labile ALP	Produced by Hepatic Cells
Alpha-2 Heat stable ALP	Produced by Placenta Inhibited by Phenyl Alanine **Most heat stable**
Pre beta ALP	Produced by Osteoblast.
Gamma ALP	Produced by Intestinal Cells
Leukocyte ALP	By leukocytes

HIGH YIELD POINTS—AMINO ACIDS

Regan Isoenzyme
- Named after the first patient from which the enzyme isolated.
- Anisoenzyme of ALP closely resemble alpha-2 heat stable ALP.
- Otherwise called Carcino Placental Isoenzyme.
- Elevated in Carcinoma of lung, liver, intestine

Nagao Isoenzyme
- A variant of ALP of germ cell origin.
- Inhibited by L-leucine
- Inappropriate expression seen in malignancy.

Kasahara Isoenzyme
- A variant of ALP of Fetal intestinal origin.
- Inappropriate expression seen in malignancy.

CARDIAC BIOMARKERS

- Creatine Kinase (CKMB)
- Cardiac Troponin T (CTnT)
- Cardiac Troponin I (CTnI)
- Brain Natriuretic Peptide (BNP)—Marker of cardiac failure not a marker of Myocardial Infarction.
- Myoglobin
- Lactate Dehydrogenase (LDH)—Not used nowadays
- Aspartate aminotransferase (AST)—Not used nowadays

Clinical Enzymology in MI
- First enzyme to rise is **CKMB**
- First cardiac biomarker to rise is **Myoglobin**.

Flipped Pattern of LDH^Q

Normally LDH-2 is present in higher concentration than LDH-1. But this pattern is reversed in MI, i.e. LDH-1> LDH-2. This limited diagnostic importance.

- Enzyme marker of myocardial reinfarction is **CK-MB**

Cardiac biomarker elevation in Myocardial Infarction			
Cardiac marker	Rises	Peaks	Return to normal
CK-MB	4–8 hrs	24 hrs	After 48–72 hrs
Cardiac Troponin (cTn)	4–6 hrs	24–36 hrs	After 3–10 days

New Cardiac Biomarkers

- Ischemia Modified Albumin
- Glycogen Phosphorylase BB Isoenzyme
- Pregnancy Associated Plasma Protein A (PAPP-A)
- Myeloperoxidase (MPO)

ENZYME PROFILE FOR LIVER DISEASES

Enzymes whose elevation in Serum reflects damage to Hepatocyte

- **Aminotransferases (transaminases)** are sensitive indicators of liver cell injury
- Are most helpful in recognizing acute hepatocellular diseases such as hepatitis.
- **Aspartate aminotransferase (AST)** (is found in liver, cardiac muscle, skeletal muscle, kidney, brain, pancreas, lungs, leukocytes, and erythrocytes)
- **Alanine aminotransferase (ALT)** (is found primarily in the **liver** and is therefore a more specific for Liver Disease than AST).

Enzymes whose elevation in serum reflects cholestasis

- Alkaline phosphatase (ALP)
- 5'-nucleotidase
- γ glutamyl Transferase (Transpeptidase)(GGT)

HIGH YIELD POINTS

GGT
- GGT elevation in serum is less specific for cholestasis than are elevations of alkaline phosphatase or 5'-nucleotidase.
- GGT is used to identify **occult alcohol use**.

ALP
- Less than threefold elevation in ALP can be seen in any type of liver disease.
- More than fourfold elevation of ALP is seen in cholestasis.
- ALP elevation is not helpful in distinguishing between intrahepatic and extrahepatic cholestasis.

5' Nucleotidase
- Specific for **cholestasis** than ALP and GGT

Non-Pathologic elevations of ALP seen in
- Patients over age 60 years
- Type O and Type B blood group
- After eating a fatty meal (due to efflux of gamma ALP)
- In children due to rapid bone growth
- Late normal pregnancy

> **HIGH YIELD POINTS**
>
> **AST/ALT Ratio^Q**
> **AST: ALT ratio < 1**
> Any condition causing hepatocellular damage ratio <1 isseenas ALT level rises above AST level. This is because ALT is more specific for hepatocellular damage than AST.
> - Chronic viral hepatitis
> - Non-alcoholic fatty liver disease
> - Toxic hepatitis
> - Paracetamol toxicity.
>
> **AST:ALT ratio > 2:1 is suggestive, while a ratio > 3:1**
> - Highly suggestive of alcoholic liver disease.

> **HIGH YIELD POINTS**
>
> **Aminotransferases (ALT and AST)**
> - **ALT** is more specific for hepatocellular damage than AST.
> - In hepatocellular disease **ALT elevation is slightly higher than or equal to AST, so AST/ALT ratio is less than 1.**
> - **If cirrhosis develop the ratio becomes more than 1.**
> - **Minimal ALT elevation** less than 300 IU/L is non-specific. Most likely explanation is **Fatty liver**
> - Level > 1000 IU/L in extensive hepatocellular injury
> - In alcoholic liver disease, alcohol-induced deficiency of Pyridoxal Phosphate causes reduced level of transaminases (ALT and AST). ALT level is often normal and AST is rarely > 300 IU/L.

ENZYME PROFILE IN PROSTATE CANCER

- **Tartarate labile acid phosphatase,**
- **Prostate specific antigen (PSA)** are enzyme markers of Prostate Cancer

Prostate Specific Antigen
- Otherwise called **Kallikrein related Peptidase 3 (KLK3)**
- It is a Serine Protease.
- Secreted by **epithelial cells of Prostate.**
- This is Prostate specific, but not prostate cancer specific.
- Commonly used cut point for Prostate cancer is **PSA level > 4 ng/mL.**
- But actually there is no PSA level below which risk of Prostate cancer is Zero.
- So **PSA level estimation should be accompanied by Prostate Biopsy.**

ENZYME PROFILE PANCREATITIS

They are:
- Amylase
- Lipase

Serum Amylase is not specific for **Pancreatic disease** as its level is increased in **Parotitis** also.

A serum lipase level measurement can be **instrumental in differentiating a pancreatic or nonpancreatic cause** for hyperamylasemia.

Apart from serum Amylase level can be estimated in **urine** also.

NOVEL BIOMARKERS OF ACUTE KIDNEY INJURY^Q

- Kidney Injury Molecule-1 (KIM-1)
- Neutrophil Gelatin associated Lipocalin (NGAL)
- IL-18
- Alanine Amino Peptidase
- Clusterin
- Alkaline Phosphatase
- α Glutathione S Transferase
- γ Glutamyl Transpeptidase
- β2 Microglobin
- α1 Macroglobin
- Retinol Binding Protein
- Cystatin C
- Microalbumin
- Osteopontin
- Liver Fatty Acid Binding Protein
- Sodium–Hydrogen Exchanger Isoform
- Exosomal Fetuin

MARKERS OF BONE DISEASES

Bone Formation
- Serum Bone-specific Alkaline Phosphatase (BAP).
- Serum Osteocalcin.
- Serum propeptide of type I Procollagen.

Bone Resorption
- Urine and serum cross linked N-telopeptide
- Urine and Serum cross linked C-telopeptide
- Urine total free deoxypyridinoline.

SERINE PROTEASES

- Proteolytic enzymes with **Serine** at their active Site
- Amino Acid triad in the active site of Serine Proteases—**Ser, His, Asp**

Examples of Serine Proteases

- Chymotrypsin
- Trypsin
- Elastase
- Thrombin
- Plasmin
- Complements
- Factor X and XI
- Prostate Specific Antigen

Serine Proteases Differ in Substrate Specificity[Q]

- Trypsin cleave **Basic Amino acid**
- Chymotrypsin cleave **hydrophobic bulky amino acid**
- Elastase cleave small neutral amino acids like Alanine, Glycine.

BI-BI REACTION

- Kinetic behavior for **two-substrate, two-product reactions** termed "Bi-Bi" reactions.
- Most Bi-Bi Reactions **follow Michaelis-Menten Kinetics.**

Bi-Bi reactions can be divided into:

- An ordered Bi-Bi reaction
  For example: NAD (P)H-dependent oxidoreductases
- A random Bi-Bi reaction
  For example: Many Kinases and some Dehydrogenases.
- A ping-pong reaction

For example: Aminotransferases and Serine proteases.[Q]

ENZYMES AS MARKERS OF ORGANELLE AND MEMBRANES[Q]

Membrane/organelle	Marker enzymes
Plasma Membrane	5'- Nucleotidase Adenylyl Cyclase Na^+-K^+ ATPase
Endoplasmic reticulum	Glucose-6-phosphatase
Golgi Complex	Galactosyl Transferase[Q]
Inner Mitochondrial Membrane	ATP Synthase
Peroxisome	Catalase, Urate Oxidase
Lysosomes	Acid Phosphatase
Cytoplasm	Lactate Dehydrogenase

ENZYMES AS DIAGNOSTIC REAGENTS

Enzyme	Diagnostic test done
Urease	Urea Estimation
Uricase	Uric Acid Estimation
Glucose Oxidase	Glucose
Hexokinase	Glucose
Peroxidase[Q]	Glucose, Cholesterol
Cholesterol Oxidase	Cholesterol
Creatininase	Creatinine
Lipase	Triglycerides

ENZYMES IN OTHER BODY FLUIDS[Q]

Enzyme	Clinical use
Lactate dehydrogenase in CSF, pleural fluid, ascitic fluid	Suggestive of malignant tumour but not confirmatory
Adenosine deaminase in pleural fluid	Suggestive of tuberculous pleural effusion
Amylase in urine	Suggestive of pancreatitis

Quick Review

- Lipoprotein lipase and Coagulation factors are functional enzymes.
- LDH, Creatine Kinase, ALP are non-functional enzymes.
- Fastest LDH in electrophoresis is LDH-1
- Fastest Creatine Kinase in electrophoresis is CK-1
- Most heat stable ALP is alpha-2 ALP in placenta
- Flipped pattern of LDH is LDH1>LDH-2
- First enzyme to rise in MI is CK-MB
- GGT is a marker of occult alchohol abuse,
- Any hepatocellular damage (Viral hepatitis, Nonalcoholic alcoholic liver disease) AST:ALT ratio is <1
- Alcoholic liver disease AST:ALT ratio >2
- PSA is a serine protease otherwise called Kallikrein related Peptidase 3 (KLK-3)

Check List for Revision

- Definition and examples of functional and non-functional enzymes.
- Names and location of Isoforms of LDH, Creatine Kinase and ALP.
- Cardiac biomarkers
- Important points of enzymes of liver disease
- Names of novel biomarkers of kidney.
- Names of Serine proteases and its substrate specificity,
- Overview of Bi Bi reaction especially ping pong reaction.
- Enzyme markers of organelle

Review Questions MCQ

1. **Biomarker of Alcoholic Hepatitis** *(Aiims Nov 2018)*
    a. ALP
    b. AST
    c. LDH
    d. GGT

2. **What happens to LDH 1 & 2 ratio in MI?**
    (Recent Question Nov 2017)
    a. LDH1 > LDH2
    b. LDH2 > LDH1
    c. LDH1 = LDH2
    d. Remains the same

3. **True about isoenzymes is:** *(AIIMS Nov 2011)*
    a. Catalyse the same reaction
    b. Same quaternary structure
    c. Same distribution in different organs
    d. Same enzyme classification with same number and name

4. **Non-functional enzymes are all except:**
    (AIIMS Nov 2008)
    a. Alkaline phosphatase
    b. Acid phosphatese
    c. Lipoprotein lipase
    d. Gamma-glutamyl transpeptidase

5. **Peroxidase enzyme is used in estimating:**
    (AIIMS Nov 2007)
    a. Hemoglobin
    b. Ammonia
    c. Creatinine
    d. Glucose

6. **Which of the following estimates blood creatinine level most accurately?** *(AIIMS May 2006)*
    a. Jaffe method
    b. Kinetic Jaffe method
    c. Technicon method
    d. Enzyme assay

7. **Not raised in liver disorder is/are:** *(PGI May 2013)*
    a. Lipase
    b. Urease
    c. ALP
    d. AST
    e. ALP

8. **LDH-5 level elevated in which cell injury?**
    a. Liver
    b. Heart
    c. Muscle
    d. RBC

9. **Which of the following LDH is having fastest electrophoretic mobility?** *(CMC Ludhiana 2014)*
    a. LDH-1
    b. LDH-2
    c. LDH-3
    d. LDH-5

Answers to Review Questions

1. **d. GGT**

    Gamma glutamyl tranferase is an inducible enzyme. It is a marker of occult alcohol abuse and alcoholic liver disease

2. **a. LDH 1 > LDH 2**

    In normal persons LDH 2 > LDH 1. In myocardial infarction LDH1 level rises, LDH1>LDH2. This is called flipped pattern of LDH.

3. **a. Catalyse the same reaction** *(Ref: Harper 31/e page 173)*
    - Isoenzymes catalyse the same reaction. For example, LDH1-5 all convert Pyruvate to lactate.
    - They have different quaternary structure. For example, the subunits in LDH-1 is different from LDH-2.
    - Tissue distribution of each isoform is different.
    - Enzyme name and number can be different

4. **c. Lipoprotein lipase**
    (Ref: Vasudevan & Sreekumari 7/e page 301)

Functional Enzymes and Nonfunctional Enzymes
- Enzymes which have specific function in the plasma.

Examples of functional enzymes:
- Coagulation Factors
- Lipoprotein Lipase

Nonfunctional enzymes
- No specific function in the serum.
- Comes out from the tissue as a result of normal wear and tear.
- Their level is very low in the serum.
- But during tissue injury their level rises in the serum.
- Hence they help to diagnose the site of tissue injury.

Examples of nonfunctional enzymes are LDH, creatine kinase, alkaline phosphatase.

5. **d. Glucose** *(Ref: Vasudevan & Sreekumari 7/e page 309)*

Enzymes as diagnostic reagents

Enzyme	Diagnostic test done
Urease	Urea estimation
Uricase	Uric acid estimation
Glucose oxidase	Glucose

Contd...

Contd...

Enzyme	Diagnostic test done
Hexokinase	Glucose
Peroxidase	Glucose, cholesterol
Cholesterol oxidase	Cholesterol
Creatininase	Creatinine
Lipase	Triglycerides

6. **d. Enzyme Assay**
   (Ref: Varley's Practical Clinical Biochemistry 6/e page 352)

   Estimation of Blood Creatinine
   Two methods:
   Chemical Method—Based on Jaffe's Test
   In alkaline medium creatinine form a red coloured tautomer of Creatinine picrate which is measured colorimetrically.
   This method can be automated in autoanalysers and Kinetic method can be used
   - Kinetic Jaffe more accurate than Jaffe's Method

   Enzymatic Method
   - By employing two enzymes Creatininase or Creatinine Deaminase
   - More specific
   - No interference by Ketones, Bilirubin or Glucose. Hence measure Creatinine accurately.

7. **a, b. Lipase, Urease** *(Ref: Harper 31/e page 566, Table 48.6)*

8. **a. Liver** *(Ref: Harper 31/e page 64)*

   Lactate dehydrogenase (LDH) is a tetrameric enzyme consisting of two monomer types: H (for heart) and M (for muscle) that combine to yield five LDH isozymes: HHHH (I1), HHHM (I2), HHMM (I3), HMMM (I4), and MMMM (I5). Tissue-specific expression of the H and M genes determines the relative proportions of each subunit in different tissues. Isozyme LDH-I predominates in heart tissue, and isozyme LDH-5 in the liver. Thus, tissue injury releases a characteristic pattern of LDH isozymes that can be separated by electrophoresis.

9. **a. LDH-1**
   - LDH having fastest electrophoretic mobility is LDH-1
   - LDH having least electrophoretic mobility is LDH-5
   - Creatine Kinase having fastest electrophoretic mobility is CK-1
   - Creatine Kinase having least electrophoretic mobility is CK-3.

CHAPTER 4

Carbohydrates

Chapter Outline

4.1 Chemistry of Carbohydrates
4.2 Major Metabolic Pathways of Carbohydrates
4.3 Minor Metabolic Pathways of Carbohydrates

CHAPTER 4

Carbohydrates

4.1 CHEMISTRY OF CARBOHYDRATES

- Classification of Carbohydrates
- Reactions of Carbohydrates
- Glycosaminoglycans
- Isomerism in Carbohydrates
- Mucopolysaccharidosis

CARBOHYDRATES

- **Carbohydrates** are the most abundant organic molecules in nature.
- Most abundant monosaccharide in nature is **D-Glucose**.
- Most carbohydrates occur in nature as **Polysaccharides**.
- The word Carbohydrate literally means hydrates of carbon.
- Main (Primary) source of energy for human beings is **Carbohydrates** (45–65% of total energy).

Definition

Aldehyde or Keto derivatives of Polyhydric Alcohols or compounds, which yield these derivatives on hydrolysis.

General Formula for Carbohydrates^Q is $C_n(H_2O)_n$, where n = no. of carbon atoms

CLASSIFICATION OF CARBOHYDRATES

Carbohydrates are classified into:	
Monosaccharides	Disaccharides
Oligosaccharides	Polysaccharides

Monosaccharides [$C_n(H_2O)_n$]

Sugars which cannot be further hydrolyzed.
They contain one sugar unit.
Building blocks of all carbohydrates.

Depending on the no. of Carbon atom, monosaccharides are subclassified into:

Number of carbon atoms	Generic name
3	Trioses
4	Tetroses

Contd...

Contd...

Number of carbon atoms	Generic name
5	Pentoses
6	Hexoses
7	Heptoses
9	Nanoses

Depending on the functional group, monosaccharides are classified into:

- Aldoses with Aldehyde group
- Ketoses with Keto group

Important monosaccharides based on functional groups are:

Generic name	Aldoses	Ketoses
Triose	Glyceraldehyde	Dihydroxy Acetone
Tetrose	Erythrose	Erythrulose
Pentose	Ribose Xylose (Epimer of Ribose) Arabinose	Ribulose Xylulose (Epimer of Ribulose)
Hexose	Glucose, Galactose, Mannose	Fructose
Heptose		Sedoheptulose

> **HIGH YIELD POINTS**
>
> - Simplest carbohydrate of biological interest are: **Glyceraldehyde and Dihydroxy Acetone**.
> - Pentoses which constitute a part of Nucleic Acid are **Riboses and Deoxyriboses**
> - Nanoses with biologic significance is **Neuraminic Acid**

The only metabolic fuel for mature erythrocytes in fed state and starving state is Glucose^Q.

Sialic Acid

- N-Acyl or O- Acyl derivative of Neuraminic Acid.
- **N-Acetyl Neuraminic Acid (NANA)** is the predominant Sialic Acid.
- Constituents of both Glycoprotein and Ganglioside.

Ring Structures of Monosaccharides

Monosaccharide molecules of 4, 5 or 6 carbon atom are quite flexible, and this flexibility brings aldehyde or keto group in close proximity to other hydroxyl groups on the same molecule.

- The reaction of Ketone with hydroxyl group form **Hemiketal Ring Structure**.
- The reaction of Aldehyde with hydroxyl group form **Hemiacetal Ring Structure**.

Image-Based Information

If the ring structure formed by cyclization is a six membered (made of 5 carbon and 1 oxygen), it is called a Pyranose ring; if it is five sided (made of four carbons and one oxygen), it is called a furanose ring.
- Fructose exist spredominantly as **Furanose** ring structure (**Fructofuranose**)
- Glucose exists predominantly as **Pyranose** ring structure (**Glucopyranose**)

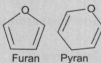

Fig. 4.1.1: Furan and pyran ring

Biologically Significant Hexoses

Glucose

- Most predominant sugar in human body.
- Main source of metabolic fuel of mammals.
- Glucose is dextrorotatory hence otherwise called **Dextrose**^{Q}
- **Universal fuel of fetus**.

 HIGH YIELD POINTS

Organs whose major energy source is glucose
- Brain
- Renal medulla
- Cornea
- Retina
- Testis
- RBC

The only metabolic fuel for mature erythrocytes in fed state and starving state is Glucose^Q.

Galactose

- Constituent of Lactose^Q (Milk sugar).
- Synthesized in the mammary gland for synthesis of lactose.
- Part of Glycoprotein, Glycosaminoglycan in Proteoglycans and Glycolipids.^{Q}

Mannose

- Isolated from plant mannans, hence the name.
- Occur in Glycoproteins and Mucoproteins.

Fructose

- Constituent if Sucrose, the common sugar.
- Present in fruit juices, honey^Q and sugarcane.
- Present in the seminal fluid^Q.

CONCEPT BOX

All the Hexoses have a free functional group, hence they are reducing sugars.

Disaccharides [$C_n(H_2O)_{n-1}$]

Two monosaccharide units are linked by a **glycosidic** bond or yield 2 monosaccharide units on hydrolysis. Depending on their reducing property they are divided into:
- Nonreducing disaccharides
- Reducing disaccharides

Nonreducing Disaccharides

The functional groups are involved in the glycosidic bond formation, hence free functional groups are not available.

Reducing Disaccharides

Free functional groups are available.

Reducing Disaccharides—Free Functional Group Present

Disaccharide	Sugar Units^Q	Linkage
1. Maltose^Q	α D-Glucose + α D-Glucose	α 1→4α Linkage (α1→4 linkage)
2. Isomaltose	α D-Glucose + α D-Glucose	α 1→6 α Linkage (α1→6 linkage)
3. Lactose^Q (Milk Sugar)	β D-Galactose + β D-Glucose	β1→4β Linkage^Q
4. Lactulose	α D-Galactose + β D-Fructose	α1→4β Linkage

Nonreducing Disaccharides^Q—No Free Functional Group

Disaccharide	Sugar units	Linkage
Trehalose (Sugar of insect hemolymph, yeast and fungi)	α D-Glucose + α D-Glucose	α1→1α Linkage
Sucrose^Q (Cane Sugar^Q)	α D-Glucose + β D-Fructose	α1→2β Linkage

> **HIGH YIELD POINTS**
>
> **LACTULOSE**
> - Osmotic Laxative
> - Mainly **Synthetic** (small amount in heated milk)
> - Not hydrolyzed by intestinal bacteria
> - But fermented by intestinal bacteria

Oligosaccharides

Condensation product of 3 to 10 monosaccharides. Blood group substances are **oligosaccharides**.
- Usually found in association with Proteins (Glycoproteins and Proteoglycans) and lipids (Glycolipids).

Polysaccharides

Condensation product of **more than 10** monosaccharide units. Also called as **Glycans**.

Depending on the type of Monosaccharide units Polysaccharides are classified into:
1. **Homoglycans (Homopolysaccharide)** contain only one type of monosaccharide unit.
2. **Heteropolysaccharides (Heteroglycans)** contain different types of monosaccharide units.

Homoglycans (Homopolysaccharide)

Glycogen
- Storage form of Glucose in animals hence called **animal Starch**.
- Made up of α **D-Glucose**.
- Branched polymer of Glucose.

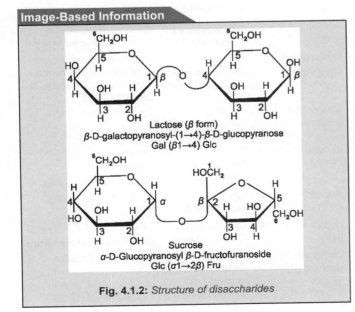

Fig. 4.1.2: *Structure of disaccharides*

- **α1, 4 Linkage** at the linear part and **α1, 6 Linkage**-at branches.
- In **Muscle** Glycogen, granules called **β particle** are present. It contains 60,000 glucose residues.
- In **Liver** apart from β particle, **rosettes of glycogen** granules, which are aggregated β-particles are also present.

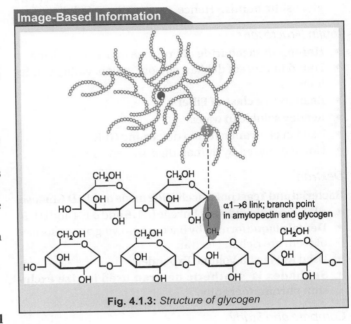

Fig. 4.1.3: *Structure of glycogen*

Starch
Major source of carbohydrate in the diet.
- Homopolysaccharides made up only of Glucose.
- Storage form of Carbohydrates in plants.
- Also called **glucan** or **Glucosan**.
- Two main constituents are:
  1. **Amylose** (13-20%) which has a **nonbranching** helical structure.
  2. **Amylopectin** (80-87%) which consist of **branched** chains with 24-30 glucose residues.

Starch [Glucosan] Mixture of Amylose 15–20% and Amylopectin 80–85%		
i. Amylose Soluble Unbranched	Glucose	α1 → 4 Linkage
ii. Amylopectin Insoluble Branched	Glucose	α1 → 4 Linkage α1 → 6 Linkage [at Branches]

Chitin
- Found in **exoskeleton** of crustaceans and insects and in mushrooms
- Made up of **N Acetyl D Glucosamine** joined by β1 → 4 linkage.

Cellulose

- Chief constituents of plant cell wall
- Homopolysaccharide of β Glucose in β1, 4 linkage
- Insoluble
- Major component of **dietary fiber** [Source of bulk in the diet]
- Humans lack enzyme, **Cellulase** that hydrolyse β1, 4 glycosidic bonds.^{Q} Hence, cannot digest Cellulose.

Inulin (Fructosan)

- **Homopolysaccharide** of Fructose in β 2 → 1 linkage.
- Found in roots of dahlias, chicory, onion, garlic, dandelions.
- Belongs to a **class of Fibers**.
- Readily soluble in water.
- Used in **clearance test** to determine GFR.
- Not hydrolyzed by human digestive enzymes.

Dextran

Bacterial and Yeast polysaccharide made up of αD Glucose.

- αD Glucose in different linkage (α1, 6 and α1, 4 and α1, 3)
- **Dental plaque** formed by bacteria growing on the surface of teeth are rich in Dextran.
- Used as Plasma Volume Expander
- **Sephadex is synthetic dextran** used in **size exclusion chromatography**.

Compare and Study

Dextrose^Q	Glucose being dextrorotatory is known as Dextrose in clinical practice.
Dextrin	Are intermediates in the hydrolysis of starch.
Dextran	Homopolysaccharide made up of Glucose.
Lactose	Disaccharide made up of Galactose and Glucose.
Lactase	Enzyme which hydrolyze Lactose to Galactose and Glucose.
Lactate	Product obtained from Pyruvate during anaerobic Glycolysis.
Lactulose	Disaccharide made up of Galactose and Fructose.

Frequently Asked Doubts

What is the difference between lectin and pectin?

Lectin refers to ant **protein that bind to polysaccharide**. They are highly selective in binding

- *For example:* Mannose binding lectin (MBL) is specific for Mannose containing polysaccharides
- Concavalin A binds to α-glucosyl and α-mannosyl residue.

Pectin is a **polysaccharide made up of Galacturonic acid** in α1 → 4 linkage.

HETEROPOLYSACCHARIDES (HETEROGLYCANS)

- Glycosaminoglycans (Mucopolysaccharides)
- Pectins
- Agarose
- Agar

MUCOPOLYSACCHARIDES (GLYCOSAMINOGLYCANS^Q)

- Glycosaminoglycans are **unbranched** hetero-polysaccharide^Q chains composed of **Disaccharide repeat units.**
- Each disaccharide repeat unit composed of an **amino sugar and uronic acid**.
- They were first isolated from mucin hence called **Mucopolysaccharides**.
- Major component of **Extracellular Matrix**.

Properties of GAG

- Carry large number of **negative charges** (COO⁻, Acetyl, Sulfate), these chains tend to repel each other.
- Hence slippery consistency of mucous secretion and Synovial fluid.
- When water squeezed out they occupy small volume. When compression released, they spring back to original hydrated volume because of repulsion of negative charges.
- Hence resilient nature of Synovial fluid and vitreous humor.
- Special ability to bind to large amounts of water. Hence forming major component of Extracellular Matrix.

Biologically Important GAGs

	Disaccharide repeat unit	Location
Hyaluronic Acid (Hyaluronan)^Q	N Acetyl Glucosamine + Glucuronic Acid^Q	Skin, Synovial fluid, bone, cartilage, vitreous humor, loose connective tissue, Umiblical Cord
Chondroitin Sulphate	N Acetyl Galactosamine + Glucuronic Acid	Cartilage, bone, CNS
Keratan Sulphate I and II	N Acetyl Glucosamine, Galactose	Cornea^Q Cartilage Loose connective tissue
Heparin	Glucosamine, Iduronic Acid	Mast cells^Q Liver, Lung, Skin
Heparan sulphate (HS)	Glucosamine, Glucuronic Acid	Skin Kidney basement membrane
Dermatan sulphate (DS)	N Acetyl Galactosamine, Iduronic Acid/ Glucuronic Acid	Skin, wide distribution

Carbohydrates

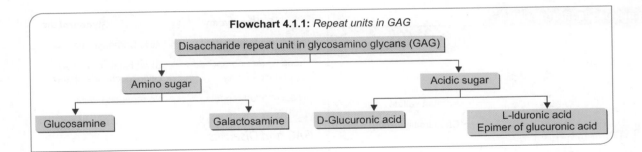

Flowchart 4.1.1: *Repeat units in GAG*

Disaccharide repeat unit in glycosamino glycans (GAG)
- Amino sugar
  - Glucosamine
  - Galactosamine
- Acidic sugar
  - D-Glucuronic acid
  - L-Iduronic acid (Epimer of glucuronic acid)

Important Points of Glycosaminoglycans

Hyaluronic acid
- Present in bacteria and ECM of nearly all animals.
- Play an important role in permitting cell migration during morphogenesis and wound repair.

Chondroitin sulphate
- Join with a protein by the Xylulose-Serine by O Glycosidic bond.
- Major component of cartilage.
- Located at sites of calcification in endochondral bone.

Keratan sulphate I and II
- Keratan sulphate I is originally isolated from **Cornea.**

Heparin
- Consist of Glucosamine and either of two uronic acid.
- Vast majority of the uronic acid residues are Iduronic acid.
- Initially all are Glucuronic acid, but 90% of Glc UA is converted to IdUA by a 5' Epimerase.
- **Heparin** is an anticoagulant.

Important Points of Glycosaminoglycans

- **Heparin** specifically binds to **Lipoprotein Lipase** present in capillary walls, causing release of this enzyme into circulation.
- Heparin is found in the granules of mast cells, also liver, lung, and skin.

Heparan sulphate
- Present on many cell surfaces as proteoglycan.
- Predominant uronic acid is GlcUA unlike Heparin.
- They act as receptors.
- Mediate cell growth and cell to cell communication.
- Found in Kidney Basement membrane along with Type IV Collagen and laminin.
- In Kidney basement membrane, it plays a role in **charges electiveness of glomerular filtration**.

Dermatan sulphate
- Widely distributed GAG.
- The main GAG of skin.

High Yield Points

- GAG with no Uronic Acid[Q]: Keratan Sulphate
- GAG with no Sulfate group: Hyaluronic Acid
- GAG found in bacteria: Hyaluronic Acid
- GAG which is an anticoagulant: Heparin
- Most abundant GAG: Chondroitin Sulphate
- Site of Synthesis of GAG: Endoplasmic Reticulum and Golgi
- Shape of Proteoglycan monomer: Bottle Brush
- Glycosaminoglycans are Polyanions.
- GAG that helps in cell migration: Hyaluronic acid
- GAG that have a role in compressibility of cartilage in weight bearing are Hyaluronic acid and Chondroitin Sulphate.
- GAG that play a role in corneal transparency. Keratan Sulphate I
- GAG that determine charge selectiveness of Renal Glomerular membrane: Heparan Sulphate.
- GAG are generally Extracellular.
- Some intracellular GAG are Heparin (in mast cells).
- GAG in synaptic and other vesicles is Heparan sulphate
- Only GAG not covalently attached to a core protien is Hyaluronic Acid

Mucin Clot Test (Rope Test)[Q]
- To detect hyaluronate in the synovial fluid.
- Normal synovial fluid forms tight ropy clot on addition of acetic acid.

Proteoglycan
- GAGs are covalently attached to a protein (termed Core Protein) to form hybrid molecules, called Proteoglycan.

Image-Based Information

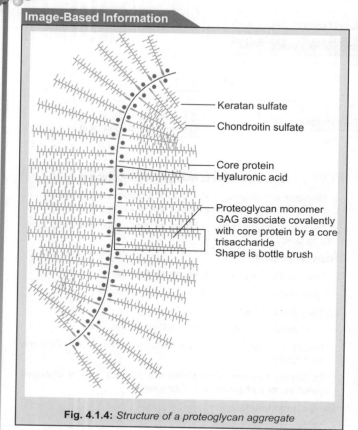

Fig. 4.1.4: Structure of a proteoglycan aggregate

Proteoglycan	Glycoprotein
>95% Carbohydrate	<5% Carbohydrate
Long linear unbranched Oligosaccharides	Short highly branched Oligosaccharide chains
Disaccharide repeats	No repeating units

GAG and Diseases

Tumor Cell Migration
- Tumor cells induce fibroblast to synthesize Hyaluronic acid.
- Hyaluronic acid permit tumor cells to migrate through ECM.
- Some tumor cells have less heparan sulfate at their surfaces, and this may play a role in the **lack of adhesiveness** that these cells display.

GAG and Atherosclerosis
- Dermatan Sulphate appears to be the major GAG synthesized by arterial smooth muscle cells.
- These cells proliferate in **atherosclerotic lesions** in arteries.
- Dermatan sulfate binds plasma low-density lipoproteins.
- Because of it, dermatan sulfate may play an important role in development of the atherosclerotic plaque.

GAG and Osteoarthritis
- In arthritis, proteoglycans may act as autoantigens.
- The amount of chondroitin sulfate in cartilage diminishes with age, whereas the amounts of keratan Sulfate and hyaluronic acid increase.
- These changes may contribute to the development of osteoarthritis.

Proteoglycan Monomer and Proteoglycan Aggregate

Proteoglycan monomer
- GAG molecules attached covalently to core protein forming Proteoglycan Monomer.
- Linking of GAG chain to core protein occurs by a Core trisaccharide, Gal-Gal-Xyl.
- GAG-Core Trisaccharide—Core Proteinis Proteoglycan monomer
- The shape of Proteoglycan monomer is Bottle brush.

Proteoglycan aggregate
- Several Proteoglycan monomer associate noncovalently to a Hyaluronic Acid by a link protein to form Proteoglycan Aggregate.

MUCOPOLYSACCHARIDOSES (MPS)

Hereditary progressive disease caused by mutation of genes coding for Lysosomal Enzymes needed to degrade GAGs results in Intralysosomal^Q accumulation of GAGs.

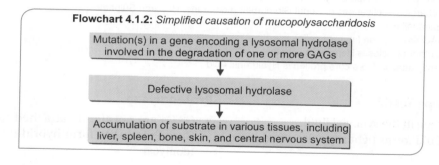

Flowchart 4.1.2: Simplified causation of mucopolysaccharidosis

Carbohydrates

	Mucopolysaccharidosis	Inheritance	Enzyme defect	Urinary metabolite
MPS I H	Hurler Disease[Q]	Autosomal Recessive	L- Iduronidase[Q]	Dermatan Sulfate Heparan Sulfate
MPS I S	Scheie Disease	Autosomal Recessive	L- Iduronidase	Dermatan Sulfate
MPS II	Hunter Disease[Q]	X-linked Recessive	Iduronate Sulfatase[Q]	Dermatan Sulfate Heparan Sulfate
MPS III A	Sanfilippo A Disease	Autosomal Recessive	Heparan Sulfate N Sulfatase	Heparan Sulfate
MPS III B	Sanfilippo B Disease	Autosomal Recessive	N- Acetyl Glucosaminidase	Heparan Sulfate
MPS III C	Sanfilippo C Disease	Autosomal Recessive	Glucosaminide N Acetyl Transferase	Heparan Sulfate
MPS III D	Sanfilippo D Disease	Autosomal Recessive	N Acetyl Glucosamine Sulfatase	Heparan Sulfate
MPS IV A	Morquio A	Autosomal Recessive	Galactosamine 6 Sulfatase	Keratan Sulfate Chondroitin 6 Sulfate
MPS IV B	Morquio B	Autosomal Recessive	Beta Galactosidase	Keratan Sulfate
MPS VI	**Maroteaux-Lamy**	**Autosomal Recessive**	**N Acetyl Galactosamine 4 Sulfatase (Aryl Sulfatase B)**	Dermatan Sulfate
MPS VII	Sly Disease	Autosomal Recessive	Beta - Glucuronidase	Dermatan Sulfate Heparan Sulfate

Note: The important MPS are given in bold letters. Please do learn them.

Recognition Pattern of Mucopolysaccharidosis

Clinical features	MPS IH	MPS IS	MPS II	MPS III	MPS IV	MPS VI	MPS VII
Common name	Hurler	Scheie	Hunter	Sanfilippo	Morquio	Maroteaux- Lamy	Sly Disease
Mental deficiency	+	−	+	+	−	−	?
Coarse facial features	+	(+)	+	−	−	+	?
Corneal clouding	+	+	−	-	(+)	+	?
Visceromegaly	+	(+)	(+)	−	−	+	+
Short stature	+	(+)	+	+	+	+	+
Joint contractures	+	+	+	−	−	+	+
Dysostosis multiplex	+	(+)	(+)	+	+	+	+
Leucocyte inclusions	+	(+)	+	+	+	+	+
Mucopolysacchariduria	+	+	+	+	+	+	+

🩺 Symptom box

Clinical Features of MPS I H (Hurler's Disease)
- Normal at birth
- Inguinal hernias
- Hepatosplenomegaly
- Coarse facial features
- Corneal clouding
- Large tongue
- Prominent forehead
- Joint stiffness
- Short stature
- Skeletal dysplasia
- Copious nasal discharge

Mucopolysaccharidosis-I H (Hurler's Disease)

Biochemical defect

Homozygous or double heterozygous nonsense mutations IDUA gene on Chr 4p encoding α-L-Iduronidase.

Scheie Disease (MPS-IS)

Biochemical Defect

A missense mutations in IDUA gene on Chr 4p encoding **α-L-Iduronidase. Partial activity of enzyme is preserved.**

Clinical Features
- Similar to MPSIH.
- Onset after the age of 5 years
- Normal intelligence and normal stature

Mucopolysaccharidosis II (Hunter Disease)

Biochemical Defect
- X-linked disorder caused by the deficiency of **iduronate-2-sulfatase (IDS)**.
- Point mutations of the *IDS* gene mapped to Xq 28 have been detected in about 80% of patients with MPSII.
- Hunter disease manifests almost exclusively in males.

Symptom box

Clinical Features
- Similar to Hurler disease
- Lack of corneal clouding.
- Grouped skin papules

EXTRA EDGE

NATOWICZ SYNDROME (MPS-IX)
- A genetic defect in hyaluronidase causes MPS-IX, a lysosomal storage disorder in which hyaluronic acid accumulates in the joints.
- Joint pains and short stature are the clinical features.

Laboratory Diagnosis of Mucopolysaccharidoses
- Urinalysis for presence GAGs.
- Assays of enzymes in white blood cells, fibroblasts, or possibly serum.
- Use of specific gene tests.
- Prenatal diagnosis in amniotic fluid cells or chorionic villus biopsy

---

CONCEPT BOX
- Impaired degradation of Heparan Sulphate associated with Mental Deterioration.
- Impaired degradation of DS, CS, KS associated with mesenchymal abnormalities.

HIGH YIELD POINTS

MPS with no Corneal Clouding
- Hunter's Disease
- Sanfilippo Disease

MPS with no Visceromegaly
- Morquio Disease
- Sanfilippo Disease

Same Enzyme-deficieincy associated with two diseases:
- Hurler Disease
- Scheie Disease

Newer Modalities of Treatment of MPS

MPS type	Stem cell transplantation (SCT)	Enzyme replacement	Remarks
I (Hurler, Scheie)	Yes	Aldurazyme	Transplantation before age 2. Enzyme replacement before and after transplantation
II (Hunter Disease)	Questionable	Elaprase	Lack of neurological improvement after SCT
III (Sanfilippo)	No	No	Experimental: Substrate reduction by Flavinoids
IVA (Morquio)	No	Preclinical	Recombinant Galactosamine Sulfatase (GALNS) in course
VI (Maroteaux-Lamy Disease)	Yes	Naglazyme	Sustained improvement
VII (Sly)	Questionable	?	Single SCT attempt without neurological improvement

DERIVED SUGARS

Monosaccharides whose structure cannot be represented by general formula or which have some unusual features. *They are*
- Acid Sugars (By Oxidation of Sugars)
- Sugar Alcohols (By reduction of Sugars)
- DeoxySugars
- AminoSugars[Q]
- Glycosides
- Furfural Derivative.

Acid Sugars

Formed by oxidation of aldehyde carbon atom, hydroxyl carbon atom or both of monosaccharides.

I. *Under Mild Oxidation Conditions*

1. Aldehyde group is oxidized to produce Aldonic Acid.
- Glucose to **Gluconic** Acid
- Mannose to Mannonic Acid
- Galactose to Galactonic Acid

Clinical application of Glucose to Gluconic acid

During Glucose oxidase method for estimation of blood glucose Gluconic Acid[Q] is formed from Glucose.

2. Last Carbon atom is oxidized to produce **Uronic Acid.**

Glucose to Glucuronic acid:
- Mannose to Mannuronic acid
- Galactose to Galacturonic acid.

Glucose to Glucuronic acid:
- Mannose to Mannuronic Acid
- Galactose to Galacturonic Acid.

Significances of Glucuronic acid
- Iduronic acid is an epimer of Glucuronic acid.
- Constituent of Glycosamino Glycans (GAGs)
- Used for conjugation of Bilirubin.

Under Strong Oxidation Condition

Both first and last carbon atom are oxidized to produce **Saccharic acid**
- Glucose to Glucosaccharic Acid.
- Mannose to Mannaric Acid
- Galactose to Mucic Acid

Significance of Galactose to Mucic acid

Mucic Acid forms *insoluble crystals* form basis for Mucic Acid test for the identification of Galactose.

Sugar Alcohols

- Monosaccharides are reduced at their carbonyl group to yield corresponding poly hydroxy alcohols.
  - Aldoses undergo reduction to form corresponding Alcohol.
  - Ketoses form two alcohols because of appearance of new asymmetric carbon atom.

Examples of Sugar Alcohols

Glucose	Sorbitol
Mannose	Mannitol
Galactose	Dulcitol/Galactitol
Fructose	Sorbitol and Mannitol

Clinical Applications of Sugar Alcohols

- Mannitol is used to reduce intracranial pressure by forced diuresis.
- Osmotic effect of Dulcitol and Sorbitol causes cataract in Galactosemia and Diabetes respectively.
- Polyol pathway is responsible for the development Diabetic cataract.

Deoxy Sugars

Hydroxyl group of sugars is replaced by hydrogen atom.

Biochemical Importance of Deoxy Sugars

Deoxy Ribose
- Oxygen is removed from 2nd position[TNPGEE04]
- It is an important component of DNA.
- *Feulgen staining* is specific for 2 deoxy sugars (and DNA) in the tissues.

L –Fucose
- Deoxy Sugar present in the Blood group antigens.

2- Deoxy Glucose
- Experimentally an inhibitor of Glucose metabolism.

Amino Sugars (Hexosamines)

Amino group replaces the hydroxyl group present in the second carbon atom of monosaccharides to form Amino Sugars.

Important Amino Sugars are:
- Glucosamine
- Galactosamine (Chondrosamine)
- Mannosamine
- Sialic acid

Sialic Acid
- An unusual Amino sugar with 9 carbon atom is Sialic Acid.
- The principal Sialic Acid found in human body is N Acetyl Neuraminic Acid (NANA)

> **HIGH YIELD POINTS**
> - Glucose is the precursor of Amino Sugar.
> - The immediate precursor of Glucosamine is Fructose 6 Phosphate[Q 2012].
> - Amino group is donated by Glutamine.
> - NANA is derived from N Acetyl Mannosamine.

Biochemical Significance of Amino Sugars
- They are components of Glycoproteins, Gangliosides and Glycolipids.
- Antibiotic which contain amino sugar is **Erythromycin.**

Glycosides

When the monosaccharide condensed with an alcohol, phenol or sterol by *O-Glycosidic linkage* to form Glycoside. The noncarbohydrate group is called Aglycone.

Clinical Importance of Glycosides

Cardiac Glycosides—Action on heart
- Digitalis (Steroid is the Aglycone)
- Quabain

Antibiotics
- Streptomycin
- Puromycin

ISOMERISM IN CARBOHYDRATES

ISOMERS

Different compounds having same molecular formula are called isomers of one another.

Asymmetric Carbon Atom

The carbon atom to which four different substituent groups are attached is called a chiral or asymmetric Carbon atom.
- In Open chain structure of Glucose, there are 4 asymmetric Carbon atom (C-2, C-3, C-4, C-5).
- But in solution 99.5% Glucose exist in Pyranose form, then first carbon atom also become an asymmetric Carbon atom.

The presence of asymmetric carbon atom imparts two important properties:
- Stereoisomerism
- Optical Isomerism

Lebervon't Hoff Rule

The relationship between the number of Asymmetric Carbon atom and the number of Stereoisomers possible.

Number of isomers = 2^n

where n is the number of Asymmetric Carbon atom.

Types of Isomers in Carbohydrates

The presence of asymmetric carbon atom imparts two important properties:
- Stereoisomerism
- Optical Isomerism

Stereoisomerism

- Compounds having the same molecular formula but different spatial configuration of H and OH group around the asymmetric carbon atoms.

Enantiomers

- A stereoisomerism in which one isomer is a mirror image of the other.
- This requires presence of an asymmetric carbon atom.
- D and L isomerism is a type of enantiomerism.

D and L Isomerism

Difference in the orientation of H and OH group around penultimate carbon atom results in two mirror images called D and L isomers.

In D and L glucose the position of H and OH varies in only the penultimate carbon atom?

Even though we say in definition of D and L isomerism at penultimate/reference carbon atom, actually the position of H and OH changes in all asymmetric carbon atoms of glucose ie C-2, C-3, C-4 and C-5. The reason is the parent carbohydrate is Glyceraldehyde, the asymmetric carbon atom in it is considered as the reference carbon atom.

The penultimate carbon atom is called **Reference Carbon atom.**

So, this is a type of Enantiomerism.
- The penultimate carbon atom in Glucose and Fructose is C-5.
- Most of the naturally occurring Monosaccharides are **D isomers** (Unlike Amino Acids, which are L isomers).

Examples of D and L Isomerism[Q]
- D Glucose and L Glucose
- D Fructose and L Fructose
- D Mannose and L Mannose
- D Glyceraldehyde and L Glyceraldehyde.

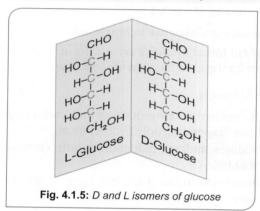

Fig. 4.1.5: *D and L isomers of glucose*

Anomerism

- Formation of ring structure in monosaccharides results in creation of an additional asymmetric carbon called anomeric Carbon atom.
- The carbon atom with functional group forms the anomeric carbon atom.
- In Glucose C-1 and in fructose C-2 form the anomeric carbon atom.
- Difference in orientation of H and OH group around the anomeric carbon atom results in Anomerism.
- The resulting isomers are called α & β anomers.
- The predominant anomeric form of Glucose is **β D Glucose.**

Examples of Anomerism

- αD Glucose and β D Glucose
- αD Fructose and β D Fructose.

Mutarotation

- The process of interconversion of α and β anomers in acqeous solution is called mutarotation.
- Mutarotation is studied by measuring the rotation of plane polarised light.
- The optical rotation of α D Glucose is +112°
- The optical rotation of β D Glucose is +19°
- Both undergo mutarotation over a period of a few hours, an equilibrium is attained
- At equilibrium the optical rotation is +52°

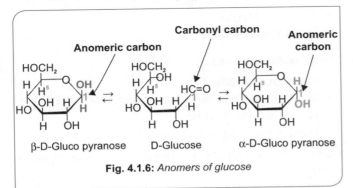

Fig. 4.1.6: *Anomers of glucose*

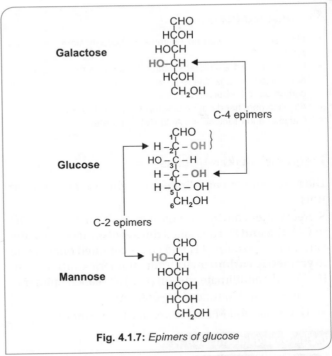

Fig. 4.1.7: *Epimers of glucose*

Epimerism

Difference in orientation of H and OH group around Carbon atoms other than Anomeric Carbon and Penultimate Carbon results in isomerism refered to as Epimerism.

The difference in configuration is confined only to one asymmetric carbon atom.

Epimers of Glucose[Q]

Difference in orientation of H and OH at C2 C3 & C4 for Glucose:
- 2nd Epimer of Glucose—Mannose
- 3rd Epimer of Glucose—Allose
- 4th Epimer of Glucose—Galactose

Diastereoisomerism

Stereoisomers that are not mirror images.
They differ in configuration of more than one asymmetric Carbon atom other than reference and functional carbon atom e.g. D Mannose and D Galactose.

Optical Isomerism

- When a beam of plane polarized light is passed through a solution of carbohydrates it rotate the light either to right or to left.

Optical Isomers

- Depends on the direction of rotation, two optical isomers possible.

Dextrorotation

- Rotate plane polarized light to right (Clockwise)
- Dextrorotatory (represented by d or+)
- D Glucose is dextrorotatory (i. e. why glucose is also called Dextrose)

Levorotation

- Rotate plane polarized light to left (Anticlockwise)
- Levorotatory (represented by l or-)
- D Fructose is levorotatory.

Racemic Mixture[AI 03]

- Equimolar mixture of optical isomers which has no net rotation of plane polarized light.

Sucrose is Otherwise Called Invert Sugar

- Sucrose is dextrorotatory. On hydrolysis of Sucrose yield a mixture of dextrorotatory Glucose and levorotatory Fructose. Because of strong levorotatory nature of fructose, Sucrose on hydrolysis is levorotatory. Hence, Sucrose is called invert sugar.

High Yield Points

- Monosaccharide with no Asymmetric Carbon atom: Dihydroxy Acetone
- Ketoses have 1 asymmetric Carbon atom less than Aldoses
- No. of Isomers possible is 2n where n is the no. of Asymmetric carbon atom (Lebervon't Hoffrule)
- All D isomers need not be dextrorotatory and vice versa
- Glucose and Fructose are Aldo-Keto Isomers

Frequently Asked Doubts

Diastereoisomers and Epimers of glucose are same or not?

Stereoisomers having variation in position of H and OH in **C-2, C-3 and C-4** are called **diastereoisomers**. But if the variation in postion of H and OH is confined only **to one asymmetric carbon** apart from anomeric Carbon atom (C-1) and penultimate carbon (C-5) are called epimers.
- Glucose and Mannose are epimers
- Galactose and Mannose are diastereoisomers.

Image-Based Information

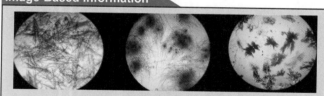

Fig. 4.1.8: *Osazones*

Sugar	Shape
Glucose, Fructose, Mannose	Needle shaped/Broomstick/Sheaves of corn
Lactose	Pincushion with pins/Hedgehog/Flower of Touch me not
Maltose	Sunflower Petal shaped

NB: Sucrose do not form Osazones

TESTS FOR CARBOHYDRATES

General test for all Carbohydrates—Molisch test

Test for Reducing Substances
- Benedict's test

Test to differentiate Monosaccharides and Disaccharides.
- Barfoed's test
- Moore's test
- Fehling's test

Test to differentiate Aldoses and Ketoses
- Seliwanoff's test
- Rapid furfural test

- Foulger's test

Test to detect Deoxy Sugar
- Feulgen staining

Test for Pentoses
- Bial's test

Test for Galactose
- Mucic acid test

Methods of Estimation of Glucose

Reductometric Methods
- Nelson Somogyi Method
- Folin WU Method
- O-Toluidine Method

Enzymatic Method
- Hexokinase Method
- Glucose Oxidase Peroxidase Method (GOD–POD) Highly Specific Method
  Used in dry analysis technique like Glucometer.

Complex Carbohydrates not digested by human digestive enzyme.
Otherwise called **Nonstarch Polysaccharide.**

Dietary ibers

Insoluble Fibres
- Cellulose
- Hemicellulose
- Lignin

Soluble Fibres
- Pectin
- Gums
- Mucilage

High Yield Points

RDA is 40 g per 2000 kcal per day is desirable. Supply of energy from dietary fibres is 2 kcal/g
Dietary fibre neither digested nor fermented-Lignin[Q]

Beneficial Effects of Dietary Fibers

- Prevents constipation
- *Maintain normal motility of GIT*
- Eliminate bacterial toxin
- Fibre absorbs large quantity of water and toxic compounds by intestinal bacteria
- *Increase bulk of the stool*
- Reduces the stool transit time
- Decreases GI Cancers—Colon and Rectum
- *Slow Gastric Emptying*

Carbohydrates

Image-Based Information

Picture A: Molisch Test
General test for all carbohydrates
Molisch reagent is alpha Naphthol in alcohol
Observation: A **purple ring** appears at the junction of two liquids
Principle: Concentrated acid dehydrates the sugar to form furfural or furfural derivatives which then condense with α-naphthol to give a purple colored complex

Picture B: Benedict's Test
Test for reducing substances Answered by
- Monosaccharides like Pentoses, Glucose, Fructose, Galactose,
- Reducing disaccharides like Lactose. Maltose
- Glucuronic acid, Homogentisic acid, Ascorbic acid
- Glutathione

Creatinine and Uric acid in high concentration

Picture C: Barfoed's Test
To distinguish monosaccharides and reducing disaccharides Barfoed's reagent is copper acetate in acetic acid
Observation: Red precipitate at the bottom of test tube.
Principle: Monosaccharides reduces Cupric ions to form red Cuprous oxide in acidic medium rapidly than disaccharides.

Picture D: Seliwanoff's Test
In picture D both negative and positive test is given.
- Test to detect Ketose sugar

Seliwanoff's reagent is resorcinol in dilute hydrochloric acid.
Principle: Ketoses are more readily dehydrated by HCl than the aldoses to form hydroxymethyl furfural which then condenses with resorcinol of Seliwanoff's reagent to form a **red colored** complex.

Picture E-Iodine Test
Test for Starch
- Reagent used is Iodine solution

Deep blue color appears which then disappears on heating and then reappears on cooling.
Principle: Starch forms an adsorption complex with iodine to give a blue color. The blue color disappears on heating due to the breaking of the Iodine starch adsorption complex and appears on cooling due to reformation of the adsorption complex.

- Improves Glucose Tolerance by decreasing rate of absorption of glucose
- *Reduces Plasma Cholesterol by binding bile salt*
- Decreases absorption of dietary cholesterol[Q]
- Binds the bile salt & decreases enterohepatic circulation of bile salts and increases excretion of bile salt, the excretory form of cholesterol
- Gives sensation of stomach fullness

A High fibre diet is associated with reduced incidence of[Q]
- Diverticulosis
- Cancer Colon
- Cardiovascular Disease
- Diabetes Mellitus

Glycemic Index

The increase in blood glucose after a test dose of a carbohydrate compared with that after an equivalent amount of glucose (as glucose or from a reference starchy food) is known as Glycemic Index.

> **HIGH YIELD POINTS**
> - Glycemic Index of **Glucose and Galactose is 1 or 100%**.
> - Glycemic index of Lactose, Maltose, Isomaltose & Trehalose is 1 or 100% (because they give rise to Glucose and Galactose).
> - **Fructose**, Sugar alcohols have **less glycemic index** as they are not absorbed completely.
> - Sucrose also low glycemic index as it is split to glucose and fructose.
> - Starch varies from 0–100% owing to variable rate of hydrolysis.
> - For **Nonstarch Polysaccharides** (Dietary fibre) it is **0**.

BIOCHEMISTRY OF ABO BLOOD GROUP ANTIGENS

ABO blood group antigens are **Glycosphingolipids (in red blood cells)** or **Glycoproteins (in secretions)**.

Basics

H blood group substance (antigen)
- H (or O) substance is the blood group substance of ABO system.
- H substance is formed by action of Fucosyl transferase
- So H substance is **Fuc-α1, 2-Gal-β-R**

A blood group substance (antigen)
- Contain an additional **N Acetyl Galactosamine (Gal Nac)**.
- **Gal Nac transferase** that add Gal Nac to H substance.

B blood group substance (antigen)
- Contain an additional **Galactose (Gal)**.
- *Galactose transferase* that add Gal to H substance.
  AB Type has both the enzymes **(Gal Nac transferase, Gal transferase)** and both **A & B substance. O type** lack A and B blood group substance.

Bombay blood group
- The h allele codes for inactive Fucosyl transferase
- hh genotype cannot generate H substance
- But they possess the enzymes to convert H substance to A and B substance.
- The RBC will be of type O.
- They are refered to as Bombay phenotype (O_h).

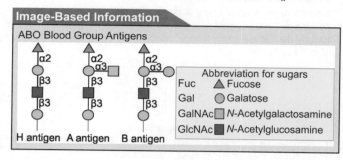

Digestion of Carbohydrates

By two set of enzymes
- Alpha Amylases from Salivary gland and Pancreas
- Disaccharidases.

Lactase Deficiency and Sucrase Deficiency[Q]

Lactase deficiency	Sucrase deficiency [AI 04]
Clinical manifestation following ingestion of milk, which contain Lactose	Clinical manifestation following ingestion of dairy products which contain Sucrose
Watery Diarrhoea Bloating Failure to thrive	Watery Diarrhoea Bloating Failure to thrive

Absorption of Carbohydrates

By two set of transporters
1. Sodium dependent Glucose transporters (SGLT)
2. Sodium independent Glucose transporters (GLUTs)

Sodium-dependent Glucose Transporter

Characteristics
- Secondary Active Transport
- Unidirectional
- SGLT is coupled with $Na^+ K^+$ ATPase pump.

Sodium dependent glucose transporters (SGLT)

SGLT1	Small Intestine, Renal Tubules	Absorption of Glucose and galactose.
SGLT2	Renal Tubules	Absorption of Glucose

Clinical Correlation: Renal Glycosuria

Isolated glycosuria in the presence of a normal blood glucose concentration is due to mutations in SLC5A2, the gene that encodes the high-capacity sodium-glucose co-transporter SGLT2 in the proximal renal tubule.

Glucose Transporters

Characteristics
- Passive process down the concentration gradient
- Bidirectional
- Facilitative Diffusion
- Ping Pong Mechanism
- Sodium Independent

Image-Based Information

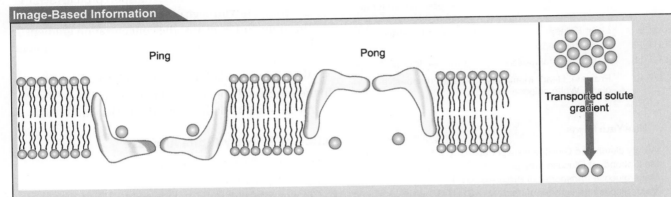

Fig. 4.1.9: *Glucose transporters—ping pong mechanism*

Two principal conformations of carrier protein.
- **"Ping"** state, it is exposed to high concentrations of solute, and molecules of the solute bind to specific sites on the carrier protein.
- **"Pong"** state: Binding induces a conformational change that exposes the carrier to a lower concentration of solute. This process is completely reversible, net flux across the membrane depends upon the concentration gradient.

Image-Based Information

Kinetics of carrier mediated diffusion of glucose transporters (GLUTS)

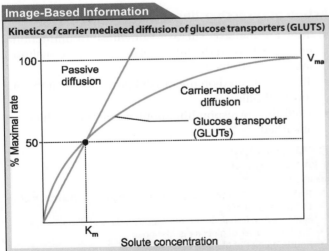

Fig. 4.1.10: *Substrate saturation curve of glucose transporters*

In carrier mediated (facilitated) diffusion (e.g. glucose transporters) the process is **saturable** as carriers are involved.
There is a maximal velocity (V_{max})
Binding Constant, K_m of carrier is solute concentration at half maximal velocity.
In Passive diffusion rate of movement of solute is directly proportional to solute concentration.

Transporter	Tissue location	Function
GLUT 1	**Brain**, Kidney, Colon, Placenta, **RBCs**, Retina	Basal Glucose Uptake
GLUT 2	**Liver Sinusoid membrane** β cells of Pancreas, Serosal side (basolateral side) of Intestinal Cell Basolateral membrane of PCT in Kidney	In liver removal of excess glucose from blood; In Pancreas regulation of insulin release. Low affinity and higher Km.

Transporter	Tissue location	Function
GLUT 3	**Neurons**, Placenta, Kidney	High affinity for Glucose
GLUT 4	Heart, Skeletal Muscle, Adipose Tissue	**Insulin Dependent** Glucose Uptake
GLUT 5	Small Intestine, Testis, Sperm	Primarily **Fructose Transporter**[Q]
GLUT 6	Spleen, Leukocytes	Possibly no transporter function
GLUT 7	Liver Endoplasmic reticulum	Glucose Transporter in the Endoplasmic Reticulum

Other Glucose transporters are

Glucose Transporter	Tissues where expressed	Functions
GLUT 8	Testis, **Blastocyst**, brain	**Insulin responsive** glucose transporter of Blastocyst. Glucose transporter to mature spermatozoa.
GLUT 9	Liver, Kidney	Urate Transporter
GLUT 10	Liver, Pancreas	
GLUT 11	Heart, Skeletal muscle	Fructose transporter
GLUT 12	Prostate, Heart, Mammary gland, White adipose tissue	Insulin responsive

High Yield Points

- Widely distributed Glucose transporter—GLUT-1
- Most abundant Glucose transporter in RBC—GLUT-1
- Major glucose transporter of Brain—GLUT-1
- Major glucose transporter of neurons—GLUT-3
- Major glucose transporter of Placenta—GLUT-1
- Glucose transporter of blastocyst—GLUT-8
- Insulin dependent Glucose transporters—GLUT-4, GLUT-8, GLUT-12
- GLUT-3 is present in neurons, whereas GLUT-1 is not present in neurons
- Urate transporter is GLUT-9

ABSORPTION OF MONOSACCHARIDES

Image-Based Information

Absorption of Glucose

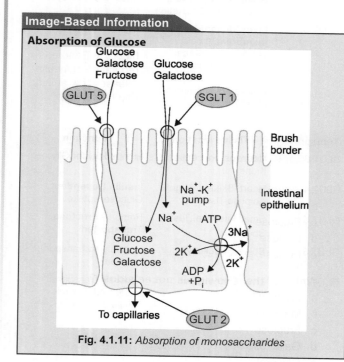

Fig. 4.1.11: Absorption of monosaccharides

In Brush Bordered Epithelium of Intestinal Mucosal Cells

- Glucose and galactose are absorbed by **a sodium-dependent process (SGLT-I)**
- They are carried by the same transport protein (SGLT 1) and compete with each other for intestinal absorption.
- This transport is against the concentration gradient.
- Fructose absorbed down their concentration gradient by GLUT5
- Absorption rate is **Galactose> Glucose >Fructose**

In the serosal side of intestinal mucosal cells
- All the sugars exit from intestinal cells via GLUT2.

Action of Salivary Alpha Amylase

- Digestion of starch starts in the mouth
- Optimal pH is 6.7
- Cleaves internal alpha 1, 4 glycosidic bond
- Produces Oligosaccharides.

Action of Pancreatic Amylase

- Cleaves internal alpha 1, 4 glycosidic bond yielding mainly maltose, and malto triose and oligosaccharides called limit dextrins and fragments of amylopectin.

Action of Disaccharidases

- They are maltase, sucrase-isomaltase (a bifunctional enzyme catalyzing hydrolysis of sucrose and isomaltose), lactose and trehalose.
- They are located on the brush borders of the intestinal mucosal cells.
- Yield the corresponding monosaccharides, which are absorbed.

Quick Review

- Primary source of energy for human beings is Carbohydrates
- Dextrose is glucose
- Sucrose and trehalose are nonreducing disaccharide.
- Second epimer of glucose is Mannose
- Fourth epimer of glucose is galactose.
- Glycosaminoglycans are negatively charged.
- GAG with no Uronic Acid[Q]—**Keratan Sulphate**
- GAG with no Sulfate group—**Hyaluronic Acid**
- GAG which is an anticoagulant—**Heparin**
- Glycosaminoglycans are **Polyanions**.
- GAG that play a role in corneal transparency—**Keratan Sulphate I**
- GAG that determine charge selectiveness of Renal Glomerular membrane.

Contd...

Contd...

Heparan Sulphate
- GAG are generally **extracellular**.
- Some intracellular GAG are **Heparin** (in mast T cells).
- Widely distributed Glucose transporter—GLUT-1
- Most abundant Glucose transporter in RBC-GLUT-1
- Major glucose transporter of Brain-GLUT-1
- Major glucose transporter of neurons-GLUT-3
- The method of absorption of glucose in intestine is secondary active transport in SGLT and facilitated diffusion in GLUTs.
- Glycemic Index of **Glucose and Galactose is 1 or 100%**.
- General test for all Carbohydrates—Molisch test

Mucopolysaccharidosis	Enzyme defect	Urinary metabolite
MPS I H Hurler Disease	L- Iduronidase^Q	Dermatan Sulfate Heparan Sulfate
MPS I S Scheie Disease	L- Iduronidase	Dermatan Sulfate
MPS II Hunter Disease	Iduronate Sulfatase^Q	Dermatan Sulfate Heparan Sulfate

Check List for Revision

- Classification and names of carbohydrates under each class.
- Properties of GAG
- Important Mucopolysaccridoses learn with clinical history
- Types of isomers of carbohydrates with examples
- Test for carbohydrtates as image based questions
- Dietary fibres
- Glucose transporters

Review Questions MCQ

1. **A patient came with lactose intolerance. He should avoid all except:** *(AIIMS May 2018)*
   a. Skimmed milk
   b. Icecream
   c. Yoghurt
   d. Condensed milk

2. **A male child presented with coarse facies, protuberant abdomen, frontal head enlargement, thickening of cardiac valve, hepatosplenomegaly, hearing impairement. What is the most probable diagnosis?**
   a. Hurler's disease
   b. Hunter's disease
   c. Fragile X syndrome
   d. Tay Sach's disease

3. **Which of the following is not dietary fibre?** *(Recent Question Jan 2019)*
   a. Cellulose
   b. Pectin
   c. Gum
   d. Inulin

4. **Which of the following are reducing sugars except?** *(PGI Nov 2018)*
   a. Glucose
   b. Maltose
   c. Isomaltose
   d. Sucrose
   e. Trehalose

5. **Method of transport of glucose in the intestine is:** *(AIIMS Nov 2018)*
   a. Primary active transport
   b. Secondary active transport
   c. Simple diffusion
   d. Counter transport

6. **The form of glucose predominantly seen is as:** *(Recent Question)*
   a. α D Glucopyranose
   b. α D Glucofuranose
   c. β D Glucopyranose
   d. β D Glucofuranose

7. **The glycemic index is highest for:** *(Recent Question)*
   a. Glucose
   b. Fructose
   c. Sucrose
   d. Sugar alcohols

8. **In Benedicts test red colour is/are produced by:** *(PGI Nov 2014)*
   a. Sucrose
   b. Inositol
   c. Fructose
   d. Lactose
   e. Maltose

9. **Which of the following is not an aldose?** *(PGI 2012)*
   a. Glucose
   b. Mannose
   c. Fructose
   d. Galactose
   e. Glycerol

10. Glucose detection can be done by the all except:
    a. Glucose oxidase
    b. Ferric Chloride test
    c. Dextrostix
    d. Folin and Wu method

11. Which of the following carbohydrate metabolism is used for liver function assessment?
    a. Galactose tolerance test
    b. Sucrose tolerance test
    c. Glucose tolerance test
    d. Lactose tolerance test

12. Which of the following are enantiomers? *(PGI May 2011)*
    a. D-galactose & L-Glucose
    b. d- Galactose & l-Glucose
    c. D-Mannose & L- Mannose
    d. d-Mannose & l- Mannose
    e. D-glucose and L -Glucose

13. Epimer combination(s) is/are: *(PGI May 2010, Nov 2009)*
    a. D-glucose & D-fructose
    b. D-mannose & D-talose
    c. D-glucose & D-mannose
    d. D-glucose & D-gulose
    e. D-galactose & D-glucose

14. True about Proteoglycans: *(PGI May 2017)*
    a. Always positively charged
    b. GAG attached to core protein covalently
    c. Always found on the cell surface
    d. Part of ECM and connective tissue
    e. Holds lot of water

15. Glycosamino Glycan present in cornea: *(Recent Question)*
    a. Keratan Sulphate
    b. Hyaluronic acid
    c. Chondroitin Sulphate
    d. Dermatan Sulphate

16. Which deposition result in cataract? *(Recent Question)*
    a. Glucose
    b. Galactose
    c. Sugar amines
    d. Sugar alcohols

17. Cellulose is: *(Recent Question)*
    a. Complex Lipoprotein
    b. Starch Polysaccharide
    c. Non Starch Polysaccharide
    d. Complex Glycoprotein

18. A 4-year-boy with mental retardation, dysostosis multiplex, coarse facial features, clear cornea. What is the diagnosis? *(Recent Question)*
    a. MPS Type IV
    b. Hunter's Disease
    c. Hurler's Disease
    d. Zellweger Syndrome

19. Mucopolysaccharide that does not contain Uronic acid residue is: *(JIPMER 2015)*
    a. Heparan Sulphate
    b. Heparin
    c. Chondroitin Sulphate
    d. Keratan Sulphate

20. Mucopolysacchridoses which is a lysosomal storage disease, occur due to abnormality in: *(PGI May 2015)*
    a. Hydrolase enzyme
    b. Dehydrogenase enzyme
    c. Lipase enzyme
    d. Phosphatase
    e. Acetyl CoA Carboxylase

21. Heparin is a: *(Recent Question)*
    a. Glycosamino glycan
    b. Polysaccharide
    c. Proteoglycan
    d. Carbohydrate

22. Glycogenin is a: *(Recent Question)*
    a. Polypeptide
    b. Polysaccharide
    c. Lipid
    d. Glycosaminoglycan

23. After overnight fasting, levels of glucose transporters reduced in: *(AIIMS May 2010)*
    a. Brain cells
    b. RBCs
    c. Adipocyte
    d. Hepatocyte

24. Glucose transporter in myocyte stimulated by insulin is: *(AIIMS Nov 2009)*
    a. GLUT-1
    b. GLUT-2
    c. GLUT-3
    d. GLUT-4

25. Defect in renal glucosuria: *(Recent Question)*
    a. GLUT 1
    b. GLUT 2
    c. SGLT 1
    d. SGLT 2

26. **Facilitated transport of glucose that is insulin insensitive (non-dependent) takes place in:**
    (Recent Question)
    a. Skeletal muscle
    b. Liver
    c. Adipose tissue
    d. Heart

27. **Glucose transporter present in the RBC:**
    (Recent Question)
    a. GLUT-1
    b. GLUT-2
    c. GLUT-3
    d. GLUT-4

28. **The monosaccharide with maximum rate of absorption in intestine is:**
    (Recent Question)
    a. Glucose
    b. Galactose
    c. Fructose
    d. Mannose

Answers to Review Questions

1. **c. Yoghurt**

   Lactobacillus present in yoghurt produce lactase enzyme which helps in the digestion of lactose in diet.

2. **b. Hunter's disease**   (Ref: Nelson 20/e Table 82.1 & 82.2)

   Clinical features of Hurlers and Hunter is almost same except for only males affected, no corneal clouding. As in the question 'a male child' and nothing mentioned in eye most probable diagnosis is Hunters disease.
   "Severe course similar to MPS I-H(Hurler) but clear corneas. Mild course: less pronounced features, later manifestation, survival to adulthood with mild or no mental deficiency"
   Nelson 20/e

3. **None of the above as all are dietary fibres.**

   But as per Park 24/e inulin is not given as dietary fibre, hence inulin is the better answer.

4. **d, e. Sucrose and Trehalose**   (Ref: Harper 31/e page 145)

   Nonreducing sugars are Trehalose and Sucrose because they do not have free functional group.

5. **Secondary active transport**   (Ref: Harper 31/e page 520)

   There are two methods for absorption of carbohydrates. For glucose and galactose Sodium dependent glucose transporter, this is a secondary active transport. Other monosaccharides are absorbed by a carrier mediated diffusion.

6. **c. β D Glucopyranose**   (Ref: Lehninger 6/e page 244)
   - The predominant anomer of glucose is β D Glucopyranose
   - The predominant anomer of Fructose is β D Fructofuranose

7. **a. Glucose**   (Ref: Harper 31/e page 520)
   - Glycemic Index of Glucose and Galactose is 1 or 100%
   - Glycemic index of Lactose, Maltose, Isomaltose & Trehalose is 1 or 100% (because they give rise to Glucose and Galactose)
   - Fructose, Sugar alcohols have less glycemic index as they are not absorbed completely.
   - Sucrose also low glycemic index as it is split to glucose and Fructose
   - Starch varies from 0-100% owing to variable rate of hydrolysis.
   - For Nonstarch Polysaccharides (Dietary fibre) it is 0.

8. **c, d, e. Fructose, Lactose, Maltose**
           (Ref: Varleys Practical Clinical Biochemistry 4/e page 110)

   Benedicts test is a test for reducing substances in urine. Carbohydrates that give positive. Benedicts test are:
   - Pentoses, Fructose, Glucose, Galactose
   - Reducing disaccharides like Lactose, Maltose, Isomaltose

   Noncarbohydrates that give Benedicts test positive are:
   - Homogentisic acid
   - Glucuronic acid
   - Salicylates
   - Ascorbic acid
   - Uric acid
   - Glutathione
   - Creatinine in very high amount

   Nonreducing disaccharides like Sucrose and Trehalose do not give positive Benedicts test.
   Test for reducing substances are Benedicts test and Fehlings test.

9. **c, e. Fructose, Glycerol**

Generic name	Aldoses	Ketoses
Triose	Glyceraldehyde	Dihydroxy Acetone
Tetrose	Erythrose	Erythrulose
Pentose	Ribose Xylose (Epimer of Ribose) Arabinose	Ribulose Xylulose (Epimer of Ribulose)
Hexose	Glucose, Galactose, Mannose	Fructose
Heptose		Sedoheptulose

10. **b. Ferric Chloride Test**

    Ferric Chloride test is the test done in Alkaptonuria and Phenyl Ketonuria.

Methods for Estimation of Glucose
- Reductometric Methods
- Nelson Somogyi Method
- Folin WU Method
- O-Toluidine Method

Enzymatic Method

Hexokinase Method

Glucose Oxidase Peroxidase Method
(GOD –POD)   (AIIMS Nov 2007)

Highly Specific Method

Used in dry analysis technique like Glucometer.

Reaction

Principle of Glucose Oxidase Peroxidase

$$Glucose + H_2O + O_2 \xrightarrow{GOD} Gluconic\ acid + H_2O_2$$

$$2H_2O_2 + 4\ aminoantipyrine + PHBS \xrightarrow{POD} Quinone\ imine\ dye + H_2O$$

11. a. Galactose Tolerance test
(Ref: Vasudevan and Sreekumari 7/e page 355)

Galactose Tolerance test assess the metabolic capacity of liver Galactose is almost entirely metabolised in the liver.

12. c, e. D-Mannose and L- Mannose, D-glucose and L-Glucose
(Ref: Harper 31/e page 145, Lehninger 6/e page 246, 247),

Enantiomerism

The isomers which are mirror images to each other are called enantiomers.

This is due to presence of asymmetric carbon atom.

D and L Isomerism

Enatiomerism at penultimate Carbon atom results in two mirror images called D and L isomers.

The penultimate carbon atom is called Reference Carbon atom.

The penultimate carbon atom in Glucose and Fructose is C-5.

Most of the naturally occurring Monosaccharides are D isomers [Unlike Amino Acids, which are L isomers.]

Examples of Enantiomers[Q]

D Glucose and L Glucose

D Fructose and L Fructose

D Mannose and L Mannose

D Glyceraldehyde and L Glyceraldehyde

13. c, e. D Glu & D Mann.., D Gal. & D Glu.
(Ref: Harper 31/e page 163)

Epimerism

Difference in orientation of H and OH group around Carbon atoms other than Anomeric Carbon and Penultimate Carbon results in isomerism refered to as Epimerism.
- In epimerism, the difference in configuration is confined to only one asymmetric carbon atom.
- In diastereoisomers, the difference in configuration is present in more than one asymmetric carbon atom.

Epimers of Glucose[Q]

Difference in orientation of H and OH at C2 C3 & C4 for Glucose:
- 2nd Epimer of Glucose – Mannose
- 3rd Epimer of Glucose – Allose
- 4th Epimer of Glucose – Galactose

14. b, d, e. GAG attached to core protein….,Part of ECM…, Holds lot of water   *(Ref: Harper 31/e page 601)*

GAG is negatively charged hence proteoglycans are negatively charged. GAG is found in ECM not cell surface.

GAG can hold lot of water.

15. a, d. Keratan Sulphate, Dermatan Sulphate
(Ref: Harper 31/e page 601)

- KS-I Originally isolated from Cornea[Q]
- KS-II from Cartilage.
- The composition of both Keratan sulphate are same (N acetyl Glucosamine and Galactose)
- No Uronic acid in Keratan Sulphate.
- In the eye, they lie between collagen fibril and play a critical role in corneal transparency.

16. d. Sugar alcohol   *(Ref: Harper 31/e page 191)*

- In Diabetes mellitus, in the lens by polyol pathway Glucose converted to Sorbitol by the enzyme Aldose reductase .
- In galactosemia, Dulcitol or Galactictol is responsible for cataract.
- Sorbitol, Dulcitol are sugar alcohols.

17. c. Non Starch Polysaccharides   *(Ref: Harper 30/e Page 153)*

- Foods contain a wide variety of other polysaccharides that are collectively known as nonstarch polysaccharides; they are not digested by human enzymes, and are the major component of dietary fiber.
- Examples are cellulose from plant cell walls (a glucose polymer) and inulin, the storage carbohydrate in some plants (a fructose polymer).

18. b. Hunter's Disease   *(Ref: Nelson 20/e Table 82.1 & 82.2)*

- Recognition Pattern of Mucopolysaccharidoses

Clinical features	MPS IH	MPS IS	MPS II
Common name	Hurler	Scheie	Hunter
Mental deficiency	+	–	+
Coarse facial features	+	(+)	+
Corneal clouding	+	+	–
Visceromegaly	+	(+)	(+)
Short stature	+	(+)	+
Joint contractures	+	+	+
Dysostosis multiplex	+	(+)	(+)
Leucocyte inclusions	+	(+)	+
Mucopolysacchariduria	+	+	+

19. d. Keratan Sulphate

GAG with no Uronic Acid: Keratan Sulphate
GAG with no Sulfate group: Hyaluronic Acid
GAG not covalently linked to Protein: Hyaluronic Acid

20. a. Hydrolase enzyme *(Ref: Nelson 20/e Chapter Defects in the metabolism of Lipids/Lysosomal storage disorder)*

The lysosomal storage diseases are diverse disorders each due to an inherited deficiency of a lysosomal hydrolase leading to the intralysosomal accumulation of the enzyme's particular substrate

21. a. Glycosamino Glycans *(Ref: Harper 31/e page 605)*

- Heparin is a Glycosamino Glycan.
- Glycosamino Glycans are heteropolysacchrides.
- Disaccharide repeat unit is Glucosamine and Iduronic acid.
- Initially uronic acid present is Glucuronic acid, 5 epimerase convert 90% of GlcUA to IdUA.
- Heparin is found in the granules of mast cells, also in lung, liver and skin. Heparin specifically bind to lipoprotein lipase present in capillary walls and cause its release into circulation.
- Heparin is an anticoagulant.

22. a. Polypeptide *(Ref: Harper 31/e page 164)*

- Glycogenin is a protein
- 37 KDa protein.
- Glycogenin catalyses transfer of 7 glucosyl residues from UDPGlucose, in alpha 1→4 linkage.
- Glucosyl residues are added on specific tyrosine residues of Glycogenin.
- Glycogenin remain at the core of Glycogen granule.

23. c. Adipocytes *(Ref: Harper 29/e page158, Harper 30/e page 192)*

GLUT-4 and Insulin

Glucose uptake into muscle and adipose tissue is controlled by insulin, which is secreted by the islet cells of the pancreas in response to an increased concentration of glucose in the portal blood.

In the fasting state, the glucose transporter of muscle and adipose tissue (GLUT-4) is in intracellular vesicles.

An early response to insulin is the migration of these vesicles to the cell surface, where they fuse with the plasma membrane, exposing active glucose transporters.

These insulin sensitive tissues only take up glucose from the bloodstream to any significant extent in the presence of the hormone.

As insulin secretion falls in the fasting state, so the receptors are internalized again, reducing glucose uptake.

24. d. GLUT-4 *(Ref: Harper 31//e page 179)*

Glucose transporter which are insulin responsive are GLUT-4, GLUT 8 and GLUT 12.

- GLUT 4 is present in the Heart, skeletal muscle, adipose tissue
- GLUT 8 is present in the Testis, blastocyst.
- GLUT 12 is present Heart, prostate, white adipose tissue, mammary gland

25. d. SGLT-2 *(Ref: Harrison 20/e page 299)*

Renal Glucosuria

- Isolated glucosuria in the presence of a normal blood glucose concentration is due to mutations in SLC5A2, the gene that encodes the high-capacity sodium-glucose co-transporter SGLT2 in the proximal renal tubule

26. b. Liver *(Ref: Harper 31/e page 191)*

Insulin responsive Glucose transporters are

- GLUT 4-Heart ,Skeletal Muscle, Adipose tissue
- GLUT 8-Testis, Blastocyst, Brain, Adipose tissue
- GLUT 12-Heart, Prostate, White adipose tissue, mammary gland

27. a. GLUT 1 *(Ref: Harper 31/e page 178t))*

- Highest level of GLUT1 is present in the RBC.
- Major Glucose transporter in brain is GLUT1 (not present in neurons)
- Major Glucose transporter in the Placenta is GLUT 1
- Major Glucose transporter in the RBC is GLUT1
- Major neuronal Glucose transporter is GLUT3
- Insulin responsive glucose transporter is GLUT4, GLUT 8 & GLUT 12
- Fructose transporter GLUT 5 (mainly) and GLUT 11
- Urate transporter is GLUT 9.

28. b. Galactose *(Ref: Harper 31/e page 538)*

In brush bordered epithelium of Intestinal mucosal cells

- Glucose and galactose are absorbed by a sodium- dependent process (SGLT-I)
- They are carried by the same transport protein (SGLT 1) and compete with each other for intestinal absorption.
- This transport is against the concentration gradient.
- Fructose absorbed down their concentration gradient by GLUT 5
- Absorption rate is **Galactose > Glucose>Fructose**

4.2 MAJOR METABOLIC PATHWAYS OF CARBOHYDRATES

- Glycolysis
- Gluconeogenesis
- Pyruvate Dehydrogenase
- Glycogen Metabolism
- Glycogen Storage Disorders

GLYCOLYSIS (EMBDEN-MEYERHOF PATHWAY)

High Yield Points

Biochemical Significances of Glycolysis
- **Principal route** for Carbohydrate metabolism.
- Pathway taking place **in all the cells** of the body.
- Only pathway which can **operate aerobically and anaerobically.**
- The ability to operate glycolysis in the absence of oxygen is important in Skeletal muscle.

High Yield Points

- Skeletal muscles can survive anoxic episodes because of anaerobic glycolysis.
- Defect in muscle Phosphofructokinase manifest as fatigue because of its significance in muscle
- **Heart muscle has relatively low glycolytic activity**, hence poor survival under conditions of ischemia.
- **Mature Erythrocyte** which lack mitochondria are completely reliant on Glucose as their metabolic fuel.^{QAIIMS May2015}
- Defect in Glycolytic enzyme like Pyruvate Kinase manifests as hemolytic anemia, because of its significance in mature erythrocytes.

Overview of Glycolysis

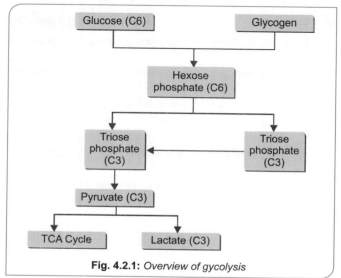

Fig. 4.2.1: Overview of gycolysis

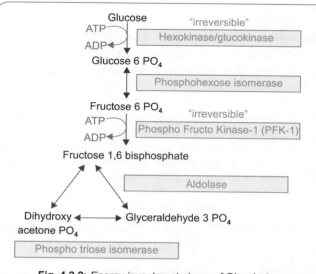

Fig. 4.2.2: Energy investment phase of Glycolysis

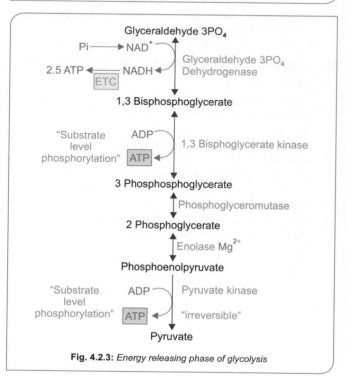

Fig. 4.2.3: Energy releasing phase of glycolysis

Steps of Glycolysis

Site–Cytoplasm^Q

Step I: Hexokinase/Glucokinase

The **flux generating reaction** of Glycolysis.

Hexokinase

- Transfer Phosphate group from ATP to Glucose. Has **high affinity** for glucose (Or Lower K_m)
- Mg^{2+} is the cofactor^Q
- Irreversible step^Q
- 1 ATP is utilized
- **Not induced by insulin**
- **Inhibited by Glucose 6 Phosphate, its product**
- **Phosphorylation of glucose commits it to metabolism within the cell**^{Q} because Glucose 6 Phosphate cannot be transported back across plasma membrane. (NB: Commits not only to glycolysis, as Glucose 6 Phosphate has other fates too)

Glucokinase (or Hexokinase IV)

- **Isoform of Hexokinase** present in **Liver** and **Pancreatic β cells**
- Has **low affinity for Glucose** (High K_m)
- Hence acts only when blood glucose is very high (**>100 mg/dL**)
- **Induced by Insulin following a meal**^{Q}. (Unlike hexokinase)
- Not inhibited by Glucose 6 Phosphate^Q. (Unlike hexokinase)
- *Play an important role in regulation of blood glucose*^{Q}
- In the **liver**, function of Glucokinase is to **remove glucose** from portal vein following a meal
- In the beta cells of **Pancreas**, function of Glucokinase is to release Insulin.
- Isoforms of Hexokinase (HK)
- Hexokinase I to IV
- Myocytic form is HK-II
- Liver and Pancreas is HK-IV or Glucokinase

> **HIGH YIELD POINTS**
>
> **FATES OF GLUCOSE 6 PHOSPHATE**^{Q}
> - HMP ShuntPathway
> - Gluconeogenesis
> - **Glycogenesis (Major fate in well fed state)** (AIIMS May 93)
> - Glycogenolysis

Step II Phospho Hexose Isomerase

- Isomerisation of Glucose 6 Phosphate to Fructose 6 Phosphate by Phospho Hexose Isomerase
- Involves **Aldose – Ketose Isomerism.**

Step III Phosphofructokinase I (PFK-I)

- Fructose 6 Phosphate is phosphorylated to Fructose 1,6 Bisphosphate.
- **Second irreversible step.**^{Q}
- **Rate limiting step of Glycolysis.**^{Q}
- **Committed**^{Q} step of Glycolysis because once Fructose 1,6 Bisphosphate is formed it should under go glycolysis.
- Otherwise called **bottle neck of Glycolysis**.
- **1 ATP is utilized**.

Step IV Aldolase

- Fructose 1,6 Bisphosphate split to two 3 Carbon compounds Glyceraldehyde 3 Phosphate and Dihydroxy Acetone phosphate.
- **Aldolase is a Lyase**^{Q}.

Step V Triose Phosphate Isomerase

- Dihydroxy Acetone Phosphate isomerized to Glyceraldehyde 3 Phosphate.

Step VI Glyceraldehyde 3 Phosphate Dehydrogenase

- Glyceradehyde 3 Phosphate is oxidized to a **high energy compound, 1,3 bisphosphoglycerate**.
- NADH is generated.
- An inorganic Phosphate is added^Q to the substrate.
- NADH generated in this step enters into mitochondria by **Malate –Aspartate Shuttle** or **Glycerophosphate shuttle** under aerobic conditions.
- But in anaerobic conditions NADH is utilized by **Lactate Dehydrogenase, NAD + is regenerated**.

Frequently Asked Doubts

What is the relationship between hypoxia and glycolysis? Hypoxia in any tissue increases Lactate production via anaerobic glycolysis.

- Because there is decreased ATP from oxidative phosphorylation.
- This results in increased AMP.
- AMP is an allosteric activator of PFK-I
- Hence Glycolysis increase, NADH level rises.
- NADH/NAD+ ratio rises, as NADH cannot be converted to NAD+ via oxidative pathways.

So Pyruvate is converted to lactate, hence lactic acidosis is the result as NADH favor LDH

Summarizing hypoxia results in

- Increased anaerobic glycolysis and lactic acidosis
- Decreased aerobic glycolysis
- Decrease Pyruvate oxidation via TCA cycle.

Step VII 1,3 Bisphospho Glycerate Kinase
- 1, 3 Bisphospho Glycerate to 3 Phospho Glycerate.
- **Only Kinase in Glycolysis which is reversible**^{Q}.
- ATP is generated.
- An example of **Substrate level Phosphorylation**^{Q}.

Step VIII Phosphoglycerate Mutase
- 3 Phosphoglycerate to 2 Phosphoglycerate
- Require **Mg^{2+}**.

Step IX Enolase
- 2 Phospho Glycerate to Phosphoenol Pyruvate
- This step involves **dehydration**.
- Enolase is dependent on **Mn^{2+} and Mg^{2+}**^{Q}
- Fluoride inhibits Enolase.

Step X Pyruvate Kinase
- Phosphoenol Pyruvate to Pyruvate^Q.
- Second Substrate level Phosphorylation^Q.
- ATP is generated.
- Irreversible step.

Anaerobic Glycolysis

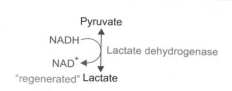

Fig 4.2.4: Lactate Dehydrogenase reaction

Lactate Dehydrogenase
- Pyruvate is to converted to lactate
- **NADH** is utilized in this step
- **NAD+ is regenerated.**

HIGH YIELD POINTS

IRREVERSIBLE STEPS OF GLYCOLYSIS^Q
- Hexokinase
- Phosphofructokinase
- Pyruvate kinase

Remember: All the Kinases are irreversible except 1,3 Bisphos-pho Glycerate Kinase which is reversible.

SUBSTRATE LEVEL PHOSPHORYLATION^Q
- **Phospho Glycerate Kinase** [1,3 Bisphospho Glycerate to 3 Phosphoglycerate]
- **Pyruvate Kinase** [Phosphoenol Pyruvate to Pyruvate]

NB: Learn the enzyme and the reaction. Question can be asked in either ways.

Inhibitors of Glycolysis^Q
- **Iodoacetate** inhibit Glyceraldehyde 3 Phosphate Dehydrogenase.
- **Fluoride** inhibit Enolase.
- Application-for Estimation of Blood Glucose Sodium Fluoride- Potassium Oxalate Mixture is used.
- **Arsenate** compete with inorganic phosphate in the step of **glyceraldehyde 3 phosphate dehydrogenase**

Clinical Correlation–Glycolysis
- Inherited Aldolase A and pyruvate Kinase deficiency in erythrocyte leads to Hemolytic anemia.
- Muscle Phosphofructokinase deficiency leads to Exercise Intolerance.

Frequently Asked Doubts
In pyruvate Kinase deficiency
1. Why there is hemolytic anemia in Pyruvate Kinase deficiency?
   In Pyruvate Kinase defect, **total ATP generated decreases to 50%.**
   Ion transporters do not function.
   RBC tend to gain Ca^{2+}, lose K$^+$ and water. This leads to hemolysis and anemia.
2. What happens to 2,3 BPG level?
   Two to three fold elevation in 2, 3 BPG level due to blockage of conversion of Phospho Enol Pyruvate to Pyruvate.
   The increase in 2, 3 BPG will moderate the effects of anemia by favoring unloading of oxygen to the tissues.

Energetics of Glycolysis^Q
Aerobic Glycolysis

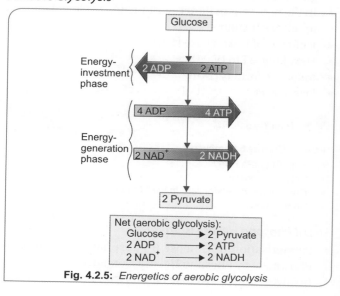

Fig. 4.2.5: Energetics of aerobic glycolysis

Enzyme	Reducing Equivalents/ATP from the step	ATP per molecule of Glucose
Glyceraldehyde 3 Phosphate Dehydrogenase	NADH = 2.5 ATPs	2 NADH = 5 ATPs
1,3 Bisphospho Glycerate Kinase	1 ATP by Substrate level Phosphorylation	2 ATPs
Pyruvate Kinase	1 ATP by Substrate level Phosphorylation	2 ATPs
The number of ATPs generated		9 ATPs
Consumption of ATPs in the Hexokinase and Phosphofructokinase		–2 ATP
No. of ATPs from Aerobic Glycolysis		9–2 = 7 ATPs

Enzyme	Reducing Equivalents/ATP from the step	ATP per molecule of Glucose
1,3 Bisphosphoglycerate Kinase	1 ATP by Substrate level Phosphorylation	2 ATPs
Pyruvate Kinase	1 ATP by Substrate level Phosphorylation	2 ATPs
The number of ATPs generated		4 ATPs
Consumption of ATPs in the Hexokinase and Phosphofructokinase		–2 ATP
No of ATPs from Anaerobic Glycolysis		4–2 = 2 ATPs

Energy Yield from 1 Molecule of Glucose under Aerobic Condition

Source	No of ATPs generated
From Aerobic Glycolysis	7 ATPs
From Pyruvate Dehydrogenase (as 2 Pyruvates from 1 mol of Glucose)	2 NADH = 5 ATPs
From TCA Cycle (as 2 Acetyl CoA from 1 mol of Glucose)	2 × 10 = 20 ATPs
Net ATPs from 1 mol of Glucose under aerobic condition	7 + 5 + 20 = 32 ATPs
Net ATPs from 1 mol of Glucose under anaerobic condition	4 – 2 = 2 ATPs

High Yield Points

- The number of ATPs produced from 1 NADH if **Malate shuttle** is used for transport of NADH into mitochondria is 2.5 ATPs
- The number of ATPs produced from 1 NADH if **Glycerophosphate shuttle** is used for transport of NADH into mitochondria is only 1.5ATPs^Q.
- If **Muscle Glycogen is used for anaerobic glycolysis**, then **3 ATPs** are produced instead of 2 ATPs because, as there is no Glucose 6 Phosphatase in muscle, Glucose 6 Phosphate directly enter into Glycolysis. Hence 1 ATP for Hexokinase step is not needed. So out of 4 ATPs produced only 1 is utilised, resulting in 3 ATPs from 1 mol of Glucose.

Concept box

Key Concept of Regulation of all Metabolic Pathways Hormonal Regulation

- **Insulin** generally favor all pathways which decrease blood glucose level by **dephosphorylating** the regulatory enzymes of these pathways.
- In other words enzymes active under the influence of insulin is active in the dephosphorylated state.
- **Glucagon** generally favor all pathways which increase blood glucose level by **phosphorylating** the regulatory enzymes of these pathways.
- In other words enzymes active under the influence of glucagon is active in the phosphorylatedstate.

Allosteric Regulation

- Substrate favor forward reaction.
- Product inhibit forward reaction

Anaerobic Glycolysis

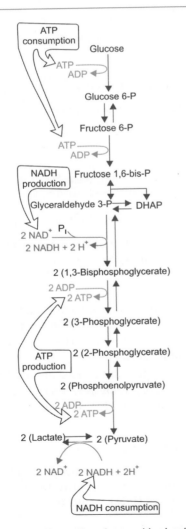

Fig. 4.2.6: *Energetics of anaerobic glycolysis*

REGULATION OF GLYCOLYSIS

Glycolysis is regulated at three physiologically irreversible steps

1. Hexokinase/Glucokinase
2. Phosphofructokinase-I (Occupies a key position in the regulation of Glycolysis)
3. Pyruvate Kinase

Hormonal Regulation

Insulin favours Glycolysis
- By dephosphorylating key enzymes of Glycolysis
- By Inducing Glucokinase.
- Glucokinase play a key role in regulating blood glucose level after a meal.

Glucagon inhibits Glycolysis
- Increasing cAMP dependent Protein Kinase A
- By Phosphorylating key Enzymes of Glycolysis.

Allosteric Regulation^Q

Enzyme	Allosteric Activator	Allosteric Inhibitor
Hexokinase		Glucose 6 Phosphate
PFK-1	5' AMP Fructose-6-Phosphate, **Fructose 2,6 Bisphosphate^Q**.	ATP^Q **Citrate^Q** Low pH
Pyruvate Kinase		ATP

- PFK-1 plays a key role in regulation of Glycolysis.
- Glucokinase play a key role in regulating blood glucose level following a meal.

Rapoport Leubering Cycle^Q (2, 3 BPG Shunt)

- **Site: Mature erythrocytes,** 10% of glucose enter this shunt pathway
- There action catalyzed by **phosphoglycerate kinase** may be by passed
- 1,3-bisphosphoglycerate is converted to 2,3-bisphosphoglycerate by Bisphosphoglycerate Mutase
- 2,3 Bisphosphoglycerate is hydrolyzed to 3-phosphoglycerate and P_i by 2,3-bisphosphoglycerate Phosphatase
- No ATP is generated by this step
- 2 ATPs at Pyruvate Kinase step is generated but that is used for Hexokinase and Phosphofructokinase
- So **no net yield of ATPs 2,3 BPG shunt pathway**
- Serve to provide **2,3-bisphosphoglycerate^Q**
- 2,3 BPG shifts the Oxygen Dissociation curve to right.

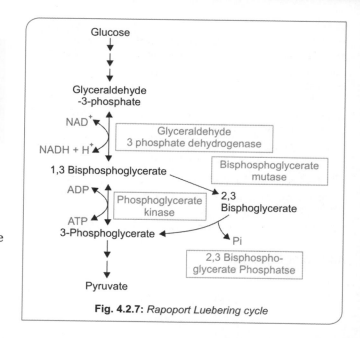

Fig. 4.2.7: Rapoport Luebering cycle

 HIGH YIELD POINTS—AMINO ACIDS

NET NO. OF ATPs PRODUCED FROM 1 MOL OF GLUCOSE BY
- Anaerobic glycolysis—2 ATPs
- Aerobic glycolysis—7 ATPs
- Aerobic oxidation—32 ATPs
- Rapoport Luebering Cycle— 0
- From aerobic glycolysis by Glycerophosphate Shuttle-5 ATPs
- From aerobic glycolysis based on old calculation-8 ATPs

Cancer Cells and Metaboilc Reprogramming

A comparison of Warburg effect and recent data about altered metabolism in cancer cells

Warburg Effect

In 1924, Otto Warburg and his colleagues observed that cancer cells take up large amount of glucose and metabolize it to lactic acid even in presence of oxygen. This observation is called Warburg effect.

The hypothesis made by him based on this data was
- Increased ratio of glycolysis when compared to aerobic respiration was likely due to defect in mitochondrial respiratory chain.

Metabolic Enzyme Reprogramming in Cancer Cells

In normal cells, the major source of ATP is oxidative phosphorylation. The major Pyruvate Kinase isozyme is Pyruvate Kinase M-I (PKM-I isoform)

In cancer cells the major isozyme is **Pyruvate Kinase-M-2 (PKM-2)**. Aerobic glycolysis is prominent. Lacate is produced via LDH.
- PKM-I is tetrameric high catalytically active form
- PKM-2 is dimeric low catalytically active form.

Consequences of metabolic enzyme reprogramming are
- Less shuttling of glucose derived energy to production of ATP.
- Shunting of glucose chemical energy for building of biomass of proteins, lipids, nucleic acid.

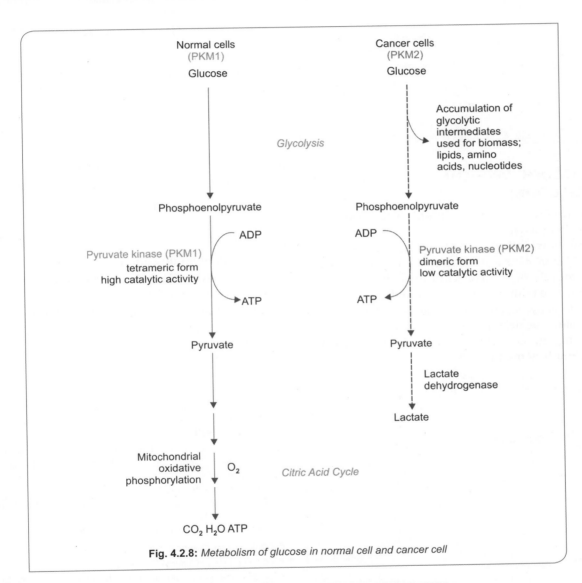

Fig. 4.2.8: *Metabolism of glucose in normal cell and cancer cell*

Clinical application of metabolic reprogramming: Increased glucose uptake by cancer cells helps in Fluorodeoxy Glucose (FDG) Positron Emission Tomography (PET) Scanning to detect tumours.

FATES OF PYRUVATE

- To Glucose (Gluconeogenesis)
- To Lactate (Lactate Dehydrogenase)
- To Oxaloacetate (Pyruvate Carboxylase)
- To Acetyl-CoA (Pyruvate Dehydrogenase)
- To Alanine (Alanine Amino Transferase)

Image-Based Information
Fates of Pyruvate

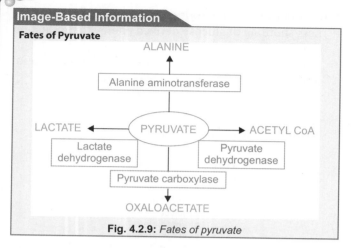

Fig. 4.2.9: Fates of pyruvate

PYRUVATE DEHYDROGENASE (PDH) COMPLEX (LINK REACTION)

- Pyruvate formed in the **cytosol enters the mitochondria by a symporter**.
- Pyruvate is **oxidatively decarboxylated** to **Acetyl-CoA**.
- The oxidation of Pyruvate to Acetyl-CoA is the irreversible route from glycolysis to the citric Acid Cycle.
- **No alternate pathway to circumvent this step.**
- Pyruvate dehydrogenase complex is analogous to the α-ketoglutarate dehydrogenase.
- Multienzyme complex associated with the **inner mitochondrial membrane**.

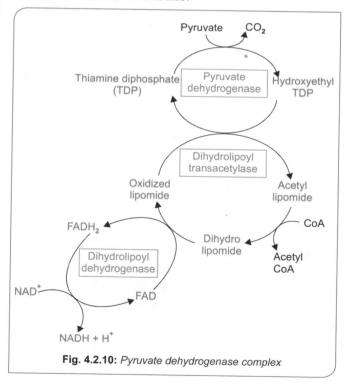

Fig. 4.2.10: Pyruvate dehydrogenase complex

PDH complex consist of 3 Enzymes and 5 Coenzymes.

Three enzymes are:
1. Pyruvate Dehydrogenase bound to Thiamine Diphosphate^Q
2. **Dihydro Lipoyl Transacetylase**, the prosthetic group is **oxidized Lipomide**
3. **Dihydro Lipoyl Dehydrogenase**, contains **FAD**.

Coenzymes^Q are:
- Thiamine Diphosphate (TDP)
- Lipomide
- Coenzyme A
- FAD
- NAD$^+$

Regulation of Pyruvate Dehydrogenase

By **end product inhibition** and **covalent modification**

End products that inhibit PDH Complex are:
1. Acetyl-CoA
2. NADH

By Covalent modification

PDH is **active** in **dephosphorylated state^Q** and inactive in phosphorylated state.

PDH is phosphorylated by a PDH Kinase.

PDH Kinase is activated by increase in:
- ATP/ADP
- Acetyl-CoA/CoA
- NADH/NAD+

HIGH YIELD POINTS

SIGNIFICANCE OF PDH COMPLEX
- Link Glycolysis and Citric acid cycle hence called link reaction
- Thiamine deficiency affects PDH. Hence complete oxidation of Glucose.
- **PDH defect can lead to Lactic Acidosis^Q.**
- Fat cannot be converted to Glucose because of the irreversible nature of PDH.
- Acetyl-CoA^Q cannot be converted to glucose.

EXCEPTION TO FAT CANNOT BE CONVERTED TO GLUCOSE
- Glycerol part of Triacyl Glycerol.
- Odd Chain Fatty Acid oxidation which forms Propionyl-CoA

HIGH YIELD POINTS

FATE OF ACETYL-CoA
- Fatty Acid Synthesis
- Ketone Body Synthesis
- Cholesterol Synthesis
- TCA Cycle

Acetyl-CoA cannot be converted to glucose^Q.

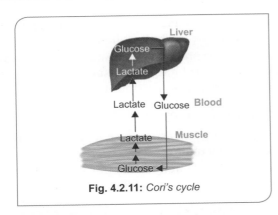

Fig. 4.2.11: *Cori's cycle*

Cori's Cycle (Glucose- Lactate Cycle) (Lactic Acid Cycle)

Lactate, formed by glycolysis in skeletal muscle and erythrocytes, is transported to the liver and kidney where it reforms glucose, which again becomes available via the circulation for oxidation in the tissues. This process is known as the Cori cycle, or the lactic acid cycle.

Uses
- Prevents Lactate accumulation in the muscle
- Reutilize lactate from muscle and erythrocyte for Gluconeogenesis.

Cori's Cycle involves
- Liver and Kidney
- Muscle
- RBC.

Glucose Alanine Cycle (Cahill Cycle)

In the fasting state, there is a considerable output of alanine from skeletal muscle formed by transamination of pyruvate produced by glycolysis of muscle glycogen, and is exported to the liver, where, after transamination back to pyruvate, it is a substrate for gluconeogenesis. This is **glucose-alanine cycle**. It provides an **indirect way of utilizing muscle glycogen to maintain blood glucose in the fasting state.**

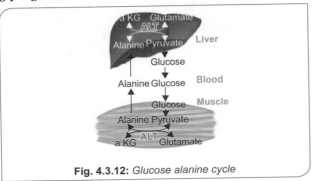

Fig. 4.3.12: *Glucose alanine cycle*

Uses of Glucose Alanine Cycle
- Carries amino group to the Liver.
- Alanine as a substrate for Gluconeogenesis during starvation.
- Amino acid increased in blood during starvation is Alanine.(AIIMS Nov 2011)

GLUCONEOGENESIS

Definition: The process of formation of *glucose* from **non-carbohydrate** precursors.

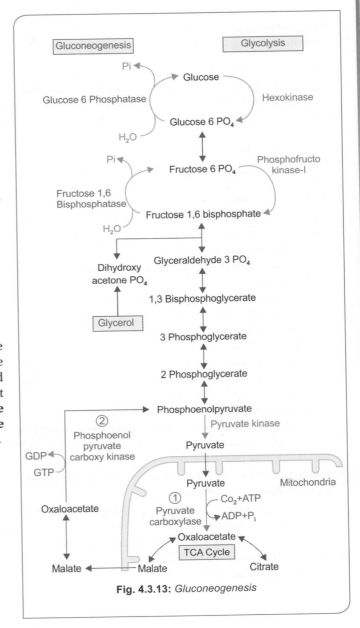

Fig. 4.3.13: *Gluconeogenesis*

High Yield Points

Substrates for Gluconeogenesis
- Glucogenic Amino Acid (Alanine is the major contributor)
- Lactate
- Glycerol
- Propionyl-CoA

Sites of Gluconeogenesis
- Liver (60–90%) Kidney (10–40%) Organelle-Cytoplasm and Mitochondria

There is a role for Smooth endoplasmic reticulum also

Key Gluconeogenic enzymes are expressed in the small intestine but it is unclear that significant Gluconeogenesis in intestine during starvation.

Biomedical Significance of Gluconeogenesis
- Provides a major contribution to **blood glucose after overnight fast**, once Glycogen stores are depleted.
- A supply of glucose is essential for **erythrocytes and brain.**
- Clear lactate from erythrocyte and muscle.

Key Enzymes of Gluconeogenesis

1. **Pyruvate to Phosphoenol pyruvate**

   Reversal of the reaction catalyzed by pyruvate kinase in glycolysis involves **two endothermic reactions.**
   - **Pyruvate carboxylase**
   - **Phosphoenol pyruvate carboxykinase (PEPCK)**

   A. **Pyruvate Carboxylase**
   - **Mitochondrial** pyruvate carboxylase catalyzes the carboxylation of Pyruvate to Oxaloacetate.
   - It is an **ATP-requiring reaction**.
   - **Biotin** is the coenzyme.
   - The resultant oxaloacetate is reduced to malate, exported from the mitochondrion into the cytosol and there oxidized back to oxaloacetate.

   B. **Phosphoenol pyruvate Carboxykinase**
   - Catalyzes the **decarboxylation and phosphorylation** of oxaloacetate to phosphoenolpyruvate using **GTP** as the phosphate donor.

2. **Fructose 1, 6-Bisphosphate and Fructose 6-Phosphate**

   C. **Fructose 1,6-Bisphosphatase**
   - The conversion of fructose1, 6-bisphosphate to fructose 6-phosphate, for the reversal of glycolysis, is catalyzed by **fructose 1, 6-bisphosphatase.**

3. **Glucose 6-Phosphate and Glucose**

   D. **Glucose 6 Phosphatase**

The conversion of glucose 6-phosphate to glucose is catalyzed by **glucose 6-phosphatase.**

This enzyme is present in liver and kidney, but **absent from muscle and adipose tissue**, which, therefore, cannot export glucose into the bloodstream.

Role of Smooth Endoplasmic Reticulum
- Glucose 6 Phosphatase is present in the **smooth endoplasmic reticulum.**

A transporter is required for the transport of Glucose 6 Phosphate from cytoplasm to SER.

Summary of Key Enzymes of Gluconeogenesis

Irreversible Steps of Glycolysis	Key Enzymes to bypass the irreversible steps in Gluconeogenesis
Pyruvate Kinase	Pyruvate Carboxylase (Mitochondria) / Phosphoenolpyruvate Carboxykinase (PEPCK)
Phosphofructo-kinase	Fructose 1,6 Bisphosphatase (Cytosol)
Hexokinase/Glucokinase	Glucose 6 Phosphatase (SER)

Entry of Propionyl-CoA to Gluconeogenesis By three Enzymes
1. Propionyl-CoA Carboxylase
2. Methyl Malonyl CoA Race mase
3. Methyl Malonyl CoA Mutase

High Yield Points
- Propionyl-CoA carboxylase require Biotin and ATP
- Methyl-Malonyl-CoA Mutase require vitamin B12

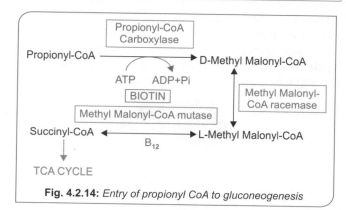

Fig. 4.2.14: Entry of propionyl CoA to gluconeogenesis

Clinical Correlation

Pyruvate Carboxylase deficiency

Biochemical reason for clinical manifestation

- **Intellectual disability**
  This is one of the anaplerotic reactions in TCA cycle. Hence TCA cycle is affected. Alpha Ketoglutarate is not formed.
  Hence no Glutamate and Glutamine. GABA cannot be formed and detoxification of ammonia affected. This leads to intellectual disability.
- **Lactic acidemia**
  Acetyl CoA has to combine with Oxaloacetate to enter into TCA cycle. So TCA cycle is decreased. Pyruvate accumulate, which enter into anaerobic glycolysis. So lactic acidemia.

> **HIGH YIELD POINTS**
>
> **ENZYMES COMMON TO GLYCOLYSIS AND GLUCONEOGENESIS^Q**
> - All the enzymes other than the irreversible enzymes in the glycolysis.
>
> **To generate 1 mol of glucose from 2 mols of lacate 6 ATPs are utilised**
> - 2 ATPs for Pyruvate Carboxylase
> - 2 ATPs for PEPCK
> - 2 ATPs for 1,3 BPG Kinase

Reciprocal Regulation of Gluconeogenesis and Glycolysis

Since Glycolysis and Gluconeogenesis share the same pathway but in opposite directions, they must be regulated reciprocally.

Three mechanisms are responsible for regulating the activity of enzymes concerned in carbohydrate metabolism:

- Changes in the rate of enzyme synthesis by induction and repression
- Covalent modification by reversible phosphorylation,
- Allosteric modification by Fructose 2,6 Bisphosphate.

Changes in the Rate of Enzyme Synthesis: By Induction and Repression

- **Insulin**, secreted in response to increased blood glucose, **enhances the synthesis of the key enzymes in glycolysis.**
- **Pyruvate Carboxylase** is repressed by **Insulin^Q**.
- Insulin also **antagonizes the effect of the glucocorticoids and glucagon**-stimulated cAMP, which induce synthesis of the key enzymes of gluconeogenesis.

Covalent Modification by Reversible Phosphorylation: By Means of Hormones

Epinephrine and Glucagon

- **Increasing the concentration of cAMP**.
- This in turn activates **cAMP-dependent protein kinase.**
- Leading to the **phosphorylation and of activation of enzymes gluconeogenesis.**

Insulin

- Decrease the concentration of cAMP.
- De phosphorylate the key enzymes of Gluconeogenesis and be come in active.

Allosteric Modification

- By **Acetyl-CoA** and **Fructose 2,6 Bisphosphate.**

Acetyl-CoA

- **Acetyl-CoA as an allosteric activator^Q of Pyruvate Carboxylase^Q**
- This ensures provision of Oxaloacetate, so that Acetyl-CoA can be oxidized by Citric acid cycle.

Fructose 2,6 Bisphosphate

First we learn about the tandem enzyme that synthesize Fructose 2,6 Bisphosphate.

Tandem Enzyme (Bifunctional Enzyme)

- Single polypeptide with two enzyme activity.
- Two enzyme activities are Phosphofructokinase–II (PFK-II) and Fructose 2, 6 Bisphosphatase (F2, 6B Pase) **Action of the Tandem Enzyme**
- PFK-II convert Fructose 6 Phosphate to Fructose 2,6 Bisphosphate
- F2, 6B Pase convert Fructose 2,6 Bisphosphate to Fructose 6 Phosphate.
- *Fructose 2,6 Bisphosphate, the product of PFK-II, is an allosteric activator of PFK-I*

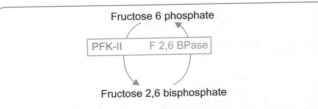

Fig. 4.2.15: *Action of bifunctional enzyme that regulate glycolysis and gluconeogenesis*

In Well Fed State

- Insulin dephosphorylate the tandem enzyme.
- PFK-II part is active and F2,6 BPase is inactive.

- Level of **Fructose 2, 6 Bisphosphate rises**.^{Q}
- **Fructose 2, 6 Bisphosphate**^{Q} favor Glycolysis.
- But Gluconeogenesis is in active.
- Decreases the blood Glucose.

In the Fasting State

- Glucagon phosphorylate the tandem enzyme by cAMP dependent Protein Kinase
- PFK-II is inactive and F2,6 B Pase is active.
- **Level of Fructose 2,6 Bisphosphate falls.**
- This favors Gluconeogenesis.
- But Glycolysis is inactive.
- Increases the blood Glucose.

Fructose 2, 6 Bisphosphate Reciprocally Regulate Glycolysis and Gluconeogenesis

Characteristics	Phosphofructo-kinase-II	Fructose 2,6 Bisphosphatase
Reaction	Fructose 6 Phosphate to Fructose 2,6 Bisphosphate	Fructose 2,6 Bisphosphate to Fructose 6 Phosphate
Hormonal regulation	Favored by Insulin	Favored by Glucagon
Covalent modification	Active in dephosphorylated state	Active in Phosphorylated state
Dietary regulation	Active in well fed state	Active in fasting state
Reciprocal regulation of Glycolysis and Gluconeogenesis	Fructose 2,6 Bisphoshate, the product of PFK-II favor Glycolysis. inhibit Gluconeogenesis	Decreases the level of Fructose 2,6 Bisphosphate, there by favor Gluconeogenesis inhibit Glycolysis

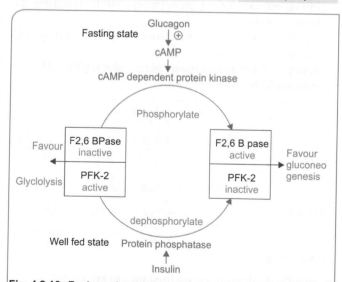

Fig. 4.2.16: Reciprocal regulation of glycolysis and gluconeogenesis

GLYCOGEN METABOLISM

- Glycogen Synthesis (Glycogenesis)
- Glycogen Degradation (Glycogenolysis)

Glycogen is the major storage carbohydrate in animals. Glycogen is present mainly in liver and muscle, with modest amount in brain.

Differences between Liver Glycogen and Muscle Glycogen

Features	Liver	Muscle
Total Glycogen content	Less (1.8 kg)	Highest (35 kg)
Percentage by tissue weight	**Highest** (5.0)	Less (0.7)
Regulation of blood glucose	Contributes to blood Glucose	Does not directly contribute to blood Glucose but serves as a source of energy to muscle it self
Glucose 6 Phosphatase	Present	**Absent**^{Q}

> **HIGH YIELD POINTS**
>
> After 12–18 h of fasting, liver glycogen is almost totally depleted.^{Q}

> **CONCEPT BOX**
>
> **Muscle Glycogen and Gluconeogenesis**
> Muscle does not contribute directly to blood Glucose but Pyruvate formed by Glycolysis, is transaminated to Alanine. This is transported to Liver, which is used for Gluconeogenesis. This is *Glucose Alanine Cycle*.

Glycogenesis

- Occurs mainly in **Muscle and Liver**
- **Organelle-Cytosol**
- Rate Limiting Enzyme-**Glycogen Synthase**^{Q}.

Glycogenesis–Steps

Synthesis of UDP Glucose

- Glucose converted to Glucose 6 Phosphate by **Hexokinase in muscle/Glucokinase in liver.**
- Glucose 6 Phosphate isomerized to Glucose 1 Phosphate **Phospho Gluco Mutase.**
- Glucose 1 Phosphate reacts with UTP to form UDP Glucose and Pyrophosphate catalyzed by UDP Glucose Pyrophosphorylase.
- **UDP Glucose** is the Glucose donor for Glycogen Synthesis.

Initiation of Glycogen Synthesis
- **Glycogen Synthase** is the enzyme that joins Glucose residues by α 1,4 Linkage (C1 of UDP Glucose and C-4 of the terminal glucose residue in the Glycogen, liberating UDP).
- But Glycogen Synthase can do this only on a preexisting
- Glycogen molecule or a primer called **Glycogenin**.
- Glycogen Synthase adds Glucose residue on this Glycogen Primer.

Glycogenin
Glycogenin is a 37 k Da protein[QNBE pattern] that is glucosylated on a specific tyrosine residue by UDPGlc.

Formation of Branch Points
- When the chain is at least 11 glucose residues long, branching enzyme acts.
- Branching enzyme transfers at least **six glucose** residues to a neighboring chain to form an α 1, 6 linkage, establishing a branch point.
- Branches grow by further addition of α 1 → 4 glucosyl units.

Glycogenolysis
- Occurs in the **Muscle and Liver**
- Organelle-Mainly **Cytoplasm** [Small proportion in the **Lysosomes**]
- Rate Limiting Enzyme-**Glycogen Phosphorylase**
- **PLP is a Coenzyme[Q] of Glycogen Phosphorylase**.

Isoenzymes of Glycogen Phosphorylase
- Present in Muscle, Liver and Brain.
- Glycogen Phosphorylase BB is a Cardiac Biomarker.

Steps of Glycogenolysis
Breaking of α1,4 Linkage
- Glycogen Phosphorylase cleave the **α1,4 linkage.**
- Releases **Glucose 1 Phosphate NOT freeGlucose.**
- Glycogen Phosphorylase stops its action when it is at least 4 glucose residues from a branch point.

Removal of Branches
By a bifunctional enzyme:
1. **First part is a α-1, 4 α1, 4 Glucantrans ferase**
   - Transfer trisaccharide residue to another forming a new α 1, 4 linkage.

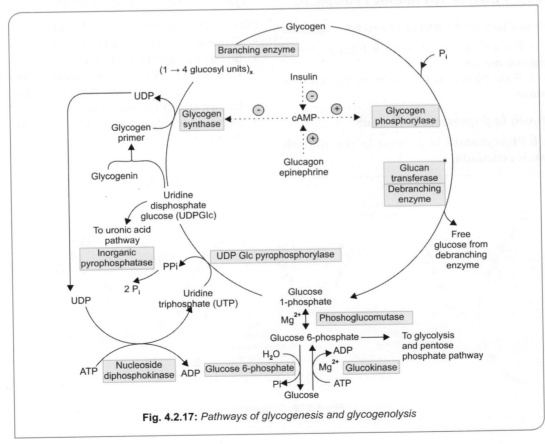

Fig. 4.2.17: *Pathways of glycogenesis and glycogenolysis*

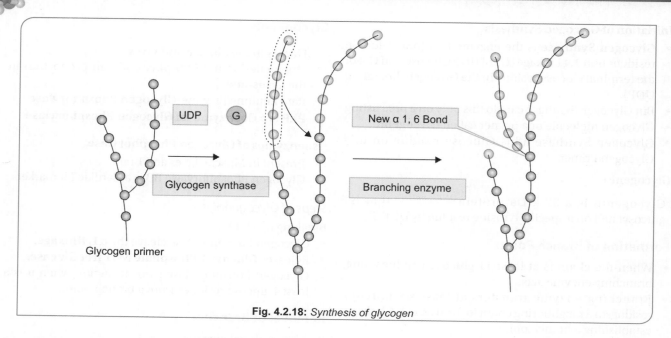

Fig. 4.2.18: *Synthesis of glycogen*

2. **Second part is an α-1, 6 Glucosidase (Amylo 1,6 Glucosidase)**
   - Hydrolyze the branching point.
   - *Releases free Glucose NOT Glucose 1 Phosphate.*

Conversion of Glucose 1 Phosphate to free Glucose

- Glucose 1 Phosphate to Glucose 6 Phosphate by **Phosphoglucomutase**
- Glucose 6 Phosphate to Glucose by **Glucose 6 Phosphatase**.

Role of Smooth Endoplasmic Reticulum

- **Glucose 6 Phosphatase is present in the smooth endoplasmic reticulum.**
- A transporter is required for the transport of Glucose 6 Phosphate from SER to cytoplasm.
- ***Defect in the Glucose 6 Phosphate transporter lead to Type Ib Glycogen Storage disorder.***

Difference between Liver and Muscle Glycogenolysis

- Glucose 6 Phosphatase is absent in the muscle[Q]
- So do not contribute free glucose.
- Hence, liver is the major contributor of blood glucose.
- Muscle utilizes Glucose 6 Phosphate for Glycolysis for its own energy during exercise.
- Muscle glycogen produces 3 ATPs by Anaerobic glycolysis

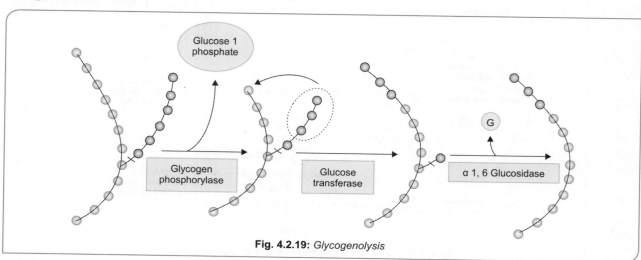

Fig. 4.2.19: *Glycogenolysis*

Minor Pathways of Glycogenolysis
- Taking place inside **lysosomes.**
- By the enzyme **Acid maltase.**
- Glycogen is hydrolyzed to Glucose.
- This is important in glucose homeostasis in **neonates.**
- Genetic lack of Lysosomal acid Maltase lead to **Type II Glycogen Storage Disorder (Pompe's Disease or Type II GSD).**

> **HIGH YIELD POINTS**
>
> Enzyme common to Glycogenesis and Glycogenolysis is Phosphoglucomutase^Q

Regulation of Glycogen Metabolism

> **CONCEPT BOX**
>
> **Basic Concepts**
> - **Insulin** favor **Glycogenesis** by **dephosphorylating, Glycogen Synthase.**
> - Glycogen Synthase active in dephosphorylated state.
> - Glucagon and Epinephrine favor **Glycogenolysis** by **phosphorylating Glycogen Phosphorylase.**
> - Glycogen Phosphorylase active in the phosphorylated state.
> - **Phosphorylase a –Active state**^{Q}
> - Phosphorylase b-Inactive state.

Rate Limiting Steps
- *Glycogenesis:* Glycogen Synthase
- *Glycogenolysis:* Glycogen Phosphorylase.

Hormonal Regulation

Insulin
- Insulin dephosphorylate Glycogen Synthase and Glycogen Phosphorylase.
- **Glycogen Synthase is active** in the dephosphorylated state.
- Glycogen Phosphorylase is inactive in the dephosphorylated state.
- So Glycogen is synthesized.

Glucagon (In Liver) and Epinephrine (In Liver and Muscle)
- Phosphosphorylate Glycogen Phosphorylase and GlycogenSynthase.
- **Glycogen Phosphorylase** active in the phosphorylated state.
- Glycogen Synthase in active in the phosphorylated state.
- So Glycogen is degraded.

> **CONCEPT BOX**
>
> - Well fed state under the influence of Insulin store excess carbohydrate as Glycogen.
> - Infasting state under the influence of Glucagon, Glycogenolysis takes place in the liver to supply Glucose.
> - In an exercising muscle, epinephrine favor Glycogenolysis for supplying energy to the muscle.

Differences between Muscle and Liver in the Regulation of Glycogen Metabolism
- Epinephrine acts in Muscle and Liver whereas Glucagon acts only in the Liver.
- In the muscle there is **cAMP-independent activation of glycogenolysis**^{Q}.
  By the stimulation of **a Ca^{2+}/calmodulin-sensitive phosphorylase kinase.**
  Phosphorylate Glycogen Phosphorylase in the muscle. Favor glycogenolysis.
- Muscle phosphorylase can be activated without phosphorylation^Q.
  Muscle Phosphorylase **has a binding site for 5' AMP. 5'AMP is an allosteric activator without phosphorylation**^{Q}.

Favor Glycogenolysis
Mechanism of Action of Glucagon and Epinephrine on Glycogen Metabolism
- Glucagon/Epinephrine bind to its receptor.
- Inactive Adenylyl Cyclase is converted to **Active Adenylyl Cyclase.**
- **Adenylyl Cyclase convert ATP to cAMP.**
- cAMP activate inactive Protein Kinase A to **active Protein Kinase A.**
- Phosphorylate Phosphorylase Kinase.
- Phosphorylase Kinase B (Inactive) is now Phosphorylase Kinase A(Active).
- This Phosphorylate Glycogen Phosphorylase.
- Glycogen Phosphorylase B (Inactive) is now Glycogen Phosphorylase A (Active)
- Glycogen is degraded (Ref Fig. 4.2.20).

Mechanism of Action of Insulin on Glycogen Metabolism
- Insulin increases the activity of Phosphodiesterase, which hydrolyzes cAMP to 5'AMP.
- Thus insulin terminates the action of cAMP.
- Increase the activity of Protein Phosphatase.
- This dephosphorylate Glycogen Synthase

Self Assessment and Review of Biochemistry

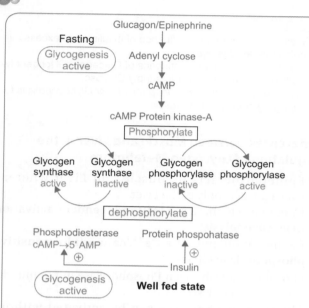

Fig. 4.2.20: cAMP dependent and cAMP independent mechanism regulation of glycogen Phosphorylase

Regulation of Glycogen

Allosteric Regulation of Glycogen Synthase

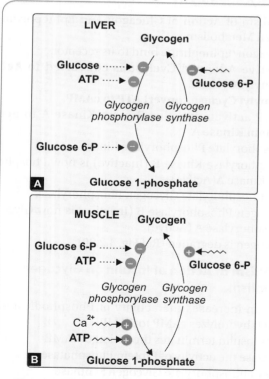

Figs. 4.2.21A and B: Allosteric regulation of glycogen metabolism in muscle and liver

Allosteric Regulators of Glycogen synthase and Glycogen phosphorylase

Organ	Enzyme	Allosteric activator	Allosteric inhibitor
Liver	Glycogen Synthase	Glucose 6 Phosphate	----
	Glycogen Phosphorylase	-----	**Glucose** Glucose 6 Phosphate ATP
Muscle	Glycogen Synthase	Glucose 6 Phosphate	-----
	Glycogen Phosphorylase	Ca^{++} AMP	Glucose 6 Phosphate ATP

GLYCOGEN STORAGE DISORDERS

Group of inherited disorders characterized by deposition of an abnormal type or quantity of glycogen in tissues, or failure to mobilize glycogen.

Liver Glycogen Storage Disorder

Type	Name	Enzyme efficiency	Characteristics
0	—	Glycogen synthase	Early morning drowsiness and **fatigue**, **fasting hypoglycemia**, and **ketosis**; early death **[No Hepatomegaly]**
Ia	Von Gierke's disease	Glucose 6-phosphatase^Q	Glycogen accumulation in liver and renal tubule cells (Kidney Enlarged) hypoglycemia; **elevated** blood lactate, cholesterol, triglyceride, and **uric acid levels**^{QPGI}
Ib	—	Endoplasmic reticulum glucose 6-phosphate transporter	Same as type Ia, with additional findings of neutropenia and impaired neutrophil function **Recurrent Bacterial Infection**, Inflammatory Bowel Disease
III	Limit dextrinosis, Forbe's or Cori's disease	Liver and muscle debranching enzyme (Amylo 1,6 Glucosidase)	**Fasting hypoglycemia** hepatomegaly in infancy accumulation of characteristic **branched polysaccharide (limit dextrin) muscle weakness**, **elevated transaminase levels**; liver symptoms can **progress to liver failure later in life**

Contd...

Contd...

Type	Name	Enzyme efficiency	Characteristics
IV	Amylopectinosis, **Andersen's disease**	Branching enzyme	Hepatosplenomegaly **Accumulation of polysaccharide with few branch points** Failure to thrive, hypotonia, hepatomegaly, splenomegaly, **progressive cirrhosis (death usually before 5th yr), elevated transaminase levels**
VI	Hers' disease	Liver phosphorylase	Hepatomegaly
VIII		Liver phosphorylase kinase	
	Fanconi Bickel Syndrome	**Glucose transporter 2 (GLUT-2)**	Failure to thrive, rickets, **hepatorenomegaly, proximal renal tubular dysfunction, impaired glucose and galactose utilization**

Type Ia GSD-Von Gierke's Disease
- Most common Glycogen Storage Disorder in childhood
- Autosomal Recessive
- Muscle not affected because Glucose 6 Phosphate absent in the muscles
- Structure of Glycogen normal

Biochemical defect
Type Ia GSD-**Glucose 6 Phosphatase absent or deficient in liver, kidney and intestinal mucosa.**
Type Ib GSD-**Translocase that transport Glucose 6 Phosphate across endoplasmic reticulum membrane is defective.**

Symptom box

Clinical Features
- Most commonly present at 3–4 months of age with
- **Doll like facies** with fat cheeks
- Relatively thin extremities
- Short stature, Protuberant abdomen
- **Massive Hepatomegaly**
- **Kidneys are also enlarged**
- **No Splenomegaly**
- Plasma may be milky due to associated hypertriglyceridemia

Type Ib has additional features of recurrent bacterial infection.

HIGH YIELD POINTS

THE BIOCHEMICAL HALLMARKS ARE
- Hypoglycemia
- Lacticacidosis
- Hyperlipidemia
- Hyperuricemia

Type III GSD (Limit Dextrinosis) (Cori's Disease)
- Autosomal Recessive.

Biochemical Defect
- **Debranching enzyme** is defective.
- **Abnormal glycogen with short outer branch chain resembling limit dextrin accumulate.**

Symptom box

Clinical Features
- Fasting hypoglycaemia, hepatomegaly, hyperlipidemia, short stature, variable skeletal muscle myopathy
- Kidneys are not enlarged
- Splenomegaly may be present
- **Progressive liver cirrhosis and failure occur**
- **Elevation of liver transaminase**
- **Fasting ketosis**
- **Blood lactate and uric acid level is normal**
- Remarkably, **hepatomegaly and hepatic symptoms in most-patients with type III GSD improve with age** and usually resolve after puberty
- The administration of glucagon 2 hr after a carbohydrate meal provokes a normal increase in blood glucose
- After an overnight fast, glucagon may provoke no change in blood glucose level.

Definitive Diagnosis
- Enzyme assay in liver, muscle, or both.
  Mutation analysis can provide a noninvasive method for diagnosis and subtype assignment in the majority of patients.

Type IV Glycogen Storage Disease (Amylopectinosis, or Andersen Disease)
- Autosomal recessive

Biochemical Defect
- **Deficiency of branching enzyme activity** results in accumulation of an abnormal glycogen with poor solubility.
- The disease is referred to as type IV GS Dora mylopectinosis because the **abnormal glycogen has a structure resembling amylopectin.**

Symptom box

Clinical Features
- This disorder is clinically variable.
- The most common and classic for mischaracterized by **progressive cirrhosis of the liver and is manifested in the 1st 18 month of life as hepatosplenomegaly and failure to thrive**.
- The cirrhosis progresses to portal hypertension, ascites, esophageal varices, and liver failure that usually leads to **death by 5 yr of age**.

Diagnosis
- The hepatic histologic findings are characterized by **micronodular cirrhosis**
- Electron microscopy shows accumulation of the **fibrillar aggregations that are typical of amylopectin**.

Muscle Glycogen Storage Disorders

Type	Name	Enzyme defect	Characteristics
II	Pompes' Disease (Belongs to **lysosomalstor age disorder**)	Lysosomal α1, 4 and α1, 6 glucosidase (acid maltase)^Q	**Cardiomegaly**, **hypotonia**, hepatomegaly; cardiorespiration failure leading to death by age 2 yr
	Danon disease	Lysosome-associated membrane protein 2 (LAMP2)	Hypertrophic cardiomyopathy Rare X linked
V	McArdle's syndrome	Muscle phosphorylase^Q	**Poor exercise tolerance** muscle glycogen abnormally high
VII	Tarui's disease	Muscle and erythrocyte phospho-fructo kinase 1	Poor exercise tolerance; Hemolytic anemia myoglobinuria

Pompe's Disease (Type II GSD)
Autosomal recessive

Biochemical Defect
- Deficiency of lysosomal enzyme, **acid α-1, 4-glucosidase (acid maltase)**.
- Lysosomal glycogen accumulation in multiple tissues likely cardiac, skeletal, and smooth muscle cells.

Symptom box

Clinical Picture
Present in the 1st few months of life
- Hypotonia
- A generalized muscle weakness with a **"floppy infant" appearance**
- Feeding difficulties
- Macroglossia
- Hepatomegaly
- **Hypertrophic cardiomyopathy**
- Followed by **death from cardiorespiratory failure or respiratory infection usually by 2 yrs of age**.

Laboratory Diagnosis
- **Serum creatine kinase, aspartate aminotransferase, lactate dehydrogenase, acid Phosphatase elevated.**
- **Chest X-ray showing massive cardiomegaly.**

Treatment
Specific enzyme replacement therapy (ERT) with re-combinant human acid α-glucosidase (alglucosidase alfa, Myozyme) is available for treatment of Pompe's disease.

Type V Glycogen Storage Disease (McCardle's Disease)
- Type V GSD is an autosomal recessive disorder.

Biochemical Defect
- Muscle Phosphorylase defect
- **Lack of this enzyme limits muscle ATP generation by glycogenolysis**, resulting in muscle glycogen accumulation.

Symptom box

Clinical Manifestations
- Exercise intolerance
- A characteristic "second wind" phenomenon. (If they slow down or pause briefly at the 1st appearance of muscle pain, they can resume exercise with more ease.)
- Burgundy-colored urine after exercise, due to myoglobinuria secondary to rhabdomyolysis.

Type VII Glycogen Storage Disease (Tarui Disease)
- Autosomal recessive disorder

Biochemical Defect
Deficiency of **muscle phosphofructokinase (MIsoenzyme form)**

Symptom box

Clinical Manifestations
- Exercise intolerance.
- Compensated hemolysis.
- **Exercise intolerance is particularly acute after meals that are rich in carbohydrates**.
- No spontaneous second-wind phenomenon.

High Yield Points

GLYCOGEN STORAGE DISORDERS (GSDs) AT A GLANCE
- Most common GSD in adolescent and adults—Type VGSD (McArdles Disease)
- Liver GSD disorder causes fasting hypoglycaemia and hepatomegaly
- GSDs associated with liver cirrhosis—Type III, Type IV, Type IX GSDs
- GSD associated with renal dysfunction—Type IGSD
- Liver GSD with myopathy—Type III GSD and Type IVGSD.
- GSD with neurological (brain and anterior horn cells) involvement—Type II GSD

Quick Review

- Glycolysis can operate both aerobically and anaerobically.
- Mature erythrocyte is reliant only on glucose as its metabolic fuel.
- Committed step of glycolysis is PFK-I
- LDH reaction regenerate NAD+.
- Iodoacetate inhibit Glyceraldehyde 3 PO4 Dehydrogenase
- Fluoride inhibit enolase.
- Arsenate compete with inorganic phosphate in the step of glyceraldehyde 3 Phospahte dehydrogenase.
- From muscle glycogen if used for anaerobic glycolysis yield 3 ATPs.
- 1 mol of glucose under aerobic condition generate 32 ATPs.
- All regulatory enzymes of glycolysis are active in dephosphorylated state.
- By Rapaport leubering cycle no net ATPs are generated
- Cancer cell utilize glucose via anaerobic glycolysis.
- PDH is the irreversible route from glycolysis to citric acid cycle.
- Inhibition of PDH lead to lactic acidosis.
- Gluconeogenesis involves glycolysis,citric acid cycle plus key enzymes of gluconeogenesis.
- Glucose lactate cycle is Cori's cycle.
- Glucose –Alanine cycle is Cahill cycle.
- To generate 1 mol of glucose from 2 mols of lactate 6 ATPs are utilised.
- The major sites of gluconeogenesis are Liver and Kidney.
- Acetyl CoA is an allosteric activator of Pyruvate Carboxylase
- Key enzymes of gluconeogenesis are active in the phosphorylated state.
- Glycogen synthesis occour in liver and muscle
- PLP is the coenzyme for glycogen phosphorylase.
- After 12-18 hours of fasting liver glycogen is almost depleted.
- Glucose 6 Phosphatase is absent in the muscle.
- Enzyme common to glycogen synthesis and glycogenolysis is Phosphoglumutase
- Epinephrine acts in liver and muscle but glucagon acts only in liver.
- Most common glycogen storage disorder is Von Gierke's disease.

Sites of Metabolic Pathways

Pathway	Organelle
Glycolysis	Cytoplasm
Gluconeogenesis	Cytoplasm & Mitochondria
Glycogen synthesis	Cytoplasm
Glycogen degradation	Cytoplasm(minimal in lysosomes)
PDH	Mitochondria

Energetics

Aerobic glycolysis	7
Anaerobic glycolysis	2
Rapoport luebering Cycle	Zero
Aerobic oxidation of I mol of glucose	32
PDH	2.5
Anaerobic oxidation of glucose	2
Utilisation of ATPs in synthesising glucose from lactate	6
Glycolysis in RBC	2
Glycolysis in cancer cell	2
Anaerobic glycolysis in muscle from glycogen	3
1 NADH by malate Aspartate shuttle	2.5
1 NADH by Glycerophosphate shuttle	1.5

Important GSD

Liver GSD	Enzyme defect
Von Gierke disease(Type I)	Glucose 6 Phosphatase
Cori's disease(Type III)	Debranching enzyme
Anderson disease(Type IV)	Branching enzyme
Her's disease(Type VI)	Liver Glycogen phosphorylase
Fanconi Bickel Syndrome	GLUT-2
Muscle GSD	**Enzyme defect**
Pompe's disease(Type II)	Acid maltase
Mc Ardles disease(Type V)	Muscle Glycogen phosphorylase
Tarui's disease(Type VII)	Muscle & Erythrocyte PFK-1

Check List for Revision

- This is a very important chapter for all exams.
- Go through each and every line in this chapter.

Review Questions MCQ

Glycolysis, GTT Curves

1. Patient with Type I Diabetes mellitus, with complains of polyuria. Which of the following will occur normally in his body. *(AIIMS Nov 2018)*
   a. Glycogenesis in muscle
   b. Increased protein synthesis
   c. Increased conversion of fatty acid to Acetyl-CoA
   d. Decreased in Cholesterol synthesis

2. Regulatory enzymes of glycolysis *(PGI Nov 2018)*
   a. PFK
   b. Pyruvate kinase
   c. PDH
   d. Hexokinase

3. Which of the following produces 3 ATP by anaerobic glycolysis? *(AIIMS May 2017)*
   a. Glucose
   b. Fructose
   c. galactose
   d. Glycogen

4. In anaerobic glycolysis, pyruvate is converted to lactate for: *(Central Institute Exam May 2017)*
   a. Removal of pyruvate
   b. Generation of NAD$^+$
   c. Generation of H$^+$
   d. Conversion of pyruvate

5. The supplement used in FSGS is: *(JIPMER Dec 2016)*
   a. Fructose
   b. Galactose
   c. Mannose
   d. Glucose

6. Which of the following is suitable test performed for diagnosis of Intestinal malabsorption? *(AIIMS Nov 2016)*
   a. D-Xylose test
   b. Stool fat estimation
   c. BT-PABA test
   d. Hydrogen breath

7. Respiratory quotient after exclusive carbohydrate meal is:
   a. 1
   b. 1.2
   c. 0.8
   d. 0.7

8. Which of the following is the normal Glucose Tolerance curve?

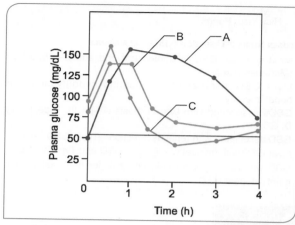

   a. A
   b. B
   c. C
   d. None

9. A 27-year lady developed severe hyperglycemia in pregnancy and it returned to normal after delivery. Her blood sugar is well under control without any medications. Her sisters and mother also have history of increased blood glucose during pregnancy, all were euglycemic after delivery. What is the enzyme defect? *(JIPMER May 2016)*
   a. Glucokinase
   b. PFK
   c. Aldolase
   d. Enolase

10. Irreversible steps of Glycolysis are catalysed by: *(AIIMS May 2013)*
    a. Hexokinase, Phosphofructokinase, Pyruvate Kinase
    b. Glucokinase, Pyruvate Kinase, Glyceraldehyde 3 Phosphate Dehydrogenase
    c. Hexokinase, Phospho Glycerate Kinase, Pyruvate Kinase
    d. Pyruvate Kinase, Fructose 1,6 Bisphosphatase, Phospho FructoKinase

11. Glycolysis occurs in: *(AIIMS May 2007)*
    a. Cytosol
    b. Mitochondria
    c. Nucleus
    d. Lysosome

12. Irreversible step(s) in Glycolysis is/are: *(PGI May 2012)*
    a. Enolase
    b. Phosphofructokinase
    c. Pyruvate Kinase
    d. Glyceraldehyde 3 Phosphate Dehydrogenase
    e. Hexokinase

13. Enzyme catalyzing reversible step in glycolysis is are: *(PGI Nov 2010)*
    a. Phosphofructokinase
    b. Enolase
    c. Pyruvate kinase
    d. Phospho-glycerate mutase
    e. Glyceraldehyde-3-Phosphate Dehydrogenase

14. In which of the following steps ATP is released?
    a. Phosphoenol pyruvate to pyruvate (Ker-2008)
    b. Glyceraldehyde 3 phosphate to 1,3 bisphosphoglycerate
    c. Fructose 6 phosphate to fructose 1,6 bisphosphate.
    d. Glucose to Glucose 6 phosphate.

15. What activate Kinases of glycolysis?
    (NBE Pattern Question)
    a. ATP
    b. cAMP
    c. Insulin
    d. Glucagon

16. About glycolysis true is: (PGI Dec 98)
    a. Occurs in mitochondria
    b. Complete breakdown of glucose
    c. Conversion of glucose to 3C units
    d. 3 ATPs are used in anaerobic pathway.

17. Compound that joins glycolysis with glycogenesis and glycogenolysis: (JIPMER 04)
    a. Glucose 1,6 bisphosphate
    b. Glucose 1 PO_4
    c. Glucose 6 PO_4
    d. Fructose 1,6 bisphosphate

18. Key glycolytic enzymes: (PGI 98)
    a. Phosphofructokinase
    b. Hexokinase
    c. Pyruvate kinase
    d. Glucose 1,6 bisphosphatase

19. In glycolysis the first committed step is catalyzed by:
    (AIIMS Dec 97)
    a. 2, 3-DPG
    b. Glucokinase
    c. Hexokinase
    d. Phosphofructokinase

20. The rate-limiting enzyme in glycolysis is: (AI 2000)
    a. Phosphofructokinase
    b. Glucose- 6-dehydrogenase
    c. Glucokinase
    d. Pyruvate kinase

21. Cancer cells derive nutrition from: (AIIMS Nov 2001)
    a. Anaerobic glycolysis
    b. Oxidative phosphorylation
    c. Increase in mitochondria
    d. Aerobic Glycolysis

22. True statements about glucokinase is/are:
    (PGI Dec 2003)
    a. Km value is higher than normal blood sugar
    b. Found in liver
    c. Glucose 6 phosphate inhibit it
    d. Has both glucose 6 phosphatase and kinase activity
    e. Glucose enter into cells through GLUT-2

23. Within the RBC, hypoxia stimulates glycolysis by which of the following regulating pathways: (AI 2007)
    a. Hypoxia stimulates pyruvate dehydrogenase by increased 2, 3-DPG
    b. Hypoxia inhibits hexokinase
    c. Hypoxia stimulates release of all glycolytic enzymes from Band 3 on RBC membrane
    d. Activation of the regulatory enzymes by high pH

24. All except occurs on decrease in blood glucose level:
    a. Inhibition of PFK-II (AI 2012)
    b. Activation of Fructose 2,6 Bisphosphatase.
    c. Increase in glucagon.
    d. Increase in Fructose 2,6 Bisphosphate.

25. The number of ATPs produced by Rapaport-leubering Cycle in RBC from Glucose: (NBE Pattern QN)
    a. 1
    b. 2
    c. 3
    d. 4

26. Enzyme responsible for complete oxidation of glucose to CO_2 and water is present in:
    (AIIMS May 2007)
    a. Cytosol
    b. Mitochondria
    c. Lysosomes
    d. Endoplasmic reticulum

27. Substrate level phosphorylation is by:
    (NBE Pattern Question)
    a. Pyruvate kinase
    b. Phosphofructokinase
    c. HexoKinase
    d. ATP Synthase

28. The enzyme not involved in substrate level phosphorylation: (JIPMER 2014)
    a. Pyruvate kinase
    b. Phosphofructokinase
    c. Succinate thiokinase
    d. Phosphoglycerate kinase

Gluconeogenesis, Fates of Pyruvate

29. The number of high energy bond require to get 1 mol of Glucose from 2 mols of lactate:
    (Central Institute Exam May 2017)
    a. 2
    b. 4
    c. 6
    d. 12

30. Which of the following is an activator of Pyruvate carboxylase?
    a. Oxaloacetate
    b. Citrate
    c. Acetyl CoA
    d. Glucose

31. All of the following amino acids forms acetyl CoA via pyruvate dehydrogenase except:
    a. Glycine
    b. Tyrosine
    c. Hydroxyproline
    d. Cysteine

32. For gluconeogenesis which of the following reaction is more effective? (AIIMS Nov 2016)

a. Citrate stimulation of Acetyl CoA Carboxylase
b. Acetyl CoA stimulation of Pyruvate Carboxylase
c. Fructose 2,6 Bisphosphate stimulates PFK-1
d. Fructose 1,6 Bisphosphate stimulation of Pyruvate Kinase

33. A baby is hypotonic and shows increased ratio of Pyruvate to Acetyl CoA. Pyruvate cannot form Acetyl CoA in fibroblast. He also shows features of lactic acidosis. Which of the following can revert the situation? *(JIPMER Dec 2016)*
    a. Biotin
    b. Pyridoxine
    c. Free fatty acid
    d. Thiamin

34. Which of the following does not contribute to glucose by gluconeogenesis? *(AIIMS Nov 2015)*
    a. Lactate
    b. Acetyl CoA
    c. Pyruvate
    d. Oxaloacetate

35. True about gluconeogenesis: *(PGI May 2013)*
    a. Prevent hypoglycemia during prolonged fasting
    b. Occur both in muscle and liver
    c. Fructose 2,6 Bisphosphate stimulate it
    d. Excess acetyl CoA causes stimulation
    e. Carbon skeleton of amino acids are involved in gluconeogenesis

36. A child with low blood glucose is unable to do glycogenolysis or gluconeogenesis. Which of the following enzyme is missing in the child? *(AIIMS Nov 2012)*
    a. Fructokinase
    b. Glucokinase
    c. Glucose 6 Phosphatase
    d. Transketolase

37. In fasted state gluconeogenesis is promoted by which enzyme? *(AIIMS May 2012)*
    a. Acetyl-CoA induced stimulation of Pyruvate Carboxylase
    b. Citrate induced stimulation of Acetyl-CoA Decarboxylase
    c. Fructose 2,6 bisphosphate induced stimulation of Phosphofructokinase-1
    d. Stimulation of Pyruvate kinase by Fructose 1,6 Bisphosphate

38. During prolonged fasting, rate of gluconeogenesis is determined by: *(AIIMS May 2012)*
    a. Essential fatty acid in liver
    b. Alanine in liver
    c. Decreased cGMP
    d. ADP in liver

39. True about gluconeogenesis is/are: *(PGI May 2013)*
    a. Prevent hypoglycemia during prolonged fasting
    b. Occur in both muscle and liver
    c. Fructose 2,6 bisphosphate stimulate it
    d. Excess of acetyl CoA stimulate it
    e. Carbon skeleton of amino acid is involved

40. Common enzyme for gluconeogenesis and glycolysis is: *(Ker-2007)*
    a. Glyceraldehyde 3 PO_4 dehydrogenase
    b. Hexokinase
    c. Pyruvate kinase
    d. Pyruvate carboxylase

41. Phosphofructokinase-I is activated by all except: *(Recent Question)*
    a. 5'AMP
    b. Fructose 2,6 Bisphosphate
    c. Fructose 6 Phosphate
    d. Citrate

42. Not a substrate for gluconeogenesis: *(NBE Pattern Q)*
    a. Acetyl-CoA
    b. Lactate
    c. Glycerol
    d. Propionyl CoA

43. Which of the following reactions takes place in two compartments? *(Recent Question)*
    a. Gluconeogenesis
    b. Glycolysis
    c. Glycogenesis
    d. Glycogenolysis

44. Glyconeogenic capability of cell is determined by the presence of: *(PGI Dec 2005)*
    a. Pyruvate dehydrogenase
    b. Glucose-6-phosphatase
    c. Pyruvate carboxylase
    d. Fructose 1,6- bisphosphatase
    e. Pyruvate carboxykinase

45. Step of Gluconeogenesis is: *(NBE Pattern QN)*
    a. Pyruvate to Lactate
    b. Glucose 6 Phosphate to Fructose 6 Phosphate
    c. Pyruvate to Acetyl CoA
    d. Oxaloacetate to Phosphoenol Pyruvate

46. Major contribution towards gluconeogenesis is by: *(AI 92)*
    a. Lactate
    b. Glycerol
    c. Ketones
    d. Alanine

47. Glucose can be synthesised from all except: *(AI 96)*
    a. Amino acids
    b. Glycerol
    c. Acetoacetate
    d. Lactic acid

48. Gluconeogenesis does not occur significantly from .... in humans: *(AIIMS 92)*
    a. Lactate
    b. Fatty acids
    c. Pyruvate
    d. Amino acid

49. Acetyl-CoA can be converted into all of the following except: *(AI 2009)*
    a. Glucose
    b. Fatty acids
    c. Cholesterol
    d. Ketone bodies

50. A genetic disorder renders fructose 1,6-bisphosphatase in liver less sensitive to regulation by fructose 2,6-biphosphate. All of the following metabolic changes are observed in this disorder except: *(AI 2004)*

a. Level of fructose 1,6-biphosphate is higher than normal
b. Level of fructose 1,6-biphosphate is lower than normal
c. Less pyruvate is formed
d. Less ATP is generated

Glycogen Metabolism and Glycogen Storage Disorders

51. A male patient came with pain in calf muscles in exercise. On biopsy glycogen present in the muscle. What is the enzyme eficiency?
    a. Branching enzyme
    b. Phosphofructokinase I
    c. Debranching enzyme
    d. Glucose 6 phosphatase

52. Limit dextrin accumulate in *(Central Institute Nov 2018)*
    a. Gaucher's disease
    b. Cori's disease
    c. VonGierke's disease
    d. Anderson disease

53. UDP glucose is used for *(PGI Nov 2018)*
    a. Glycogen synthesis
    b. Galactose metabolism
    c. Heparin synthesis
    d. Bilirubin metabolism

54. A four-year-old child with exercise intolerance. On investigation Blood pH 7.3, FBS 60 mg%, hypertriglyceridemia, ketosis and lactic acidosis. The child had hepatomegaly and renomegaly. Biopsy of liver and kidney showed increased glycogen content. What is the diagnosis? *(AIIMS Nov 2017)*
    a. McCardle's disease
    b. Cori's disease
    c. Von Gierke's disease
    d. Pompe's disease

55. Glycogen Phosphorylase, coenzyme is: *(AIIMS Nov 2017)*
    a. Pyridoxal Phosphate
    b. Thiamin
    c. Biotin
    d. Pantothenic acid

56. Glycogenin primer is glucosylated by: *(Recent Question)*
    a. UDP Glucose
    b. Glucose 1 PO_4
    c. UDP Glucose 1 PO_4
    d. UDP Glucose 6 PO_4

57. A female infant appeared normal at birth but developed signs of liver disease one month of age and muscle weakness at 3 months and severe hypoglycemia on early morning awakening. Examination revealed hepatomegaly, laboratory analysis showed ketoacidosis, pH 7.2, increased AST and ALT over 1000 IU. Intravenous administration glucagon followed by meals normalised blood levels, but glucose levels did not rise when glucagon was administered overnight fast. Liver biopsy was done and glycogen constituted (8%) of wet weight. With the above clinical picture which of the following enzyme is deficient? *(AIIMS Nov 2016)*
    a. Debranching enzyme
    b. Glucose 6 phosphatase
    c. Muscle phosphorylase
    d. Branching enzyme

58. Why Glucose 6 Phosphate in the cytoplasm of hepatocyte is not acted upon by Glucose 6 Phosphatase as soon as it is formed? *(AIIMS Nov 2015)*
    a. Thermodynamically possible only when gluconeogenesis occur
    b. Need Protein Kinase for its activation
    c. Enzyme is present in SER, Glucose 6 Phosphate need to be transported into SER
    d. Steric inhibition of Phosphatase by albumin

59. The reason for ketosis in von Gierke's Disease are all except: *(AIIMS Nov 2013)*
    a. Hypoglycemia
    b. Oxaloacetate is necessary for gluconeogenesis
    c. Low blood glucose less than 40 mg%
    d. Fatty acid mobilisation is low

60. A child with low blood glucose is unable to do glycogenolysis or gluconeogenesis. Which of the following enzyme is missing in the child? *(AIIMS Nov 2012)*
    a. Fructokinase
    b. Glucokinase
    c. Glucose 6 Phosphatase
    d. Transketolase

61. In which of the following tissues, is glycogen incapable of contributing directly to blood glucose: *(AI-2008)*
    a. Liver
    b. Muscle
    c. Both
    d. None

62. In humans carbohydrates are stored as: *(Ker-2006)*
    a. Glucose
    b. Glycogen
    c. Starch
    d. Cellulose

63. Glycogen is released from the muscle due to increased cAMP due to:
    a. Epinephrine
    b. Thyroxine
    c. Glucgon
    d. Growth hormone

64. Pancreatic alpha amylase:
    a. Convert starch to glycogen
    b. Hydrolyses starch to limit dextrin
    c. Hydrolyses Starch to Monosaccharides.
    d. Convert maltose to glucose

65. A 5 years old boy presents with hepatomegaly, hypoglycaemia, ketosis. The diagnosis is:
    a. Mucopolysaccharidosis
    b. Glycogen storage disorder
    c. Lipopolysaccharidosis
    d. Diabetes mellitus

66. Glycogen Phosphorylase can be regulated by all following EXCEPT: *(AIIMS Nov 2015)*
    a. cAMP
    b. Calmodulin
    c. Protein Kinase A
    d. Glycogenin

Self Assessment and Review of Biochemistry

67. Cofactor for Glycogen Phosphorylase:
(AIIMS Nov 2015)
   a. Thiamine Pyrophosphate   b. Pyridoxal Phosphate
   c. Citrate   d. FAD

68. Pompe's disease is due to deficiency of:
   a. Debranching enzyme   (Recent Question)
   b. Muscle Phosphorylase
   c. Acid Maltase
   d. Branching enzyme

69. Glycogen storage disorder is/are: (PGI Nov 2014)
   a. Niemann pick disease   b. Gaucher disease
   c. Taysach's disease   d. Pompe's disease
   e. McCardles disease

70. How many hours for depletion of glycogen?
(NBE Pattern Question)
   a. 9   b. 18
   c. 24   d. 48

71. In the fed state, major fate of glucose-6-phosphate in tissues is: (AIIMS May 93)
   a. Storage as fructose
   b. Storage as glyceraldehyde-3-phosphate
   c. Enters HMP shunt via ribulose-5-phosphate
   d. Storage as glycogen

72. Which of the following is a debranching enzyme?
(AIIMS 90)
   a. Glycogen synthetase
   b. Glucose-6-phosphatase
   c. Amylo (1,6) glucosidase
   d. Amylo 1,4-1,6 transglycosylase

73. Sequence of events in glycogenolysis:
(PGI June 97, Dec 96)
   a. Phosphorylase, glucan transferase, debranching, phosphorylase
   b. Debranching, phosphorylase, transferase, phosphorylase
   c. Transferase, phosphorylase, debranching, phosphorylase
   d. Any of the above

74. Muscles are not involved in which glycogen storage disease?   (PGI Dec 97)
   a. I   b. II
   c. III   d. IV

75. An infant has hepatosplenomegaly, hypoglycaemia, hyperlipidemia, acidosis & normal structured glycogen deposition in liver. What is the diagnosis: (PGI June 01)
   a. Her's disease   b. Von Gierke's disease
   c. Cori's disease   d. Anderson's disease
   e. Pompe's disease

76. Glycogen storage diseases include all the following except: (PGI Dec 01)
   a. Von Gierke's disease   b. Fabry's disease
   c. McArdle's disease   d. Fragile X syndrome
   e. Krabbe's disease

77. The cause of hyperuricemia and gout in glucose-6-phosphatase deficiency is: (AIIMS Nov 01)
   a. More formation of pentose
   b. Decreased availability of glucose to tissues
   c. Increased accumulation of sorbitol
   d. Impaired degradation of free radicals

Answers to Review Questions

Glycolysis, GTT Curves

1. **c > d**

Explanation:
In Type I DM....Insulin decreased...decreased Acetyl-CoA carboxylase activity hence decreased Malonyl CoA, an allosteric inhibitor of CPT-I the RLE of beta oxidation...so uninhibited beta oxidation...i.e. FA conversion to Acetyl CoA.
No insulin and increased Glucagon activity Increased AMP Kinase activity ....So HMG CoA reductase is phosphorylated hence less cholesterol synthesis .But a study given here is saying decreased cholesterol synthesis in Type I Diabetes mellitus when compared to Type II DM.
Decreased cholesterol synthesis in Type I Diabetes mellitus when compared to Type II
http://diabetes.diabetesjournals.org/content/53/9/2217

The ratios of the absorption marker sterols in serum were higher, and those of the synthesis markers were lower in type 1 diabetic than in control subjects. The increased cholestanol ratios were seen in all lipoproteins, and those of free and total plant sterols were mainly in LDL, whereas the decreased free and total synthesis markers were mainly in all lipoproteins. In conclusion, high absorption and low synthesis marker sterols seem to characterize human type 1 diabetes. These findings could be related to low expression of ABC G/5 G/8 genes, resulting in high absorption of cholesterol and sterols in general and low synthesis of cholesterol compared with type 2 diabetes

2 **a, b, d. PFK, Pyruvate Kinase; Hexokinase**
(Ref: Harper 31/e page 159)

These three are the irreversible enzymes of glycolysis also.

3. **d. Glycogen**
(Ref: Harper 29/e page 171-174, Harper 31/e page 170, 172)
In Muscle Glucose 6 Phosphatase is absent, hence Glucose 6 Phosphate directly enter into Glycolysis, which spare 1

ATP utilized by Hexokinase. Hence out of 4 ATPs generated by anaerobic glycolysis only 1 is utilised, hence 3 ATPs.

In Fructose no PFK I step hence no ATP utilized but Glyceraldehyde formed by Aldolase reaction require ATP to get converted to Glyceraldehyde 3 PO_4 hence 4- 2 ATPs itself.

4. **b. Generation of NAD+** (Ref: Harper 31/e page 157)

In anaerobic glycolysis, the NADH generated in Glyceraldehyde 3 Phosphate Dehydrogenase has to be converted to NAD+. But in organs where anaerobic glycolysis operate as in RBC no mitochondria to regenerate NAD+. The only way to regenerate NAD + is LDH reaction that convert Pyruvate to Lactate.

5. **b. Galactose** (Ref: Journal Tranlational research, Galactose binds to focal segmental glomerulosclerosis permeability factor and inhibits its activity.)

"We propose testing galactose as a novel nontoxic therapy for nephrotic syndrome in FSGS to determine whether galactose slows progression and whether pretransplant therapy decreases rates of recurrence"

6. **b. Stool fat estimation** (Ref: Varley 6/e page 699)

Tests used to diagnose malabsorption

This involves tests to detect malabsorption of Fat, Carbohydrate and protein. But fat digestion and absorption, being more complex is often first to be disturbed. The resultant increase in faecal fat, steatorrhea, is present in generalised malabsorption. So in the given option Test for fat malabsorption i.e. Stool fat estimation is the answer.

Tests for fat malabsorption

- Determination of total faecal fat—Fat Balance test
- Breath tests using ^{14}C labelled triglycerides
- Tests using 131 I-Triolein

7. **a. 1** (Ref: Harper 31/e page 136)

	Energy Yield (kJ/g)	O_2 Consumed (L/g)	CO_2 Produced (L/g)	RQ (CO_2 Produced/ O_2 Consumed)	Energy (kJ)/L O_2
Carbohydrate	16	0.829	0.829	1.00	~20
Protein	17	0.966	0.782	0.81	~20
Fat	37	2.016	1.427	0.71	~20
Alcohol	29	1.429	0.966	0.66	~20

8. **b. B** (Ref: Ganong 24/e page 448)

- Curve A is Excessive Glucose absorption
- Curve B is normal glucose tolerance
- Curve C is Liver disease
- Horizontal line represent the approximate plasma glucose value at which hypoglycemia appear

9. **a. Glucokinase**

10. **a. Hexokinase, Phosphofructokinase, Pyruvate Kinase** (Ref: Harper 29/e page 171-174, Harper 31/e page 160,161)

Remember: *All the Kinases are irreversible except 1,3 Bisphospho Glycerate Kinase which is reversible.*

11. **a. Cytosol** (Ref: Harper 29/e page 171, Harper 31/e page 157)

Metabolic	Pathway site
Glycolysis	Cytoplasm
Gluconeogenesis	Cytoplasm & Mitochondria
Glycogen Synthesis	Cytoplasm
Glycogenolysis	Cytoplasm & some in Lysosomes
HMP Pathway	Cytoplasm
Pyruvate Dehydrogenase	Mitochondria
Krebs Cycle	Mitochondria

12. **b, c, e. Phospho.., Pyru…, Glyceral…** (Ref: Harper 29/e Page 171,172, Harper 31/e Page 160,161)

Irreversible Steps of Glycolysis[Q]

- Hexokinase
- Phosphofructokinase
- Pyruvate Kinase

Remember: All the Kinases are irreversible except 1,3 Bisphospholycerate Kinase which is reversible.

Substrate: Level Phosphorylation[Q]

- Phosphoglycerate kinase [1,3 Bisphospho Glycerate to 3 Phosphoglycerate]
- Pyruvate Kinase [Phosphoenol Pyruvate to Pyruvate] NB: Learn the enzyme and the reaction. Question can be asked in either ways.

13. **b, d, e. Eno…, Phos…, Glycer…** (Ref: Harper 29/e page 171, 172, Harper 31/e page 157-159)

14. **a. Phosphoenol pyruvate to pyruvate** (Ref: Harper 31/e page 158,159)

Steps releasing ATP at the level of substrate

1, 3 Bisphosphoglycerate to 3 Phosphoglycerate (1, 3 Bisphosphoglycerate Kinase.

Phospho enol Pyruvate to Pyruvate. (Pyruvate Kinase)

Succinyl CoA to Succinate (Succinate Thiokinase)

15. **c. Insulin** (Ref: Harper 31/e page 175 Table 19-1)

Regulation of Carbohydrate Metabolism (NB: This table is an important topic for all exams)

Enzyme	Inducer	Repressor	Activator	Inhibitor
Glycogen synthase	Insulin	Glucagon	Insulin, glucose 6-phosphate	Glucagon
Hexokinase		Glucagon		Glucose 6-phosphate
Glucokinase	Insulin	Glucagon		

Contd...

Self Assessment and Review of Biochemistry

Contd...

Enzyme	Inducer	Repressor	Activator	Inhibitor
Phosphofructokinase-1	Insulin	Glucagon[Q]	5'AMP, fructose 6-phosphate, fructose 2,6-bisphosphate[Q] Inorganic Phosphate	Citrate, ATP, glucagon
Pyruvate kinase	*Insulin*	Glucagon[Q]	Fructose 1,6-bisphosphate, insulin	ATP Alanine Glucagon norepinephrine
Pyruvate dehydrogenase	*Insulin*	Glucagon[Q]	CoA, NAD+, insulin[Q], ADP, pyruvate	Acetyl-CoA[Q], NADH, ATP (fatty acids, ketone bodies)

Gluconeogenesis				
Enzyme	Inducer	Repressor	Activator	Inhibitor
Pyruvate carboxylase	Glucocorticoids, Glucagon, Epinephrine	Insulin	Acetyl-CoA[Q]	ADP
Phosphoenolpyruvate carboxykinase	Glucocorticoids, Glucagon, Epinephrine	Insulin		
Glucose 6-phosphatase	Glucocorticoids, Glucagon, Epinephrine	Insulin		

16. **c. Conversion of Glucose to 3 C units** *(Ref: Harper 31/e page 157)*

    Option a. Glycolysis occur in cytosol

    Option b. Complete breakdown of Glucose happens when Pyruvate formed by Glycolysis undergo Pyruvate Dehydrogenase reaction, followed by TCA Cycle.

    Option d. In anaerobic Glycolysis
    - Number of ATPs produced is 4
    - Number of ATPs used is 2
    - Net ATP yield by anaerobic glycolysis is 2

17. **c. Glucose 6 Phosphate** *(Ref: Harper 31/e page 158,159)*

    Fates of Glucose 6 Phosphate
    - Can undergo Glycolysis
    - Can enter into Glycogenesis
    - Can be used for gluconeogenesis
    - Is an intermediate in Glycogenolysis
    - Can enter into HMP Pathway

18. **a,b,c. Phosphofruct..., Hexoki..., Pyruvate Kin...** *(Ref: Harper 31/e page 160,161)*

    Irreversible steps of Glycolysis are
    - Hexokinase/Glucokinase
    - Phosphofructokinase
    - Pyruvate Kinase

19. **d. Phosphofructokinase** *(Ref: Harper 31/e page 159)*
    - First Committed step is catalysed by Phosphofructo Kinase.
    - This is otherwise called the bottle neck of Glycolysis.

    Hexokinase/Glucokinase
    - Convert Glucose to Glucose 6 Phosphate.
    - Glucose 6 Phosphate has different fates, not only Glycolysis.

    Fates of Glucose 6 Phosphate
    - Can undergo glycolysis
    - Can enter into glycogenesis
    - Can be used for gluconeogenesis
    - Is an intermediate in Glycogenolysis
    - Can enter in to HMP Pathway

20. **a. Phosphofructokinase** *(Ref: Harper 31/e page 161, 176)*

    Regulatory steps of Glycolysis are
    - Hexokinase/Glucokinase
    - Phosphofructokinase
    - Pyruvate Kinase

    Harper says Phosphofructokinase occupy a key position in regulating Glycolysis and is also subject to feedback control.

21. **d. Aerobic glycolysis**
    (Ref: Harper 31/e page 695) (Harrison 19/e page 102 e-13)

    Many cancer cells use aerobic glycolysis (Warburg effect) to metabolize glucose leading to increased lactic acid production, where as normal cells use oxidative phosphorylation in mitochondria under aerobic conditions, a much more efficient process.

22. **a,b,e. Km value..., Found in liver, Glucose enter in to...** *(Ref: Harper 31/e page 158, 178-179)*
    - **Glucokinase is important in regulating blood Glucose after a meal**
    - Glucokinase has a considerably higher *Km* (lower affinity) for glucose, so that its activity increases
    - With increases in the concentration of glucose in the hepatic portal vein.
    - Found in liver cells and Pancreatic Beta islet cells.
    - Glucose 6 Phosphate inhibit Hexokinase but not Glucokinase
    - Glucose enter into liver cells and Pancreatic beta cells through GLUT2.

23. **c. Hypoxia stimulates release of all glycolytic enzymes from Band 3 on RBC membrane**

24. **d. Increase in Fructose 2,6 Bisphosphate**
(Ref: Harper 31/e page 178, 179)

On Decreasing Blood Glucose Level *Glucagon is released from β cells of Pancreas* **Increases the Blood Glucose Level by Inhibiting Glycolysis**

By phosphorylating the key enzymes of Glycolysis By decreasing the level of Fructose 2,6 Bisphosphate, the product of PFK-II

Favour Gluconeogenesis

By phosphorylating Key enzymes of Gluconeogenesis.
By favouring Fructose 2,6 Bisphosphatase, which decreases the level of Fructose 2,6 Bisphosphate, a potent activator of PFK-I.

25. **b. 2** *(Ref: Harper 31/e page 161)*

Rapaport Leubering cycle (2,3 BPG Cycle.takes place in the erythrocytes
- The reaction catalyzed by phosphoglycerate kinase may be bypassed
- 1,3-bisphosphoglycerate is converted to 2,3- bisphosphoglycerate by bisphosphoglycerate 2,3-bisphosphoglycerate
- 2,3 Bisphosphoglycerate is hydrolysed to 3-phosphoglycerate and P_i by 2,3-bisphosphoglycerate phosphatase mutase.
- No ATP is generated by this step.
- But 2 ATPs are generated by Pyruvate Kinase.
- As 2 ATPs are utilised by Hexokinase and PFK-1, No net ATPs are generated by this pathway.

26. **b. Mitochondria**
(Ref: Harper 29/e page 173; Harper 31/e page 160)
- Under aerobic conditions, pyruvate is taken up into mitochondria, and after oxidative decarboxylation to acetyl-CoA is oxidized to CO_2 by the citric acid cycle.
- Under anaerobic conditions, pyruvate is reduced by the NADH to lactate, catalyzed by lactate dehydrogenase.

27. **a. Pyruvate Kinase** *(Ref: Harper 30/e page 159)*

28. **b. Phosphofructokinase** *(Ref: Harper 31/e page 159)*

Gluconeogenesis, Fates of Pyruvate

29. **c. 6** *(Ref: Harper 31/e page 188)*

When lactate is converted to Glucose ATPs utilised are
- 2 for Pyruvate Carboxylase
- 2 for PEPCK
- 2 for 1,3 BPG Kinase

30. **c. Acetyl CoA** *(Ref: Harper 31/e page 188)*
- Allosteric Activator of Pyruvate Carboxylase is Acetyl CoA
- Allosteric inhibitor of Pyruvate Carboxylase is ADP

31. **b. Tyrosine**
- Tyrosine enter TCA cycle as Fumarate
- This is a must learn topic

Amino acid that enter as Pyruvate are
- Hydroxyproline
- Serine
- Cysteine
- Threonine
- Glycine

Amino acids that enter as Succinyl-CoA are
- Valine
- Isoleucine
- Methionine
- Threonine

Amino acid that enter as Glutamate to Alpha Ketoglutarate by transamination
- Histidine
- Proline
- Glutamine
- Arginine

Amino acid that enter as Fumarate
- Phenyl Alanine
- Tyrosine

Amino acid that enter as Alanine to Pyruvate is
- Tryptophan

32. **b. Acetyl-CoA stimulation of Pyruvate Carboxylase**
(Ref: Harper 31/e page 175)

Other options
- Citrate stimulation of Acetyl-CoA Carboxylase is for Fatty acid synthesis
- Fructose 2,6 Bisphosphate stimulation of PFK-1 favour Glycolysis
- Fructose 1,6 Bisphosphate stimulation of Pyruvate Kinase favour Glycolysis

33. **d. Thiamin** *(Ref: Harper 31/e page 163)*

Increased Pyruvate to Acetyl-CoA, with lactic acidosis, inability to convert Pyruvate to Acetyl-CoA are all suggestive of a Pyruvate Dehydrogenase deficiency. Thiamine Pyrophosphate is one of the coenzymes of PDH. Hence Thiamine is the answer.

34. **b. Acetyl-CoA** *(Ref: Harper 31/e page 172,173)*

Lactate from muscle and RBC are converted to glucose in the liver (Coris Cycle)
Lactate and Alanine is converted to Pyruvate which can enter into Gluconeogenesis.
Oxaloacetate is converted to Phospho enol Pyruvate by PEPCK and enter in to Gluconeogenesis.

35. **a, d, e. Prevent hypo….,Excess acetyl CoA…,Carbon skeleton of…**
- Gluconeogenesis occur in liver and Kidney.
- Fructose 2,6 Bisphosphate activate PFK-I ,hence stimulate Glycolysis.

36. **c. Glucose 6 Phosphatase** *(Ref: Harper 29/e page 179, 181)*
Table 19.2, 189, Harper 31/e page 166 Table 18.2)

The answer should be an enzyme common to Glycogenolysis and Gluconeogenesis.

37. **a. Acetyl CoA induced stimulation of Pyruvate Carboxylase**
(Ref: Harper 29/e page 190 Table 20-1, Harper 31/e page 175 Table 19.1)

Regulation of Carbohydrate Metabolism (NB: This table is an important topic for all exams)

Enzyme	Inducer	Repressor	Activator	Inhibitor
Glycogen synthase	Insulin	Glucagon	Insulin, glucose 6-phosphate	Glucagon
Hexokinase		Glucagon		Glucose 6-phosphate
Glucokinase	Insulin	Glucagon		
Phosphofructokinase-1	Insulin	Glucagon[Q]	5' AMP, fructose 6-phosphate, fructose 2,6-bisphosphate[Q], Pi	Citrate, ATP, glucagon
Pyruvate kinase	Insulin	Glucagon[Q]	Fructose 1,6-bisphosphate, insulin	ATP, alanine, glucagon, norepinephrine
Pyruvate dehydrogenase	Insulin	Glucagon[Q]	CoA, NAD+, insulin[Q], ADP, pyruvate	Acetyl-CoA[Q], NADH, ATP (fatty acids, ketone bodies)

Gluconeogenesis				
Enzyme	Inducer	Repressor	Activator	Inhibitor
Pyruvate carboxylase	Glucocorticoids, Glucagon, Epinephrine	Insulin	Acetyl-CoA[Q]	ADP
Phosphoenol pyruvate carboxy kinase	Glucocorticoids, Glucagon, Epinephrine	Insulin		
Glucose 6-phosphatase	Glucocorticoids, Glucagon, Epinephrine	Insulin		

38. **b. Alanine in the liver**
    (Ref: Harpers 29/e page 160, Harper 31/e page 172)

    Major Substrates for Gluconeogenesis[Q (AI 97)]
    - Glucogenic Amino Acid [Alanine[Q] is the major contributor]
    - Lactate
    - Glycerol

    Propionate (Major contributor in Ruminants)

39. **a, d, e. Prevent Hypo…, Excess Acetyl CoA…, Carbon skeleton of…** *(Ref: Harper 29/e page 187-191 Harper 31/e page 172-175)*
    - Gluconeogenesis occur in liver and kidney not in the muscle.
    - Gluconeogenesis prevent hypoglycemia in prolonged fasting.
    - Fructose 2,6 Bisphosphate stimulate Glycolysis
    - Excess Acetyl CoA is an allosteric activator of Pyruvate Carboxylase, a key enzyme of Gluconeogenesis.
    - Carbon skeleton of gluconeogenic amino acid are involved in gluconeogenesis.

40. **a. Glyceraldehyde 3 Phosphate dehydrogenase**
    - An enzyme that catalyse a reversible step in Glycolysis is common to both Glycolysis and Gluconeogenesis.

41. **d. Citrate** *(Ref: Harper 31/e page 175 Table 19-1)*

Enzyme	Inducer	Repressor	Activator	Inhibitor
Hexokinase		Glucagon		Glucose 6-phosphate
Glucokinase	Insulin	Glucagon		
Phosphofructokinase-1	Insulin	Glucagon	5' AMP, fructose 6-phosphate, fructose 2,6-bisphosphate, Inorganic Phosphate	Citrate, ATP, glucagon
Pyruvate kinase	Insulin	Glucagon	Fructose 1,6-bisphosphate, insulin	ATP, alanine, glucagon, norepinephrine

42. **a. Acetyl-CoA** *(Ref: Harper 29/e page 172)*

    Substrates for Gluconeogenesis[Q]
    - Glucogenic Amino Acid [Alanine[Q] is the major contributor]
    - Lactate
    - Glycerol
    - Propionyl-CoA

43. **a. Gluconeogenesis** *(Ref: Harper 31/e page 172)*
    - Gluconeogenesis takes place in cytosol and mitochondria.
    - The Mitochondrial step is Pyruvate Carboxylase reaction

    Pathways taking place in two compartments are
    - Heme Synthesis
    - Urea Cycle
    - Gluconeogenesis

44. **b, c, d. Glucose 6 Pho…, Pyruvate Carbo…., Fructose 1,6 Bis…,** *(Ref: Harper 31/e page 172-174)*

- Glyconeogenic capacity is determined by the presence of key enzymes of Gluconeogenesis.
- Gluconeogenesis is the process of synthesizing glucose or glycogen from noncarbohydrate precursors
- Pyruvate Dehydrogenase and Pyruvate carboxykinase are not enzymes of gluconeogenesis.

45. **d. Oxaloacetate to PEP** (Ref: Harper 31/e page 172-174)
- Pyruvate to lactate is a step in Anaerobic Glycolysis
- Glucose 6 Phosphate to Fructose 6 Phosphate is a step in Glycolysis
- Pyruvate to Acetyl Co A is a step in aerobic oxidation of Glucose.
- Oxaloacetate to PEP catalysed by PEPCK is a step in gluconeogenesis.

46. **d. Alanine** (Ref: Harper 31/e page 172)

Alanine is the principal gluconeogenic amino acid

47. **c. Acetoacetate** (Ref: Harper 31/e page 172)

Substrates for Gluconeogenesis
- Glucogenic amino acids
- Lactate
- Glycerol

Acetyl-CoA, Acetoacetate are not substrates for gluconeogenesis

48. **b. Fatty acids** (Ref: Harper 31/e page 172)

Glycerol part of fat and Propionyl-CoA from odd chain fatty acid oxidation are Gluconeogenic part of fat.

49. **a. Glucose** (Ref: Harper 31/e page...)
- Acetyl-CoA is NEVER a substrate for gluconeogenesis
- Acetyl-CoA is the starting material for Fatty acid and Cholesterol synthesis.
- Acetyl-CoA is an intermediate in Ketone body synthesis.

50. **a. Level of fructose 1,6-biphosphate is higher than normal**
- The action of Fructose 2,6 Bisphosphate on Fructose 1,6 Bisphosphatase is decreasing its activity.
- Here as the control of Fructose 2,6 BP on Fructose 1,6 Bisphosphatase is lost.
- The enzyme is more active.
- So the level of Fructose 1,6 Bisphosphate is lower than normal.
- Less Pyruvate as more gluconeogensis
- Less ATP as gluconeogenesis utilise ATPs

Glycogen Metabolism and Glycogen Storage Disorders

51. **b. Phosphofructokinase I**

NB: If Muscle phosphorylase is there then it's the best answer/ as McArdles is MC GSD in adolescent age. if its not there then PFK-1 is the answer

52. **b. Cori's disease** (Ref: Harper 31/e page 167)

Accumulating substances in
- Gaucher's disease-Glucocerebroside
- Cori's disease-Limit dextrin
- Anderson disease-Amylopectin

53. **a, b. Glycogen synthesis, Galactose metabolism** (Ref: Harper 31/e page 321)

UDP Glucose is used for sugar interconversions like galactose metabolism and Glycogen synthesis

54. **c. Von Gierke's Disease**
(Ref: Nelson 20/e page 715, Defects in metabolism of Carbohydrates)

Clinical Picture of Von Gierke's Disaese Clinical presentation
- Most commonly present at 3-4 months of age
- Doll like facies with fat cheeks
- Relatively thin extremities
- Short stature, Protuberant abdomen
- Massive Hepatomegaly
- Kidneys are also enlarged
- No Splenomegaly
- Plasma may be milky due to associated hypertrigly-ceridemia

55. **a. Pyridoxal Phosphate** (Ref: Harper 31/e page 166)

80% of PLP is in the muscle .It is a coenzyme for Glycogen Phosphorylase.

56. **a. UDP Glucose** (Ref: Harper 31/e page 164)
- Glycogenin, a 37 kDa protein is glucosylated on specific tyrosine residue by UDP glucose.
- Glycogenenin catalyses transfer of 7 glucose residue from UDP-Glc, in 1→4 linkage to form Glycogen primer.
- Further Glucose on glycogen primer are added by Glycogen Synthase to nonreducing end till growing chain is at least 11 glucose residue long.

57. **a. Debranching enzyme**
(Ref: Nelson 20e Chapter 715 Defects in metabolism of Carbohydrates)

Cori's Disease (Type III Glycogen Storage Disorder)
Enzyme deficient is Debranching enzyme

Clinical Features
- Fasting Hypoglycaemia, Hepatomegaly, hyperlipidemia, short stature, **variable skeletal muscle myopathy**
- Kidneys are not enlarged
- Splenomegaly may be present.
- **Elevation of liver transaminase**
- **Fasting Ketosis**
- **Blood lactate and Uric acid level is normal.**
- **Progressive liver cirrhosis and failure occurs.**
- Remarkably, hepatomegaly and hepatic symptoms in most patients with type III GSD improve with age and usually resolve after puberty.

- The administration of glucagon 2 hr after a carbohydrate meal provokes a normal increase in blood glucose
- After an overnight fast, glucagon may provoke no change in blood glucose level.

58. c. Enzyme is present in SER, Glucose 6 phosphate need to be transported in to SER *(Ref: Harper 31/e page 166)*

- Glucose-6-phosphatase is in the lumen of the smooth endoplasmic reticulum, Glucose 6 Phosphate is transported to SER by a transporter called translocase to be acted by Glucose 6 Phosphatase.
- Genetic defects of the glucose-6-phosphate transporter can cause a variant of type I glycogen storage disease

Other options
- Glycogenolysis provide blood glucose before Gluconeogenesis sets in fasting state.
- Glucose 6 Phosphatase does not need Protein Kinase, but Glycogen Phosphorylase need Protein Kinase for its activation.

59. d. Fat mobilisation is low

Glucose 6 Phosphatase deficiency leading to hypoglycemia. Glucose 6 Phosphate converted to Pyruvate. This is converted to Acetyl-CoA. As oxaloacetate is depleted because of using up of Oxaloacetate for Gluconeogenesis, Acetyl-CoA enter into Ketone body Synthesis. Hence Ketosis.

Glucose 6 Phosphate also enter into HMP shunt pathway leads to more production of Pentoses. Therefore more purine synthesis. Purines degraded to Uric Acid. Hence there is Hyperuricemia.
As there is hypoglycemia, *fat is mobilised*. This also leads to more Acetyl-CoA by Fatty acid oxidation. This increases Ketone body synthesis. Hence Ketosis.

60. c. Glucose 6 Phosphatase *(Ref: Harper 29/e page 179, 181 Table 19.2, 189 ,Harper 31/e page 167)*

The answer should be an enzyme common to Glycogenolysis and Gluconeogenesis.

Glycogen Storage Disorders
Inborn errors of metabolism of Glycogen associated with accumulation or altered function of Glycogen in various organs concerned with Glycogen metabolism.

Liver Glycogen Storage Disorder

Type	Name	Enzyme efficiency	Characteristics
Ia	Von Gierke's disease	Glucose 6-phosphatase^Q	Glycogen accumulation in liver and renal tubule cells (Kidney Enlarged. hypoglycemia; elevated blood lactate, cholesterol, triglyceride, and uric acid levels
Ib	—	Endoplasmic reticulum glucose 6-phosphate transporter/translocase	Same as type Ia, with additional findings of neutropenia and impaired neutrophil function. Recurrent Bacterial Infection, Inflammatory Bowel Disease

61. b. Muscle *(Ref: Harper 29/e page 180, Harper 31/e page 164)*

Differences between liver Glycogen and Muscle Glycogen

Features	Liver	Muscle
Total Body Glycogen content	Less	Highest
Percentage by tissue weight	Highest	Less
Regulation of blood glucose	Contributes to blood Glucose	Does not contribute to blood Glucose.

62. b. Glycogen *(Ref: Harper 30/e page 164)*

- Glycogen is the storage polysaccharide in animals and is sometimes called animal starch.

63. a. Epinephrine *(Ref: Harper 31/e page 166,167)*

Differences between Muscle and Liver in the regulation of Glycogen Metabolism
- Epinephrine acts in Muscle and Liver whereas Glucagon acts only in the Liver.
- In the muscle there is cAMP-independent activation of glycogenolysis^Q
  - By the stimulation of a Ca^{2+}/calmodulin-sensitive phosphorylase kinase

  - Phosphorylate Glycogen Phosphorylase in the muscle.
  - Favour glycogenolysis.
- Muscle phosphorylase can be activated without phosphorylation.
  - Muscle Phosphorylase has a binding site for 5'AMP.
  - 5'AMP is an allosteric activator without phosphorylation.
  - Favour Glycogenolysis.

64. b. Hydrolyses Starch to limit dextrins *(Ref: Harper 31/e page 520)*

65. b. Glycogen Storage Disorder *(Ref: Nelson 20e Chapter 715 Defects in metabolism of Carbihydrates)*

- Patients with type I GSD may present in the neonatal period with hypoglycemia and lactic acidosis
- These children often have doll-like faces with fat cheeks, relatively thin extremities, short stature, and a protuberant abdomen that is due to massive hepatomegaly; the kidneys are also enlarged, whereas the spleen and heart are normal.
- The biochemical hallmarks of the Type Ia GSD (Von Gierke's) disease are hypoglycemia, lactic acidosis, hyperuricemia, and hyperlipidemia

66. d. Glycogenin *(Ref: Harper 31/e page 167,168)*

Regulation of Glycogen metabolism at Glycogen Phosphorylase
- cAMP activates Glycogen Phosphorylase by cAMP dependent Protein Kinase
- cAMP independent Calcium /Calmodulin sensitive Phosphorylase Kinase also activates Glycogen Phosphorylase.

Glycogenin is a protein on which initial glucose is added in the synthesis of Glycogen

67. b. Pyridoxal Phosphate *(Ref: Harper 31/e page 166)*

68. c. Acid Maltase *(Ref: Harper 31/e page 166)*

	Muscle glycogen storage disorder	
II	Pompes Disease (Belongs to lysosomal storage disorder)	Lysosomal α1, 4 and α 1, 6 glucosidase (acid maltase^Q)
	Danon disease	Lysosome-associated membrane protein 2 (LAMP2)
V	McArdle's syndrome	Muscle phosphorylase^Q
VII	Tarui's disease	Muscle and erythrocyte phosphofructokinase 1

69. d,e. Pompes disease, McCardles disease *(Ref: Nelson 20/e page 715 Chapter Defects in the metabolism of Carbohydrates)*

70. b. 18 hours *(Ref: Harper 31/e page 149)*
- Liver and Muscle Glycogen exhausted by 18 hours of fasting.

71. d. Stored as Glycogen *(Ref: Harper 31/e page 136,137)*
- The uptake of glucose into the liver GLUT 2 is independent of insulin.
- In the well fed state, the concentration of glucose entering the liver increases, so does the rate of synthesis of glucose-6-phosphate.
- This is in excess of the liver's requirement for energy-yielding metabolism. So it is used mainly for synthesis of **glycogen**.
- In both liver and skeletal muscle, insulin acts to stimulate glycogen synthetase and inhibit glycogen phosphorylase.

72. c. Amylo 1, 6 Glucosidase *(Ref: Harper 31/e page 166)*

73. a. Phosphorylase, Glucan transferase, debranching, phosphorylase *(Ref: Harper 31/e page 166)*

Steps of Glycogenolysis
Breaking of α 1,4 linkage
- Glycogen Phosphorylase cleave the α **1,4 linkage**.
- Glycogen Phosphorylase stops its action when it is at least 4 glucose residues from a branch point.

Removal of Branches
By a bifunctional enzyme:
- First part is a α-1, 4 α 1,4 Glucan transferase
  Transfer trisaccharide residue to another forming a new α 1, 4 linkage.
- Second part is a α1, 6 Glucosidase (Amylo 1,6 Glucosidase. Hydrolyse the branching point.

Conversion of Glucose 1 Phosphate to free Glucose
- Glucose 1 Phosphate to Glucose 6 Phosphate by Phospho glucomutase
- Glucose 6 Phosphate to Glucose by Glucose 6 Phosphatase.

Defect in the Glucose 6 Phosphate transporter lead to Type Ib Glycogen Storage disorder

74. a. I *(Ref: Nelson 20/e Defects in the metabolism of Carbohydrates)*

Glycogen Storage Disorder II (Limit dextrinoses) and GSD IV (Anderson disease. are liver GSD with muscle involvement.

75. b. Von Gierkes Disease
(Ref: Nelson 20/e Defects in the metabolism of Carbohydrates)

76. b,d,e. Fabrys…,Fragile X…,Krabbes …
(Ref: Nelson 20/e Defects in the metabolism of Carbohydrates)
- Fabrys Disease and Krabbes Disease are Sphingolipidoses.

77. a. More formation of Pentoses
(Ref: Nelson 20/e Defects in the metabolism of Carbohydrates)

Biochemical defect in Von Gierkes

Glucose 6 Phosphatase defect hence Glucose 6 Phosphate accumulate. It is chanelled to HMP Pathway. Hence more pentoses formed. These are shunted to Purine synthesis. Hence Hyperuricemia.

4.3 MINOR METABOLIC PATHWAYS OF CARBOHYDRATES

- ☞ HMP Shunt Pathway
- ☞ Galactose Metabolism Polyol Pathway
- ☞ Uronic Acid Pathway
- ☞ Fructose Metabolism
- ☞ Glucose Tolerance Curves

HMP SHUNT PATHWAY

- **Other names:** Pentose Phosphate Pathway, Dicken Horecker Pathway, Phosphogluconate Pathway.
- Organelle—**Cytosol**.
- Rate limiting step-**Glucose 6 Phosphate Dehydrogenase**.

Biochemical Significances of HMP Pathway

- Alternative route for metabolism of Glucose.
- **Complete oxidation of Glucose.**
- More complex pathway than Glycolysis.
- Major function is to generate **NADPH and Riboses**.
- **No ATP is generated by this pathway.**

HMP Shunt Pathway has Two Phases

Three molecules of Glucose 6 phosphate give rise **to three molecules of CO_2** and three 5 Carbon sugars. They are rearranged to regenerate two molecules of Glucose 6 Phosphate and one molecule of Glyceraldehyde 3 Phosphate. This is taking place in two phases.

1. Oxidative phase
2. Nonoxidative phase.

Characteristics of the two Phases of HMP Shunt Pathway

Oxidative Phase—Characteristics and Organs

Sites: Liver, adipose tissue, adrenal cortex, erythrocytes, gonads, thyroid, lactating mammary glands (Not in non-lactating mammary glands)

- Irreversible
- Takes place in sites where NADPH is required.
- Glucose 6 Phosphate undergoes **dehydrogenation and decarboxylation** to Ribulose 5 Phosphate.
- **Oxidation is achieved by dehydrogenation using NADP+, not NAD+, as the hydrogen acceptor.**
- **Produce NADPH**^{Q}

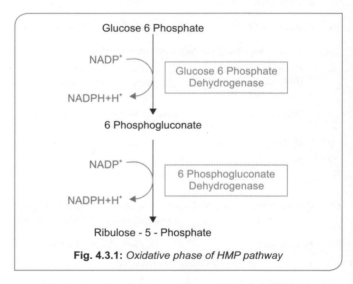

Fig. 4.3.1: *Oxidative phase of HMP pathway*

HIGH YIELD POINTS

PATHWAYS THAT REQUIRE NADPH^Q

Detoxification
- Reduction of oxidized glutathione
- Cytochrome P450 mono-oxygenases

Reductive Synthesis
- Fatty acid synthesis
- Fatty acid chain elongation
- Cholesterol synthesis
- Steroid synthesis
- Deoxyribonucleotide synthesis

Nonoxidative Phase Regenerate Glucose 6 Phosphate

Sites: In rapidly dividing cells, bone marrow, skin, intestinal mucosa and virtually in all tissues.

- Reversible
- **Produce Pentoses**
- Ribulose 5-phosphate is converted back to glucose 6-phosphate mainly by **two transketolase reactions and one transaldolase.**

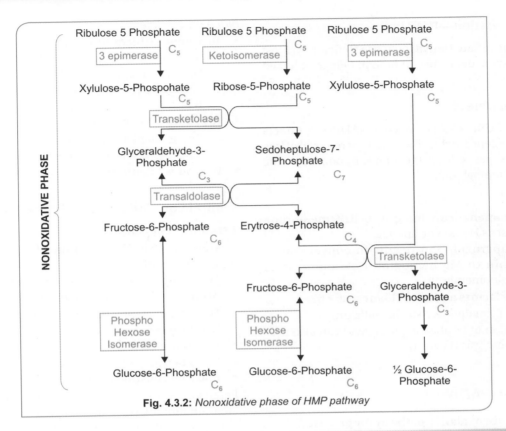

Fig. 4.3.2: *Nonoxidative phase of HMP pathway*

Transketolase

Two Transketolase Reactions

- Transfers the two-carbon unit comprising carbons 1 and 2 of a ketose on to the aldehyde carbon of an aldose sugar.
- It therefore affects the conversion of a ketose sugar into an aldose with two carbons less and an aldose sugar into a ketose with two carbons more.
- Require **thiamine as coenzyme**.
- Erythrocyte transketolase[Q] is a measure thiamine status of the body.

Transaldolase

One Transaldolase Reaction

Transfer the three carbon unit of a 7 carbon ketosugar (sedoheptulose 7 phosphate) to a 3 C aldosugar (Glyceraldehyde 3 Phosphate) to form a 6 carbon ketosugar (Fructose 6 Phosphate) and 4 carbon aldosugar.

No cofactor for this enzyme.

 HIGH YIELD POINTS

METABOLITES IN HMP SHUNT PATHWAY
- Glucose 6 Phosphate
- 6 Phosphogluconate
- Ribulose 5 Phosphate
- Xylulose 5 Phosphate
- Ribose 5 Phosphate
- Glyceraldehyde 3 Phosphate
- Sedoheptulose 7 Phosphate
- Fructose 6 phosphate
- Erythrose 4 Phosphate

 HIGH YIELD POINTS

KEY POINTS TO REMEMBER IN HMP (PENTOSE PHOSPHATE) PATHWAY
- Main source of NADPH[Q2012] and Pentoses.
- **Two transketolation and one transaldolation** reactions are involved.
- **No ATP is produced**[Q].
- CO_2 is produced in this pathway and not in glycolytic pathway[Q].
- Deficiency of Glucose 6 phosphate dehydrogenase is a major cause of acute hemolysis in erythrocytes.
- NADPH is used for **reductive biosynthetic pathways,** like fatty acid synthesis, steroid synthesis, Amino acids by Glutamate dehydrogenase.
- NADPH is required for regeneration of reduced Glutathione, that clears free radicals from erythrocytes and lens.
- Wernicke's Korsakoff's syndrome is exacerbated by defect in **trans ketolase**[Q].

Clinical Correlation—HMP Pathway

Glucose 6 Phosphate Dehydrogenase Deficiency: Most common enzyme deficiency in human beings. X-linked recessive.

Biochemical Defect

Deficiency of G6PD → Decreased NADPH → Affects Glutathione defence system necessary for removal of free radicals (esp Superoxide ion) generated by nonenzymatic oxidation of Hemoglobin.

This results in:

- **Loss of maintenance of integrity of RBC membrane** → manifest as Hemolytic anemia
- Excess Superoxide ion results in conversion of hemoglobin to Methemoglobin → manifest as methemoglobinemia

Precipitating factors are any oxidant stress like
1. Drugs like Primaquine, Aspirin, Sulfa drugs
2. Consumption of favabeans (vicia fava) can also precipitate hemolysis.(Favism)
3. Infections.

URONIC ACID PATHWAY

- An alternative **oxidative pathway for glucose**
- UDP Glucose converted to UDP Glucuronic Acid (Glucuronate) by using NAD
- No ATP is formed
- Site—liver
- Organelles—cytosol.

Biochemical Role

- **To produce Glucuronic Acid**
  - Source of Glucuronate for the synthesis of Glycosaminoglycans and Proteoglycans.
  - For Phase II Conjugation reaction (Conjugation of Bilirubin and many drugs).
- **To produce Ascorbic Acid**
  - But not in higher primates and mammals
  - Because they lack **L Gulonolactone Oxidase**^{Q}.
- **To Produce Pentoses**.

Clinical Correlation—Uronic Acid Pathway

Essential Pentosuria

- **Defect in Uronic Acid Pathway**^{Q}
- Benign condition
- One among the **Garrod's tetrad (Pentosuria, Albinism, Cystinuria, Alkaptonuria)**
- Due to deficiency of **Xylitol Dehydrogenase or D Xylulose reductase**
- **L Xylulose** excreted in urine
- Gives Benedict's test positive
- **Bial's test positive.**

Pentosuria also occurs after consumption of relatively large amounts of fruits such as pears that are rich sources of pentoses (**alimentary pentosuria**).

POLYOL/SORBITOL PATHWAY

- To produce **Fructose from Glucose**.
- Responsible for occurrence of Fructose in seminal fluid.
- Sorbitol is responsible for Diabetic cataract because, it cannot pass through cell membrane so accumulate, causing osmotic damage.

Pathway

- Glucose is reduced to **Sorbitol by Aldose Reductase.**
- Sorbitol is oxidized to Fructose by Sorbitol dehydrogenase.

METABOLISM OF GALACTOSE

Galactose is seen in:

- Lactose
- Glycolipids
- Glycoproteins
- GAG (Keratan Sulphate).

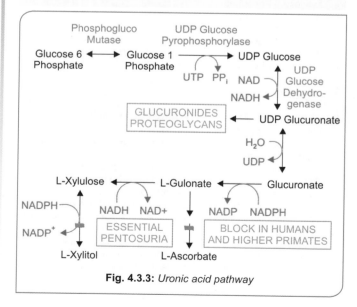

Fig. 4.3.3: Uronic acid pathway

Carbohydrates

Galactose is derived from intestinal hydrolysis of the disaccharide **lactose,** the sugar of milk.

It is readily converted in **the liver to glucose.**

Conversion of Galactose to Glucose

Rate limiting Step—**Galactose 1 Phosphate uridyl Transferase.**

Site: Galactose is metabolized **exclusively in the Liver**, hence **Galactose Tolerance Test is a Liver Function Test.**[Q]

Pathway

- Galactokinase catalyzes the phosphorylation of galactose, using ATP as phosphate donor.
- Galactose 1-phosphate reacts with UDPGlc to form uridine diphosphate galactose (UDP Gal) and glucose 1-phosphate, in a reaction catalyzed by galactose 1-phosphate uridyl transferase.
- The conversion of UDPGal to UDPGlc is catalyzed by UDPGal 4-epimerase.
- The UDPGlc is then incorporated into glycogen.
- In the synthesis of lactose in the mammary gland, UDPGal condenses with glucose to yield lactose, catalyzed by lactose synthase.

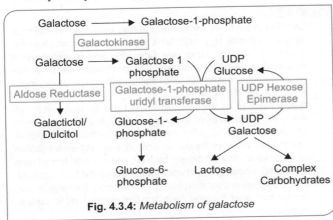

Fig. 4.3.4: *Metabolism of galactose*

Galactosemia[Q]

Biochemical Defect

Enzyme deficiencies are:
- Galactose 1 Phosphate uridyl transferase (Classic Galactosemia)
- Galactokinase
- UDP Hexose 4 Epimerase.

Galactose-1-phosphate, the accumulation of which results in injury to kidney, liver, and brain.

Symptom Box

Clinical Features
- Classic galactosemia is a serious disease with onset of symptoms typically by **the 2nd half of the 1st week of life.**
- With jaundice, vomiting, seizures, lethargy, irritability, feeding difficulties, poor weight gain or failure to regain birth weight.
- Hepatomegaly **Oil drop cataracts**, Hepatic failure.

- Mental retardation.
- Patients with galactosemia are at increased risk for *Escherichia coli* **neonatal sepsis.**

Diagnosis

- Demonstrating a reducing substance in several urine specimens collected while the patient is receiving human milk, cow's milk, or any other formula containing lactose.
- **Benedict's Test Positive**[Q].
- Chromatography for presence of Galactose in urine.
- Clinistix urine test results are usually negative because the test materials rely on the action of glucose oxidase, which is specific for glucose and is nonreactive with galactose (**Glucose Oxidase Test is negative**).
- **Mucic Acid Test Positive**.
- Galactose Tolerance Test is contraindicated.
- Direct enzyme assay using erythrocytes establishes the diagnosis.

Treatment

- **Lactose free diet till 4–5 years of age.**
- Because **Galactose 1 Phosphate Pyrophosphorylase becomes active by 4–5 years.** This enzyme reduces the level of Galactose 1 Phosphate.
- *Breast milk is avoided*[Q].

METABOLISM OF FRUCTOSE

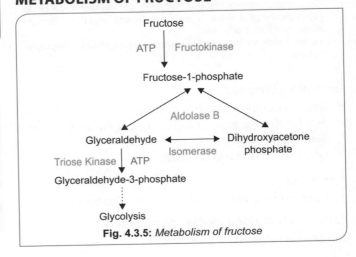

Fig. 4.3.5: *Metabolism of fructose*

Key Points in the Metabolism of Fructose

Fructose undergoes more rapid glycolysis in the liver than does glucose, because it bypasses the regulatory step catalyzed by phosphofructokinase.

This allows fructose to flood the pathways in the liver, leading to **enhanced fatty acid synthesis**, increased esterification of fatty acids, and increased VLDL secretion, which may raise serum triacylglycerols and ultimately raise LDL cholesterol concentrations.

Fructokinase, in liver, kidney, and intestine, catalyzes the phosphorylation of fructose to fructose 1-phosphate.

Unlike glucokinase, its activity is not affected by fasting or by insulin, which may explain why fructose is **cleared from the blood of diabetic patients** at a normal rate.

Fructose 1-phosphate is cleaved to D-glyceraldehyde and dihydroxyacetone phosphate by aldolase B.

Clinical Correlations

Hereditary Fructose Intolerance^Q
- Autosomal recessive.

Biochemical Defect
- Deficiency **of Fructose-1, 6-Bisphosphate Aldolase (Aldolase B)**
- Deficiency of this enzyme activity causes a rapid **accumulation of fructose-1-phosphate** and initiates severe toxic symptoms when exposed to fructose.

Symptom Box

Clinical Manifestations
- Patients with HFI are asymptomatic until fructose or sucrose (table sugar) is ingested (usually from fruit, fruit juice, or sweetened cereal).
- Symptoms may occur **early in life, soon after birth if foods or formulas containing these sugars** are introduced into the diet.
- Early clinical manifestations resemble galactosemia and include jaundice, hepatomegaly, vomiting, lethargy, irritability, convulsions, and hypoglycemia.
- **Acute fructose ingestion produces symptomatic hypoglycemia.**

Laboratory Diagnosis
- Urine reducing substance positive.
- Test for Ketose sugar positive (Rapid Furfural test and Seliwanoff's test)
- Glucose oxidase test is negative

Treatment
- Complete exclusion of all sources of Fructose.

Essential Fructosuria (Benign Fructosuria)
Autosomal recessive.

Benign Condition

Biochemical Defect
- **Fructokinase defect**
- Fructose increased in the blood, **is excreted in urine because there is practically no renal threshold for fructose.**

Laboratory Diagnosis
- **Rapid Furfural Test and Seliwanoff's Test Positive.**
- Clinistix for reducing sugars positive or **Benedict's test is positive in urine.**
- Urine chromatography for fructose.

Frequently Asked Doubts

Why Fructokinase defect is called as Fructosuria NOT Fructosemia?
- Some of ingested fructose is slowly metabolized by Hexokinase in nonhepatic tissue and metabolized by glycolysis, Fructose level need to be elevated always.
- No renal threshold for Fructose, so appearance of fructose in urine does not need a high fructose concentration in blood.

Extra Edge

- **Loading of the Liver with Fructose may potentiate Hyper triacylglycerolemia, Hypercholesterolemia, and Hyperuricemia.**
- In the liver, fructose **increases fatty acid and triacylglycerol synthesis and VLDL secretion**, leading to hypertriacylglycerolemia—and increased LDL cholesterol—which can be regarded as potentially atherogenic.
- **Acute loading of the liver with fructose,** as can occur with intravenous infusion or following very high fructose intakes, causes sequestration of inorganic phosphate in fructose **diminished ATP synthesis**. As a result, there is **less inhibition of de novo purine synthesis by ATP**, and uric acid formation is increased, **causing hyperuricemia**, which is the cause of gout.
- Since fructose is absorbed from the small intestine by (passive) carrier-mediated diffusion, high oral doses may lead to osmotic diarrhea

GLUCOSE TOLERANCE CURVES

Oral Glucose Tolerance Tests

Glucose Load
- Adults—75 g anhydrous glucose (82.5 g of glucose monohydrate) in 250–300 mL of water
- Children—1.75 g/kg body weight.

Samples Taken

Classical GTT
- Fasting Urine and blood sample called "0" hr sample.
- Then ½ hrly sample for next 2½ hrs.

Modern GTT (WHO recommendation) called Mini GTT
- Only 2 samples
- Fasting and 2 hr post-glucose load urine and blood samples.

Glucose Challenge Test
- Only 2-hour blood post-glucose load value is taken.

Glucose Tolerance Test in Pregnancy
- Three blood samples are taken, Fasting, 1 hour and 2 hours.

Plasma Sugar Value in Normal and Diabetes and Impaired Glucose Tolerance

	Normal	Diabetes mellitus	Impaired glucose tolerance
Fasting	<100 mg/dL (<5.6 mmol/L)	>126 mg/dL >7 mmol/dL	110–125 mg/dL
1 hr post-glucose load	<160 mg/dL <9 mmol/L	—	—
2 hr post-glucose load	<140 mg/dL <7.8 mmol/L	>200 mg/dL >11.1 mmol/L	140–199 mg/dL
HbA1C	<5.6%	>6.5%	5.6–6.4%

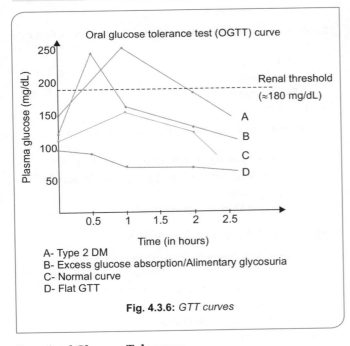

A- Type 2 DM
B- Excess glucose absorption/Alimentary glycosuria
C- Normal curve
D- Flat GTT

Fig. 4.3.6: *GTT curves*

Impaired Glucose Tolerance
- Blood sugar value above normal but less than diabetic level.

Impaired Fasting Glycemia
- Fasting blood sugar is above normal but less than diabetic level.

Alimentary Glycosuria or Excessive Uptake of Glucose
- Fasting and 2 hour value is normal. But excessive glucose uptake just after ingestion of glucose. Urine sugar is also positive at that time.

Renal Glucosuria
- Blood glucose is normal but urine sugar is positive as renal threshold is lowered (Normal Renal threshold = 170–180 mg/dL). SGLT-2 is affected.

Quick Review

- Pentose phosphate pathway forms NADPH and pentoses.
- Oxidative phase generate NADPH
- Nonoxidative phase generate pentoses.
- Oxidative phase utilises NADP+ but glycolysis utilises NAD+
- CO_2 is produced in HMP pathway
- No ATP generated by pentose phosphate pathway.
- The pentose phosphate pathway and glutathione peroxidise scavenges free radicals.
- Uronic acid pathway produces glucuronic acid.
- Ascorbic acid cannot be synthesised in humans due to lack of L-Gulanolactone oxidase.
- Most common enzyme deficiency in humans is Glucose 6 phosphate dehydrogenase.
- Metabolic disorders and enzyme deficiency.

Metabolic disorder	Enzyme defect
Essential fructosuria	Fructokinase
Essential pentosuria	Xylulose reductase
Hereditary fructose intolerance	Aldolase B
Classic Galactosemia	Galactose 1 Phosphate Uridyl transferase
Galactosemia	UDP hexose 4 epimerase Galactokinase

Check List for Revision

- All boxes and tables in HMP pathway
- Uronic acic pathway all boxes and learn functions
- Polyol pathway important points in boxes
- Metabolic disorders of galactose
- Metabolic disorders of fructose
- Starve feed cycle
- Metabolic fuels
- GTT as image-based question.

Review Questions MCQ

HMP Shunt Pathway

1. **True about HMP shunt:** *(PGI May 2017)*
   a. Takes place in cytosol
   b. Does not produce ATP
   c. NADH is produced in oxidative phase
   d. Found in liver, adipose tissue, gonads
   e. Pyruvate is produced in nonoxidative phase

2. **Metabolites in HMP shunt are all except:** *(AI 95)*
   a. Glycerol-3-phosphate
   b. Sedoheptulose-7-phosphate
   c. Glyceraldehyde-3-phosphate
   d. Xylulose-5-phosphate

3. **NADPH is produced by:** *(AIIMS Nov 2003)*
   a. Glycolysis
   b. Citric acid cycle
   c. HMP shunt
   d. Glycogenesis

4. **Reduced NADPH produced from which pathway?**
   a. Krebs cycle *(Recent Question)*
   b. Anaerobic glycolysis
   c. Uronic acid pathway
   d. Hexose monophosphate pathway

5. **Which of the following metabolic pathways does not generate ATP?** *(AI-2008)*
   a. Glycolysis
   b. TCA cycle
   c. Fatty acid oxidation
   d. HMP pathway

6. **Severe thiamine deficiency is associated with:** *(JIPMER 2013)*
   a. Decreased RBC transketolase activity
   b. Increased clotting time
   c. Decreased RBC transaminase activity
   d. Increased xanthic acid excretion

Other Metabolic Pathways of Glucose

7. **Vitamin C cannot be produced in humans due to lack of:** *(AIIMS May 2018)*
   a. L Gulonolactone oxidase
   b. Phosphoglucomutase
   c. UDPGlc dehydrogenase
   d. Glucose 6 Phosphate Dehydrogenase

8. **Product of uronic acid pathway in human-beings are all except:** *(Recent Qestion)*
   a. Vitamin C
   b. Glucuronic acid
   c. Pentoses
   d. NADH

9. **Uronic acid pathway is not involved in:** *(Recent Question)*
   a. Conjugation of bilirubin
   b. GAG synthesis
   c. Vitamin C synthesis
   d. Biotransformation

Fructose Metabolism and Disorders, Metabolism of Amino Sugars

10. **A baby boy 10-month-old comes with vomiting severe jaundice, hepatomegaly and features of irritability on starting weaning with fruit juice. Which of the following enzymes is defective?** *(JIPMER May 2016)*
    a. Aldolase B
    b. Fructokinase
    c. Glucose 6 phosphatase
    d. Galactose 1 Phosphate Uridyl transferase

11. **Fate of Fructose 6 Phosphate:** *(Recent Question)*
    a. Glucuronic acid
    b. N-Acetyl glucosamine
    c. Hyaluronic acid
    d. Heparan sulphate

12. **Hereditary fructose intolerance is due to deficiency of:** *(JIPMER 2014)*
    a. Aldolase B
    b. Aldolase A
    c. Fructokinase
    d. Sucrase

13. **False about hereditary fructose intolerance:** *(AIIMS June 98)*
    a. Deficiency of fructose 1-phosphate aldolase
    b. Accumulation of fructose 1-phosphate in tissues
    c. Hyperglycaemia
    d. Liver and kidneys are involved

Galactose Metabolism and Disorders

14. **Enzyme deficiency in Galactosemia:** *(Recent Question Nov 2017)*
    a. Galactose 1 phosphate uridyl transferase
    b. Aldolase B
    c. UDP galactose 4 epimerase
    d. Fructokinase

15. **E. coli sepsis commonly seen in:** *(Recent Question)*
    a. Urea cycle disorder
    b. Galactosemia
    c. Glycogen storage disorder
    d. Lysosomal storage disorder

16. A patient has normal blood glucose level as estimated by Glucose-oxidase Peroxidase method, shows positive Benedict's test in urine. Which of the following is likely cause? *(AIIMS Nov 2016)*
    a. Fructosemia
    b. Galactosemia
    c. Latent diabetes mellitus
    d. Glucose intolerance

17. Galactosemia enzyme defect:
    a. Fructokinase
    b. Glucokinase
    c. Galactose 1 Phosphate Uridyl transferase
    d. Glucose 6 Phosphatase

18. A newborn baby refuses breast milk since the second day of birth, vomits on force-feeding but accepts glucose-water, develops diarrhea on third day, by fifth day she is jaundiced with liver enlargement and eyes show cataract. Urinary reducing sugar was positive but blood glucose estimated by glucose oxidation method was found low. The most likely cause is deficiency of: *(AIIMS May 03)*
    a. Galactose 1-phosphate uridyl transferase
    b. Beta galactosidase
    c. Glucose 6-phosphate
    d. Galactokinase

19. A child presents with hepatomegaly and bilateral lenticular opacities. Deficiency of which of the following enzymes will not cause such features?
    a. Galactose-1-phosphate uridyl transferase
    b. UDP galactose 4-epimerase
    c. Galactokinase
    d. Lactase

20. True regarding galactosemia: *(PGI Dec 01)*
    a. Mental retardation occurs
    b. Absent disaccharidase in intestine
    c. Defect in epimerase
    d. Defect in galactose 1-phosphate uridyl transferase

21. Which of the following test shows positive result in prolonged fasting? *(AIIMS Nov 2017)*

    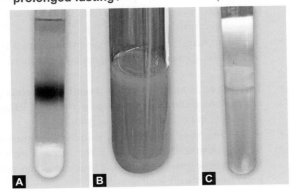

    a. A
    b. B
    c. C
    d. None

22. Fatty acid is not utilized by:
    a. RBC
    b. Skeletal muscle
    c. Liver
    d. Heart

23. Which is used for energy? *(PGI May 2013)*
    a. Ketone bodies
    b. Glucose
    c. Free fatty acids
    d. Creatinine
    e. Collagen

24. All the following are increased in fasting except: *(PGI Nov 2012)*
    a. Lipolysis
    b. Ketogenesis
    c. Gluconeogenesis
    d. Glycogenesis
    e. Glycogenolysis

25. Which enzyme is active when insulin: glucagon ratio is low? *(AIIMS Nov 2013)*
    a. Glucokinase
    b. Hexokinase
    c. Glucose 6 phosphatase
    d. Pyruvate carboxylase

26. Substrate used by RBC in fasting state is: *(May AIIMS 2015)*
    a. Glucose
    b. Amino acids
    c. Ketone body
    d. Fatty acid

27. Lactic acidosis in thiamine deficiency is due to which enzyme dysfunction? *(AIIMS May 2015)*
    a. Phosphoenol pyruvate carboxykinase
    b. Pyruvate dehydrogenase
    c. Pyruvate carboxylase
    d. Aldolase

28. During exercise, most rapid way to synthesize ATP is: *(AIIMS Nov 2016)*
    a. Glycogenolysis
    b. Glycolysis
    c. Phosphocreatine
    d. TCA cycle

Answers to Review Questions

HMP Shunt Pathway

1. **a,b,d. Takes place in…, Does not produce …, Found in liver…**
   (Ref: Harper 31/e page 182)
   - NADPH is produced in oxidative phase
   - Pentoses are produced in nonoxidative phase

   Key points to remember in HMP (Pentose Phosphate Pathway)
   - Main source of NADPH & Pentoses.
   - Two transketolation and one transaldolation reactions are involved.
   - No ATP is produced^Q
   - CO_2 is produced in this pathway and not in glycolytic Pathway^Q *(DNB 01)*
   - Deficiency of Glucose 6 phosphate dehydrogenase is a major cause of acute hemolysis in erythrocytes.
   - NADPH is used for **reductive biosynthetic pathways**, like fatty acid synthesis, steroid synthesis, amino acids by glutamate dehydrogenase.
   - NADPH is required for regeneration of reduced Glutathione, that clears free radicals from erythrocytes and lens.

2. **a. Glycerol 3 Phosphate** *(Ref: Harper 31/e page 182-185)*

   Metabolites in HMP Shunt Pathway
   - Glucose 6 phosphate
   - 6 Phosphogluconate
   - Ribulose 5 phosphate
   - Xylulose 5 phosphate
   - Ribose 5 phosphate
   - Glyceraldehyde 3 phosphate
   - Sedoheptulose 7 phosphate
   - Fructose 6 phosphate
   - Erythrose 4 phosphate

3. **c. HMP shunt** *(Ref: Harper 31/e page 182)*

4. **d. HMP pathway** *(Ref: Harper 31/e page 182)*

5. **d. HMP pathway** *(Ref: Harper 29/e page199, Harper 31/e page 185)*

Metabolic pathway	Net ATP yield
Anaerobic glycolysis	2
Aerobic glycolysis	7
Aerobic oxidation of 1 mol of glucose	32
Palmitic acid oxidation	106
Citric acid cycle	10
Rappaport leubering cycle	0
HMP pathway	0

6. **a. Decreased RBC transketolase activity**
   (Ref: Harper 31/e page 185)
   Thiamine is a coenzyme of Erythrocyte transketolase so its activity is decreased if Thiamine deficiency.

Other Metabolic Pathways of Glucose

7. **a. L Gulanolactone oxidase** *(Ref: Harper 31/e page 186)*
   In human beings and other primase, guinea pigs, bats, fishes and some birds ascorbic acid cannot be synthesized because of absence L gulanolactone oxidase.

8. **a. Vitamin C** *(Ref: Harper 31/e page 186)*
   - Uronic acid pathway cannot synthesize vitamin C in humans and higher primates because of lack of L-Gulano lactone oxidase

9. **c. Vitamin C synthesis** *(Ref: Harper 31/e page 186)*
   - Glucuronic acid is involved in the conjugation of Bilirubin.
   - Uronic acid and amino sugar are the repeating disaccharide unit in the glycosamino glycans.
   - Glucuronidation is a Phase II Xenobiotic reactions (Biotransformation)
   - Uronic acid pathway cannot synthesize vitamin C in humans and higher primates because of lack of L-Gulano lactone oxidase

Fructose Metabolism and Disorders, Metabolism of Amino Sugars

10. **a. Aldolase B** *(Ref: Nelson 20/e Chapter Defects in Metabolism of Carbohydrates)*

    Biochemical Defect of Hereditary Fructose Intolerance
    - Deficiency of Fructose-1,6-Bisphosphate Aldolase (Aldolase B)
    - The gene for aldolase B is on chromosome 9q22.3
    - Deficiency of this enzyme activity causes a rapid accumulation of fructose-1-phosphate and initiates severe toxic symptoms when exposed to fructose.

    Clinical manifestations
    - Patients with HFI are asymptomatic until fructose or sucrose (table sugar) is ingested (usually from fruit, fruit juice, or sweetened cereal).
    - Symptoms may occur early in life, soon after birth if foods or formulas containing these sugars are introduced into the diet.
    - Early clinical manifestations resemble galactosemia and include jaundice, hepatomegaly, vomiting, lethargy, irritability, and convulsions, hypoglycemia.
    - Acute fructose ingestion produces symptomatic hypoglycaemia. If the intake of the fructose persists, hypoglycemic episodes recur, and liver and kidney failure progress, eventually leading to death.
    - Chronic ingestion results in failure to thrive

11. **b. N-Acetylglucosamine** *(Ref: Harper 31/e page 190)*

   Fructose 6 Phosphate is converted to N-Acetyl glucosamine and various other amino sugars which can be used for the synthesis of Glycosaminoglycans. But as it is initially converted to N Acetyl glucosamine, hence it is the answer.

12. **a. Aldolase B.** *(Ref: Nelson 20/e Chapter Defects in Metabolism of Carbohydrates)*

   - Heriditary fructose intolerance due to deficiency of Aldolase B
   - Essential fructosuria due to deficiency of fructokinase

13. **c. Hypergly…** *(Ref: Nelson 20/e Chapter Defects in Metabolism of Carbohydrates)*

   Biochemical Defect of Hereditary Fructose Intolerance
   - Deficiency of Fructose-1,6-Bisphosphate Aldolase (Aldolase B)
   - Deficiency of this enzyme activity causes a rapid accumulation of fructose-1-phosphate and initiates severe toxic symptoms when exposed to fructose.

   Clinical manifestations
   - Patients with HFI are asymptomatic until fructose or sucrose (table sugar) is ingested **(usually from fruit, fruit juice, or sweetened cereal).**
   - Symptoms may occur early in life, soon after birth if foods or formulas containing these sugars are introduced into the diet.
   - Early clinical manifestations resemble galactosemia and include jaundice, hepatomegaly, vomiting, lethargy, irritability, and convulsions, hypoglycemia.
   - Acute fructose ingestion produces symptomatic hypoglycaemia. If the intake of the fructose persists, hypoglycemic episodes recur, and liver and kidney failure progress, eventually leading to death.
   - Chronic ingestion results in failure to thrive

Galactose Metabolism and Disorders

14. **a. Galactose 1 Phosphate uridyl transferase** *(Ref: Harper 31/e page 191)*

   Classic galactosemia due to Galactose 1 Phosphate uridyl transferase is most common cause of galactosemia enzyme deficiency
   - **Galactose 1 Phosphate uridyltransferase (Classic Galactosemia)**
   - Galactokinase
   - UDP Hexose 4 Epimerase

15. **b. Galactosemia** *(Ref: Nelson 20/e Chapter Defects in 5 Metabolism of Carbohydrates)*

   Classic galactosemia is a serious disease with onset of symptoms typically by the 2nd half of the 1st wk of life. With jaundice, vomiting, seizures, lethargy, irritability, feeding difficulties, poor weight gain or failure to regain birth weight.

   Hepatomegaly oil drop cataracts due to accumulation of Galactictol/dulcitol

   Hepatic failure

   Mental retardation

   Patients with galactosemia are at increased risk for Escherichia coli neonatal sepsis.

16. **b>a. Galactosemia>Fructosemia** *(Ref: Nelson 20/e Chapter Defects in Metabolism of Carbohydrates)*

   In this patient, Blood glucose level is normal, but reducing sugar is present in urine.

   Galactose and Fructose are reducing sugars
   - Conditions causing excretion of Galactose in urine is Galactosemia
   - Condition causing excretion of fructose in urine are Essential fructosuria & Hereditary fructose intolerance.

   From the above list, Galactosemia is the single best answer for the following reason:
   - The option given is Fructosemia NOT fructosuria.
   - Even in Essential fructosuria and Hereditary fructose intolerance, Fructosemia is unlikely because Fructose has no renal threshold and Fructose can be metabolised by hexokinase even in the absence of Fructokinase or Aldolase B.

17. **c. Galactose 1 Phosphate uridyl transferase** *(Ref: Harper 31/e page 191)*

   Galactosemia

   Enzyme Deficiency
   - **Galactose 1 Phosphate uridyl transferase (Classic Galactosemia)**
   - Galactokinase
   - UDP Hexose 4 Epimerase

18. **a. Galactose 1 phos…..** *(Ref: Harper 31/e page 191)*

   - Classic galactosemia is a serious disease with onset of symptoms typically by the 2nd half of the 1st week of life.
   - With jaundice, vomiting, seizures, lethargy, irritability, feeding difficulties, poor weight gain or failure to regain birth weight.
   - Hepatomegaly oil drop cataracts, Hepatic failure
   - Mental retardation
   - Patients with galactosemia are at increased risk for *Escherichia coli* neonatal sepsis.

   Other options
   - Beta Galactosidase is lactase.
   - In Galactokinase deficiency the sole manifestation is Cataract.
   - Glucose 6 Phosphate deficiency is Von Gierke's disease.

19. **d. Lactase….** *(Ref: Nelson 20/e Chapter Defects in Metabolism of Carbohydrates)*

20. **a, c, d. Mental reta…,Defect in Epi…,Defect in Galact….**
    (Ref: Nelson 20/e Chapter Defects in Metabolism of Carbohydrates)

21. **a. A Rothera's test**

    In prolonged fasting, ketone bodies level rises. Hence Rothera's test is positive.

    Test A is Rothera's test

    Test B is Benedict's test for reducing substances. Test C is Heat and acetic acid test for Urine Protein.

22. **a. RBC** *(Ref: Harper 31/e page 136)*

    The sole fuel for RBC is glucose.

    Metabolic Fuels for Different Organs

Organ	Major metabolic fuels
Liver	Free fatty acids, glucose (in fed state., lactate, glycerol, fructose, amino acids, alcohol
Brain	Glucose, amino acids, ketone bodies in prolonged starvation
Heart^Q	Ketone bodies, free fatty acids, lactate, chylomicron and VLDL triacylglycerol, some glucose
Adipose tissue	Glucose, chylomicron and VLDL triacylglycerol
Fast twitch muscle	Glucose, glycogen
Slow twitch muscle	Ketone bodies, chylomicron and VLDL triacylglycerol
Kidney	Free fatty acids, lactate, glycerol, glucose
Erythrocyte	Glucose

23. **a, b, c. Ket, Glu, Free Fa** *(Ref: Harper 29/e page161 Table 16.3)*
    (Harper 31/e page 136 Table 14.3)

24. **d. Glycogenesis** *(Ref: Harper 31/e page 138)*

    Glycogenesis is increased in well fed state.

    In Fasting state, Glycogenolysis increase, then gluconeogenesis, lipolysis and Ketogenesis are favoured. These pathways are favoured depending on the duration of fasting.

25. **c, d. Glucose 6 Phosphatase, Pyruvate Carboxylase**
    (Ref: Harper 31/e page 175 Table 19.1)

    Insulin

    Glucagon ratio is more when body is in the fasting state. Key Enzymes of Gluconeogenesis will be active. So Pyruvate Carboxylase and Glucose 6 Phosphatase is active.

26. **a. Glucose** *(Ref: Harper 31/e page 138 Table 14.3)*

27. **b. Pyruvate dehydrogenase** *(Ref: Harper 31/e page 138)*

    In Thiamine deficiency, PDH reaction is defective. So Pyruvate is converted to lactic acid.

    Causes of inhibition PDH leading to lactic acidosis
    - Inherited PDH deficiency.
    - Thiamin deficiency.
    - Alcoholics due to thiamine deficiency.
    - Arsenite and Mercury poisoning.

28. **c. Phosphocreatine** *(Ref: Harper 31/e page 663)*
    - Creatine phosphate represent a major energy reserve in the muscle.
    - Creatine phosphate prevent rapid depletion of ATP by providing readily available high energy phosphate that can regenerate ATP from ADP.
    - Creatine phosphate is formed from Ceatine and ATP at times when the muscle is relaxed and demand for ATP is low.
    - This is done by the enzyme muscle specific Creatine Kinase.

CHAPTER 5

Lipids

Chapter Outline

5.1 Chemistry of Lipids
5.2 Phospholipids and Glycolipids
5.3 Metabolism of Lipids
5.4 Lipoprotein Metabolism

CHAPTER 5

Lipids

5.1 CHEMISTRY OF LIPIDS

- Definition of Lipids
- Classification of Lipids
- Fatty Acids
- Triacylglycerol

DEFINITION OF LIPIDS

Lipids are heterogeneous group of compounds relatively insoluble in water and freely soluble in nonpolar organic solvents–ether, chloroform, etc.

Unlike Carbohydrates and Proteins they are not chemically related, but they are physically related.
- **Carbohydrates:** Polymer of Monosaccharide
- **Proteins:** Polymer of Amino Acids

Lipids are not true polymers, but mixture of chemically unrelated substances.

CLASSIFICATION OF LIPIDS

Bloor's Classification

1	Simple Lipids
2	Complex (Compound) Lipids
3	Precursor Lipids or Derived Lipids
4	Miscellaneous Lipids

Simple Lipids

They are esters of **fatty acid with alcohol**.

Fatty Acid + Alcohol = Esters of fatty acid = Simple lipid

They are divided into:
- Fats and oils
- Waxes

Fats and Oils

They are esters of fatty acid with alcohol, glycerol.

Fatty Acid + Glycerol = Fat

Fats and oils are the same except that fats are solid at room temperature, and oils are liquid at room temperature.

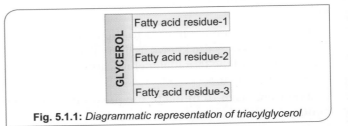

Fig. 5.1.1: *Diagrammatic representation of triacylglycerol*

Waxes

They are esters of fatty acid with **higher molecular weight monohydric alcohol** other than glycerol, e.g. Bee waxes, Lanolin.

Fatty Acid + High mol weight alcohol other than glycerol

Complex Lipids

They are esters of fatty acid with alcohols having **additional groups** like Phosphoric acid, Carbohydrates, Proteins, etc.

Fatty Acid + Alcohol (Glycerol/Sphingosine) + **additional groups = Complex lipid**

Examples
Phospholipids, Glycolipids, Lipoproteins, Sulfolipids
- Glycolipids
- Other complex lipids like Sulfolipid, Lipoproteins, Amino lipids.

Glycolipids or Glycosphingolipids[Q]

- Lipids containing carbohydrate apart from Fatty acid and Alcohol (usually Sphingosine)
- **Fatty Acid + Alcohol (Sphingosine) + Carbohydrate** e.g. **Cerebroside, Ganglioside**

Derived Lipids or Precursor Lipids

Compounds which are derived from the above group of Lipids, e.g. Fatty acids, Glycerol, Cholesterol

Miscellaneous Lipids

Vast number of lipids which are not classified under any of the above groups, e.g. Squalene, Carotenoids, Vitamin E, Vitamin K

FATTY ACIDS

They are aliphatic carboxylic acid.

General Structural Formula of Fatty acid— R-COOH
- R is the Aliphatic Hydrocarbon chain.
- R group accounts for the non-polar nature of fat.

Numbering the carbon atoms in fatty acids

Carbon atoms are numbered from the carboxyl carbon (carbon no. 1).

The carbon atoms adjacent to the carboxyl carbon (nos. 2, 3, and 4) are also known as the α, β, and γ carbons, respectively, and the terminal methyl carbon is known as the ω– or n–Carbon.

Conventions used to indicate the positions of double bonds

Various conventions use for indicating the number and position of the double bonds:
- Δ^9 indicates a double bond between carbons 9 and 10 of the fatty acid counting from the carboxyl end.
- ω indicates a double bond on the ninth carbon counting from the ω-carbon.

$$\underset{n\ \ \ \ \ \ \ \ 17}{\overset{\omega\ \ \ 2\ \ \ 3\ \ \ 4\ \ \ 5\ \ \ 6\ \ \ 7\ \ \ 8\ \ \ 9\ \ \ \ \ \ 10\ \ \ \ \ \ \ \ \ \ \ \ 18}{CH_3CH_2CH_2CH_2CH_2CH_2CH_2CH_2CH = CH(CH_2)_7COOH}}\ \underset{10\ \ \ \ \ \ \ \ \ \ 9\ 1}{}$$

Fig. 5.1.2: Nomenclature of number and position of double bond of unsaturated fatty acid

Classification of Fatty Acid

Depending on the chain length
1. Short chain fatty acid (C2-C6)
2. Medium chain fatty acid (C8-C14)
3. Long chain fatty acid ($\geq$ C16)

Depending on the presence of double bond
1. **Saturated fatty acid:** No double bond in the Hydrocarbon chain.
2. **Unsaturated fatty acid:** Double bonds are present in the Hydrocarbon chain.

Depending on the number of double bonds present in unsaturated fatty acid is again classified into:
A. Monounsaturated fatty acid—Only one double bond is present.
B. Polyunsaturated fatty acid—More than one double bonds are present.

Common Saturated Fatty Acids and their Sources

Saturated fatty acid	Source
Acetic Acid (2C)	Vinegar
Butyric Acid (4C)	Butter
Valeric Acid (5C)	Butter
Caproic Acid (6C)	Butter and Coconut milk
Lauric Acid (12C)	Coconut milk
Myristic Acid (14C)	Coconut milk
Palmitic Acid (16C)	Body fat
Stearic Acid (18C)	Body fat

Common Unsaturated Fatty Acids

Number of C atoms and number and position of common double bonds	Family	Common name	Occurrence[Q]
Monoenoic acids (one double bond)			
16:1;9	ω7	Palmitoleic	In nearly all fats.
18:1;9	ω9	Oleic	Possibly **the most common fatty acid in natural fats**; particularly high in olive oil.
18:1;9	ω9	Elaidic	Hydrogenated and ruminant fats.
Dienoic acids (two double bonds)			
18:2;9,12	ω6	Linoleic	Corn, peanut, **cotton seed**, soybean, and many plant oils.
Trienoic acids (three double bonds)			
18:3;6,9,12	ω6[Q]	γ-Linolenic (GLA)	Some plants, e.g. **oil of evening primrose**, borage oil; minor fatty acid in animals Linseed oil

Contd...

Contd...

Number of C atoms and number and position of common double bonds	Family	Common name	Occurrence[Q]
18:3;9,12,15	ω3[Q]	α-Linolenic	Frequently found with linoleic acid but particularly in **linseed oil**.
Tetraenoic acid (four double bonds[Q])			
20:4;5,8,11,14	ω6	Arachidonic	Found in animal fats; important component of phospholipids in animals.
Pentaenoic acids (five double bonds)			
20:5;5,8,11,14,17	ω3	Timnodonic (Eicosapentaenoic)	Important component of fish oils, e.g. codliver, mackerel, menhaden, salmon oils.
Hexaenoic acids (six double bonds)			
22:6;4,7,10,13,16,19	ω3	Cervonic (Docosahexaenoic) (DHA)	**Fish oils**, phospholipids in brain, **Breast milk[Q]**

NB: *This table is very important, Numerous questions can be asked Bold letters are very important*

Significance of Medium Chain Fatty Acid

Absorbed directly into the Blood
Do not need Carnitine for transport into Mitochondria
No effect on Atherosclerosis

Essential Fatty Acid

The fatty acids that are required by humans, **but are not synthesized in the body** hence need to be supplied in the diet are known as essential fatty acid (EFA). Humans lack the enzymes that can introduce double bond **beyond 9th Carbon**. (Remember the numbering is done from the Carboxyl end of the fatty Acid).

> **HIGH YIELD POINTS**
>
> Essential fatty acids are Polyunsaturated Fatty Acid namely:
> 1. Linoleic Acid[Q]
> 2. Alpha Linolenic Acid[Q]

Arachidonic Acid is considered as semi essential fatty acid as it can be synthesized from Linoleic Acid[Q].

Functions of Essential Fatty Acid

- They are integral components of membrane structure, often in the 2 position of phospholipids.
- Eicosanoids are synthesized from Arachidonic Acid.
- Essential fatty acids are needed for the synthesis of Arachidonic Acid.
- Lower the risk of Cardiovascular Diseases.
- Lower the risk of Fatty liver.

Deficiency of Essential Fatty Acid

- **Skin-Acanthosis and Hyperkeratosis**
- Fatty liver
- Swelling of mitochondrial membrane and reduction in efficiency of oxidative phosphorylation
- Decrease in fibrinolytic activities.

Omega Classification of Fatty Acid

In this type of classification position of double bond is counted from the methyl end (ω carbon atom).

Depending on the position of first double bond from the terminal end, the fatty acids are again classified into:

ω3 Series	• α **Linolenic Acid** • **Timnodonic Acid (Eicosapentaenoic Acid)** • **Cervonic Acid (Docosahexaenoic Acid) (DHA)**
ω6 Series	• **Linoleic Acid** • γ Linolenic Acid **(GLA)** • Arachidonic Acid
ω9 Series	• Oleic Acid • Elaidic Acid

Significance of ω3 Fatty Acid

- Decrease the risk of Cardiovascular Disease.
- Appear to replace arachidonic acid in platelet membranes
- Lower the production of Thromboxane and tendency of the platelet aggregation.
- Promote the synthesis of less inflammatory prostaglandin and leukotrienes, hence they **are anti-inflammatory**.

- **Decrease Serum Triglycerides**
- **Fish oils and certain plant oils are rich in** ω 3 Fatty Acid are hence used in familial **hyperlipoproteinemia with hypertriglyceridemia**[Q]
- Important for Infant Development
- Lower the risk of various mental illness (Depression, ADHD)
- Lower the risk of chronic degenerative diseases such as Cancer, Rheumatoid Arthritis, and Alzheimer's Disease.

> **HIGH YIELD POINTS**
>
> **Docosahexaenoic Acid (DHA)**
> - Sources: **Human milk, Fish liver oils, Algaloils**
> - Synthesized in the body from a **Linolenic acid.**
> - **Highest concentration of DHA found in retina, cerebral cortex, sperms.**
> - Functions: Needed for the development of **foetal brain and retina.**
> - DHA is **supplied transplacentally** and through breast milk.
> - Clinical significance: Low DHA is associated with increased risk of **Retinitis Pigmentosa.**

Isomerism in Fatty Acids

Image-Based Information—cis-Trans Isomerism of Fatty Acid

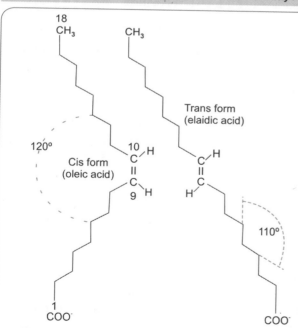

Fig. 5.1.3: *Cis and trans isomer-oleic acid and elaidic acid*

A type of geometrical isomerism that occur in unsaturated fatty acids, depending on the orientation of groups around the axis of double bond. If the acyl chains are **on the same side of the double bond** it is cis form (oleic acid). If the acyl chain on the **opposite sides of double bond** it is trans-form (elaidic acid)

Biological Significance of Cis-Isomers

All naturally occurring fatty acids are cis-isomers.
Cis-isomers increases the fluidity of biological membranes.
Profound effect in molecular packing of cell membranes

Trans Fatty Acids (TFA)[Q]

- Present in dairy products and **partially hydrogenated** edible oils (e.g. Margarine)
- They are used in food industry to improve the shelf life.
- Trans fatty acids are present in high amounts in processed foods, fast foods and bakery items and fried foods
- Transfatty acid compete with essential fatty acid, hence exacerbate essential fatty acid deficiency.
- **Trans fatty acids raise the level of LDL, TAG and lower the level of HDL.**
- Consumption of trans fatty acid for long-terms may **raise the risk factor for cardiovascular** diseases like Atherosclerosis and Coronary Artery disease and Diabetes mellitus.
- It also **increases the body's inflammatory response**.
- Deleterious effect of TFA is seen if intake is more than **2–7 g/day.**

Dietary Sources of Fatty Acid

Fats/Oil	SFA (%)	MUFA (%)	Linoleic acid (%)	Alpha Linolenic acid (%)
High medium chain and short chain fatty acid				
Coconut	92 (Highest SFA)	6	2	–
Palm kernel	83	15	2	–
Butter/Ghee	68	29	2	1
High SFA and MUFA				
Palmolein	39	46	11	<0.5
High MUFA and moderate linoleic acid				
Ground nut	19	41	32	<0.5
Rice bran	17	43	38	1
Sesame	16	41	42	<0.5
High linoleic acid				
Cotton seed	24	29	48	1
Corn	12	35	50	1
Safflower	9	13	75 (Highest)	–
Sunflower	12	22	62	–
High linoleic acid and alpha linolenic acid				

Contd...

Contd...

Fats/Oil	SFA (%)	MUFA (%)	Linoleic acid (%)	Alpha Linolenic acid (%)
Soybean	14	24	53	7
Canola	6	60	22	10
Mustard/Rapeseed	4	65 (Highest)	15	14
Flax-Seed	10	21	16	53 (Highest)
High trans fatty acid				
Vanaspati	46	49	4	–

HIGH YIELD POINTS

Dietary Sources of Fatty Acids—at a Glance
Highest content of MUFA in **Mustard/Rapeseed oil** Highest content of Medium chain fatty acid in **Coconut oil** Highest content Linoleic acid in **Safflower oil**
Highest content alpha linolenic acid in **Flax seed oil** Highest amount of PUFA is present in **Safflower oil**. Second highest source of PUFA is Sunflower oil.
Least source of PUFA is Coconut oil.
Fatty acid present in human milk **is Docosahexaenoic Acid (Cervonic Acid)**
Fatty acid present in the fish oils are:
- Timnodonic Acid
- Clupadonic Acid
- Cervonic Acid

TRIACYLGLYCEROL (TAG)

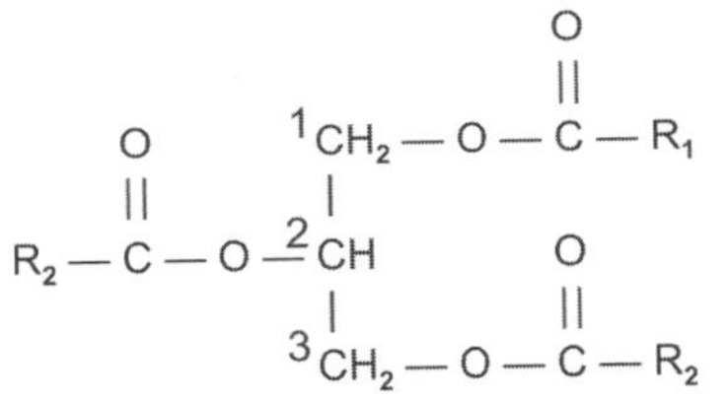

Fig. 5.1.4: *Structure of triacylglycerol*

- **Main storage form of lipids in the body.**
- Stored in the adipose tissues.
- Otherwise called **Neutral fat**.

Physical Properties of Triacylglycerol

Saponification and Saponification Number
The hydrolysis of triacylglycerol by alkali into glycerol and soap is called Saponification.

Saponification number
The mg of KOH required to saponify 1 g of fat or oil completely.

This is a measure of average molecular weight and chain length of the fatty acid present.
- *It is inversely proportional to the chain length of fatty acid present in the fats.*

Type of fat	Saponification number
Human Fat	195–200
Butter	230–240
Coconut oil	250–260

Iodine Number
- The number of grams of Iodine absorbed by 100 g of fat or oil.
- Iodine number is used to assess the degree of unsaturation of fat.
- *It is directly proportional to the degree of unsaturation of fatty acid.*

Type of fat/oil	Iodine number
Butter	25–28
Human fat	65–70
Linseed oil	170–200

Reichert-Meissl (RM) Number
The number of 0.1 NKOH required to neutralize the volatile fatty acids distilled from 5 g of fat.

Assess the purity of fats having more volatile fatty acid.

Rancidity of Fat
Rancidity refers to unpleasant smell and taste for fats and oils.

Hydrolytic rancidity due to partial hydrolysis of Triacylglycerol due to hydrolytic enzymes present in naturally occurring fats and oils.

Oxidative rancidity due to partial oxidation of unsaturated fatty acid.

Vegetable oils with high content of Polyunsaturated Fatty Acid are easily oxidised.

Hence vegetable oils are preserved with antioxidants.

Self Oriented Structures Formed by Amphipathic Lipids
Amphipathic lipids self-orient at Oil: Water interfaces. They form:
1. **Membranes (lipid bilayer)**
2. **Micelles**
3. **Liposomes**
4. **Emulsions.**

Lipid bilayer

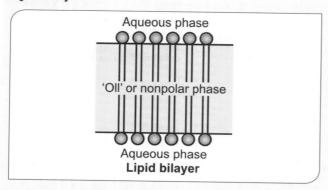

Lipid bilayer

- A bilayer of such amphipathic lipids is the basic structure in biologic membranes

Micelle

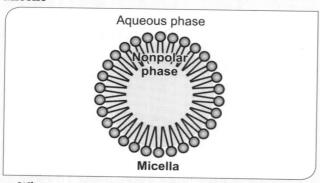

Micelle

- When a critical concentration of these lipids is present in an aqueous medium, they form micelles.

Aggregation of bile salts into micelles and liposomes and the formation of mixed micelles with the products off at digestion are important in facilitating absorption of lipids from the intestine.

Liposomes

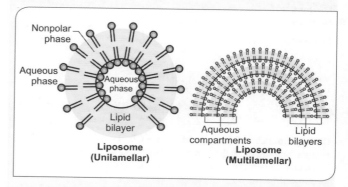

A reformed by sonicating an amphipathic lipid in an aqueous medium.

They consist of spheres of lipid bilayers that enclose part of the aqueous medium.

Clinical uses of liposomes

- As carriers of drugs in the circulation, targeted to specific organs, for example, in cancer therapy.
- They are used for gene transfer into vascular cells.
- As carriers for topical and transdermal delivery of drugs and cosmetics.

Quick Review

- Lipids are physically related than chemically.
- MCFA are Lauric acid and Myristic acid.
- Richest source of MCFA is coconut milk
- Richest source of PUFA is safflower oil.
- Fish oils and breast milk are rich in cervonic acid or Docosa Hexaenoic acid
- Omega 3 fatty acids are anti-inflammatory
- Low DHA is associated with Retinitis pigmentosa
- Trans fatty acids raises level of TAG, LDL-C and lower HDL-C
- Essential fatty acids are Linoleic acid and gamma linolenic acid.

Check List for Revision

- Classification of Fatty acids with examples
- Important unsaturated fatty acids and its sources
- Omega classification with examples.
- Essential fatty acids
- Significance of omega 3 and DHA
- All high yield boxes and bold letters

Review Questions MCQ

1. Retinitis pigmentosa is associated with deficiency of which of the following: *(Recent Question Jan 2019)*
   a. Timnodonic acid
   b. Eicopentaenoic acid
   c. Arachidonic acid
   d. Docosahexaenoic acid

2. Which are omega 6 fatty acid? *(PGI Nov 2018)*
   a. Gamma linolenic acid
   b. Alpha linolenic acid
   c. Arachidonic acid
   d. Palmitic acid
   e. Linoleic acid

3. Essential fatty acid is/are: *(PGI May 2013)*
   a. Palmitic acid
   b. Linoleic acid
   c. Linolenic acid
   d. Oleic acid
   e. Free fatty acid

4. True statement about fatty acid: *(PGI Nov 2011)*
   a. PUFA is essential for membrane structure
   b. Biologically arachidonic acid is essential to life
   c. Hydrogenated vegetable oils contain trans fatty acid
   d. Most of the naturally occurring unsaturated FA exist as transisomer

5. True about trans fatty acid: *(PGI Nov 2010)*
   a. Fried rice have high content of trans fatty acid
   b. Partial hydrogenation increases trans fatty acid
   c. Refining decreased TFA
   d. ↑LDL
   e. ↓HDL

6. PUFA (more than 50%) content is seen in: *(PGI Nov 2009)*
   a. Groundnut oil
   b. Safflower oil
   c. Corn oil
   d. Sunflower oil

7. Which among the following is a cardioprotective fatty acid? *(Kerala 2011)*
   a. Palmitic acid
   b. Stearic acid
   c. Oleic acid
   d. Omega-3 fatty acid

8. Which among the following is not a saturated fatty acid? *(Recent Question 2016)*
   a. Myristic acid
   b. Stearic acid
   c. Palmitic acid
   d. Linoleic acid

9. Most essential fatty acid is: *(Kerala 2008)*
   a. Linolenic acid
   b. Linoleic acid
   c. Arachidonic acid
   d. Eicosapentaenoic acid

10. All are true *except*: *(PGI Dec 2002)*
    a. Linoleic acid is found in soyabean oil
    b. Linolenic and linoleic acids are cis derivatives containing double bonds
    c. Arachidonic acid contains five double bonds
    d. Monoenoic acids contain one double bond at 9th position

11. Maximum source of linoleic acid is: *(AIIMS Jun 97)*
    a. Coconut oil
    b. Sunflower oil
    c. Palm oil
    d. Vanaspati

12. Which of these fatty acids is found exclusively in breast milk? *(AI 2001)*
    a. Linoleate
    b. Linolenic
    c. Palmitic
    d. Docosahexaenoic acid

13. The following fatty acid does not be long to ω6 series: *(PGI June 2000; AIIMS Dec 90)*
    a. Linoleic acid
    b. Arachidonic acid
    c. Gamma linoleic acid
    d. Alpha linolenic acid
    e. Timnodonic acid

14. An example of Omega 6 fatty acid is: *(CMC Ludhiana 2014)*
    a. Cervonic acid
    b. α Linolenic acid
    c. Arachidonic acid
    d. Timnodonic acid

15. Which is not present in plants? *(CMC Ludhiana 2014)*
    a. Cholesterol
    b. Linolenic acid
    c. Linoleic acid
    d. Lauric acid

16. True about Long chain fatty acids is: *(PGI May 2017)*
    a. Hydrogenation convert unsaturated fatty acid to saturated fatty acid
    b. Linoleic acid is an essential fatty acid
    c. Refining oils lead to removal of trans fatty acid
    d. Arachidonic acid can be synthesized from linoleic acid

Answers to Review Questions

1. **d, Docosa hexaenoic acid** *(Harper 31/e page 199)*

2. **a, c, e. Gamma linolenic acid, Arachdonic acid, Linoleic acid**
   (Ref: Harper 31/e page 197 Table 21-2)

   Omega 6 fatty acids are Gamma Linolenic acid, Linoleic acid and Arachdonic acid

   Omega 3 fatty acids a Alpha linolenic acid, Timnodonic acid, Cervonic acid,

3. **b, c. Linoleic acid, Linolenic acid**

   Essential Fatty Acid

   The fatty acids that are required by humans, but are not synthesized in the body hence need to be supplied in the diet are known as essential fatty acid (EFA). Humans lack the enzymes that can introduce double bond beyond 9th Carbon.

   They are Polyunsaturated fatty acid namely
   - Linoleic acid
   - Linolenic acid

   Linoleic acid is the most essential fatty acid

   Arachidonic acid is considered as *semiessential fatty acid* as it can be synthesized from linoleic acid.

   Functions of Essential Fatty Acid
   - They are integral components of membrane structure, often in the 2 position of phospholipids.
   - Eicosanoids are synthesized from arachidonic acid. Essential fatty acids are needed for the synthesis of arachidonic acid.
   - Lower the risk of cardiovascular diseases.

4. **a, b, c. PUFA is essential for membrane structure, Biologically arachidonic acid is essential to life, Hydrogenated vegetable oils contain trans fatty acid** *(Ref: Harper 31/e page 222)*

   Most of the naturally occurring UFA exist in cis form.

5. **a, b, d, e. Fried rice have high content of trans fatty acid, Partial hydrogenation increases trans fatty acid, ↑LDL, ↓HDL**
   (Ref: Harper 31/e page 198)

   Trans Fatty Acids
   - Present in dairy products and **partially hydrogenated** edible oils.
   - They are used in food industry to improve the shelf life.
   - Trans fatty acids are present in high amounts in processed foods, fast foods and bakery items.
   - Consumption of trans fatty acid for long-terms may raise the risk factor for cardiovascular diseases like atherosclerosis, coronary artery disease and diabetes mellitus.
   - Exacerbate essential fatty acid deficiency by competing for essential fatty acid.

6. **b, c, d. Safflower oil, Corn oil, Sunflower oil**
   (Ref: Park 23/e page 611, Table 2)
   - Highest content of MUFA in mustard/rapeseed oil
   - Highest content of Medium chain fatty acid in Coconut oil
   - Highest content linoleic acid in safflower oil
   - Highest content alpha linolenic acid in flax seed oil.

7. **d. Omega-3 fatty acid** *(Ref: Harper 31/e page 199)*

   Significance of ω3 Fatty Acid
   - Decrease the risk of cardiovascular disease.
   - Appear to replace arachidonic acid in platelet membranes
   - Lower the production of thromboxane and tendency of the platelet aggregation.
   - Decrease serum triglycerides
   - Important for infant development
   - Lower the risk of various mental illness [depression, ADHD]
   - Lower the risk of chronic degenerative diseases, such as cancer, rheumatoid arthritis, and Alzheimer's disease.

8. **d. Linoleic acid** *(Ref: Harper 31/e page 197)*

9. **b. Linoleic acid** *(Ref: Harper 31/e page 211)*
   - Linoleic acid is the most essential fatty acid

   Arachidonic Acid is considered as *semi essential fatty acid* as it can be synthesized from linoleic acid.

   Functions of Essential Fatty Acid
   - They are integral components of membrane structure, often in the 2 position of phospholipids.
   - Eicosanoids are synthesized from arachidonic acid. Essential fatty acids are needed for the synthesis of Arachidonic acid.
   - Lower the risk of cardiovascular diseases.

10. **c. Arachidonic acid contains five double bonds**
    (Ref: Harper 31/e page 212, 238)
    - 53% of soyabean oil is linoleic acid.
    - Almost all unsaturated fatty acids have cis configuration
    - Arachidonic acid has 4 double bonds
    - First double bond is usually added in 9th position by a delta 9 desaturase.

11. **b. Sunflower oil**
    - Highest content of MUFA in mustard/rapeseed oil
    - Highest content of medium chain fatty acid in coconut oil
    - Highest content linoleic acid in safflower oil
    - Highest content alpha linolenic acid in flax seed oil.

12. **d. Docosahexaenoic acid** *(Ref: Harper 31/e page 197, Table 21.2)*
    - Docosahexaenoic acid is present in fish oils, phospholipids of brain, algal oils.

- DHA is otherwise called cervonic acid.
- It belongs to ω3 fatty acid.

13. d. Alphalinolenic acid (Ref: Harper 31/e page197)

ω3 FA	ω6 FA	ω7 FA	ω9 FA
Alpha linolenic acid	Linoleic acid	Palmitoleic acid	Oleic acid
Timnodonic acid (eicosa-pentaenoic acid)	Gamma linolenic acid		
Cervonic acid (docosa-hexaenoic acid)	Arachidonic Acid		

14. c. Arachidonic acid (Ref: Harper 31/e page197)

15. a. Cholesterol

16. a, b, d. Hydrogenation convert unsaturated fatty acid to saturated acid, linoleic acid is an essential fatty acid Arachidonic acid can be synthesized from linoleic acid

- Hydrogenation of double bonds convert unsaturated fatty acid to saturated fatty acid
- Essential fatty acids are linoleic acid and alpha linolenic acid
- Linoleic acid can be the precursor of arachidonic acid
- Refining of oils cannot remove trans fatty acids

5.2 PHOSPHOLIPIDS AND GLYCOLIPIDS

- Classification of Phospholipids
- Glycolipids (Glycosphingolipids)
- Sphingolipidosis

CLASSIFICATION OF PHOSPHOLIPIDS

Definition

Compound lipids composed of fatty acid, alcohol, phosphoric acid and a nitrogenous base.

Based on the alcohol present phospholipids are divided into:

Lecithin (Phosphatidyl Choline)

Glycerophospholipid with Choline as nitrogenous base, i.e. Phosphatidic Acid + Cholinsse.

Most abundant phospholipid of cell membrane.

Dipalmitoyl Lecithin (Di Palmitoyl Phosphatidyl Choline) **is a major constituent of lung surfactant.**

Largest body store of choline.

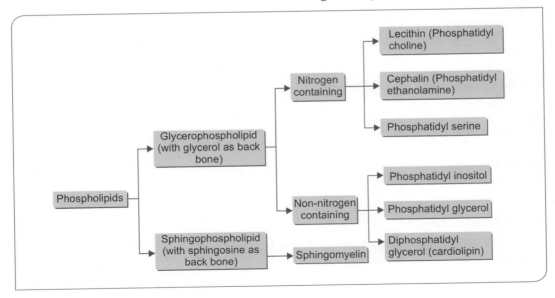

Glycerophospholipids

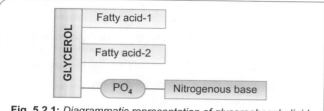

Fig. 5.2.1: Diagrammatic representation of glycerophospholipid

Glycerophospholipid Contains

1. Glycerol
2. Fatty acid esterified to the first two carbon atoms
3. Nitrogenous base
4. Phosphoric acid.

Glycerophospholipid

Phosphatidic Acid

Simplest phospholipid

Does not contain any nitrogenous base.

Phosphatidic acid contains:
- Glycerol
- Fatty acid esterified to the first two carbon atoms.
- Phosphoric Acid.
  All glycerophospholipids are derived from Phosphatidic acid, *i.e. Phosphatidic acid + Nitrogenous base*

Importance of Choline
- Acetyl Choline in nerve transmission.
- Transmethylation reaction

> **EXTRA EDGE**
>
> **Lung Surfactants**
> Consist of Dipalmitoyl Lecithin, Phosphatidyl Glycerol, Cholesterol and Surfactant protein A, B, C.
>
> **Lecithin Sphingomyelin Ratio (L/S ratio)**
> Before 28 weeks fetal lung synthesizes Sphingomyelin. But as lung matures more Lecithin is synthesized. L/S ratio indicates lung maturity. Ratio of 2 indicate full lung maturity.
>
> **Respiratory Distress Syndrome**
> In premature infants due to decreased lung surfactants.

Cephalin (Phosphatidyl Ethanolamine)

Glycerophospholipid with Ethanolamine as nitrogenous base, i.e. Phosphatidic Acid + Ethanolamine
- Component of cell membrane
- Play a role in Blood coagulation.

Cardiolipin
- First isolated **from Cardiac muscle** and hence the name.
- Made up of two molecules of Phosphatidic Acid linked by a molecule of Glycerol, i.e. **Diphosphatidyl Glycerol**.
- Is a **major lipid of Inner Mitochondrial Membranes**.
- Only Phospholipid which possess **antigenicity**.
- Recognized by antibodies raised against *Treponema pallidum*, that causes syphilis.
- Decreased Cardiolipin levels or alteration in its structure or metabolism cause mitochondrial dysfunction.

 HIGH YIELD POINTS

Cardiolipin Associated Mitochondrial Dysfunction is Seen in:
- Aging
- Heart failure
- Barth Syndrome (Cardioskeletal Myopathy)
- Hypothyroidism

Phosphatidyl Serine
- Nitrogenous base is Serine
- Plays an important role in **Programmed cell death.**

Phosphatidyl Inositol
- Phosphatidyl Inositol present in the cell membrane.
- The inositol is present as its stereo isomer, Myoinositol.
- Play an important role in **cell signalling and membrane trafficking.**
- Phosphatidyl Inositol is a precursor of **second messenger in hormonal pathways**.

Action of Phospholipase C on Phosphatidyl Inositol 4,5 Bisphosphate (PiP$_2$)
- Diacyl Glycerol and Inositol Tris phosphate IP3 are formed and both act as second messengers.

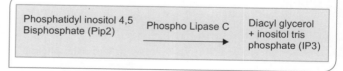

Image-Based Information

Ether Lipids
When fatty acid is attached to C1 of Glycerol in Glycerophospholipids by an **ether linkage instead of usual Ester linkage**, they are called Ether lipids.
Two biologically important Ether Lipids are:
1. **Plasmalogen**
2. **Platelet Activating Factor (PAF)**

Contd...

Contd...

Plasmalogen
On C1 instead of fatty acid an **unsaturated alkyl** is attached by an ether linkage.

Glycerol + Unsaturated Alkyl residue + Fatty Acid + Phosphoric Acid + Ethanolamine

- Plasmalogens occur in Brain and heart
- The Plasmalogen may have a protective effect against reactive oxygen species.

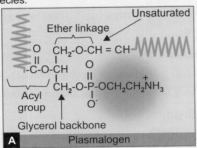

A Plasmalogen

Platelet Activating Factor
It has an ether linkage at C1 to an Alkyl residue and Ester linkage at C2 to Acetyl group
Components are
Glycerol + Alkyl residue + Acetic Acid + Phosphoric Acid + Choline.

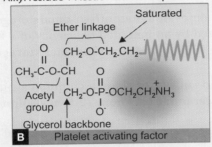

B Platelet activating factor

Sphingophospholipid

The second group of phospholipid which contain **sphingosine** as the backbone alcohol.

Only Sphingophospholipid is Sphingomyelin^Q.

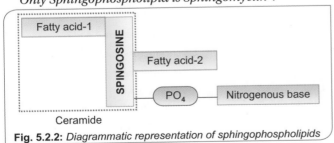

Fig. 5.2.2: *Diagrammatic representation of sphingophospholipids*

Sphingomyelin

Structure

- The **Amino Alcohol, Sphingosine** is attached to a fatty acid by an amide linkage forming Ceramide.
- Ceramide is further linked to a phosphoryl and nitrogenous base (Choline) to form Sphingomyelin.

Image-Based Information
Structure of Sphingosine and Ceramide

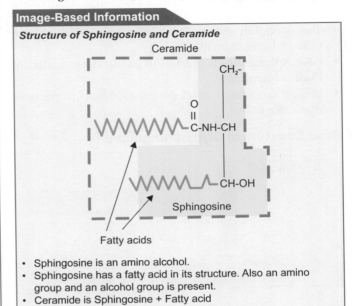

- Sphingosine is an amino alcohol.
- Sphingosine has a fatty acid in its structure. Also an amino group and an alcohol group is present.
- Ceramide is Sphingosine + Fatty acid

Sphingosine + Fatty Acid = Ceramide

Ceramide + Phosphoryl group + Nitrogenous base = Sphingomyelin

- Sphingomyelins are found in **outer leaflet of cell membrane bilayer**.
- Particularly abundant in specialised areas of plasma membrane called **lipid rafts.**
- They are also particularly abundant in the myelin sheath that surrounds nerve fibres.
- They play a role in cell signalling and apoptosis

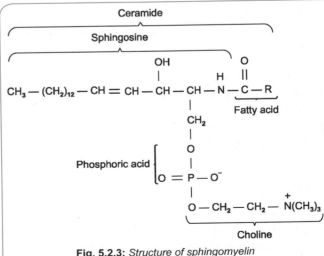

Fig. 5.2.3: *Structure of sphingomyelin*

GLYCOLIPIDS (GLYCOSPHINGOLIPIDS)

- Complex lipids which contain carbohydrates, but no phosphoric Acid.
- The alcohol in glycolipid is always Sphingosine hence called Glycosphingolipid.
- Glycosphingolipids are found in the outer leaflet of Plasma membrane.
- Widely distributed in all tissues especially in nervous tissue like brain.

Basic Structure of Glycosphingolipid

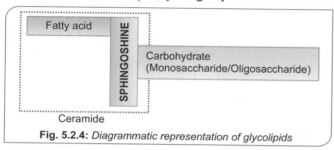

Fig. 5.2.4: Diagrammatic representation of glycolipids

Three Types of Glycolipids

Ceramide (Sphingosine + Fatty Acid) attached to different carbohydrate to form three types of Glycolipids

Glycolipids or Glycosphingolipids
1. Cerebroside
2. Globoside
3. Ganglioside

Cerebroside
- **Ceramide + Monosaccharide**
- Glucocerebroside is Ceramide + Glucose
- Galactocerebroside is Ceramide + Galactose
- Galactocerebroside found in the **brain and other neural tissues**
- The fatty acid present in Galactocerebroside is characteristically **Cerebronic Acid (C-24)**
- **Sulfogalactosyl ceramide** present in high amounts in the myelin
- Glucocerebroside found in the **non-neural tissues and some amount in brain.**

Globoside
- **Ceramide + Oligosaccharide.**

Ganglioside
- ***Ceramide + Oligosaccharide containing N Acetyl Neuraminic Acid [NANA] or Sialic Acid)***

Ganglioside is named as GMn where
- G represent Ganglioside
- M represent Monosialo, as it contain Sialic Acid.
- n stands for number assigned on the basis of chromatographic migration.
- Gangliosides are present in the brain in high concentration.
- They act as receptors for bacterial toxins and for hormones.

GM3 Ganglioside
- *The simplest ganglioside found in tissues is GM3*
- **Ceramide + Glucose + galactose + NANA**

GM1 Ganglioside
- More complex than GM3 Ganglioside.
- Derived from GM3 Ganglioside.
- This is known to be receptor for Cholera toxin in human intestine.

SPHINGOLIPIDOSIS

Sphingolipidoses are a group of **lysosomal storage disorder** characterized by an inherited deficiency of a **lysosomal hydrolase** leading to **intralysosomal storage of lipid** substrates.

The lipid substrates share a common structure that includes a **ceramide backbone (2-N-acylsphingosine).**

| Enzyme deficiencies in sphingolipidosis ||
Sphingolipidoses	Enzyme deficiency
Farber disease	Ceramidase
Fabry disease	α-Galactosidase
GM1 gangliosidosis	β-Galactosidase
GM2 gangliosidosis	
1. Tay-Sachs disease	β-Hexosaminidases A
2. Sandhoff disease	β-Hexosaminidases A and B
Gaucher's disease	Glucocerebrosidase/β Glucosidase
Niemann-Pick disease	Sphingomyelinase
Metachromatic leukodystrophy	Arylsulfatase A Sphingolipid Activator Protein (SAP-1)
Krabbe's disease	β-Galactosidase/ β-Galactocerebrosidase

Important Sphingolipidosis

GM1 Gangliosidosis
Autosomal recessive trait.

Biochemical defect
- Deficient activity of β-**galactosidase**
- Accumulation of **GM1 gangliosides** in the lysosomes of both neural and visceral cells.
- A mucopolysacchride accumulate in GM1
- A mucopolysacchride **Keratan sulphate** is accumulated in liver and excreted in urine.

Symptom box

Clinical presentation
- Developmental delay
- Hepatosplenomegaly
- Skin eruptions (**angio keratoma**).
- **A typical facies** is characterized by low-set ears, frontal bossing, a depressed nasal bridge, and an abnormally long philtrum.
- 50% of patients have a **macular cherry red spot.**
- Blindness and deafness with severe neurologic impairment

GM_2 Gangliosidoses

This include:
1. **Tay-Sachs Disease**
2. **Sandhoff Disease**

Biochemical defect
- Autosomal recessive
- Deficiency of β-**Hexosaminidase**
- Lysosomal accumulation of **GM2 Ganglioside**, particularly in CNS.

Frequently asked doubts:
Sandhoff's disease is due to a defect in beta subunit of Hexosaminidase or Hexosaminidase A and B defect? The answer lies in the concept of enzyme subunits
Concept of enzyme Beta Hexosaminidase
- β Hexosaminidase has two isoforms
- β Hexosaminidase A, composed of 1 α and 1 β subunit.
- β Hexosaminidase B composed of 2 β subunits.
- Mutation in α subunit causes Tay-Sachs Disease, so only β Hexosaminidase A
- Mutation in β subunit causes Sandhoff's Disease, so both β Hexosaminidase A and B defectas β subunit is common to both isoforms.

Symptom box

Clinical features
Tay-Sachs disease
- Loss of motor skills
- **Increased startle reaction (hyperacusis)**
- Macular pallor and **retinal cherry red spots**
- Macrocephaly
- Seizures
- Neurodegeneration

Sandhoff disease
Similar to Tay-Sachs but infants with Sandhoff disease have **hepatosplenomegaly**, cardiac involvement, and mild bony abnormalities.

Gaucher's Disease^Q

Most common lysosomal storage disorder.
Autosomal recessive.

Biochemical defect
Deficient activity of the **lysosomal** hydrolase, acid β-**glucosidase** (β **Glucocerebrosidase**)

Symptom box

Clinical manifestations Type I Gaucher disease
- Hematological features—Pancytopenia, bleeding manifestation
- Hepatosplenomegaly
- Bone pain (In a pseudo-osteomyelitis pattern)
- Pathological fractures of long bones
- Bone crises with severe pain and swelling can occur
- **No cherry red spot in the macula (But pseudocherry red spot is present)**
- **No mental deterioration.**

Diagnosis
1. X-ray Femur-**Erlenmeyer Flask Deformity**
2. Bone marrow examination
   - The **pathologic hallmark of Gaucher disease is the Gaucher cell particularly in the bone marrow.**

Gaucher cell
- They have characteristic **wrinkled paper appearance resulting from the presence of intracytoplasmic substrate inclusions.**

High Yield Points

Two other disorders with Gaucher cells
1. Granulocytic Leukemia
2. Myeloma

Treatment

Enzyme replacement therapy (ERT)^Q

1. Mannose terminated Recombinant **Human Acid Beta Glucosidase (Imiglucerase)**
   - Most symptoms including organomegaly, hematological manifestation, bone pain improves.
2. Two additional Enzyme Preparation approved by FDA are:
   i. **Velaglucerase alfa**, which is produced from human fibrosarcoma cells
   ii. **Taliglucerase alfa**, which is produced in carrot cells.

Oral substrate reduction agents

1. **Miglustat**
   - To decrease glucosylceramide by chemical inhibition of glucosylceramide Synthase.

> **HIGH YIELD POINTS**
>
> Bone marrow transplantation
> Concept of clinical manifestations and diagnosis of Gaucher Disease
> - Accumulation of Glucocerebroside in the reticuloendothelial system, so there is Hepatosplenomegaly
> - Infiltration into the bone marrow, so pancytopenia which includes thrombocytopenia, which causes bleeding manifestation and anaemia
> - Skeletal manifestation like bone pain and pathological fractures.
>
> Diagnosis
> - Gaucher disease should be considered in the differential diagnosis of patients with unexplained organomegaly, who bruise easily, have bone pain, or have a combination of these conditions.

Niemann-Pick Disease

Autosomal recessive.

Biochemical Defect

- Deficient activity of **acid sphingomyelinase**, a lysosomal enzyme encoded by a gene on chromosome 11.
- Accumulation of **sphingomyelin and other lipids in the monocyte-macrophage system.**

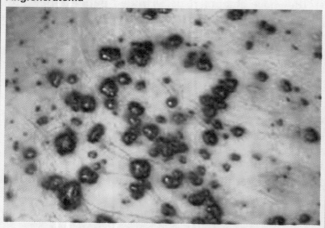

Telangiectatic skin lesion (Angiokeratoma) seen in GM1 Gangliosidoses and Fabry's Disease

Fabry's Disease

X-linked recessive condition.

Biochemical Defect

- Mutations in the α-galactosidase. Accumulation of neutral glycosphingolipids, primarily globotriaosylceramide.

Clinical Features

- **Angiokeratomas** (telangiectatic skin lesions characteristically, the lesions are most dense between the umbilicus and knees, in the "bathing trunk area").
- **Hypohidrosis**
- **Corneal and lenticular opacities**
- **Fabry crises**, -agonizing, burning pain in the hands, feet, and proximal extremities.
- Red cell casts and lipid inclusions with characteristic birefringent **"Maltese crosses"** appear in the urinary sediment

Enzyme Replacement Therapy^Q

- Recombinant α-galactosidase (**Agalzidase β or Fabrazyme**) is approved by FDA
- **Agalzidase α (Replagal).**

Krabbe Disease

- Also called **globoid cell leukodystrophy**
- Autosomal recessive.

Biochemical Defect

- Deficiency of the enzymatic activity of **galactocerebrosidase (Beta Galactosidase)**
- Accumulation of **galactosylceramide** in the white matter of brain

> **Symptom box**
>
> Clinical Features
> - Mental development is severely impaired

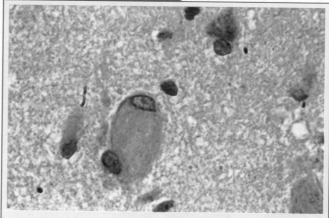

A unique and diagnostic feature of Krabbe disease is the aggregation of **engorged macrophages (globoid cell)** in the parenchyma and around blood vessels.

Farbers Disease

Biochemical Defect
- Deficiency of the lysosomal enzyme **acid ceramidase**
- The accumulation of ceramide in various tissues, especially the joints.

Symptom box

Clinical Features
- Painful joint swelling and nodule formation over the joints.
- Mimicks **Rheumatoid Arthritis.**

Wolman Disease and Cholesterol Ester Storage Disease (CESD)

- Autosomal recessive condition
- Lysosomal storage diseases.

Biochemical Defect
- Deficiency of acid lipase.
- Accumulation **of cholesterol esters and triglycerides in histiocytic foam cells of most visceral organs.**

Symptom box

Clinical Features
- Presents in 1st week of life
- Failure to thrive
- Relentless vomiting,
- Abdominal distention,

Steatorrhea
- Hepatic dysfunction and fibrosis may occur
- **Calcification of the adrenal glands is pathognomonic for the disorder**

Image-Based Information
Cherry Red Spot in the Macula

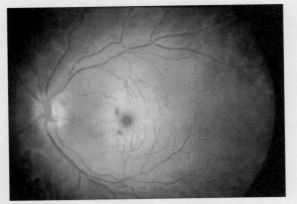

Seen in all Sphingolipidosis except
1. Gaucher's Type I (Only Pseudocherry red spot seen)
2. Fabry's Disease
3. Krabbe's Disease its presence is variable

Image-Based Information
Calcification of Adrenals-Wolman's Disease

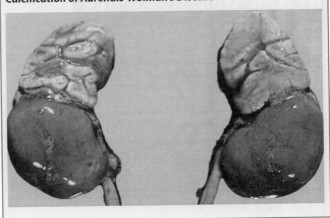

Quick Review

Most abundant phospholipid present in cell membrane is lecithin

Most abundant lung surfactant is lecithin

Phospholipid involved in programmed cell death is phosphatidyl serine.

Only antigenic phospholipid is cardiolipin

Only sphingosine containing phospholipid is sphingomyelin.

All sphingolipidosis are lysosomal storage disorders

All are Autosomal recessive EXCEPT
- Fabry's Disease [X linked Recessive]

Sphingolipidoses with no cherry red spot on the macula
- Fabry's Disease
- Gauchers Disease (Only a rare Type II has pseudocerry red spot)

Sphingolipidoses with no mental retardation
- Fabry's Disease, Gaucher's Type I

Zebra Body Inclusion seen in
- Niemann Pick's Disease

Globoid Cell Inclusion seen in
- Krabbe's Disease

Check List for Revision

- Name of all phospholipids and glycolipids
- All points in boxes and bold letters.
- Name of all sphingolipidoses with its enzyme deficiency.
- Important clinical features i.e. written in bold letters

Review Questions MCQ

1. **True about role of phospholipids:** *(PGI Nov 2016)*
   a. Cell to cell recognition
   b. Cell signaling
   c. Precursor of second messenger
   d. Mediators of inflammation
   e. Regulate membrane permeability

2. **Gangliosides contain:** *(PGI May 2015)*
   a. Phosphate
   b. Galactose
   c. Sulfate
   d. Serine
   e. Sialic acid

3. **Glycosphingolipid is made up of:** *(PGI Nov 2010)*
   a. Glucose
   b. Glycerol
   c. Sphingosine
   d. Fatty acids
   e. Thromboxane A2

4. **Alcoholic group is found in:** *(PGI June 98)*
   a. Ganglioside
   b. Sphingomyelin
   c. Cerebroside
   d. Ceramide

5. **Which of the following is a glycolipid?** *(JIPMER 2012)*
   a. Cerebroside
   b. Plasmalogen
   c. Sphingomyelin
   d. Lecithin

6. **Second messenger is produced from:**
   a. Phosphatidylinositol
   b. Phosphatidylserine
   c. Phosphatidylcholine
   d. None

Sphingolipidosis

7. **A child presents with hepatosplenomegaly and pancytopenia. Bone marrow shows "crumbled tissue paper appearance". It is due to accumulation of:** *(AIIMS May 2013)*
   a. Glucocerebroside
   b. Sphingomyelin
   c. Ganglioside
   d. Galactocerebroside

8. **Which is/are sphingolipidosis?** *(PGI May 2013)*
   a. Tay-Sachs disease
   b. Fabry's disease
   c. Krabbe's disease
   d. Sandhoff's disease
   e. Wolman's disease

9. **Sphingomyelinase deficiency is seen in:** *(AI 2010)*
   a. Niemann-Pick disease
   b. Farber's disease
   c. Tay-Sach's disease
   d. Krabbe's disease

10. **Deficiency of phosphorylating enzymes for the formation of which of the following recognition marker leads to I-cell disease?** *(Ker-2008)*
    a. GM2 ganglioside
    b. Mannose 6 phosphate
    c. Galactose
    d. Globoside

11. **Which of the following disease occurs due to the deficiency of glucocerebrosidase?** *(Ker-2008)*
    a. Gaucher's disease
    b. Pompe's disease
    c. Fabry's disease
    d. Krabbe's disease

12. **Accumulation of sphingomyelin in phagocytic cells is feature of:** *(AIIMS Nov 02)*
    a. Tay-Sach's disease
    b. Gaucher's disease
    c. Niemann-Pick disease
    d. Down's syndrome

13. **Tay-Sachs disease is due to accumulation of:** *(JIPMER 2012)*
    a. GM2 ganglioside
    b. GM1 ganglioside
    c. Glucocerebroside
    d. Galactocerebroside

14. **The enzyme defect in Wolmans disease** *(Central institute Nov 2018)*
    a. Acid lipase
    b. Acid maltase
    c. Sphingomyelinase
    d. Sucrase

Answers to Review Questions

1. **b, c, d, e. Cell signalling, Precursor of second messenger, Mediators of..., Regulate membrane...**
   - Phosphatidylinositol is involved in signal cascade pathway.
   - Phosphartidylinositol is the precursor of second messenger DAG and Ca^{2+}
   - Lipid composition regulate membrane permeability

2. **b, e. Galactose, Sialic acid** *(Ref: Harper 31/e page 202)*

   Ganglioside
   - Contain Sphingosine+Fatty acid+Oligosaccharide that contain one or two molecules of sialic acid.

3. **a, c, d. Glucose, Sphingosine, Fatty acids**
   (Ref: Harper 31/e page 196, page 202)

   Phospholipids
   Lipids containing phosphoric acid and nitrogenous base in addition to glycerol and fatty acid.
   Fatty acid + Alcohol (Glycerol/Sphingosine)+ Phosphate + Nitrogenous base

   Glycerophospholipids
   The alcohol in this group is glycerol.
   e.g. Lecithin, Cephalin

   Sphingophospholipids
   The alcohol in this group is sphingosine.
   e.g. sphingomyelin

   Glycolipids or glycosphingolipids
   Lipids containing carbohydrate apart from fatty acid and alcohol (usually sphingosine)
   Fatty acid + Alcohol (Sphingosine) + Carbohydrate
   e.g. Cerebroside, Ganglioside

4. **a, b, c, d. Ganglio.., Sphingomye..., Cerebro..., Ceram...**
   - Ceramide = Sphingosine (Amino alcohol) + Fatty acid
   - Ganglioside = Ceramide + Oligosaccharide that contains N Acetyl Neuraminic acid
   - Cerebroside = Ceramide + Monosaccharide
   - Sphingomyelin = Glycerol + 2 fatty acid + PO_4 + Choline

5. **a. Cerebroside** *(Ref: Harper 31/e page 202)*

   Glycolipids or glycosphingolipids
   1. Cerebroside 2. Globoside 3. Ganglioside

6. **a. Phosphatidylinositol** *(Ref: Harper 31/e page 202)*

   Phosphatidylinositol
   - Phosphatidylinositol presents in the cell membrane.
   - The inositol is present as its stereoisomer, myoinositol.
   - Plays an important role in cell signalling and membrane trafficking.
   - Phosphatidyl inositol is a precursor of second messenger in hormonal pathways.

   Action of Phospholipase C on Phosphatidyl Inositol 4,5 Bisphosphate (PiP_2)
   Diacyl Glycerol and Inositol Tris phosphate IP3 are formed and both act as second messengers.

Sphingolipidosis

7. **a. Glucocerebroside** *(Ref: Nelson 20/e, Chapter 86.4 Lipidoses)*

Sphingolipidoses	Enzyme deficiency	Sphingolipid accumulated
Farber disease	Ceramidase	Ceramide
Fabry disease	α-Galactosidase	Globotriaosyl Ceramide
GM1 gangliosidosis	β-Galactosidase	GM1 Ganglioside
GM2 gangliosidosis		
Tay-Sachs disease,	β-Hexosaminidases A	GM2 Ganglioside
Sandhoff disease	β-Hexosaminidases A and B	GM2 Ganglioside
Gaucher disease	Glucocerebrosidase/ β Glucosidase	Glucosylceramide
Niemann-Pick Disease	Sphingomyelinase	Sphingomyelin
Metachromatic leukodystrophy	Arylsulfatase A Sphingolipid activator Protein (SAP-1)	Sulfogalactosyl Ceramide
Krabbe disease	β-Galactosidase β-Galactocerebrosidase	Galactosyl Ceramide

Gaucher's Disease
- Most common lysosomal storage disorder
- Glucocerebrosidase defect
- Lysosomes filled with glucocerebroside
- No cherry red spot in the macula
- No mental deterioration (Type I)
- Hematological features-Pancytopenia, bleeding manifestation
- Hepatosplenomegaly
- Bone pain and pathological fractures of long bones
- X-ray Femur-Erlenmeyer flask deformity
- Bone marrow biopsy-Gaucher cell with Wrinkled paper appearance/crumbled tissue paper appearance.

8. **a, b, c, d. Tay-Sachs disease, Fabry's disease, Krabbe's disease, Sandhoff's disease**

(Ref: Nelson 20/e, Chapter 86.4 Lipidoses)

9. **a. Niemann-Pick disease**

(Ref: Nelson 20/e, Chapter 86.4 Lipidoses)

Other options

Farber's disease-Ceramidase
Tay-Sachs disease-β Hexosaminidase A
Krabbe's disease-β Galactosidase

10. **b. Mannose 6 phosphate**

(Ref: Harper 31/e page 554)

Disorders associated with defect in import of proteins to lysosomes.

I Cell Disease

- Due to defective synthesis of recognition marker– Mannose 6 phosphate
- Serum level of lysosomal enzymes raised.

11. **a. Gaucher disease**

(Ref: Nelson 20/e, Chapter 86.4 Lipidoses)

12. **c. Niemann-Pick disease**

Niemann-Pick disease

Autosomal recessive

Biochemical Defect

- Deficient activity of acid sphingomyelinase, a lysosomal enzyme encoded by a gene on chromosome 11
- Accumulation of sphingomyelin and other lipids in the monocyte-macrophage system.

Clinical Features

- Failure to thrive
- Hepatosplenomegaly
- Rapidly progressive neurodegenerative course

Treatment

- Orthotopic liver transplantation
- Amniotic cell transplantation
- Bone marrow transplantation
- Miglustat
- A phase I trial of enzyme replacement therapy for type BNPD.

13. **a. GM2 Ganglioside**

(Ref: Nelson 20/e, Chapter 86.4 Lipidoses)

14. **a. Acid Lipase** *(Ref: Nelson 20/e, Chapter 86.4 Lipidoses)*

- Acid Lipase-Wolman's disease
- Acid Maltase-Pompe's disease
- Sphingomyelinase-Niemann Pick disease
- Sucrase-Sucrose intolerance

5.3 METABOLISM OF LIPIDS

- Digestion of Lipids
- Absorption of Lipids
- Metabolism of Triacylglycerol
- Fatty Acid Synthesis
- Ketone Body Metabolism
- Fatty Acid Oxidation
- Cholesterol Synthesis

DIGESTION OF LIPIDS

The major lipids in the diet are **triacylglycerols (>90%)** and the rest is made of phospholipids, cholesterol, cholesterol esters and free fatty acids.

Enzymes for Digestion of Lipids

Lingual Lipase and Gastric Lipase

- Hydrolysis of triacylglycerols is initiated by lingual and gastric lipases
- Site: Stomach
- Action: Hydrolysis **of 3rd ester bond** of triacyl glycerol forming 1,2-Diacylglycerol.
- **30%** of total triacylglycerol digestion takes place in the stomach.

Pancreatic Lipase

- Requires a further pancreatic protein, colipase, for activity.
- **Colipase prevents the inhibition lipases by bile acids**.
- This is **the major enzyme of triacylglycerol hydrolysis**
- It is specific for the **primary ester** links—i.e., positions 1 and 3 in triacylglycerol
- Resulting in 2-monoacylglycerols and free fatty acids
- **2-Monoacylglycerol** and **fatty acids** are the major end products of luminal triacylglycerol digestion.

Pancreatic Esterase

- 25% of monoacylglycerols are hydrolysed to glycerol and fatty acids.
- Cholesterol esters and other lipid esters are also hydrolysed to some extent.

Phospholipases
- Specifically **Phospholipases A2** is present in pancreatic juice.

ABSORPTION OF LIPIDS

Role of Bile Salt/Bile Acids
- Synthesized in the **liver**.
- They are **biological detergents**.
- At physiological pH bile acids are seen ionized (anions) so bile acids and bile salts (anions) are interchangeably used.
- The products of lipid digestion are hydrophobic molecules.
- Bile salt/bile acids helps in **emulsification of products of lipid digestion to micelle**.
- The minimal concentration of bile acids necessary for micelle formation is called **critical micellar concentration. It is about 5 mm.**
- Micelles are less than 1μm in diameter and are soluble.
- They allow the products of digestion, to be transported through the aqueous environment of the intestinal lumen to come into close contact with the brush border of the mucosal cells, allowing uptake into the epithelial cells.
- The fat-soluble **vitamins, A, D, E and K** also transported in the micelles.

Within the Intestinal Epithelial Cells
- 1-monoacyglycerols are hydrolyzed to **fatty acids and glycerol**.
- **Short chain fatty acid and medium chain fatty acids are passed into portal vein** without any modification.
- Glycerol released in the intestinal lumen is absorbed into the hepatic portal vein.
- Long-chain fatty acid and 2-monoacylglycerol are transported into endoplasmic reticulum.
- Long-chain fatty acids and 2-monoacylglycerols are **reacylated to triacylglycerols** via the **monoacylglycerol pathway.**
- Cholesterol is esterified by Cholesterol Acyltransferase
- The newly synthesized **triacylglycerol, cholesterol ester and phospholipids are incorporated into chylomicrons.**
- Chylomicrons are secreted into the lymphatics, enter blood stream via thoracic duct.

METABOLISM OF TRIACYLGLYCEROL

Synthesis of Triacylglycerol
Triacylglycerol are the predominant form of simple lipids, which are composed of glycerol esterified to three fatty acids.

Site: Almost all tissues but predominantly in **Liver and Adipose tissue, Intestinal mucosal cells**

Organelle: Majority **Endoplasmic Reticulum**, a few in mitochondria.

Three Steps for TAG Synthesis
1. Activation of fatty acid.
2. Activation of glycerol.
3. Esterification of fatty acid to glycerol.

Activation of Fatty Acid

Enzyme: Acyl-CoA synthetase or thiokinase
Reaction: Fatty acid converted to CoA derivative.
Activation of Glycerol Enzyme: Glycerol kinase
Reaction: Glycerol is phosphorylated to glycerol 3-phosphate.

In Muscle and Adipose Tissue
Glycerol kinase is absent **in muscle and white adipose tissue**.

Glycerol 3-phosphate is formed **from dihydroxy acetone phosphate**, an intermediate in glycolysis.

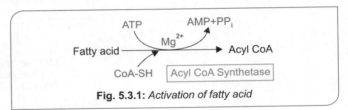

Fig. 5.3.1: *Activation of fatty acid*

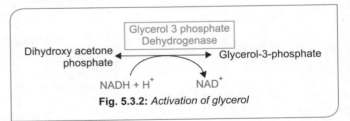

Fig. 5.3.2: *Activation of glycerol*

Esterification of Activated Fatty Acid to Triacylglycerol

Enzyme: Acyltransferase
Reaction: Transfer of 3 Acyl moiety to glycerol successively.
Synthesis of Triacylglycerol in the Intestine- Monoacylglycerol Pathway

In the **intestinal mucosa**, Monoacylglycerol transferase converts

Monoacylglycerol to 1,2-Diacylglycerol in the **Monoacylglycerol pathway.**

Degradation of Triacylglycerol (Lipolysis)

- Triacylglycerols are hydrolyzed by a **lipase** to their constituent fatty acids and glycerol
- Much of this hydrolysis (lipolysis) occurs in **adipose tissue.**
- The free fatty acids are released into the plasma, where they are found combined with **serumalbumin.**
- The free fatty acid uptake into tissues (including liver, heart, kidney, muscle, lung, testis, and adipose tissue, but **not readily by brain**), where they are oxidized or re-esterified.
- The utilization of glycerol depends upon whether such tissues have the enzyme **glycerol kinase**.
- The Glycerol kinase is found in significant amounts in **liver, kidney, intestine, brown adipose tissue, and the lactating mammary gland.**

NB: Brown adipose tissue contain glycerol kinase unlike white adipose tissue which lack glycerol kinase.

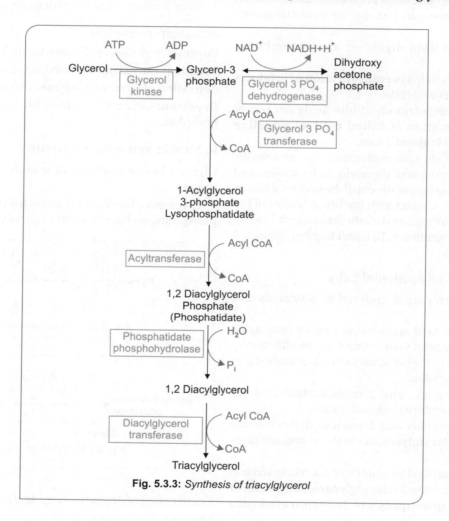

Fig. 5.3.3: *Synthesis of triacylglycerol*

Lipolysis in Adipose Tissue

- Stored fat, triacylglycerol (TAG) is degraded.
- By special enzyme: **hormone sensitive Lipase (hSL)**
  Reaction: Hormonesensitive lipase removes fatty acid from **1st and 3rd carbon** of TAG to form diacylglycerol, and Monoacylglycerol sequentially.

The hormone sensitive lipase **cannot remove fatty acid from 2nd carbon atom** which is removed by 2-monoacyl glycerol lipase.

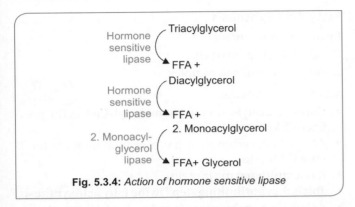

Fig. 5.3.4: *Action of hormone sensitive lipase*

- It is **active in phosphorylated state**
- It is **inactive in dephosphorylated state**

Hormone Sensitive Lipase is Activated by
- **Glucagon**^{Q}
- Catecholamines (epinephrine and norepinephrine)
- ACTH
- TSH
- Glucocorticoids
- Thyroid Hormones
- Growth hormone
- α and β melanocyte stimulating hormone (MSH)
- Vasopressin

Hormone Sensitive Lipase is Inactivated by
- **Insulin**^{Q}
- Nicotinic acid
- Prostaglandin E1

Regulation of Hormone Sensitive Lipase
- The activity of this enzyme is under the control of hormones, hence the name.
- **It is present in the adipose tissue.**

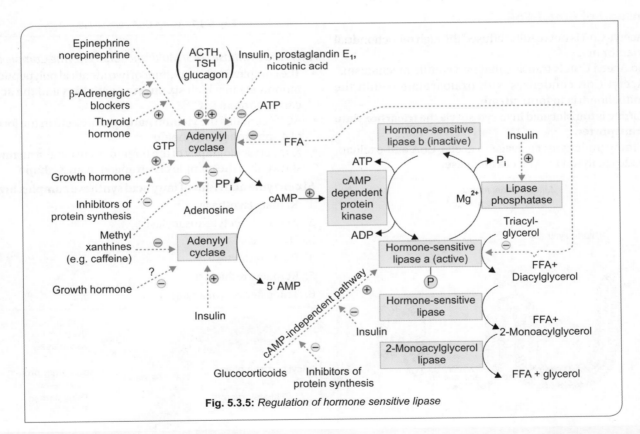

Fig. 5.3.5: *Regulation of hormone sensitive lipase*

Clinical correlation
In Diabetes mellitus hormone sensitive lipases are activated

DE NOVO FATTY ACID SYNTHESIS (LIPOGENESIS)

- Major contribution to discovery by Feodor Lynen, hence the pathway is otherwise called **Lynen's Spiral**
- Site^Q: Liver, kidney, brain, lung, lactating mammary gland, and adipose tissue
- **Organelle**^{Q}: By an extramitochondrial system [in the cytosol]
- Cofactor^Q requirements include
- **NADPH, ATP, Mn2$^+$, Biotin, and HCO3$^-$** (as a source of CO_2)
- **Acetyl-CoA**^{Q} is the principal building block of fatty acid.

Sources of Acetyl-CoA

- Aerobic glycolysis (pyruvate to acetyl-CoA by PDH in the mitochondria)
- Fatty acid oxidation (in the mitochondria)

Transport of Acetyl-CoA

- Acetyl-CoA is not readily diffused through mitochondrial membrane.
- So Acetyl CoA is translocated by a shuttle mechanism.
- Acetyl-CoA condenses with oxaloacetate within the mitochondria to form **citrate.**
- Citrate is translocated in to cytosol via the **tricarboxylate transporter.**
- Citrate undergoes cleavage to acetyl-CoA and oxaloacetate catalysed by **ATP citrate lyase.**

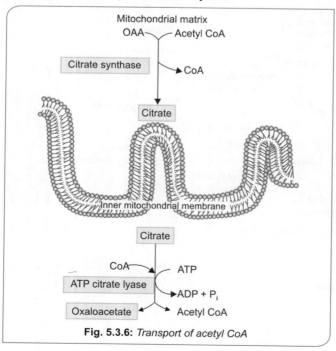

Fig. 5.3.6: Transport of acetyl CoA

Fatty Acid Synthesis

By two enzyme system

1. Acetyl-CoA carboxylase
2. Fatty acid synthase complex

Acetyl CoA Carboxylase

- Converts acetyl-CoA (2C) to **malonyl-CoA** in the presence of ATP.
- Acetyl-CoA carboxylase has a requirement for the **B vitamin biotin.**
- It is a **multienzyme** protein.
- This is the **rate limiting step**^{Q} **in the fatty acid synthesis**.
- *Acetyl-CoA carboxylase is not a part FAS complex*^{Q}

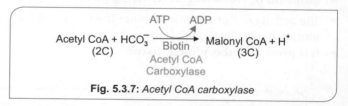

Fig. 5.3.7: Acetyl CoA carboxylase

Fatty Acid Synthase (FAS) Multifunctional Enzyme Complex^Q

- The complex is a homodimer of two identical polypeptide monomers in which **six enzyme activities** and the acyl carrier protein (ACP)
- ACP^Q contains the vitamin **pantothenic acid** in the form of 4'-phosphopantetheine
- X-ray crystallography of the three-dimensional structure, shown that the complex is arranged in an **X-shape**

Six enzyme activities of fatty acid synthase complex are

1. Ketoacylsynthase
2. Malonyl-acetyltransacylase
3. Hydratase
4. Enoyl reductase
5. Ketoacyl reductase
6. Thioesterase (deacylase)

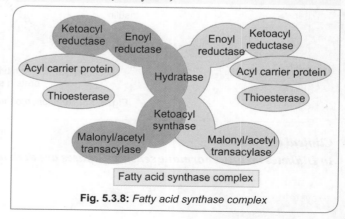

Fig. 5.3.8: Fatty acid synthase complex

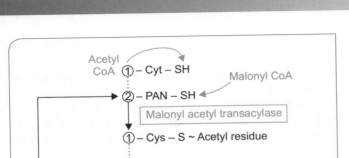

Fig. 5.3.9: *De novo fatty acid synthesis*

While malonyl-CoA combines with the adjacent—SH on the 4'-phosphopantotheine of ACP of the other monomer. These reactions are catalyzed by **malonyl-acetyl transacylase**, to form **acetyl (acyl)-malonyl enzyme**

2. **Ketoacyl synthase**

   The acetyl group attacks the methylene group of the malonyl residue, catalyzed by 3-ketoacyl synthase, and **liberates CO_2**, forming 3-ketoacyl enzyme (acetoacetyl enzyme) (reaction2), freeing the cysteine—SH group.

Reduction Reactions

1. **Ketoacyl reductase**

   The 3-ketoacyl group is reduced to hydroxy acyl group by Ketoacyl reductase

2. **Hydratase**
   - Hydroxy acyl group is dehydrated to unsaturated acyl (enoyl) group by hydratase

3. **Enoyl Reductase**
   - Unsaturated acyl (enoyl) group is reduced to Acyl group by enoyl reductase

Releasing of Fatty Acid

These reactions of condensation and reduction repeated several times till the desired acyl group is assembled on the enzyme.

Thioesterase
- Fatty acid (acyl group) is liberated from the enzyme complex by the activity of the sixth enzyme in the complex, thioesterase (deacylase).

Regulation of Fatty Acid Synthesis

Rate limiting step: Acetyl-CoA carboxylase

Short-term Regulation

1. Allosteric regulation
2. Covalent modification

Long-term Regulation

1. Control of enzyme synthesis by regulation of gene expression

Short-term Regulation

1. **Allosteric regulation:**
   i. **Positive Allosteric regulation or allosteric activation of acetyl-CoA carboxylase by citrate^Q**
   ii. Citrate promotes the conversion of Acetyl-CoA carboxylase from an **inactive dimer to active polymeric form.**

Sources of NADPH

1. The main Source of NADPH for lipogenesis is the **HMP Pathway (pentose phosphate pathway)**
2. **Malic enzyme** (NADP malate dehydrogenase)
3. The **extramitochondrial isocitrate dehydrogenase** reaction.

Reactions of Fatty Acid Synthase Complex

By three stages

1. Condensation
2. Reduction
3. Release of fatty acid

Condensation Reactions

1. **Malony/Acetyl transacylase**

   A priming molecule of acetyl-CoA combines with a cysteine—SH group.

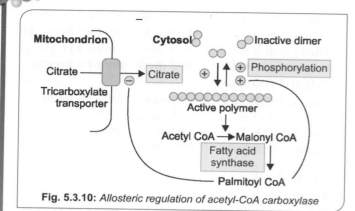

Fig. 5.3.10: Allosteric regulation of acetyl-CoA carboxylase

Long chain acyl-CoA inactivates acetyl-CoA carboxylase by
i. Favoring phosphorylation of Acetyl-CoA carboxylase
ii. **By inhibiting the tricarboxylic transporter** that transport citrate from mitochondria to cytosol.

Covalent Modification

By phosphorylation-dephosphorylation by hormones
Acetyl-CoA carboxylase is **active** in **dephosphorylated state**^{Q} and **inactive** in **phosphorylated** state.

Glucagon and Epinephrine

Glucagon and epinephrine inactivate Acetyl-CoA carboxylase by phosphorylating the enzyme.

Insulin

Insulin activate acetyl-CoA carboxylase, by dephosphorylating the enzyme.

Activation of Acetyl-CoA Carboxylase	Inactivation of Acetyl-CoA Carboxylase
Citrate	Acyl-CoA
Insulin	Glucagon, Epinephrine
Dephosphorylation	Phosphorylation

Fates of Acyl-CoA

1. Esterified into triacylglycerol
2. Chain elongation to produce very long chain fatty acid
3. Desaturation to produce unsaturated fatty acid
4. Esterified into cholesterol Ester.

Elongation of Fatty Acid Chains

- Occurs in the **endoplasmic reticulum (the "microsomal system")** and some in mitochondria also.
- By fatty acid elongase system
- Elongates saturated and unsaturated fatty acyl-CoAs (from C10 upward) by two carbons
- Malonyl-CoA donates 2 carbon atoms in step-wise manner.
- In the same manner as fatty acid synthase complex in the cytosol.
- **NADPH** is required at the two reductase step
- Elongation reaction is particularly increased in brain during myelination to provide C22 and C24 fatty acids for sphingolipids.

Synthesis of Unsaturated Fatty Acids

- Occurs in **endoplasmic reticulum**.
- By an enzyme called fatty acyl-CoA desaturase.
- The most common desaturase is Δ^9 **Desaturase, a monoxygenase**.
- The first double bond introduced **is always in the Δ^9 position**.
- Some unsaturated fatty acids are essential fattyacids.

Humans cannot introduce additional double bond in the fatty acid chain beyond Δ^9 (i.e. between C-10 and terminal methyl group). Hence linoleic acid ($\Delta^{9,12}$) and linolenic acid ($\Delta^{9,12,15}$) become essential fatty acid.

OXIDATION OF FATTY ACID

The process by which fatty acids are successively cleaved to two carbon compounds, acetyl-CoA and release energy.

Different types of fatty acid oxidation are:
1. β Oxidation of fatty acid
2. Oxidation of very long chain fatty acids
3. Oxidation of unsaturated fatty acids
4. Oxidation of odd chain fatty acid
5. α Oxidation of fatty acid
6. ω Oxidation of fatty acid.

β Oxidation of Fatty Acid

Most common type of fatty acid oxidation.^{Q}

Two carbon at a time are cleaved from carboxyl end of activated fatty acid as **acetyl-CoA**.^{Q}

The cut is between α and β carbon atom and hence the name Beta oxidation.
Site-Organs-Liver, Adipose tissue, Muscle
Organelle-**Mitochondria**^{Q}.

Steps of Fatty Acid Oxidation

1. Activation of fatty Acids.
2. Transport of activated fatty acid from cytosol to mitochondria
3. Reactions of betaoxidation

Activation of Fatty Acids

- **Site:** Cytoplasm
- **Enzyme:** Acyl-CoA synthetase/thiokinase is present in outer mitochondrial membrane.
- The only step in the complete degradation of fatty acid that require energy.
- **Two inorganic phosphates**[Q] are used.

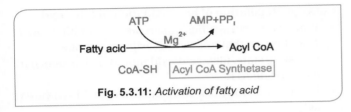

Fig. 5.3.11: *Activation of fatty acid*

Transport of Acyl-CoA into Mitochondria

- Long-chain fatty acids penetrate the innermitochondrial membrane as **Carnitine**[Q] derivatives.
- Short-chain and medium-chain fatty acids do not need carnitine for transport.

Carnitine

- **Beta hydroxy gamma trimethyl ammonium butyrate**.
- Store house is the **muscle**.
- Synthesized from **Lysine** in the liver.
- **S. Adenosyl methionine** is the methyl donor.
- **Ascorbic acid** is the vitamin needed for its synthesis

Steps of Transport of Activated Fatty Acid

1. **Carnitine Acyltransferase I**
   - Also called as Carnitine Palmitoyl transferase-I (CPT-I) as most common fatty acid translocated is palmitic acid.
   - Located in the **outer mitochondrial membrane**.
   - Transfer acyl group present in the Acyl-CoA to carnitine to formacylcarnitine.
2. **Carnitine Acylcarnitine Translocase**
   - Acylcarnitine is translocated across the inner mitochondrial membrane.
3. **Carnitine Acyltransferase II**
   - Also called Carnitine Palmitoyl transferase-II (CPT-II)
   - Located in the Inner mitochondrial membrane.
   - Converts Acyl Carnitine to Acyl-CoA
4. **Carnitine Acyl Carnitine translocase**
   - Returns the carnitine back to cytosol

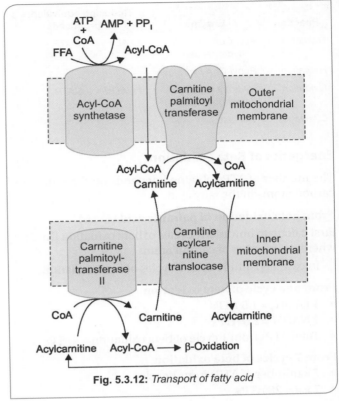

Fig. 5.3.12: *Transport of fatty acid*

Reactions of Beta Oxidation

Involves sequence of four reaction successively cleave 2 carbons as **Acetyl-CoA**

See Figure 5.3.13 for reactions of Beta oxidation.

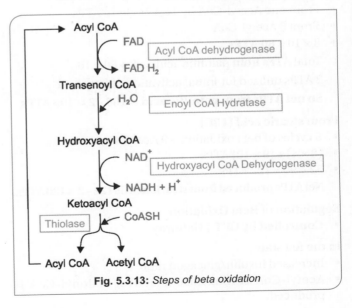

Fig. 5.3.13: *Steps of beta oxidation*

Reaction	Enzyme	Reducing equivalents produced
Oxidation	Acyl-CoA dehydrogenase	1 $FADH_2$ = 1.5 ATP
Hydration	Enoyl-CoA hydratase	—
Oxidation	Hydroxy acyl CoA dehydrogenase	1 NADH = 2.5 ATP
Cleavage	Thiolase	—

Energetics of Beta Oxidation

The number of ATPs obtained depends on the number of carbon atoms in the fatty acid.

From beta oxidation of palmitic acid (C-16)

First calculate how many cycles of beta oxidation [(n/2)–1] where n = number of carbon atoms

In case of palmitic acid, 7 cycles of beta oxidation

From one cycle of beta oxidation
- 1 $FADH_2$ = 1.5 ATPs
- 1 NADH = 2.5 ATPs
- Total ATPs from 1 cycle of beta oxidation = 4 ATPs

From 7 cycles of beta oxidation
- 7 × number of ATPs from 1 cycle
- 7 × 4 = **28 ATPs**

Second calculate how many 2 carbon acetyl-CoA from 16 carbon palmitic acid
- (n/2) where n = number of carbon atoms

So in case of palmitic acid
- (16/2) = 8 Acetyl-CoA
- From one Acetyl-CoA by TCA Cycle 10 ATPs
- From 8 Acetyl-CoA
- 8 × 10 = **80 ATPs**

Total ATPs from palmitic acid = 28 + 80 = 108
2 ATPs utilized for initial activation of fatty acid

So net ATPs from palmitic acid = 108 – 2 = 106 ATPs

From stearic acid (18C)
- 8 cycles of beta oxidation + 9 Acetyl-CoA
- (8 × 4) + (9 × 10) ATPs
- 32 + 90 = 122 ATPs

Net ATPs produced from stearic acid is 122 – 2 = 120 ATPs.

Regulation of Beta Oxidation
- **Controlled by CPT-I Gateway**

In the fed state
- Increased insulin/glucagon ratio.
- Acetyl-CoA carboxylase is active, Malonyl-CoA is produced.
- Malonyl-CoA^Q is an inhibitor of CPT-I.
- So decreased beta oxidation.

In the fasting state
- Decreased insulin/glucagon ratio.
- Acetyl-CoA carboxylase is inactive, Malonyl-CoA is not produced
- So CPT-I is active
- So increased beta oxidation

> **Concept–Regulation of beta oxidation of fatty acid**
> - When fatty acid synthesis takes place it inhibit its own oxidation, so that futile cycles will not operate.
> - In fasting state, fatty acid oxidation is active, hence it can provide ATP.
> - In fed state, body can store fat as fatty acid synthesis active.

CLINICAL CORRELATIONS

Defects in β Oxidation of Fatty Acids

Medium Chain Acyl-CoA Dehydrogenase Deficiency (MCAD deficiency)
- Fasting hypoglycemia
- No ketone bodies
- Vomiting, coma and death
- C8-C10 acyl carnitine in the blood
- Episodes may be provoked by overnight fast in an infant
- Primary treatment is IV glucose
- Prevention is by frequent feeding with high carbohydrate low fat diet.

Fasting hypoglycemia in MCAD deficiency
- Lack of ATP to support gluconeogenesis, as Fatty acid oxidation is not taking place.
- Acetyl-CoA, the end product of beta oxidation is the allosteric activator of pyruvate carboxylase, the enzyme of gluconeogenesis
- Hence fasting hypoglycemia.

Sudden infant death syndrome (SIDS) common in MCAD deficiency
- Infants with MCAD deficiency if not fed for 12 hours or more especially in the nights, hypoglycemia sets in due to lack of gluconeogenesis and death results.

Jamaican Vomiting Sickness
- Ackee fruit that grows in Jamaica and West Africa contain a toxin called hypoglycin.
- Hypoglycin is an inhibitor of fatty Acyl-CoA dehydrogenase.
- Severe hypoglycemia if ingested as it inhibit beta oxidation of fatty acids.

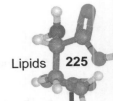

- Charaterised by sudden onset of vomiting 2-6 hrs after ingestion, followed by convulsion, coma and death.

Carnitine Deficiency

It can occur particularly in the newborn—and especially in preterm infants and in hemodialysis.

Clinical Features
- Hypoglycemia, which is a consequence of impaired fatty acid oxidation.
- Muscular weakness, due to lipid accumulation.
- Hence they have a vitamin-like dietary requirement for carnitine.
- Treatment is by oral supplementation with carnitine

CPT-I Deficiency
- Affects only the liver, resulting in reduced fatty acid oxidation and ketogenesis, with hypoglycemia.

CPT-II Deficiency
- Affects primarily skeletal muscle and, when severe, the liver.

Sulfonylurea in Type II Diabetes Mellitus
- The sulfonylurea drugs **(glyburide [glibenclamide]** and **tolbutamide)**, used in the treatment of type II diabetes mellitus, reduce fatty acid oxidation by inhibiting CPT-I
- Hence it reduces gluconeogenesis, there by preventing hyperglycemia.

Acute Fatty Liver of Pregnancy
- Defects in long chain 3-hydroxyacyl-CoA dehydrogenase.

Oxidation of Very Long Chain Fatty Acid (VLCFA)
- By a modified β oxidation pathway
- For fatty acids > C_{20}, C_{22}
- Activation of very long chain fatty acid is needed. It takes place within the peroxisomes.
- Oxidation in peroxisome produces acetyl-CoA and $H_2O_2^Q$ (instead of $FADH_2$)
- Oxidation of VLCFA takes place in **Peroxisomes** till Octanoyl-CoA
- Further oxidation of acetyl-CoA and octanoyl-CoA takes place in the **mitochondria.**
- Peroxisomes do not attack shorter chain fatty acids beyond octanoyl-CoA.
- **No ATP is generated**.
- Peroxisomal oxidation **shorten the side chains of cholesterol** in bile acid formation.
- Peroxisome takes part in the synthesis of **cholesterol, and dolichol and ether glycerolipids.**

Clinical Correlation Defects in the Oxidation of VLCFA in the Peroxisomes

Peroxisomal Disorders
The peroxisomal diseases are genetically determined disorders caused either by the failure to form or maintain the peroxisome or by a defect in the function of a single enzyme that is normally located in this organelle.

Basic Defect
- The proteins that are destined to the peroxisomes has a specific targeting sequence called peroxisome targeting sequence (PTS).
- Defects in the PTS leads to peroxisomal disorders

Peroxisomal Ghost
- Absence or reduction in the number of peroxisome is pathognomonic for disorders of peroxisome biogenesis.
- In most disorders there are membranous sacs that contain peroxisomal integral membrane proteins, which lack the normal complement of matrix proteins; these are peroxisome "ghosts".

Some Peroxisomal Disorders
- **Zellweger syndrome (Cerebero hepatorenal disease)**
- Neonatal adrenoleukodystrophy (NALD)
- Infantile Refsum's disease (IRD)
- Rhizomelic chondrodysplasia punctata (RCDP)

Lorenzo's Oil Therapy
- Treatment for **Adrenoleuko dystrophy**^{Q}
- **Lorenzo's oil** (4:1 mixture of glyceryl trioleate and glyceryltrierucate).

Oxidation of Unsaturated Fatty Acids
- Occurs by a modified β-oxidation pathway.
- In the mitochondria.
- Till the double is reached, normal beta oxidation will take place.
- An additional **isomerase and a reductase** helps to shifts the double bond.
- The first step, **FAD dependent Acyl-CoA dehydrogenase is bypassed**.
- **$FADH_2$ is not formed**.
- The energy yield by oxidation of unsaturated Fatty Acid is **1.5 ATP less per double bond**^{Q}.

Oxidation of Odd Chain Fatty Acid
- Takes place in the **mitochondria.**
- Oxidation of a fatty acid with an odd number of carbon atoms yields acetyl-CoA and a molecule of **propionyl- CoA**

- The propionyl residue from an odd-chain fatty acid is the only part of a fatty acid that is glucogenic.

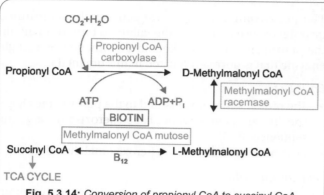

Fig. 5.3.14: Conversion of propionyl CoA to succinyl CoA

Quick Review—Sites of Oxidation of Fatty Acids	
Beta oxidation of fatty acid	Mitochondria
Beta oxidation of unsaturated fatty acids	Mitochondria
Beta oxidation of very long chain fatty acid	Peroxisomes up to Octanoyl-CoA, then rest in mitochondria
Alpha oxidation of fatty acid	Peroxisomes, smooth endoplasmic reticulum
Omega oxidation of fatty acid	Smooth endoplasmic reticulum/microsomes
Activation of fatty acid	Cytosol

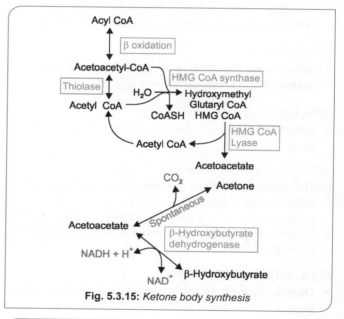

Fig. 5.3.15: Ketone body synthesis

Minor Oxidation Pathways of Fatty Acid

α Oxidation of Fatty Acids
- Site: **Endoplasmic reticulum and peroxisome**.
- Removal of **one carbon at a time from the α carbon atom**.
- For oxidation of **branched chain fatty acid** to remove methyl group at the branch points.
- Used for oxidation of **phytanic acid, a major dietary methylated fatty acid seen in dairy products**.

Refsum's Disease
Defect in alpha oxidation of phytanic acid in peroxisomes.

Enzyme defect is phytanoyl-CoA hydroxylase^Q in the peroxisome.

The manifestation of classic Refsum's disease includes impaired vision **from retinitis pigmentosa, ichthyosis, peripheral neuropathy, ataxia, and, occasionally, cardiac arrhythmias.**

Classic Refsum's disease often does not manifest until young adulthood, but visual disturbances such as night blindness, ichthyosis, and peripheral neuropathy may already be present in childhood and adolescence.

Restrict dietary dairy products and green leafy vegetables.

ω Oxidation of Fatty Acid
- Occur in the **endoplasmic reticulum/microsomes**.
- Oxidation involves methyl group at the ω end.
- Involves hydroxylation at the terminal methyl group at the ω end by Cyt P450 (mixed function oxidase).
- Resulting in short-chain Dicarboxylic acid (double headed fatty acid).

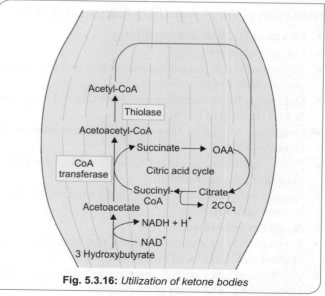

Fig. 5.3.16: Utilization of ketone bodies

KETONE BODIES

Ketogenesis occurs in metabolic conditions associated with high rate of fatty acid oxidation.

- Primary ketone body is **acetoacetate**
- Secondary ketone bodies are **acetone and beta hydroxybutyrate.**
- Concentration of ketone bodies in the blood does not normally exceed **0.2 mmol/L**.
- **In normal persons the ratio of beta hydroxybutyrate to acetoacetate is 1:1.**
- In ketosis the ratio of beta hydroxybutyrate to acetoacetate is **6:1.**

Ketone Body Synthesis

Site: Exclusively liver mitochondria^Q

- **Acetoacetyl-CoA** from beta oxidation is the starting material.
- **HMG CoA synthase**^{Q} is the rate limiting step.
- HMG CoA synthase is common to cholesterol synthesis and ketone body synthesis.

Steps of Ketone Body Synthesis

1. Two acetyl-CoA molecules formed in β-oxidation condense to form acetoacetyl-CoA by a reversal of the **thiolase** reaction.
2. Condensation of acetoacetyl-CoA with another molecule of acetyl-CoA by **hydroxy-3-methylglutaryl-CoA synthase** forms 3-hydroxy-3-methylglutaryl-CoA **(HMG-CoA).**
3. **Hydroxy-3-methylglutaryl-CoA lyase** then causes acetyl-CoA to split off from the HMG-CoA, leaving free acetoacetate.
4. Acetoacetate continually undergoes spontaneous decarboxylation to yield acetone.
5. Acetoacetate converted to β hydroxybutyrate by the mitochondrial enzyme β **hydroxybutyrate dehydrogenase.**

High Yield Points

The pathways where HMG-CoA is an intermediate are
- Ketone body Synthesis
- Cholesterol synthesis
- Leucine catabolism

Utilization of Ketone Bodies

Ketone bodies serve as a fuel for extrahepatic tissues.

Almost all the organs utilize ketone bodies with the exception of **liver and RBCs.**^{Q}

Steps of utilisation of ketone bodies

- In extrahepatic tissues, acetoacetate is activated to **acetoacetyl-CoA by succinyl-CoA-acetoacetate CoA transferase or thiophorase**^{Q}.
- CoA is transferred from succinyl-CoA to form acetoacetyl-CoA.
- The acetoacetyl-CoA is split into two acetyl-CoAs by thiolase.
- Acetyl-CoA is oxidized in the citric acid cycle.
- Acetone is difficult to oxidize in vivo and to a large extent is volatilized in the lungs.

Extra Edge

Energy yield from ketone bodies from acetoacetate-19 ATPs
Acetoacetate yield two mols of acetyl-CoA. Two acetyl-CoA yields 20 ATPs via TCA cycle
   But activation of acetoacetate to acetoacetyl-CoA, by CoA transferase, Succinyl-CoA is converted to succinate without generation of GTP.
   So 20–1 = 19 ATPs are generated

From beta hydroxybutyrate-21.5 ATPs Generate 1 NADH = 2.5 ATPs
2 Acetyl-CoA = 20 ATPs
Activation of acetoacetate, needs 1 ATP
Hence 22.5–1 = 21.5 ATPs

High Yield Points

Test for ketone bodies in urine
- Gerhardt's ferric chloride Test: Answered by **acetoacetate**
- Rothera's Nitroprusside test: Answered by **acetoacetate and acetone**
- NB: None of these tests answer beta hydroxybutyrate the predominant ketone body in ketosis.

Quick review: Ketone bodies
- Primary ketonebody-Acetoacetate
- Secondary ketone bodies: Acetone, Beta hydroxybutyrate
- Neutral ketonebody-Acetone
- Ketone body excreted through lungs: Acetone
- Site of synthesis of ketone bodies: Liver mitochondria
- Organs which do not utilise ketone bodies: Liver, RBCs
- Rate limiting step of ketone body synthesis-HMG CoA synthase
- Ketone bodies that do not answer Gerhadt's test: beta hydroxybutyrate and acetone.
- Ketone body that does not answer neither Rothera test nor Gerhadt's test: Beta hydroxybutyrate

Clinical Correlation: Fatty Liver^Q

Lipid mainly as **triacylglycerol**^{Q} can accumulate in the liver, called fatty liver.

Nonalcoholic Fatty Liver Disease (NAFLD)
Stages of Progression of NAFLD

Nonalcoholic steatohepatitis (NASH), *which can progress to liver diseases including* cirrhosis, hepatocarcinoma, and liver failure.

Causes of Fatty Liver[Q]

- The basic cause of fatty liver is imbalance between the rate of **Triacyl Glycerol synthesis and its export** from the liver

This can be due to:

a. **Starvation**
b. The feeding of high-fat diets.
c. **In uncontrolled diabetesmellitus**
d. Twin lamb disease
e. Ketosis in cattle

2. **Due to a metabolic block in the production of plasma lipoproteins, thus allowingtriacylglycerol to accumulate.**

This may be due to:

- A block in apolipoprotein synthesis as in Kwashiorkar
- A block in the synthesis of the lipoprotein from lipid and apolipoprotein
- A failure in provision of phospholipids that are found in lipoproteins, or
- A failure in the secretory mechanism itself
- **Orotic acid** also causes fatty liver
- The antibiotic **puromycin, ethionine, carbon tetrachloride, chloroform, phosphorus, lead, and arsenic** all cause fatty liver
- **Lack of lipotropic factors**[Q]

HIGH YIELD POINTS

Lipotropic factors are
- Choline, betaine
- **Vitamin E**-supplemented diets
- Methionine in S Adenosyl Methionine trap available adenine and prevent synthesis of ATP
- **Selenium**
- Essential fatty acid. e.g. linoleic acid
- Pyridoxine and pantothenic acid

Alcoholic Fatty Liver

Alcoholic fatty liver is the first stage in **alcoholic liver disease (ALD)** which is caused by **alcoholism** and ultimately leads to **cirrhosis**.

Fatty liver and Gout in Alcoholic Liver Disease

Oxidation of ethanol by **alcohol dehydrogenase** and **aldehyde dehydrogenase** leads to excess production of NADH.

This results in increased NADH/NAD+ ratio. This result in:

- Increased esterification of fatty acids to form triacylglycerol, resulting in the fatty liver.
- Increased (lactate)/(pyruvate), resulting in **hyperlactic acidemia,** which decreases excretion of uric acid, aggravating **gout.**

CHOLESTEROL

- **Exclusive animal sterol** never seen in plants.
- Major component of plasma membrane.
- Made up of steroid nucleus called **"Cyclopentanoperhydrophenanthrene (CPP)"**.
- Total number of carbon atom in cholesterol is **27**.
- **Amphipathic** in nature.

Cholesterol Synthesis

Major sites—**All tissues containing nucleated cells are capable of cholesterol synthesis** especially in liver, adrenal cortex, testes, ovaries, intestine

Orgenelle—**Smooth endoplasmic reticulum and Cytoplasm**

Starting Material–**Acetyl-CoA**

Steps of Cholesterol Synthesis

Formation of HMG-CoA

- Two molecules of Acetyl-CoA (2C) condense to form acetoacetyl-CoA (4C) by the enzyme **thiolase.**
- Acetoacetyl-CoA condenses with a third molecule of acetyl-CoA to form HMG-CoA by the enzyme **HMG CoA synthase.**

Synthesis of mevalonate

- HMG-CoA (6C) converted to mevalonate (6C) by HMG-CoA reductase
- *This is the rate limiting step*[Q]
- Takes place in the **Endoplasmic reticulum**.
- **NADPH** is required.
- Statins are competitive inhibitor of HMG-CoA reductase.

Generation of isoprenoid units (5C)

- Mevalonate on decarboxylation and phosphorylation to form isoprenoid units.

Condensation of 5 carbon isoprenoid units to form squalene (30C)

- Two 5C unit condense to form **10C compound–Geranyl pyrophosphate**
- 10C unit, geranyl pyrophosphate condense with a 50 isoprenoid unit to form **15C compound-Farnesyl diphosphate.**

- Two farnesyl diphosphate (15C) condense to form **30C compound-Squalene**.

Formation of cholesterol
- Linear 30C molecule cyclises to form a structure that closely resembles steroid nucleus called lanosterol.
- Lanosterol undergo further modification to form cholesterol (27C)
- The intermediates in the conversion of Lanosterol to Cholesterolare
  - 14-desmethyl-lanosterol
  - Zymosterol
  - Desmosterol

> **HIGH YIELD POINTS**
>
> **Lanosterol**
> - First Cyclical Compound formed
> - First Steroid Compound formed

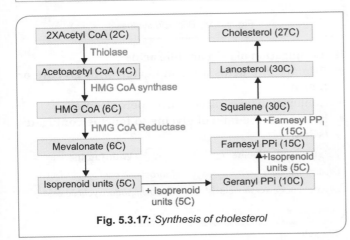

Fig. 5.3.17: *Synthesis of cholesterol*

Uses of Isoprenoid Units in Farnesyl and Geranyl Diphosphate

1. Polyisoprenoid compounds **dolichol and ubiquinone** are formed from farnesyl diphosphate.
2. **Prenylation** of proteins:
   - It is a post translational modification.
   - GTP binding proteins are prenylated.
   - Facilitate anchoring of proteins to lipid membranes.
   - Facilitate protein trafficking.

Regulation of Cholesterol Synthesis
Rate limiting enzyme is HMG-CoA reductase.
1. Feedback inhibition
2. Feedback regulation
3. Hormonal regulation
- HMG-CoA reductase is inhibited by **mevalonate and cholesterol.**

Feedback Regulation
- Cholesterol represses the transcription of genes for HMG-CoA reductase.
- This acts **via sterol regulatory element binding protein (SREBP.)**

Hormonal Regulation
By covalent modification—Phosphorylation dephosphorylation
- HMG-CoA reductase is active in **dephosphorylated** state and vice versa.
- **Insulin and thyroxine increase the activity of HMG-CoA reductase**
- **Glucagon and flucocorticoids decrease the activity of HMG-CoA reductase**

Insulin has a dominant role in regulation than glucagon

> **HIGH YIELD POINTS**
>
> **Key Points Cholesterol Regulation**
> - Dietary cholesterol inhibit its own synthesis by repressing HMG-CoA reductase enzyme.
> - Decrease of 100 mg of dietary cholesterol causea decrease of approximately 0.13 mmol/L of serum cholesterol.
> - HMG-CoA reductase is active in dephosphorylated state.
> - Insulin and thyroxine favour cholesterol synthesis.
> - Glucagon and glucocorticoids inhibit cholesterol synthesis.

Tests for Cholesterol
1. Liebermann Burchard test.
2. Salkowski's Test.
3. Zlatki-Zak's reaction.

Compare–Cholesterol Synthesis and Ketone Body Synthesis

Characteristics	Ketone Body Synthesis	Cholesterol Synthesis
Site	Mitochondria	Cytoplasm/Smooth endoplasmic reticulum
HMG-CoA as an intermediate	Yes	Yes
HMG-CoA synthase	Yes, the regulatory step	Yes
HMG-CoA reductase	No	Yes, the rate limiting step
HMG-CoA lyase	Yes	No

> **HIGH YIELD POINTS**
>
> - Cytoplasmic HMG-CoA synthase is for cholesterol synthesis
> - Mitochondrial HMG-CoA synthase is for ketone body synthesis

Cholesterol cannot Generate Energy
- Unlike other biomolecules, cholesterol does not degrade to generate energy.

The Fates of Cholesterol
- Nearly half converted to bile acids.
- Some excreted in faeces as cholestanol and coprostanol.
- Coprostanol is the principal sterol in the faeces.
- Rest serve as precursors of vitamin D and sex hormones, corticosteroids.

> **HIGH YIELD POINTS**
>
> **Specialised Products of Cholesterol[Q]**
> 1. Bile acids (excretory form of cholesterol)
> 2. Vitamin D
> 3. Sex hormones
> 4. Corticosteroids

BILE ACID SYNTHESIS

Starting material—**Cholesterol[Q]**

Bile Acids and Its Site of Synthesis

Primary bile acids—Liver
They are:
1. **Cholic acid (mostabundant** bile acids in mammals)
2. Chenodeoxycholic acid or chenicacid.

Secondary bile acids-Intestine
They are:
1. Deoxycholic acid
2. Lithocholic acid.

Steps of Bile Acid Synthesis

I. Synthesis of primary bile acids—In the liver

First step:

Cholesterol converted to 7 hydro cholesterol by **7 α Hydroxylase, a microsomal Cyt P450** enzyme designated as CYP7A1.

A typical monooxygenase **requires oxygen, NADPH and vitamin C.**

- This is the **rate limiting step.[Q]**

Further multiple steps:

7 Dehydrocholesterol is divided in to two subpathways leading to synthesis of cholic acid and chenodeoxy cholic acid.

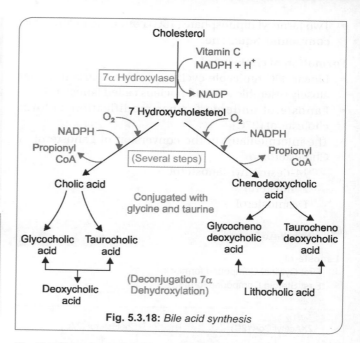

Fig. 5.3.18: *Bile acid synthesis*

II. Conjugation of primary bile acids
- Primary bile acids are conjugated with **glycine or taurine**.
- Conjugation takes place in **liver peroxisomes.**
- In humans the ratio of **glycine to taurine conjugates is 3:1.**

With Glycine	With Taurine
Glycocholic acid	Taurocholic acid
Glycochenodeoxycholic acid	Taurochenodeoxycholic acid.

> **CONCEPT BOX**
>
> **Bile Salts**
> - In alkaline bile (pH 7.6 to 8.4) usually conjugated bile acids exists as sodium and potassium salts.
> - So they are called bile salts.
> - Bile salts enter in to liver through bile.

III. Synthesis of secondary bile acids—In the intestine

Intestinal bacterial enzymes deconjugate and dehydroxylate primary bile acids to form secondary bile acids.
- Cholic acid to deoxycholic acid
- Cheno deoxycholic acid to lithocholic acid.

IV. Enterohepatic circulation
- Primary and secondary bile acids are absorbed exclusively in theileum.
- **98-99%** of absorbed bile acids are returned to the liver via portal circulation (enterohepatic circulation).
- **Lithocholic acid** is the bile acid which undergo **least** enterohepatic circulation.

Regulation of Bile Acid Synthesis

- The principal rate-limiting step is cholesterol 7α-hydroxylase (CYP7A1) reaction.
- The activity of the enzyme is feedback regulated via the nuclear bile acid-binding receptor, **farnesoid X receptor (FXR)**.
- When the size of the bile acid pool in the enterohepatic circulation increases, FXR is activated, and transcription of the cholesterol 7α-hydroxylase gene is suppressed.
- **Chenodeoxycholic acid** is particularly important in activating FXR.

Methods of Separation of Lipids and Its Analysis

1. **Thin layer chromatography**
2. **Adsorption chromatography by simple Column method or a high performance liquid chromatography**
3. **Gas liquid chromatography**

Quick Review

- TAG synthesis takes place in Smooth endoplasmic reticulum
- White adipose tissue lack Glycerol kinase.
- Hormone sensitive lipase is inhibited by Insulin
- In diabetes mellitus hormone sensitive lipase is activated and lipoprotein lipase is activated.
- Building block of Fatty acid synthesis is Acetyl CoA.
- Acetyl CoA is transported as Citrate via Tricarboxylic acid transporter.
- ATP citrate lyase releases acetyl CoA in Cytoplasm for Fatty acid synthesis.
- Acetyl CoA Carboxylase require Biotin, CO_2 and ATP.
- Fatty acid synthase complex require NADPH, Mn^{2+}
- Pantothenic acid is present in Fatty acid Synthase Complex.
- Allosteric activator of Acetyl CoA carboxylase is Citrate
- Acetyl CoA Carboxylase is active in dephosphorylated state.
- Tricarboxylate transporter is inhibited by long chain acyl CoA.
- From 1 palmitic acid 106 ATS are generated by beta oxidation.
- Jamaican Vomiting sickness is due to a toxin hypoglycin which inhibit Acyl CoA dehydrogenase
- Zellweger syndrome (Cerebrohepatorenal Syndrome) is a peroxisomal targeting disorder,
- VLCFA accumulate in Zellweger Syndrome.
- Zellweger syndrome resembles Downs Syndrome.
- Thiophorase is the enzyme that utilise ketone bodies.
- First steroid compound that is synthesised in cholesterol synthesis is Lanosterol
- Bile acid, Vitamin D and steroid hormones are synthesised from Cholesterol

Sites of metabolic pathway

Metabolic pathway	Enzyme deficiency
TAG synthesis	Smooth endoplasmic reticulum
Denovo fatty acid synthesis (Lipogenesis)	Cytoplasm
Elongation of Fatty acid	Smooth endoplasmic reticulum (Microsomes) Mitochondria (Minor)
Synthesis of unsaturated fatty acid	Smooth endoplasmic reticulum
Beta oxidation of fatty acid	Mitochondria
Beta oxidation of unsaturated fatty acids	Mitochondria
Beta oxidation of very long chain fatty acid	Peroxisomes up to Octanoyl-CoA, then rest in mitochondria
Alpha oxidation of fatty acid	Peroxisomes, smooth endoplasmic reticulum
Omega oxidation of fatty acid	Smooth endoplasmic reticulum/ microsomes
Activation of fatty acid	Cytosol
Ketone body synthesis	Exclusively mitochondrial
Cholesterol synthesis	Cytosol and Smooth endoplasmic reticulum
Bile acid synthesis	Smooth endoplasmic reticulum

Rate limiting enzyme

Pathway	RLE
Fatty acid synthesis	Acetyl CoA carboxylase
Beta oxidation of fatty acid	CAT-I or CPT-I
Ketone body synthesis	HMG CoA Synthase
Cholesterol synthesis	HMG CoA Reductase
Bile acid synthesis	7 alpha Hydroxylase

Check List for Revision

- This is a very important chapter.
- Learn each and every points of major pathways Fatty acid synthesis, beta oxidation, Ketone body synthesis Cholesterol synthesis.
- Other pathways learn the points given in bold letters.

Review Questions MCQ

1. **True about acetyl-CoA:** *(PGI Nov 2011)*
   a. Precursor for synthesis of cholesterol and other steroids
   b. Form ketone bodies
   c. Starting material for synthesis of fatty acid
   d. Arise from glycolysis

2. **Regarding synthesis of triacylglycerol in adipose tissue, all of the following are true except:** *(AI 07)*
   a. Synthesis from dihydroxy acetone phosphate
   b. Enzyme glycerol kinase plays an important role
   c. Enzyme glycerol 3-phosphate dehydrogenase plays an important role
   d. Phosphatidate is hydrolysed

3. **The storage triacylglycerol are hydrolysed by:** *(JIPMER 2012)*
   a. Pancreatic lipase
   b. Lipoprotein lipase
   c. Lysosomal lipase
   d. Hormone sensitive lipase

4. **Hormone sensitive lipase acts on:** *(Recent Question 2016)*
   a. Triglycerides
   b. Cholesterol ester
   c. Phospholipids
   d. Gangliosides

Fatty Acid Synthesis

5. **Most abundantly synthesised Fatty acid in the body is:** *(Recent Question Nov 2017)*
   a. Palmitic acid
   b. Oleic acid
   c. Arachidonic acid
   d. Stearic acid

6. **Which of the following is not a part of fatty acid synthase Complex?** *(AIIMS Nov 2013)*
   a. Ketoacyl reductase
   b. Enoyl reductase
   c. Acetyl-CoA carboxylase
   d. Ketoacyl synthase

7. **Mitochondria is involved in A/E:** *(AI 2012)*
   a. Fatty acid synthesis
   b. DNA synthesis
   c. Fatty acid oxidation
   d. Protein synthesis

8. **Fatty acid synthase complex contain the following enzymes except:** *(Kerala 2010)*
   a. Enoyl reductase
   b. Ketoacyl reductase
   c. Acetyl: CoA carboxylase
   d. Dehydratase

9. **NADPH is required for:** *(AI 1998)*
   a. Gluconeogenesis
   b. Glycolysis
   c. Fatty acid synthesis
   d. Glycogenolysis

10. **The first step in fatty acid synthesis involves:**
    a. Acetyl-CoA carboxylase
    b. Hydroxyl-CoA dehydrogenase
    c. Acetyl dehydrogenase
    d. Pyruvate kinase

11. **In fatty acid synthesis CO_2 step loss occurs in which?** *(PGI Dec 2006)*
    a. Hydration
    b. Dehydration
    c. Condensation reaction
    d. Reduction

12. **Carbon atoms added in fatty acid synthesis:** *(AIIMS Nov91)*
    a. 2 in Ist cycle and 4 in IInd cycle
    b. 4 in Ist cycle and 2 in IInd cycle
    c. 2 in Ist cycle and 2 in IInd cycle
    d. 4 in Ist cycle and 4 in IInd cycle

13. **True about mitochondrial chain elongation of fatty acid is/are** *(PGI June 1999)*
    a. Operates under anaerobic conditions
    b. Operates aerobically
    c. Common pathway
    d. Not a common pathway
    e. Pyridoxal-Phosphate and NADPH is required

14. **PAN-SH site of fatty acid synthase complex accepts:** *(PGI Dec 2000, AIIMS Dec 94)*
    a. Acetyl-CoA
    b. Malonyl-CoA
    c. Propionyl-CoA
    d. All

15. **In which organelle(s) of hepatocyte, the elongation of long chain fatty acid takes place?** *(PGI Dec 08)*
    a. Endoplasmic reticulum
    b. Golgi body
    c. Mitochondria
    d. Lysosomes
    e. Ribosome

16. Acetyl-CoA acts as a substrate for all the enzymes except: (AIIMS May 03)
    a. HMG-CoA synthetase
    b. Malic enzyme
    c. Malonyl CoA synthetase
    d. Fatty acid synthetase

17. Acetyl CoA Carboxylase is activated by: (Recent Question 2016)
    a. Malonyl-CoA
    b. Citrate
    c. Palmitoyl-CoA
    d. Acetoacetate

Oxidation of Fatty Acid and Disorders

18. In well fed state, the activity of Carnitine Palmitoyl Transferase-1 in outer mitochondrial membrane is inhibited by: (AIIMS Nov 2011)
    a. Glucose
    b. Acetyl-CoA
    c. Malonyl-CoA
    d. Pyruvate

19. Number of ATP formed by oxidation of one molecule of palmitic acid (16c): (Kerala 2009)
    a. 146
    b. 106
    c. 135
    d. 34

20. Beta oxidation in peroxisome generate: (Recent Question 2016)
    a. NADPH
    b. H_2O_2
    c. Long chain fatty acid
    d. $FADH_2$

21. All are features of Refsum's disease except: (PGI Nov 2014)
    a. Deficiency of alpha oxidation
    b. Defect of beta oxidation
    c. Accumulation of phytanic acid
    d. Peripheral neuropathy
    e. Treated by removing phytanic acid precursors from diet

22. Enzyme defect in Refsum's disease: (NBE Pattern Question)
    a. Phytanoyl alpha oxidase
    b. Acyl-CoA dehydrogenase
    c. Thiolase
    d. Thiokinase

23. Adrenoleukodystrophy is associated with: (CMC Vellore 2014)
    a. Accumulation of very long chain fatty acids
    b. Accumulation of medium chain fatty acid
    c. Increased plasmalogen
    d. Decreased pipecolic acid

24. β-oxidation of palmitic acid yields: (PGI Dec 05)
    a. 3-acetyl CoA
    b. 129 ATP net
    c. 131 ATP net
    d. 16-acetyl CoA
    e. 96 ATP from citric acid cycle

25. β–oxidation in peroxisome is differentiated from that occurring in mitochondria by: (PGI June 03)
    a. Acetyl CoA
    b. H_2O_2 formed
    c. Different enzymes are found in different site
    d. NADH is required

26. One of the following is obtained in the by beta oxidation of odd chain fatty acids: (JIPMER 2013)
    a. Acetyl-CoA + Acetyl-CoA
    b. Acetyl-CoA + Propionyl-CoA
    c. Propionyl CoA + Propionyl-CoA
    d. Acetyl-CoA alone

27. Zellweger Syndrome is associated with: (Recent Question Jan 2019)
    a. Lysosome
    b. Peroxisome
    c. Mitochondria
    d. Golgi apparatus

Ketone Bodies

28. Which of the following takes place in low insulin/glucagon ratio? (AIIMS May 2017)
    a. Cholesterol synthesis
    b. Glycogen synthesis
    c. Ketogenesis
    d. Fatty acid synthesis

29. Which of the following organs do not utilise ketone bodies? (PGI May 2014)
    a. Brain
    b. RBC
    c. Muscle
    d. Heart
    e. Liver

30. Ketone bodies can be utilised by all, except: (AIIMS May 2013)
    a. RBC
    b. Brain
    c. Skeletal muscle
    d. Renal cortex

31. Rothera's test used for detection of: (Kerala 2010)
    a. Proteins
    b. Glucose
    c. Fatty acid
    d. Ketones

32. Which organ does not utilize ketone bodies? (AIIMS Sep 96)
    a. Liver
    b. Brain
    c. Skeletal muscle
    d. Cardiac muscle

33. The immediate precursor in the formation of aceto-acetate from acetyl-CoA in the liver is:
    (PGI June 99)
    a. Mevalonate
    b. HMG-CoA
    c. Acetoacetyl-CoA
    d. 3-hydroxyl-butyryl-CoA

34. In a well fed state, acetyl-CoA obtained from diet is least used in the synthesis of: *(AI 2002)*
    a. Palmitoyl-CoA
    b. Citrate
    c. Acetoacetate
    d. Oxalosuccinate

35. The major fuel in the brain after several weeks of starvation is: *(JIPMER 2014)*
    a. Glucose
    b. Fatty acid
    c. β-Hydroxybutyrate
    d. Glycerol

Cholesterol Synthesis

36. Common enzyme in cholesterol and ketone body metabolism: *(AI 2012)*
    a. HMG-CoA reductase
    b. HMG-CoA lyase
    c. HMG-CoA synthase
    d. Thiolase

37. All are derived from cholesterol except:
    (Kerala 2011)
    a. Vitamin D
    b. Bile salt
    c. Bile pigment
    d. Steroid

38. Which of the following does not have cholesterol?
    (Kerala 2006)
    a. Vitamin D
    b. Estrogen
    c. Adrenaline
    d. Progesterone

39. Which coenzyme act as reducing agent in anabolic reaction?
    a. FADH2
    b. FMNH2
    c. NADPH
    d. NADH

Bile Acids

40. Bile acids are derived from: *(AI 1994)*
    a. Fatty acids
    b. Cholesterol
    c. Bilirubin
    d. Proteins

41. Bile acids synthesised in liver (primary bile acids):
    (PGI Dec 2000)
    a. Lithocolic acid
    b. Cholic acid
    c. Chenodeoxycholic acid
    d. Deoxycholic acid
    e. Taurocholic acid

Answers to Review Questions

1. **a, b, c, d. Precursor..., Form..., Starting..., Arise...,**
   (Ref: Harper 31/e page 250,252)

   Fates of Acetyl-CoA
   1. Synthesis of cholesterol and other steroids.
   2. Synthesis of fatty acid.
   3. Synthesis of ketone bodies.
   4. Enter into TCA cycle.

2. **b. Enzyme glycerol kinase plays an important role**

   In Muscle and Adipose Tissue
   - Glycerol kinase is absent in muscle and white adipose tissue.
   Glycerol 3-phosphate is formed from dihydroxy acetone phosphate, an intermediate in glycolysis.

3. **d. Hormone sensitive lipase** *(Ref: Harper 31/e page 244)*
   - Pancreatic lipase to hydrolyse dietary TGs
   - Lipoprotein lipase to hydrolyse TGs in lipoprotein in the blood
   - Lysosomal hydrolase to act on TGs in lysosomes
   - Hormone sensitive lipase hydrolyse stored TGs in adipose tissue

4. **a. Triglycerides** *(Ref: Harper 31/e pag 244)*
   - Stored triacylglycerol in adipose tissue is converted to 2-Monoacyl glycerol and fatty acid by hormone sensitive lipase.

Fatty Acid Synthesis

5. **a. Palmitic acid** *(Ref: Harper 31/e page 217)*

6. **c. Acetyl-CoA carboxylase** *(Ref: Harper 31/e page 217)*

   Fatty Acid Synthase (FAS) Multienzyme Complex
   - The complex is a homodimer of two identical polypeptide monomers in which **six enzyme activities and the acyl carrier protein (ACP)**
   - ACP contains the vitamin **pantothenic acid** in the form of 4'-phosphopantetheine
   - X-ray crystallography of the three-dimensional structure, shown that the complex is arranged in an X shape
   - Acetyl-CoA carboxylase is not a part FAS Complex^Q.

7. **a. Fatty Acid Synthesis** *(Ref: Harper 31/e page 216)*

Mitochondria is involved in mitochondrial DNA synthesis and protein synthesis. (described in Molecular genetics chapter)

De Novo Fatty Acid Synthesis (Lipogenesis)
- Major contribution to discovery by Feodor Lynen, hence the pathway is otherwise called **Lynen's spiral**
- Site: Liver, kidney, brain, lung, lactating mammary gland, and adipose tissue
- Organelle: By an **extra mitochondrial system** [in the cytosol]

Cofactor requirements include NADPH, ATP, Mn^{2+}, Biotin, and HCO_3^- (as a source of CO_2)
- Acetyl-CoA is the principal building block of fatty acid.

8. **c. Acetyl: CoA carboxylase** *(Ref: Harper 31/e page 217)*

Six enzyme activities of fatty acid synthase complex are:
- Ketoacyl Synthase
- Malonyl-acetyl transacylase
- Hydratase
- Enoyl reductase
- Ketoacyl reductase
- Thioesterase (Deacylase)

9. **c. Fatty acid synthesis** *(Ref: Harper 31/e page 217)*

10. **a. Acetyl-CoA carboxylase** *(Ref: Harper 31/e page 217)*

Fatty Acid Synthesis

By two enzyme system
1. Acetyl-CoA carboxylase
2. Fatty acid synthase complex

Acetyl-CoA Carboxylase
- Converts acetyl-CoA (2C) to malonyl-CoA in the presence of ATP.
- Acetyl-CoA carboxylase has a requirement for the B vitamin biotin.
- It is a multienzyme protein.
- This is the rate limiting step^Q in the fatty acid synthesis.
- Acetyl-CoA carboxylase is not a part FAS Complex^Q.

11. **c. Condensation reaction** *(Ref: Harper 31/e page 218)*

Reactions of Fatty Acid Synthase Complex by three stages
- Condensation
- Reduction
- Release of fatty acid

Condensation Reactions
1. **Malony/Acetyl transacylase**
   - A priming molecule of acetyl-CoA combines with a cysteine—SH group.
   - While malonyl-CoA combines with the adjacent— SH on the 4'-phosphopantetheine of ACP of the other monomer.
   - These reactions are catalyzed by **malonyl** acetyl **transacylase**, to form **acetyl (acyl)-malonyl enzyme**

2. **Ketoacyl synthase**
   - The acetyl group attacks the methylene group of the malonyl residue, catalyzed by 3-ketoacyl synthase, and **liberates CO_2**, forming 3-ketoacyl enzyme (acetoacetyl enzyme) (reaction2), freeing the cysteine—SH group.

12. **b. 4 in Ist cycle and 2 in IInd cycle**
    (Ref: Harper 31/e page 218)

First cycle of FAS Complex 2 C from Acetyl-CoA condenses with 2 carbon atoms of Malonyl-CoA by liberating 1 CO_2, So 4 Carbon atoms added.

In the second cycle to the existing 4 carbon fatty acyl residue, 2 carbon is added by ketoacyl synthase

13. **b, d. Operates aerobically, Not a common pathway**

Elongation of Fatty Acid Chains
- Occurs in the Endoplasmic Reticulum (the "microsomal system") and some in mitochondria also.
- By fatty acid elongase system
- Elongates saturated and unsaturated fatty acyl-CoAs (from C10 upward) by two carbons.
- Malonyl-CoA donates 2 carbon atoms in step wise manner.
- In the same manner as fatty acid synthase complex in the cytosol.
- NADPH is required at the two reductase step
- Elongation reaction are **particularly increased in brain during myelination to provide C22 and C24 fatty acids for sphingolipids. (So not common).**

14. **b. Malonyl-CoA** *(Ref: Harper 31/e page 218)*
- Cys –SH group accept Acetyl-CoA
- Pan-SH group accepts Malonyl-CoA

15. **a, c. Endoplasmic reticulum, Mitochondria**
    (Ref: Harper 31/e page 220)

16. **b. Malic enzyme** *(Ref: Harper 31/e page 220)*
- Malic enzyme converts malate to pyruvate by liberating CO_2.

17. **b. Citrate** *(Ref: Harper 31/e page 221)*

Allosteric regulation of acetyl-CoA carboxylase
- Positive allosteric regulation or allosteric activation of acetyl-CoA carboxylase by citrate^Q
- **Palmitoyl-CoA is an inhibitor of Acetyl CoA Carboxylase.**

Oxidation of Fatty Acid and Disorders

18. **c. Malonyl-CoA** *(Ref: Harper 31/e page 214)*

Fatty acid oxidation is inhibited in the well fed state because, CPT-I activity is low in fed state because of inhibition by malonyl-CoA, the initial intermediate in fatty acid synthesis formed by acetyl-CoA carboxylase.

19. **b. 106** *(Ref: Harper 31/e page 209, 210)*

Energetics of Beta Oxidation

The number of ATPs obtained depends on the number of carbon atoms in the fatty acid.

From Beta oxidation of palmitic acid (C-16)

First calculate how many cycles of beta oxidation [(n/2)–1] where n = number of carbon atoms

- In case of palmitic acid, 7 cycles of beta oxidation
- From one cycle of beta oxidation
- 1 FADH2 = 1.5 ATPs
- 1 NADH = 2.5 ATPs
- Total = 4 ATPs
- **From 7 cycles of beta** oxidation
- 7 × number of ATPs from 1 cycle
- **7 × 4 = 28 ATPs**

In case of palmitic acid (16/2) = 8 Acetyl-CoA

- From one Acetyl-CoA by TCA cycle 10 ATPs
- From 8 Acetyl-CoA
- 8 × 10 = **80 ATPs**
- **Total ATPs from palmitic acid = 28 + 80 = 108**
- 2 *ATPs utilised for initial activation of fatty acid*
- *So net ATPs from palmitic acid = 108-2 = 106 ATPs*

20. b. H_2O_2 *(Ref: Harper 31/e page 210)*

A modified form of β-oxidation is found in peroxisomes and leads to the formation of acetyl-CoA and H_2O_2 (from the flavoprotein-linked dehydrogenase step), which is broken down by catalase. Thus, this dehydrogenation in peroxisomes is not linked directly to phosphorylation and the generation of ATP. The system facilitates the oxidation of very long chain fatty acids (e.g., C_{20}, C_{22}). The enzymes in peroxisomes do not attack shorter chain fatty acids; the β-oxidation sequence ends at octanoyl-CoA.

21. b. Defect of beta oxidation

Refsum's Disease

- Defect in **Alpha oxidation of phytanic acid (phytanic acid oxidase) (phytanoyl-CoA hydroxylase)** in the peroxisome.
- The manifestation of classic Refsum's disease includes impaired vision from retinitis pigmentosa, ichthyosis, peripheral neuropathy, ataxia, and, occasionally, cardiac arrhythmias.
- Classic Refsum's disease often does not manifest until young adulthood, but visual disturbances such as night blindness, ichthyosis, and peripheral neuropathy may already be present in childhood and adolescence.
- Restrict dietary dairy products and green leafy vegetables.

22. a. Phytanoyl alpha oxidase

(Ref.: Dinesh Puri 3/e page 210; Nelson 20/e Chapter Defects in Metabolism of Lipids)

23. a. Accumulation of very long chain fatty acids

Abnormal laboratory findings common to disorders of peroxisome biogenesis

- Peroxisomes absent to reduced in number
- Catalase in cytosol
- Deficient synthesis and reduced tissue levels of plasmalogens
- Defective oxidation and abnormal accumulation of very long chain fatty acids
- Deficient oxidation and age-dependent accumulation of phytanic acid
- Defects in certain steps of bile acid formation and accumulation of bile acid intermediates
- Defects in oxidation and accumulation of l-pipecolic acid
- Increased urinary excretion of dicarboxylic acids

24. b, e. 129 ATP net, 96 ATP from citric acid cycle

This question is based older calculation, i.e. 1 NADH = 3 ATPs, 1 $FADH_2$ = 2 ATPs

- Palmitic acid oxidation has 7 cycles and 8 Acetyl-CoA
- I cycle of beta oxidation yield 3 + 2 = 5 ATPs
- 7 cycles yield 35 ATPs
- 8 Acetyl-CoA by TCA cycle yield 8 × 12 = 96 ATPs
- So total ATPs = 96 + 35 = 131 ATPs
- 2 ATPs for initial activation
- So net ATPs produced from palmitic acid = 131–2 = 129 ATPs
- According to new calculation
- 106 ATPs from 1 mol of Palmitic acid by beta oxidation.

25. b. H_2O_2 formed *(Ref: Harper 31/e page 210)*

Oxidation of Very Long Chain Fatty Acid

- By a modified β oxidation pathway
- For fatty acids >C20, C22
- Takes place in the **Peroxisomes** till Octanoyl-CoA
- Oxidation in peroxisome produces Acetyl-CoA and $H_2O_2^Q$ (instead of FADH2)
- No ATP is generated.

Further oxidation of acetyl-CoA and octanoyl-CoA takes place in the **mitochondria**.

26. b. Acetyl-CoA + Propionyl-CoA

(Ref: Harper 31/e page 209)

Oxidation of Odd Chain Fatty Acid

- Takes place in the mitochondria.
- Oxidation of a fatty acid with an odd number of carbon atoms yields acetyl-CoA and a molecule of propionyl-CoA.
- The propionyl residue from an odd-chain fatty acid is the only part of a fatty acid that is glucogenic.

27. b. Peroxisome *(Ref: Harper 31/e page 210)*

In Zellweger syndrome (Cerebro hepatorenal syndrome) Peroxisomal oxidation is affected as it is a peroxisomal targeting disorder

Ketone Bodies

28. c. Ketogenesis

Here Glucagon level is high, so it is regarding fasting state. Ketogenesis is in fasting state. Rest all in fed state.

29. b, e. RBC, Liver

30. a. RBC *(Ref: Harper 31/e page 210)*

- Ketone bodies serve as a fuel for extrahepatic tissues

31. d. Ketones

Test for Ketone Bodies

- Gerhardt's ferric chloride Test—Detect acetoacetate
- Rothera's Nitroprusside test—Detect acetoacetate and acetone
- None of the above tests detect beta hydroxy butyrate

Name of the test	Compound detected
Rothera's test	Ketone bodies, Branched chain ketoacids
Hay's test	Bile salt
Liebermann Burchard reaction	Cholesterol
Salkowski's reaction	Cholesterol

32. a. Liver *(Ref: Harper 31/e page 210)*

33. b. HMG-CoA *(Ref: Harper 31/e page 210)*

Hydroxy-3-methylglutaryl-CoA lyase then causes acetyl-CoA to split off from the HMG-CoA, leaving free acetoacetate.

34. c. Acetoacetate

- In well fed state fatty acid (Palmitic acid) synthesis is active
- TCA cycle also takes place, citrate and oxalosuccinate are intermediates in TCA cycle
- Oxalosuccinate is an intermediate in the reaction of isocitrate dehydrogenase of TCA cycle
- Acetoacetate, a primary ketone body is synthesized in fasting state.

35. c. β-Hydroxybutyrate *(Ref: Harper 31/e page 210)*

In the prolonged fasting stage, 20% of brains energy is met by ketone bodies. Rest by available glucose. But after several weeks of starvation, ketone bodies provide the major metabolic fuel.

Cholesterol Synthesis

36. b. c HMG-CoA Synthase *(Ref: Harper 31/e page 250)*

Compare–Cholesterol synthesis and ketone body synthesis

Characteristics	Ketone Body synthesis	Cholesterol synthesis
Site	Mitochondria	Cytoplasm
HMG-CoA as an intermediate	Yes	Yes
HMG-CoA synthase	Yes, the regulatory step	Yes
HMG-CoA reductase	No	Yes, the rate limiting step
HMG-CoA lyase	Yes	No

- Cytoplasmic HMG-CoA synthase for cholesterol synthesis
- Mitochondrial HMG CoA synthase for KB synthesis

37. c. Bile pigment *(Ref: Harper 31/e page 249)*

Derivatives of Cholesterol
- Bile acids
- Vitamin D
- Corticosteroids
- Sex Hormones

38. c. Adrenaline *(Ref: Harper 31/e page 241)*

Specialised Products of Cholesterol
- Bile acids [Excretory form of cholesterol]
- Vitamin D
- Sex hormones
- Corticosteroids

39. c. NADPH

NADPH helps in the reductive biosynthesis of fatty acids, cholesterol and steroid hormones.

Bile Acids

40. b. Cholesterol *(Ref: Harper 31/e page 249)*

- Bile acids are the excretory form of cholesterol.

41. b, c, e. Cholic…, Chenodeoxycholic…, Taurocholic…, *(Ref: Harper 31/e page 249)*

Primary Bile Acids–Liver
They are:
1. Cholic acid (Most abundant bile acids in mammals)
2. Chenodeoxycholic acid or chenic acid

Secondary Bile Acids-Intestine
They are:
1. Deoxycholic acid
2. Lithocholic acid

5.4 LIPOPROTEIN METABOLISM

- Introduction
- Composition of Lipoproteins
- Major Classes of Apolipoproteins
- Lipoprotein Metabolism
- Dyslipoproteinemia

INTRODUCTION

Definition

Lipoproteins are compound lipids formed as a combination of lipids with proteins.

The protein part of lipoprotein is called apolipoprotein.

Helps to transport lipids in the plasma.

Structure of Lipoprotein

- Lipoproteins consist of a nonpolar core and a single surface layer of amphipathic lipids.
- The nonpolar lipid core consists of mainly **triacylglycerol^Q and cholesteryl ester^Q**.
- It is surrounded by a single surface layer of amphipathic phospholipid and cholesterol molecules.
- These are oriented so that their polar groups face outward to the aqueous medium, as in the cell membrane.

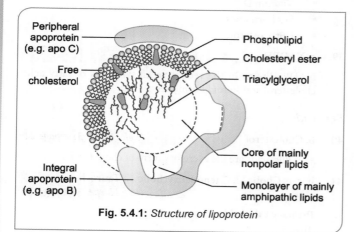

Fig. 5.4.1: *Structure of lipoprotein*

MAJOR CLASSES OF LIPOPROTEINS

Based on ultracentrifugation, in the ascending order of density is:

1. Chylomicrons (least density)
2. Very low density lipoproteins (VLDL)
3. Low density lipoproteins (LDL)
4. Intermediate density lipoproteins (IDL)
5. High density lipoproteins (HDL)

Based on electrophoretic separation

From cathode to anode the order of lipoprotein in an electrophoretogram is (see Figure 5.4.2)

1. Chylomicron
2. LDL (β lipoprotein)
3. VLDL (pre β lipoprotein)
4. IDL (broad β lipoprotein)
5. HDL (α lipoprotein)

> **CONCEPT BOX**
>
> **Concept of lipoprotein separation by electrophoresis**
> Separation of lipoproteins in an electric field depends on the *protein content*
> - Higher the protein content faster the mobility of lipoprotein in the electric field.
> - **Chylomicron^Q** with least protein content remains at the origin and HDL with highest protein content moves fastest.
> - An exception is VLDL and IDL with less protein content moves ahead of LDL.

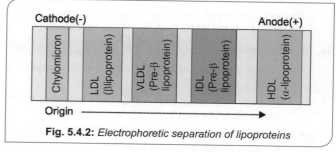

Fig. 5.4.2: *Electrophoretic separation of lipoproteins*

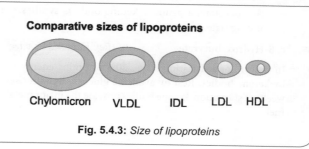

Fig. 5.4.3: *Size of lipoproteins*

Lipoproteins–Composition and Apolipoproteins

Lipoprotein	Apolipoproteins	Composition		Density (g/mL)
		Protein (%)	Lipid (%)	
Chylomicron	apo B48 apo C-I, C-II, C-III apo E apo A-I, A-II, A-IV	1–2 Least	98–99 Maximum	<0.95 Least
VLDL	apo B100 apo C-I, C-II, C-III apo E	7–10	90-93	0.95–1.006
IDL (VLDL remnant)	apo B100 apo E	11	89	1.006–1.019
LDL	apo B100	21	79	1.019–1.063
HDL	apo A-I, A-II, A-IV apo C-I, C-II, C-III apo D and E	32–57 Maximum	43–68	1.063–1.210 Maximum

COMPOSITION OF LIPOPROTEINS

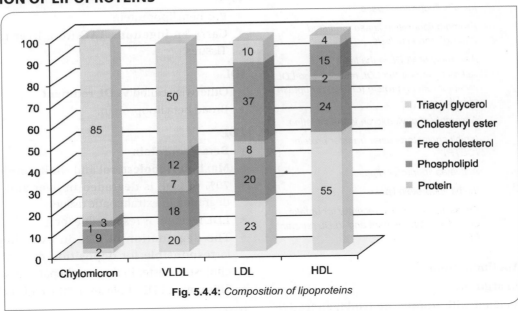

Fig. 5.4.4: *Composition of lipoproteins*

Lipoproteins and Its Function

Lipoprotein	Function
Chylomicron	Formed in the intestine. **Carry dietary triacylglycerol (Exogenous TAGs) to the liver.**
VLDL	Formed in the liver. **Carry endogenous triacylglycerol**
IDL (VLDL remnant)	Formed from VLDL. LDL is formed from IDL.
LDL	Derived from VLDL remnant. Deliver cholesterol and cholesterol ester to extrahepatic tissues and to liver.
HDL	Formed in the **liver and intestine**. Deliver cholesterol from **peripheral tissues to liver and other steroidogenic tissues.**[Q]

APOLIPOPROTEINS

- The protein part of lipoprotein is apo lipoprotein abbreviated as apo.
- They are integral protein (e.g. apo B) which cannot be removed to other lipoprotein or peripheral proteins (e.g. apos C and E)
- The major apolipoprotein in HDL is **apo A**
- The major apolipoprotein in LDL and VLDL is **apo B100**.
- Chylomicron contain a **truncated apolipoprotein apo B48**.

Apolipoproteins and Its Function

Apolipoprotein	Function
Apo A-I	Activates lecithin cholesterol acyl transferase (LCAT)
Apo A-II	Inhibits lipoprotein lipase
Apo A-V	Promote lipoprotein lipase mediated triacylgycerol lipolysis.
Apo B-100	Assembly of VLDL in the liver. Act as ligand for the LDL receptor and LDL receptor related protein (LRP-1) for uptake of LDL
Apo B-48	Assembly of chylomicron in the intestine.
Apo C-I	Inhibit cholesterol ester transfer protein (CETP)
Apo C-II	Activates lipoprotein lipase
Apo C-III	Inhibit lipoprotein lipase
Apo E	Act as ligand for LDL receptor for uptake of chylomicron remnant and VLDL remnant (IDL)

More about Apolipoproteins

Apo E is rich in **arginine**.

Apo D is associated with **human neurodegenerative disease**.

Isoforms of apo E
- Apo E gene is polymorphic in sequence.
- This result in expression of three common isoforms.
- Apo E2, Apo E-3, Apo E-4
- **Apo-E3 is most common**.
- Individuals carry **two apo E-4 alleles** are prone to develop **late onset Alzheimer's disease.**
- Apo E-2 has a low affinity for LDL-receptor.
- Individuals carrying **two apo E-2 allele** is prone to develop **type III hyperlipoproteinemia (familial dysbetalipoproteinemia)**

LIPOPROTEINS AT A GLANCE

Chylomicron
- **Maximum diameter**
- **Least density**
- Most buoyant lipoprotein
- Least protein content
- **Maximum triacylglycerol**
- Least electrophoretic mobility [Remain at the point of application]
- Least plasma half-life
- Carry **exogenous (dietary triacylglycerol from intestine to peripheral tissues)**
- Carry **dietary cholesterol and cholesterol** ester into the liver.

VLDL
- Pre-beta lipoprotein
- Carry **endogenous TAG** from liver to peripheral Tissues.

IDL
- Otherwise called **VLDL remnant**
- Broad beta lipoprotein.

LDL
- **Beta lipoprotein**
- **Maximum cholesterol and cholesterolester**
- 70% of LDL is degraded in liver and 30% of LDL degraded in extrahepatic tissue
- LDL receptor is responsible for it
- The degradation of LDL in extra hepatic tissue is responsible for deposition of cholesterol and cholesterol ester in the extrahepatict issues.
- This makes LDL cholesterol "the bad cholesterol".

HDL
- **Alpha lipoprotein**
- **Least diameter**
- **Maximum electrophoretic mobility.**
- **Maximum protein content.**^{Q}
- Carry cholesterol from peripheral tissues to liver and other steroidogenic tissues
- This is called **reverse cholesterol transport**.
- This makes HDL Cholesterol "the good cholesterol"
- The major role of HDL is to acts as the repository for apo C and apo E required for the metabolism of VLDL and chylomicron.

Contd...

Contd...

Lp (a)
- Almost similar to LDL.
- **Apo(a) is attached to apo B 100 by disulphide bond.**
- Major site of clearance of Lp(a) is liver.
- Strongly associated with Atherosclerosis and myocardial infarction.
- **Apo(a) has significant homology with plasminogenQ.**
- It interferes with activation of Plasminogen to plasmin.
- Hence fibrin clot is not lysed.
- Susceptible to Intravascular thrombosis

Lpx^Q
- Cholesterol is excreted as bile acids in the bile.
- In cholestasis, cholesterol combines with phospholipid and form lipoprotein X.
- Hence it is an index of cholestasis.

LIPOPROTEIN METABOLISM

Chylomicron Metabolism

Step I: Formation of Nascent Chylomicron

Assembly of nascent chylomicron in the intestine transported by lymphatics.

Step II: Formation of Mature Chylomicron

Remodelled to mature chylomicron by receiving apo C-II and apo E from HDL.
- *Remember: HDL acts as the repository of apo C and apo E.*

Step III: Formation of Remnant Chylomicron
- Apo C-II activates lipoprotein lipase
- **Lipoprotein lipase** that is located on the walls of blood capillaries, anchored to the endothelium by negatively charged proteoglycan chains of **heparan sulphate.**
- Lipoprotein lipase hydrolyses TAG in mature chylomicron to fatty acid and glycerol, to form **remnant chylomicron.**
- Fatty acid delivered to these tissues for storage/utilization.
- Thus chylomicron remnant is formed.

Step IV: Uptake of Remnant Chylomicron
- Chylomicron remnant is taken up in the liver by **receptor mediated endocytosis.**
- Uptake is mediated by **apo E** via two apo E dependent receptors, **LDL receptor and LDL receptor related protein-I (LRP-I)**
- Hepatic lipase hydrolyses remnant triacylglycerol and phospholipid.

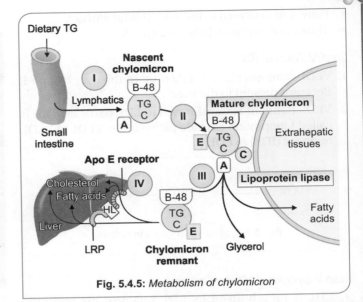

Fig. 5.4.5: Metabolism of chylomicron

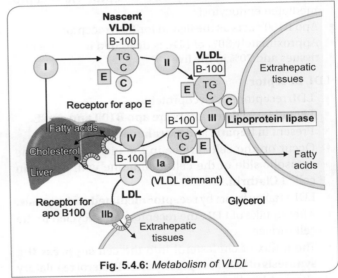

Fig. 5.4.6: Metabolism of VLDL

VLDL Metabolism

Step I: Formation of Nascent VLDL
- Assembly of nascent VLDL from the liver which carry **endogenous triacylglycerol.**

Step II: Formation of Mature VLDL
- Nascent VLDL remodelled to mature VLDL by receiving **apo C-II and apo E** from HDL.

Step III: Formation of Remnant VLDL (IDL)
- **apo C-II activates lipoprotein lipase**
- Lipoprotein lipase that is present in the walls of the capillaries that lines the extrahepatic tissues hydrolyses TAG to fatty acid and glycerol.

- Fatty acid delivered to these tissues for storage.
- Thus VLDL remnant (IDL) is formed.

Step IV: Fates of IDL

1. VLDL remnant (IDL) is taken up in the liver by apo E Receptor present in the liver.
2. VLDL remnant is transformed to LDL particles. This is called Lipoprotein cascade pathway, i.e. VLDL to VLDL remnant (IDL) to LDL

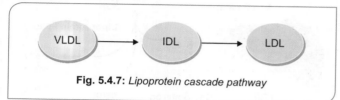

Fig. 5.4.7: *Lipoprotein cascade pathway*

Step V: Uptake of LDL by Tissues

- LDL is metabolised by LDL receptor via receptor mediated endocytosis.
- Apo B100^Q acts as the ligand for LDL receptors.
- Approximately 30% of LDL is degraded in extrahepatic tissues and 70% in the liver.

LDL Receptor

- LDL receptor is a glycoprotein.
- Ligand for LDL receptors are **apo B100 and apo E.**
- Present in **hepatic** and **extrahepatic tissues**^{Q}.
- Occur on the cell surface in the pits coated on the cytosolic side of the cell membrane with a protein called **Clathrin.**
- LDL is taken intact by **receptor mediated endocytosis.**
- After uptake of LDL, the receptors are recycled to the cell surface.
- The influx of cholesterol into the cell **suppress the synthesis of LDL receptor** by SREBP (sterol regulatory element-binding protein) pathway.
- Cholesterol lowering effect of **PUFA and MUFA** is thought to be due to **upregulation of LDL-receptor**, that increases the catabolic rate of cholesterol laden LDL.

HDL Metabolism and Reverse Cholesterol Transport

- Nascent HDL is synthesized and secreted from intestine and liver. Nascent HDL is discoidal in shape, consist of phospholipid bilayer, cholesterol and apo A.
- Lecithin cholesterol acyltransferase (LCAT) binds to nascent HDL.
- **Apo A-I** activates LCAT.
- LCAT converts cholesterol to nonpolar cholesterol ester.

Lecithin cholesterol acyltransferase (LCAT)

Cholesterol + Lecithin ⟶ Cholesterol ester + Lysolecithin

- Cholesterol ester is nonpolar. So a nonpolar core is generated forming a spherical HDL (HDL3) with surface film of amphipathic lipids and apolipoproteins.
- HDL3 accepts cholesterol from tissues by **Class B Scavenger Receptor B-I (SR-BI) and ATP-binding cassette transporters A1 (ABCA1) and G-I (ABCG-I).**
- LCAT acts on the cholesterol in HDL3, convert it into cholesteryl esters.
- Thus **less dense HDL$_2$** is formed.
- HDL2 delivers Cholesterol and cholesterol ester to liver via SR-B1 receptor or transport it to steroidogenic tissues or acted upon by hepatic lipase or endothelial Lipase.
- Thus HDL3 is reformed and free apo A-I is released.
- Free apo A-I forms poorly lipidated pre beta HDL.
- HDL3 and pre beta HDL again carry out cholesterol efflux from tissues.
- Thus *HDL collects excess cholesterol from tissues and transport it to liver and steroidogenic tissues^Q. This is called reverse cholesterol transport.*

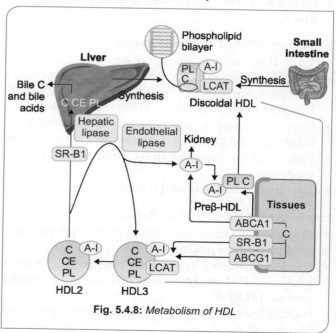

Fig. 5.4.8: *Metabolism of HDL*

Enzymes Responsible Reverse Cholesterol Transport

Lecithin cholesterol acyl transferase (LCAT)
- Apo AI activates LCAT
- Seen **associated with HDL**
- Responsible for virtually all plasma cholesteryl esters in humans.
- **Esterify cholesterol** in HDL

Cholesterol Ester Transfer Protein
- **Seen associated with HDL.**
- Facilitate transfer of cholesteryl ester from HDL to other lipoproteins, VLDL, IDL, LDL in exchange of Triacylglycerol.
- Cholesteryl ester, that is transferred to other lipoproteins find its way to liver by IDL or LDL.

Receptors Responsible for Reverse Cholesterol Transport

Class B scavenger receptor B1 (SR-B1)
- Considered as **HDL receptor.**
- Apo A-I acts as the ligand for the receptor.
- SR-BI has **dual role in HDL metabolism**
1. In liver and steroidogenic tissues it helps in the delivery of cholesterol and cholesterol ester to the cells from HDL.
2. Whereas from peripheral organs SR-BI facilitates cholesterol efflux to HDL particles.

ATP-binding cassette transporters
- ATP-binding cassette transporters AI (ABCAI) and ATP-binding cassette transporters G1 (ABCG1).
- Present in the extrahepatic tissues to facilitate efflux of cholesterol from peripheral tissues.
- **ABCAI preferentially transport cholesterol to poorly lipidated pre beta HDL or apo AI**
- ABCGI transport cholesterol to HDL.

> **HIGH YIELD POINTS**
>
> **HDL Fractions**
> **Nascent HDL or Discoidal HDL**
> - Contain phospholipid bilayer, cholesterol and apo AI
>
> **Spherical HDL or HDL3**
> - Contain cholesterol, cholesteryl ester, phospholipid, Apo AI, LCAT
>
> **Spherical less dense HDL or HDL2**
> - Contain cholesterol, cholesteryl ester, phospholipid, Apo AI
>
> **Pre beta HDL**
> - Most potent HDL in efflux of cholesterol from tissues.
> - Contain apo AI, cholesterol and phospholipid.
> - ABCAI preferentially transfer cholesterol to pre beta HDL.
>
> **HDL cycle**
> - The interchange of HDL2 and HDL3 is called HDL cycle.

Transport of Cholesterol between Tissues
- Dietary cholesterol is incorporated into chylomicrons.
- 95% of chylomicron cholesterol is delivered to liver in **chylomicron remnant.**
- The cholesterol is secreted from liver in VLDL.
- Most of the cholesterol secreted in VLDL is retained in IDL and ultimately LDL.
- LDL cholesterol is taken up by **liver and extrahepatic tissues**.

DISORDERS OF LIPOPROTEIN METABOLISM (DYSLIPOPROTEINEMIAS) (DYSLIPIDEMIAS)
- Hyperlipoproteinemias
- Hypolipoproteinemias

Hyperlipoproteinemias

Fredrickson Classification of Hyperlipoproteinemia

Frederickson and Levy classified hyperlipoproteinemias according to the lipoprotein particles that accumulate in the blood.

Frederickson type	Nomenclature	Molecular defect	Genetic transmission	Estimated incidence
Type I	Familial chylomicronemia syndrome (FCS)	Lipoprotein lipase or Apo CII	AR	1/1000000
Type IIa	Familial hypercholesterolemia (FH)	LDL receptor	AD	1/250-1/500 (Most common[Q])
	Familial Defective Apo B (FDB) Autosomal dominant hypercholesterolemia type II (**ADH Type II**)	Apo B100	AD	<1/1500
	Autosomal dominant hypercholesterolemia Type III (**ADH Type III**)	PCS K9	AD	<1/1000000
	Autosomal recessive hypercholesterolemia (**ARH**)	LDL receptor adapter protein (LDLRAP)	AR	<1/1000000

Contd...

Contd...

Frederickson type	Nomenclature	Molecular defect	Genetic transmission	Estimated incidence
	Sitosterolemia	ABCG5 or ABCG8	AR	<1/1000000
Type II b	Familial combined hyperlipidemia (FCHL)	-----	AR	<1/1000000
Type III	Familial dysbeta lipoproteinemia (FDBL)	Apo E	AR	1/10000
Type IV	Familial hypertriglyceridemia (FHTG)	Apo A-V	AR	<1/1000000
Type V	Familial hypertriglyceridemia (FHTG)	Apo A-V and GPIHBP 1	AR	<1/1000000

Lipoproteins Accumulated and Clinical Presentation of Different Hyperlipoproteinemias

Phenotype	I	IIa	IIb	III	IV	V
Lipoprotein, elevated	Chylomicrons (predominant) VLDL	LDL	LDL and VLDL	Chylomicron and VLDL remnants	VLDL	Chylomicrons and VLDL
Triglycerides	↑↑↑	N	↑	↑↑	↑↑	↑↑↑
Cholesterol (total)	↑	↑↑↑	↑↑	↑↑	N/↑	↑↑
LDL-cholesterol	↓	↑↑↑	↑↑	↓	↓	↓
HDL-cholesterol	↓↓↓	N/↓	↓	N	↓↓	↓↓↓
Plasma appearance	Lactescent	Clear	Clear	Turbid	Turbid	Lactescent
Xanthomas	Eruptive	Tendon, tuberous	None	Palmar, tuberoeruptive	None	Eruptive
Pancreatitis	+++	0	0	0	0	+++
Coronary atherosclerosis	-	+++	+++	+++	+/–	+/–
Peripheral atherosclerosis	-	+	+	++	+/–	+/–

Primary Hyperlipoproteinemias Causing Hypertriglyceridemia

I. **Familial Chylomicronemia Syndrome (Type I Hyperlipoproteinemia)**

Biochemical abnormalities

- **Lipoprotein lipase (LPL) or Apo CII defect.**
- Lipoprotein lipase is required for hydrolysis of TGs in chylomicrons and VLDL. Apo CII is the cofactor for lipoprotein lipase.
- **Lipoprotein accumulated is chylomicron and VLDL,** but chylomicron predominates.
- **Fasting triglycerides is >1000 mg/dL**
- Fasting cholesterol is elevated to a lesser degree.

Symptom box

Clinical presentation
- Present in childhood with recurrent **abdominal pain** due to acute pancreatitis
- On fundoscopic examination opalescent retinal blood vessels **(lipemia retinalis)**
- **Lactescent plasma**^{Q}
- **Eruptive xanthoma** (small yellowish white papules appear in clusters on backs, buttocks, extensor surfaces of arms and legs. These are painless skin lesions may become pruritic).
- Hepatosplenomegaly
- Premature CHD is not a feature of FCS.

Diagnosis

1. **Assaying triglyceride lipolytic activity in post heparin plasma (IV heparin injection to release the endothelial-bound LPL)**

Observation
- LPL activity is profoundly reduced in both LPL and apo C-II deficiency
- In patients with apo C-II deficiency, it normalizes after the addition of normal plasma (providing a source of apo C-II).

2. **Molecular sequencing of gene**

New Advances in Treatment of Familial Chylomicronemia Syndrome

A gene therapy approach—Alipogene tiparvovec
- Multiple intramuscular injection of an adeno associated viral vector encoding a gain of function LPL variant, leading to skeletal myocyte expression of LPL.
- Approved in Europe.

II. Familial Hypertriglyceridemia (FHTG) (Types IV and V Hyperlipoproteinemia)

Biochemical abnormalities

Apo A-V deficiency
- Apo A-V facilitate association of VLDL and chylomicrons with LPL.
- Loss of function mutation of Apo A-V causes accumulation of Chylomicrons and VLDL.

GPIHBP-I deficiency
- GPIHBP-I is glycosylated phosphatidyl Inositol HDL binding protein-I
- Mutation in GPIHBP-I causes defect in transport of LPL to vascular endothelium.
- Chylomicron is elevated
- Triglycerides are also markedly elevated.

Primary Hyperlipoproteinemias Causing Hypercholesterolemia

I. Familial Hypercholesterolemia (FH)

Also known as autosomal dominant hypercholesterolemia Type I (ADH Type I)

Biochemical abnormalities
- Loss of function mutation in **LDL receptor gene**.
- Can be
  1. Homozygous or receptor negative called as homozygous FH
  2. Heterozygous or receptor defective
- Reduced clearance of LDL from circulation.
- As LDL receptor is responsible for clearance of IDL, LDL production from IDL is also increased.
- **So lipoprotein accumulated is LDL and lipid accumulated is cholesterol.**
- LDL-Cholesterol >400–1000 mg/dL in homozygous cases
- Triglycerides are usually normal.

> **Symptom box**
> **Clinical presentation**
> - Family history of **premature CHD**.
> - **Corneal arcus**
> - **Plasma is clear**
> - No pancreatitis.
> - **Tendon xanthomas** particularly dorsum of hands and Achilles tendon.
> - Increased risk of cardiovascular disease.

> **EXTRA EDGE**
> **Recent Advances in the Treatment of Homozygous Familial Hypercholesterolemia (FH)**
> Two orphan drug developed are
> 1. **Lomitapide**
> Small molecule inhibitor of microsomal triglyceride transfer protein, which decreases the production of VLDL. There by LDL production from VLDL.
> 2. **Mipomersen**
> Antisense oligonucleotide to apo B.

II. Familial Defective Apo B100 (FDB)

Also known as autosomal dominant hypercholesterolemia Type II (ADH Type II)

Biochemical abnormalities
- Mutation in gene encoding apo B100, specifically in the LDL receptor–binding domain of apo B100.

Clinical presentation
- Almost like heterozygous FH.
- Plasma level of LDL-C is lower than heterozygous FH.
- This is because; IDL clearance is not impaired unlike FH as apo B100 binding domain of LDL receptor is only affected.

Diagnosis
- Sequencing receptor binding region of apo B gene.

III. Autosomal Dominant Hypercholesterolemia Type III (ADH Type III)

Biochemical abnormalities
- **Gain of function mutation of PCSK9**
- PCSK9 is a secreted protein that binds to the LDL receptor and is redirected to the lysosome and degraded.
- So in ADH 3 there is accelerated degradation of LDL receptor.
- So LDL is not cleared.

IV. Autosomal Recessive Hypercholesterolemia (ARH)

Biochemical abnormalities

- Due to mutations in a **protein LDL Receptor adaptor protein, LDL RAP** involved in LDL receptor-mediated endocytosis in the liver.

V. Sitosterolemia

Biochemical abnormalities

- Mutations ATP-binding cassette (ABC) half transporter family, **ABCG5 and ABCG8,** expressed in enterocytes and hepatocytes which pumps plant sterols such as sitosterol and campesterol, and animal sterols, predominantly cholesterol, into the gut lumen and into the bile.
- Hence intestinal absorption of sterols is increased and biliary excretion of the sterols is reduced, resulting in increased plasma and tissue levels of both plant sterols and cholesterol.
- Increase in hepatic sterol level results in transcriptional suppression of the expression of LDL receptors, which results in reduced uptake of LDL and substantially increased LDL-C

> **Symptom box**
>
> **Clinical presentation**
> - Usual presentation of familial hypercholesterolemia like tendon xanthoma, premature atherosclerotic cardiovascular disease.
> - Anisocytosis, poikilocytosis of erythrocytes, megathrombocytes due to incorporation of plant sterols into cell membrane.
> - Episodes of hemolysis and splenomegaly are distinctive features.
> - Severe hypercholesterolemia not responding to statins but dramatic response to dietary therapy or ezetemibe.

Diagnosis

- Substantial increase in plant sterols level.

Treatment

- Bile acid sequestrants
- Cholesterol absorption inhibitors.

Primary Hyperlipoproteinemias Causing Both Hypertriglyceridemia and Hypercholesterolemia

Type III Hyperlipoproteinemia

Familial Dysbetalipoproteinemia (FDBL) or Familial Broad Beta Disease or Remnant Removal Disease

Autosomal Recessive

Biochemical abnormalities

- Due to genetic variations in apo E, especially apo E2 that interfere with its ability to bind lipoprotein receptors.
- Lipoprotein elevated is chylomicron and VLDL remnant.
- Lipid elevated is cholesterol and triacylglycerol

> **Symptom box**
>
> **Clinical features**
> - Tuberoeruptive xanthoma
> - Palmar Xanthoma
> - Both xanthomas are pathognomonic of FDBL
> - Risk of coronary atherosclerosis

> **High Yield Points**
>
> **Xanthomas in FDBL**
> - Two distinctive types of xanthomas, tuberoeruptive and palmar, are seen in FDBL patients.
> - **Tuberoeruptive xanthomas** begin as clusters of small papules on the elbows, knees, or buttocks and can grow to the size of small grapes.
> - **Palmar xanthomas** (alternatively called xanthomata striata palmaris) are orange-yellow discolorations of the creases in the palms and wrists.

Hypolipoproteinemias

Abetalipoproteinemia

Autosomal recessive disease

Biochemical defect

- Loss-of-function mutations in the gene encoding **microsomal triglyceride transfer protein (MTP)** the gene name MTTP.
- MTP transfers lipids to nascent chylomicrons and VLDLs in the intestine and liver, respectively.
- So **Chylomicron^Q, VLDL^Q and hence LDL^Q** not produced.
- Plasma levels of cholesterol and triglyceride are extremely low in this disorder.
- Chylomicrons, VLDLs, LDLs, and apo B are undetectable in plasma.

Clinical presentation

- Presents in early childhood with **diarrhea and failure to thrive due to fat malabsorption.**
- Neurologic manifestations—loss of deep-tendon reflexes, followed by decreased distal lower extremity vibratory and proprioceptive sense, dysmetria, ataxia, and the development of a spastic gait.
- Progressive **pigmented retinopathy**.
- **Acanthocytes**
- **Bleeding manifestation**
- **Marked deficient in vitamin E and are also mildly to moderately deficient in vitamins A and K (due to defective fat absorption and hence the fat soluble vitamins.)^Q**

Tangier Disease

Biochemical defect
- Mutations in the gene encoding **ABCA1**, a cellular transporter that facilitates efflux of unesterified cholesterol and phospholipids from cells to apo A-I.
- In the absence of ABCA1, the nascent HDL, poorly lipidated.
- Apo A-I is immediately cleared from the circulation.
- **Extremely low circulating plasma levels of HDL-C and apo A-I.**
- Cholesterol accumulates in the reticuloendothelial system.

Clinical presentation
- Hepatosplenomegaly
- **Enlarged, greyish yellow or orange tonsils (pathognomonic clinical feature)**
- An intermittent **peripheral neuropathy (mononeuritis multiplex)**

LCAT Deficiency

Two genetic forms of LCAT deficiency.

Complete deficiency (*also called classic LCAT deficiency*) or Norum disease

Clinical presentation
- Progressive corneal opacification due to the deposition of free cholesterol in the cornea.
- Hemolytic anemia
- Progress to ESRD.

Biochemical abnormalities
- Very low plasma levels of HDL-C
- **Rise in free cholesterol**
- Rise in lecithin
- Fall in lysolecithin, cholesterol ester, HDL.

Partial LCAT deficiency (also called fish-eye disease)

Same features as classic LCAT deficiency, but no hemolytic anemia and do not progress to ESRD.

Diagnosis
- Assay plasma LCAT activity.
- Sequencing LCAT gene.

SECONDARY HYPERLIPOPROTEINEMIA AT A GLANCE

- In high carbohydrate diet excess carbohydrate converted to fatty acid in the liver, esterified to form **Triacyl glycerols**, secreted from liver as **VLDL**^{Q}.
- The most common effect of **alcohol** is to **increase plasma triglyceride** levels.
- Increased Insulin resistance causes **reduced LPL** activity which causes **reduced clearance of Chylomicron and VLDL.** This results in **hypertriglyceridemia in Diabetes mellitus.**
- **Nephrotic Syndrome is a classic** cause of **excessive VLDL production.**
- In **Cushing's Syndrome** exogenous or endogenous glucocorticoids lead to increased **VLDL production.**

Hypothyroidism and Hypercholesterolemia
- Thyroid hormone increases the hepatic expression of LDL receptor.
- So hypothyroidism **reduction in hepatic LDL receptor**, reduced clearance of LDL.
- **Elevated LDL-C**

HIGH YIELD POINTS

Calculation of lipid fractions
In fasting, cholesterol is carried primarily on three lipoproteins—the VLDL, LDL, and H
1. Total Cholesterol = HDL Cholesterol + VLDL Cholesterol + LDL Cholesterol
2. VLDL contains five times as much triglyceride by weight as cholesterol.

$$\text{So VLDL Cholesterol} = \frac{\text{Triglycerides}}{5}$$

This formula will work when triglyceride level <400 mg/dL

3. **Friedwalds formula for LDL-cholesterol (all values in mg/dL)**

$$\text{LDL Cholesterol} = \text{Total Cholesterol} - \text{HDL Cholesterol} - \frac{\text{Triglycerides}}{5}$$

Pharmacologic Treatment of Lipoprotein Disorders

For Severe Hypertriglyceridemia
1. Fibrates
2. Omega-3 Fatty acids^Q
3. Nicotinic acid or Niacin

For Hypercholesterolemia
1. HMG CoA Reductase Inhibitors
2. Cholesterol Absorption Inhibitors
3. Bile Acid Sequestrants

LDL Apheresis
- Treatment of hypercholesterolemia
- Patients blood is passed through a column that selectively removes the LDL.

Specialised Drugs for Homozygous Familial Hypercholesterolemia.
- MTP inhibitor—Lomitapide
- Apo B inhibitor—Mipomersen

ATP IV Guidelines for Ideal Level of LDL, HDL and Cholesterol

Biochemical parameter	Values mg/dL	Risk
LDL- Cholesterol	<100	Optimum
	100–129	Near or above Optimum
	130–159	Borderline high
	160–189	High
	>190	Very High
Total - Cholesterol	<200	Desirable
	200–239	Borderline High
	>240	High
HDL-Cholesterol	<40	Low
	>60	High

High Yield Points

Predictors of Coronary Artery Diseases
- HsCRP
- Total Cholesterol/HDL ratio
- Apo B/Apo A ratio
- LDL
- Non HDL Cholesterols
- HDL-Cholesterol

Quick Review

- Inner core of lipoprotein is Choelesterol ester and triacyl glycerol
- Fastest moving lipoprotein is HDL
- Least density is for Chylomicron.
- Apo B48 is unique for chylomicron.
- Apo E4 is associated with late onset Alzheimer's disease.
- Maximum protein/apoprotein content in HDL
- Maximum TAG content in Chylomicron
- Maximum phospholipid in HDL.
- Maximum cholesterol content in LDL

Important Dyslipoproteinemias and its Biochemical Defects

Frederickson type	Name	Biochemical defect
Type I	Familial chylomicronemia syndrome (FCS)	Lipoprotein lipase or Apo CII
Type IIa	Familial hypercholesterolemia (FH)	LDL receptor
	Familial Defective Apo B (FDB) Autosomal dominant hypercholesterolemia type II (**ADH Type II**)	Apo B100
	Autosomal dominant hypercholesterolemia Type III (**ADH Type III**)	PCS K9
	Autosomal recessive hypercholesterolemia (**ARH**)	LDL receptor adapter protein (LDLRAP)
	Sitosterolemia	ABCG5 or ABCG8
Type II b	Familial combined hyperlipidemia (FCHL)	-----
Type III	Familial dysbeta lipoproteinemia (FDBL)	Apo E
Type IV	Familial hypertriglyceridemia (FHTG)	Apo A-V
Type V	Familial hypertriglyceridemia (FHTG)	Apo A-V and GPIHBP 1

Hypolipoproteinemias

Abetalipoproteinemia	MTP
Tangier's disease	ABCA-1
Fish eye disease	LCAT

Check List for Revision

- Names of lipoproteins
- All bold letters and boxes are very important
- Dysbetalipoproteinemia learn the biochemical defect and its clinical features.

Review Questions MCQ

1. Alzheimer's disease is associated with which apolipoprotein? *(Recent Question Jan 2019)*
   a. Apo E1
   b. Apo E4
   c. Apo E3
   d. Apo E4

2. Apo integrated with HDL: *(PGI Nov 2018)*
   a. Apo A1
   b. Apo A2
   c. Apo D
   d. Apo B100
   e. Apo B48

3. Scavenger receptor is used in the metabolism of: *(Recent Question 2016)*
   a. HDL
   b. LDL
   c. IDL
   d. VLDL

4. HDL has highest content of: *(AIIMS Nov 2016)*
   a. Saturated fatty acid
   b. Triglycerides
   c. Cholesterol
   d. Apolipoproteins

5. Lipase that is regulated by glucagon: *(AIIMS Nov 2016)*
   a. Lipoprotein lipase
   b. Hormone sensitive lipase
   c. Gastric lipase
   d. Pancreatic lipase

6. A patient has total cholesterol 300, TG 150, and HDL 25. What would be the LDL value? (All values in mg/dL) *(May 2015)*
   a. 245
   b. 125
   c. 55
   d. 35

7. Regarding LDL receptors, all are true except: *(AIIMS Nov 2011)*
   a. Found in Clathrin coated pits of cell membrane
   b. Found only in extrahepatic tissue
   c. Internalized by endocytosis
   d. High levels of cellular cholesterol down regulate LDL receptors

8. Which is the ligand for receptors present in liver for uptake of LDL? *(May 2009)*
   a. apo E
   b. apo A and apo E
   c. apo E and apo B100
   d. apo B100

9. Triglycerides are maximum in: *(AIIMS May 2007)*
   a. Chylomicrons
   b. VLDL
   c. LDL
   d. HDL

10. Increased level of lipoprotein (a) predisposes to: *(AIIMS May 2007)*
    a. Liver cirrhosis
    b. Atherosclerosis
    c. Nephritic syndrome
    d. Pancreatitis

11. Main transporter of cholesterol to peripheral tissue: *(PGI Nov 2011)*
    a. HDL
    b. LDL
    c. VLDL
    d. IDL
    e. Chylomicron

12. Which of the following lipoproteins does not move towards charged end in electrophoresis: *(AI 2010)*
    a. VLDL
    b. LDL
    c. HDL
    d. Chylomicrons

13. All of the following statements about lipoprotein Lipase are true, except: *(AI 2009)*
    a. Found in adipose tissue
    b. Found in myocytes
    c. Deficiency leads to hypertriacylglycerolemia
    d. Does not require CII as cofactor

14. All of the following statements about apoproteins are true except: *(AI 2008)*
    a. Apoprotein A-I activates LCAT
    b. Apoprotein C-I activates lipoprotein lipase
    c. Apoprotein C-II inhibits lipoprotein lipase
    d. Apoprotein C-II activates lipoprotein lipase

15. Which of the following types of hypertriglyceridemia is associated with an increase in chylomicron and VLDL remnants? *(AI 2007)*
    a. Type I
    b. Type IIa
    c. Type III
    d. Type IV

16. The human plasma lipoprotein containing the highest percentage of triacylglycerol by weight is: *(AI 2006)*
    a. VLDL
    b. Chylomicrons
    c. HDL
    d. LDL

17. Cholesterol from dietary sources is transported to the peripheral tissue by: *(Kerala 2012)*
    a. Chylomicron
    b. VLDL
    c. LDL
    d. HDL

18. Action of lipoprotein lipase is: *(Recent Question 2016)*
    a. To form remnant lipoprotein
    b. Promote lipolysis in adipose tissue
    c. To form mature chylomicron
    d. To form HDL

19. In coronary artery disease the cholesterol level (mg/dL) recommended is: *(Recent Question 2016)*
    a. Below 200
    b. <250
    c. <220
    d. <280

20. Lipoprotein X is an indicator of: *(Recent Question 2016)*
    a. Atherosclerosis
    b. Cholestasis
    c. Hepatitis
    d. Myocardial infarction

21. Which is the lipoprotein with lowest density?
    a. HDL   b. LDL
    c. VLDL   d. Lpa

22. Which of the following has highest electrophoretic mobility and least lipid content? *(PGI June 01)*
    a. Chylomicrons b. HDL
    c. LDL   d. VLDL
    e. IDL

23. Which helps in the transport of chylomicrons from intestine to liver? *(AI 2000)*
    a. Apoprotein B b. Apoprotein A
    c. Apoprotein C d. Apoprotein E

24. Which of the following is an activator of LCAT? *(JIPMER 2002)*
    a. Apo B100   b. Apo B48
    c. Apo E   d. Apo A-I

25. Cholesterol present in LDL: *(AIIMS May 03)*
    a. Represents primarily cholesterol that is being removed from peripheral cells
    b. Binds to a receptor and cholesterol diffuses across the cell membrane
    c. On accumulation in the cell inhibits replenishment of LDL receptors
    d. When enters a cell, suppresses activity of acyl-CoA: cholesterol acyl transferase ACAT

26. A person on a fat free carbohydrate rich diet continues to grow obese. Which of the following lipoproteins is likely to be elevated in his blood? *(AI 2004)*
    a. Chylomicrons b. VLDL
    c. LDL   d. HDL

27. Which of the following is false about heparin? *(JIPMER May 2014)*
    a. Releases lipoprotein lipase
    b. Releases hormone sensitive lipase
    c. It is an anticoagulant
    d. It is a glycosaminoglycan

28. Lipoprotein A resembles: *(JIPMER 2014)*
    a. Plasminogen   b. Plasmin
    c. Thrombin   d. Prothrombin

Dysbetalipoproteinemia

29. Which of the following is increased in lipoprotein lipase deficiency? *(Recent Question Jan 2019)*
    a. LDL   b. Chylomicron
    c. HDL   d. VLDL

30. In uncontrolled diabetes mellitus what is the cause of high level of VLDL and TAG *(AIIMS May 2017)*
    a. Increased hepatic lipase
    b. Increased LDL receptors
    c. Increased activity of lipoprotein lipase and decreased activity of hormone sensitive lipase
    d. Increased activity of hormone sensitive lipase and decreased lipoprotein lipase activity

31. Defect in familial hypercholesterolemia *(Central Institute Exam May 2017)*
    a. LDL receptor defect
    b. Lipoprotein lipase defect
    c. Increased HDL
    d. Defect in apo E

32. Secondary hypertriglyceridemia is associated with: *(PGI May 2017)*
    a. Diabetes mellitus   b. Alcoholism
    c. Nephritic syndrome d. Cholestasis
    e. Congestive heart disease

33. A patient with eruptive xanthomas, drawn blood milky in appearance. Which lipoprotein is elevated in the plasma? *(AIIMS May 2015)*
    a. Chylomicron   b. Chylomicron remnants
    c. LDL   d. HDL

34. Very high total cholesterol, elevated LDL, normal level of LDL receptors. What is the probable cause? *(AIIMS May 2015)*
    a. Apo B100 mutation
    b. Complete deficiency of lipoprotein lipase
    c. Cholesterol acyltransferase deficiency
    d. Apo E defect

35. Fish oil is not used in the treatment of: *(JIPMER Dec 2016)*
    a. Type 2A Hyperlipoproteinemia
    b. Type 2B Hyperlipoproteinemia
    c. Type 3 Hyperlipoproteinemia
    d. Type 5 Hyperlipoproteinemia

36. Which of the following is increased in lipoprotein lipase deficiency? *(AIIMS Nov 2000)*
    a. VLDL   b. LDL
    c. HDL   d. Chylomicrons

37. Familial hypercholesterolemia is: *(PGI Dec 98)*
    a. Deficient LDL receptors
    b. Deficient HDL receptors
    c. HMG-CoA reductase deficiency
    d. Deficient VLDL receptors

38. Hypertriglyceridemia not seen in: *(CMC Vellore 2014)*

a. Hypothyroidism
b. Type 2 Diabetes Mellitus
c. Cushing's syndrome
d. Hepatitis

39. A patient was diagnosed with isolated increase in LDL. His father and brother had the same disease with increased cholesterol. The likely diagnosisis: *(AIIMS May 2009)*
    a. Familial type III hyperlipoproteinemia
    b. Abetalipoproteinemia
    c. Familial LPL deficiency (type1)
    d. LDL receptor mutation

40. Both Triglycerides and HDL increased: *(JIPMER Nov 2015)*
    a. Smoking
    b. Athletes
    c. Statin/Anabolic steroid abusers
    d. Alcoholism

41. Abetalipoproteinemia result in absence of: *(PGI Dec 02)*
    a. Chylomicron
    b. LDL
    c. VLDL
    d. HDL
    e. TG

42. Absence of this apo lipoprotein is responsible for the genetic disorder, familial type III hyperlipoproteinemia: *(Recent Question 2016)*
    a. Apo B100
    b. Apo B48
    c. Apo E
    d. Apo CII

43. Patient with abetalipoproteinemia frequently manifests with delayed blood clotting. This is due to in-ability to: *(JIPMER 2012)*
    a. Produce chylomicrons
    b. Produce VLDL
    c. Synthesise clotting factors
    d. Synthesise fatty acids

44. Apolipoprotein of chylomicron is: *(JIPMER Nov 2015)*
    a. Apo B100
    b. Apo B48
    c. Apo E
    d. Apo AI

45. Both triglycerides and HDL increased: *(JIPMER Nov 2015)*
    a. Smoking
    b. Athletes
    c. Statin
    d. Alcoholism

46. In prolonged fasting glycerol formed from triglyceride. Which of the following statement is true regarding glycerol? *(PGI 2013)*
    a. Used in synthesis of chylomicron
    b. It is directly used by tissues for energy needs
    c. It is formed due to increased activity of hormone sensitive lipase
    d. Glycerol acts as a substrate for gluconeogenesis
    e. It is formed by increased activity of lipoprotein lipase

47. Full form of LCAT: *(PGI May 2014)*
    a. Lecithin cholesterol acyl-transferase
    b. Lecithin choline acyl-transferase
    c. Lecithin cholesterol alkyl-transferase
    d. Lecithin choline alcohol-transferase
    e. Lecithin CoA transferase

Answers to Review Questions

1. **e. Apo E4** *(Ref: Harper 31/e page 238)*

   Late onset Alzheimer's disease is associated with Apo E 4 isoform

2. **a, b, c. Apo A1, Apo A2, Apo D**

   (Ref: Harper 31/e page 237 & Table 25-1)

   The apolipoproteins linked to HDL are Apo AI, Apo AII. Apo A IV, CI, CII, CIII, D & E

3. **a. HDL** *(Ref: Harper 31/e page 241)*

   Scavenger receptor B1 (SRB-1) is used in the metabolism HDL.

4. **d. Apolipoproteins** *(Ref: Harper 31/e page 237)*

Lipoprotein	Source	Protein content (%)	Lipid content (%)	Main lipid component	Apolipoproteins
Chylomicrons	Intestine	1–2 (Minimum)	98–99 (Maximum)	Triacylglycerol	A-I, A-II, A-IV, B-48, C-I, C-II, C-III, E
VLDL	Liver (intestine)	7–10	90–93	Triacylglycerol	B-100, C-I, C-II, C-III
IDL	VLDL	11	89	Triacylglycerol, cholesterol	B-100, E
LDL	VLDL	21	79	Cholesterol	B-100
HDL	Liver, intestine	32 (Maximum)	68 (Minimum)	Phospholipids, cholesterol	A-I, A-II, A-IV, C-I, C-II, C-III, D, E

5. **b. Hormone sensitive lipase** *(Ref: Harper 31/e page 242)*

   Hormone Sensitive Lipase
   - The activity of this enzyme is under the control of hormones, hence the name.
   - Present in the adipose tissue.
   - Involved in the metabolism of triacylglycerol stored in the adipose tissue.
   - Hence mobilisation of fatty acids from adipose tissue during starvation and diabetes mellitus

6. **a. 245**

   According to Friedewald's equation to calculate LDL-C
   LDL C = TC - HDLC - TG/5
   = 300 – 25 – 150/5
   = 245

   Also remember VLDL-C = TG/5

7. **b. Found only in extra hepatic tissues**

   (Ref: Harper 31/e page 240)

   LDL Receptor
   - Function uptake of LDL and remnant lipoprotein containing apo E.
   - Ligand for LDL receptor is apo B100 and apo E
   - Present in the liver and extrahepatic tissues (adipose tissue, Heart etc.).
   - **High level of cholesterol upregulate LDL receptor, causing an increase in the uptake LDL.**
   - Mechanism of uptake is **receptor mediated uptake or absorptive pinocytosis.**
   - Vesicles formed during absorptive pinocytosis are derived from invaginations (pits) are coated on the cytoplasmic side with a filamentous material clathrin, hence the pits are named coated pits.
   - For uptake of LDL, apo B100 acts as the ligand.
   - **Clathrin**
   - Three-limbed structure (called a triskelion).
   - With each limb being made up of one light and one heavy chain of clathrin.

8. **d. apo B100** *(Ref: Harper 31/e page 240)*
   - LDL receptor is used for the uptake of LDL and other remnant lipoprotein (VLDL remnant).
   - LDL receptor has ligand binding site for both apo E and apo B100.
   - For uptake of LDL apo B100 acts as the ligand for LDL receptor.
   - For uptake of remnant lipoproteins apo E acts as the ligand for LDL receptor.

9. **a. Chylomicrons** *(Ref: Harper 31/e 237 Table 25-1)*

10. **b. Atherosclerosis** *(Ref: Vasudevan and Sreekumari 7/e page 179)*

    Lp (a)
    - Almost similar to LDL.
    - Attached to B100 by disulphide bond.
    - Strongly associated with atherosclerosis and Myocardial infarction.
    - Significant homology with plasminogen.
    - It interferes with activation of plasminogen to plasmin.
    - Hence fibrin clot is no lysed.
    - Susceptible to Intravascular thrombosis.

    Lpx
    - Cholesterol is excreted as bile acids in the bile.
    - In cholestasis, cholesterol combines with phospholipid and form lipoprotein X.
    - Hence it is an index of cholestasis.

11. **b. LDL** *(Ref: Harper 31/e page 240)*

    Transport of cholesterol between tissues
    - Dietary cholesterol is incorporated into chylomicrons.
    - 95% of chylomicron cholesterol is delivered to liver in chylomicron remnant.
    - The cholesterol is secreted from liver in VLDL.
    - Most of the cholesterol secreted in VLDL is retained in IDL and ultimately LDL.
    - LDL cholesterol is taken up by liver and extrahepatic tissues (peripheral tissues)

12. **d. Chylomicrons** *(Ref: Lippincott 7/e page 228, Fig.18.5)*

    Electrophoretic Separation of Lipoproteins

    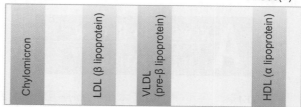

    Depends on the charge (protein content is one of the major determinants of charge) content
    - Higher the protein content faster the mobility of lipoprotein in the electric field.
    - Chylomicron remains at the origin
    - HDL moves fastest.

    From Cathode to Anode the Order of Lipoprotein
    - Chylomicron
    - LDL (β lipoprotein)
    - VLDL (pre β lipoprotein)
    - IDL (broad β lipoprotein)
    - HDL (α lipoprotein)

13. **d. Does not require CII as cofactor** *(Ref: Harper 31e page239)*

Option a and b
- Lipoprotein lipase is not present in the myocyte and adipose tissue.
- It is actually present in the capillaries of these organs.
- But option d, definitely wrong so it is given as the answer.

Lipoprotein Lipase
- It is located on the walls of blood capillaries.
- It is anchored to the endothelium by negatively charged proteoglycan chains of heparan sulfate.
- It has been found in heart, adipose tissue, spleen, lung, renal medulla, aorta, diaphragm, and lactating mammary gland, although it is not active in adult liver.
- It is not normally found in blood; however, following injection of heparin, lipoprotein lipase is released from its heparan sulfate binding sites into the circulation.

Both **phospholipids** and **apo C-II** are required as cofactors for lipoprotein lipase activity, while **apo A-II** and **apo C-III** act as inhibitors.

Reaction Catalysed by Lipoprotein Lipase
- Triacylglycerol is hydrolyzed progressively through a diacylglycerol to a monoacylglycerol and finally to FFA plus glycerol.
- Bulk of FFA is directed into the tissues.
- Thus it helps in the **hydrolysis of chylomicron and VLDL** to chylomicron remnant and VLDL remnant respectively.

Lipoprotein Lipase in Heart during Starvation

Heart lipoprotein lipase has a low K_m for triacylglycerol, about one-tenth of that for the enzyme in adipose tissue.

This enables the delivery of fatty acids from triacylglycerol to be **redirected from adipose tissue to the heart in the starved state** when the plasma triacylglycerol decreases.

Lipoprotein Lipase in Lactating Mammary Gland

A similar redirection to the mammary gland occurs during lactation.

Allowing uptake of lipoprotein triacylglycerol fatty acid for **milk fat** synthesis.

Lipoprotein Lipase and Insulin

In adipose tissue, **insulin enhances lipoprotein lipase synthesis in adipocytes** and its trans location to the luminal surface of the capillary endothelium.

Deficiency of Lipoprotein lipase lead to Type I Hyperlipoproteinemia

Hepatic Lipase
- It is bound to the sinusoidal surface of liver cells
- It is also released by heparin
- This enzyme, however, does not react readily with chylomicrons or VLDL
- But is involved in chylomicron remnant and HDL metabolism.

Hormone Sensitive Lipase
- The activity of this enzyme is under the control of hormones, hence the name.
- Present in the adipose tissue.
- Involved in the metabolism of triacylglycerol stored in the adipose tissue.
- Hence mobilisation of fatty acids from adipose tissue during starvation and diabetes mellitus.

14. **c. Apoprotein C-II inhibits lipoprotein lipase**

(Ref: Harper 31/e page 240)

Apolipoproteins and its function

Apolipoprotein	Function
Apo A-I	Activates lecithin cholesterol acyltransferase (LCAT)
Apo A-II	Inhibits Lipoprotein lipase
Apo A-V	Promote lipoprotein lipase mediated Triacyl glycerol lipolysis.
Apo B-100	Assembly of VLDL in the liver. Act as ligand for the LDL receptor and LDL receptor related protein (LRP-1) for uptake of LDL
Apo B-48	Assembly of Chylomicron in the intestine
Apo C-I	Inhibit cholesterol ester transfer protein (CETP)
Apolipoprotein	Function
Apo C-II	Activates lipoprotein Lipase
Apo C-III	Inhibit lipoprotein lipase
Apo E	Act as ligand for LDL receptor for uptake of chylomicron remnant and VLDL remnant (IDL)

15. **c. Type III** *(Ref: Harrison 18/e, Chapter 356, Table 356-4)*

16. **b. Chylomicrons** *(Ref: Harper 31/e 237 Table 25-1)*
- Lipoprotein with highest TAG content is chylomicron
- Lipoprotein with highest cholesterol and cholesterol ester content is LDL

17. **c. LDL** *(Ref: Harper 31/e page 272)*

Transport of Cholesterol between Tissues
- Dietary cholesterol is incorporated into chylomicrons.
- 95% of chylomicron cholesterol is delivered to liver in chylomicron remnant.
- The cholesterol is secreted from liver in VLDL.
- Most of the cholesterol secreted in VLDL is retained in IDL and ultimately LDL.
- LDL cholesterol is taken up by liver and extrahepatic tissues (peripheral tissues).

18. **a. To form remnant lipoprotein** *(Ref: Harper 31/e page239)*

The Action of Lipoprotein Lipase Forms Remnant Lipoproteins

Reaction with lipoprotein lipase results in the loss of 70–90% of the triacylglycerol of chylomicrons and in the loss of apo C (which returns to HDL) but not apo E, which is retained. The resulting chylomicron remnant is about half the diameter of the parent chylomicron and is relatively enriched in cholesterol and cholesteryl esters because of the loss of triacylglycerol. Similar changes occur to VLDL, with the formation of VLDL **remnants** [also called **intermediate-density lipoprotein (IDL)**.

19. a. Below 200
(Ref: Tietz Fundamentals of Clinical Chemistry 6/e page 419)

Biochemical parameter	Values mg/dL	Risk
LDL - Cholesterol	<100	Optimum
	100–129	Near or above Optimum
	130–159	Borderline high
	160–189	High

Contd...

Contd...

Biochemical parameter	Values mg/dL	Risk
	>190	Very high
Total - Cholesterol	<200	Desirable
	200–239	Borderline high
	>240	High
HDL - Cholesterol	<40	Low
	>60	High

20. b. Cholestasis

Incholestasis, free cholesterol, coupled with phospholipids, is secreted into the plasma as a constituent of a lamellar particle called LP-X. The particles can deposit in skin folds, producing lesions resembling those seen in patients with FDBL (xanthomata strata palmaris). Planar and eruptive xanthomas can also be seen in patients with cholestasis.

21. c. VLDL (Ref: Harper 31/e page 237, Table 25.1)

Composition of Different Lipoproteins

Lipoprotein	Source	Diameter (nm)	Density (g/mL)	Protein (%)	Lipid (%)	Main Lipid Components	Apolipoproteins
Chylomicrons	Intestine	90–1000	< 0.95	1–2	98–99	Triacylglycerol	A-I, A-II, A-IV, B-48, C-I, C-II, C-III, E
Chylomicron remnants	Chylomicrons	45–150	< 1.006	6–8	92–94	Triacylglycerol, phospholipids, cholesterol	B-48, E
VLDL	Liver (intestine)	30–90	0.95–1.006	7–10	90–93	Triacylglycerol	B-100, C-I, C-II, C-III
IDL	VLDL	25–35	1.006–1.019	11	89	Triacylglycerol, cholesterol	B-100, E
LDL	VLDL	20–25	1.019–1.063	21	79	Cholesterol	B-100
HDL	Liver, intestine, VLDL, chylomicrons	20–25	1.019–1.063	32	68	Phospholipids, cholesterol	A-I, A-II, A-IV, C-I, C-II, C-III, D, 2 E

22. b. HDL (Ref: Harper 31/e page 237, Table 25-1)

HDL
- Alpha lipoprotein
- Least diameter
- Maximum electrophoretic mobility.
- Maximum protein content.
- Least lipid content
- Carry cholesterol from peripheral tissues to liver and other steroidogenic tissues
- This is called reverse cholesterol transport
- This makes HDL cholesterol "the good cholesterol"
- The major role of HDL is to acts as the repository for apo C and apo E required for the metabolism of VLDL and chylomicron.

23. d. Apoprotein E (Ref: Harper 31/e page 237)
- Transport of chylomicron remnant to liver by Apo E
- Transport of IDL or VLDL remnant to liver by Apo E
- Transport of LDL to liver and extrahepatic tissues by apo B100.

24. d. Apo A-1 (Ref: Harper 31/e page 241)

25. c. On accumulation in the cell inhibits replenishment of LDL receptors
(Ref: Harper 31/e page 271)

Option a: LDL cholesterol is primarily from other lipoproteins whereas HDL cholesterol is from peripheral organs

Option b:

Mechanism of uptake of LDL by LDL receptor

- LDL receptors occur on the cell surface in pits that are coated on the cytosolic side of the cell membrane with a protein called clathrin.
- The LDL receptor spans the membrane, the B-100 binding region being at the exposed amino terminal end.
- After binding, LDL is taken up intact by endocytosis.
- The apoprotein and cholesteryl ester are then hydrolyzed in the lysosomes, and cholesterol is translocated into the cell.
- The receptors are recycled to the cell surface.

Option b: LDL is taken up intact, and not cholesterol alone diffuses across the membrane

Cholesterol balance in tissues

This influx of cholesterol inhibits the transcription of the genes encoding

- HMG-CoA synthase, HMG-CoA reductase, and other enzymes involved in cholesterol synthesis
- The LDL receptor itself, via the SREBP pathway

Thus coordinately suppresses cholesterol synthesis and uptake.

- In addition, ACAT activity is stimulated, promoting cholesterol esterification.

Recent advance

In addition, recent research has shown that the protein **proprotein convertase subtilisin/kexin type 9 (PCSK9)** regulates the recycling of the receptor to the cell surface by targeting it for degradation.

By these mechanisms, Cholesterol balance is maintained within normal limits in the cell.

26. **b. VLDL**

Excess carbohydrates are converted to acetyl-CoA, which is converted to fatty acid, esterified to produce endogenous TGs. VLDL carry the endogenous TGs VLDL level rises.

27. **b. Releases hormone sensitive lipase**

- Heparin is an anticoagulant and it is a glycosamino glycan.
- It releases Lipoprotein lipase not hormone sensitive lipase.

28. **a. Plasminogen** *(Ref: Chatterjea and Shinde 8/e page 448)*

Lpa

- Almost silmilar to LDL.
- Apo (a) is attached to apo B100 by disulphide bond.
- Major site of clearance of Lp(a) is liver.
- Strongly associated with Atherosclerosis and myocardial infarction.
- Apo A has significant homology with plasminogen^Q.
- It interferes with activation of plasminogen to plasmin.
- Hence fibrin clot is not lysed.
- Susceptible to Intravascular thrombosis.

Dysbetalipoproteinemia

29. **b. Chylomicron**

(Ref: Harrison 19/e Chapter Disorders Associated with Metabolism of Lipoproteins)

Familial Chylomicronemia Syndrome (Type I Hyperlipoproteinemia)

Biochemical abnormalities

- Lipoprotein Lipase (LPL) or Apo-CII defect.
- Lipoprotein accumulated is chylomicron and VLDL, but chylomicron predominates.
- Fasting triglycerides is >1000 mg/dL
- Fasting cholesterol is elevated to a lesser degree.

30. **d. Increased activity of hormone sensitive lipase and decreased lipoprotein lipase activity**

Insulin increases activity of Lipoprotein Lipase but inhibit Hormone sensitive lipase. So in Diabetes mellitus, due to decreased insulin, Lipoprotein lipase activity is decreased which increase VLDL. Hormone sensitive lipase activity is increased which releases stored TAG from adipose tissue, hence TAG level rises.

31. **a. LDL receptor defect**

32. **a, b, d. Diabetes mellitus, Alcoholism, Cholestasis**

33. **a. Chylomicron**

(Ref: Harrison 19/e Chapter Disorders Associated with Metabolism of Lipoproteins)

This is a case of Type I hyperlipoproteinemia (familial chylomicronemia syndrome)

Familial Chylomicronemia Syndrome (Type I Hyperlipoproteinemia)

Biochemical abnormalities

- Lipoprotein Lipase (LPL) or Apo-CII defect.
- Lipoprotein accumulated is chylomicron and VLDL, but chylomicron predominates.
- Fasting triglycerides is >1000 mg/dL
- Fasting cholesterol is elevated to a lesser degree.

Clinical presentation

- Present in childhood with recurrent **abdominal pain** due to acute pancreatitis
- On fundoscopic examination opalescent retinal blood vessels (**lipemia retinalis**)
- **Lactescent plasma**^{Q}
- **Eruptive xanthoma** (small yellowish white papules appear in clusters on backs, buttocks, extensor surfaces of arms and legs. These are painless skin lesions may become pruritic)
- Hepatosplenomegaly
- Premature CHD is not a feature of FCS.

34. a. Apo B100 mutation *(Ref: Harrison 19/e Chapter Disorders Associated with Metabolism of Lipoproteins)*

Familial defective apo B100 (FDB)
Also known as Autosomal Dominant Hypercholesterolemia Type II (ADH Type II)
Biochemical abnormalities
- Mutation in gene encoding apo B100, specifically in the LDL receptor–binding domain of apo B100.
- Cholesterol level is high.
- LDL receptor level is not decreased.

35. a. Type 2A Hyperlipoproteinemia *(Ref: Harrison 18/e, Table 356.3)*

Fish oil is used in the treatment of hypertriglyceridemia. The hyperlipoproteinemia with hyper triglyceridemia are all except Type IIA.

Hyperliporoteinemia Types	I	IIa	IIb	III	IV	V
Lipoprotein, elevated	Chylomicrons (predominant) VLDL	LDL	LDL and VLDL	Chylomicron and VLDL remnants	VLDL	Chylomicrons and VLDL
Triglycerides	↑↑↑	N	↑	↑↑	↑↑	↑↑↑

36. d >> a. Chylomicron >> VLDL *(Ref: Harrison 19/e, Table 356.3)*

Phenotype	I	IIa	IIb	III	IV	V
Lipoprotein, elevated	Chylomicrons (predominant) VLDL	LDL	LDL and VLDL	Chylomicron and VLDL remnants	VLDL	Chylomicrons and VLDL

37. a. Deficient LDL receptors *(Ref: Harrison 19/e, Table 356.3)*

Frederickson Type	Nomenclature	Molecular Defect
Type IIa	Familial hypercholesterolemia (FH)	LDL receptor
	Familial defective Apo B (FDB) Autosomal dominant hypercholesterolemia Type II (ADH Type II)	Apo B100
	Autosomal dominant hypercholesterolemia Type III (ADH Type III)	PCS K9
	Autosomal recessive hypercholesterolemia (ARH)	LDL receptor adapter protein (LDLRAP)
	Sitosterolemia	ABCG5 or ABCG8

38. a. Hypothyroidism *(Ref: Harrison 19/e, Table 356-5)*

Conditions associated with increased LDL (Increased Cholesterol)
- Hypothyroidism
- Nephrotic syndrome
- Cholestasis
- Acute intermittent porphyria
- Anorexia nervosa
- Hepatoma
- Drugs: thiazides, cyclosporin, tegretol

Conditions associated with decreased LDL
- Severe liver disease
- Malabsorption
- Malnutrition
- Gaucher's disease
- Hyperthyroidism
- Drugs: niacin toxicity

39. d. LDL receptor mutation *(Ref: Harrison 19/e, Table 356.3)*

Frederickson's Classification of Hyperlipoproteinemia

Phenotype	Type I	Type II	Type III
Molecular defect	Lipoprotein lipase apo CII	LDL receptor defect apo B100 defective	Apo E defect
Genetic nomenclature	Familial chylomicronemia syndrome	Familial Hypercholesterolemia Familial Defective apo B	Familial dysbetalipoproteinemia

Contd...

Contd…

Phenotype	Type I	Type II	Type III
Clinical features	Eruptive Xanthoma Pancreatitis Lactescent Plasma No Coronary/ Peripheral Atherosclerosis	Tendon Xanthoma Tuberous Xanthoma No Pancreatitis No Lactescent plasma Coronary atherosclerosis +++ Peripheral Atherosclerosis +	Tuberoeruptive Xanthoma No Pancreatitis Coronary atherosclerosis +++ Peripheral Atherosclerosis +
Lipid elevated	Triacylglycerol NB: Cholesterol normal	Cholesterol NB: Triacylglycerol normal	Cholesterol triacylglycerol
Lipoprotein elevated	Chylomicron	LDL	Chylomicron remnant VLDL remnant

40. **d. Alcoholism**

    (Ref: Harrison 19/e Chapter Disorders Associated with Metabolism of Lipoproteins)

    - Regular alcohol consumption has a variable effect on plasma lipid levels. *The most common effect of alcohol is to increase plasma triglyceride levels.*
    - Alcohol consumption stimulates hepatic secretion of VLDL, possibly by inhibiting the hepatic oxidation of free fatty acids, which then promote hepatic triglyceride synthesis and VLDL secretion.
    - The usual lipoprotein pattern seen with alcohol consumption is Type IV (increased VLDLs), but persons with an underlying primary lipid disorder may develop severe hypertriglyceridemia (Type V) if they drink alcohol.
    - Regular alcohol use also **raises plasma levels of HDL-C.**

41. **a, b, c, e. Chylomicron, LDL, VLDL, TG**

    (Ref: Harrison 19/e Chapter Disorders Associated with Metabolism of Lipoproteins)

    Abetalipoproteinemia
    - Autosomal recessive disease
    - Biochemical defect
    - Loss-of-function mutations in the gene encoding microsomal triglyceride transfer protein (MTP) the gene name MTTP
    - MTP transfers lipids to nascent chylomicrons and VLDLs in the intestine and liver, respectively.
    - So Chylomicron[Q], VLDL[Q] and hence LDL[Q] not produced.
    - Plasma levels of cholesterol and triglyceride are extremely low in this disorder.
    - Chylomicrons, VLDLs, LDLs, and apo B are undetectable in plasma.

42. **c. Apo E** *(Ref: Harrison 18/e, Table 356.3)*

43. **a>>>b. Produce chylomicrons>>Produce VLDL**

    Fat soluble vitamins are absorbed in Chylomicrons, from liver. It is secreted out by VLDL.

Vitamin deficiency in Abetalipoproteinemia[Q]

- Most clinical manifestations of abetalipoproteinemia result from defects in the absorption and transport of fat-soluble vitamins.
- Vitamin E and retinyl esters are normally transported from enterocytes to the liver by chylomicrons.
- Vitamin E is dependent on VLDL for transport out of the liver and into the circulation.
- As apo B containing lipoproteins are not formed these patients, there is, marked deficiency of vitamin E, mild to moderate deficiency of vitamins A and K.
- Bleeding manifestation due to defective absorption of fat and hence fat soluble vitamins.

44. **b. Apo B48** *(Ref: Harper 31/e page 238)*

    - Apo B48 is found exclusively in chylomicron.
    - The only one apolipoprotein in LDL is apo B100.

45. **d. Alcoholism**

    (Ref: Harrison 19/e Disorders of Lipoprotein Metabolism)

    - Regular alcohol consumption has a variable effecton plasma lipid levels.
    - The most common effect of alcohol is to increase plasma triglyceride levels.
    - Alcohol consumption stimulates hepatic secretion of VLDL, possibly by inhibiting the hepatic oxidation of free fatty acids, which then promote hepatic triglyceride synthesis and VLDL secretion.
    - Regular alcohol use also raises plasma levels of HDL-C.

46. **c, d. It is formed due to increased activity of hormone sensitive lipase, Glycerol acts as a substrate for gluconeogenesis**

    In prolonged fasting gluconeogenesis sets in and glycerol liberated from stored TGs in adipose tissue by the action of hormone sensitive lipase.

47. **a. Lecithin cholesterol acyl transferase**

CHAPTER 6

Bioenergetics

Chapter Outline

6.1 TCA Cycle
6.2 Electron Transport Chain
6.3 Integration of Metabolism
6.4 Shuttle Systems

CHAPTER 6

Bioenergetics

6.1 TCA CYCLE

- Definition
- Steps of TCA Cycle
- Energetics
- Regulation of TCA Cycle

DEFINITION

- Sequence of reactions in **mitochondria** that oxidizes the acetyl moiety of acetyl-CoA and reduces coenzymes, that are reoxidized through the electron transport chain, linked to the formation of ATP.
- The citric acid cycle is the **final common pathway**[Q] for the oxidation of carbohydrate, lipid, and protein.
- **Site-Mitochondria.**

Other Names

- Citric acid cycle because citric acid is involved in the cycle.
- Hans Krebs elucidated this cycle hence also called Krebs cycle

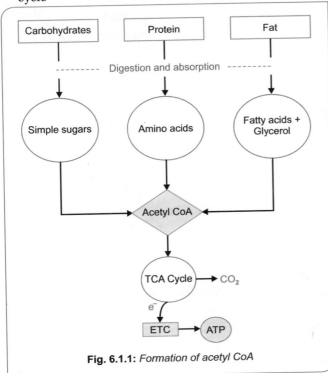

Fig. 6.1.1: *Formation of acetyl CoA*

Overview of TCA Cycle

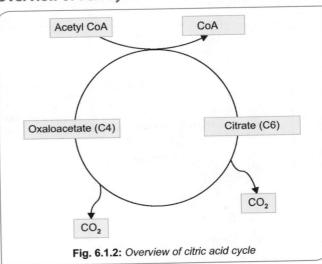

Fig. 6.1.2: *Overview of citric acid cycle*

STEPS OF TCA CYCLE

Citrate Synthase

- Acetyl CoA + Oxaloacetate → Citrate
- **Oxaloacetate** is the first substrate of TCA cycle.
- First tricarboxylic acid formed is Citrate (6C)
- **Irreversible step**[Q]
- Citrate can cross the mitochondrial membrane and release Acetyl CoA for the synthesis of fatty acid by **ATP Citrate Lyase**.

Aconitase (Aconitate Hydratase)

- Citrate isomerised to Isocitrate.
- Reversible reaction.
- The reaction occurs in two steps: **dehydration to cis-aconitate and rehydration to isocitrate**.
- Inhibited noncompetitively by **Fluoroacetate**[Q].
- **Aconitase is a Lyase**.

Isocitrate Dehydrogenase

- Isocitrate undergoes **dehydrogenation** catalyzed by isocitrate **dehydrogenase** to form, initially, oxalosuccinate
- Oxalosuccinate is **decarboxylated** to α-ketoglutarate (5C)
- The decarboxylation requires Mg^{2+} or Mn^{2+} ions
- **First Oxidative decarboxylation**
- **1 NADH** is formed
- **Reversible** reaction
- *Remember Cytoplasmic ICDH generate NADPH unlike NADH from mitochondrial ICDH.*

α-Ketoglutarate Dehydrogenase

- α-Ketoglutarate (5C) **oxidised and decarboxylated** to Succinyl CoA(4C)
- **Second Oxidative Decarboxylation**
- **1 NADH** is formed.
- **Physiologically unidirectional step**
- Alpha Ketoglutarate Dehydrogenase noncompetitively inhibited by **Arsenite**^{Q}.
- Multienzyme complex similar to Pyruvate Dehydrogenase.

5 Coenzymes of this enzyme are:

1. Lipomide
2. Thiamine Pyrophosphate
3. NAD^+
4. FAD
5. Coenzyme A

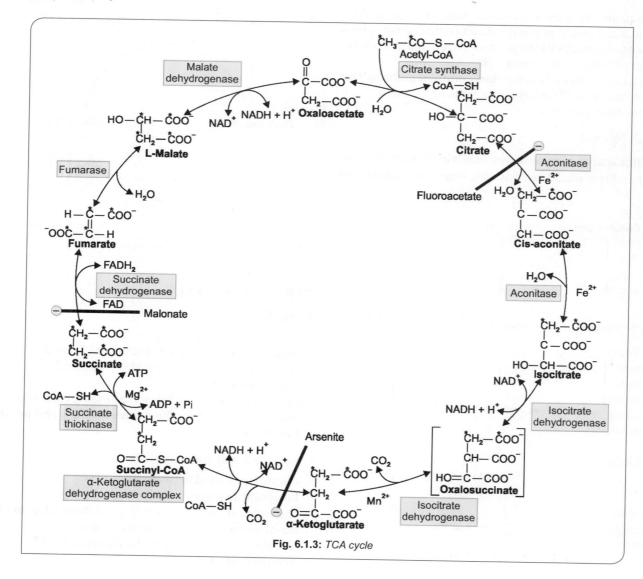

Fig. 6.1.3: *TCA cycle*

Succinate Thiokinase[Q] (Succinyl-CoA Synthetase)

- Convert Succinyl CoA to Succinate[Q]
- **1 ATP/GTP is generated**.
- GTP is generated in Gluconeogenic tissues like **Liver and Kidney**.
- **Substrate Level Phosphorylation.**

Succinate Dehydrogenase

- Succinate undergo **dehydrogenation** reaction, forming fumarate.
- The enzyme contains **FAD and iron–sulfur (Fe:S) protein**
- The enzyme directly reduce subiquinone in the electron transport chain **and is a part of Complex II**
- **Only enzyme in TCA cycle attached to Inner mitochondrial membrane**.
- All other enzymes are in the mitochondrial matrix.
- **FADH2 is formed**.
- Succinate dehydrogenase is competitively inhibited by **Malonate.**[Q]

Fumarase (Fumarate Hydratase)

- Catalyzes the addition of water across the double bond of fumarate, yielding malate.
- **Fumarase is a Lyase**.

Malate Dehydrogenase

- Final step in TCA cycle.
- Malate is **dehydrogenated** to Oxaloacetate.
- Oxaloacetate is regenerated.
- **1 NADH** generated.
- Oxaloacetate regenerated, hence **Oxaloacetate has a catalytic role like Ornithine in Urea Cycle**.

Inhibitors of TCA Cycle[Q]

- Aconitase noncompetitively inhibited by **Fluoroacetate**[Q].
- **Fluoroacetyl-CoA condenses with Oxaloacetate to form fluorocitrate which inhibit Aconitase**.
- Alpha Ketoglutarate dehydrogenase noncompetitively inhibited by **Arsenite**[Q].
- Succinate Dehydrogenase is competitively inhibited by **Malonate.**[Q] (Inhibitor of complex II of ETC).

ENERGETICS OF TCA CYCLE[Q]

Reaction	Method of ATP Production	No. of ATP Generated
Isocitrate dehydrogenase	1 NADH enter ETC	2.5 ATPs
α-Ketoglutarate Dehydrogenase	1 NADH enter ETC	2.5 ATPs
Succinate thiokinase	Substrate level Phosphorylation	1 ATP
Succinate dehydrogenase	1 FADH2 enter ETC	1.5 ATPs
Malate dehydrogenase	1 NADH enter ETC	2.5 ATPs
Total number of ATP per turn of TCA cycle		10 ATPs

Three molecules of NADH[Q] *and one molecule of FADH2 are produced for each molecule of acetyl-CoA catabolized in one turn of the cycle.*

Based on old calculation, 12 ATPs are generated by 1 turn of TCA cycle.

REGULATION OF TCA CYCLE

Regulatory steps are:

1. Citrate synthase
2. Isocitrate dehydrogenase
3. α-Ketoglutatrate dehydrogenase
4. Pyruvate dehydrogenase is also considered as the regulatory step of TCA cycle.

> **CONCEPT BOX**
>
> **Concept of regulation of TCA cycle**
> - High energy states inhibit TCA cycle and vice versa
> - High ATP/ADP ratio and High NADH/NAD+ ratio are inhibitors of TCA Cycle.
> - High ADP and High NAD+ are activators of TCA Cycle
> - Products of the pathway inhibit the regulatory enzymes

Allosteric Activators and Inhibitors of Individual Enzymes

- **Long Chain Acyl CoA and ATP** inhibit Citrate Synthase.
- Isocitrate dehydrogenase is inhibited by **ATP and NADH**
- Succinate dehydrogenase is inhibited by **Oxaloacetate**.
- In **Muscle**, the dehydrogenases of TCA cycle are **activated by Ca^{2+}**, which increases during muscle contraction.
- Mitochondrial isocitrate dehydrogenase is activated by **ADP**.
- In tissue such as **brain**, which is largely dependent on carbohydrate to supply acetyl-CoA, control of the citric acid cycle may occur at **pyruvate dehydrogenase**.

> **CONCEPT BOX**
>
> To answer whether a compound is an activator or inhibitor of an enzyme, think whether they are substrate or products of that enzyme/pathway.

High Yield Points

TCA Cycle is Truly an Amphibolic Pathway[Q]
A pathway with both catabolic and an abolic role is called edamphibolic pathway.

Anabolic Role of TCA Cycle
- Citrate to Fatty Acid Synthesis
- Alphaketoglutarate to GABA and Glutamate
- Succinyl CoA to Heme
- Oxaloacetate to Gluconeogenesis

Catabolic role of TCA cycle
- Acetyl CoA is completely oxidised to CO_2.

Image-Based Information

Anaplerotic Reactions of TCA Cycle[Q]
The 6 Carbon, 5 Carbon and 4 Carbon intermediates are used for various synthetic or anabolic reactions mentioned above. So these intermediates get depleted.
To replenish these compounds filling reactions takes place. These filling up reactions are called **Anaplerotic reactions**.

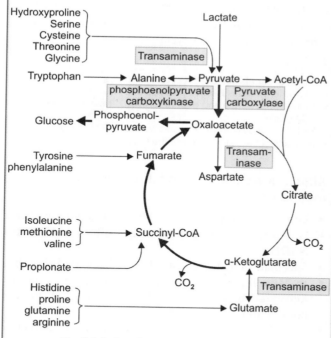

Fig. 6.1.4: Anaplerotic reactions of TCA cycle

Filling up (Anaplerotic) Reactions are:
1. As Pyruvate to Oxaloacetate
   - Hydroxy Proline, Serine, Cysteine, Threonine, Glycine to Pyruvate
   - Lactate to Pyruvate

Contd...

Contd...

Remember
- Pyruvate is converted to Oxaloacetate by **Pyruvate Carboxylase**.
- **Acetyl CoA** is an allosteric activator of Pyruvate Carboxylase.
- **This is the major filling up (anaplerotic) reaction**
2. As Alanine to Pyruvate to Oxaloacetate
   - Tryptophan to Alanine to Pyruvate
3. **Directly to Oxaloacetate**
   - Aspartate
4. **As Aspartate to Oxaloacetate**
   - Asparagine
5. **As Glutamate to Alpha Ketoglutarate(5C)**
   - Glutamine and Glutamate are the major anaplerotic substrates of Alpha Ketoglutarate
   - Histidine, Proline, Glutamine and Arginine
6. **At the level of Succinyl CoA (4C)**
   Valine, Isoleucine and Methionine
   Threonine
   Compounds that form Propionyl CoA
7. **At the level of Fumarate (4C)**
   Tyrosine and Phenyl Alanine

High Yield Points

Vitamins in TCA Cycle
1. Pantothenic Acid as a part of CoA
2. Riboflavin as FAD
3. Thiamine
4. Niacin as NAD^+

Quick Revision

- Final common oxidative pathway for carbohydrates lipids and proteins is TCA Cycle.
- Aconitase and Fumarase belongs to Lyase.
- ICDH and Alpha KGDH are two oxidative decarboxylation.
- 3 NADH and 1 FADH2 are the reducing equivalents generated.
- Succinate Thiokinase is the substrate level phosphorylation.
- Total 10 ATPs from 1 cycle of TCA cycle.
- Citrate Synthase and Alpha KGDH are the irreversible steps
- Fluoroacetate inhibit Aconitase.
- Arsenite inhibit Alpha KGDH.
- Cytoplasmic ICDH generate NADPH.
- Major anaplerotic reaction is Pyruvate Carboxylase.
- TCA cycle has both catabolic and anabolic roles, hence it is truly an amphibolic pathway.
- Citrate Synthase, ICDH and Alpha KGDH are the regulatory enzymes
- Valine Isoleucine Methionine and Threonine enter TCA cycle T the level of succinyl CoA.

Check List for Revision

- Each and every line in this chapter is important for exams.

Review Questions MCQ

1. Fluoride released from fluoroacetate inhibit which metabolic pathway in cancer metabolism?
   (AIIMS May 2018)
   a. Citric acid cycle
   b. Glycolytic pathway
   c. Oxidative phosphorylation
   d. ETC

2. An example of Anaplerotic reaction is:
   (AIIMS May 2017)
   a. Pyruvate to acetaldehyde
   b. Pyruvate to oxaloacetate
   c. Pyruvate to lactate
   d. Pyruvate to acetyl CoA

3. A chronic alcoholic have low energy production because of Thiamine deficiency as it is:
   (AIIMS May 2017)
   a. Acting as a cofactor for alpha ketoglutarate dehydrogenase and pyruvate dehydrogenase
   b. Acting as cofactor for transketolase in pentose phosphate pathway
   c. Interferes with energy production from amino acids
   d. Act as cofactor for oxidation reduction

4. In which step of TCA cycle ATP is generated:
   (Central Institute Exam, May 2017)
   a. Succinate dehdrogenase
   b. Succinate thiokinase
   c. Fumarase
   d. Malate dehydrogenase

5. All of the following amino acids forms Acetyl CoA via pyruvate dehydrogenase except:
   (AIIMS May 2016)
   a. Glycine
   b. Tyrosine
   c. Hydroxyproline
   d. Alanine

6. A child ingested cyanide and rushed to the emergency room. Which of the following of citric acid cycle is inhibited at the earliest?
   (AIIMS Nov 2016)
   a. Citrate synthase
   b. Aconistase
   c. Acetyl CoA production
   d. NAD + donor

7. Which of the following is not an intermediate of TCA cycle?
   (AIIMS May 2014)
   a. Acetyl CoA
   b. Citrate
   c. Succinyl CoA
   d. Alphaketoglutarate

8. Which of the following is true about Krebs cycle?
   (JIPMER May 2015)
   a. Pyruvate condenses with oxaloacetate to form citrate
   b. Alpha ketoglutarate is a five carbon compound
   c. Oxidative phosphorylation occurs in the cytoplasm only
   d. Krebs cycle can operate in anaerobic condition

9. Which of the following substance binds to CoA and condenses oxaloacetate to inhibit the TCA cycle?
   (AIIMS Nov 2010)
   a. Malonate
   b. Arsenite
   c. Fluoroacetate
   d. Fumarate

10. First substrate of Krebs cycle is: *(AIIMS May 2007)*
    a. Oxaloacetate
    b. Acetyl CoA
    c. Pyruvate
    d. Lipoprotein

11. Hyperammonemia inhibits TCA cycle by depleting:
    (PGI June 2009)
    a. Oxaloacetate
    b. Alpha-ketoglutarate
    c. Citrate
    d. Succinyl CoA
    e. Fumarate

12. What is liberated when Citrate converted to Cis Aconitate? *(Recent Question 2016)*
    a. H_2O
    b. H_2
    c. H_2O_2
    d. CO_2

13. False about reducing equivalents is:
    (Recent Question 2016)
    a. They are NADH and NADPH
    b. Only produced during primary metabolic pathway
    c. Formed in TCA cycle
    d. Formed in mitochondria

14. High energy phosphate is not produced in:
    a. TCA cycle
    b. Hexose monophosphate pathway
    c. Glycolysis
    d. Beta oxidation of fatty acid

Answers to Review Questions

1. **a. Citric acid cycle** *(Ref: Harper 31/e page 155)*
   - The poison fluoroacetate is found in some of plants, and their consumption can be fatal to grazing animals.
   - Some fluorinated compounds used as anticancer agents and industrial chemicals (including pesticides) are metabolized to fluoroacetate.
   - It is toxic because fluoroacetyl-CoA condenses with oxaloacetate to form fluorocitrate, which inhibits aconitase, causing citrate to accumulate.

2. **b. Pyruvate to oxaloacetate** *(Ref: Harper 31/e page 151-155)*

   Anaplerotic reactions are those reactions which replenish the depleted intermediates of TCA cycle, to have functional TCA cycle. Major filling-up or anaplerotic reaction is Pyruvate to Oxaloacetate.

3. **a. Acting as cofactor for alpha ketoglutarate Dehydrogenase...** *(Ref: Harper 31/e page 151-155)*

   Alpha ketoglutarate dehydrogenase (TCA cycle) and PDH (link between aerobic glycolysis and TCA) require Thiamine, hence affect energy production, in alcoholic patients thiamin deficiency is present.

4. **b. Succinate thiokinase** *(Ref: Harper 31/e page 151-155)*

5. **b. Tyrosine** *(Ref: Harper 31/e page 151-155)*

   Amino acid that enter via Pyruvate to TCA cycle are:
   - Hyrdoxy Proline, Serine, Cysteine, Threonine, Glycine
   - Tryptophan, Alanine

6. **d. NAD + donor**
   - Two closely related answers are here, NAD+ donor and Acetyl CoA production.

   As there is a word, earliest NAD+ donor is a better answer. Because as Cyanide inhibit Cyt c Oxidase, it inhibits the electron transfer of ETC, hence cannot accept electrons from reducing equivalents like NADH, so NAD+ donor is affected. This in turn affect PDH reaction, as NAD+ is less, hence acetyl CoA production is affected. But the earliest affected is NAD + donor. In Cyanide poisoning Pyruvate is converted to Lactate, instead of PDH. This leads to Metabolic Acidosis.

7. **a. Acetyl CoA** *(Ref: Harper 31/e page 151-155)*

   Acetyl CoA and Oxaloacetate are the starting materials of TCA cycle

8. **b. Alpha ketoglutarate is a five carbon compound** *(Ref: Harper 31/e page 151-155)*
   - Acetyl CoA condenses with Oxaloacetate to form Citrate
   - Oxidative phosphorylation occurs in mitochondria by ETC
   - Krebs cycle cannot operate in anaerobic condition.

9. **c. Fluoroacetate** *(Ref: Harper 31/e page 151-155)*

10. **a. Oxaloacetate** *(Ref: Harper 31/e page 151-155)*

    Oxaloacetate binds to the enzyme Citrate synthase, which induces conformational changes in the enzyme that facilitate the binding of Acetyl CoA. Hence, first substrate is Oxaloacetate.

11. **b. Alpha-ketoglutarate** *(Ref: Harper 31/e page 396)*

    Hyperammonemia, as occurs in advanced liver disease and a number of (rare) genetic diseases of amino acid metabolism, leads to loss of consciousness, coma and convulsions, and may be fatal. This is because of the withdrawal of α-ketoglutarate to form glutamate (catalyzed by glutamate dehydrogenase) and then glutamine (catalyzed by glutamine synthetase), leading to reduced concentrations of all citric acid cycle intermediates, and hence reduced generation of ATP.

12. **a. H_2O** *(Ref: Harper 31/e page 151-155)*
    - Citrate isomerised to Isocitrate by Aconitase
    - Reversible reaction.
    - The reaction occurs in two steps: dehydration to cis-aconitate and rehydration to isocitrate.

13. **b. Only produced during primary metabolic pathway**

14. **b. Hexose monophosphate pathway**

    Pathways which do not synthesize ATP are:
    - HMP pathway
    - Rapoport Luebering cycle
    - Uronic acid pathway
    - Alpha oxidation of fatty acid
    - Omega Oxidation of fatty acid

6.2 ELECTRON TRANSPORT CHAIN

- Basics of Biological Oxidation
- Electron Transport Chain
- Oxidative Phosphorylation
- Inhibitors of ETC
- Components of ETC
- High Energy Compounds

BASICS OF BIOLOGICAL OXIDATION

- Oxidation is loss of electrons
- Reduction is gain of electrons
- Redox potential is the ability to transfer electrons.

Redox couple or redox pair are compounds that can exist both in reduced as well as oxidised state Each redox couple has a particular redox potential. Electrons are transferred in the ascending order of redox potential. The redox pair with high redox potential receive electron from redox pair with low redox potential.

Eg:- NAD + /NADH, FAD/FADH2

Redox Potential of Common Redox Couples

Redox couple	Redox potential
H^+/H_2	−0.42
$NADP^+/NADPH$	−0.34
$NAD^+/NADH$	−0.32
NADH Dehydrogenase (FMN/FMNH2)	−0.30
Lipoate	−0.29
Acetoacetate/β Hydroxybutyrate	−0.27
FAD/FADH2	−0.219
Pyruvate/Lactate	−0.19

Contd...

Contd...

Redox couple	Redox potential
Oxaloacetate/Malate	−0.19
Fumarate/Succinate	+0.03
$Cyt b; Fe^{3+}/Fe^{2+}$	+0.08
Ubiquinone (CoQ)	+0.10
$Cyt c_1; Fe^{3+}/Fe^{2+}$	+0.22
$Cyt a; Fe^{3+}/Fe^{2+}$	+0.29
Oxygen/water	+0.82

Remember this table is important for national board pattern of exams. It is important to learn the order in which they are arranged, not the value of redox potential.

ELECTRON TRANSPORT CHAIN

Site : In the Inner Mitochondrial Membrane Components of the Electron Transport Chain^Q are contained in four large protein complexes.

Concept of ETC

ETC is a series of redox couples arranged in the ascending order of redox potential. So electrons jump from one redox couple to the next redox couple.

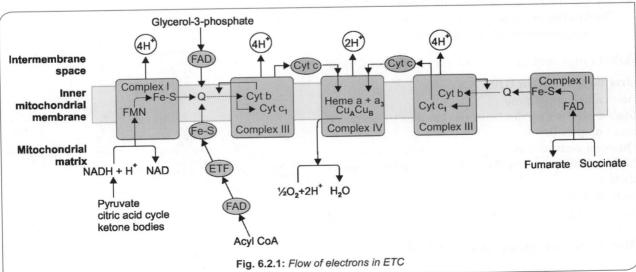

Fig. 6.2.1: *Flow of electrons in ETC*

COMPONENTS OF ELECTRON TRANSPORT CHAIN

1. **Complex I NADH Coenzyme° Oxidoreductase:**
   Contain FMN and Fe-S (Iron- Sulfur) Complex
   Pumps 4 H+^Q to Intermembrane space
2. **Complex II Succinate Q Reductase**
   Contain FAD and FeS Complex
   No H+ pumped to intermembrane space

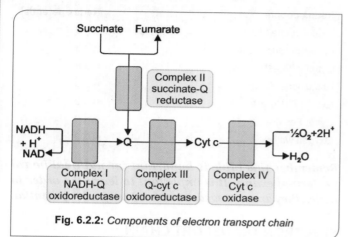

Fig. 6.2.2: *Components of electron transport chain*

3. **Complex III Q Cytochrome c Oxidoreductase:**
   Contain Cyt b and Cyt c1
   Contain **Reiske**Fe-S Complex

   Pumps $4H^{+Q}$ to Intermembrane Space (PGI Nov 09, May10)
4. **Complex IV Cytochrome c Oxidase:**
   – Contain Cyt a and a3 (now known as Heme a a3 and Copper A and Copper B centre
   – **2H+** pumped to Intermembrane Space
   – The **final electron acceptor of ETC is oxygen.**
   – This is the **irreversible complex.**

Mobile Complexes in the Electron Transport Chain

Coenzyme Q or CoQ or Q10
- Also called ubiquinone.
- Mobile electron carrier between Complex I/Complex II and Complex III
- Quinone derivative with a polyisoprenoid side chain.
- Lipid solubility and small size make it a mobile electron carrier.

Cytochrome c
- Mobile electron carrier between Complex III and Complex IV.
- Also play a role in **programmed cell death**

OXIDATIVE PHOSPHORYLATION

- The flow of electrons through the respiratory chain generates ATP by the process of oxidative phosphorylation.
- Oxidation is coupled with phosphorylation.
- The theory behind the oxidative phosphorylation is the **Chemiosmotic theory.**

The Chemiosmotic Theory^Q

- Proposed by **Peter Mitchell** in 1961
- Postulates that the two processes, oxidation and Phosphorylation are coupled by a **proton gradient across the inner mitochondrial membrane.**
- The **proton motive force^Q** caused by the electrochemical potential difference (negative on the matrix side) drives the mechanism of ATP synthesis.

Complex V-ATP Synthase Complex

- Also called as **the Fifth Complex** of Electron transport chain.
- **The smallest molecular motor present in the human body**.
- Location–ATP synthase is embedded in the **inner mitochondrial membrane.**

Divided into two Subcomplexes:
- F0 Subcomplex
- F1 Subcomplex

F0 Subcomplex
- Hydrophobic innature.
- F0 spans the **inner mitochondrial membrane**
- Forms a **proton channel**.
- Made up of a disk of 10 "C"protein subunits.

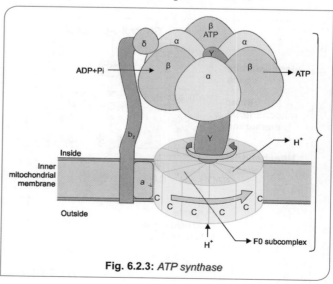

Fig. 6.2.3: *ATP synthase*

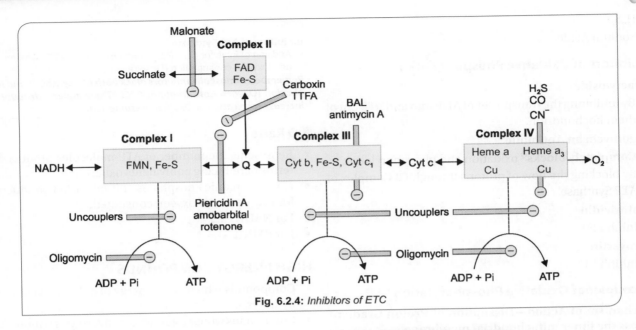

Fig. 6.2.4: *Inhibitors of ETC*

F1 Subcomplex
- Hydrophilic in nature.
- Projects into the mitochondrial matrix
- F1 is attached to FO Subcomplex.
- Made up 9 Subunits ($\alpha 3\beta 3\gamma\delta\varepsilon$)
- γ subunit in the form of a **"bentaxle"**.
- γ subunit is surrounded by 3α and 3 β subunit alternatively.
- The flow of protons through F0 causes, rotation of F0 Complex along with γ subunit of F1 complex to rotate.
- This causes the **production of ATP in the F1 complex**.
- β-**subunit of F1 Complex is called Catalytic Subunit**.

ATP Synthase Complex as a Rotor-Stator Molecular Motor

Because it has two functional unit.

A rotating subunit
- Consist of F0 Complex and γ Subunit of F1 Complex.

A stationary subunit
- F1 Complex other than γ subunit.

Binding Change Mechanism
- The theory behind the ATP production in the β subunit of F1 Subcomplex
- Proposed by **Paul Boyer**.
- States that re-entry of protons through F0 Subcomplex causes **rotation of γ subunit** which in turn causes conformational changes in the β subunits of F1 Sub-complex.

INHIBITORS OF ELECTRON TRANSPORT CHAIN

Divided into
1. Inhibitors of Electron transfer
2. Inhibitors of Oxidative Phosphorylation.
3. Uncouplers of Oxidative Phosphorylation
4. Ionophores

Inhibitors of Electron Transfer

Between NADH and CoQ [At Complex I]
1. An insecticide and a fish Poison Rotenone
2. Amobarbital which is abarbiturate.
3. PiericidinA.

Inhibitor of Complex II
1. TTFA (Trienoyl Tri Fluoro Acetone) a Fe^{2+} Chelating agent
2. Carboxin
3. Malonate, a competitive inhibitor of Succinate Dehydrogenase.

Between Cyt b and Cyt c [At Complex III]
1. Antimycin A
2. British Antilevisite (Dimercaprol)

Inhibitor at Cytochrome C Oxidase (Complex IV)
1. CO
2. Cyanide

3. H_2S
4. Sodium Azide.

Inhibitors of Oxidative Phosphorylation

Atractyloside
- By inhibiting the transporter of ADP into and ATP out of them itochondrion.

Oligomycin an Antibiotic
- **Completely blocks**^{Q} oxidation and phosphorylation
- By blocking the flow of protons through F0 Complex of ATP Synthase.

Venturicidin
- Inhibit F0

Aurovertin
- Inhibit F1

Uncoupler of Oxidative Phosphorylation

Mechanism of Action—Disruption of Proton Gradient across the inner mitochondrial membrane
1. 2,4 Dinitrophenol
2. Dinitrocresol
3. FCCP (Fluoro Carbonyl Cyanide Phenyl hydrazine)
4. Aspirin in high dose.

Physiological Uncoupler
1. Thermogenin (uncoupling protein 1) in Brown Adipose tissue
2. Thyroxine
3. Long chain free fatty acid
4. Unconjugated bilirubin.

Ionophores

- Ionophores or Channel formers, create an ion channel which permit specifications to penetrate membranes.
- Hence dissipate Proton Gradient.
  1. Valinomycin
  2. Gramicidin
  3. Nigericin

CONCEPT BOX

Effects of Inhibitors of Electron Transport Chain
Inhibitors of Electron Transfer
- Addition of inhibitors which block electron transfer inhibit ATP synthesis, which in turn block electron transfer also.
- Oxygen consumption (respiration) and ATP synthesis come to a stand still

Uncouplers of Oxidative phosphorylation
- Addition of an inhibitor which is an uncoupler (e.g.: DNP) respiration (oxygen consumption) increases but ATP is not synthesized.

Contd...

Contd...

Inhibitors of ATP Synthase
- Addition of inhibitor of ATP Synthase block both ATP Synthesis and Oxygen consumption (Respiration)

Remember: Almost all inhibitors block both ATP synthesis and respiration (Oxygen Consumption) EXCEPT uncouplers. Uncouplers increase respiration or Oxygen consumption.

P:O Ratio

- Represents the number of ATP molecules produced in terms of reducing equivalents oxidised.
- No. of inorganic Phosphates utilised for ATP production for every atom of oxygen consumed.
- For NADH -2.5
- For $FADH_2$ -1.5

HIGH ENERGY COMPOUNDS

- Compounds which yield energy of at least 7 kcal/m on hydrolysis.
- Compounds whose free energy of hydrolysis is more than that of ATP is called High energy phosphates.
- Compounds whose free energy of hydrolysis is less than that of ATP is called low energy phosphates.

Classification of high energy Phosphates
1. Pyrophosphate, e.g.: ATP
2. Acyl Phosphate, e.g.: 1,3 Bisphosphoglycerate
3. Enol Phosphate, e.g.: Phosphoenol pyruvate
4. Thioester e.g.: Acetyl CoA, Succinyl CoA
5. Phosphagen, e.g.: Phosphocreatine, Phosphoarginine

- All high energy compounds given yield energy higher than ATP.
- Most of the compound contain Phosphate group (hence also called High Energy Phosphates) except Acetyl CoA.

Compound	Free energy kJ/mol	Free energy Kcal/mol
Phosphoenolpyruvate	−61.9	−14.8
Carbamoyl phosphate	−51.4	−12.3
1,3-Bisphosphoglycerate (to 3-phosphoglycerate)	−49.3	−11.8
Creatine phosphate	−43.1	−10.3
ATP→AMP + PPi	−32.2	−7.7
ATP → ADP + Pi	−30.5	−7.3
Glucose-1-phosphate	−20.9	−5.0
PPi	−19.2	−4.6
Fructose-6-phosphate	−15.9	−3.8
Glucose-6-phosphate	−13.8	−3.3
Glycerol-3-phosphate	−9.2	−2.2

Image-Based Information

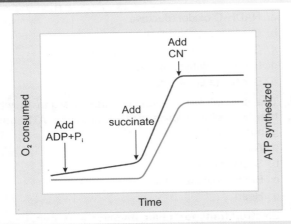

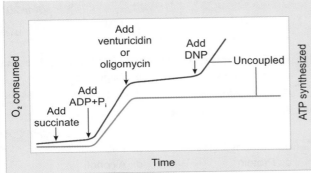

Fig. 6.2.5: Effect of inhibitors of respiratory chain

In an experiment mitochondria is suspended in a buffered medium with O_2 electrode to monitor Oxygen consumption and ATP production. At intervals various compounds are added and examined for Oxygen consumption and ATP production.

Experiment 1
First added ADP and Pi
- Result No ATP produced and No Oxygen consumption
- Reason: There is no electron donors.
- Second added Succinate
- Result ATP is produced, Oxygen consumption increased.
- Reason: Electron donor is added, which donated electron to Complex II.
- Thirdly Cyanide is added
- Result: Both ATP and Oxygen consumption stopped
- Reason: Cyanide is an inhibitor of electron transfer in Complex IV

Experiment 2
Succinate, ADP and Pi added like previous experiment
Added Oligomycin or Venturicidin
- Result: Both ATP production and O_2 consumption stopped
- Reason: Oligomycin/Venturicidin is and inhibitor of ATP Synthase
  Added Dinitrophenol
- Result: Oxygen increase but no ATP is produced
- Reason: DNP is an uncoupler, hence electron transfer is increased but no ATP is produced

NB: Based on this experiment you can expect some NBE model questions

Clinical Correlation

MELAS (Mitochondrial Encephalopathy, Lactic Acidosis, Stroke)
An inherited condition due to deficiency of **NADH-Q Oxido reductase (Complex I)** or **Cytochrome Oxidase (Complex IV)**.

Quick Revision

- Fe-S complex is present in all complexes except Complex IV
- The complex which pump no protons to intermembrane space is Complex II.
- Mobile complexes in ETC are CoQ and Cyt c
- The chemiosmotic theory explains the mechanism of oxidative phosphorylation.
- The Complex V is ATP Synthase Complex.
- The irreversible complex is Cytochrome c oxidase.

Inhibitors of ETC

Site of inhibition	Compound
Complex I	Rotenone
	Amobarbital
	Piericidin A
Complex II	TTFA
	Carboxin
	Malonate
Complex III	Antimycin A
	British Anti-Lewisite
Complex IV	CO
	Cyanide
	H_2S
	Sodium azide
ATP-ADP Transporter	Atractyloside
Fo subcomplex	Oligomycin
	Venturicidin
F1 subcomplex	Aurovertin
Chemical uncouplers	2,4 Dinitrophenol
	Dinitrocresol
	FCCP
	Aspirin in high dose
Physiological uncouplers	Thermogenin
	Thyroxine
	Long chain Fatty acid
	Unconjugated bilirubin
Ionophores	Valinomycin
	Gramicidin
	Nigericin

Check List for Revision

- Learn the names of all complexes
- Component of ETC
- Inhibitors of ETC is the most important topic in this chapter

Review Questions

1. Oxidative phosphorylation is not inhibited by:
   (PGI Nov 2016)
   a. Fluoride
   b. 2,4 Dinitrophenol
   c. Oligomycin
   d. Carboxin
   e. Quabain

2. Which of the following is a physiological uncoupler?
   (Recent Question 2016)
   a. Thyroxine
   b. Insulin
   c. Glucagon
   d. Norepinephrine

3. True about effect of 2,4 Dinitrophenol is:
   (Recent Question 2016)
   a. Oxygen consumption is increased
   b. ATP is produced
   c. Respiration decreased
   d. Electron transfer is decreased

4. MELAS is an inherited disorder affect which complex of ETC:
   (Recent Question 2016)
   a. Complex I
   b. Complex II
   c. Complex III
   d. Complex IV

5. Transport of ADP in and ATP out of mitochondria is inhibited by:
   (AIIMS Nov 2010)
   a. Atractyloside
   b. Oligomycin
   c. Rotenone
   d. Cyanide

6. The electron flow in cytochrome Coxidase can be blocked by:
   (AIIMS May 2006)
   a. Rotenone
   b. Antimycin-A
   c. Cyanide
   d. Actinomycin

7. Cytosolic cytochrome C mediates: (AIIMS May 2006)
   a. Apoptosis
   b. Electron transport
   c. Krebs cycle
   d. Glycolysis

8. High energy compounds is/are: (PGI May 2012)
   a. ATP
   b. Creatine phosphate
   c. Glucose1 phosphate
   d. Glycerol 3 phosphate
   e. ADP

9. In ETC, oxidative phosphorilation (ATP formation) is regulated by:
   (PGI May 2011)
   a. NADH Co-Q reductase
   b. Cytochrome Coxidase
   c. Glutathione reductase
   d. Isocitrate dehydrogenase
   e. Co-Q cytochrome C reductase

10. Which component transfer four protons?
    (PGI Nov 2009)
    a. NADH-Q oxidoreductase
    b. Cytochrome-Coxidase
    c. Cytochrome C–Q oxidoreductase
    d. Isocitrate dehydrogenase
    e. Succinate Q reductase

11. The specialized mammalian tissue/organ in which fuel oxidation serves not to produce ATP but to generate heat is: (AI 2006)
    a. Adrenal gland
    b. Skeletal muscle
    c. Brown adipose tissue
    d. Heart

12. Electron transport chain involves all except:
    (Ker 2011)
    a. NADP
    b. NAD
    c. Coenzyme^Q
    d. FAD

13. FO-F1 complex, ATP synthase inhibitor is: (Ker 2007)
    a. Atractyloside
    b. Oligomycin
    c. Antimycin
    d. Rotenone

14. Respiratory Quotient 0.7 is seen in:
    (Recent Question 2016)
    a. Carbohydrates
    b. Fat
    c. Protein
    d. Alcohol

15. Phenobarbitone inhibits which complex of ETC?
    (Recent Question 2016)
    a. Complex I
    b. Complex II
    c. Complex III
    d. Complex IV

16. Dinitrophenol inhibits the electron transport chain by:
    (Recent Question 2016)
    a. Cytochrome b
    b. Inhibits ATP synthesis and electron transport chain
    c. Inhibits ATP synthesis but not electron transport chain
    d. Inhibits electron transport chain but not ATP synthesis

17. Mechanism of Cyanide poisoning:
    (Recent Question 2016)
    a. Inhibition of cytochrome oxidase
    b. Inhibition of carbonican hydrase
    c. Inhibition of cytochrome c
    d. Inhibition of ATP synthase

18. Final acceptor of electrons in ETC is:
    (AIIMS May 2014)
    a. Cyt c
    b. Oxygen
    c. $FADH_2$
    d. Co^Q

Answers to Review Questions

1. **a, d, e. Fluoride, Carboxin, Quabain**

   Fluoride is not an inhibitor of respiratory chain. Carboxin is an inhibitor of complex II hence it is not an inhibitor of oxidative phosphorylation.
   Quabain is an inhibitor of $Na^+ K^+$ ATPase.

2. **a. Thyroxine** (Ref: Harper 31/e page 123)

   Uncoupler of Oxidative Phosphorylation
   Mechanism of Action – Disruption of Proton Gradient across the inner mitochondrial membrane.
   1. 2,4 Dinitrophenol
   2. Dinitrocresol
   3. FCCP (Flouro Carbonyl Cyanide Phenylhydrazine)
   4. ? Aspirin in high dose

   Physiological Uncoupler
   1. Thermogenin (uncoupling Protein 1) in Brown Adipose Tissue
   2. Thyroxine
   3. Long chain free fatty acid
   4. Unconjugated bilirubin.

3. **a. Oxygen consumption is increased**
   (Ref: Lehninger 6/e page 695)

   Addition of an inhibitor which is a nuncoupler (e.g.: DNP) to mitochondria, then electron transfer is increased. Hence, respiration (oxygen consumption increases) but ATP is notsynthesized.

4. **a, d. Complex I, Complex IV** (Ref: Harper 31/e page 123)

5. **a. Atractyloside** (Ref: Harper 31/e page 123)
   a. Atractyloside inhibits oxidative phosphorylation by inhibiting the transporter of ADP into and ATP out of the mitochondrion
   b. The antibiotic oligomycin completely blocks oxidation and phosphorylation by blocking the flow of protons through ATP synthase
   c. Barbiturates such as amobarbital, Rotenone & Piericidin A inhibit electron transport via Complex I
   d. Antimycin A and dimercaprol inhibit the respiratory chain at Complex III. The classic poisons H_2S, carbon monoxide, and cyanide inhibit Complex IV and can therefore totally arrest respiration. Malonate is a competitive inhibitor of Complex II.

6. **c. Cyanide** (Ref: Harper 31/e page 123)

   Cytochrome Oxidase is inhibited by CO, HCN, H_2S and Na Azide

7. **a. Apoptosis** (Ref: Harper 31/e page 123)

   Mitochondrial cytochrome c is a mobile electron carrier in Electron Transport Complex. This also mediates Apoptosis.

8. **a, b. ATP, Creatine phosphate** (Ref: Harper 31/e page 107)

High energy compound	Free energy kJ/mol	Free energy kcal/mol
Phosphoenol pyruvate	–61.9	–14.8
Carbamoyl phosphate	–51.4	–12.3
1, 3-Bisphosphoglycerate (to 3-phosphoglycerate)	–49.3	–11.8
Creatine phosphate	–43.1	–10.3
ATP → AMP + PPi	–32.2	–7.7
ATP → ADP + Pi	–30.5	–7.3

9. **a, b, c, e. NADH-CoQ reductase, Cytochrome c oxidase, Glutathione reductase, Co-Q Cytochrome C reductase**
   (Ref: Harper 31/e page 123)

   Components of the Electron Transport Chain^Q are contained in four large protein complexes.
   1. Complex I NADH CoQ Oxidoredutase-
   2. Complex II CoQ Succinate Reductase
   3. Complex III CoQ Cytochrome c Oxidoreductase
   4. Complex IV Cytochrome c Oxidase

10. **a, c. NADH-Q Oxidoreductase, Cytochrome c—Q oxidoreductase** (Ref: Harper 31/e page 123)
    a. Complex I and III pumps $4H^+$
    b. Complex II pumps noprotons
    c. Complex IV pumps $2H+$

11. **c. Brown adipose tissue**

    Brown adipose tissue contains thermogenin, which is a physiological uncoupler of oxidative phosphorylation. This process is called Nonshivering Thermogenesis.

12. **a. NADP**

    NADP is involved in reductive biosynthesis, not in ETC.

13. **b. Oligomycin** (Ref: Harper 31/e page 123)
    - Atractyloside inhibits oxidative phosphorylation by inhibiting the transporter of ADP into and ATP out of the mitochondrion.
    - The antibiotic oligomycin completely blocks oxidation and phosphorylation by blocking the flow of protons through ATP synthase.

14. **b. Fat**

15. **a. Complex I** (Ref: Harper 31/e page 123)

    Inhibitors of ETC at Complex I
    1. An insecticide and a fish Poison Rotenone
    2. Amobarbital which is a barbiturate.
    3. Piericidin

16. **c. Inhibits ATP synthesis but not electron transport chain**

    Dinitrophenol is a nuncoupler of Oxidative Phosphorylation. Sono ATP synthesis but electron transfer and oxidation of reducing equivalents takes place.

17. a. Inhibition of Cytochrome Oxidase *(Ref: Harper 31/e page 123)*

Inhibitors of Complex IV are CO, Cyanide, H_2S, Sodium Azide

18. b. Oxygen *(Ref: Harper 31/e page 123)*

Electrons are transferred in the ascending order of redox potential, the final oxygen electron acceptor is oxygen.

6.3 INTEGRATION OF METABOLISM

- Metabolic Fuels for Major Organs
- Steps of TCA Cycle
- Energetics
- Regulation of TCA Cycle

METABOLIC FUELS FOR MAJOR ORGANS

Metabolic Fuel of Major Organs[Q]

Organ	Major metabolic fuels
Liver	**Free fatty acids, glucose (in fed state)**, lactate, glycerol, fructose, amino acids, alcohol
Brain	**Glucose**, amino acids, **ketone bodies** in prolonged starvation
Heart[Q]	**Ketone bodies, free fatty acids**, lactate, chylomicron and VLDL triacylglycerol, some glucose
Adipose tissue	Glucose, chylomicron and VLDL triacylglycerol
Fast twitch muscle	Glucose, glycogen
Slow twitch muscle	**Ketone bodies**, chylomicron and VLDL triacylglycerol
Kidney	**Free fatty acids**, lactate, glycerol, glucose
Erythrocyte	Glucose

Starve Feed Cycle

Divided into 5 stages:
1. Well fed state (1–4 hours after food)
2. Early fasting (4–16 hours after food)
3. Fasting (16–48 hours after food)
4. Starvation (2–4 days without food)
5. Prolonged starvation (>5 days without food)

Stage I Well Fed State (Post-prandial Stage)
- **Exogenous dietary supply of Fuel from the Intestine**
- **Glucose** is the major fuel
- **Insulin** is the hormone—Most of the enzyme regulated by covalent modification is in the **Dephosphorylated State.**
- **Favor Glucose utilisation and storage of excess glucose.[Q]**

In the Liver
On Carbohydrate Metabolism
- Favor Glycolysis
- Favor Glycogenesis
- **Uptake of Glucose by GLUT-2 is insulin independent** But **Glucokinase is an inducible enzyme by insulin,** when Glucose is excess
- Decreased Gluconeogenesis.

Fat Metabolism
- Favor **Lipogenesis[Q]**
- Increased fatty acid synthesis
- Increased triacylglycerol synthesis.

In Adipose Tissue
- Favor the transport of **GLUT-4 from intracellular vesicle to the cell surface[Q]**
- Favor the uptake of glucose
- **Inhibit hormone sensitive Lipase[Q]—Inhibit Lipolysis**
- Favor fatty acid and triacylglycerol synthesis.

In the Muscle
- **Glucose uptake by insulin dependent GLUT-4**
- Favor glycolysis
- Favor glycogenesis.

Post-absorptive Phases
- No fuel from gut.
- Plasma insulin decreases,
- **Glucagon level begin to rise**
- **Increased cAMP level—Increased cAMP dependent Protein Kinase A**
- **Most of the enzyme active in post-absorptive phase are in Phosphorylated state.**

Glucagon
- Activate **Adenylyl Cyclase.**
- Increase cAMP level.

Insulin
- Favor Phosphodiesterase which convert cAMP to 5'AMP.
- Decrease cAMP level. (DNB 2011, AIIMS May 95)

Stage II Early Fasting (4–16 Hours after Food)
Fuel from the gut stopped.

On Carbohydrate Metabolism
- Increased **breakdown of liver glycogen (not Muscle)**
- Decreased glycogenesis.

On Lipid Metabolism
- Decreased lipogenesis.

Stage III Fasting (16–48 Hours after Food)
- No fuel from the gut
- **By 12-18 hours liver Glycogen^Q stores are depleted.**
- **Hepatic Gluconeogenesis favored**
- Muscle proteins are degraded to supply amino group to Pyruvate to form Alanine.
- This Alanine reaches the liver used for gluconeogenesis. This is Glucose Alanine Cycle.

Stage IV Starvation or Prolonged Fasting (2-4 Days without Food)
- No fuel from the gut.
- Decreased Gluconeogenesis
- **Increased activity of hormone sensitive lipase.**
- **Increased lipolysis, free fatty acid level rises.**
- **Increased ketone bodies synthesis.**
- **Ketone bodies inhibit muscle proteolysis.**

Stage V Prolonged Starvation (After 5 days)
Increased Proteolysis—Muscle Wasting Cachexia

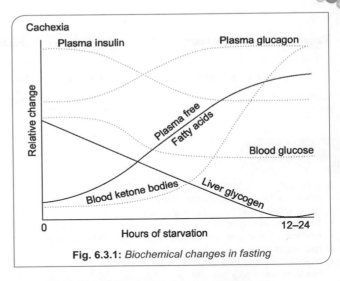

Fig. 6.3.1: Biochemical changes in fasting

Integration of Metabolism—At a Glance
Major Metabolic Pathways in Different Stages of Fast/Fed Cycle

Well fed	1–4 hrs after food	Dietary metabolic fuels are available. All anabolic pathways are active.
Early fasting	4–16 hrs after food	Liver **Glycogenolysis**
Fasting	16–48 hrs after food	**Gluconeogenesis** mainly from Glycerol and Alanine from muscle TAG undergo lipolysis to provide Glycerol for Gluconeogenesis Glycogen stores are depleted by 18 hrs.
Starvation (Prolonged starvation)	2–5 days	Gluconeogenic substrates are decreased like Glycerol from TAG, Alanine, etc. Fatty acid oxidation **Ketone body synthesis**
Prolonged starvation	>5 days	Muscle Proteolysis

Metabolic Fuels in Various Stages of Fed Fast Cycle^Q

Organ	Fed state	Early Fasting	Fasting	Prolonged fasting/ starvation	Prolonged starvartion
Liver	Free fatty acid/Glucose	Free Fatty acid/Glucose		Amino acids/Free fatty acids	
	Other fuels are Lactate and Glycerol in fasting via **Gluconeogenesis** Fructose (fed state)				
Heart	**Free fatty acids^Q(preferred fuel)**/Glucose	Free fatty acids/Glucose		Ketone bodies	
	Other fuels are Amino acids, Lactate, Chylomicron and VLDL TAG				
Brain	Glucose	Glucose		**Glucose (80%)^Q** Ketone bodies (20%)	Ketone bodies
	Other fuels are Branched amino acids				
Skeletal muscle	Glucose/Free fatty acids/ Glucose from Glycogen	Free fatty acids		Ketone bodies	
RBC	**Glucose only^Q**				

Contd...

Contd...

Organ	Fed state	Early Fasting	Fasting	Prolonged fasting/ starvation	Prolonged starvartion
Kidney	Free fatty acids/Glucose		Ketone bodies		
	Other fuels Lactate and Glycerol esp in Fasting via **Gluconeogenesis**				
Adipose tissue	Glucose		Chylomicron and VLDL TAG	Ketone bodies	

In fed state
- Most organs use glucose than free fatty acids
- Heart utilises free fatty acid > glucose
- Brain uses only glucose
- Erythrocyte can use only glucose in fed, fast and starvation.

In fasting state
- Most organs use free fatty acid than glucose
- But brain uses only glucose
- Erythrocyte can use only glucose

In starvation state
- Liver uses free fatty acids and amino acids
- Heart uses ketone bodies

Contd...

- In brain 80% energy need by glucose itself and only 20% by Ketone bodies.
- Skeletal muscle, Kidney and Adipose tissue uses Ketone bodies.
- Liver and RBC cannot utilize ketone bodies.
- Eythrocyte can use only glucose even in starvation.

Check List for Revision

- Learn different stages of fed fast cycle and major events that occur in each stage.
- After going through tables given in the chapter learn the quick review points.

Contd...

Review Questions MCQ

1. Which of the following organs utilises glucose, fatty acid and ketone bodies as fuel?
   a. Heart
   b. Liver
   c. RBC
   d. Brain

2. In fasting state which of the following is false?
   a. Insulin level falls
   b. Level of cAMP is high
   c. Glycogenesis is active
   d. Lipolysis active

3. Which of the following pathway is active in high insulin glucagon ratio?
   a. Glycogenolyis
   b. Lipolysis
   c. Pyruvate dehydrogenase
   d. Gluconeogenesis

4. After 48 hours of fasting which of the following do not be active?
   a. Lipolysis
   b. Ketone body synthesis
   c. Gluconeogenic substrates are depleted
   d. Glycogenolysis

5. Which of the following enzyme activity decrease in fasting? (AIIMS May 2018)
   a. Hormone sensitive lipase
   b. Glycogen Phosphorylase
   c. Acetyl CoA Carboxylase
   d. CPS I

1. **a. Heart**
   - Liver can not utilise Ketone bodies
   - RBC can utilise only glucose
   - Brain cannot utilise fatty acid

2. **c. Glycogenesis is active**

Early fasting	4–16 hrs after food	Liver **glycogenolysis**
Fasting	16–48 hrs after food	**Gluconeogenesis** mainly from Glycerol and Alanine from muscle TAG undergo lipolysis to provide Glycerol for Gluconeogenesis Glycogen stores are depleted by 18 hrs
Starvation (Prolonged starvation)	2–5 days	Gluconeogenic substrates are decreased like Glycerol from TAG, Alanine, etc. Fatty acid oxidation Ketone body synthesis
Prolonged starvation	>5 days	Muscle proteolysis

3. **c. Pyruvate dehydrogenase**

High insulin glucagon ratio means Well fed state hence PDH, the link reaction is active.

4. **d. Glycogenolysis**

Glycogen stores are depleted by 18 hours of fasting. Hence, Glycogenolysis is not active

5. **c. Acetyl CoA Carboxylase**
   - Option a, Hormone sensitive lipase increase in fasting.
   - Option b, Glycogen phosphorylase, the rate limiting step of glycogenolysis will be active.
   - But Acetyl CoA Carboxylase, the rate limiting step of fatty acid synthesis is inactive in fasting state.
   - CPSI is not related to fed fast cycle.

6.4 SHUTTLE SYSTEMS

☞ Shuttle Systems

SHUTTLE SYSTEMS

Concept

The inner mitochondrial membrane (IMM) is selectively permeable to various compounds. This necessitates exchange transporters and shuttle mechanisms

Molecules which are freely permeable to IMM are uncharged small molecules like

- Oxygen
- Water
- CO_2
- NH_3
- Monocarboxylic acid like acetic acid
- NADH cannot penetrate the mitochondrial membrane, but it is produced continuously in the cytosol by 3-phosphoglyceral dehyde dehydrogenase, an enzyme in the glycolysis sequence.
- The transfer of reducing equivalents is carried out by using the various substrate shuttle systems.

Glycerophosphate Shuttle

Site :- **brain, white muscle**
- But absent in **heart tissue.**

Sequence of events are:

In the cytosol
- **NADH to NAD+,**
- **DHAP to Glycerol 3 PO4**

In the mitochondria
- **Glycerol 3 PO4 to DHAP**
- **FAD to FADH2**
- **By enzyme Glycerol 3 PO4 dehydrogenase**

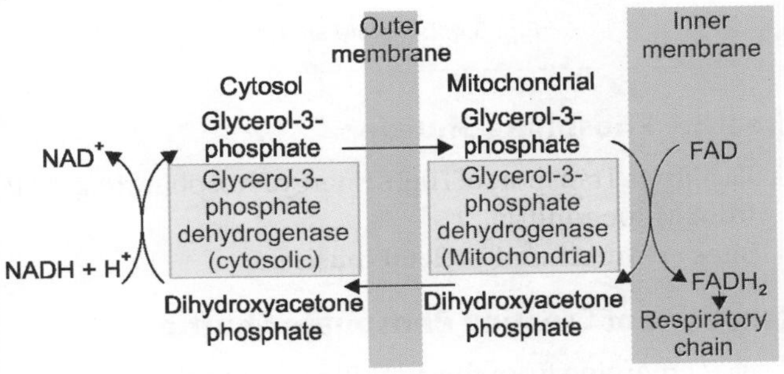

Fig. 6.4.1: *Glycerophosphate shuttle*

The number of ATPs from 1 NADH transported to mitochondria by Glycerophosphate Shuttle is 1.5 NOT 2.5. Why?

Since the mitochondrial glycerol phosphate dehydrogenase is linked to the respiratory chain via a flavo protein (FAD) rather than NAD, **only 1.5 mol** rather than 2.5 mol of ATP are formed per atom of oxygen consumed.

Malate Shuttle

- **Malate shuttle** system is of more universal utility.
- **Used to transport NADH from Cytosol to Mitochondria.**

Reactions involved in Malate Shuttle

- NADH converted to NAD+, Oxaloacetate to Malate
- Malate enter mitochondria via α Ketoglutarate Transporter.
- Malate converted to Oxaloacetate, NADH is released.
- But there is no transporter for Oxaloacetate.
- Oxaloacetate react with glutamate to form aspartate and (α-ketoglutarate by transamination).
- Aspartate and αKetoglutarate are transported to cytosol and Oxaloacetate is reconstituted.

Self Assessment and Review of Biochemistry

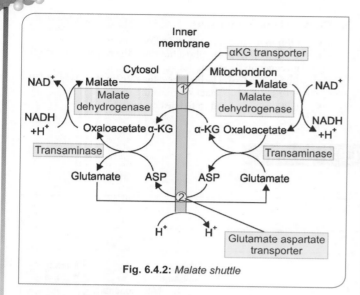

Fig. 6.4.2: *Malate shuttle*

Via protein pores in the outer mitochondrial membrane, Creatine Kinase generate extramitochondrial ATP.

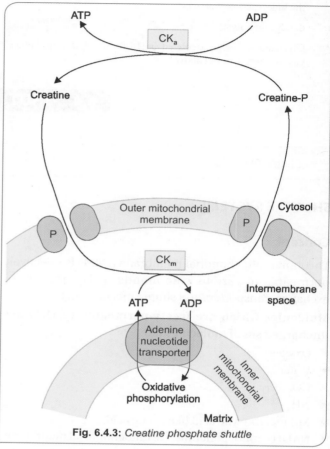

Fig. 6.4.3: *Creatine phosphate shuttle*

Creatine Phosphate Shuttle

- Facilitates Transport of High-Energy Phosphate (e.g. ATP) from Mitochondria
- Sites are **heart and skeletal muscle**.

Reactions of Creatine Phosphate Shuttle

- ATP emerging from the adenine nucleotide transporter to intermembrane space.
- CK_m (An isoenzyme of **creatine kinase** [CK_m] is found in the mitochondrial intermembrane space) transfer high-energy phosphate from ATP to creatine form Creatine Phosphate.
- The creatine phosphate is transported into the cytosol

- IMM is selectively permeable.
- Uncharged molecules are freely permeable to IMM.
- From 1 NADH via Glycerophosphate shuttle generate only 1.5 ATPs
- Malate Aspartate shuttle is the universal transporter of NADH to mitochondria.
- Glycerophosphate shuttle operate in Brain and white muscle.
- The organ that lack Glycerophosphate shuttle is heart.
- Malate Aspartate shuttle transport Oxaloacetate to cytosol in gluconeogenesis.
- Creatine phosphate shuttle transport ATP from mitochondria in heart and skeletal muscle.

Check List for Revision

- This is not a very frequently asked topic for exams, so an overall idea is only needed.
- Just concentrate on quick revision points.

Review Questions MCQ

1. 1 mol of glucose generate how many net ATPS if NAD is transported via Glycerophosphate shuttle:
   a. 6
   b. 7
   c. 5
   d. 9

2. Malate shuttle is involved in:
   a. Glycolysis
   b. Gluconeogenesis
   c. Both glycolysis and gluconeogenesis
   d. HMP Pathway

Answers to Review Questions

1. **c. 5**

   1 NADH by glycerophosphate shuttle generate only 1.5 ATPs as mitochondrial Glycerol 3 Phosphate dehydrogenase generate FADH2. So from 2 NADH by Glyceraldehyde 3 Phosphate via Glycerol Phosphate shuttle produce only 3 ATPs.

   Energetics of 1 mol of glucose is
   - Glyceraldehyde 3 Phosphate Dehydrogenase = 2 NADH = 3 ATPs
   - 1, 3 BPG Kinase = 2 ATPS
   - Pyruvate Kinase = 2 ATPs
   - Utilise 2 ATPs
   - Hence, net ATPs = 9–2 = 7ATPs

2. **c. Both glycolysis and gluconeogenesis**
   - Malate shuttle transport NADH to mitochondria so it is involved in glycolysis to transport NADH to mitochondria.
   - In gluconeogenesis Oxaloacetate is transported from mitochondria to Cytosol via Malate Aspartate shuttle.
   - In HMP Pathway Malate shuttle has no role.

CHAPTER 7

Heme and Hemoglobin

Chapter Outline

7.1 Heme Synthesis and Porphyrias
7.2 Heme Catabolism and Hyperbilirubinemia
7.3 Hemoglobin

CHAPTER 7

Heme and Hemoglobin

7.1 HEME SYNTHESIS AND PORPHYRIAS

- ☞ Heme Synthesis and Porphyrias
- ☞ Heme Catabolism and Hyperbilirubinemia
- ☞ Hemoglobin

STRUCTURE OF HEME

Heme is a metalloporphyrin.

Porphyrin

Porphyrins are cyclic compounds formed by the linkage of **four pyrrole** rings through **methyne or methenyl bridges** (= CH –).

Special Features of Porphyrins

- Porphyrinogens are colorless whereas **porphyrins are coloured** compounds. The **conjugated double bond** in the pyrrole ring and linking methenyl bridges is responsible for characteristic absorption and fluorescence spectra.
- Sharp absorption band near **400 nm called Soret band** is distinguishing feature of all porphyrins after its discoverer, the French physicist Charles Soret.
- When porphyrins dissolved in strong mineral acids or in organic solvents are illuminated by ultraviolet light, they emit a strong **red fluorescence**.
- Used in **cancer phototherapy** because tumors often take up more porphyrins than normal tissue. The tumor is then exposed to an argon laser, which excites the porphyrins, producing cytotoxic effects. This called **Photodynamic therapy**.
- Porphyrins cause **photosensitivity**.

Image-Based Information

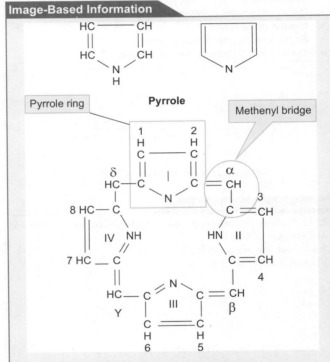

Fig. 7.1.1: *Structure of porphyrin*

- 4 pyrrole rings joined by methyne or methenyl bridge.
- Pyrrole ring—heterogeneous five membered ring with a nitrogen atom. They are labeled I, II, III, IV
- Methyne or methenyl bridge is = C–. They are labeled α, β, γ, δ.

Property of Porphyrins

A characteristic property of porphyrins is the formation of complexes with metal ions bound to the nitrogen atom of the pyrrole rings.
- Iron porphyrins—**Heme** of hemoglobin
- **Magnesium**-containing porphyrin—**Chlorophyll**, the photosynthetic pigment of plants.

284 Self Assessment and Review of Biochemistry

The *porphyrins* found in nature are compounds in which various *side chains* are substituted for the eight hydrogen atoms numbered in the porphyrin nucleus. The side chains are

    M—Methyl   A—Acetate
    V—Vinyl   P—Propionate

The distribution of side chain in each of the above porphyrin is given in the Figure 7.1.2 below.

HIGH YIELD POINTS

- The three important porphyrins are **Uroporphyrin, Coproporphyrin** and **Protoporphyrin**.
- Most water soluble porphyrin is **Uroporphyrin**.
- Least water soluble is **Protoporphyrin**.
- Coproporphyrin is having intermediate water solubility.
- Heme is metal and Porphyrin
- Porphyrin in heme is **Protoporphyrin**
- Metal in heme is **iron**
- So Heme is **Ferroprotoporphyrin**.
- The iron in the Heme is in **Ferrous state**.

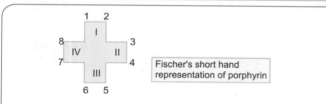

Fig. 7.1.2: *Representation of porphyrin*

Image-Based Information

The side chain distribution of major porphyrins are shown here. A is Acetate (-CH$_2$COO-)
P is Propionate (-CH$_2$CH$_2$COO-)
V is Vinyl (-CH=CH$_2$)

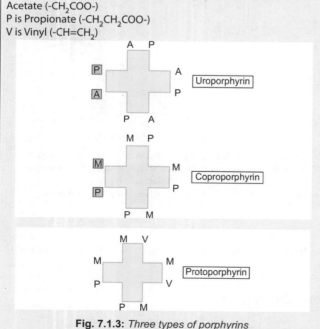

Fig. 7.1.3: *Three types of porphyrins*

Image-Based Information

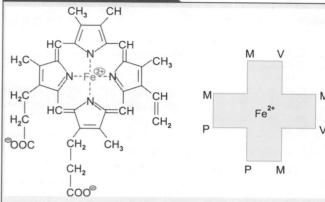

Fig. 7.1.4: *Structure of heme. Heme is Ferroprotoporphyrin. Protoporphyrin side chain distribution is Methyl (M) and Vinyl (V)*

Important heme containing proteins	
Hemoglobin	Transport of oxygen in blood
Myoglobin	Storage of oxygen in muscle
Cytochrome C	Involvement in electron transport chain
Cytochrome P450	Hydroxylation of xenobiotics
Catalase	Degradation of hydrogen peroxide
Tryptophan pyrrolase	Oxidation of tryptophan

BIOSYNTHESIS OF HEME

- Site: Synthesized in almost **all tissues** in the body EXCEPT in mature erythrocytes.
- 85% in **erythroid Precursor** cells in the bone marrow and majority of remainder in **hepatocyte**
- Organelle: **Partly cytoplasmic and partly mitochondrial**
- Starting materials: **Succinyl CoA** and **Glycine**

Steps of Synthesis of Heme

Can be divided into:
1. Synthesis of Porphobilinogen (Monopyrrole)
2. Synthesis of Uroporphyrinogen (Tetrapyrrole)
3. Conversion of Uroporphyrinogen to Protoporphyrin
4. Formation of Heme by incorporation iron.

Synthesis of Porphobilinogen (Monopyrrole)

ALA Synthase (ALAS)

Catalyse condensation reaction between succinyl-CoA and glycine to form α-**amino**-β-**ketoadipic acid,** which is rapidly decarboxylated to form δ-**aminolevulinate (ALA)**
Synthesis of ALA occurs in mitochondria.

Heme and Hemoglobin

High Yield Points

Comparison of two isoforms of ALAS
- ALAS-I is expressed throughout the body.
- ALAS-II is expressed in erythrocyte precursorcells.
- ALAS-1 is the rate limiting step of hepatic hemesynthesis. Heme acts as the negative regulator of ALAS-1.
- ALAS-1 is induced by the drugs whose metabolism require Cytochromes.
- ALAS-2 is not feedback regulated by heme.
- ALAS-2 is not induced by drugs.

ALA Dehydratase

Two molecules of ALA are condensed by the enzyme **ALA dehydratase** to form two molecules of water and one mol of **porphobilinogen** (PBG). Thus the first precursor **monopyrrole** is formed.

Takes place in the cytosol.

ALA dehydratase is a zinc-containing enzyme.

This enzyme is sensitive to inhibition by **lead,** as can occur in lead poisoning.

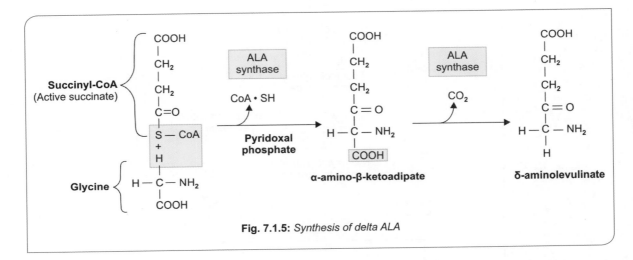

Fig. 7.1.5: Synthesis of delta ALA

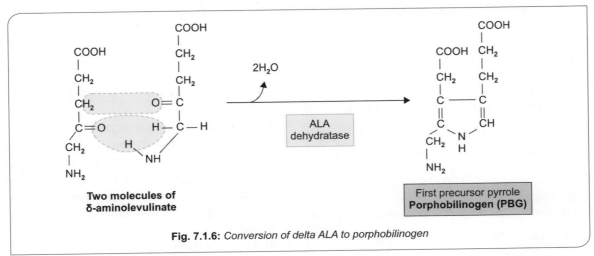

Fig. 7.1.6: Conversion of delta ALA to porphobilinogen

Synthesis of Uroporphyrinogen (Tetrapyrrole)

The formation of a cyclic tetrapyrrole — i.e., aporphyrin — occurs by condensation of four molecules of PBG.

Uroporphyrinogen—I Synthase or HMB Synthase or PBG Deaminase

Four molecules of PBG condense in a head-to-tail manner to form a line artetrapyrrole, **hydroxy methylbilane** (HMB).

4 mols of NH_3 is released

Takes place in the cytosol.

There action is catalyzed by **uroporphyrinogen I synthase,** also named PBG deaminase or HMB synthase.

Uroporphyrinogen III Synthase

- HMB is converted to Uroporphyrinogen III by Uroporphyrinogen III synthase.
- Uroporphyrinogen is thus the first porphyrin precursor formed.
- Under normal conditions, the uroporphyrinogen formed is almost exclusively the III isomer.
- But in certain porphyrias, HMB cyclizes spontaneously to form **uroporphyrinogen I**

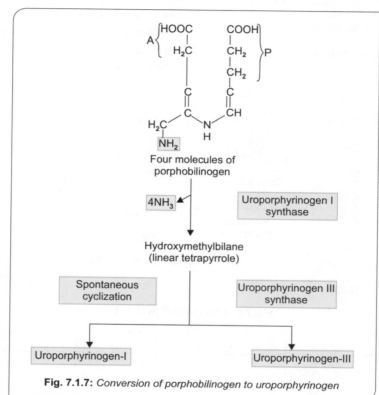

Fig. 7.1.7: Conversion of porphobilinogen to uroporphyrinogen

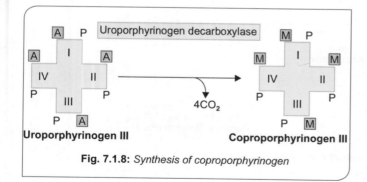

Fig. 7.1.8: Synthesis of coproporphyrinogen

Conversion of Uroporphyrinogen to Protoporphyrin Uroporphyrinogen Decarboxylase

- Uroporphyrinogen III is converted to **coproporphyrinogen III** by decarboxylation of all of the acetate (A) groups, which changes them to methyl (M) substituent.
- The reaction is catalyzed by **uroporphyrinogen decarboxylase.**
- This also takes place in the cytosol
- Coproporphyrinogen III then enters the mitochondria, where it is converted to **protoporphyrinogen III**
- The **mitochondrial** enzyme **coproporphyrinogen oxidase** catalyzes the decarboxylation and oxidation of two propionic side chains to form protoporphyrinogen.
- This enzyme is able to act only on type-III coproporphyrinogen, which would explain why type I protoporphyrins do not generally occur in nature.

Protoporphyrinogen Oxidase

The oxidation of protoporphyrinogen to **protoporphyrin** is catalyzed by another **mitochondrial** enzyme, **protoporphyrinogen oxidase.**

Heme and Hemoglobin

Formation of Heme by Incorporation of Iron
- This is the final step in hemesynthesis
- It involves the incorporation of ferrous iron into protoporphyrin in a reaction
- This step is catalyzed by **ferrochelatase (hemesynthase)**
- Takes place in the mitochondria.

Regulation of Heme Synthesis

ALA Synthase[Q] is the key regulatory enzyme in hepatic biosynthesis of heme.

ALA synthase occurs in both **hepatic** (ALAS-1) and **erythroid** (ALAS-2) forms.

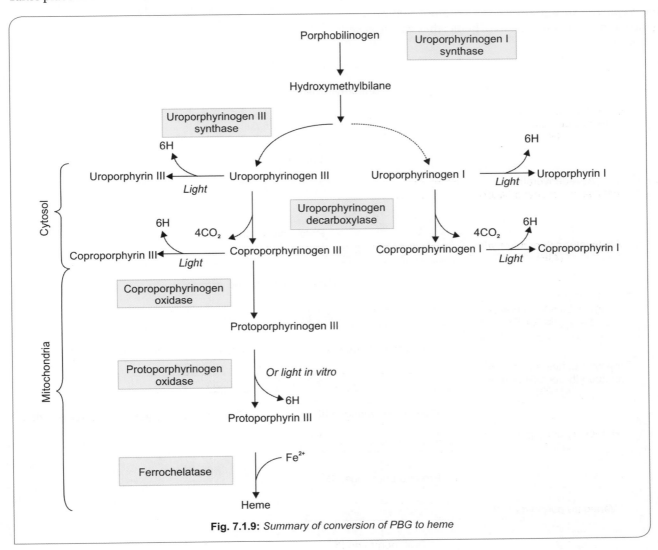

Fig. 7.1.9: *Summary of conversion of PBG to heme*

The rate-limiting reaction in the synthesis of heme in liver is **ALAS-1**.

It appears that **heme**, probably acting through an aporepressor molecule, acts as **a negative regulator** of the synthesis of ALAS-1 by repression-derepression mechanism.

Thus, the rate of synthesis of ALAS-1 increases greatly in the absence of heme and is diminished in its presence.

Factors that Affect Heme Synthesis

- **Drugs**—That induce Hepatic Cytochromes, e.g. Barbiturates[Q], Griseofulvin
- **Lead**[Q]—Inhibit steps catalyzed by **ALA dehydratase and ferrochelatase**
- **INH**—Decrease availability of PLP

Extra Edge

Basis of administration of glucose to relieve acute attacks of porphyrin
High cellular concentration of glucose prevents induction of ALA Synthase

High Yield Points

Most of the porphyrias are inherited autosomal dominant EXCEPT
- ALAD enzyme deficiency [ADP]
- Congenital erythropoietic porphyria [CEP]
- Erythropoietic protoporphyria [EPP]
- X Linked Protoporphyria [XLP]

PORPHYRIAS

The *porphyrias* are a group of disorders due to abnormalities in the pathway of biosynthesis of heme; they can be *genetic* or *acquired*.

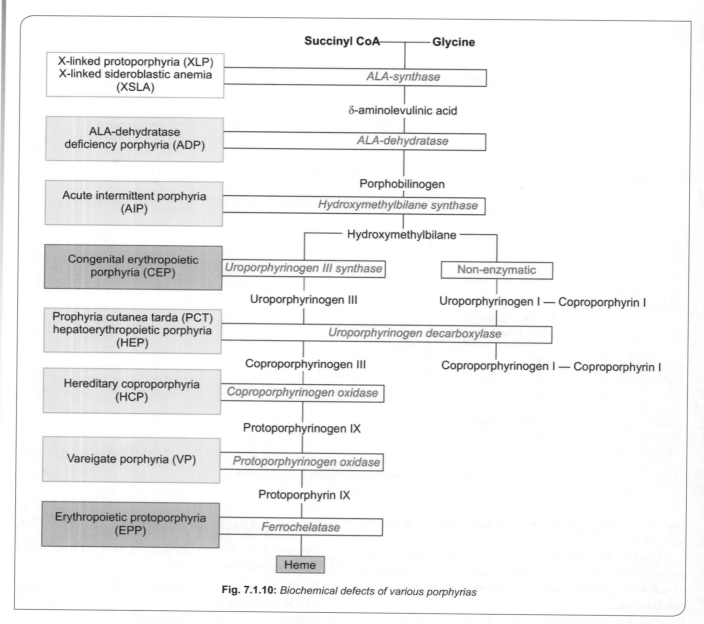

Fig. 7.1.10: *Biochemical defects of various porphyrias*

Heme and Hemoglobin

Concept of Clinical features of Porphyria

CONCEPT BOX

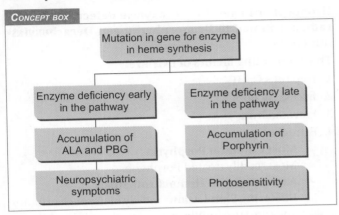

HIGH YIELD POINTS

Porphyrias at a Glance
- Erythropoietic porphyrias usually present with **Cutaneous photo sensitivity**.

Contd…

- Hepatic Porphyria with Cutaneous photosensitivity is **Porphyria CutaneaTarda (PCT)**
- **Hepatic Porphyria** is symptomatic in adults.
- **Erythropoietic Porphyria** usually present at birth or early childhood.
- Porphyria that presents as **Nonimmune Hydrops Fetalis** is Congenital Erythropoietic Porphyria.
- Most common Porphyria is **PCT**.
- Most common acute Porphyria is **Acute Intermittent Porphyria (AIP)**
- Most common Porphyria in children is **Erythropoietic Proto Porphyria (EPP)**
- Porphyria which can be sporadic is **Porphyria Cutanea Tarda (PCT)**
- Porphyria that is frequently seen in countries where Hepatitis C and HIV is prevalent is **PCT**
- Confirmatory diagnostic test for all Porphyrias are **Enzyme analysis and Mutation analysis**.
- First line investigation of Porphyrias with neuro visceral symptoms is **Spot Urine ALA and PBG**.
- First line investigation of Porphyrias with Photosensitivity is **Plasma Porphyrins**.

Major Clinical Features and Laboratory Features of Porphyrias

Porphyria	Deficient Enzyme	Inheritance	Principal Symptoms NV or CP+	Results of Laboratory Tests
5-ALA dehydratase-deficient porphyria (ADP)	ALA-dehydratase	AR	NV	Zn Protoporphyrin in erythrocytes
Acute intermittent porphyria (AIP)	HMB-synthase or Uroporphyrinogen I synthase	AD	NV	Urinary ALA and PBG increased
Porphyria cutaneatarda (PCT)	Uroporphyrinogen decarboxylase	AD	CP	Urinary uroporphyrin I increased
Hereditary coproporphyria (HCP)	Coproporphyrinogen oxidase	AD	NV and CP	Urinary ALA, PBG, and coproporphyrin III and fecal coproporphyrin III increased
Variegate porphyria (VP)	Protoporphyrinogen oxidase	AD	NV and CP	Urinary ALA, PBG, and coproporphyrin III and fecal protoporphyrin IX increased
Congenital erythropoietic porphyria (CEP)	Uroporphyrinogen III synthase	AR	CP	Urinary, fecal, and red cell uroporphyrin I increased
Erythropoietic protoporphyria (EPP)	Ferrochelatase	AR	CP	Fecal and red cell protoporphyrin IX increased
X-linked protoporphyria (XLP)	ALA-synthase 2	XL	CP	Protoporphyrin in Erythrocytes and stool

ALAD Deficient Porphyria (ADP)

- Aminolevulinic Acid Dehydratase Deficient Porphyria (ADP)
- Otherwise called **Doss Porphyria**
- Rare Porphyria.

Differential Diagnosis of ADP

- **Lead intoxication resembles ADP** as lead inhibit ALA Dehydratase and there is increased excretion of ALA.
- Hereditary Tyrosinemia Type I resembles ADP, as Succinyl Acetone resembles ALA.

Acute Intermittent Porphyria (AIP)

Defect in PBG Deaminase/HMB Synthase/Uroporphyrinogen I Synthase

- Most Common **Acute Porphyria**
- Most Common symptom—**Abdominal Pain**
- Most common Physical Sign—**Tachycardia**
- **Neurovisceral**
- Levels of ALA and PBG increased
- Urine Red coloured
- Currently liver directed gene therapy with Adeno Associated Viral vector (AAV-HMBS) has been proven to prevent drug induced AIP.
- **Hepatic targeted RNA interference (RNAi) therapy** directed to inhibit markedly elevated ALAS-1 mRNA, reduced the ongoing attack.

Congenital Erythropoietic Porphyria (CEP) Gunther's Disease

Uroporphyrinogen III Synthase defect

- Presents shortly **after birth or in utero as non-immune hydrops**
- Presents with severe **Cutaneous Photosensitivity**
- Porphyrins deposit in teeth and bones
- So **brownish discoloration of teeth**
- Hemolysis due to erythrocyte porphyrins, hence splenomegaly
- **Uroporphyrin I and Coproporphyrin I** accumulate in bone marrow, erythrocyte, plasma, urine and feces.
- **Coproporphyrin I** is the predominant porphyrin in feces.
- **Portwine urine or Pink stain in diapers**
- **Erythrodontia**—Reddish Fluorescence of teeth when illuminated with long UV light
- Prenatal diagnosis by Porphyrin in amniotic fluid, **URO Synthase enzyme activity** in chorionic villus and cultured amniotic cells.
- **Beta Carotene**^{Q} to protect from sunlight
- **Gene therapy** using Human cDNA UROS retroviral vectors.

Image-Based Information
Congenital Erythropoietic Porphyria

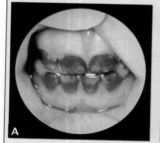 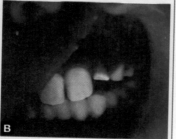

Figs. 7.1.11A and B: A: Brownish discoloration of teeth; B: Erythrodontia

Porphyria Cutanea Tarda

Uroporphyrinogen Decarboxylase defect 80% is sporadic attributed to Uroporphyrinogen Decarboxylase inhibitors.

The aggravating factors of PCT are:

1. Hepatitis C, HIV
2. Excess Alcohol
3. Elevated Iron
4. Estrogen
   - **Most Common Porphyria**
   - Most readily treated Porphyria
   - Associated with **Hemochromatosis**
   - **Blistering Skin Lesions** mostly in the back of hands.
   - They are susceptible to develop chronic liver disease and are at risk for Hepatocellular Carcinoma.

Treatment

- Repeated Phlebotomy to reduce hepatic iron.
- Low dose regimen of Chloroquine or Hydroxychloroquine.
- In patients with End stage renal Disease, administer Erythropoietin.

Image-Based Information
Cutaneous Photosensitivity in Porphyria

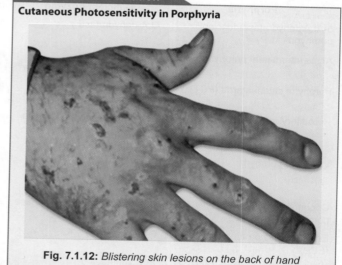

Fig. 7.1.12: Blistering skin lesions on the back of hand

Erythropoietic Protoporphyria

Due to defect in Ferrochelatase (FECH Mutation)

- Most common porphyria in children and second most common in adults.
- **Non-Blistering Photosensitivity**
- Characterised by pain, swelling redness within minutes of sunlight exposure, resembling angioedema.
- Vesicular lesions are uncommon.

Diagnosis
- A substantial increase in **erythrocyte protoporphyrin**, which is predominantly free and not complexed with Zn is the hallmark of EPP.
- Erythrocytes exhibits **red fluorescence under fluorescence microscopy at 620 nm.**
- FECH mutation analysis.

Treatment
- Oral Beta Carotene may improve tolerance to sunlight.
- **Afamelanotide, an alfa Melanocyte—stimulating hormone** has completed phase III trials for patients with EPP and XLP.

X-linked Protoporphyria
- *Due to increased* activity of ALA Synthase-2 due to gain of function mutation.

X-linked Sideroblastic Anemia
- Not a porphyria
- Due to decreased activity of ALA Synthase-2.

Quick Revision

- Porphyrins can be used for Photodynamic therapy.
- Porphyrins emit red fluorescence.
- A sharp absorption band near 400 nm is called Soret band.
- Porphyrin present in heme is Protoporphyrin.
- Heme is ferroprotoporphyrin.
- Heme is not synthesised in mature erythrocyte
- Rate-limiting step of heme synthesis is ALA Synthase.
- PLP is the coenzyme for ALA Synthase.
- First pyrrole compound synthesised is PBG.
- First porphyrinogen synthesised is Uroporphyrinogen.
- ALA and PBG accumulate causes neuropsychiatric and Visceral symptoms.
- Porphyrins accumulate causes Cutaneous photosensitivity.
- Lead inhibit ALA Dehydratase and Ferrochelatase.
- Almost all porphyrias are autosomal dominant.

Porphyria	Deficient Enzyme	Inheritance	Principal Symptoms Neurovisceral (NV) /Cutaneous Photosensitivity (CP)	Important remarks
ALA dehydratase-deficient porphyria (ADP)	ALA-dehydratase	**AR**	NV	Also called Doss Porphyria Resemble lead poisoning
Acute intermittent porphyria (AIP)	HMB-synthase or Uroporphyrinogen I synthase	AD	NV	MC acute porphyria. No cutaneous photosensitivity
Porphyria cutanea tarda (PCT)	Uroporphyrinogen decarboxylase	AD	CP	Most common porphyria
Hereditary coproporphyria (HCP)	Coproporphyrinogen oxidase	AD	NV and CP	–
Variegate porphyria (VP)	Protoporphyrinogen oxidase	AD	NV and CP	–
Congenital erythropoietic porphyria (CEP)	Uroporphyrinogen III synthase	**AR**	CP	Nonimmune hydrops Also called Gunther's disease
Erythropoietic protoporphyria (EPP)	Ferrochelatase	**AR**	CP	MC porphyria in children. Nonblistering photosensitivity.
X-linked protoporphyria (XLP)	ALA-synthase 2	**XL**	CP	–

Check List for Revision

- In structure of heme mainly concentrate on boxes, tables and text written in bold letters.
- Structure of heme as an IBQ.
- Heme synthesis pathway is important as all enzymes are associated with porphyrias.
- Porphyrias is a must learn topic for almost all exams.

Review Questions MCQ

1. A girl licks paint that is peeled of from the toys, develop acute abdominal pain, tingling sensation of hands and legs and weakness. Which enzyme is inhibited in this child? *(AIIMS May2017)*
   a. ALA synthase
   b. Heme oxygenase
   c. Coproporphyrinogen oxidase
   d. ALA dehydratase

2. Heme biosynthesis does not occur in: *(AIIMS Nov 2015)*
   a. Osteocyte
   b. Liver
   c. RBC
   d. Erythroid cells of bone marrow

3. In lead poisoning which of the following is seen in urine? *(AIIMS NOV2015)*
   a. Delta ALA
   b. Uroporphyrin
   c. Coproporphyrin
   d. Protoporphyrin

4. In HbS, Glutamic acid replaced by valine. What will be its electrophoretic mobility? *(AIIMS Nov 2015)*
   a. Increased
   b. Decreased
   c. No change
   d. Depends on level of concentration of HbS

5. Which of the following porphyrias does not present with photosensitivity? *(AIIMS May 2012)*
   a. Uroporphyrin decarboxylase
   b. HMB synthase
   c. Protoporphyrinogen oxidase
   d. Coproporphyrinogen oxidase

6. A boy with staining of teeth and raised Coproporphyrin-I levels and increased risk of photosensitivity, the enzyme deficient is: *(Recent Question 2016)*
   a. Uroporphyrinogen synthase
   b. Uroporphyrinogen III synthase
   c. Uroporphyrinogen decarboxylase
   d. Coproporphyrinogen oxidase

7. Acute Intermittent Porphyria is caused by: *(Recent Question 2016)*
   a. ALA synthase
   b. ALA dehydratase
   c. Ferrochelatase
   d. Uroporphyrinogen I synthase

8. Variegate porphyria enzyme defect is: *(Recent Question 2016)*
   a. Protoporphyrinogen oxidase
   b. Coproporphyrinogen oxidase
   c. Uroporphyrinogen decarboxylase
   d. Uroporphyrinogen synthase

9. No. of iron in transferrin: *(Recent Question 2016)*
   a. 1
   b. 2
   c. 3
   d. 4

10. No of iron in ferritin: *(Recent Question 2016)*
    a. 4
    b. 40
    c. 400
    d. 4000

11. No. of pyrrole rings in Porphyrins: *(Recent Question 2016)*
    a. 2
    b. 3
    c. 4
    d. 5

Answers to Review Questions

1. **d. ALA dehydratase**

   Paints contain lead hence it is most probably lead poisoning. Biochemical effects of lead poisoning are:
   - Inhibition of ALA dehydratase >>Ferrochelatase
   - Increased ALA excretion in urine. Increased ALA synthase activity.

2. **c. RBC** *(Ref: Harper 31/e page 310)*

   Site: Synthesised in almost all tissues in the body EXCEPT in mature erythrocytes. 85% in erythroid precursor cells in the bone marrow and majority of remainder in hepatocyte

   Organelle: Partly cytoplasmic and partly mitochondrial Starting Materials: Succinyl CoA and Glycine

3. **a. Delta ALA**

   Lead inhibit ALA dehydratase. So delta ALA increases, hence found in urine.

4. **b. Decreased**

   In HbS, Glutamic acid is replaced by Valine. Glutamic acid is negatively charged and Valine is neutral. Movement of Hb in electrophoresis depends on charge. More the negative charge, fast the mobility towards the anode. So, HbS with less negative charge lags behind HbA1.

5. **b. HMB synthase**

- Porphyrias that present with Neurov is ceral symptoms are Acute Intermittent Porphyria and ALA Dehydratase deficient Porphyria.
- Porphyrias that present with Cutaneous Photosensitivity are Congenital Erythropoietic Porphyria, Porphyria Cutanea Tarda, Erythropoietic Porphyria, X linked Protoporphyria.
- Porphyria that presents with neuro visceral Symptoms and Cutaneous Photosensitivity is Hereditary Coproporphyria, Variegate Porphyria.

6. **b. Uroporphyrinogen III synthase**

   This is a case of congenital Erythropoietic porphyria Enzyme defect in CEP is Uroporphyrinogen III Synthase.

7. **d. Uroporphyrinogen I synthase**

8. **a. Protoporphyrinogen oxidase**

9. **b. 2**

   Transferrin
   Transport form of Iron
   Has two iron binding state

10. **d. 4000**

    Ferritin
    Storage Form of Iron
    Poly nuclear complex of Hydrous ferric oxide carry about 4500 iron atoms
    Seen in Intestinal cells, Liver, Spleen Bone marrow Sensitive index of Body iron stores

11. **c. 4**

7.2 HEME CATABOLISM AND HYPERBILIRUBINEMIA

- The Fate of Hemoglobin
- Steps of Catabolism of Heme
- Hyperbilirubinemias

THE FATE OF HEMOGLOBIN

- Under normal conditions in human adults, some **200 billion erythrocytes** are destroyed per day.
- A 70 kg human turns over approximately **6 g of hemoglobin** daily.
- **1 g of hemoglobin** yields **35 mg of bilirubin^Q**.
- The daily bilirubin formation in human adults is approximately **250–350 mg**.
- When hemoglobin is destroyed in the body
- Globin is degraded to its constituent amino acids, which are reused.
- Iron of heme enters the iron pool, also forreuse.
- The iron-free porphyrin portion of heme is degraded, mainly in the reticuloendothelial cells of the liver, spleen, and bone marrow.

STEPS OF CATABOLISM OF HEME

Site: Microsomal fraction of reticuloendothelial cells of the liver, spleen, and bone marrow.

Hemoxygenase

- Oxygen is added to the α-methynebridge between pyrroles I and II of the porphyrin.
- Splitting of tetrapyrrole ring.
- Green pigment, **Biliverdin^Q** is produced.
- **Carbon monoxide** is produced by this reaction.
- This is the only source of endogenous CO^Q in the body.

Biliverdin Reductase

- Reduces the **methyne bridge between pyrrole III and pyrrole IV** of Biliverdin to a methylene group.
- **Yellow pigment, bilirubin** is produced.
- This takes place in the **cytosol**.

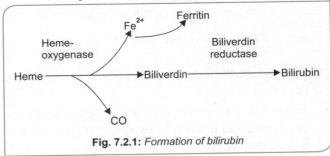

Fig. 7.2.1: *Formation of bilirubin*

Transport of Bilirubin

- Bilirubin formed in peripheral tissues is transported to the liver by **plasma albumin.**
- Bilirubin is only sparingly soluble in water, but its solubility in plasma is increased by **noncovalent** binding toalbumin.

- Each molecule of albumin appears to have **one high-affinity site and one low-affinity site for bilirubin**.
- In 100 mL of plasma, approximately 25 mg of bilirubin can be tightly bound to albumin at its high-affinity site.
- Bilirubin in excess of this quantity can be bound only loosely and thus can easily be detached and diffuse into tissues.

Metabolism of Bilirubin

Occurs Primarily in the Liver.
It can be divided into three processes.
1. Uptake of bilirubin by liver parenchymal cells
2. Conjugation of bilirubin with glucuronate in the endoplasmic reticulum
3. Secretion of conjugated bilirubin into the bile.

Uptake of Bilirubin in the Liver

- In the liver, the bilirubin is removed from albumin
- Taken up at the sinusoidal surface of the hepatocytes by a carrier-mediated saturable system **(facilitated transport system)**
- This has a very large capacity, so that even under pathologic conditions the system does not appear to be rate limiting in the metabolism of bilirubin.
- Once bilirubin enters the hepatocytes, it can **bind to certain cytosolic proteins. This is called intracellular binding**.

> **Extra Edge**
>
> The proteins help in intracellular binding are:
> 1. **Ligandin** (a member of the family of glutathione S-transferases)
> 2. **Protein Y**.
>
> **Functions of intracellular binding**
> They may also help to prevent efflux of bilirubin back into the bloodstream.
> They help to keep bilirubin solubilized prior to conjugation.

Conjugation of Bilirubin with Glucuronic Acid Occurs in the Liver

- Hepatocytes convert bilirubin to a **polar** form, which is readily excreted in the bile, by adding glucuronic acid molecules to it.
- This process is called **conjugation** and can employ polar molecules other than glucuronic acid.
- The conjugation of bilirubin is catalyzed by a specific **glucuronyl transferase**.
- The enzyme is mainly located in the **endoplasmic reticulum**^{Q}, uses UDP-glucuronic acid as the glucuronyl donor, and is referred to as bilirubin-UGT.

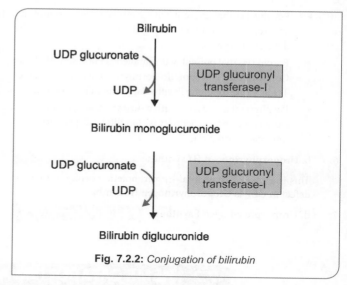

Fig. 7.2.2: *Conjugation of bilirubin*

- Bilirubin monoglucuronide is an intermediate and is subsequently converted to the **diglucuronide**
- Most of the bilirubin excreted in the bile of mammals is in the form of bilirubin diglucuronide.
- However, when bilirubin conjugates exist abnormally in **human plasma** (e.g. in obstructive jaundice), they are predominantly **monoglucuronides.**
- Bilirubin-UGT activity can be **induced** by a number of clinically useful drugs, including **phenobarbital.**

Secretion of Conjugated Bilirubin into the Bile

- This occurs by an **active transport** mechanism.
- This is **the rate-limiting** for the entire process of hepatic bilirubin metabolism.
- The protein involved is **MRP-2** (multidrug-resistance-like protein 2), also called **multispecific organic anion transporter (MOAT).**
- It is located in the **plasma membrane** of the bile canalicular membrane.

> **Extra Edge**
>
> - Recently it is found that apart from secretion into biliary canaliculi, a portion of bilirubin diglucuronide is transported into portal circulation by MRP-3 (Multi-drug Resistance associated protein-3)
> - **They are subjected to reuptake into hepatocyte by transporters, Organic anion Transporter 1B1 (OATP 1B1), and 1B3 (OATP 1B3).**
> - It is a member of the family of ATP-binding cassette transporters.

- The hepatic transport of conjugated bilirubin into the bile is **inducible** by those same drugs that are capable of inducing the conjugation of bilirubin.

Conjugated Bilirubin is Reduced to Urobilinogen by Intestinal Bacteria

- The conjugated bilirubin reaches the **terminal ileum** and the large intestine
- The glucuronides are removed by specific bacterial enzymes (β-**glucuronidases**)
- This is subsequently reduced by the fecal flora to a group of colorless tetrapyrrolic compounds called **urobilinogens**.

Fates of Urobilinogen

- 80–90% of the urobilinogen is converted to stercobilinogen and stercobilin by intestinal flora and excreted through feces.
- 10–20% enterohepatic circulation reaches the liver. This is called **enterohepatic urobilinogen cycle**.
- A small fraction <3 mg/dL escape hepatic uptake, filters across renal glomerulus and is excreted through urine.

HYPERBILIRUBINEMIAS

Jaundice

- Scleral icterus indicate serum bilirubin >3 mg/dL
- Carotenoderma can be distinguished from Icterus by sparing of sclera.

Congenital Hyperbilirubinemias

Unconjugated hyperbilirubinemias
1. Gilbert's disease
2. Crigler-Najjar syndrome

Principal Differential Characteristics of Crigler-Najjar and Gilbert Syndrome

Features	Crigler- Najjar Syndrome		Gilbert Syndrome
	Type-I	Type-II	
Total serum bilirubin, mol/L (mg/dL)	310–755 [18–45 (usually >20)]	100–430 (usually 345) [6–25 (usually 20)]	Typically 70 mol/L (4 mg/dL)
Routine liver tests	Normal	Normal	Normal
Response to Phenobarbital	None	Decreases bilirubin by >25%	Decreases bilirubin to normal
Kernicterus	Usual	Rare	No
Hepatic histology	Normal	Normal	Usually normal; **increased lipofuscin pigment**
Bile characteristics			
Color	Pale or colorless	Pigmented	Normal dark color
Bilirubin fractions	>90% unconjugated	Largest fraction (mean: 57%) monoconjugates	Mainly diconjugates but monoconjugates increased (mean: 23%)
Bilirubin UDP-glucuronosyltransferase activity	**Typically absent**; traces in some patients	Markedly reduced: 0–10% of normal	Reduced: typically 10–33% of normal
Inheritance (all autosomal)	**Recessive**	**Predominantly recessive**	7 of 8 dominant; 1 reportedly recessive

Physiological Jaundice

- Predominantly **unconjugated hyperbilirubinemia**.

Causes
- Incompletely developed hepatic system
- **Low UDP Glucuronyl Transferase** enzyme. Unconjugated hyperbilirubinemia that **develop 2nd to 5th day of birth**
  Peak level of Serum Bilirubin is 5–10 mg/dL
  Decline to normal adult concentration in 2 weeks.

Breast Milk Jaundice

- Bilirubin conjugation is inhibited by certain **fatty acids** that are present in the breast milk and not in serum.
- A correlation between **epidermal growth factor content of breast milk** and elevated bilirubin level is noted in these infants.

Lucey Driscoll Syndrome

- Transient familial neonatal hyperbilirubinemia.
- **UGT1A1 inhibitor is in maternal serum**, and not in breast milk unlike breast milk jaundice.

Conjugated Hyperbilirubinemias

1. Dubin Johnson's syndrome
2. Rotor syndrome
3. Benign Recurrent intrahepatic Cholestasis (BRIC)
4. Progressive Familial intrahepatic Cholestasis (FIC)

Dubin Johnson's Syndrome
- Autosomal recessive
- Cause—Mutation of **gene encoding MRP2**
- A cardinal feature is accumulation in the lysosomes of centrilobular hepatocytes of dark, coarsely granular pigment. **Liver is grossly black in appearance**. This pigment is thought to be an **epinephrine** metabolite that are not excreted properly.
- **Black liver jaundice**
- Bromsulphthalein test (BSP test) shows 2° Peak.

Rotor Syndrome
- Autosomal recessive
- Defective bilirubin excretion
- Recent studies indicate that deficiency of plasma transporters **OATP1B1 and OATP1B3** is the cause.
- This result in reduced reuptake of conjugated bilirubin pumped into portal circulation.

Benign Recurrent Intrahepatic Cholestasis (BRIC)
- Rare disorder characterised by recurrent attacks of jaundice and pruritus.
- There are two types BRIC-1 and BRIC-2.
- In BRIC-1, agenenamed **FIC-1** is mutated, that encodes a protein that play a role in biliary canalicular excretion of various compounds.
- In BRIC 2 mutationis in **Bile salt Excretory Protein (BSEP)**
- This protein is defective in Familial Intrahepatic Cholestasis type 2 (FIC-type 2)

Progressive Familial Intrahepatic Cholestasis (FIC)

FIC Type 1 (Byler's disease)
- Due to **FIC-1 mutation**.
- Present in early infancy as cholestasis, initially episodic.
- Unlike BRIC, Byler's disease progress to malnutrition, growth retardation and End stage Liver Disease.

FIC Type II
- Due to mutation of **Bile Salt Excretory Protein (BSEP)**

FIC Type III
- Due to mutation of **MRP-3**

Differentiating features of important Conjugated Hyperbilirubinemia

	DJS	Rotor	PFIC1	BRIC1	PFIC2	BRIC2	PFIC3
Gene	ABCCA	SLCO1B1/SLCO1B3	ATP8B1	ATP8B1	ABCB11	ABCB11	ABCB4
Protein	MRP2	OATP1B1,OATP1B3	FIC1	FIC1	BSEP	BSEP	MDR3
Cholestasis	No	No	Yes	Episodic	Yes	Episodic	Yes
Serum γ GT	Normal	Normal	Normal	Normal	Normal	Normal	↑↑
Serum Bile acids	Normal	Normal	↑↑	during episodes	↑↑	during episodes	↑↑

NB: Learn BRIC1 and PFIC1 together, then BRIC 2 and PFIC 2 together.

Acquired Hyperbilirubinemias
1. Hemolytic Jaundice
2. Hepatic Jaundice
3. Obstructive Jaundice

Laboratory Tests Done to Differentiate Jaundice

1. **Serum Bilirubin**

   By van den Bergh test
   - Chemical method to estimate bilirubin in serum.
   - Bilirubin + Ehrlich's Diazo reagent (Diazotised Sulfanilic Acid).
   - Reddish purple azopigment is formed.
   - They are analyzed by photometry at 540 nm.

   Two types of reaction in this test: Direct and Indirect
   - **Conjugated bilirubin:** Direct positive, so conjugated bilirubin is otherwise Direct bilirubin.
   - **Unconjugated bilirubin:** Indirect positive, so unconjugated bilirubin is otherwise indirect bilirubin.

   Tests in Urine
   - **Urine and fecal Urobilinogen—By Ehrlich's test**
   - **Urine Bilirubin (Bile pigment)—By Modified Fouchet's test**
   - **Urine Bile salt by Hay's test, Pettenkofer test.**

   Liver Enzyme Panel:
   - Transaminases (AST and ALT) elevated in Hepatitis
   - Alkaline Phosphatase elevated in Obstructive Jaundice
   - 5' Nucleotidase in Obstructive Jaundice.

Laboratory Tests in three Different Types of Jaundice

Condition	Serum Bilirubin	Urine Urobilinogen	Urine Bilirubin
Normal	Direct: 0.1–0.4 mg/dL Indirect: 0.2–0.7 mg/dL	0–4 mg/24 h	Absent
Hemolytic anemia	Indirect	Increased	Absent
Hepatitis	Direct and indirect	Decreased if micro-obstruction is present	Present if micro-obstruction occurs
Obstructive jaundice	Direct	Absent	Present

Extra Edge

Delta Bilirubin or Biliprotein
- **Conjugated bilirubin that is covalently linked to albumin.**
- Half life of delta bilirubin is 12–14 days [Half life of unbound Bilirubin is only 4hrs]
- In case of conjugated hyperbilirubinemia bilirubinuria starts late.

Quick Revision

- 1 g of Hb yield 35 mg of bilirubin.
- Hemoxygenase is an enzyme present in microsomes of liver, spleen & bone marrow.
- The only endogenous source of Carbon monoxide in the body is Microsomal hemoxygenase.
- Bilirubin is transported to liver by plasma albumin.
- UDP Glucuronyl transferase is the enzyme that conjugate bilirubin.
- Conjugation of bilirubin takes place in Endoplasmic reticulum of hepatocytes.
- Rate limiting step of catabolism of heme is secretion of conjugated bilirubin into the bile.
- The protein involved in secretion of conjugated bilirubin is MRP-2 (Multi drug Ressistance like Protein-2) or MOAT (Multispecific Organic Anion Transporter)
- Dubin Johnson's syndrome is due to mutation in MRP-2 or MOAT

Important hyperbilirubinemias and its causes

Congenital Hyperbilirubinemias	Etiology
Unconjugated Hyperbilirubinemia	
Crigler- Najjar Syndrome	UDP Glucuronyl Transferase is absent/reduced
Gilbert Syndrome	UDP Glucuronyl Transferase is reduced
Physiological Jaundice	Low UDP Glucuronyl transferase enzyme activity
Lucey Driscoll Syndrome	UGT1A1 inhibitor in maternal serum
Conjugated Hyperbilirubinemia	
Dubin Johnson's Syndrome (Black Liver Jaundice)	Mutation in MRP-2
Rotor Syndrome	Mutation in OATP1B1 and OATP1 B3
Benign Recurrent Intrahepatic Cholestatasis	Mutation in FIC-1 in Type I Mutation in Bile salt excretory Protein (BSEP) in Type II
Progressive Familial Intrahepatic Cholestasis	FIC-1 mutation in Type I BSEP mutation in Type II MRP-3 mutation in Type III

Check List for Revision

- In heme catabolism and Fate of hemoglobin bold letters and boxes are most important points.
- Different type of jaundice with its etiology given in tables is important

Review Questions MCQ

1. A 10-year-old boy presents with increase bilirubin, increased bilirubin in urine and no urobilinogen. Diagnosis: *(Recent Question 2016)*
   a. Gilbert syndrome
   b. Hemolytic jaundice
   c. Viral hepatitis
   d. Obstructive jaundice

2. MRP-2 deficiency is seen in: *(Recent Question Jan 2019)*
   a. Dubin Johnson syndrome
   b. Crigler-Najjar syndrome-I
   c. Rotor's syndrome
   d. Crigler-Najjar syndrome-II

3. Severity of jaundice is high in: *(Recent Question Jan 2019)*
   a. Dubin Johnson syndrome
   b. Crigler-Najjar syndrome-I
   c. Gilbert's disease
   d. Crigler-Najjar syndrome-II

Answers to Review Questions

1. **d. Obstructive jaundice**

Laboratory tests in three different types of Jaundice

Condition	Serum Bilirubin	Urine Urobilinogen	Urine Bilirubin
Normal	Direct: 0.1–0.4 mg/dL	0–4 mg/24 h	Absent
	Indirect: 0.2–0.7 mg/dL		
Hemolytic anemia	Indirect	Increased	Absent
Hepatitis	Direct and indirect	Decreased if micro-obstruction is present	Present if micro-obstruction occurs
Obstructive jaundice	Direct	Absent	Present

2. **a. Dubin Johnson syndrome**
3. **b. Crigler-Najjar syndrome**

7.3 HEMOGLOBIN

- Structure of Hemoglobin
- Interaction with 2,3 BPG
- Genetics of Alpha and Beta Chains
- Glycated Hb
- Types of Hemoglobin
- Myoglobin
- Function and Characteristics of Hemoglobin
- Hemoglobinopathies

STRUCTURE OF HEMOGLOBIN

Primary Structure

- Hemoglobin, a **tetrameric protein** of erythrocytes.
- Consists of **a pair of α-like chains** 141 amino acids long and a **pair of β-like chains** 146 amino acids long.
- Each globin chain enfolds a single **heme** moiety.
- A single heme moiety, consists of a **protoporphyrin IX** ring complexed with a single iron atom in the **ferrous state (Fe^{2+})**.
- Each heme moiety can bind a single oxygen molecule.
- A molecule of hemoglobin can transport up to **four oxygen molecules**.

Higher Order Structure

- **Secondary structure**: Various globin chains are predominantly **alpha helix**.
- **Tertiary structure** is globular *tertiary structures* with the exterior surfaces to be rich in polar (hydrophilic) amino acids that enhance solubility, and the interior lined with nonpolar groups, forming a hydrophobic pocket into which heme is inserted.
- **Quaternary structure**: It is a tetramer that contains **two α β dimers.**
- The hemoglobin tetramer is highly soluble but individual globin chains are insoluble.
- Unpaired globin precipitates, forming inclusions that damage the cell.

GENETICS OF ALPHA AND BETA CHAINS

The human hemoglobins are encoded in two tightly linked gene clusters; the α-like globin genes are clustered on chromosome 16 and the β-like genes on chromosome 11.

Types of Hemoglobin

Embryonic Hemoglobins[Q]
- Hb Portland - $\zeta_2\gamma_2$
- Hb Gower I - $\zeta_2\epsilon_2$
- Hb Gower II - $\alpha_2\epsilon_2$

Fetal Hemoglobin[Q]
- Hb F - $\alpha_2\gamma_2$

Characteristics of Fetal Hemoglobin[Q]
1. Slower Electrophoretic Mobility
2. Increased resistance to Alkali Denaturation
3. Decreased Interaction with 2,3 BPG

Adult Hemoglobins
- Hb A_1 - $\alpha_2\beta_2$
- HbA_2 - $\alpha_2\delta_2$

CHARACTERISTICS OF HEMOGLOBIN

Hemoglobin is an Allosteric Protein

Binding of one molecule of oxygen to one heme residue increases the affinity of binding of oxygen to other heme residue called **Heme-heme interaction or Positive Cooperativity.**

Hence, Oxygen Dissociation Curve is **Sigmoidal**.

FUNCTIONS OF HEMOGLOBIN

- The functions of hemoglobins are to transports O_2 to the tissues and returns CO_2 and protons to the lungs.

Functional Histidine Residues in Hemoglobin

- Proximal Histidine [F8] and Distal Histidine [E7] are the functional His residues in Hemoglobin.
- Fifth coordination position of iron is occupied by a nitrogen of imidazole ring of **the proximal histidine, His F8.**
- Distal Histidine, **His E7** lies on the side of heme ring opposite, His F8.

HIGH YIELD POINTS

Conformational States of Hemoglobin
- Two states are T (taut) state and R (relaxed) state
- **T state** is the **low affinity to Oxygen state**, so it is the deoxygenated state.
- T to R state is formed by breaking the salt bridges.
- **R state** is the **high affinity to oxygen state**, it is the oxygenated state.
- T state favour Oxygen delivery to tissues.
- R state favour binding of Oxygen to Hemoglobin.
- Binding of first Oxygen to deoxy Hb shifts the conformation from **T to R state.**
- Binding **of 2,3 BPG stabilises the T structure of Hb**, so shifts ODC to right.

Image-Based Information

Factors affecting Oxygen dissociation curve of Hemoglobin

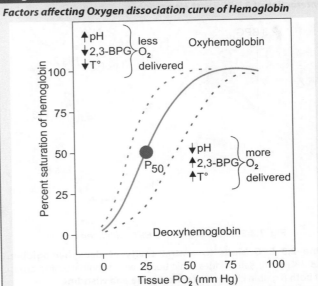

Fig. 7.3.1: *Factors affecting ODC of hemoglobin*

Factors that shift the ODC to right (decreases Oxygen affinity)
1. Decreased pH
2. Interaction with 2,3 BPG
3. Increased Temperature

Factors that shift the ODC to left (increases Oxygen affinity)
1. Increased pH
2. No interaction with 2,3 BPG
3. Decreased temperature

INTERACTION OF HEMOGLOBIN WITH 2, 3 BPG[Q]

- 2, 3 BPG is an intermediate of Glycolysis
- Synthesized with the help of 2,3 BPG Shunt or Rapoport Leubering Cycle.
- The Hemoglobin tetramer binds to **one mol of 2,3 BPG** in the central cavity formed **by four subunits**.
- 2,3 BPG forms salt bridges with terminal amino group of **both β globin** chain via, Valine, Lysine and Histidine.
- Binding of 2,3 BPG **decreases the affinity of Hb to Oxygen, by stabilising the T state.**
- **Shifts the ODC to right.**
- 2, 3 BPG has decreased interaction with HbF.
- Because, **Histidine that forms salt bridges with 2,3 BPG is not present in γ chain of HbF.**
- Instead of Histidine it is Serine, so 2,3 BPG has less interaction with Fetal Hb.
- This accounts for high affinity of HbF towards Oxygen.

Image-Based Information
Interaction of 2,3 BPG with Hemoglobin

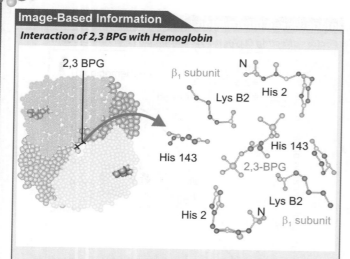

Fig. 7.3.2: *Interaction of 2,3 BPG with haemoglobin*

One 2,3 BPG binds to central cavity of deoxyhemoglobin. 2,3 BPG forms salt bridges/ionic bonds with terminal amino group of both β globin chain via, **Valine, Lysine and Histidine**

Name of Glycated Hemoglobin	Components
HbA_{1a1}	HbA with fructose 1,6 bisphosphate attached to the N terminal of β chain
HbA_{1a2}	HbA with Glucose 6 Phosphate attached to the N terminal of the β chain
HbA_{1a}	HbA_{1a1} and HbA_{1a2} together
HbA_{1b}	HbA with Pyruvic acid attached to the N terminal of the β chain
HbA_{1c}	HbA with glucose attached to the N **terminal Valine of the β chain**
HbA_1	HbA_{1a}, HbA_{1b}, HbA_{1c} together

HbA_{1c} Derived Average Glucose Value

HbA_{1c} level	Average whole blood glucose
5.5%	110 mg/dL
6%	126 mg/dL
7%	154 mg/dL
8%	183 mg/dL
9%	212 mg/dL
10%	240 mg/dL

GLYCATED HEMOGLOBINS (GHB)

- These are formed **nonenzymatically** by condensation of glucose or other reducing sugars, called Glycation with alpha- and beta-chains of hemoglobin A.
- N terminal amino group of beta chain interact with Glucose to form Pre HbA_{1c} (with Aldimine linkage). This undergo Amadori rearrangement to form HbA_{1c}, which is a stable Keto amine.
- Overall reaction called **Maillard Reaction**.
- A stands for Aldimine linkage.
- HbA_{1a}, HbA_{1b}, HbA_{1c} are collectively called as Glycated Hemoglobin or Glycohemoglobin.
- **HbA_{1c} is the major fraction constituting 80% of glycated Hb (GHb)**
- They migrate faster than adult Hb, hence called as fast hemoglobins.
- Best index of long-term control of Blood Glucose
- Formation GHb is irreversible, the concentration of GHb depends on life span of RBC (120 days) and the blood glucose concentration.
- GHb concentration represents integrated values for glucose over preceding **8 to 12 weeks**.
- Normal level of HbA_{1c} is <5.5%
- **5.5% to 6%** is good control

Methods of Assay of Glycated Hemoglobin
1. High Performance Liquid Chromatography which employ Cation Exchange Chromatograpy
2. Immunoassay
3. Affinity Chromatography.

HEMOGLOBIN ELECTROPHORESIS

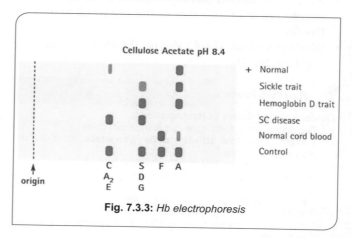

Fig. 7.3.3: *Hb electrophoresis*

Heme and Hemoglobin

MYOGLOBIN

- Myoglobin in red muscle **store Oxygen**.
- Myoglobin is a **monomeric** protein.
- Its MW is 17,000
- **Sperm whale myoglobin** is the first protein whose three dimensional structure is elucidated.
- Amino acid sequence of α and β chains of human hemoglobin and Myoglobin shows 25% and 24% similarity respectively.
- Myoglobin and the β **subunit of hemoglobin**[Q] share almost identical secondary and tertiary structure.
- Oxygen is stored in red muscle myoglobin is released during severe exercise for use in muscle mitochondria for aerobic synthesis of ATP.
- Rich **in alpha helix**.
- Bind only **one mol of Oxygen**.
- **High affinty** for Oxygen.
- Exhibit no Bohr effect.
- Exhibit **no Cooperativity**.
- **No interaction with 2, 3 BPG**.
- ODC is **Hyperbolic** and not sigmoidal.

Image-Based Information
ODC of Myoglobin and Hemoglobin

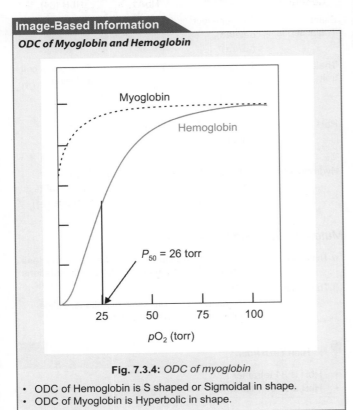

Fig. 7.3.4: *ODC of myoglobin*

- ODC of Hemoglobin is S shaped or Sigmoidal in shape.
- ODC of Myoglobin is Hyperbolic in shape.

Image-Based Information
Oxygen Affinity of Maternal and Fetal Haemoglobin

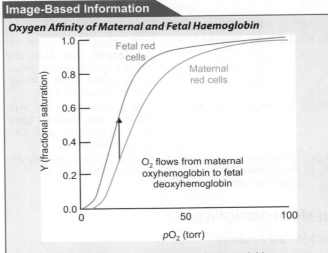

Fig. 7.3.5: *Oxygen affinity of haemoglobin*

Oxygen affinity of fetal Hb is higher than maternal Hb. ODC of Fetal Hb is seen left to ODC of maternal Hb. So Oxygen flows from maternal blood to fetal blood.

Classification of Hemoglobinopathies

	Name of Hemoglobinopathy		Altered functional or physical and chemical properties
I	Structural hemoglobinopathies		Amino acid sequences is altered
IA (1)	HbS, Hemoglobin Sickling	β 6 Glu→Val	Abnormal Polymerisation
IA (2)	HbC	β 6 Glu→Lys	
IB Altered Oxygen affinity			
IB (1)	Hb Yakima		Increased O_2 affinity— hence polycythemia
IB (2)	Hb Kansas		Low O_2 affinity— cyanosis, pseudoanemia
IC Hemoglobin that oxidize readily			
	Hb Philly		Unstable hemoglobins
	Hb Genova		
	Hb Koln		
	HbM Boston		M hemoglobins Prox or Distal Histidine mutated to Tyrosine
	HbM Iwata		
	HbM Saskatoon		
	HbM Hyde park		
	HbM Milwaukee		
II	Thalassemias		Defective biosynthesis of globin chains
II (A)	α Thalassemias		Defective biosynthesis of α globin chains

Contd...

Contd...

Classification of Hemoglobinopathies			
III (B)	Hb Constant Spring		In the alpha chain C terminal **termination codon is mutated to a coding codon** hence an elongated alpha chain of 173 aminoacid is formed.
III (C)	Hb Lepore	$\alpha(\delta\beta)2$	Unequal cross over and recombination even that fuses proximal end of δ gene with distal end of β gene.
IV	Hereditary persistence of fetal hemoglobin		Persistence of high levels of HbF into adult life

HEMOGLOBINOPATHIES

Sickle Cell Anemia

Molecular defect in Sickle cell anemia

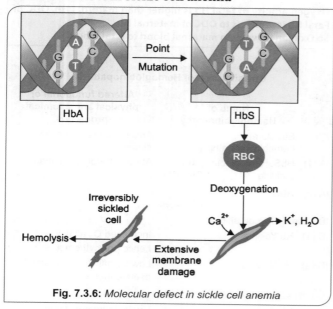

Fig. 7.3.6: Molecular defect in sickle cell anemia

HIGH YIELD POINTS

Mutations in Sickle cell syndromes are:
- A **partially acceptable missense** mutation in sequence of codon 6 of β globin chain.
- **GAG to GTG**, So A to T mutation hence an example of **transversion**.
- The sixth amino acid in β globin chain is changed from glutamic acid to valine.
- A polar amino acid is replaced by an on polar amino acid, so an example of **Nonconservative mutation**.

Pathological Changes due to Mutation are:

- HbS **polymerizes reversibly** when deoxygenated to form age latinous network of fibrous polymers that stiffen the RBC membrane,
- There is increased viscosity
- There is **dehydration** due to **potassium leakage and calcium influx**
- These changes also produce the sickle shape or holly leaf shape.
- Sickled cells **lose the pliability** needed to traverse small capillaries.
- They possess altered "**sticky**" membranes that are abnormally adherent to the endothelium of small venules.
- These abnormalities provoke unpredictable episodes of microvascular vaso-occlusion and premature RBC destruction (hemolytic anemia).

Thalassemia Syndromes

Alpha Thalassemia Syndromes

As alpha chain is encoded by two tightly linked gene clusters and not by a single gene as in the case of β globin gene. Deletion in each gene loci is associated with a syndrome.

Alpha Thalassemia Syndromes

Condition		HbA1, %	HbH (β4), %
Silent Thalassemia	1 α gene loci deleted ($-\alpha/\alpha\alpha$)	98–100	0
Thalassemia trait	2 α gene loci deleted ($-\alpha/-\alpha$) or ($-/\alpha\alpha$)	85–95	Rare blood cell inclusion
HbH disease	3 α gene loci deleted ($-/-\alpha$)	70–95	5–30
Hydrops fetalis	All the four loci deleted	0	5–10 NB: 90–95% is HbB arts (γ4)

Mutations in α and β Thalassemia

α Thalassemia	Unequal crossing-over large deletions less commonly nonsense and frame shift mutations
β Thalassemia	Deletions Nonsense and frameshift mutations, Splice sites and promoter mutations

HIGH YIELD POINTS

- HbH is β4 tetramer
- HbB arts is γ4 tetramer

Extra Edge

Novel treatments for Hemoglobinopathies

For Sickle Cell Anemia:
- Direct gene correction in situ (Genomic editing) using Zn finger nucleases or CRISPR C as 9
- Derepressing HbF by interfering with Bcl 11a.

For Sickle cell and Thalassemia Syndromes
- Bone marrow transplantation to provide stem cells used in large number of β Thalassemia and a few cases of sickle cell anemia
- Gene therapy
- Re-establishing high level of HbF by stimulating proliferation of primitive HbF producing progenitor cells (i.e. F cell progenitors) by Cytarabine, Hydroxy Urea, Pulse dorintermittent administration of Butyrates.

Quick Review

- Hemoglobin is an allosteric protein.
- ODC is sigmoidal due to positive cooperativity.
- Fifth coordination position of iron is occupied by the proximal histidine.
- 2,3 BPG shifts the ODC to right.
- Fetal hemoglobin has less interaction with HbF.
- HbA1c is glucose attached to N terminal valine.
- Average glucose level for 6% of HbA1c is 126 mg/dL.

Check List for Revision

- These topics are dealt in other subjects also, hence only the summary is given here.
- Learn all bold letters and images.

Review Questions MCQ

1. **HbA_{1c} is:**
   a. Glucose to N terminal β globin
   b. Glucose to lysine residue of β globin
   c. Glucose to valine residue of β globin
   d. Glucose to glutamine residue of β globin

2. **Structure of Hemoglobin and Myoglobin are similar in:**
   a. Primary structure
   b. Secondary structure
   c. Tertiary structure
   d. Both secondary and tertiary structure

3. **Identify the structure given below:**
   (Recent Question 2016)

   a. Porphyrin
   b. Heme
   c. Chlorophyll
   d. Pyrrole

4. **2,3 DPG binds to sites in hemoglobin and causes in its oxygen affinity:**
   (AIIMS May 2014)
   a. Four, increases
   b. Four, decreases
   c. One, increases
   d. One, decreases

Answers to Review Questions

1. **c. Glucose to Valine residue of β globin**
   (Ref: Teitz text book of Clinical Chemistry)

Name of Glycated Hemoglobin	Components
HbA_{1a}	HbA_{1a1} & HbA_{1a2} together
HbA_{1b}	HbA with Pyruvic acid attached to the N terminal of the β chain
HbA_{1c}	HbA with glucose attached to the N terminal Valine of the β chain

2. **d. Both secondary and tertiary structure**
   (Ref: Harper 30/e page 58)

   Amino acid sequence of α and β chains of human hemoglobin and Myoglobin shows 25% and 24% similarity respectively.
   Myoglobin and the β subunit of **hemoglobin**[Q] share almost identical secondary and tertiary structure.

3. **a. Porphyrin**

4. **d. One, decreases**

   The Hemoglobin tetramer binds to one mol of 2,3 BPG in the central cavity formed by four subunits.
   2,3 BPG forms salt bridges with terminal amino group of both β globin chain via, Valine, Lysine and Histidine.
   Considering these two sentences, If central cavity is taken into account 2,3 BPG binds to one site.
   But actually 2,3 BPG is forming salt bridges with two beta subunit, so correctly it is binding to two β sites.
   So two is a better answer than four, for the number of binding site. As it is not there in the option, better answer is one.
   Binding of 2,3 BPG decreases the affinity of Oxygen towards Hb
   - Hb F has low affinity towards 2, 3 BPG.

CHAPTER 8

Nutrition

Chapter Outline

8.1 Fat Soluble Vitamins
8.2 Water Soluble Vitamins
8.3 Minerals
8.4 Basics of Nutrition

CHAPTER 8

Nutrition

8.1 FAT SOLUBLE VITAMINS

- Vitamins
- Classification
- Vitamin A
- Vitamin D
- Vitamin E
- Vitamin K

VITAMINS

Definition: Vitamins are organic compounds occurring in small quantities in different natural foods and necessary for growth and maintenance of good health.

Vitamins are mainly classified into:
- Fat soluble vitamins—Vitamins A, D, E and K
- Water soluble vitamins—B Complex Vitamins and Vitamin C.

Endogenously Synthesised Vitamins[Q]

Vitamins are generally not synthesized by the humans, but some vitamins can be synthesized endogenously. They are:
- **Vitamin D** from precursor steroids
- **Niacin** from tryptophan, an essential amino acid.
- **Vitamin K, Biotin, and pantothenic acid** by the intestinal microflora.

VITAMIN A

- Ring structure present in vitamin A is β *ionone ring*
- Pro vitamin A, β carotene contain 2 β ionone ring
- Cleaved in the intestine by a **dioxygenase.**

Retinoids

All compounds chemically related to retinol are called retinoids. They are:
- **Retinal: 11 cis retinal** for normal Vision
- **Retinoic Acid:** Normal morphogenesis, growth and cell differentiation

- **Retinol:** Reproduction.
  Vitamin A, in the strictest sense, refers to Retinol.

Carotenoids
- They are provitamins of Vitamin A present in plants.
- More than 600 carotenoids in nature, and approximately 50 of them can be metabolized to vitamin A.
- β **Carotene** is the most prevalent carotenoid in the food supply that has provitamin A activity.

> **EXTRA EDGE**
> **Non-Provitamin A Carotenoids**
> - **Lutein and Zeaxanthin:** Protect against macular degeneration
> - **Lycopene[Q]:** Protect against prostate cancer

Vitamin A Metabolism

Absorption and Transport of Vitamin A
- Beta Carotene from plant sources is absorbed and cleaved to two molecules of Retinal by Beta Carotene
- Dioxygenase: Retinal is reduced to retinol by Retinol Reductase.
- Retinol ester from animal sources is hydrolysed in the intestinal lumen to Retinol and absorbed into the intestinal cells.
- Retinol from animal and plant sources is reesterified to retinol esters and transported in **Chylomicrons**[Q] to Liver
  - Uptake takes place in liver cells by means of **apo E receptors.**

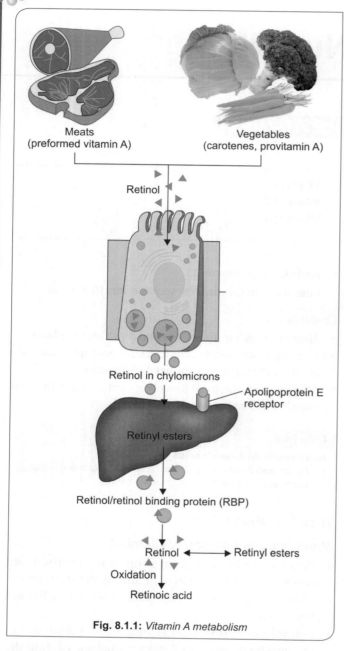

Fig. 8.1.1: *Vitamin A metabolism*

Storage of Vitamin A

Stored in the **Liver Perisinusoidal Stellate (Ito) cells** as Retinyl Ester (Retinol Palmitate).

Transport of Vitamin A from Liver to Target Organs

Carried to target sites in the plasma as trimolecular complex bound to **Retinol Binding Protein (RBP)** and **Transthyretin**.

Functions of Vitamin A

- **Vision**

  Visual process involves 3 forms of Vitamin A containing pigments

  Rhodopsin
  – Most light sensitive pigment present in rods
  – Formed by covalent association between **11 cis retinal** and 7-transmembrane rod protein called opsin.

  Three iodopsin each responsive to specific colours in cones in bright light.

- Regulation of gene expression and differentiation.
  – **Retinoic Acids** are involved in this function.
  – Biologically important retinoic acids are all- trans retinoic acid and 9 -cis retinoic acid.
  – They act like steroid hormones.
  – They bind to **nuclear receptors**.

> **EXTRA EDGE**
>
> **Retinoic Acid Receptors**
> Retinoic receptors regulate transcription by binding to specific DNA site.
> - Retinoic Acid Receptors (RARs) bind with high affinity to All-trans–retinoic acid and 9-cis retinoic acid
> - Retinoic X receptor (RXRs) binds only to 9-cis retinoic acid

- Normal reproduction
  – Retinol is necessary for this function.
- Maintenance of normal epithelium of skin and mucosa.
- Antioxidant Properties and photoprotective property is attributed to Beta Carotenes.
- Host resistance to infection.

Vitamin A Deficiency Manifestations

- **Most common vitamin deficiency**
- **Most common cause of preventable blindness**.

1. Eyes
   – The most characteristic and specific signs of vitamin A deficiency are eye lesions.
   – Loss of sensitivity to green light is the earliest manifestation.
   – An early symptom is delayed adaptation to the dark.
   – Later, **night blindness or impairment in adaptation to dim light or nyctalopia** due to the absence of retinal in the visual pigment, rhodopsin.
   – Dryness of Conjunctiva or Conjunctival Xerosis.
   – The conjunctiva keratinizes and develops plaques (**Bitot spots**).
   – Cornea keratinizes, becomes opaque (Corneal Xerosis)

- Later it degenerates irreversibly (keratomalacia), resulting in blindness.
- All the ocular manifestations are collectively called as Xerophthalmia.

2. **Skin and Mucosa**
   - **Epithelial metaplasia and keratinization**
   - Hyperplasia and hyperkeratinisation of the epidermis with plugging of ducts of **adnexal gland** produce **Follicular Hyperkeratosis**^{Q} or Papular dermatosis. This is called as **Phrynoderma** or Toad Skin
   - **Squamous Metaplasia** in the mucus secreting epithelium of upper respiratory tract and urinary tract
   - Loss of **taste sensation.**

- **Concurrent Zinc deficiency** can interfere with mobilization of Vitamin A from liver stores.
- **Alcohol** interferes with conversion of Retinol to retinaldehyde in the eyes.

EXTRA EDGE

Vitamin A as Therapeutic Agent
- β Carotene used in Cutaneous Porphyria
- **All transretinoic acid** in Acute Promyelocytic Leukemia (called as Differentiation therapy)
- **13-cis retinoic acid (Isotretinoin)** in Cystic Acne.
- 13-cis retinoic acid in childhood neuroblastoma.

Hypervitaminosis A

- Common in Arctic Explorers who eat Polar bear liver.
- Organelle damaged in hypervitaminosis is **Lysosomes.**
- Acute Toxicity: **Pseudotumor cerebri**^{Q} (headache, dizziness, vomiting, stupor, and blurred vision, symptoms that may be confused with a brain tumor) and exfoliative dermatitis. In the liver, hepatomegaly and hyperlipidemia.
- Chronic Toxicity: If intake of > 50,000 IU/day for >3 months
- Weight loss, anorexia, nausea, vomiting, bony exostosis, bone and joint pain, decreased cognition, hepatomegaly progresses to cirrhosis.
- **Retinoic acid stimulates osteoclast production and activity leading to increased bone resorption and high risk of fractures, especially hip fractures.**
- **In pregnancy retinoids causes teratogenic effects.**

Carotenemia

- Persistent excessive consumption of foods rich in Carotenoids.
- Causes yellow staining of skin but not Sclera (Unlike Hyperbilirubinemia which stain both skin and sclera).

Required Daily Allowance of Vitamin A^Q (μg of Retinol) (ICMR 2010)

Children (1–6 yrs)	= 400 μg/day
Men	= 600 μg/day
Women	= 600 μg/day
Pregnancy	= 800 μg/day
Lactation	= 950 μg/day

Units of Vitamin A

Vitamin A in food is expressed as micrograms of retinol equivalent.

6 μg of beta Carotene = 1 μg of Preformed retinol.

Pure Vitamin A for pharmaceutical uses is expressed International Units (IU)

1 IU = 0.3 μg of Retinol

1 μg of Retinol = 3.33 IU

Sources of Vitamin A

1. Animal food (mainly as Retinol)
2. Plant food as Carotenes

Animal sources
- *Fish liver oils^Q are the rich sources of Vitamin A.*
- **Hali but liver oil** is the richest source (900,000 μg/100 g) followed by cod liver oil.
- *Other animal sources are liver, egg, butter, cheese, whole milk, fish and meat.*

Plant Sources
- Richest plant source is **Carrot.**
- Others are GLV like Spinach, Amaranth, Green andyellow fruits like papaya, mango, pumpkin.

Treatment of Vitamin A Deficiency

- 200,000 IU or 110 mg of Retinol Palmitate orally in two successive days.

Prevention of Vitamin A Deficiency

- Single massive dose 200,000 IU to children (1–6 years) once in 6 months.
- Single massive dose 100,000 IU to children (6 mo–1 years) once in 6 months.

Assays of Vitamin A

- Dark adaptation time.
- Serum Vitamin A by **Carr and Price reaction.**

VITAMIN D

Group of sterols having a **hormone like function:**
- **Ergocalciferol (Vit D$_2$)**—Commercial Vitamin D obtained from the fungus, ergot.
- **Cholecalciferol (Vit D$_3$)**—Endogenous synthesis from 7 Dehydrocholesterol.

Vitamin D Metabolism

- **Sources of Vitamin D**
  - The major source of vitamin D for humans is its **endogenous synthesis** in the skin by photochemical conversion of a precursor, **7-dehydrocholesterol,** to Cholecalciferol or Vitamin D$_3$ via the energy of solar or artificial **UV light in the range of 290 to 315 nm (UV B radiation)** in the **stratum corneum of the epidermis** of skin.
  - Absorption of vitamin D from foods and supplements in the gut.
- Binding of vitamin D from both of these sources to plasma α1-globulin (D-binding protein or DBP) and transport into the liver.
- Conversion of vitamin D into 25-hydroxy cholecalciferol **(25-OH-D)** in the **liver**, through the effect of **25-Hydroxylases. Most abundant** circulatory form of Vitamin D. This is because there is little regulation of this liver hydroxylation. The measurement of 25-OH-D is the standard method for determining patient's Vitamin D status.
- Conversion of 25-OH-D into **1, 25-dihydroxy vitamin D, (1, 25 (OH)$_2$D$_3$) or Calcitriol** in the **kidney^Q, the biologically most active^Q form of vitamin D,** through the activity of α1-hydroxylase. This is the rate limiting step. PTH and Hypophosphatemia upregulate 1α Hydroxylase.
- When Ca^{2+} level is high, kidney produces the relatively inactive metabolite **24, 25 Dihydroxy Cholecalciferol (Calcitroic acid)** excreted through urine.

Functions of Vitamin D

- Regulation of Calcium and Phosphorus homeostasis
  Action on Intestine
  - Vitamin D increases Ca^{2+} absorption
  - By increasing the transcription of **TRPV6** (a member of the transient receptor potential vanilloid family), which encodes acritical calcium transport channel. This increases Calcium absorption from **duodenum**.

  Action on Kidney
  - Vitamin D- increases Ca^{2+} and Phosphorus reabsorption

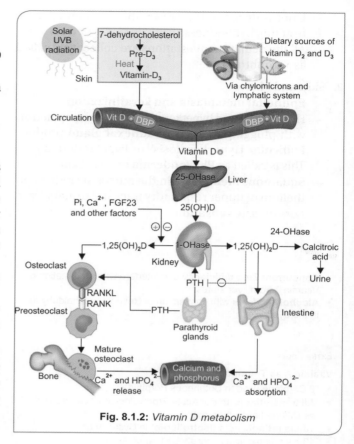

Fig. 8.1.2: *Vitamin D metabolism*

  - Increases calcium influx in distal tubules of the kidney through the increased expression of **TRPV5**, another member of the transient receptor potential vanilloid family.

  Action on Bones
  - 1, 25-dihydroxy vitamin D and parathyroid hormone enhance the expression of RANKL (receptor activator of NF-κB ligand) on osteoclasts.
  - **RANKL** binds to its receptor (RANK) located in preosteoclasts, inducing the differentiation of these cells into mature osteoclasts.
  - They dissolve bone and release calcium and phosphorus into the circulation.
- **Immunomodulatory and antiproliferative effects.** Prevent infection by **Mycobacterium tuberculosis.** Within macrophages, synthesis of 1,25-dihydroxyvitamin D occurs through the activity of CYP27B located in the mitochondria.

Pathogen-induced activation of Toll-like receptors in macrophages causes a transcription-induced increase in vitamin D receptor and CYP27B.

The resultant production of 1,25-dihydroxyvitamin D then stimulates the synthesis **of cathelicidin**, an antimicrobial peptide from the defense in family, which is effective against infection by *Mycobacterium tuberculosis*.

EXTRA EDGE

Anti-Proliferative Role of Vitamin D

1, 25 (OH) 2 D level less than 20 ng/mL is associated with increase in incidence of:
- Colon Cancer
- Breast Cancer
- Prostate Cancer

- **Mineralisation of bones**
  - Vitamin D contributes to mineralisation of osteoid matrix and epiphyseal cartilage in both flat and long bones.
  - It stimulates osteoblast to synthesise calcium binding protein osteocalcin involved in deposition of calcium during bone development.

Vitamin D Deficiency

- The normal reference range for circulating 25-(OH)Dis 20 to 100 ng/mL
- The concentration circulating 25-(OH) D < 20 ng/mL is called Vitamin D deficiency.

Causes inadequate mineralization of bone osteoid
- Before closure of epiphysis—**Rickets in children**
- After closure of epiphysis—**Osteomalacia in adults.**

Clinical Manifestations of Rickets

Symptom Box

General
- Failure to thrive
- Listlessness
- Protruding abdomen
- Muscle weakness (especially proximal)
- Fractures

Head
- Craniotabes
- Frontal bossing
- Delayed fontanel closure
- Delayed dentition; caries
- Craniosynostosis

Chest
- Rachitic rosary Widening of the costochondral junctions results in a **rachitic rosary**
- **Harrison groove** occurs from pulling of the softened ribs by the diaphragm during inspiration
- Respiratory infections and atelectasis.

Spine
- Scoliosis
- Kyphosis
- Lordosis.

Extremities
- Enlargement of wrists and ankles—**Growth plate** widening is also responsible for the enlargement at the wrists and ankles
- Valgus or varus deformities
- Windswept deformity (combination of valgus deformity of 1 leg with varus deformity of the other leg)
- Anterior bowing of the tibia and femur
- Coxa vara
- Leg pain

Biochemical Defect of Different Types of Rickets

Nutritional Vitamin D Deficiency

Most common cause of rickets globally.

CONCEPT BOX

Concept of biochemical changes that occur in nutritional Vitamin D deficiency
- Due to Vitamin D deficiency, Serum Calcium level and Phosphorus level is low.
- This causes Secondary Hyperparathyroidism, so PTH level is high
- This increases the 1α hydroxylation in kidney, so 1, 25 D level increases.
- This will increase the Serum Calcium level, but Phosphorus level remain at low level.
- So Serum Calcium level is variable, Serum Phosphorus is low, S PTH increase, 25 D is decreased, 1, 25 D is low initially but later increase due to secondary hyperparathyroidism.

- Serum calcium need not be always low in Rickets
- 1,25 D level also need not be always low in Rickets
- Serum Phosphorus remain low

Vitamin D-dependent Rickets Type 1 (Pseudo-Vitamin D-resistant Rickets)

- An autosomal recessive disorder.
- Mutations in the gene encoding **renal 1α-hydroxylase**.
- Prevent conversion of 25 D to 1,25D.
- Even with high PTH, as 1 α Hydroxylase is defective, 1,25 D is low.

- Usually presents in first 2 years of life.
- With classic features of rickets.

> **Concept box**
> **Concept of biochemical changes in Vitamin D Dependent Rickets Type I**
> - In spite of secondary hyperparathyroidism, 1, 25 D will remain decreased as 1 α hydroxylase gene is mutated.

Vitamin D-dependent Rickets Type 2 (True Vitamin D-resistant Rickets)

- An autosomal recessive disorder.
- Due to mutations in the **gene encoding the vitamin D receptor** causing end-organ resistance to the active metabolite 1,25D
- Presents in infancy with less severe manifestation.
- 50–70% of children have **alopecia**.
- Epidermal cyst is also a common manifestation.

> **Concept box**
> **Concept of biochemical changes in Vitamin D Dependent Rickets Type II**
> - 1,25 D level is extremely elevated because of end-organ resistance.

X-Linked Hypophosphatemic Rickets

- X-linked dominant disorder.
- The **most common hypophosphatemic rickets.**
- The defective gene is called **PHEX** (PHosphate-regulating gene with homology to Endopeptidases on the X chromosome).
- The product of this gene have either a direct or an indirect role in inactivating a phosphatonin or phosphatonins (FGF-23).
- Mutation of PHEX gene lead to **increased level of FGF-23.**
- **Hypophosphatemia** with normal PTH, normal calcium and low or inappropriately normal 1,25 D are the laboratory findings.

> **Extra Edge**
> **Phosphatonins (FGF-23)**
> - Humoral mediator that decreases renal tubular reabsorption of phosphate, therefore decreases serum phosphorus.
> - This also decreases the activity of 1α hydroxylase, resulting in deficiency of 1,25 D.
> - Fibroblast Growth Factor-23 (FGF-23) is the most well characterised phosphatonin.
> - Increased level of phosphatonins causes increased excretion of phosphorus in urine.
> - So serum Phosphorus is decreased.

Autosomal Dominant Hypophosphatemic Rickets

An autosomal dominant condition.

Due to a mutation in the gene encoding **FGF-23** which prevents the degradation of FGF-23 by proteases.

So there is increased levels of phosphatonins.

- Hypophosphatemia with normal PTH, normal calcium and low or inappropriately normal 1,25 D are the lab findings.

> **Concept box**
> - Biochemical findings of X linked and autosomal dominant Hypophosphatemic rickets is same as phosphatonins is excess in both.
> - Hypophosphatemia is due to increased excretion of phosphates through kidney by phosphatonins.
> - Low or normal 1,25 D is due to decreased activity of 1α Hydroxylase.

Autosomal Recessive Hypophosphatemic Rickets

- Extremely rare disorder due to mutation in the gene encoding **dentin matrix protein 1**, which results in elevated level of FGF-23.

> **Extra Edge**
> **Hereditary Hypophosphatemic Rickets with Hypercalciuria (HHRH)**
> - Autosomal recessive disorder due to mutation in the gene for a **sodium phosphate cotransporter** in the proximal renal tubules.
> - Hypophosphatemia, stimulates production of 1,25D.
> - This causes increased intestinal absorption of calcium.
> - Symptoms of rickets, along with muscle pain, bone pain short stature with disproportionate decrease in length of lower extremities, kidney stones.

Chronic Renal Failure

- There is decreased activity of 1α-hydroxylase in the kidney, leading to diminished production of 1,25-D.
- Unlike the other causes of vitamin D deficiency, patients have **hyperphosphatemia** as a result of decreased renal excretion.

Requirement of Vitamin D

- Children—10 μg/day (400 IU)
- Adults—5 μg/day (200 IU)
- Pregnancy, Lactation—10 μg/day (400 IU)

> **Extra Edge**
> **Vitamin D is Toxic in Excess**
> Upper limit of Vitamin D intake has been set 4000 IU/day.
> Some *infants are sensitive to intakes of vitamin D as low as 50 μg/day*[Q], resulting in an elevated plasma concentration of calcium.

This can lead to contraction of blood vessels, high blood pressure, and **calcinosis**—the calcification of soft tissues. Although excess dietary vitamin D is toxic, excessive exposure to sunlight does not lead to vitamin D poisoning, because there is a limited capacity to form the precursor, 7-dehydrocholesterol, and prolonged exposure of pre-vitamin D to sunlight leads to formation of inactive compounds.

> **HIGH YIELD POINTS**
>
> **Beneficial Effects of Vitamin D**
> - Protective against the cancer of Prostate, Colorectal cancer.
> - Protective against Prediabetes, and metabolic Syndrome.

Laboratory Findings in Disorders Causing Rickets

Disorder	Serum calcium	S phosphorus	PTH	25(OH)D	1,25(OH)D	ALP
Vitamin D deficiency	N/ Decrease	Decrease	Increase	Decrease	Decrease, N, Increase	Increase
Vitamin D Dependent rickets type I	N/ Decrease	Decrease	Increase	N	Decrease	Increase
Vitamin D dependent rickets type II	N/ Decrease	Decrease	Increase	N	Increase	Increase
Chronic renal failure	N/ Decrease	Increase	Increase	N	Decrease	Increase
X Linked hypophosphatemic rickets	N	Decrease	Normal	N	Relatively decrease	Increase
Autosomal dominant hypophosphatemic rickets	N	Decrease	Normal	N	Relatively decreased	Increase

Sources of Vitamin D
- Sunlight.
- Foods: Only animal sources Liver, Egg yolk, butter and liver oils. Out of the food sources Fish liver oils are the richest source.
- The richest source of Vitamin D is also **Halibut Liveroil**.

Assays of Vitamin D
- The release into the circulation of **osteocalcin** provides an index of vitamin D status.
- **25 (OH) Vitamin D** level is measured in the serum indicate Vitamin D status.

VITAMIN E
- *Vitamin E is a collective name for all stereo isomers of* **tocopherols and to cotrienols.**
- The **most powerful** naturally occurring antioxidant^Q.
- Ring Structure Present in Vitamin E
- Chromane (Tocol) ring with isoprenoid side chain.
- Vitamin E is carried to liver in **Chylomicron**.

Biochemical Functions of Vitamin E
- Biologically most potent form of Vitamin E is α **Tocopherol**^{Q}
- **Chain-breaking antioxidant**^{Q} and is an efficient pyroxyl radical scavenger that protect slow-density lipoproteins (LDLs) and polyunsaturated fats in membranes from oxidation.
- Lipid soluble antioxidant.

Relationship with Selenium
- Selenium decreases the requirement of Vitamin E^Q.

Vitamin E Deficiency
- **Axonal degeneration** and of the large myelinated axons and result in posterior column and spinocerebellar symptoms.
- **Hemolytic anaemia:** The erythrocyte membranes are abnormally fragile as a result of poor lipid peroxidation, leading to haemolytic anemia.
- **Peripheral neuropathy** initially characterized by Areflexia with progression to ataxic gait.
- Decreased position and vibration sense.
- Spinocerebellar ataxia (So all cerebellar signs are positive).
- Skeletal myopathy.
- Pigmented retinopathy
- Ophthalmoplegia, nystagmus.

> **EXTRA EDGE**
>
> **Vitamin E in high doses may protect against:**
> - Oxygen-induced retrolental fibroplasia
> - Bronchopulmonary dysplasia
> - Intraventricular hemorrhage of prematurity
> - Treat intermittent claudication
> - Slow the aging process.

Toxicity of Vitamin E
- Reduce platelet aggregation and interfere with Vitamin K.

Required Daily Allowance
- Males 10 mg/day
- Females 8 mg/day
- Pregnancy 10 mg/day
- Lactation 12 mg/day.

Sources of Vitamin E
- Vegetable oils like Wheat germ oil, sunflower oil, Cotton seed oil, etc.

VITAMIN K
- **Naphthoquinone derivative with long isoprenoid side chain.**
- **Letter K is the abbreviation of German word, *K*oagulation Vitamin.**

Three Forms of Vitamin K
1. Vitamin K_1: Phylloquinone from dietary sources
2. Vitamin K_2: **Menaquinone–Synthesized by Bacterial Flora**
3. Vitamin K_3: **Menadione** (and Menadiol diacetate)–Synthetic, **Water Soluble.**

Functions of Vitamin K

Vitamin K is required for the post translational carboxylation of glutamic acid (Gamma Carboxylation), which is necessary for calcium binding to γ carboxylated proteins.
- Prothrombin (factor II)
- Factors VII, IX, and X
- Protein C, protein S
- Proteins found in bone (osteocalcin)
- Matrix Gla protein
- Nephrocalcin in kidney
- Product of growth arrest specific gene Gas6.

Drugs Causing Vitamin K Deficiency
- Warfarin and Dicoumarol inhibit γ carboxylation by competitively inhibiting the enzyme that convert vitamin K to its active hydroquinone form.
- Antiobesity drug or list at.

Vitamin K Deficiency
- Bleeding manifestations
- Elevated prothrombin time, bleeding time
- Newborns, especially premature infants are particularly susceptible to Vitamin K deficiency because of
- Low Fat stores in newborns
- Low level of Vitamin K in breast milk
- Sterility of the infantile intestinal tract
- Liver immaturity
- Poor placental transport.

Hypervitaminosis K
- Hemolysis
- Hyperbilirubinemia
- Kernicterus and brain damage.

Quick Review
- Fat soluble vitamins are carried to storage organs by chylomicrons.
- Only Fat soluble vitamin with coenzyme activity is Vitamin K in gamma carboxylation.
- 11-cis retinal is the form of vitamin present in rhodopsin.
- Retinoic acid is like steroid hormone in its mechanism of action.
- Most prevalent provitamin A is beta carotene.
- Vitamin D is group of sterols having hormone like function.
- The active form of Vitamin D is Calcitriol or 1,25 Dihydroxycholecalciferol.
- The rate limiting step of vitamin D metabolism is 1 alpha hydroxylase.
- Vitamin D is immunomodulatory and antiproliferative.
- 25OH Vitamin D is the indicator of Vitamin D status of the body.
- Water soluble Vitamin K is Menadione.
- Endogenously synthesised Vitamin K is Vitamin K_2, i.e. Menaquinone.
- Gamma carboxylation is function of vitamin K.
- Most potent naturally occurring chain breaking antioxidant is alpha tocopherol (Vitamin E).

Check List for Revision
- This is a must learn topic for all exams and thoroughly revise it several times.
- All important points are given in quick review.

ANNEXURE

Fat Soluble Vitamins: Deficiency and Toxicity

Vitamin	Deficiency	Toxicity (if any)
Vitamin A	**EYES** • Loss of sensitivity towards green light • Night blindness • Conjunctival xerosis, • Bitot's spots • Corneal xerosis, • Keratomalacia **SKIN & MUCOSA** • Epithelial metaplasia keratinisation • Follicular hyperkeratosis • Squamous metaplasia • Loss of taste sensation	• Pseudotumour cerebri • Exfoliative dermatitis • Hepatomegaly • Hyperlipidemia • Bony exostosis • Decreased cognition • Cirrhosis • Increased bone resorption causing # of bones • In pregnancy teratogenicity
Vitamin D	• Rickets • Osteomalacia	• Contraction of blood vessel • High blood pressure • Calcinosis
Vitamin E	• Axonal degeneration • Peripheral neuropathy • Decreased position and vibration sense • Spinocerebellar ataxia • Skeletal myopathy • Hemolytic anaemia • Pigmented retinopathy • Ophthalmoplegia • Nystagmus	• Reduced platelet aggregation • Interfere with Vit K
Vitamin K	• Bleeding manifestation • Elevated PT & BT	• Hemolysis • Hyperbilirubinemia • Kernicterus & Brain damage

Fat Soluble Vitamins: Active Form and RDA

Vitamin	Active form (if any)	RDA
Vitamin A	Retinal Retinoic acid Retinol	**In µg/day** Men–400 Women–600 Pregnancy–800 Lactation–950
Vitamin D	1, 25 Dihydroxycholecalciferol	**In µg/day (IU in brackets)** Children–10 (400) Adults–5 (200) Pregnancy–10 (400) Lactation–10 (200)
Vitamin E	Tocopherol	**In mg/day** Males–10 Females–8 Pregnancy–10 Lactation–12–15
Vitamin K	_____	50–100 **µg/day**

Review Questions MCQ

1. Vitamin K as coenzyme form is regenerated by ? *(AIIMS May 2018)*
   a. Pyruvate carboxylase
   b. Epoxide reductase
   c. Glutathione reductase
   d. Dihydrofolate reductase

2. Most abundant form of Pro Vit A is: *(Recent Question Nov 2017)*
   a. Alpha carotene
   b. Beta carotene
   c. Cryptoxanthine
   d. Lycopene

3. Which of the following is true about Vitamin K? *(Recent Question Nov 2017)*
   a. It is a water soluble vitamin
   b. It helps in the carboxylation of factor VIII
   c. Chronic use of antibiotics lead to deficiency of Vitamin K
   d. Vitamin K deficiency manifest as multiple thrombotic episodes

4. Which vitamin is synthesized in the body? *(Recent Question 2016)*
   a. Thiamine
   b. Vitamin B_3
   c. Vitamin B_6
   d. Riboflavin

5. Tocopheryl radical is converted to Tocopherol by which vitamin? *(Recent Question 2016)*
   a. Vitamin D
   b. Niacin
   c. Vitamin E
   d. Vitamin C

6. In the crystalline lens, level of tocopherol and Ascorbate is maintained by: *(AIIMS May 2014)*
   a. Glutathione
   b. Glycoprotein
   c. Fatty acid
   d. Glucose

7. All are true about vitamin D metabolism, except: *(AIIMS Nov 2011)*
   a. 1-alpha hydroxylation occurs in kidney
   b. 25-alpha hydroxylation occurs in liver
   c. In absence of sun light, the daily requirement is 400- 600 IU per day
   d. Williams syndrome is associated with mental retardation, precocious puberty and obesity

8. Vitamin K is required for: *(AIIMS Nov 2007)*
   a. Hydroxylation
   b. Chelation
   c. Transamination
   d. Carboxylation

9. Vitamin A intoxication cause injury to: *(AIIMS Nov 2006)*
   a. Lysosomes
   b. Mitochondria
   c. Endoplasmic reticulum
   d. Microtubules

10. Active form of Vitamin D is: *(AIIMS Nov 2006)*
    a. Cholecalciferol
    b. 24,25$(OH)_2$ vit-D
    c. 1,25$(OH)_2$ vit-D
    d. 25-OH vit-D

11. Which of these has antioxidant properties? *(PGI Nov 2012)*
    a. Tocopherol
    b. Reduced glutathione
    c. Citrulline
    d. Lycopene

12. Vitamin K is involved in the post-translational modification of: *(AIIMS Nov 08, May 01, AIPGME 2011)*
    a. Glutamate
    b. Aspartate
    c. Lysine
    d. Proline

13. Which Vitamin is required for carboxylation of clotting factors?
    a. Vitamin A
    b. Vitamin D
    c. Vitamin E
    d. Vitamin K

14. All the following have antioxidant action except: *(Ker 2011)*
    a. Vitamin A
    b. Vitamin E
    c. Selenium
    d. Vitamin D

15. Which of the following is true about vitamin K? *(Recent Question 2016)*
    a. Vit K dependent factors undergo post-transcriptional modification
    b. Prothrombin is a vitamin K dependent factor
    c. Stuart-Prower factor is not vitamin K dependent
    d. Menadione is a natural water insoluble vitamin K used in clinical practice

16. Vitamin E deficiency causes all except: *(Recent Question 2016)*
    a. Ataxia
    b. Areflexia
    c. Ophthalmoplegia
    d. Neuropathy

17. Which coenzyme acts as reducing agent in anabolic reaction? *(Recent Question 2016)*
    a. FADH2
    b. FMNH2
    c. NADPH
    d. NADH

18. Most powerful chain breaking antioxidant: *(Recent Question 2016)*
    a. Glutathione peroxidase
    b. Alpha tocopherol
    c. Superoxide dismutase
    d. Vitamin C

Answers to Review Questions

1. **b. Epoxide reductase** (Ref: Harper 31/e page 529-533)

   Coenzyme role of Vitamin K is Gamma carboxylation. During the reaction Vitamin K is converted to its epoxide form. The epoxide form is converted back to hydrquinone form by Vit K Epoxide reductase

2. **b. Beta carotene**

3. **c. Chronic use of antibiotics lead to deficiency of Vitamin K** (Ref: Harper 31/e page 529-533)

   Vitamin K is a fat soluble vitamin. It helps in the gamma carboxylation of factors VII, IX, X. Vitamin K deficiency causes bleeding manifestations. Vitam in K2 is synthesized by bacterial flora. Hence chronic use of antibiotics can cause deficiency of VitaminK.

4. **b. VitaminB_3**

   Endogenously synthesised Vitamins^Q

   Vitamins are generally not synthesized by the humans, but some vitamins can be synthesized endogenously. They are:
   - Vitamin D from precursor steroids
   - Vitamin K, Biotin, and pantothenic acid by the intestinal microflora
   - Niacin, Vitamin B_3 from tryptophan, an essential amino acid.

5. **d. Vitamin C** (Ref: Harper 31/e page 529-533)
   - The main function of vitamin E is as a chain-breaking, free-radical-trapping antioxidant in cell membranes and in plasma lipoproteins
   - In plasma lipoproteins, peroxidation of polyunsaturated fatty acids forms Lipid peroxide radicals, is scavenged by Tocopherol
   - Tocopheroxyl radical, a relatively unreactive compound is formed by above reaction
   - Commonly, the tocopheroxyl radical is reduced back to tocopherol by reaction with vitamin C from plasma.

6. **a. Glutathione** (Ref: Harper 31/e page 529-533)

7. **d. Williams Syndrome is associated with mental retardation, precocious puberty and obesity**

8. **d. Carboxylation** (Ref: Harper 31/e page 529-533)

9. **a. Lysosomes**

10. **c. 1,25 (OH)$_2$vit D**

    This is otherwise called calcitriol

11. **a, b, d. Toco...., Reduced Glut...., Lycopene**

12. **d. Proline > a. Glutamate** (Ref: Harper 31/e page 529-533)

13. **d. Vitamin K** (Ref: Harper 31/e page 529-533)

    Vitamin K is required for the post-translational carboxylation of glutamic acid (Gamma Carboxylation), which is necessary for calcium binding to γ carboxylated proteins.

14. **d. Vitamin D**
    - Adequate Selenium intake maximize the antioxidant action of Glutathione Peroxidase.
    - So Selenium is also considered as a compound having antioxidant action.
    - Now a days Vitamin D is also considered as powerful natural membrane antioxidant.
    - But as it is an old question, Vitamin D is the best answer.

15. **b. Prothrombin is a vitamin K dependent factor**
    - Vitamin K helps Post-translational modification.
    - Stuart Prower factor is Factor X, which is not Vitamin K dependent.
    - Prothrombin is factor II, so it is Vitamin K dependent.
    - Menadione is Synthetic Vitamin K.

16. **None** (Ref: Harper 31/e page 529-533)

    Manifestations of Vit E deficiency

    Patients may have cerebellar disease, posterior column dysfunction, and retinal disease. Loss of deep tendon reflexes, Peripheral neuropathy is usually the initial finding. Subsequent manifestations include limb ataxia (intention tremor, dysdiadochokinesia), truncalataxia (wide-based, unsteady gait), dysarthria, ophthalmoplegia (limited upward gaze), nystagmus, decreased proprioception (positive Romberg test), decreased vibratory sensation, and dysarthria. Some patients have pigmentary retinopathy. Visual field constriction can progress to blindness. Cognition and behavior can also be affected. Myopathy and cardiac arrhythmias are less common findings.

17. **c. NADPH**

    NADPH is used in reductive biosynthesis of Fatty acids, Steroids, etc. So it is used for anabolic reactions.

18. **b. Alpha tocopherol**

    Antioxidants fall into two classes:
    - Preventive antioxidants, which reduce the rate of chain initiation

      They are Glutathione Peroxidase, Catalase
    - Chain-breaking antioxidants, which interfere with chain propagation

    They are Superoxide Dismutase, Uric Acid, Vitamin E (Most powerful).

8.2 WATER SOLUBLE VITAMINS

- B Complex Vitamins
- Vitamin C

THIAMINE (VITAMIN B₁)

Thiamine is also called Aneurine.

Sources

Aleurone layer of cereals. Hence whole wheat flour and unpolished hand pound rice has better nutritive value. Yeast is also a good source of thiamine.

Active form of Thiamine

Thiamine Pyrophosphate (TPP) also called Thiamine diphosphate (TDP).

> **Extra Edge**
> **Thiamine and nerve conduction**
> Thiamine triphosphate has a role in nerve conduction; it phosphorylates, and so activates, a chloride channel in the nerve membrane.

> **High Yield Points**
> **Coenzyme Role of Thiamine Pyrophosphate^Q**
> Thiamine generally function in the decarboxylation reaction of alpha ketoacids and branched chain amino acids
> - **Pyruvate Dehydrogenase^Q** which convert Pyruvate to Acetyl CoA.
> - **α Keto Glutarate Dehydrogenase^Q** in Citric Acid Cycle which convert α Keto Glutarate to Succinyl CoA.
> - **Branched Chain Ketoacid Dehydrogenase^Q** which catalyses oxidative decarboxylation of branched chain amino acids.
> - **Trans Ketolase^Q** in **Pentose Phosphate Pathway^Q**. This is the biochemical basis of assay of Thiamine status of the body.

Deficiency of Vitamin B₁ (Thiamine)

Beriberi^Q

Two types

1. **Wet beriberi:** Marked peripheral vasodilatation, resulting in high output cardiac failure with dyspnoea, tachycardia, cardiomegaly, pulmonary and peripheral oedema.
   The right heart is enlarged.
   Increased QT interval. Inverted T wave
2. **Dry beriberi**—Involves both peripheral and central nervous system.

Peripheral Nervous System

- Typically a symmetric motor and sensory neuropathy with pain, paraesthesia and loss of reflexes.
- The legs are affected more than the arms.
- Tenderness and cramping of leg muscles.
- Muscle Atrophy.

Central Nervous System

Wernicke's encephalopathy—in alcoholics with chronic thiamine deficiency
- Horizontal Nystagmus
- Ophthalmoplegia (Ptosis of eyelids, Atrophy of optic nerve)
- Truncal ataxia
- Confusion.

Wernicke-Korsakoff Syndrome

- Along with features of Wernicke's encephalopathy
- Amnesia
- Confabulatory psychosis.

Acute pernicious (fulminating) beriberi (shoshin beriberi), in which heart failure and metabolic abnormalities predominate.

Biochemical Assessment of Thiamine Deficiency

- Erythrocyte Transketolase activity is reduced.
- Urinary Thiamine excretion

Thiamine Toxicity

There is no known toxicity of thiamine

> **Recommended Daily Allowance (RDA) of Vitamin B₁**
> - 1–1.5 mg/day.

RIBOFLAVIN (VITAMIN B₂)

- Is called **Warburg Yellow enzyme^Q** of cellular respiration.
- Riboflavin is heatstable.
- Enzymes containing riboflavin are called Flavoproteins.
- Act as respiratory coenzyme and an electron donor.

Nutrition

> **HIGH YIELD POINTS**
>
> **Active Forms of Riboflavin**
> - They are FAD (Flavin Adenine Dinucleotide) and FMN (Flavin Mononucleotide)
>
> **Coenzyme Role of Riboflavin**
> **FMN Dependent Enzymes**^{Q}
> - L- Amino Acid Oxidase
> - NADH Dehydrogenase (Complex I of ETC)
> - Monoamine Oxidase
>
> **FAD Dependent Enzymes**
> - Complex II (Succinate Dehydrogenase) of ETC
> - D Amino Acid Oxidase
> - Acyl CoA Dehydrogenase
> - Alpha Ketoglutarate Dehydrogenase
> - Pyruvate Dehydrogenase
> - Xanthine Oxidase

Deficiency Manifestation of Vitamin B$_2$ (Riboflavin)

- Cheilosis, glossitis (Magenta coloured tongue).
- In glossitis, the tongue becomes smooth, with loss of papillary structure.
- Cheilosis begins with pallor at the angles of the mouth and progresses to thinning and maceration of the epithelium, leading to **fissures** extending radially into the skin.
- Keratitis, conjunctivitis, photophobia, lacrimation, **corneal vascularization**.
- **Seborrheic dermatitis**.
- **Normochromic, normocytic anemia** may also be seen because of the impaired erythropoiesis.

Biochemical Assessment of Nutritional Status of Riboflavin

- Measurement of activation of erythrocyte **Glutathione Reductase by FAD** added *in vitro*.
- Urinary excretion of Riboflavin.

Riboflavin Toxicity

Riboflavin toxicity is not reported yet because of limited absorption capacity of GIT.

RDA of Riboflavin

- 1.5 mg/day

NIACIN OR NICOTINIC ACID (VITAMIN B$_3$)

- **Not strictly a Vitamin**.
- Can be synthesized from **Tryptophan**.
- **60 mg of Tryptophan** yield 1 mg of Niacin.

Active Form of Niacin

- Two Coenzyme forms are NAD$^+$ (Nicotinamide Adenine Dinucleotide) and NADP$^+$ (Nicotinamide Adenine Dinucleotide Phosphate).

> **HIGH YIELD POINTS**
>
> **Coenzyme Role of Niacin**
> - Important in numerous oxidation reduction reactions.
>
> **NAD$^+$ Linked Enzymes**
> - Lactate Dehydrogenase
> - Pyruvate Dehydrogenase
> - α-Ketoglutarate Dehydrogenase
> - Isocitrate Dehydrogenase
> - Malate Dehydrogenase
> - β-Hydroxy Acyl CoA Dehydrogenase
> - Glycerol 3 Phosphate Dehydrogenase (cytoplasmic)
> - Glutamate Dehydrogenase
> - Glyceraldehyde 3 phosphate Dehydrogenase
>
> **NADP$^+$ Utilizing Enzymes**
> Mainly for Reductive Biosynthesis^Q of steroids and Cholesterol^Q, Fre eradical Scavenging^Q, Formation of deoxyribonucleotides, One carbon metabolism.
> - 3 Keto acyl reductase
> - Enoyl reductase
> - HMG CoA Reductase
> - Folate reductase
> - Glutathione Reductase
> - Ribonucleotide Reductase
>
> **NADPH Generating Reactions**^{Q}
> - Glucose 6 Phosphate Dehydrogenase in HMP shunt pathway.
> - 6 Phosphogluconate Dehydrogenase in HMP Shunt Pathway.
> - Cytoplasmic Isocitrate Dehydrogenase.
> - Malic Enzyme. (NADP Malate Dehydrogenase)
>
> **Other Function of NAD**
> NAD is the **source of ADP-ribose** for the ADP-ribosylation of proteins and poly ADP-ribosylation of nucleoproteins involved in the DNA repair mechanism.

Deficiency of Niacin

Pellagra

- The early symptoms of pellagra are vague: anorexia, lassitude, weakness, burning sensations, numbness, and dizziness. After along period of deficiency, the classic triad of **dermatitis, diarrhea**, and **dementia** appears.
- Diarrhoea can be severe resulting in malabsorption due to atrophy of intestinal villi.
- The cutaneous lesions may be preceded by or accompanied by stomatitis, glossitis.
- Nervous symptoms include **Depressive Psychosis**, disorientation, insomnia, and delirium.
- Advanced Pellagra can result in **death**.

Image-Based Information

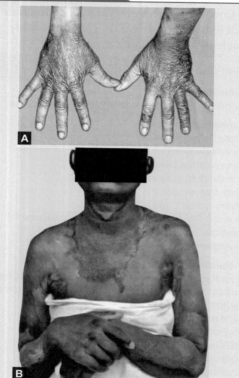

Figs. 8.2.1A and B: *Photosensitive dermatitis in pellagra*
The most characteristic manifestation of pellagra. Symmetric are as of erythematous, dry scaly lesions on exposed surfaces. They are sharply demarcated from the surrounding healthy skin
- Around the neck (Casal necklace)
- On the hands and feet often have the appearance of a glove or stocking

High Yield Points

4D s of Pellagra
- Dermatitis (Photosensitive Dermatitis)
- Dementia
- Diarrhoea
- Death

High Yield Points

Conditions Associated with Pellagra Like Symptoms
- **Hartnup Disease** (Due to intestinal malabsorption and renal reabsorption of Tryptophan)
- **Carcinoid Syndrome** (Over production of serotonin leads to diversion of Tryptophan from NAD^+ pathway)
- **Vitamin B_6 deficiency** (Defective Kynureninase that lead to defective synthesis of Niacin)
- Pellagra is common in people whose staple diet is **maize and jowar**.
Maize-Niacin present in unavailable form Niacytin
Sorghum vulgare (Jowar)-High Leucine content inhibit QPRTase, rate limiting enzyme in Niacin synthesis.

Recommended Daily Allowance of Niacin (RDA)
20 mg/day

Toxicity of Niacin
- **Prostaglandin mediated cutaneous flushing** due to binding of vitamin to a G Protein coupled receptor.
- Gastric irritation.
- **Hepatic toxicity** is the most serious toxic reaction with sustained release niacin. Presents with jaundice, elevated liver enzymes (AST and ALT) even **fulminant hepatitis**.
- Other toxic reactions include **glucose intolerance**, hyperuricemia, macular oedema and cysts.

Treatment of cutaneous flushing
- Laropiprant, a selective Prostaglandin D2 receptor 1 antagonist.
- Premedication with Aspirin.

Therapeutic uses of Niacin (Nicotinic Acid)
- Used as Lipid modifying Drug.
- Niacin reduces plasma triglyceride and LDL-C levels and raises the plasma concentration of HDL-C.

PYRIDOXINE (VITAMIN B_6)

Family of 3 related Pyridine Derivatives
1. Pyridoxine
2. Pyridoxal
3. Pyridoxamine

High Yield Points

Some 80% of the body's total vitamin B_6 is pyridoxal phosphate in muscle^Q, mostly associated with glycogen phosphorylase.

Active Form of Pyridoxine
- Pyridoxal phosphate (PLP)
- Mainly used for Amino Acid metabolism^Q.

High Yield Points

Coenzyme Role of Pyridoxal Phosphate (PLP)^Q
Transamination
- Alanine Amino Transferase (ALT)
- Aspartate Amino Transferase (AST)
- Alanine Glyoxalate Amino Transferase

Decarboxylation of Amino Acids
This results in the formation of Biogenic Amines
- Glutamate → GABA
- 5-Hydroxy → Tryptophan Serotonin
- Histidine → Histamine
- Cysteine → Taurine

Contd...

Contd...
- Serine→Ethanolamine
- DOPA→Dopamine

Transulfuration
- Involved in the metabolism of Sulfur containing amino acids.
- Synthesis of Cysteine from methionine.
- Enzymes are Cystathionine Beta Synthase and Cystathioninase.

Tryptophan Metabolism
- Coenzyme of Kynureninase involved in the synthesis of niacin from Tryptophan.
- In Pyridoxine deficiency Xanthurenic acid is excreted because of defective Kyneureninase in Niacin synthesis

Heme Synthesis
- ALA Synthase that catalyse condensation of Succinyl CoA and Glycine.

Glycogenolysis
- Glycogen phosphorylase (Only enzyme in Carbohydrate metabolism).

Deficiency of Vitamin B_6 (Pyridoxine)

- Neurological manifestation—Due to deficiency of Catecholamines
- Peripheral neuropathy
- Personality changes that include **depression and confusion**
- **Convulsions**—Due to decreased synthesis of GABA
- **Microcytic hypochromic Anemia**—Due to decreased hemesynthesis
- Pellagra due defective niacin synthesis.

> **EXTRA EDGE**
>
> **Other Conditions Caused by PLP Deficiency**
> - **Oxaluria**—Due to defective Alanine—Glyoxylate Amino Transferase. Glyoxylate converted to Oxalic acid.
> - **Homocystinuria**—Due to defective Cystathionine Beta Synthase.
> - **Xanthurenic Aciduria**—Due to defective Kynureninase.
> - **Cardiovascular risks** because of homocysteinemia.

Drugs that Interact with Carbonyl Group and Causes PLP Deficiencies are
- L-Dopa, Pencillamine, Cycloserine.

Pyridoxine Dependency Syndromes that need Pharmacological Dose of PLP
- Classic homocystinuria (due to cystathionine beta synthase deficiency)
- Sideroblastic anemia (due to ALA Synthase deficiency)
- Gyrate atrophy of retina and choroid in δ-ornithine aminotransferase.

High Doses of Pyridoxine Given in
- Carpal Tunnel syndrome
- Premenstrual syndrome
- Schizophrenia
- Diabetic neuropathy.

> **EXTRA EDGE**
>
> **Pyridoxine and Hormone dependent cancer**
> - Pyridoxine is important in steroid hormone action.
> - Pyridoxal phosphate removes the hormone-receptor complex from DNA binding, terminating the action of the hormones.
> - In vitamin B_6 deficiency, there is increased sensitivity to the actions of low concentrations of estrogens, androgens, cortisol, and vitamin D.
> - Increased sensitivity to steroid hormone action may be important in the development of hormone-dependent cancer of the **breast, uterus**, and **prostate**, and vitamin B_6 status may affect the prognosis.

Biochemical Assays of Vitamin B_6
- Erythrocyte Transaminase activity.
- **Tryptophan load test**—measurement of Xanthurenic acid following Tryptophan load.
- Measurement of PLP in the blood.

Toxicity of Vitamin B_6
- Excess Pyridoxine may lead to *Sensory Neuropathy*.

RDA of Pyridoxine
- 1–2 mg/day.
- RDA of Pyridoxine depends on Protein intake.

PANTOTHENIC ACID (VITAMIN B_5)

- Derived from the Greek word panto smeans everywhere.
- Endogenously synthesised by bacterial flora in the intestine.
- Vitamin that contains **Beta Alanine**.
- Vitamin present in **Coenzyme A (CoA)** and **Acyl Carrier Protein (ACP)** in **Fatty Acid Synthase Complex**.

The important CoA derivatives are:
- Acetyl CoA
- Succinyl CoA
- HMG CoA
- Acyl CoA

Pantothenic acid as a part of CoA take part in:
- Fatty acid Oxidation
- Acetylation
- Citric acid cycle
- Cholesterol synthesis

Deficiency of Pantothenic Acid

- Gopalan's Burning feet Syndrome or **Nutritional Melalgia** or Peripheral nerve damage.

RDA of Pantothenic acid
10 mg/day.

> **EXTRA EDGE**
>
> **Pantothenate Kinase Associated Neurodegeneration (PKAN) (Formerly Hallervorden-Spatz Syndrome)**
> - Rare autosomal recessive neurodegenerative disorder
> - Chorea, dystonia, parkinsonian features, pyramidal tract features and MR
> - MRI-decreased T2 signal in the globus pallidus and substantia nigra, "**eye of the tiger**" sign (hyperintense area within the hypointense area);
> - Some times acanthocytosis
> - Neuropathologic examination indicate sex successive accumulation of iron-containing pigments in the globus pallidus and substantia nigra.
> - Similar disorders are grouped as neurode generation with brain iron accumulation (NBIA).

BIOTIN OR VITAMIN H OR VITAMIN B_7

Also known as anti-egg white injury factor.
Endogenously synthesized by intestinal flora.
Reactive form is the enzyme bound **Carboxybiocytin**.

> **HIGH YIELD POINTS**
>
> **COENZYME ROLE OF BIOTIN**
> **Play a role in**
> - Gene expression,
> - Fatty acid synthesis,
> - Gluconeogenesis
> - Serve as a CO_2 carrier for Carboxylases enzymes
> - Gene regulation by histone biotinylation.
>
> **Coenzyme for ATP dependent Carboxylation reaction (Carbon Dioxide Fixation)**
> - Pyruvate Carboxylase (Pyruvate to Oxaloacetate)
> - Propionyl CoA Carboxylase (Propionyl CoA to Methyl Malonyl CoA)
> - Acetyl CoA Carboxylase (Acetyl CoA to MalonylCoA)
> - Methyl Crotonyl CoA Carboxylase

> **HIGH YIELD POINTS**
>
> **Biotin Independent Carboxylation Reaction**
> - Carbamoyl Phosphate Synthetase-I and II
> - Addition of CO_2 to C_6 in Purine ring (AIR Carboxylase)
> - Malic Enzyme (Pyruvate to Malate)
> - Gamma Carboxylation (Vitamin K dependent)

> **EXTRA EDGE**
>
> **Biotin Antagonist Avidin**
> - Protein present in the raw egg white.
> - Eating raw egg is harmful because of Avidin present in raw egg inhibit biotin.
> - Affinity of Avidin to Biotin is stronger than most of the Antigen antibody reaction.
>
> **This property is used in:**
> - ELISA test
> - Labelling of DNA.
>
> **Streptavidin**
> - Purified from Streptomyces avidinii
> - Bind 4 molecules of Biotin.

Deficiency of Biotin

- Mental changes (Depression, hallucination) paresthesia, anorexia, and nausea.
- A scaling, seborrheic and erythematous rash around nose, eyes and mouth.

Biochemical Tests to Diagnose Biotin Deficiency
- Decreased concentration of Urinary biotin
- Increased urinary excretion of 3-hydroxyvaleric acid after leucine challenge.
- Decreased activity of biotin dependent enzymes in lymphocytes.

FOLIC ACID OR VITAMIN B_9

- Derived from Latin word **folium**, which means leaf of vegetable.
- Folic Acid is abundant in leafy vegetables.
- Folic Acid is absorbed from upper part of **Jejunum**^{Q}.

Functions of Folic Acid

Active form of Folic acid is **TetraHydro Folic Acid (THFA)**
- THFA is the carrier of One Carbon group.

One Carbon Metabolism
- One Carbon units are:
  - Methyl (CH_3)
  - Methylene (CH_2)
  - Methenyl (CH)
  - Formyl (CHO)
  - Formimino (CH = NH)

One carbon groups bind to THF through:
- N_5 are Formyl, Formimino or methyl
- N_{10} are Formyl
- Both N_5 and N_{10} are Methylene and Methenyl.

Sources of One Carbon Groups

- The major point of entry of one carbon unit is **Methylene THF**[Q]
- **Serine**[Q] is the most important source of One Carbon units.
- **Serine**[Q] **Hydroxy Methyl Transferase**[Q] is the enzyme involved in this pathway.

Important Sources of One Carbon Groups

Source of Methylene THF
- Serine to Glycine by Serine Hydroxy Methyl Transferase
- Glycine
- Choline

Source of Formimino THF
- Histidine→FIGLU →Formimino THF

Utilization of One Carbon Groups
- Glycine to Serine
- Homocysteine to Methionine.
- Synthesis of Purine Nucleotides.
- Synthesis of TMP.
- Synthesis of Choline

Biochemical Assessment of Folate Deficiency

- Serum Folate (Normal level is 2–20 ng/mL)
- Red Cell Folate
- Histidine Load test[Q] or FIGLU excretion test
- **AICAR (Amino Imidazole Carboxamide Ribose 5 Phosphate)** Excretion Test
- Serum Homocysteine
- Peripheral Blood Smear (Macrocytes, tear drop cells, hypersegmented neutrophils, anisopoikilocytosis)

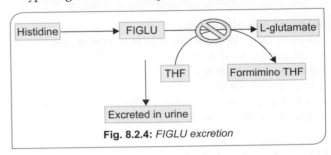

Fig. 8.2.4: *FIGLU excretion*

Deficiency of Folic Acid

- **Reduced DNA Synthesis** because of THF derivatives are involved in purine synthesis and thymidylate Synthesis.
- Megaloblastic anemia
- Vitamin B_{12} deficiency and Folated deficiency can lead to this condition.
- In Vitamin B_{12} deficiency Megaloblastic anemia is due to folate trap.
- **Homocysteinemia** due to decreased conversion of homocysteine to Methionine. This is because Methyl THFA is the methyl donor for this reaction.
- **Neural tube defects (like Spinabifida, Anencephaly)** during pregnancy.
- Non hematologic manifestations include glossitis, listlessness, and growth retardation.
- Depression.

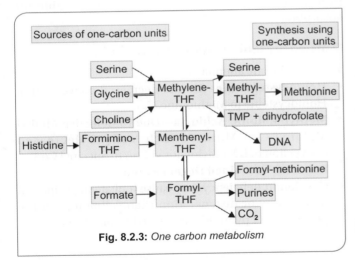

Fig. 8.2.3: *One carbon metabolism*

High Yield Points

Pharmaceutically Used THFA Derivative
- **5-Formyl-tetrahydrofolate**[Q2013] **is more stable than folate and is therefore used pharmaceutically (known as folinic acid)**, and the synthetic (racemic) compound **(leucovorin)**.
- It is given orally or parenterally to overcome the toxic effects of methotrexate or other DHF reductase inhibitors.

Extra Edge

Folic Acid and Cancer
- Protein present in the raw egg white.
- Low folate status results in impaired methylation of CpG islands in DNA, which is a factor in the development of **colorectal and other cancers**.
- Prophylactic Folic Acid during pregnancy reduce chance of Acute Lymphoblastic Lymphoma.
- But, folate supplements increase the rate of transformation of preneoplastic colorectal polyps into cancers.
- Folic acid "feed" tumours by increasing thymidine pools and "better" quality DNA.
- So Folic Acid should be avoided in established tumors.

VITAMIN B₁₂ (COBALAMIN)

- Other name is Extrinsic factor of castle.
- Contain **4.35% cobalt by weight**.
- Contain 4 pyrrole rings coordinated with a cobalt atom, called Corrin ring.

> **HIGH YIELD POINTS**
>
> **Active forms of Vitamin B₁₂ (Cobamide Enzyme)**
> - Methyl Cobalamin and Adenosyl Cobalamin (Ado B₁₂)
>
> **Coenzyme Role of Methyl B₁₂**
> Methionine Synthase or Homocysteine Methyl Transferase
> - Homocysteine →Methionine
>
> **Coenzyme role of Ado B₁₂**
> Methyl Malonyl CoA Mutase
> - L Methyl Malonyl CoA → Succinyl CoA
> - Leucine Amino Mutase

Vitamin B₁₂ Metabolism

Absorption of Cobalamin

- 99% of absorption Cobalamin are active
- Active mechanism-site is **Ileum**ᵠ
- 1% passive occurs equally in Buccalcavity, Duodenum, Ileum.

Cobalamin Binding Proteins

- Cobalamin binding proteins in the saliva are called
- Haptocorrins or Cobalophilin or R Binders.
- **Intrinsic Factor of Cast**le from **parietal cells of body and fundus of the stomach.**

Vitamin B₁₂ is freed from binding proteins in food through the action of pepsin in the stomach and binds to salivary proteins called **cobalophilins, or R-binders.**

In the **duodenum**, bound vitamin B₁₂ is released by the action of pancreatic proteases

It then associates with **intrinsic factor**.

Actively absorbed from the **ileum**ᵠ by binding to IF receptor.

IF receptor in the ileum is called **CUBULIN.**

Transport of Cobalamin to the target tissues

- Major Cobalamin transport protein in plasma is
- Transcobalamin II (TC II)ᵠ
- Transcobalamin I [TC I] plays a role in the transport of Cobalamin analogues.
- At the target tissues by receptor mediated endocytosis involving TC II receptor.

Causes of Vitamin B₁₂ Deficiency

Nutritional

Vitamin B₁₂ is found only in foods of animal origin, there being no plant sources of this vitamin. *This means that strict vegetarians (vegans) are at risk of developing B₁₂ deficiency.*

Malabsorption-Pernicious Anemia

- Pernicious anemia is a specific form of megaloblastic anemia caused by autoimmune gastritis and an attendant failure of intrinsic factor production, which leads to vitamin B₁₂ deficiency.

Gastric causes

- Congenital absence of intrinsic factor or functional abnormality
- Total or partial gastrectomy.

Intestinal causes

Intestinal stagnant loop syndrome: jejunal diverticulosis, ileocolic fistula, anatomic blind loop, intestinal stricture, etc.

Ileal resection and Crohn's disease.

Fish tapeworm

The fish tapeworm (*Diphyllobothrium latum*) lives in the small intestine of humans and accumulates cobalamin from food, rendering the cobalamin unavailable for absorption.

Deficiency Manifestation of Vitamin B₁₂

- Megaloblastic anemia
- **Homocysteinemia**—Due to decreased conversion of Homocysteine to Methionine.
- **Methyl Malonic Aciduria**—Due to defective Methyl Malonyl CoA Mutase which leads to decreased conversion of L Methylmalonyl CoA to Succinyl CoA.
- **Subacute Combined Degeneration.**
- Cobalamin deficiency may cause a bilateral peripheral neuropathy or degeneration (demyelination) of the posterior and pyramidal tracts of the spinal cord.

Biochemical Assessment of Cobalamin Deficiency

- Serum Cobalamin
- **Serum Methyl** Malonate (This helps to distinguish between Megaloblastic anemia due to Cobalamin deficiency and Folate deficiency.)
- Serum homocysteine
- Schilling Test using Radioactive labelled Cobalt-60
- Urine Homocysteine and MMA
- Bone marrow and Peripheral Blood Smear

Image-Based Information

Megaloblastic Anaemia
Megaloblastic anaemia is seen in Folate deficiency and Cobalamin deficiency

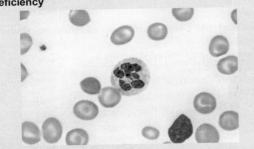

Fig. 8.2.5: *Peripheral smear showing hypersegmented neutrophils*

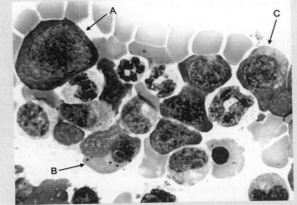

Fig. 8.2.6: *Bone marrow aspirate showing megaloblast in various stages*

VITAMIN C (ASCORBIC ACID)

- Other name is anti-**Scorbutic factor**.
- Most animals synthesize Vitamin C from Glucose by **uronic Acid Pathway^Q**.
- Humans and higher Primates cannot due to absence of **Gulonolactone Oxidase^Q**.

Biochemical Functions of Ascorbic Acid

- Acts as a good reducing agent and a scavenger of free radicals **(Antioxidant)**
- In **Collagen Synthesis**—Vitamin C is required for the post-translational modification, Hydroxyl at ionoflysine and Proline.
- **Hydroxylation** of Tryptophan
- Tyrosine Metabolism—Oxidation of P hydroxyl Phenyl Pyruvate to Homogentisic Acid.
- **Bile Acid Synthesis** in 7 alpha Hydroxylase
- **Iron Absorption**—Favour Iron absorption by conversion of Ferric ions to Ferrous ions.

- **Folate Metabolism**—Conversion of Folate to its active form.
- Adrenal steroid synthesis.

Vitamin C Deficiency

Scurvy

- Petechiae, ecchymosis, coiled hairs, inflamed and bleeding gums, joint effusion, poor wound healing, fatigue.
- Perifollicular hemorrhages.
- Perifollicular hyperkeratotic papules, petechiae, purpura.
- Splinter hemorrhage, bleeding gums, hemarthroses, subperiosteal hemorrhage.
- A "Scorbutic rosary" at the costochondral junctions and depression of the sternum are other typical features.
- Pseudoparalysis and pithed frog position in infants.
- Anaemia.
- Late stage are characterised by edema, oliguria, neuropathy, intracerebral hemorrhage and death.

Barlow's Syndrome (Infantile Scurvy)

- In infants between 6 and 12 months, the diet if not supplemented with Vitamin C then deficiency will result.

Vitamin C Toxicity

- Gastric irritation, flatulence, diarrhoea
- Oxalate stones are of theoretic concern.

Image-Based Information

Scurvy

- A "rosary" at the costochondral junctions called Scorbutic rosary
- Gum changes are bluish purple, spongy swellings of the mucous membrane, especially over the upper incisors.
- Hemorrhagic manifestations of scurvy include petechiae, purpura, and ecchymoses at pressure points; epistaxis; gum bleeding; and the characteristic perifollicular hemorrhages.

Quick Review

Vitamin Deficiencies Causing Dementia
- Thiamin
- Niacin
- Cobalamin

Sulphur containing Vitamins
- Biotin
- Thiamin

Antioxidant Vitamin
- Vitamin E
- Vitamin C
- Beta Carotene

Antioxidant Vitamins are also Pro-oxidants
- Vitamin C
- Beta Carotene
- Vitamin E

Contd...

B complex Vitamins with Toxicity
- Niacin
- Pyridoxine

Redox Vitamins
Vitamins that take part in Oxidation reduction reaction
Niacin and Riboflavin

Endogenously Synthesized Vitamins
- Niacin (Vitamin B_3)
- Biotin
- Vitamin D
- Pantothenic Acid
- Vitamin K

Check List for Revision

- This is a must learn chapter.
- Do revise the quick revision tables and boxes

Contd...

ANNEXURE

Ring Structures of B-Complex Vitamins

Vitamin	Ring structure
Vitamin B_1 [Thiamine]	Pyrimidine + Thiazole
Vitamin B_2 [Riboflavin]	Isoalloxazine
Vitamin B_3 [Niacin]	Pyridine
Vitamin B_6 [Pyridoxine]	Pyridine
Vitamin B_{12} [Cobalamin]	Corrin [Tetrapyrrole with Co at its center]
Folic acid	Pteridine + PABA
Biotin	Imidazole + Thiophene
Pantothenic acid	**No ring Structure** Contain Pantoic Acid and **Beta Alanine**^{Q} in amide linkage

B Complex Vitamins – Active Form and RDA

Thiamin	Thiamine Pyrophosphate (TPP) or Thiamine Diphosphate (TDP)	1-1.5 mg/day
Riboflavin	FAD FMN	1.5 mg/day
Niacin	NAD+ NADP+	20 mg/day
Pantothenic acid	----------------	10 mg/day
Pyridoxine	Pyridoxal Phosphate (PLP)	1-2 mg/day RDA depends on Protein intake
Biotin	Carboxybiotin	Adults 20-30 μg/day
Folic acid	THF (Tetra Hydro Folate)	1-1.5 mg/day
Cobalamin	Methyl Cobalamin Adenosyl Cobalamin	Adults 2-3 μg/day Pregnancy & lactation 4 μg/day
Vitamin C		Children 50-60 mg/day Adults 500 mg/day Preg & lact 500 +20-40 mg/day
Thiamin	• Wet beriberi-High output cardiac failure • Dry beri beri affects PNS & CNS **PNS** Symmetric motor and sensory neuropathy **CNS** 1. Wernicke's encephalopathy (Horizontal nystagmus, Ophthalmolegia, Trunal ataxia, Confusion) 2. Wernicke's Korsakoffs Syndrome (Above clinical features + Amnesia & Confabulatory psychosis) • Shoshin beriberi (Cardiac manifestation and metabolic abnormalities predominate)	Not known
Riboflavin	• Cheilosis, Glossitis, Fissuring, Magenta tongue • Keratitis, Conjunctivitis, Photophobia, Lacrimation, Corneal vascularisation • Sebhorrheic dermatitis • Normochromic normocytic anaemia	Not known

Vitamin	Deficiency	Toxicity
Niacin	• Pellagra with classic triad of diarrhoea, photosensitive dermatitis, dementia • Depressive psychosis in somnia, delirium • Death in advanced cases	• PG mediated cutaneous flushing • Gastric irritation • Hepatic toxicity • Glucose intolerance • Hyperuricemia • Macular oedema
Pantothenic acid	• Burning foot syndrome or nutritional melalgia	Not known
Pyridoxine	• Peripheral neuropathy • Personality changes (Depression & Confusion) • Convulsion • Microcytic hypochromic anaemia • Pellagra like symptoms • Oxaluria, Homocystinuria, Xanthurenic aciduria	• Sensory neuropathy
Folic acid	• Megaloblastic anaemia • Homocysteinemia • Neural tube defects in pregnancy • Depression	Not known
Cobalamin	• Megaloblastic anaemia • Homocysteinemia • Methyl Malonic Aciduria • SACD • Bilateral peripheral neuropathy • Degeneration of posterior & pyramidal tracts of spinal cord	Not known
Vitamin C	• Scurvy (Petechiae, bleeding gums, poor wound healing, perifollicular hemorrhage, splinter hemorrhage) • Follicular hyperkeratosis) • Scorbutic rosary, Pseudoparalysis • Barlow's Syndrome	Gastric irritation Oxalate stones

Coenzyme Role of B Complex Vitamins

Vitamins	Enzymes
Thiamin	• **Pyruvate dehydrogenase** • **α Keto Glutarate Dehydrogenase** • **Branched Chain Ketoacid Dehydrogenase** • **Trans Ketolase**
Riboflavin	**FMN Dependent Enzymes**[Q] • L- Amino Acid Oxidase • NADH Dehydrogenase (Complex I of ETC) • Monoamino Oxidase **FAD Dependent Enzymes** • Complex II (Succinate Dehydrogenase) of ETC • D Amino Acid Oxidase • Acyl CoA Dehydrogenase • Alpha Ketoglutarate Dehydrogenase • Pyruvate Dehydrogenase
Niacin	**NAD^+ Linked Enzymes** • Lactate Dehydrogenase • Pyruvate Dehydrogenase • α-Ketoglutarate Dehydrogenase • Isocitrate Dehydrogenase • Malate Dehydrogenase • β-Hydroxy Acyl CoA Dehydrogenase • Glycerol 3 Phosphate Dehydrogenase (cytoplasmic) • Glutamate Dehydrogenase • Glyceraldehyde 3 phosphate Dehydrogenase **$NADP^+$ Utilizing Enzymes** • 3 Keto acyl reductase • Enoyl reductase • HMG CoA Reductase • Folate reductase • Glutathione Reductase • Ribonucleotide Reductase **NADPH generating enzymes** • Glucose 6 Phosphate Dehydrogenase } HMP Pathway oxidative phase • 6 Phospho gluconate Dehydrogenase • Cytoplasmic Isocitrate Dehydrogenase • Malic enzyme
Pantothenic acid	**Pantothenic acid as a part of CoA take part in:** • Fatty acid Oxidation • Acetylation • Citric acid cycle • Cholesterol synthesis
Biotin	**Coenzyme for ATP dependent Carboxylation reaction (Carbon Dioxide Fixation)** • Pyruvate Carboxylase • Propionyl CoA Carboxylase • Acetyl CoA Carboxylase Methyl Crotonyl CoA Carboxylase
Pyridoxine	**Transamination** • Alanine Amino Transferase (ALT) • Aspartate Amino Transferase (AST) • Alanine Glyoxalate Amino Transferase **Decarboxylation of Amino Acids** Transsulfuration • Enzymes are Cystathionine Beta Synthase and Cystathioninase. **Tryptophan Metabolism** Kynureninase **Heme Synthesis** • ALA Synthase Glycogenolysis • Glycogen phosphorylase
Folic acid	
Cobalamin	**Coenzyme Role of Methyl B_{12}** • Methionine Synthase or Homocysteine Methyl Transferase **Coenzyme role of Ado-B_{12}** • Methyl Malonyl CoA Mutase • Leucine Amino Mutase
Vitamin C	Hydroxylation of proline and Lysine in Collagen 7 Alpha hydroxylase Dopamine β Hydroxylase

Review Questions MCQ

1. A middle aged woman with fissures in tongue angular stomatitis, tingling and numbness in hands. Investigation showed reduced glutathione reductase activity in RBC. Which vitamin deficiency causes this? *(AIIMS May 2018)*
   a. Vitamin B_1
   b. Vitamin B_2
   c. Vitamin B_6
   d. Vitamin B_{12}

2. Fe absorption increased by which vitamin? *(AIIMS May 2018)*
   a. Vitamin A
   b. Vitamin C
   c. Thiamin
   d. Riboflavin

3. An alcholoic malnourished patient present to hospital with respiratory distress. His pulse rate is high, pedal oedema, hypertension and systolic murmur along with bilateral crepitation. A diagnosis of congestive high output cardiac failure is made. Which vitamin deficiency can cause this? *(Recent Question Jan 2019)*
   a. Vitamin B_1
   b. Vitamin C
   c. Vitamin B_2
   d. Vitamin B_6

4. Cause of thiamine induced nerve weakness: *(Recent Question 2016)*
   a. Hypocalcimia
   b. Hypomagnesmia
   c. Inactivation of chloride channel
   d. Difficult to produce ACh molecules

5. Fulminant Hepatitis is associated with which vitamin toxicity? *(Central Institute Exam May 2017)*
   a. Vitamin B_1
   b. Vitamin B_2
   c. Vitamin B_3
   d. Vitamin B_6

6. Vitamin deficiency causing mental disorder? *(Recent Question 2016)*
   a. Thiamine
   b. Riboflavin
   c. Niacin
   d. Biotin

7. Site of absorption of Vitamin B_{12}: *(Recent Question 2016)*
   a. Ileum
   b. Jejunum
   c. Duodenum
   d. Stomach

8. Vitamin B_{12} deficiency causes all except: *(Recent Question 2016)*
   a. Neural tube defect
   b. Peripheral neuropathy
   c. Megaloblastica naemia
   d. Demyelination

9. Isoniazid toxicity can be prevented by intake of: *(Recent Question 2016)*
   a. Vitamin B_6
   b. Vitamin B_3
   c. Vitamin B_{12}
   d. Vitamin B_1

10. Which among the following causes generalized oedema? *(Recent Question 2016)*
    a. Vitamin B_1
    b. Vitamin B_2
    c. Vitamin B_{12}
    d. Vitamin A

11. A 50-year-male with symptoms of fatigue and he has swelling of feet and loss of sensations in legs and anaemia. He also has dilatation of ventricle and high cardiac output state. What is the vitamin deficiency associated with this presentation? *(JIPMER May 2016)*
    a. Vitamin B_1
    b. Vitamin B_2
    c. Vitamin B_{12}
    d. Vitamin B_3

12. A mineral which can generate free radical are all except: *(Recent Question 2016)*
    a. Copper
    b. Cobalt
    c. Selenium
    d. Nickel

13. Cobalt is present in which vitamin? *(Recent Question 2016)*
    a. Vitamin B_1
    b. Vitamin B_2
    c. Vitamin B_3
    d. Vitamin B_{12}

14. Antioxidant in Vitamin is: *(Recent Question 2016)*
    a. Beta carotene
    b. Thiamine
    c. Niacin
    d. Riboflavin

15. The proxidant action of Vitamin C is potentiated by: *(Recent Question 2016)*
    a. Selenium
    b. Copper
    c. Calcium
    d. Iron

16. Biotin act as a coenzyme for all except: *(AIIMS Nov 2015)*
    a. Pyruvate to oxaloacetae
    b. Acetyl CoA to malonyl CoA
    c. Propionyl CoA to methyl malonyl CoA
    d. Glutamate to gamma carboxy glutamate

17. Vitamin B12 is not required for: *(AIIMS Nov 2015)*
    a. Glycogen phosphorylase
    b. Methionine synthase
    c. Methyl malonyl CoA mutase
    d. Leucine amino mutase

18. A vitamin derived from amino acid is: *(JIPMER Nov 2015)*
    a. Biotin
    b. Pantothenic acid
    c. Niacin
    d. Folic acid

19. Vitamin for which RDA is based on protein intake is: *(JIPMER Nov 2015)*
    a. Niacin
    b. Riboflavin
    c. Pyridoxine
    d. Thiamine

20. Megaloblastic anaemia seen in: *(JIPMER Nov 2015)*
    a. Ornithine transcarbamoylase defect
    b. MSUD
    c. Citrullinemia
    d. Oroticaciduria

21. In one carbon metabolism when Serine converted to Glycine, which carbon atom is added to THFA?
    (Recent Question 2016)
    a. Alpha carbon
    b. Beta carbon
    c. Delta carbon
    d. Gamma carbon

22. Vitamin deficiency that causes or ooculo genital syndrome: *(APPG 2012)*
    a. Vitamin B_2
    b. Vitamin B_{12}
    c. Zinc
    d. Vitamin B_3

23. NADPH is produced by: *(PGI May 2014)*
    a. Pyruvate dehydrogenase
    b. Isocitrate dehydrogenase
    c. Succinate dehydrogenase
    d. Malate dehydrogenase
    e. α Keto glutarate dehydrogenase

24. Vitamin deficiency causing circum corneal vascularisation is: *(AIIMS May 2014)*
    a. Biotin
    b. Riboflavin
    c. Thiamine
    d. Vitamin D

25. False about folic acid: *(AIIMS Nov 2013)*
    a. It is present in all the green leafy vegetables
    b. It is proven to decrease the occurrence of neural tube defects when taken preconceptionally
    c. Wheat flour in India is fortified with folate as in USA
    d. Methylfolate trap is because of methionine synthase defect

26. Which of the vitamin deficiency lead to lactic acidosis?
    a. Riboflavin
    b. Thiamine
    c. Niacin
    d. Panthothenic acid

27. Thiamin requirement increases in excessive intake of:
    a. Carbohydrate
    b. Amino acid
    c. Fat
    d. Lecithin

28. Which of the following statement about Thiamine true? *(AIIMS Nov 2008)*
    a. It is a coenzyme of lactate dehydrogenase
    b. Its deficiency is associated with scurvy
    c. Its coenzyme function is done by thiamine monophosphate
    d. It is coenzyme for pyruvate dehydrogenase and α-ketoglutarate dehydrogenase

29. Vitamin which is excreted in urine is? *(AIIMS Nov 2006)*
    a. Vitamin A
    b. Vitamin C
    c. Vitamin D
    d. Vitamin K

30. Which of the following cannot be synthesised in the body? *(PGI Nov 2012)*
    a. Vit K
    b. Vit C
    c. Thiamine
    d. Riboflavin
    e. Cyanocobalamin

31. Thiamine deficiency causes decreased energy production because: *(AIPGMEE 2010)*
    a. It is required for the process of transamination
    b. It is a cofactor in oxidative reduction
    c. It is a coenzyme for transketolase in pentose phosphate pathway
    d. It is a coenzyme for pyruvate dehydrogenase and alpha ketoglutarate dehydrogenase

32. Vitamin B12 acts as coenzyme to which one of the following enzymes?
    a. Isocitrate dehydrogenase
    b. Homocysteine methyltransferase
    c. Glycogen synthase
    d. Glucose-6- Phosphate dehydrogenase

33. Biotin is a cofactor of:
    a. Carboxylase
    b. Oxidase
    c. Hydrolase
    d. Decarboxylase

34. Post-translation modification of hydroxylysine and hydroxyproline is by: *(Ker 2009)*
    a. Vit C
    b. Vit K
    c. Vit E
    d. Vit D

35. Pantothenic acid containing coenzyme is involved in: *(Recent Question 2016)*
    a. Decarboxylation
    b. Dehydrogenation
    c. Acetylation
    d. Carboxylation

36. Vitamin given in pregnant women to prevent neural tube defect: *(Recent Question 2016)*
    a. Folic acid
    b. Vitamin B12
    c. Vitamin C
    d. Vitamin A

37. Not needed in TCA cycle: *(Recent Question 2016)*
    a. Pyridoxine
    b. Thiamine
    c. Riboflavin
    d. Niacin

38. Identify the vitamin deficiency: *(Recent Question 2016)*

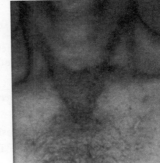

a. Riboflavin b. Ascorbic acid
c. Niacin d. Biotin

39. **Neurological worsening with anemia what is the treatment to be given?** *(Recent Question 2016)*
    a. Folic acid alone
    b. Folic acid along with hydroxycobalamin
    c. Iron
    d. Pyridoxine

40. **Vitamin deficiency causing dementia:** *(Recent Question 2016)*
    a. Biotin b. Thiamine
    c. Pyridoxine d. Vitamin B_{12}

41. **Pantothenate Kinase associated neurodegenaration is:** *(Recent Question 2016)*
    a. Wiison's disease
    b. Hallervorden-Spatz syndrome
    c. McLeod syndrome
    d. Lesch Nyhan syndrome

42. **The form of THFA used in treatment is:**
    a. N_5 Formyl THFA *(Recent Question 2016)*
    b. N_{10} Formyl THFA
    c. N_5 Formimino THFA
    d. N_5 Methyl THFA

43. **Excess of avidin causes deficiency of:** *(Recent Question 2016)*
    a. Biotin b. Choline
    c. Vitamin B_{12} d. Folate

44. **Thiamine act as a cofactor in:** *(Recent Question 2016)*
    a. Pyruvate to oxaloacetate
    b. Malonate to oxaloacetate
    c. Succinate to fumarate
    d. Pyruvate to acetyl CoA

45. **Sebhorreic dermatitis is produced by deficiency of:** *(Recent Question 2016)*
    a. Vitamin A b. Vitamin B_1
    c. Vitamin B_2 d. Vitamin C

46. **Severe thiamine deficiency is associated with:** *(Recent Question 2016)*
    a. Decreased RBC transketolase activity
    b. Increased clotting time
    c. Decreased RBC transaminase activity
    d. Increased xanthurenic acid excretion

Answers to Review Questions

1. **b. Vitamin B_2** *(Ref: Harper 31/e page 533-540)*

2. **b. Vitamin C** *(Ref: Harper 31/e page 533-540)*

   Vitamin C has Ferrireductase activity which convert Fe^{3+} to Fe^{2+}. Iron is absorbed in reduced Ferrous state from proximal duodenum.

3. **a. Vitamin B_1** *(Ref: Harper 31/e page 533-540)*

   In Alcoholic patients thiamine deficiency is possible. The given clinical features are consistent with wet beri beri hence answer is Vitamin B1.

4. **c. Inactivation of chloride channel** *(Ref: Harper 31/e page 534)*

   Thiamine triphosphate has a role in nerve conduction; it phosphorylates, and so activates, achloride channel in the nerve membrane

5. **c. Vitamin B_3** *(Ref: Harper 31/e page 533-540)*

   Toxicity of Niacin
   - Prostaglandin mediated cutaneous flushing due to binding of vitamin to a G Protein coupled receptor.
   - Gastricirritation.
   - Hepatic toxicity is the most serious toxic reaction with sustained release niacin. Presents with jaundice, elevated liver enzymes (AST and ALT) even fulminant hepatitis.
   - Other toxic reactions include glucose intolerance, hyperuricemia, macular edema and cysts.

6. **a, c. Thiamine; Niacin** *(Ref: Harper 31/e page 533-540)*
   - Thiamine deficiency can cause Confabulatory psychosis
   - Niacin deficiency can cause Depressive Psychosis
   - But Niacin is strictly not a vitamin as it can be endogenously synthesized from Tryptophan. Hence Thiamine is a better answer.

7. **a. Ileum** *(Ref: Harper 31/e page 533-540)*
   - Site of absorption of Iron—Proximal Duodenum
   - Site of absorption of Folic acid—ProximalJejunum
   - Site of absorption of Cobalamin—Ileum

8. **a. Neural tube defect** *(Ref: Harper 30/e page 560)*

   Deficiency manifestation of Vitamin B_{12}
   - Megaloblastic anemia
   - **Homocysteinemia**-Due decreased conversion of **Homocysteine to Methionine.**
   - **Methyl Malonic Aciduria**-Due to defective Methyl Malonyl CoA Mutase which leads to decreased con- version of L Methyl malonyl CoA to Succinyl CoA.
   - Subacute Combined Degeneration.

- Cobalamin deficiency may cause a bilateral peripheral neuropathy or degeneration (demyelination) of the posterior and pyramidal tracts of the spinal cord.

9. **a. Vitamin B_6**

10. **a. Vitamin B_1** (Ref: Harper 31/e page 533-540)

    Wet beriberi caused by Vitamin B_1 deficiency there is generalised oedema

11. **a. Vitamin B_1** (Ref: Nelson 20/e, Chapter 46.1 Vitamin B_1)

12. **c. Selenium** (Ref: Harper 30/e page 566)
    - Transition metal which can generate free radical are Cu^+, Co^{2+}, Ni^{2+}, Fe^{2+}
    - Selenium dependent enzyme Glutathione Peroxidase is a free radical trapping enzyme.

13. **d. Vitamin B_{12}** (Ref: Harper 30/e page 558)

    Cobalamin is a Corrinoid, i.e. Cobalt containing compound possessing the corrin ring having biological activity of the vitamin.

14. **a. Beta carotene** (Ref: Harper 30/e page 567)

    Antioxidant Vitamins are
    - Beta Carotene
    - Vitamin C
    - Vitamin E

15. **b. Copper** (Ref: Harper 31/e page 533-540)

    Ascorbate is antioxidant, but it can be a source of hydroxyl radical by reacting with Cu^{2+} ions. Antioxidant vitamins that can act as prooxidants are:
    - Beta carotene
    - Ascorbate
    - Tocopherol

16. **d. Glutamate to gamma carboxy glutamate** (Ref: Harper 31/e page 533-540)

    Biotin transfer CO_2 in:
    - Acetyl CoA Carboxylase
    - Pyruvate Carboxylase
    - Propionyl CoA carboxylase
    - Methyl Crotonyl CoA carboxylase

17. **a. Glycogen phosphorylase** (Ref: Harper 31/e page 533-540)

    Coenzyme Role of Cobalamin
    Methyl Malonyl CoA Mutase
    - L MethylMalonyl CoA → Succinyl CoA
      Methionine Synthase or Homocysteine Methyl Transferase
    - Homocysteine → Methionine
    - Leucine Amino Mutase

18. **c. Niacin** (Ref: Harper 31/e page 533-540)

    Niacin is strictly not a vitamin as it can be synthesized from Tryptophan.

19. **c. Pyridoxine** (Ref: Harper 31/e page 533-540)

    The RDA of Pyridoxine is dependent on Protein intake.

    The RDA of Thiamine is dependent of Carbohydrate intake.

20. **d. Oroticaciduria**

21. **b. Beta carbon**

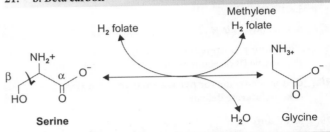

22. **a. Vitamin B_2**

23. **b, d. Isocitrate dehydrogenase, Malate dehydrogenase**

    Sources of NADPH are:
    - HMP Shunt Pathway (Major source)
    - Cytoplasmic Isocitrate Dehydrogenase
    - Malic Enzyme (NADP Malate Dehydrogenase)

24. **b. Riboflavin**

Principal Clinical Findings of Vitamin Malnutrition	
Nutrient	Clinical finding
Thiamine	Peripheral nerve damage (beriberi) or central nervous system lesions (Wernicke-Korsakoff syndrome)
Riboflavin	Magenta tongue, angular stomatitis, cheilosis, seborrheic dermatitis, circumcorneal Vascularisation
Niacin	Pellagra: pigmented rash of sun-exposed areas (photosensitive dermatitis), bright red tongue, diarrhea, apathy, memory loss, disorientation, depressive psychosis
Vitamin B_6	Seborrhea, glossitis, convulsions, neuropathy, depression, confusion, microcytic anemia
Folate	Megaloblastic anemia, atrophic glossitis, depression, homocysteine
Vitamin B_{12}	Pernicious anemia = megaloblastic anemia with degeneration of the spinal cord, loss of vibratory and position sense, abnormal gait, dementia[a], impotence, loss of bladder and bowel control, homocysteine, methyl malonic acid
Pantothenic Acid	Peripheral nerve damage (nutritional melalgia or "burning foot syndrome")
Vitamin C	Scurvy: petechiae, ecchymosis, coiled hairs, inflamed and bleeding gums, joint effusion, poor wound healing, fatigue
Vitamin A	Xerophthalmia, night blindness, Bitot's spots, follicular hyperkeratosis, impaired embryonic development, immune dysfunction
Vitamin D	Rickets: skeletal deformation, rachitic rosary, bowed legs; osteomalacia
Vitamin E	Peripheral neuropathy, spinocerebellar ataxia, skeletal muscle atrophy, retinopathy
Vitamin K	Elevated prothrombin time, bleeding

25. **c. Wheat flour in India is fortified with folate as in USA**

26. **b. Thiamine**

    Thiamine deficiency affect Pyruvate Dehydrogenase, so it causes Lactic acidosis.

27. **a. Carbohydrate**

    Thiamine requirement increases in carbohydrate intake. Pyridoxine (B_6) requirement increases in protein intake.

28. **d. It is coenzyme for pyruvate dehydrogenase and α-ketoglutarate dehydrogenase**

29. **b. Vitamin C**

    Water soluble vitamins are excreted in urine.

30. **b, c, d, e. Vit C, Thiamine, Riboflavin, Cyanocobalamin**

31. **d. It is a coenzyme for pyruvate dehydrogenase and alpha ketoglutarate dehydrogenase**
    (Ref: Harper 30/e page, Harper 30/e page 555)

32. **b. Homocysteine methyltransferase**

33. **a. Carboxylase** *(Ref: Harper 30/e page 556)*

    Coenzyme Role of Biotin
    Play a role in gene expression, fatty acid synthesis, gluconeogenesis and serve as a CO_2 carrier for Carboxylases enzymes and gene regulation by histone biotinylation.

    Coenzyme for ATP dependent Carboxylation reaction (Carbon Dioxide Fixation)
    - Pyruvate Carboxylase (Pyruvate to Oxaloacetate)
    - Propionyl CoA Carboxylase (Propionyl CoA to Methyl Malonyl CoA)
    - Acetyl CoA Carboxylase (Acetyl CoA to Malonyl CoA)
    - Methyl Crotonyl CoA Carboxylase

    Biotin independent Carboxylation reaction
    - Carbamoyl Phosphate Synthetase –I and II
    - Addition of CO_2 to C_6 in Purine ring (AIR Carboxylase)
    - Malic Enzyme (Pyruvate to Malate)

34. **a. Vit C**

    Apart from Vitamin C, Alpha Ketoglutarate is also acting as a coenzyme in Prolyl and Lysyl Hydroxylase reaction.

35. **c. Acetylation** *(Ref: Harper 30/e page 561)*

    Pantothenic acid as a part of CoA take part in
    - Fatty acid Oxidation
    - Acetylation
    - Citric acid cycle
    - Cholesterol synthesis

    As a part of ACP in fatty acid Synthesis

36. **a. Folic acid** *(Ref: Harper 30/e page 558)*

37. **a. Pyridoxine**

    Vitamins required for TCA Cycle are Riboflavin, Niacin, Thiamine, and Pantothenic acid

38. **c. Niacin**

    The given diagram is Casal's Necklace in Pellagra.

39. **b. Folic acid along with hydroxy cobalamin**

40. **d. Vitamin B_{12}**

    Vitamin deficiencies associated with dementia are Vitamin B_1, Niacin, Vitamin B_{12}.

41. **b. Hallervorden-Spatz syndrome**

    Pantothenate kinase associated neuro degeneration (PKAN) (formerly Hallervorden-Spatz syndrome)
    - Rare autosomal recessive neuro degenerative disorder
    - Chorea, dystonia, parkinsonian features, pyramidal tract features and MR
    - MRI-decreased T2 signal in the globus pallidus and substantia nigra, "eye of the tiger" sign (hyperintense area within the hypointense area);
    - Sometimes acanthocytosis
    - Neuropathologic examination indicates excessive accumulation of iron-containing pigments in the globus pallidus and substantia nigra.
    - Similar disorders are grouped as neurodegeneration with brain iron accumulation (NBIA).

42. **a. N_5 Formyl THFA** *(Ref: Harper 30/e page 559)*
    - 5 Formyl THFA is more stable than Folate, therefore used pharmaceutically.
    - Known as Folinic acid.
    - The synthetic racemic compound of Folinic acid is Leucovorin

43. **a. Biotin** *(Ref: Harper 30/e page 560)*

44. **d. Pyruvate to Acetyl CoA** *(Ref: Harper 30/e page 555)*

    Thiamine generally function in the decarboxylation reaction of alpha keto acids and branched chain amino acids
    - **Pyruvate Dehydrogenase**^{Q} which convert Pyruvate to Acetyl CoA.
    - Alpha **Ketoglutarate Dehydrogenase**^{Q} in Citric Acid Cycle which convert α Ketoglutarate to Succinyl CoA.
    - **Branched Chain Ketoacid Dehydrogenase**^{Q} which catalyses oxidative decarboxylation of Branched Chain Aminoacids.
    - **Trans Ketolase**^{Q} in **Pentose Phosphate Pathway**^{Q}. This is the biochemical basis of assay of Thiamine status of the body.

45. **c. Vitamin B_2**

Riboflavin	Magenta tongue, angular stomatitis, cheilosis, **seborrheic dermatitis**

46. **a. Decreased RBC transketolase activity**
    - Thiamine is a cofactor of Transketolase.

8.3 MINERALS

- Classifiaction
- Iron
- Copper
- Zinc
- Selenium
- Fluoride
- Chromium
- Other Important Minerals

CLASSIFICATION

- **Macrominerals (Major elements)**
  Daily requirement >100 mg
  Calcium, Magnesium, Phosphorus, Sodium, Potassium, Chloride, Sulphur
- **Micromineral (Trace elements)**
  Daily requirement <100 mg
  Iron, Copper, Chromium, Cobalt, Mangenese, Zinc, and Fluorine
- **Ultra trace elements**
  Daily requirement < 1 mg/day
  Iodine, Molybdenum, Selenium

IRON

Body Distribution of Iron^Q

Body distribution	Iron content, mg	
	Adult male	Adult female
Hemoglobin	2500	1700
Myoglobin/Enzymes	500	300
Transferrin	3	3
Iron stores	600–1000	0–300
Total body iron content	3603–4003	2003–2303

HIGH YIELD POINTS

Heme Containing^Q—Iron Containing Proteins
- Hemoglobin
- Myoglobin
- Cytochrome C
- Tryptophan pyrrolase
- Catalase
- Nitric Oxide Synthase

Nonheme–Iron containing Proteins
- Aconitase
- Transferrin
- Ferritin
- Hemosiderin

Iron-Sulphur Complex
- Complex I of ETC
- Complex II of ETC
- Complex III of ETC
- Xanthine oxidase

Proteins that has Role in Iron Metabolism

Storage Form—Ferritin and Hemosiderin

Ferritin
- The human body can typically store up to 1g of iron, the vast majority of which is bound to **ferritin**
- MW 440 kDa
- **Ferric iron + Apoferritin = Ferritin**
- Polynuclear complex of **Hydrous ferric oxide**
- Ferritin is composed of 24 identical subunits, which surround as many as **3000 to 4500 ferric atoms**.
- The subunits may be of the **H (heavy)** or the **L (light)** type.
- The H-subunit possesses ferroxidase activity, which is required for iron-loading of ferritin.
- The function of the L subunit is not clearly known but is proposed to play a role in ferritin nucleation and stability.
- Seen in **intestinal** cells, **liver, spleen** and **bone marrow**.
- Plasma ferritin levels thus are considered to be an **indicator of body iron stores.**

Hemosiderin
- A **partly degraded form of ferritin** that contains iron is Hemosiderin
- Iron is **not easily mobilised** from Hemosiderin unlike ferritin.
- It can be detected in tissues by histological stains (e.g., Prussian blue), under conditions of iron overload (**hemosiderosis**).
- Hemosiderin is an Index of **Iron Overload^Q**.

Transport form-Transferrin

Transferrin and Transferrin receptors
- Iron is transported in plasma in the Fe^{3+} form by the transport protein, **transferrin.**

- Ferric iron combines with apotransferrin to form transferrin
- Synthesized in the **Liver**
- Transferrin is a β1 globulin.
- Transferrin is a **bilobed glycoprotein** with **two iron binding sites**.
- Transferrin that carries iron exists in two forms—*monoferric* (one iron atom) or *diferric* (two iron atoms).
- The turnover (half-clearance time) of transferrin-bound iron is very rapid—typically **60–90 min.**
- Normal **1/3rd transferrin** saturated with Iron
- The iron-transferrin complex circulates in the plasma until it interacts with specific *transferrin receptors* on the surface of marrow erythroid cells.
- **Diferric transferrin** has the highest affinity for transferrin receptors.
- The greatest number of transferrin receptors (300,000 to 400,000/cell) is the **developing erythroblast.**
- **The Transferrin receptor1 (TfR1)** can be found on the surface of most cells.
- **Transferrin receptor 2 (TfR2)**, by contrast, is expressed primarily on the surface of hepatocytes and also in the crypt cells of the small intestine.
- The affinity of TfR1 for Tf-Fe is much higher than that of TfR2.
- The major role of TfR2 is sensing iron level, rather than internalizing iron.

Concept box

Reciprocal Regulation of TfR1 and Ferritin
- The rates of synthesis of TfR1 and ferritin are reciprocally linked to intracellular iron levels.
- When iron is low, TfR1 synthesis increases and that of ferritin declines.
- The opposite occurs when iron is abundant.
- Control is exerted through the binding of iron regulatory proteins (IRPs) called **iron response elements (IREs)** located in the 5' and 3' untranslated regions of mRNA.

Concept
When iron level is low, tissue demand for iron is high, increased transferrin receptors, help to internalize the available iron in the plasma. Decreased ferritin will help to mobilize the maximum iron stores to meet the demand of iron.

Extra Edge

Carbohydrate Deficient Transferrin (CDT)
- Glycosylation of transferrin is impaired in **congenital disorders of glycosylation** as well as in **chronic** alcoholism.
- The presence of **carbohydrate-deficient transferrin (CDT)**, which can be measured by isoelectric focussing (IEF).
- This is used as a biomarker of chronic alcoholism and Congenital Disorders of Glycosylation (CDGs).

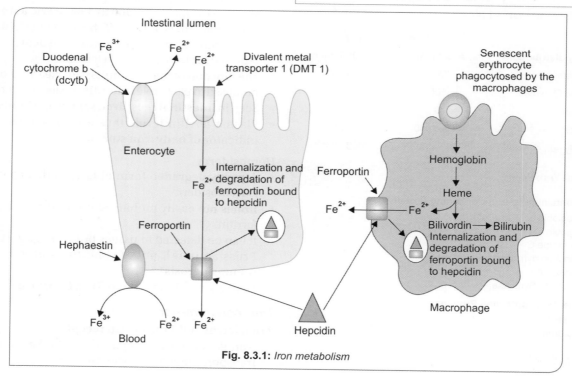

Fig. 8.3.1: *Iron metabolism*

Iron Metabolism

- Site of absorption—Enterocytes in the **proximal duodenum**.
- Heme iron is absorbed by a **hemetransporter**.
- Iron is absorbed in the **ferrous** form^Q.
- **Inorganic dietary iron** in the ferric state (Fe^{3+}) is reduced to its ferrous form (Fe^{2+}) by a brush border membrane-bound ferrireductase, **duodenal cytochrome b (Dcytb)**.
- **Vitamin C** in food also favors reduction of ferric iron to ferrous iron.
- The transfer of iron from the apical surfaces of enterocytes into their interiors is performed by a proton-coupled **divalent metal transporter (DMT1)**.
- This protein is not specific for iron, as it can transport a wide variety of divalent cations ($Co^{2+}, Zn^{2+}, Pb^{2+}, Cu^{2+}$).
- Once inside the enterocytes, iron can either be stored as **ferritin** or transferred across the basolateral membrane into the circulation by the iron exporter protein, **ferroportin** or **iron-regulated protein 1 (IREG1** or **SLC40A1)**.
- This protein may interact with the copper-containing protein **hephaestin**, a protein similar to **ceruloplasmin**.
- **Hephaestin** is thought to have a ferroxidase activity, which is important in the release of iron from cells.
- Thus, Fe^{2+} is converted back to Fe^{3+}, the form in which it is transported in the plasma by transferrin.

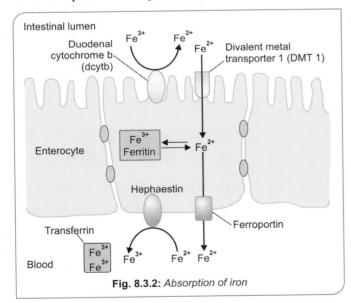

Fig. 8.3.2: *Absorption of iron*

Dietary Regulation of Iron by Mucosal Block at the Level of Enterocyte Hepcidin

- **Hepcidin** is the Chief Regulator of Systemic Iron Homeostasis
- It is a 25-amino acid peptide.
- Synthesized in the **liver** as an 84-amino acid precursor (prohepcidin).

Mechanism of Iron Regulation by Hepcidin

Function of Hepcidin

- Binds to the cellular iron exporter, ferroportin, triggering its internalization and degradation.

Function of Ferroportin

- Decreased export of iron into circulation
- Depressed iron recycling by macrophages.

Result of combined action of Hepcidin and Ferroportin

Reduction in circulating iron levels (hypoferremia)

Mechanism of Iron Regulation

- When plasma iron levels are high, hepatic synthesis of hepcidin increases, thus reducing circulating iron level.
- The opposite occurs when plasma iron levels are low.

Regulation of Expression of Hepcidin

The hepcidin level is influenced by:

- Circulatory level of iron
- Bone Morphogenic Proteins (BMPs) and Hemojuvelin
- Erythropoietic signals
- Inflammation
- Hypoxia

> **CONCEPT BOX**
>
> **Concept** is increased level of iron → increased expression of hepcidin → which in turn decreases circulating iron.

Conservation of Iron

- Extra corpuscular haemoglobin is bound by **haptoglobin**
- **Hemopexin** is a β1 globin that binds Heme.
- **Albumin** will bind some metheme (ferric heme) to form methemalbumin, which then transfers the metheme to hemopexin.
- Transferrin bind **free Iron (Fe^{3+})** in plasma.

> **EXTRA EDGE**
>
> **Haptoglobin**
>
> - Human haptoglobin exists in **three polymorphic forms**, known as Hp 1-1, Hp 2-1, and Hp 2-2.
> - Haptoglobin is an acute phase protein, and its plasma level is elevated in a variety of inflammatory states.
> - Haptoglobin scavenges hemoglobin that has escaped recycling.
>
> **Haptoglobin protects the kidneys from damage by extracorpuscular hemoglobin**
>
> - During the course of red blood cell turnover, approximately 10% of an erythrocytes hemoglobin is released into the circulation.

Contd...

Contd...

- This free, **extracorpuscular** hemoglobin is sufficiently small at = 65 kDa to pass through the glomerulus of the kidney into the tubules, where it tends to form damaging precipitates.
- **Haptoglobin** (Hp) is a plasma glycoprotein that binds extra-corpuscular hemoglobin (Hb) to form a tight noncovalent complex (Hb-Hp).
- Since the Hb-Hp complex is too large (≥155 kDa) to pass through the glomerulus, this protects the kidney from the formation of harmful precipitates and reduces the loss of the iron associated with extracorpuscular hemoglobin.

Haptoglobin Level in Hemolytic Anemia
- Patients suffering from **hemolytic anemias** exhibit low levels of haptoglobin.
- The half-life of haptoglobin is approximately 5 days.
- The Hb-Hp complex is removed rapidly by the hepatocytes (half-life 90 minutes).
- Thus, when haptoglobin is bound to hemoglobin, it is cleared from the plasma about 80 times faster than normally.
- So the level of haptoglobin falls rapidly in situations where hemoglobin is constantly being released from red blood cells, such as occurs in hemolytic anemias.

Haptoglobin-related protein and cancer
- A plasma protein that has a high degree of homology to haptoglobin
- It is elevated in some patients with cancers, although the significance of this is not understood.

Iron Deficiency Anaemia

Stages of Iron Deficiency

The progression to iron deficiency can be divided into three stages.

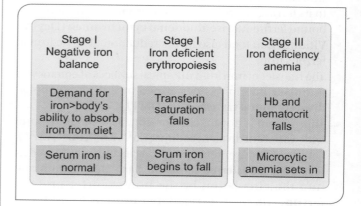

Laboratory Iron Studies in Normal and Different Stages of Evolution of Iron Deficiency

	Normal	Negative iron balance	Iron deficient erythropoiesis	Iron deficiency anemia
Iron stores Erythron iron				
Marrow iron stores	1–3 +	0–1 +	0	0
Serum ferritin (µg/L)	50–200	< 20	< 15	< 15
TIBC (µg/dL)	300–360	> 360	> 380	> 400
SI (µg/dL)	50–150	NL	< 50	< 30
Saturation (%)	30–50	NL	< 20	< 10
Marrow sideroblasts (%)	40–60	NL	< 10	< 10
RBC protoporphyrin (µg/dL)	30–50	NL	> 100	> 200
RBC morphology	NL	NL	NL	Microcytic/hypochronic

Lab parameters that increase in iron deficiency anemia
- TIBC
- RBC Protoporphyrin
- TR (TRP) (Transferrin Receptor Protein)
- RBC Distribution Width (RDW)

Diagnosing Microcytic anemia					
Tests	Iron deficiency	Inflammation	Thalassemia	Sideroblastic anemia
Peripheral Smear	Microcytic Hypochromic	Normal/Micro/Hypo	Microcytic Hypochromic with targeting	Variable
S Iron (µg/dL)	<30	<50	Normal to high	Normal to high
TIBC (µg/dL)	>360	<300	Normal	Normal
Transferrin saturation (%)	<10	10–20	30–80	30–80
Ferritin (µg/L)	<15	30–200	50–300	50–300
Hb electrophoresis pattern	Normal	Normal	Abnormal pattern in beta Thalassemia	Normal

Iron Overload Conditions

TYPE-I Hereditary Hemochromatosis (HFE related)
- Mutation in HFE gene located on Chr6p
- Tightly linked to the HLA-A locus
- Most Common Hemochromatosis (80–90%)

Non-HFE related Hereditary Hemochromatosis
- Juvenile hemochromatosis (type 2A) (hemojuvelin mutations)
- Juvenile hemochromatosis (type 2B) (hepcidin mutation)
- Mutated transferrin receptor 2 *TFR 2* (type 3)
- Mutated ferroportin 1 gene, *SLC11A 3* (type 4)

Secondary Hemochromatosis
- Anemia characterized by ineffective erythropoiesis (e.g., thalassemia major)
- Repeated blood transfusions
- Parenteral iron therapy
- Dietary iron overload (Bantusiderosis)

Miscellaneous Conditions Associated with Iron Overload
- Alcoholic liver disease
- Nonalcoholic steatohepatitis
- Hepatitis C infection

Hemochromatosis
Inherited disorder of iron metabolism that lead to iron overload, leading to deposition of iron in the parenchymal cells leading to fibrosis and organ failure.

Hemosiderosis
- **Acquired** condition
- Presence of stainable iron in tissues

High Yield Points

Hemochromatosis at a glance
The first organ to be affected in Hemochromatosis **Liver** Maximum deposition of Hemosiderin is seen in **Liver**. Least Hemosiderin deposition is seen in **Skin**.

Classical Triad of Hemochromatosis is:
- Cirrhosis with Hepatomegaly,
- Skin Pigmentation [Bronzing],
- Due to the epidermis of the skin is thin, and melanin is increased in the cells of the basal layer and dermis
- Diabetes Mellitus
  First joint to be affected in hemochromatosis — 2nd and 3rd MCP joint
  .Most common cause of death in treated patients—Hepatocellular **Carcinoma**

Role of HFE Mutations in other diseases
- Nonalcoholic Steatohepatitis
- Porphyria Cutanea Tarda

COPPER

High Yield Points

Cofactor Role of Copper
- Amine oxidases
- Ferroxidase (ceruloplasmin) (Iron metabolism, Copper Transport)
- Cytochrome-c oxidase (in Complex IV of Electron Transport Chain)
- Cytosolic Superoxide dismutase (Free Radical Scavenging enzyme)
- Tyrosinase (Melanin Synthesis)
- Component of ferroportin (Iron Metabolism)
- Lysyl Oxidase (Cross linking in Collagen)
- Dopamine β hydroxylase (Catecholamine synthesis)

High Yield Points

Copper Deficiency Anemia is a microcytic hypochromic type of Anemia.

Wilson's Disease

Autosomal recessive

Biochemical Defect

- **ATP7 B mutation**, a gene encoding for Copper transporting ATPase in the cells.
- Defective Biliary Copper Excretion from liver cells.
- Defective Copper incorporation into Apoceruloplasmin
- Copper accumulate in cells leading to copper deposits in the liver and brain.

> **HIGH YIELD POINTS**
>
> **Quick glance—Wilson's disease**
> - The most common presentation in Wilson's disease Acute or Chronic Liver Disease
> - Neuro psychiatric manifestation in Wilson's resembles Parkinson's Disease like Syndrome
> - Most Sensitive test in Wilson's disease is or gold standard investigation is liver biopsy quantitative copper assay
> - False positive is liver biopsy quantitative copper assay in obstructive liver disease.
> - The most specific Screening Test—Urinary Excretion of Copper

Diagnosis of Wilson's disease

- **99% of cases Kayser–Fleischer ring** is present, but absence of KF ring does not excludes the disease.
- Serum Ceruloplasmin (18–35 mg/dL) decreased.
- But normal in 10% of affected individuals and decreased in 20% of carriers
- 24 hour Urinary Copper >100 µg/24hour
- *3 methyl histidine excretion is reduced in urine*
- **Gold Standard investigation**^{Q} is **Liver Biopsy**^{Q} with quantitative Copper assays (>200 mg/g dry weight of Liver).

Test	Useful-ness	Normal value	Wilson's disease
Serum Ceruloplasmin	+	180–350 mg/L (18–35 mg/dL)	Low in 90%
Kayser-Fleischer ring	++	Absent	Present in >99% if neurologic or psychiatric symptoms are present. Present 30–50% in hepatic presentation and presymptomatic patients
Urine Copper (24 h)	+++	0.3–0.8 µmol (20–50 µg)	>100 µg in symptomatic patients 60-100 µg in presymptomatic
Liver Copper	++++	0.3–0.8 µmol/g (20–50 µg/g of tissue)	>3.1 µmol (>200 µg)

Treatment of Wilson's Disease

Disease status	First line	Second line
Hepatitis or Cirrhosis without decompensation	Zinc	Trientene
Hepatic decompensation		
Mild	Trientene and Zinc	Penicillamine and Zinc
Moderate	Trientene and Zinc	Hepatic Transplantation
Severe	Hepatic Transplantation	Trientene and Zinc
Initial neurologic/psychiatric	Tetrathiomolybdate and Zinc	Zinc
Maintenance/Presymptomatic/Pregnant/Paediatric	Zinc	Trientene

Method to Assess Severity of Hepatic Decompensation in Wilson's Disease

Nazer's Prognostic Index

- Serum Bilirubin
- Serum Aspartate Transferase [AST]
- Prolongation of Prothrombin Time
  - Score <7 Medical management
  - Score >9 Liver Transplantation

Menke's (Kinky or Steely) Hair Syndrome

- **Mutation in *ATP7A* gene**
- X linked recessive condition
- Defective Copper binding P-type ATPase
- Copper is not mobilised from **Intestine.**
- **Xanthine oxidase is defective as it is copper dependent**

> **EXTRA EDGE**
>
> **MEDNIK Syndrome**
> - A rare multisystem disorder of copper metabolism with features of both Wilson's and Menke's disease.
> - Caused by mutation in *AP1S1* gene, which encodes an adaptor protein necessary for intracellular trafficking of ATP7A and ATP7B.
> - MEDNIK stands for Mental retardation, Enteropathy, Deafness, Neuropathy, Iethyosis, Keratodermia

ZINC

- Zinc is an integral component of many metallo enzymes in the body;
- It is involved in the synthesis and stabilization of proteins, DNA, and RNA and plays a structural role in ribosomes and membranes.
- Zinc is necessary for the binding **of steroid hormone receptors** and several other transcription factors to DNA.
- Zinc is absolutely required **for normal spermatogenesis,** fetal growth, and embryonic development.

Zn Deficiency

- Mild chronic zinc deficiency can cause stunted growth in children, **decreased taste sensation (hypogeusia), impaired immune function**.
- Severe chronic zinc deficiency can cause hypogonadism, dwarfism, hypopigmented hair.

Acrodermatitis Enteropathica

- Rare **autosomal recessive** disorder characterized by abnormalities in zinc absorption.
- Clinical manifestations include **diarrhea, alopecia, muscle wasting, depression, irritability, and a rash involving the extremities, face, and perineum**.
- The rash is characterized by vesicular and pustular crusting with scaling and erythema.
- The diagnosis of zinc deficiency is usually made by a serum zinc level <12 mol/L (<70 g/dL).

Zn Toxicity

- Acute zinc toxicity after oral ingestion causes nausea, vomiting, and fever.
- Zinc fumes from welding may also be toxic and cause fever, respiratory distress, excessive salivation, sweating, and headache.

Selenium

- Selenium, in the form of selenocysteine, is a component of all the enzymes that contain Selenocysteine.
- Selenium is being actively studied as a chemopreventive agent against certain cancers, such as prostate cancer.

> **HIGH YIELD POINTS**
>
> **Keshan disease**
> An endemic cardiomyopathy found in children and young women residing in regions of China where dietary intake of selenium is low (<20 g/d).
>
> **Kashinbeck Disease**
> The etiology and pathogenesis of Kashin-Beck disease (KBD) remain uncertain at present.
> A deficiency of selenium and iodine is considered common in KBD-affected areas
> It is a chronic joint disease

CHROMIUM

- Chromium potentiates the action of insulin in patients with impaired glucose tolerance, by increasing insulin receptor-mediated signalling.

> **HIGH YIELD POINTS**
>
> **Chromium-6**
> - Chromium in the trivalent state is found in supplements and is largely nontoxic;
> - Chromium-6 is a product of **stainless steel welding** and is a known pulmonary carcinogen as well as a cause of liver, kidney, and CNS damage.

FLUORIDE

- An essential function for fluoride in humans has not been described, although it is useful for the maintenance of structure in teeth and bone.
- **Adult fluorosis** results in mottled and pitted defects in tooth enamel as well as brittle bone (skeletal fluorosis).

> **HIGH YIELD POINTS**
>
> **Minerals at a Glance**
> - Zinc containing protein present in the **Saliva-Gusten**^{Q}
> - Mineral stabilize **hormone insulin** is Zinc
> - Mineral that potentiates action of **Insulin-Chromium**^{Q}
> - Mineral deficiency that leads to impaired **Glucose tolerance-Chromium**^{Q}
> - Highest concentration of Zn seen in **Hippocampus and Prostatic Secretion**
> - The mineral deficiency leads to impaired Spermatogenesis—Zinc
> - Garlicky odor in breath is seen in—Selenosis (Due to Dimethyl selenide)
> - Selenium toxicity lead to Kaschinbeck Disease.
> - Low Selenium level leads to Keshan disease (Endemic Cardiomyopathy)
> - Calcium dependent Cysteine Protease are called Calpain
> - Calpain associated with Type II Diabetes Mellitus—Calpain 10
> - Normal Blood Calcium level—9–11 mg/dL
> - Total **Calcium level in the body**^{Q} is 1.5 kg

Recommended Daily Allowances (RDA) of Important Minerals

Mineral	RDA
Calcium^Q	Adult-0.5 g Children-1 g
	Pregnancy and Lactation-1.5 g
Iron^Q	Males-15–20 mg Females-20–25 mg
	Pregnancy-40–50 mg
Iodine^Q	150–200 µg
	200–250 µg
Phosphorus	500 mg
Magnesium	400 mg
Mangenese	5–6 mg
Sodium	5–10 g
Potassium	3–4 g
Copper	1.5–3 mg
Zinc^Q	8–10 mg
Selenium^Q	50–200 µg

OTHER IMPORTANT MINERALS—FUNCTIONS AND DEFICIENCY MANIFESTATION

Mineral	Function	Deficiency
Cobalt	Core component of Adenosyl cobalamin is the only known function.	Macrocytic Anemia
Chromium	Potentiate the action of Insulin is called Glucose tolerance factor	Impaired Glucose Tolerance
Fluoride	Constituent of Bone and teeth	Dental caries
Iodine	Thyroid Hormone Synthesis	Thyroid enlargement, ⁻T4, cretinism
Molybdenum	Cofactor for Xanthine Oxidase and Sulfite Oxidase, Aldehyde oxidase	Severe neurologic abnormalities, Xanthinuria
Selenium	Cofactor for Glutathione Peroxidase Deiodinase, Thioredoxin Reductase Antioxidant along with Vitamin E	**Keshan's Disease (Cardiomyopathy)**, heart failure, striated muscle degeneration
Zinc Is Redox inert as it can exist only in divalent state	Cofactor for **Carbonic Anhydrase Carboxy Peptidase** Lactate Dehydrogenase Alcohol Dehydrogenase Alkaline Phosphatase Cytosolic SOD	Growth retardation, **taste** and smell, alopecia, dermatitis, diarrhea, **immune dysfunction**, failure to thrive, **gonadal atrophy**, congenital malformation Impaired wound healing
Manganese	Cofactor for Arginase, Carboxylase, **Kinase**, Enolase, Glucosyl Transferase, Phosphoglucomutase Required for RNA Polymerase Mitochondrial SOD Ribonucleotide reductase Mn Catalase	Impaired growth and skeletal development, reproduction, lipid and carbohydrate metabolism; upper body rash
Nickel	Urease Absorbed through lungs	
Vanadium	No vanadium containing cofactors identified till date.	Plasma proteins Immunoglobulin G, Albumin & transferin bind to oxides of vanadium.

Transition Metals

Nutritionally essential transition metals are:
- Iron, Manganese, Molybdenum, Chromium, Cobalt, Copper, Nickel & Zinc

Check List for Revision
- Must learn topic is Iron and copper
- This is a highly volatile topic hence multiple revision is a must.
- Learn all new updates and bold letters.

Review Questions MCQ

1. **Which enzyme defect is seen in Menke's disease?**
   (Recent Question Jan 2019)
   a. Lysyl hydroxylase   b. Lysyl oxidase
   c. Prolyl hydroxylase   d. Prolyl oxidase

2. **Copper containing enzyme is:**
   (Central Institute Nov 2018)
   a. Lysyl oxidase   b. Lysyl hydroxylase
   c. Prolyl hydroxylase   d. Prolyl oxidase

3. **RDI of Iodine (µg/day) for pregnant woman:**
   (Recent Question Nov 2017)
   a. 150   b. 220
   c. 290   d. 120

4. **All are involved in iron metabolism except:**
   (AIIMS Nov 2017)
   a. Hepcidin   b. Ferroportin
   c. Transthyretin   d. Ceruloplasmin

5. **A child with kinky hair, intellectual disability, seizures. What is the diagnosis?** *(AIIMS Nov 2017)*
   a. Down's syndrome   b. Menke's disease
   c. Iron deficiency anemia   d. Lesch Nyhan syndrome

6. Identify the deficiency disorder:
   (Recent Question 2016)

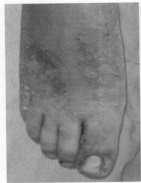
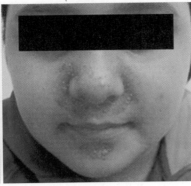

   a. Zinc deficiency
   b. Calcium deficiency
   c. Iron deficiency
   d. Copper deficiency

7. Iron absorption is inhibited by all except:
   (Recent Question 2016)
   a. Vitamin C
   b. Phytates
   c. Caffeine
   d. Milk

8. Minky Kinky hair disease is due to defect in:
   (Recent Question 2016)
   a. Cu transporter
   b. Fe transporter
   c. Zn transporter
   d. Mg transporter

9. Which of the following is wrongly matched?
   (JIPMER Nov 2015)
   a. Folate—Anemia
   b. Zinc—Immunodeficiency
   c. Iodine –Dry Skin
   d. Iron—Anemia

10. Which of the following is a non-essential metal/mineral?
    a. Sodium
    b. Manganese
    c. Iron
    d. Lead

11. The 40 nm gap in between the tropocollagen molecule in collagen which serve as the site of bone formation is occupied by:
    (AIIMS Nov 06)
    a. Carbohydrates
    b. Ligand moiety
    c. Calcium
    d. Ferric ion

12. Zinc is a cofactor for:
    (AIIMS Nov 2009)
    a. Pyruvate dehydrogenase
    b. Pyruvate decarboxylase
    c. Alpha ketoglutarate dehydrogenase
    d. Alcohol dehydrogenase

13. Cardiomyopathy is due to deficiency of:
    (PGI Nov 2012)
    a. Selenium
    b. Phosphorus
    c. Boron
    d. Zinc
    e. Iron

14. Selenium deficiency causes:
    (PGI May 2012)
    a. Dermatitis
    b. Cardiomyopathy
    c. Diarrhea
    d. Alopecia
    e. Gonadal atrophy

15. Copper containing enzymes are:
    (PGI May 2012)
    a. Superoxide dismutase
    b. Cytochrome oxidase
    c. Myeloperoxidase
    d. Tyrosinase
    e. Amino acid oxidase

16. Which of the following is considered the active from of calcium?
    a. Ionized calcium
    b. Albumin bound calcium
    c. Phosphate bound calcium
    d. Protein bound calcium

17. Copper involves collagen synthesis by:
    (Ker 2010)
    a. Lysyl oxidase
    b. Lysyl hydroxylase
    c. Cytochrome oxidase
    d. Tyrosinase

18. Zinc is present in:
    (Recent Question 2016)
    a. Carbonic anhydrase
    b. Xanthine oxidase
    c. Glutathione reductase
    d. Glutathione synthetase

19. Selenium is a cofactor in the following enzyme:
    (Recent Question 2016)
    a. Glutathione peroxidase
    b. Cytochrome oxidase
    c. Cytochrome reductase
    d. Xanthine oxidase

Answers to Review Questions

1. **b. Lysyl oxidase** (Ref: Harper 31/e page 595)

   Menke's disease affect collagen crosslinking of Collagen & Elastin

   Collagen disorders due to enzyme defect
   - Lysyl oxidase—Menke's disease
   - Lysyl hydroxylase—EDS Kyphoscoliotic, Scurvy

2. **a. Lysyl oxidase** (Ref: Harper 31/e page 595, 596)
   - Lysyl hydroxylase and Prolyl hydroxylase are alpha ketoglutarate containing iron linked hydroxylases that require Vit C also.

3. **b. 220**

   RDI of Iodine for pregnant women is 220 µg, lactating mothers is 290 µg.

4. **c. Transthyretin**

Hepcidin is the regulator of iron homeostasis. Ferroprotin is their onexporter protein into circulation from intestinal cells. Ceruloplasmin has ferroxidase which convert Fe^{2+} to Fe^{3+} so it can serve the purpose of hephaestin in iron metabolism. Transthyretin is a transport protein for Thyroxine and Retinol.

5. **b, Menke's Disease**

   Menke's (Kinky or Steely) Hair Syndrome
   - Mutation in ATP7A gene
   - X-linked recessive condition
   - Defective Copper binding P type ATPase
   - Copper is not mobilised from **Intestine**.
   - Deficiency of Xanthine oxidase.

6. **a. Zinc deficiency Acrodermatitis enteropathica**
   - Rare autosomal recessive disorder characterized by abnormalities in zinc absorption.
   - Clinical manifestations include diarrhea, alopecia, muscle wasting, depression, irritability, and a rash involving the extremities, face, and perineum.
   - The rash is characterized by vesicular and pustular crusting with scaling and erythema.
   - The diagnosis of zinc deficiency is usually made by a serum zinc level <12 mol/L (<70 g/dL).

7. **a. Vitamin C** (Ref: Harper 31/e page 533-540)

   Iron absorption is enhanced by:
   - Vitamin C, Fructose, Alcohol iron absorption is inhibited by
   - Phytates, Oxalates, Caffeine, Calcium

8. **a. Cu transporter**

   Menke's (kinky or steely) hair Syndrome
   - Mutation in *ATP7A* gene
   - X-linked recessive condition
   - Defective Copper binding P-typeATPase
   - Copper is not mobilised from **Intestine**

9. **c. Iodine—Dry skin**

Mineral	Function	Deficiency
Cobalt	Constituent of Vitamin B_{12}	Macrocytic anemia
Chromium	Potentiate the action of insulin	Impaired glucose tolerance
Fluoride	Constituent of bone and teeth	Dental caries
Iodine	Thyroid hormone synthesis	Thyroid enlargement, ¯T4, cretinism
Molybdenum	Cofactor for Xanthine Oxidase and Sulfite Oxidase, Aldehyde oxidase	Severe neurologic abnormalities, Xanthinuria
Selenium	Cofactor for Glutathione Peroxidase Deiodinase, Thioredoxin Reductase Antioxidant along with Vitamin E	**Keshan's Disease (Cardiomyopathy),** heart failure, striated muscle degeneration
Zinc	Cofactor for **Carbonic Anhydrase Carboxy Peptidase** Lactate Dehydrogenase; Alcohol Dehydrogenase; Alkaline Phosphatase	Growth retardation, ¯**taste** and smell, alopecia, dermatitis, diarrhea, **immune dysfunction**, failure to thrive, **gonadal atrophy**, congenital malformation Impaired wound healing
Manganese	Cofactor for Arginase, Carboxylase, **Kinase**, Enolase, Glucosyl Transferase, Phosphoglucomutase Required for RNA Polymerase	Impaired growth and skeletal development, reproduction, lipid and carbohydrate metabolism; upper body rash

10. **d. Lead**

    Lead is a toxic mineral.

11. **c. Calcium**

12. **d. Alcohol dehydrogenase**

Zinc	Cofactor for **Carbonic Anhydrase Carboxy Peptidase** Lactate Dehydrogenase; Alcohol Dehydrogenase **Alkaline Phosphatase**

13. **a. Selenium**

    Nutritional Causes of Cardiomyopathy are:
    Deficiency of thiamine, selenium, calcium and magnesium.
    Excess of iron (Hemochromatosis)

14. **b. Cardiomyopathy**

15. **a, b, d, e.** Superoxide dismutase, Cytochrome oxidase, Tyrosinase, Amino acid oxidase

16. **a. Ionised calcium**

17. **a. Lysyl oxidase**

Cofactor Role of Copper
- Amine oxidases
- Ferroxidase (ceruloplasmin) (Iron metabolism, Copper Transport)
- Cytochrome-c oxidase (in Complex IV of Electron Transport Chain)
- Superoxide dismutase (Free Radical Scavenging enzyme)
- Tyrosinase (Melanin Synthesis)
- Component of ferroportin (Iron Metabolism)
- Lysyl Oxidase (Cross linking in Collagen).

18. a. Carbonic anhydrase
19. a. Glutathione peroxidase   (Ref: Harper 31/e page 534)

Selenium	Cofactors for: • Glutathione peroxidase • Deiodinase, • Thioredoxin reductase • Antioxidant along with vitamin E

8.4 BASICS OF NUTRITION

☞ Energy Content of Food   ☞ Respiratory Quotient

ENERGY CONTENT OF FOOD

The energy content of food is calculated from the heat released by the total combustion of food in a calorimeter. It is expressed in Kcal (Kilocalories) or (Cal) Calories.

Average energy available from macronutrients is as follows

- Carbohydrate—4 kcal/g
- Protein—4 kcal/g
- Fat—9 kcal/g
- Alcohol—7 kcal/g

In 24 hours total kilocalories expended by these processes is called total energy expenditure (TEE).

The energy released from by metabolism of macronutrients are used for:

1. **Resting Metabolic Rate (RMR)**—Energy expenditure by an individual in resting post absorptive state. It reperesents energy for respiration, blood flow, ion transport and other body functions.
   It is determined by measuring O_2 consumed or CO_2 produced. It is around 10% of TEE. Basal metabolic rate is 10% higher than RMR. BMR is calculated in a stringent environmental condition.
2. Physical activity provides the greatest variation in TEE.
3. **Thermic effect of food.** The production of heat by the body during digestion and absorption of food is called thermic effect of food. Its 5-10% of the TEE. This is also called **Specific dynamic action (SDA)**. It depends on the food types in the diet.
   - **Carbohydrates—5-10% of energy consumed**
   - **Proteins—20-35% of energy is consumed**
   - **Fats—0-5% of energy is consumed**
   - **So in descending order Proteins> Carbohydrates> Fats.**

RESPIRATORY QUOTIENT[Q]

Measurement of the ratio of the volume of carbon dioxide produced: volume of oxygen consumed in the oxidation of metabolic fuels in unit time. It gives us the information regarding substrate being used by an organism for respiration. It is a ratio hence no units are there.

For carbohydrates by aerobic oxidation

$$C_6H_{12}O_6 + 6 O_2 \longrightarrow 6 CO_2 + 6 H_2O$$

So for carbohydrates RQ=$6CO_2/6O_2$=1.00

But for **anaerobic oxidation RQ is infinity** as oxygen consumed is zero.

For exclusive carbohydrate meal RQ is 1.

For excess carbohydrate meal RQ is 1 or even less than 1 as excess carbohydrate is used for glycogen synthesis initially

If excess carbohydrates is used for fat synthesis then **RQ is more than 1**

For fatty acid oxidation

$C_{16}H_{32}O_2 + 23 O_2 \longrightarrow >16CO_2 + 16 H_2O + ENERGY$

So for fatty acid RQ=$16CO_2/23 O_2$ =0.7

Metabolic fuel	Energy Yield (kJ/g)	RQ (CO_2 Produced/ O_2 Consumed)	Energy (kJ)/L O_2
Carbohydrate	16	1.00	20
Protein	17	0.81	20
Fat	37	0.71	20
Alcohol	29	0.66	20
Mixed diet		0.80 -0.85	

- The organ with maximum respiratory quotient is organs which exclusively use glucose as metabolic fuel RQ is 1 e.g.: Brain
- In RBC even though depending solely on glucose RQ is infinity as oxygen consumed is zero.

Check List for Revision

- Must learn topic is Respiratory quotient and Specific dynamic action.

Review Questions MCQ

1. Respiratory quotient after exclusive carbohydrate meal is: *(AIIMS Nov 2016)*
   a. 1
   b. 1.2
   c. 0.8
   d. 0.7

2. Specific dynamic action of which of the following nutrient is maximum?
   a. Fats
   b. Carbohydrates
   c. Alcohol
   d. Proteins

3. Respiratory quotient after heavy carbohydrate meal is: *(AIIMS Nov 2018)*
   a. 1
   b. 1.2
   c. 0.8
   d. 0.7

4. Respiratory quotient of mixed diet is:
   a. 1
   b. 1.2
   c. 0.85
   d. 0.7

5. Which of the following is having maximum thermic effect food? *(AIIMS May 2017)*
   a. Fat
   b. Protein
   c. Carbohydrate
   d. Does not depend on type of food/macronutrient content

Answers to Review Questions

1. **a. 1**

   Respiratory Quotient

   Measurement of the ratio of the volume of carbon dioxide produced: volume of oxygen consumed.

   (Respiratory Quotient, RQ) *is an indication of the mixture of metabolic fuels being oxidized.*

Metabolic fuel	Energy Yield (kJ/g)	RQ (CO_2 Produced/ O_2 Consumed)	Energy (kJ)/L O_2
Carbohydrate	16	1.00	20
Protein	17	0.81	20
Fat	37	0.71	20
Alcohol	29	0.66	20

2. **d. Proteins**

3. **a. 1** (Ref. https://academic.oup.com/nutritionreviews/article/22/4/105/1916688, Harper 31/e page 104)
   - Carbohydrate taken in diet is primarily used for oxidation i.e. Glycolysis then to to TCA and ETC for which RQ is 1. If excess is used mainly for glycogen metabolism, then only fat conversion can be considered.
   - Harper if carbohydrate in excess liver's requirement will go for glycogen synthesis. for that RQ be 1 itself.

4. **c. 0.85**

5. **b. Protein**
   - Specific dynamic action (thermic effect) is energy expenditure above resting metabolic energy. This includes energy expenditure for digestion and absorption of food, etc.
   - Maximum is for protein (20%) then fat or lipid or carbohydrates. Hence, for weight loss dieticians advise high protein diet.

CHAPTER 9

Special Topics

Chapter Outline

9.1 Metabolism of Alcohol
9.2 Free Radicals
9.3 Xenobiotics
9.4 Biomembranes and Cell Organelle

CHAPTER 9

Special Topics

9.1 METABOLISM OF ALCOHOL

- Metabolic Pathways of Alcohol
- Acute Effects of Alcohol
- Chronic Effects of Alcohol

METABOLIC PATHWAYS OF ALCOHOL

Major site of absorption of alcohol is small intestine.
- The energy released by alcohol is 7 kcal/g.

I. Ethanol is converted to Acetaldehyde by three routes:
  1. **Cytoplasmic NAD+ dependent Alcohol dehydrogenase**

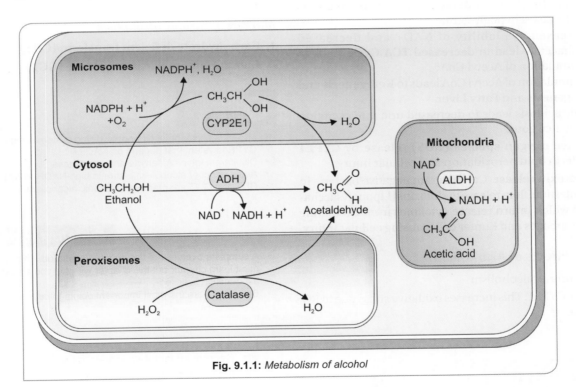

Fig. 9.1.1: *Metabolism of alcohol*

  2. **Microsomal NADP+ dependent CYP2E1** a type of CytP450 called Microsomal ethanol Oxidising system (MEOS)
  3. Minor route is **Peroxisome by Catalase**

II. The Acetaldehyde is converted to Acetate by
  - Mitochondrial acetaldehyde dehydrogenase

Concept Box

1. Why metabolic tolerance to alcohol intoxication in some persons?
Protective Mechanism of Microsomal Ethanol Oxidizing System (MEOS)
- Some metabolism of ethanol takes place via a cytochrome P450-dependent microsomal ethanol oxidizing system (MEOS), CYP2E1 involving **NADPH and O_2**.

Contd...

Contd...

- So NADH/NAD+ ratio is not altered hence account for metabolic tolerance in chronic alcoholics.
- This system is inducible hence increases in activity in **chronic alcoholism**.

2. In some Asian populations and Native Americans, alcohol consumption results in increased adverse reactions to acetaldehyde. Why?
This is due to a genetic defect of mitochondrial aldehyde dehydrogenase in 50% of Asians. Such persons experience tachycardia, flushing & hyperventilation.

Acute Effects of Alcoholism

1. Metabolic Changes following Ingestion of Alcohol
High concentration of NADH leads to high NADH/NAD+ ratio. This is the basic cause of all metabolic alteration in alcoholism.

- Favour conversion of pyruvate to lactate (this leads to **Lactic acidosis**)
- Deficiency of pyruvate leads to deficiency of oxaloacetate. This leads to **decrease in Gluconeogenesis** which causes hypoglycemia.
- Decreased availability of NAD+ and decreased Oxaloacetate lead to **decreased TCA Cycle** leads to accumulation of Acetyl CoA.
- Accumulation of Acetyl CoA leads to **Ketogenesis and Lipogenesis** and **Fatty Liver.**
- Lactic acidosis leads to decreased uric acid excretion and hence **gout**.

2. Rective oxygen species (ROS) release by **CYPE1** results in lipid peroxidation and cellular injury.

3. Endotoxin release - Certain gram negative bacteria in the intestinal flora release endotoxins (Lipopolysaccharide) which inturn release cytokines from circulating macrophages and Kupfer cells causing cellular injury.

Chronic Effects of Alcoholism

Cause: Chronic alcoholism

1. Induce CYPE1: This increases oxidative stress due to ROS release.

2. Accumulation of acetaldehyde that adduct glutathione and other amino acids

Liver is the major site of injury
- Alcoholic hepatitis, Fibrosis and Cirrhosis with portal hypertension

GI Tract:
- Gastric bleeding and esophageal varices Neurological deficits This is due to thiamine deficiency

Cardiovascular effects:
- Moderate amount of alcohol raises the HDL and inhibit platelet aggregation, hence cardioprotective.
- But heavy ingestion has opposite effects.
- Myocardial injury causes dilated congestive cardiomyopathy (Alcoholic cardiomyopathy)

Effect on Fetus-Fetal alcohol syndrome
- Microcephaly, growth retardation and facial abnormalities.

Markers of chronic alcoholism

1. **Carbohydrate deficient transferin (CDT)**
2. **Gamma glutamyl transpeptidase (GGT)**

Quick Review

- MEOS is Microsomal ethanol oxidising system is inducible CYPE1 system that depend on NADP+
- Disulfuram inhibit Aldehyde dehydrogenase
- Acute effect of alcoholism is due to high NADH to NAD+ ratio
- CDT and GGT are markers of chronic alcoholism.

Check List for Revision

- A complete overview of this chapter is needed. This is not a must learn chapter but this chapter will give you an extra edge if a question is asked.
- Bold letters are the most important points to be learnt.

Review Questions MCQ

1. The energy generated from alcohol is:
   a. 4 kcal/g
   b. 0 kcal/g
   c. 7 kcal/g
   d. 9 kcal/g

2. Toxicity of ethanol is due to: *(JIPMER 2012)*
   a. Increased NADH/NAD+ ratio
   b. Decreased lactate/pyruvate ratio
   c. Inhibition of gluconeogenesis
   d. Stimulation of fatty acid oxidation

3. Best explained pathogenes is of fatty liver in alcoholic liver disease: *(JIPMER 2013)*
   a. Increased hydrolysis of fat from adipocytes
   b. Decreased synthesis of fatty acids
   c. Decreased [NADH]/[NAD+] ratio
   d. Impaired beta oxidation of fatty acids

Answers to Review Questions

1. c. 7 kcal/g

Average energy available from macronutrients is as follows:
- Carbohydrate—4 kcal/g
- Protein—4 kcal/g
- Fat—9 kcal/g
- Alcohol—7 kcal/g

2. a. Increased NADH/NAD⁺ ratio

High concentration of NADH leads to high NADH/NAD+ ratio. This is the basic cause of all metabolic alteration in alcoholism.

3. d. Impaired beta oxidation of fatty acids

Impaired beta oxidation, leads to increased TAG in the liver, which leads to fatty liver.

9.2 FREE RADICALS

☞ Generation of Oxygen Free Radicals (OFR) or Reactive Oxygen Species (ROS)

☞ Free Radical Scavenging System

DEFINITION

Free radical is molecule or molecular fragment that contains one or more unpaired electrons in its outer orbit and has an independent existence.

A free radical is designated by a superscript dot (R*).

GENERATION OF OXYGEN FREE RADICALS (OFR) OR REACTIVE OXYGEN SPECIES (ROS)

Incomplete Reduction of Oxygen

In the body oxidative reactions normally ensures that molecular oxygen is completely reduced to water. Normally, four electrons are transferred to molecular oxygen so that it is completely reduced to form a watermolecule.

$$4e^- + O_2 \xrightarrow{4H^+} 2H_2O$$

Incomplete reduction of oxygen generates **Oxygen free radicals or Reactive Oxygen Species**. They are Superoxide, Hydrogen Peroxide and Hydroxy radical.

1. **Superoxide Radicals** are produced when a single electron is transferred to oxygen. It is both an anion and free radical.

$$O_2 + e^- \longrightarrow O_2^- \text{ (Superoxide)}$$

2. **Hydrogen Peroxide:** The two electron reduction product of oxygen is hydrogen peroxide (H_2O_2)

$$O_2^- + 2e^- \xrightarrow{2H^+} H_2O_2 \text{ (Hydrogen peroxide)}$$

This reaction is called **Dismutation reaction**. This can occur spontaneously or as enzyme catalysed reaction.

 High Yield Points

Hydrogen Peroxide is not a free radical as it does not have an unpaired electron, but it is a Reactive Oxygen Species.

3. **Hydroxy radical:** *The three electron reduction product of Oxygen is* **Hydroxy radical (OH•)**

This is the most powerful oxygen free radical.

Extra Edge

Free Radicals generated by Iron
The electron is transferred from a Ferrous ion to Hydrogen Peroxide. This reaction is called **Fenton reaction.**

$$H_2O_2 + Fe^{2-} \longrightarrow OH_2^- + OH^• + Fe^{3+}$$

Another iron catalysed reaction for generation of hydroxyl radical is **Haber Weiss Reaction.**

$$O_2^- + H_2O_2 \xrightarrow{Fe^{2+}} O_2 + OH^- + OH^-$$

Sequential univalent reduction of oxygen

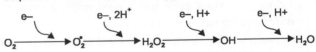

Fig. 9.2.1: *Reactive oxygen species (ROS)*

High Yield Points

- Most powerful oxygen free radical is Hydroxy radical (OH•)Q
- Least powerful Reactive oxygen species is Hydrogen peroxide (H_2O_2).
- Precursor of all reactive oxygen species is Superoxide radical (•O_2^-)
- H_2O_2 is not a free radical but a ROS

Common Sources of Free Radicals in the Body

I. Electron leakage in mitochondrial Electron Transport Chain.

II. Normal Oxidation reduction reactions in the body
   1. Xanthine Oxidase, Aldehyde Oxidase, Dihydroorotate Dehydrogenase

2. Flavin Coenzymes in Peroxisomes generate H_2O_2.
   L- Amino Acid Oxidase (Coenzyme–FMN)
   D-Amino Acid Oxidase (Coenzyme–FAD)
III. Respiratory burst

Respiratory Burst

NADPH Oxidase in the inflammatory cells (**neutrophils, eosinophils, monocytes and macrophages**) produces superoxide anion by a process of respiratory burst during phagocytosis.

Steps of Respiratory Burst
- The enzyme NADPH Oxidase catalyses the formation of Superoxide radical from **oxygen and NADPH**.
- By Dismutation Hydrogen Peroxide is generated.
- Hydrogen Peroxide generate Hydroxyradical spontaneously
- Hydrogen peroxide generate Hypochlorous acid by the action of enzyme Myeloperoxidase **exclusively present in the neutrophil granules.**
- They mediate killing of bacteria. This is called respiratory burst.

Damages Produced by Free Radicals

- Free radicals are highly reactive molecular species with an unpaired electron.
- They persist for only a very short time (of the order of **10–9 to 10–12 sec**) before they collide with another molecule and either abstract or donate an electron in order to achieve stability.
- In doing so, they generate a new radical from the molecule with which they collided.
- The main way in which a free radical can be quenched, so terminating this chain reaction, is if two radicals react together, when the unpaired electron scan become paired in one or other of the parent molecules.
- They cause damage to **nucleic acids, proteins, and lipids in cell membranes and plasma lipoproteins.**
- This can cause **cancer, atherosclerosis and coronary artery disease, and autoimmune diseases.**

Lipid Peroxidation

Lipids are most susceptible to damaging effects of free radicals. PUFA present in cell membrane and plasma lipoproteins are especially prone to damage.

Radical damage to unsaturated fatty acids in cell membranes and plasma lipoproteins leads to the formation of lipid peroxides, then highly reactive dialdehydes^Q that can chemically modify proteins and nucleic acid bases.

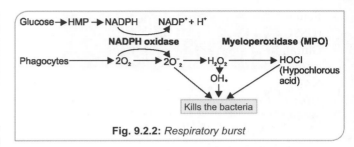

Fig. 9.2.2: *Respiratory burst*

Transition Metals that Generate Free Radicals

Transition metals can react nonenzymically with oxygen or hydrogenperoxide, leading to the formation of hydroxyl radical. They are **Cu^+, Co^{2+}, Ni^{2+}, Fe^{2+}**

Nitric Oxide (Endothelium Derived Relaxation Factor)

- It is a free radical.
- It can also yield other free radicals peroxynitrite, which decays to form hydroxyl radical.

Measurement of Body Free radical Burden

> **EXTRA EDGE**
>
> **The total body radical burden can be estimated by measuring the products of lipid per oxidation.**
> 1. **FOX (Ferrous Oxidation in Xylenol) Assay**
>   Lipid peroxides can be measured by the ferrous oxidation in xylenolorange (FOX) assay. Under acidic conditions, they oxidize Fe^{2+} to Fe^{3+}, which forms a chromophore with xylenolorange.
> 2. **Estimation of Dialdehydes**
>   The dialdehydes formed from lipid per oxide scan be measured by reaction with thio barbituric acid, when they form a red fluorescent adduct. The results of this as say are generally reported as total **t**hio **b**arbituric **a**cid **r**eactive **s**ubstances, **TBARS**.
> 3. **Measurement of Pentane and Methane in Exhaled air.**
>   Peroxidation of ω-6 polyunsaturated fatty acids leads to the formation of pentane, and of ω-3 polyunsaturated fatty acids to ethane, both of which can be measured in exhaled air.

FREE RADICAL SCAVENGING SYSTEM

Antioxidants fall into two classes:

1. **Preventive antioxidants**, which reduce the rate of chain initiation
   They are **Glutathione Peroxidase, Catalase.**
2. **Chain-breaking antioxidants**, which interfere with chain propagation
   They are superoxide dismutase, uric acid, vitamin E.

Preventive Antioxidants

Glutathione Peroxidase
- Eliminates hydrogen peroxide and organic hydroperoxides by reaction with reduced glutathione (GSH).

- Reduced glutathione is converted to oxidized glutathione (GSSG).
- Glutatione reductase convert oxidised glutathione back to reduced glutathione using the reducing equivalent NADPH.
- **Glutathione peroxidase is a selenium containing enzyme.**

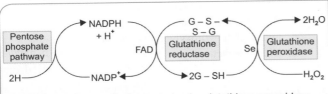

Fig. 9.2.3: *Free radical scavenging by glutathione peroxidase*

Catalase

It is a hemoprotein with four heme groups. Causes decomposition of peroxides to yield water and oxygen. **Highest concentration of catalase is present in Peroxisomes.**

$$2H_2O_2 \xrightarrow{\text{Catalase}} 2H_2O + O_2$$

H_2O_2 is not free radical but can generate free radical.

Chain Breaking Antioxidant

Superoxide Dismutase

It is the only enzyme that takes a free radical a sits substrate, hence a scavenger. Two isoenzymes are there for Super Oxide Dismutase (SOD).

Cytosolic SOD is copper dependent and mitochondrial SOD is manganese dependent^Q.

$$O_2^- + O_2^- + 2H + \xrightarrow{\text{Superoxide dismutase}} H_2O_2 + 2H +$$

Other Non-Enzymic Antioxidants

Vitamin E (α-Tocopherol)
- Most potent chain breaking **natural antioxidant**.
- Being lipophilic it acts on biological membranes.
- It terminates **lipid per oxidation.**

Other Antioxidants are:

β Carotene and Ubiquinone are **lipid soluble radical-trapping antioxidants** in membranes and plasma lipoproteins.

Ascorbate, uric acid and a variety of **polyphenols** derived from plant foods act as **water-soluble radical trapping** antioxidants, forming relatively stable radicals that persist long enough to undergo reaction to nonradical products.

Antioxidants can be Pro-oxidants Called as Antioxidant Paradox

- **Ascorbate**, can also be a source of superoxide radicals by reaction with oxygen, and hydroxyl radicals by reaction with **Cu^{2+} ions**.
- β-carotene is indeed a radical-trapping antioxidant

Under conditions of low partial pressure of oxygen, as in most tissues, a thigh partial pressures of oxygen (as in the lungs) and especially in high concentrations, β-carotene is an autocatalytic pro-oxidant.

— Increased mortality among those taking supplements of **Vitamin E.**

Quick Review

- Most powerful oxygen free radical is Hydroxy radical (OH•)^Q
- Respiratory Burst via NADPH Oxidase in neutrophils, eosinophils, monocytes and macrophages) produces superoxide anion
- The total body radical burden can be estimated by FOX (Ferrous Oxidation in Xylenol) Assay
- Preventive antioxidants are Glutathione Peroxidase, Catalase.
- Chain-breaking antioxidants, are superoxide dismutase, uric acid, vitamin E.
- Glutathione peroxidase is a selenium containing enzyme.
- Ascorbate, uric acid are water-soluble radical trapping antioxidants
- Vitamin E (α-Tocopherol) β Carotene and Ubiquinone are lipid soluble radical- trapping antioxidants
- Most potent chain breaking natural antioxidant.
- Antioxidants can be pro-oxidants called as Antioxidant Paradox

Check List for Revision

- An overall understanding is enough.
- Do learn bold letters, boxes and quick review in the last.

Review Questions MCQ

1. **Assay for lipid peroxidationis:** *(Recent Question 2016)*
   a. MTT assay
   b. Ame's test
   c. Guthrie test
   d. FOX assay

2. **Pro-oxidant action of vitamin A is potentiated by:** *(Recent Question 2016)*
   a. Copper
   b. Selenium
   c. Iron
   d. Cobalt

3. **Free radical with highest activity:**
   a. O_2^-
   b. OH•
   c. Hypochlorite
   d. Peroxynitrite

4. **Most powerful chain breaking antioxidant:**
   a. Glutathione peroxidase
   b. Alpha tocopherol
   c. Superoxide dismutase
   d. Vitamin C

5. **Enzyme which catalyse the reaction H_2O_2 give $H_2O + O_2$:**
   a. Catalase
   b. Glutathione reductase
   c. Glutathione peroxidase
   d. Glutathione s-transferase

6. **Which of the following is not a free radical?**
   a. Hydroxyl radical
   b. Hydrogen peroxide
   c. Superoxide
   d. $O_2^{\bullet}$

7. **Cytochrome p450 is involved in:** *(PGI Nov 2016)*
   a. Hydroxylation of xenobiotics
   b. Methylation of xenobiotics
   c. Deamination reaction
   d. Involved in hydroxylation of steroids
   e. Drug interaction

Answers to Review Questions

1. **d. FOX assay**

   Measurement of Body Free Radical Burden

   The total body radical burden can be estimated by measuring the products of lipid peroxidation:
   - FOX (Ferrous Oxidation in Xylenol) Assay
     Lipid peroxides can be measured by the ferrous oxidation in xylenolorange (FOX) assay. Under acidic conditions, they oxidize Fe^{2+} to Fe^{3+}, which forms a chromophore with xylenolorange.
   - Estimation of Dialdehydes
     The dialdehydes formed from lipid peroxides can be measured by reaction with thiobarbituric acid, when they form a red fluorescent adduct. The results of this assay are generally reported as total **t**hio**b**arbituric **a**cid **r**eactive substances, **TBARS**.
   - Measurement of Pentane and Methane in Exhaled Air.
     Peroxidation of ω-6 poly unsaturated fatty acids leads to the formation of pentane, and of ω-3 polyunsaturated fatty acids to ethane, both of which can be measured in exhaled air.

2. **a. Copper**

3. **b. OH•**

   Most powerful oxygen free radical is **Hydroxyradical (OH•)^Q**
   Least powerful reactive oxygen species is **Hydrogen peroxide (H_2O_2)**·
   Precursor of all reactive oxygen species is **Superoxide radical (O_2^-)**

4. **b. Alpha tocopherol**

5. **a. Catalase**

   Is a **hemoprotein** *with four hemegroups. Causes decomposition of peroxides to yield water and oxygen. Highest concentration of Catalase is present in* **Peroxisomes.**

   $$2H_2O_2 \xrightarrow{\text{Catalase}} 2H_2O + O_2$$

   H_2O_2 is not free radical but can generate free radical.

6. **b. Hydrogen peroxide**

   It is a reactive oxygen species, not a free radical.

7. **a, d, e. Hydroxylation of xenobiotics, involved in hydroxylation of steroids, drug interaction** *(Ref: Harper 30/e)*

9.3 XENOBIOTICS

- Definition
- Phase I Reactions
- Phase II Reactions

DEFINITION

A xenobiotic (Gk xenos "stranger") is a compound that is foreign to the body.

Metabolism of Xenobiotics in two phases
- Phase I reactions
- Phase II reactions

Phase I Reactions

In phase 1 reactions, xenobiotics are generally converted to more *polar, hydroxylated derivatives.*

In phase 2 reactions, these derivatives are conjugated with molecules such as glucuronic acid, sulfate, or glutathione. This renders them even more water-soluble, and they are eventually excreted in the urine or bile.

Phase II Reactions

- **Hydroxylation,** catalyzed mainly by members of a class of enzymes referred to as **monooxygenases** or **cytochrome** P450s.
- Deamination
- Dehalogenation
- Desulfuration
- Epoxidation
- Peroxygenation
- Reduction.

Biotransformation to Toxic Compounds

In certain cases, phase 1 metabolic reactions convert xenobiotics from **inactive** to **biologically active** compounds. In these instances, the original xenobiotics are referred to as **"prodrugs"** or **"procarcinogens."**

Example:
- Vinyl Chloride to Vinyl Chloride Epoxide which can bind to DNA and RNA.
- Mercury methylated, making them neurotoxic.
- Methanol to formic acid
- Benzopyrene converted to its epoxide by epoxidation.

Cytochrome P450

The main reaction involved in phase 1 metabolism is **hydroxylation,** catalyzed by a family of enzymes known as **monooxygenases** or **"mixed-function oxidases."**

- There are at least 57 cytochrome P450 genes in the human genome.
- Cytochrome P450 is a heme enzyme.
- They exhibit an absorption peak at **450 nm**.
- Approximately 50% of the common drugs that humans ingest are metabolized by isoforms of cytochrome P450.
- They also act on steroid hormones, carcinogens, and pollutants.
- In addition to their role in metabolism of xenobiotics, cytochromes P450 are important in the metabolism of a number of physiological compounds—for example, the synthesis of steroid hormones and the conversion of vitamin D to its active metabolite, calcitriol.
- NADPH is required to reduce cytochrome P450
- Lipids which are components of Cyt P450 is **Phosphatidylcholine**.

Sites where Cyt P450 is present

- In mammals, cytochromes P450 are present in highest amount in **liver cells** and **enterocytes** but are probably present in all tissues.
- In liver and most other tissues, they are present mainly in the membranes of the smooth endoplasmic reticulum **(microsomal fraction)**.
- In the **adrenal gland**, they are found in **mitochondria** as well as in the endoplasmic reticulum.
- In the adrenal gland are involved in cholesterol and steroid hormone biosynthesis.
- The mitochondrial cytochrome P450 system differs from the microsomal system in that it uses an NADPH-linked flavoprotein, **adrenodoxin reductase**, and a nonheme iron-sulfur protein, **adrenodoxin**.

EXTRA EDGE

Properties of Human Cytochromes P450
- Involved in phase I of the metabolism of a large number of xenobiotics, including perhaps 50% of the clinically used drugs;
- Involved in the metabolism of many endogenous compounds (e.g. steroids).
- All are hemoproteins.
- Often exhibit broad substrate specificity, thus acting on many compounds; consequently, different P450s may catalyze formation of the same product.
- Basically they catalyze reactions involving introduction of one atom of oxygen into the substrate and one into water.

Contd...

Contd...

- Their hydroxylated **products are more water-soluble than** their generally lipophilic substrates, facilitating excretion.
- Liver contains highest amounts, but found in most if not all tissues, including small intestine, brain, and lung.
- Located in the smooth endoplasmic reticulum or in mitochondria (steroidogenic hormones).
- In some cases, their products are mutagenic or carcinogenic.
- Many have a molecular mass of about 55kDa.
- Many are **inducible**, resulting in one cause of drug interactions.
- Some exhibit genetic polymorphisms, which can result in a typical drug metabolism.
- Their activities may be altered in diseased tissues (e.g., cirrhosis), affecting drug metabolism.

Phase 2 Reactions

Most abundant Phase 2 reaction is Conjugation.

Glucuronidation

- The glucuronidation of bilirubin by UDP-glucuronic acid
- *Glucuronidation is probably the most frequent conjugation reaction.*

Sulfation

- The **sulfate donor** in these and other biologic sulfation reactions) is **adenosine 3'-phosphate-5'-phosphosulfate (PAPS)**, this compound is called "active sulfate" e.g., sulfation of steroids, glycosaminoglycans, glycolipids, and glycoproteins

Conjugation with Glutathione

- Glutathione (γ-glutamyl cysteinyl glycine) is a **tripeptide** consisting of glutamic acid, cysteine, and glycine
- Glutathione is commonly abbreviated GSH (because of the sulfhydryl group of its cysteine, which is the business part of the molecule)

- The enzymes catalyzing these reactions are called glutathione S-transferases and are present in high amounts in liver cytosol and in lower amounts in other tissues.

Acetylation

Acetyl-CoA (active acetate) is the acetyl donor. These reactions are catalyzed by **acetyl transferases**.

Methylation

A few xenobiotics are subject to methylation by methyl-transferases, employing S-adenosyl methionine as the methyl donor.

Conjugation with Glycine

Benzoic acid conjugated with Glycine to form Hippuric Acid (Benzoyl Glycine).

Conjugation with Glutamine

Phenyl acetic acid is conjugated with glutamine to form phenyl acetyl glutamine.

Quick Revision

- Hydroxylation is the most common phase I reaction
- Cyt P450 family catalyses most phase I hydroxylation reactions
- Cyt P450 absorption peak is at 450 nm.
- Most common phase II reaction is conjugation.
- Glycosylation is probably the most common conjugation reaction

Check List for Revision

- An overall idea is enough.
- Bold letters are the must learn points

Review Questions

1. Hippuric acid is:
   a. Benzoyl Glutamine
   b. Benzoyl Glycine
   c. Benzoyl acetate
   d. Benzoyl lactate
2. Phase I Xenobiotic reaction include all except:
   a. Hydroxylation
   b. Reduction
   c. Peroxygenation
   d. Methylation
3. Lipid present in Cyt P450 is:
   a. Phosphatidyl choline
   b. Ceramide
   c. Triacyl glycerol
   d. Cholesterol

Answers to Review Questions

1. **b. Benzoyl Glycine**

   Conjugation with glycine is a phase II Xenobiotic reaction.

2. **d. Methylation**

   Methylation is Phase II reaction. The methyl donor is S Adenosyl Methionine.

3. **a. Phosphatidyl choline**

9.4 BIOMEMBRANES AND CELL ORGANELLE

- Properties of Biomembranes
- Components of Membranes
- Specialised Region of Plasma Membrane
- Cell Organelle

PROPERTIES OF BIOMEMBRANES

- Membranes are *sheet like structures*, that form *closed boundaries* between different compartments. The thickness of most membranes is between **60 Å (6 nm) and 100 Å (10 nm).**
- Membranes consist mainly **of *lipids* and *proteins***. The mass ratio of lipids to proteins ranges from 1:4 to 4:1.
- Membranes also contain **carbohydrates** that are linked to lipids and proteins.
- Membrane lipids are small molecules that have both *hydrophilic* and *hydrophobic* moieties. These lipids spontaneously form **closed bimolecular sheets** in **aqueous media**. These *lipid bilayers* are barriers to the flow of polar molecules.
- ***Specific proteins mediate distinctive functions of membranes.*** Proteins serve as pumps, channels, receptors, energy transducers, and enzymes.
- Membranes are **noncovalent assemblies**. The constituent protein and lipid molecules are held together by many noncovalent interactions, which act cooperatively.
- Membranes are *asymmetric.* The two faces of biological membranes always differ from each other.
- Membranes are *fluid structures*.
- **Lipid molecules** diffuse rapidly in the plane of the membrane, as **do proteins**, unless they are anchored by specific interactions. This is **lateral diffusion.**
- In contrast, **lipid molecules and proteins do not readily rotate across the membrane. This is transverse diffusion or flip flop movement**) (*see* Fig.18.5).
- Most cell membranes are *electrically polarized,* such that the inside is negative [typically 260 millivolts (mV)].

Membrane potential plays a key role in transport, energy conversion, and excitability.

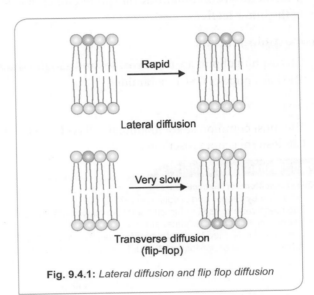

Fig. 9.4.1: *Lateral diffusion and flip flop diffusion*

COMPONENTS OF MEMBRANES

Membranes are complex structures composed of lipids, proteins, and carbohydrate-containing molecules.

Ratio of Protein to Lipid in Different Membranes

Proteins equal or exceed the quantity of lipid in nearly all membranes.

The outstanding exception is myelin (protein to lipid ratio is 0.23).

The Major Lipids in Mammalian Membranes
- **Phospholipids**
- **Glycosphingolipids**
- Sterols

Membrane lipids are **amphipathic**

Phospholipid

Two major phospholipid classes present in membranes are glycerophospholipid and sphingomyelin.

- **Glycerophospholipid are the most common phospholipid (Lecithin is the most common glycerophospholipid)**
- Choline-containing phospholipids (phosphatidylcholine and sphingomyelin) are located mainly in the **outer molecular layer**
- Aminophospholipids (**phosphatidyl serine and phosphatidyl ethanolamine**) are preferentially located in the **inner leaflet.**

Glycosphingolipids

- Glycosphingolipids are **cerebrosides and ganglio sides.**
- The back bone of GSL is ceramide.

Sterol

- The most common sterol in animal cell is **cholesterol**
- Cholesterol is not present in plants.

Image-Based Information

Fluid Mosaic Model of Biomembranes[Q]
- Proposed by Singer and Nicolson in 1972.
- The membrane consist of bimolecular lipid bilayer with proteins inserted in it or bound to either surface.
- This model is likened to integral membrane protein "icebergs" floating in a sea of predominantly fluid phospholipid molecules.

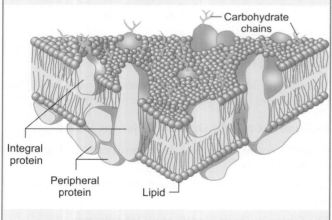

Fig. 9.4.2: Fluid mosaic model of plasma membrane

Membrane Proteins

Membranes contain integral and peripheral proteins.

Integral Protein
- **Deeply embedded** (to both hydrophilic and hydrophobic portions) in the lipid bilayer
- Usually **globular** and are themselves **amphipathic.**
- Transmembrane integral protein are the proteins that span the whole lipid bilayer
- They are held in membranes by hydrophobic interaction between membrane lipid and hydrophobic domain of protein.

Peripheral Proteins
- Attached to the hydrophilic portions of plasma membrane.
- **By electrostatic interaction and hydrogen bonding**
- Do not interact directly with the hydrophobic cores of the phospholipids in the bilayer.

Image-Based Information

Integral and Peripheral Proteins

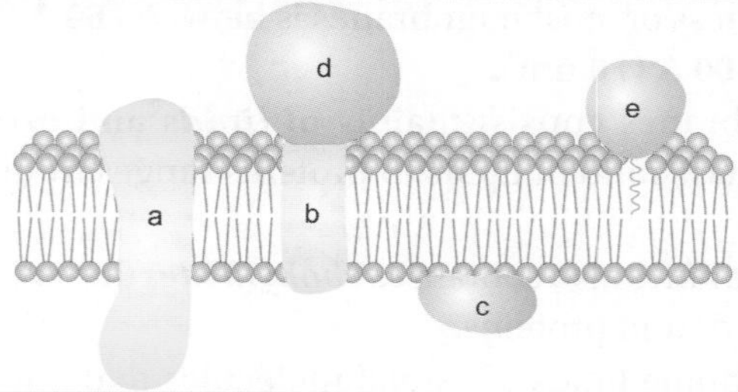

Fig. 9.4.3: *Integral and peripheral proteins*

Fluidity of Membranes

The temperature at which the structure undergoes the transition from ordered to disordered (i.e., melts) is called the "transition temperature" (Tm).

Factors Affecting the Melting Temperature

The properties of fatty acids and of lipids derived from them are markedly dependent on chain length and degree of saturation.

- The longer and more saturated fatty acid chains cause higher values of Tm. So decreases the fluidity of membranes.
- **Unsaturated fatty acid increases the fluidity** (i.e. they lower Tm)
- **Cis fatty acids increases the fluidity of membranes**
- **Cholesterol modifies the fluidity of membranes.**
  – At temperatures below the Tm increases fluidity
  – At temperatures above the Tm decreases fluidity

At normal body temperature (37°C) the lipid bilayer is in a fluid state.

Specialised Region of Plasma Membrane

1. **Lipid Rafts**
2. **Caveolae**
3. **Intercellular connections**

Image-Based Information
Lipid Rafts

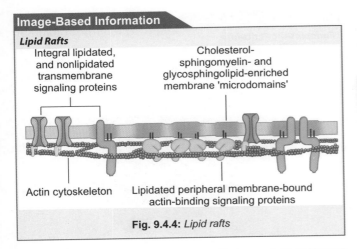

Fig. 9.4.4: Lipid rafts

Lipid Rafts[Q]

- These are specialized are as of the exoplasmic leaflet of the lipid bilayer:
- Enriched **in cholesterol, sphingolipids**
- Contain **certain GPI-linked proteins (outer leaflet)** and acylated and prenylated proteins (inner leaflet)
- Important for signal transduction and other processes

Image-Based Information
Caveolae

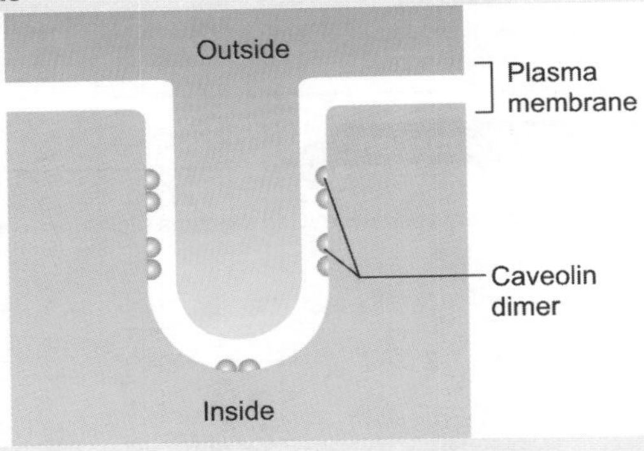

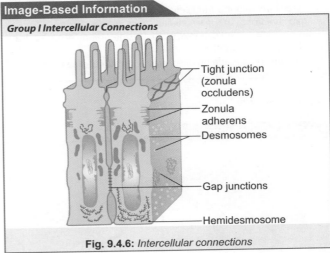

Fig. 9.4.5: Caveolae

Caveolae
Flask-shaped indentation of the cell membrane facing the cytosol. Contain the protein caveolin-1

Functions
- Signal transduction system (e.g. the insulin receptor and some G proteins), the folate receptor, and endothelial nitric oxide synthase (eNOS)
- Transport of macromolecules [IgA]
- Endocytosis of Cholesterol containing lipoprotein.

Intercellular Connections

Inter cellular junctions that form between the cells in tissues can be broadly split into two groups:

1. Junctions that fasten the cells to one another and to surrounding tissues
2. Junctions that permit transfer of ions and other molecules from one cell to another.

Image-Based Information
Group I Intercellular Connections

Fig. 9.4.6: Intercellular connections

Group I—The Types of Junctions that Tie Cells Together

1. **Tight junctions [the zonula occludens]**
   Prevent the diffusion of macromolecules between cells. Those are composed of various proteins, including occludin, various claudins, and junctional adhesion molecules [JAMs].
   Absence of Tight junction implicated in loss of contact inhibition
2. **Desmosome and zonula adherens** also help to hold cells together
3. **Hemidesmo some and focal adhesions** attach cells to their basal laminas

Image-Based Information
Gap Junction

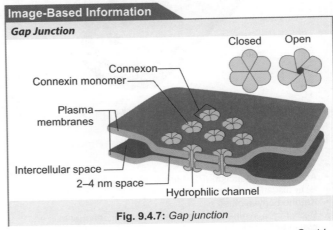

Fig. 9.4.7: Gap junction

Contd...

Contd...

Gap Junctions
Belongs to Group II— Junctions that permit transfer of ions and other molecules from one cell to another.

Cytoplasmic "tunnel" for diffusion of small molecules (<1000 Da) between two neighbouring cells.

At gap junctions, the intercellular space narrows from 25 nm to 3 nm, and units called **connexons**[Q] in the membrane of each cell are lined up with one another.

Each connexon is made up of six protein subunits called connexins.

Cell Organelle

Image-Based Information

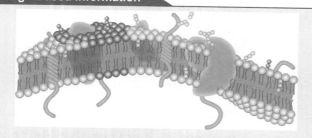

Fig. 9.4.8: *Plasma membrane*

Plasma Membrane—Marker Enzymes
- 5'Nucleotidase
- Adenylyl cyclase
- $Na^+ K^+$ ATPase

Image-Based Information
Mitochondria

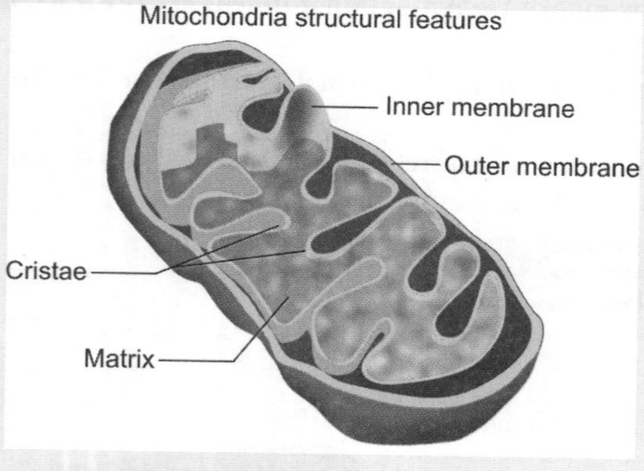

Mitochondria structural features
- Inner membrane
- Outer membrane
- Cristae
- Matrix

Fig. 9.4.9: *Mitochondria*

Marker Enzyme
Inner mitochondrial membrane is ATP Synthase Functions
- Electron transport chain
- TCA Cycle
- Acetyl CoA production (PDH reaction)
- Beta oxidation of fatty acids
- Ketone body synthesis
- Part of urea synthesis, heme synthesis, gluconeogenesis and pyrimidine synthesis

Image-Based Information
Endoplasmic Reticulum

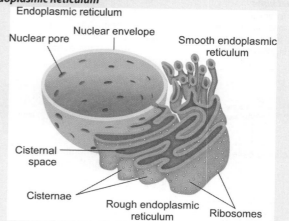

Fig. 9.4.10: *Endoplasmic reticulum*

Marker Enzyme
- Glucose 6 Phosphatase

Functions
- Synthesis of proteins
- Glycosylation of proteins
- Lipoprotein assembly
- Synthesis of cholesterol
- Steroid hormone synthesis
- Collagen synthesis
- Phospholipid synthesis
- Elongation of fatty acids
- Minor pathways of oxidation of fatty acids.
- Desaturation of fatty acid

Image-Based Information
Golgi Apparatus

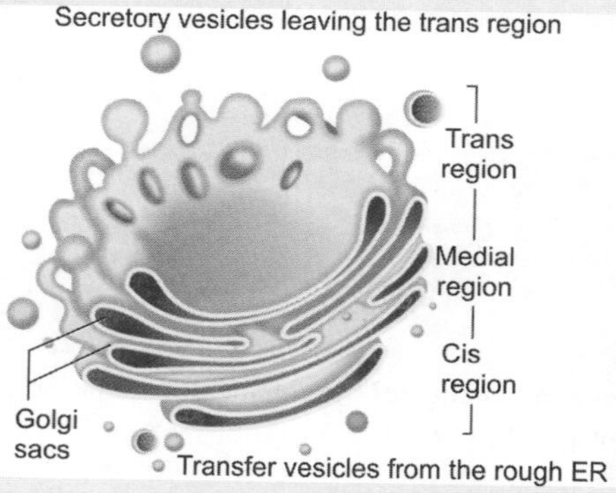

Secretory vesicles leaving the trans region
- Trans region
- Medial region
- Cis region
- Golgi sacs
- Transfer vesicles from the rough ER

Fig. 9.4.11: *Golgi apparatus*

Marker Enzyme of parts of Golgi apparatus
- Cis –Glc NAc transferase I
- Medial-Golgi mannosidase II
- **Trans-Galactosyl Transferase**
- Trans Golgi Network-Sialyl transferase

Functions
- Glycosylation of proteins
- Packing and sorting of proteins

Special Topics 359

Image-Based Information
Peroxisomes

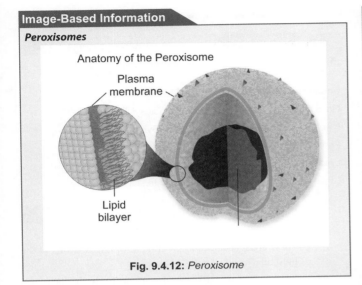

Fig. 9.4.12: *Peroxisome*

Image-Based Information

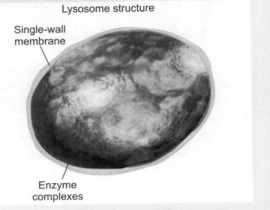

Fig. 9.4.13: *Lysosome*

Lysosome
- Marker enzyme—Acid Phosphatase

Functions
- Contain hydrolysing enzymes of Proteins, Lipids, Carbohydrates etc.

Marker Enzyme
- Catalase, urate oxidase

Functions
- Free radical scavenging (catalase, peroxidase)
- Oxidation of very long chain fatty acids
- Oxidation of fatty acids which leads to H_2O_2 production.

HIGH YIELD POINTS

Functions of Cytosol
Marker enzyme—Lactate Dehydrogenase
Functions
- Glycolysis
- Glycogen metabolism
- HMP pathway
- Transaminations
- Fatty acid synthesis
- Purine synthesis
- Cholesterol synthesis (Partly)
- Heme Synthesis, Urea synthesis, Pyrimidine synthesis (partly)

Check List for Revision

- Markers of organelle and composition of biomembranes is high yield topic.
- Overall this chapter is an extra edge topic not must learn

REVIEW QUESTIONS MCQ

1. **Correct statement about membrane:** *(PGI Nov 2016)*
   a. Phospholipids undergo rapid lateral diffusion
   b. Transverse movement of lipids across the membrane is faster than protein
   c. Hydrophobic core of phospholipid bilayer remains constantly in motion because of rotation around the bonds of lipid tail
   d. Phospholipids that have one fatty acyl group cannot form the bilayer.
   e. Phospholipid span the whole bilayer

2. **Marker enzyme for golgi complex is:**
   a. Glucose 6 Phosphatase
   b. Catalase
   c. Galactosyl Transferase
   d. Lactate Dehydrogenase

3. **Marker enzyme for plasma membrane is:**
   a. 5' Nucleotidase
   b. Catalase
   c. Acid Phosphatase
   d. GGT

4. **GPI anchored glycoproteins are found in:**
   a. Caveoli
   b. Lipid rafts
   c. Connexons
   d. Intercellular connections

Answers to Review Questions

1. **a, c, d, e.** Phospholipids undergo…, Hydrophobic core of phospholipid bilayer…, Phospholipids that have one fatty acyl…, Phospholipid span the whole… *(Ref: Harper 30/e)*

   - Transverse movement of lipids across the membrane is slower for lipid than protein.
   - In fluid mosaic model, phospholipids are dynamic and in motion especially because of cis fatty acid increase fluidity because of rotation around cis double bond.
   - Phospholipid in the membrane should have at least one unsaturated fatty acid and atleast 1 cis double bond. Phospholipids have two fatty acyl group.

2. **c. Galactosyl Transferase**

3. **a, 5' Nucleotidase**

Membrane/organelle	Marker enzymes
Plasma Membrane	5'- Nucleotidase Adenylyl Cyclase Na^+-K^+ ATPase
Endoplasmic reticulum	Glucose-6-phosphatase
Golgi Complex	Galactosyl Transferase[Q]
Inner Mitochondrial Membrane	ATP Synthase
Peroxisome	Catalase, Urate Oxidase
Lysosomes	Acid Phosphatase
Cytoplasm	Lactate Dehydrogenase

4. **b. Lipid rafts**

   Lipid rafts
   - These are specialized areas of the exoplasmic leaflet of the lipid bilayer.
   - Contain **certain GPI-linked proteins (outer leaflet)**

CHAPTER 10

Molecular Genetics

Chapter Outline

10.1 Chemistry of Nucleic Acids
10.2 Metabolism of Nucleotides
10.3 Organization and Structure of DNA
10.4 DNA Replication and Repair
10.5 Transcription
10.6 Different Classes of RNA
10.7 Translation
10.8 Regulation of Gene Expression
10.9 Mutations
10.10 Mitochondrial DNA
10.11 Patterns of Inheritance
10.12 DNA Polymorphism

CHAPTER 10

Molecular Genetics

10.1 CHEMISTRY OF NUCLEIC ACIDS

- Nucleic Acid
- Composition of Nucleotides

NUCLEIC ACIDS

Nucleic Acids are polymers of nucleotides joined by **3'-5' phosphodiester bond**.

Two Types of Nucleic Acid
1. Deoxy Ribonucleic Acid (DNA)
2. Ribonucleic Acid (RNA)

General Properties of Nucleotides
1. Nucleotides **are poly functional acids**. Nucleotides bear a **negative** charge at physiological pH.
2. Absorbs UV light at a wavelength **260 nm**, at pH 7.0. The conjugated double bond of Purine and Pyrimidine nucleotide is responsible for it.

> **HIGH YIELD POINTS**
>
> **Ultraviolet rays (UV-C) are mutagenic**
> The mutagenic effect of ultraviolet light is due to its absorption of UV light by nucleotides in DNA that result in chemical modifications.

Composition of Nucleotides

The components of nucleotides are
Nitrogenous base + Pentose Sugar + Phosphate Group

Nitrogenous Base
They are nitrogen containing heterocyclic ring structures.

Two types of nitrogenous bases
1. Purine
2. Pyrimidine

1. **Purine bases**^{Q} are adenine and guanine
   - Adenine-6 Amino Purine
   - Guanine-2-amino, 6-oxopurine.

Adenine and Guanine are present in both DNA and RNA.

Minor purine bases are
   - Hypoxanthine-6 oxopurine.
   - Xanthine-2, 6 dioxopurine.
   - Uric Acid-2, 6, 8 trioxopurine.

2. **Pyrimidine bases are** cytosine, uracil and thymine
   - Cytosine-2 oxo 4 amino Pyrimidine
   - Uracil-2, 4 dioxo Pyrimidine
   - Thymine-2, 4 dioxo 5 methyl Pyrimidine

Modified nitrogenous bases
   - Dihydrouracil
   - Pseudouridine
   - 5-Methyl cytosine
   - Dimethyl amino adenine
   - 7-methyl Guanine.

> **HIGH YIELD POINTS**
>
> - Cytosine is present in both DNA and RNA
> - Uracil present only in the RNA
> - Thymine is present only in the DNA
> - Most common modified bases are methylated forms of major bases

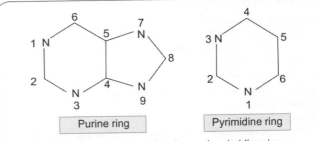

Fig. 10.1.1: *Structure of purine and pyrimidine ring*

Methylated Xanthine derivative present in the
Coffee-Caffeine (Trimethyl Xanthine)
Tea-Theophylline (Dimethyl Xanthine)
Cocoa-Theobromine (Dimethyl Xanthine)

> **HIGH YIELD POINT**
>
> Vitamin that contains pyrimidine ring is Thiamin

Pentose Sugar

Two types of Pentose Sugar
1. D-Ribose Sugar in RNA
2. 2'deoxy D-Ribose Sugar in DNA

Both pentose sugar their b Furanose form

Composition of Nucleoside

- **Nitrogenous base + Pentose Sugar**
- C1 of ribose or deoxyribose sugar joined covalently to **N1 of Pyrimidine or N9 of Purine by β N Glycosidic linkage**.
- So Nucleosides are **N Glycosides**

Composition of Nucleotides

- Nucleoside + Phosphoryl groups
- Usually phosphoryl group is attached to **5' hydroxyl group of pentose sugar by an ester bond.**
- Additional phosphoryl group are attached by an **acid anhydride** bond to form Nucleoside diphosphates and triphosphates.

Nitrogenous Base	Pentose Sugar	Nucleoside = Nitrogenous Base + Pentose Sugar	Nucleotide = Nucleoside + Phosphate group
Ribonucleotides			
Adenine	Ribose	Adenosine	Adenosine Monophosphate (AMP)
Guanine	Ribose	Guanosine	Guanosine Monophosphate (GMP)
Cytosine	Ribose	Cytidine	Cytidine Monophosphate (CMP)
Uracil	Ribose	Uridine	Uridine Monophosphate (UMP)
Hypoxanthine	Ribose	Inosine	Inosine Monophosphate (IMP)
Xanthine	Ribose	Xanthosine	Xanthosine Monophosphate (XMP)
Deoxyribonucleotides			
Adenine	Deoxy Ribose	dAdenosine	dAdenosine Monophosphate (dAMP)
Guanine	Deoxy Ribose	dGuanosine	dGuanosine Monophosphate (dGMP)
Cytosine	Deoxy Ribose	dCytidine	dCytidine Monophosphate (dCMP)
Thymine	Deoxy Ribose	Thymidine	Thymidine Monophosphate

> **EXTRA EDGE**
>
> **Spontaneous Alterations in the Covalent Structure of Purines and Pyrimidines**
>
> **Spontaneous deamination of**
> 1. Cytosine to Uracil (Most common alteration)
> 2. 5 Methyl Cytosine to Thymine
> 3. Adenine to Hypoxanthine
> 4. Guanine to Xanthine
>
> **Think:** What is the base formed by deamination and Methylation of Cytosine?
> Answer:
> Cytosine deaminated to Uracil, Uracil methylated to Thymine

Nucleic Acid

Polymers of nucleotides joined by 3'-5' phosphodiester bond.

They are RNA and DNA.

3'-5' Phosphodiester Bond

3'hydroxyl group of sugar of first mononucleotide linked to 5'phosphoryl group of the second mononucleotide by a 3'-5' phosphodiester bond.

Nucleic acid exhibit polarity

- Nucleotides are linked by 3'-5' phosphodiester bond, there is a free phosphoryl group at 5' of the first nucleotide and free hydroxyl group at the 3' end of last nucleotide.
- Hence nucleic acid exhibit polarity.
- The base sequence is usually written from 5' end to 3' end. (see Figure 10.1.2)

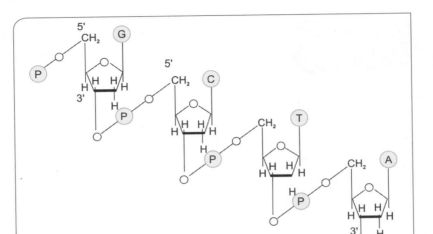

Fig. 10.1.2: *Nucleic acid*

HIGH YIELD POINTS

Functions of Nucleotides and Nucleotide Derivative
Most abundant free nucleotide in mammalian cell is ATP
- Building blocks of Nucleic acid is **nucleotides**.
- The principal biological transducer of free energy is **Adenosine Triphosphate (ATP)**
- Nucleotide derivative that are second messengers in hormonal pathways are **cAMP and cGMP**
- Nucleotide derivative that is important methyl donor is **S Adenosyl Methionine (SAM)**
- An important Sulfate donor is **Adenosine 3'Phosphate -5'phosphosulphate (PAPS)**
- Nucleotide derivative in Glycogen synthesis is **UDP Glucose**
- Nucleotide derivative that conjugate bilirubin is **UDP–Glucuronic acid**
- Nucleotide derivative that act as coenzymes are **NAD+, NADP, FMN, FAD+**
- **FMN has no Adenine in its structure**

HIGH YIELD POINTS

Pseudouridine in tRNA
- Seen in the **Ribothymidine Pseudouridine Cytidine (TψC) arm of tRNA.**
- Formed by modification of **UMP** on preformed tRNA.
- D Ribose of pseudouridine linked to **C-5 of Uracil**, by **a carbon to carbon bond** rather than by β N Glycosidic bond.
- Pseudouridine is excreted unchanged in urine

HIGH YIELD POINTS

Thymidine in tRNA
- UMP in preformed tRNA is methylated by S-adenosyl Methionine (SAM) to form TMP.
- **tRNA is the RNA with thymidine**.
- Ribothymidine is also seen in **Pseudouridine Cytidine (TψC) arm of tRNA**.

Quick Review

- Nucleotide is nitrogenous base + Pentose sugar + Phosphate group.
- Nucleic acid absorb UV light at 260 nm.
- Purines are Adenine and Guanine
- Pyrimidines are Cytosine, Uracil and Thymine
- Least oxygen content in Adenine.
- Uric acid is 2,6,8 Trioxopurine.
- Nucleotides are joined by 3' to 5' phosphodiester bond.
- Pseudouridine is C1' of pentose sugar joined to C5 of Uracil to form a C-C bond.

Comparison of DNA and RNA

DNA	RNA
Mostly seen in the Nucleus	Mostly seen in the Cytoplasm
Pyrimidine bases are Thymine and Cytosine	Pyrimidine bases are Uracil and Cytosine
Sugar is deoxy ribose	Sugar is Ribose
Pyrimidines = Purines	Pyrimidines not equal to Purines
Not destroyed by Alkali	Destroyed by Alkali
Usually Double Stranded	Single Stranded

Check List for Revision

- This is a simple chapter so you can easily learn this chapter.
- Boxes and tables are very important.

Review Questions MCQ

1. **Pyrimidine is a part of:** *(PGI May 2017)*
   a. Adenosine
   b. Cytidine
   c. Uridine
   d. Cysteine
   e. Guanosine

2. **Nucleoside is made up of:** *(PGI Nov 2010)*
   a. Pyrimidine
   b. Histone
   c. Sugar
   d. Purine
   e. Phosphate

3. **The followings correctly arranged:** *(PGI June 2009)*
   a. GMP-Guanine monophosphate
   b. UMP-Uracil monophosphate
   c. TMP-Thymine monophosphate
   d. CMP-Cytidine monophosphate
   e. AMP-Adenine monophosphate

4. **Apart from occurring in nucleic acid, pyrimidines are also found in:** *(AIIMS Nov 05)*
   a. Theophylline
   b. Theobromine
   c. Flavin mononucleotide
   d. Thiamin

5. **Which of the following is not a nitrogenous base?**
   a. Adenine
   b. Guanosine
   c. Cytosine
   d. Thymine

6. **Which is not found in DNA?** *(AI 1994)*
   a. Adenine
   b. Thymine
   c. Guanine
   d. Uracil

7. **At the physiological pH the DNA molecules are:** *(AIIMS Nov 02)*
   a. Positively charged
   b. Negatively charged
   c. Neutral
   d. Amphipathic

Answers to Review Questions

1. **b, c. Cytidine, Uridine** *(Ref: Harper 31/e page 322)*

   Pyrimidine base present in Cytidine is Cytosine, Uridine is Uracil

2. **a, c, d. Pyrimidine, Sugar, Purine** *(Ref: Harper 31/e page 322)*

   Composition of Nucleoside
   - Nitrogenous base + Pentose Sugar
   - C1 of ribose or deoxyribose sugar to N1 of Pyrimidine or N9 of Purine by β N Glycosidic linkage

   Composition of Nucleotides
   - Nucleoside + Phosphoryl groups
   - Usually phosphoryl group is attached to 5'hydroxyl group of pentose sugar by an ester bond.
   - Additional phosphoryl group are attached by an Acid anhydride bond to form Nucleoside diphosphates and triphosphates

3. **d. CMP-Cytidine monophosphate** *(Ref: Harper 31/e page 322)*

Nitrogenous Base	Pentose Sugar	Nucleoside = Nitrogenous Base + Pentose Sugar	Nucleotide = Nucleoside + Pentose Sugar
Ribonucleotides			
Adenine	Ribose	Adenosine	Adenosine Monophosphate (AMP)
Guanine	Ribose	Guanosine	Guanosine Monophosphate (GMP)
Cytosine	Ribose	Cytidine	Cytidine Monophosphate (CMP)
Uracil	Ribose	Uridine	Uridine Monophosphate (UMP)
Hypoxanthine	Ribose	Inosine	Inosine Monophosphate (IMP)
Xanthine	Ribose	Xanthosine	Xanthosine Monophosphate (XMP)
Deoxyribonucleotides			
Adenine	Deoxy Ribose	dAdenosine	dAdenosine Monophosphate (dAMP)
Guanine	Deoxy Ribose	dGuanosine	dGuanosine Monophosphate (dGMP)
Cytosine	Deoxy Ribose	dCytidine	dCytidine Monophosphate (dCMP)
Thymine	Deoxy Ribose	Thymidine	Thymidine Monophosphate

4. **d. Thiamin**
   - Vitamin with pyrimidine ring is Thiamin
   - Theophylline and Theobromine has Purine ring
   - FMN also contain Purine ring
5. **b. Guanosine** (Ref: Harper 31/e page 322, Table 32-1)
   - Guanosine is nucleoside.

6. **d. Uracil** (Ref: Harper 31/e page 328)
   - Uracil is found only in the RNA
7. **b. Negatively charged** (Ref: Harper 31/e page 321)
   - DNA is negatively charged because of Phosphate group.

10.2 METABOLISM OF NUCLEOTIDES

☞ Metabolism of Purines. ☞ Metabolic of Pyrimidines

METABOLISM OF PURINES

Metabolic Pathways of Purine
- De novo synthesis of Purine Nucleotides.
- Salvage pathways of Purine Nucleotides.
- Catabolism of Purines

De Novo synthesis of Purine nucleotides
- Site of synthesis: Most of the tissues but majority in the **Liver.**
- **Organelle: Cytoplasm**

Sites where de novo synthesis do not takes place:
- Brain
- Erythrocytes
- Polymorphonuclear Leukocytes
- Bone marrow

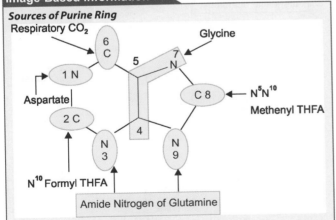

Fig. 10.2.1: *Source of purine ring*

Sources of Purine Ring
- C4, C5, N7 by Glycine
- N3, N9 by Amide nitrogen of Glutamine

Contd...

Contd...
- N1 by Aspartate
- C2 by N^{10} Formyl THFA
- C8 by $N^5 N^{10}$ Methenyl THFA
- C6 by Respiratory CO_2

> **HIGH YIELD POINTS**
> - Rate limiting Step in Purine Synthesis: **PRPP Glutamyl Amidotransferase**
> - Purine ring is built up on a ribose-5-phosphate, hence nucleotides are the products of the de novo synthesis.
> - 1st Nucleotide formed in Purine Synthesis: Inosine Monophosphate (IMP)
> - From IMP Adenosine Monophosphate (AMP) and Guanosine monophosphate (GMP) is formed.
> - Committed step is **PRPP Glutamyl Amidotransferase**

Conversion of IMP to AMP and GMP

- Amino group of AMP is contributed by Aspartic Acid
- Amino group of GMP is contributed by Glutamine

Mnemonic
A for A (Aspartic Acid for AMP)
G for G (Glutamine for GMP)

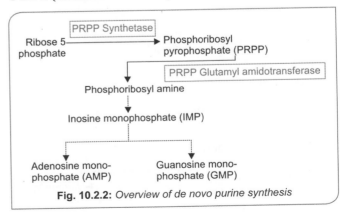

Fig. 10.2.2: *Overview of de novo purine synthesis*

Regulation of De Novo Purine Synthesis

- The regulatory enzymes of purine synthesis are **PRPP Synthetase and PRPP Glutamyl Amido transferase**.
- The rate limiting step of De novo purine synthesis is **PRPP Glutamyl amido Transferase**
- The overall determinant of de novo purine synthesis is level of PRPP.
- This activate PRPP Synthetase.
- The rate of PRPP Synthesis depends on Ribose 5 Phosphate.
- AMP, ADP, GMP and GDP the end products of De Novo Purine synthesis feedback regulate PRPP Synthase and PRPP Glutamyl Amido transferase.

Salvage Pathway of Purine Nucleotides

Definition

Recycling of degraded purine and purine nucleosides to their corresponding mononucleotides.

Significances
- Less energy consuming.
- Effective recycling of degraded nucleotides.
- Important in organs where de novo synthesis do not takes place

Salvage pathway reactions are
1. Phosphoribosylation of Purine bases (Pu) by PRPP.
2. Phosphorylation of Purine Nucleosides (PuR) by ATP

Salvage Pathway Reaction of Purines

1. Phosphorylation of Purine (Pu) by PRPP

   a. Adenine + PRPP → (Adenine phosphoribosyl transferase (APRTase)) → Adenosine monophosphate + PPi

   b. Hypoxanthine + PRPP → (Hypoxanthine Guanine Phosphoribosyl Transferase (HGPRTase)) → Inosine monophosphate + PPi

   c. Guanine + PRPP → Guanosine Monophosphate + PPi

2. Phosphorylation of Purine Nucleoside

   Adenosine + ATP → (Adenosine kinase) → Adenosine monophosphate + ADP

Catabolism of Purine Nucleotides

Site :
- Liver (Endogenously synthesized purines are catabolised majority in liver.)
- GIT (Dietary purines are catabolised in the intestinal mucosal cells)
- The end product of Purine catabolism in humans is **Uric Acid.**^{Q}
- The end product of purine catabolism in mammals other than higher primates is **Allantoin.**
- Uricase enzyme convert uric acid to water soluble, Allantoin

Steps of Catabolism of Purine Nucleotides

Adenosine Deaminase
- Adenosine converted to inosine by Adenosine Deaminase.

Purine Nucleoside Phosphorylase
- Inosine is converted to Hypoxanthine
- Guanosine is converted to Guanine.

Xanthine Oxidase
- Convert Hypoxanthine and Guanine to Xanthine.
- Same enzyme catalyse conversion of Xanthine to Uric acid

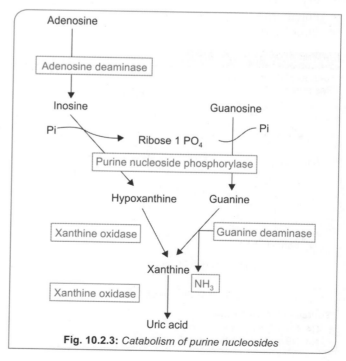

Fig. 10.2.3: *Catabolism of purine nucleosides*

Clinical Correlations—Purine Metabolism

Lesch Nyhan Syndrome[Q]
X-linked Recessive Disorder[Q]

Biochemical Defect
- Complete deficiency of **HGPRTase.**
- Purine accumulates.
- Purines degraded to Uric Acid.
- Uric acid level increases.

> **Symptom box**
>
> *Clinical Features*
> - **Hyperuricemia,** intellectual disability
> - Dystonic movement, choreoathetosis, dysarthric speech.
> - **Compulsive Self mutilation**[Q]
> - Megaloblastic anemia also can occur in the Lesch-Nyhan syndrome, in which regeneration of purine nucleotides is blocked

Diagnosis
- Hyperuricemia.
- HGPRTase enzyme activity in the Erythrocytes is deficient.

Treatment
- Allopurinol.
- Alkalinization of urine.
- High fluid intake.

> **Extra Edge**
>
> **Kelley-Seegmiller Syndrome**
> - **Partial deficiency of HGPRTase** with > 1.5-2% enzyme activity.
> - Associated with hyperuricemia and variable neurologic dysfunction.

> **Extra Edge**
>
> Adenine Phosphoribosyl transferase (APRTase) deficiency autosomal recessive
>
> ### Biochemical Defect
> - Deficiency APRTase
> - Adenine accumulates.
> - Adenine oxidized by xanthine dehydrogenase to **2,8-dihydroxyadenine**, which is extremely insoluble.
>
> ### Clinical manifestations
> - Urinary calculus formation with crystalluria,
> - The presence of brownish spots on the infant's diaper or of yellow-brown crystals in the urine is suggestive of the diagnosis.

Gout
Group of disorders presented with
1. Hyperuricemia.
2. Uric acid nephrolithiasis.
3. Acute inflammatory arthritis.

Biochemical defect
1. **Primary Gout:** Due to defect in the enzymes that lead to overproduction of purine nucleotides
   - Superactivity PRPP Synthetase (X linked Disorder)
   - Superactivity PRPP Amido Transferase
   - Deficiency of HGPRTase (Lesch Nyhan Syndrome)
   - Glucose 6 Phosphatase Deficiency (Type I Glycogen Storage Disorder)
2. **Secondary Gout**
   Increased production of uric acid
   Leukemia, Lymphoma
   Decreased excretion rate
   Renal failure, Thiazide diuretics, Lactic acidosis

> **Symptom box**
>
> *Clinical Features*
> 1. **Acute Gouty Arthritis**
>   Typically in the metatarsophalangeal joint of the big toe.
> 2. **Chronic Cases**
>   Tophi[Q] deposits of monosodium urate crystals in the subcutaneous tissue

Diagnosis
Aspiration and examination of synovial fluid
- Negatively birefringent[Q] needle shaped **monosodium urate crystals** using polarized light microscopy

Treatment
- Colchicine, an anti-inflammatory agent
- Uricosuric agents, such as probenecid or sulfinpyrazone.
- Allopurinol

Severe Combined Immunodeficiency (SCID)

Adenosine Deaminase (ADA) defect is one of the causes.
- Both **B cells and T cells** are affected
- First disorder to be treated by **Gene Therapy**[Q]
- Enzyme Replacement therapy with **polyethylene glycol modified bovine adenosine deaminase (PEG-ADA).**

HIGH YIELD POINT

Father of Gene therapy — French Anderson

EXTRA EDGE

Immunodeficiency in SCID is due to
- Deficiency of ADA.
- Adenosine accumulate
- Adenosine converted to its ribonucleotides and deoxyribo-nucleotides (dATP).
- dATP inhibit ribonucleotide reductase
- Decreases production of all deoxyribose containing nucleotides.
- Hence DNA synthesis decreased.
- Decrease in both T-cells and B-cells, hence immunodeficiency.

Purine Nucleoside Phosphorylase Defect
- Severe deficiency of T cells but apparently normal B-cell function[Q].
- **This is an enzyme in Purine Catabolism.**

Comparison of Immunodeficiency in SCID and Purine Phosphorylase Deficiency

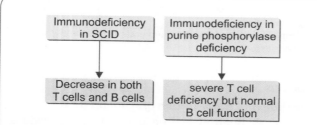

Fig. 10.2.4: *Immunodeficiency in SCID and purine phosphorylase deficiency*

Xanthine Oxidase Deficiency

- Genetic defect in Xanthine Oxidase enzyme.
- Associated with Hypouricemia.
- Xanthine crystals in urine.
- Xanthine lithiasis.

METABOLIC PATHWAYS OF PYRIMIDINES

Pyrimidine Synthesis
- Site mainly liver
- Organelle **Cytoplasm and Mitochondria**

Image-Based Information
Sources of Pyrimidine Ring

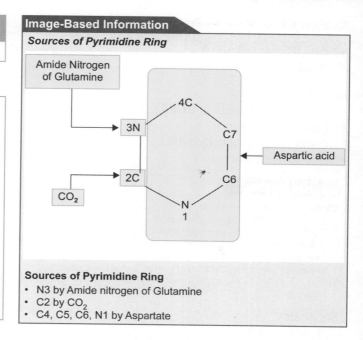

Sources of Pyrimidine Ring
- N3 by Amide nitrogen of Glutamine
- C2 by CO_2
- C4, C5, C6, N1 by Aspartate

Steps of Pyrimidine Biosynthesis

This pathway is similar to Urea Cycle

Enzymes of Pyrimidine Synthesis

Catalysed by multifunctional enzymes (Single polypeptide with more than one enzyme activity).

1. **C**PS II, **A**spartate Transcarbamoylase, **D**ihydro Orotase (CAD) in the Cytoplasm
2. Dihydro Orotate Dehydrogenase
   – **Only mitochondrial step in Pyrimidine Synthesis**
3. Orotate Phosphoribosyl Transferase and OMP Decarboxylase
   – Together called UMP Synthase
   – Seen in the Cytoplasm

Synthesis of Uridine Nucleotides
- OMP is decarboxylated to UMP, the first Pyrimidine nucleotide by OMP Decarboxylase

Synthesis of Cytidine Nucleotides
- UTP converted to CTP by CTP synthase
- Amino group is donated by Glutamine.

Synthesis of Thymine Nucleotides
- First UDP is converted to dUDP by ribonucleotide reductase (explained later).
- dUDP is converted to dUMP

- dUridine Mono Phosphate^Q (dUMP) to Thymidine Mono Phosphate (TMP) by Thymidylate Synthetase
- $N^5 N^{10}$ Methylene THFA donates the methyl group for thymidine Monophosphate (TMP)
- The only reaction in pyrimidine Synthesis where a Tetrahydrofolate derivative is needed.

Comparison of Purine and Pyrimidine Synthesis

	Purine synthesis	Pyrimidine synthesis
Amphibolic intermediates added	On Ribose 5 Phosphate	Not on Ribose 5 Phosphate
End product	Purine nucleotide	Pyrimidine base, then Ribose and Phosphate are added by OPRTase results in OMP
Site	Cytosol alone	Both Cytosol and Mitochondria
Starting material	Ribose 5 Phosphate	CO_2, Glutamine

EXTRA EDGE

Two Anticancer Drug Inhibit the Synthesis of TMP

Methotrexate
- Inhibit Dihydrofolate Reductase
- Dihydrofolate Reductase convert Dihydrofolate to Tetrahydrofolate.
- Hence TMP Synthesis is affected.

5 Fluorouracil
- Competitively inhibit Thymidylate Synthase.

Regulation of Pyrimidine Synthesis

- **C**PS II, **A**spartate Transcarbamoylase, **D**ihydro Orotase (CAD) is the primary focus of regulation of Pyrimidine Synthesis.
- Expression of CAD gene is controlled at genetic level.
- CPS II is activated by PRPP and is feedback inhibited by UTP.

Differences between CPS-I and CPS-II

	CPS-I	CPS-II
Cellular Location	Mitochondria	Cytosol
Pathway involved	Urea Cycle	Pyrimidine Synthesis
Source of Nitrogen	Ammonia	Amide nitrogen of Glutamine
Allosteric Regulators	Activator-N-Acetyl Glutamate No inhibitors	Activator-PRPP Inhibitor-UTP

Conversion Ribonucleotides to Deoxyribonucleotides

- Forms deoxyribonucleoside diphosphates (dNDPs) from ribonucleoside diphosphate^Q
- Ribonucleotide reductase complex catalyses the reaction.
- Reduction requires thioredoxin,^{Q} thioredoxin reductase, and NADPH.
- Thioredoxin reductase is a selenocysteine containing enzyme

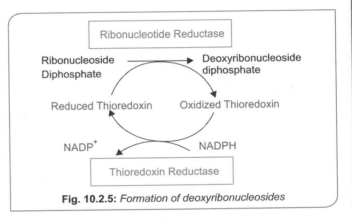

Fig. 10.2.5: *Formation of deoxyribonucleosides*

Catabolism of Pyrimidines

HIGH YIELD POINTS

- Important Point-Catbolism of Pyrimidines
- Cytosine and Uracil is catabolised to Beta Alanine, CO_2 & NH_3
- But Thymine to β–aminoisobutyrate, CO_2 & NH_3
- β–aminoisobutyrate to Succinyl CoA
- End products of Purine catabolism is water insoluble.
- But Pyrimidine catabolic end products are water soluble.
- No clinical symptoms due to pyrimidine over production as its end products are water soluble.

Clinical Correlation: Pyrimidine Metabolism

Orotic Acidurias

- Rare autosomal recessive condition
- The most common metabolic error in the de novo synthesis of pyrimidines

Biochemical Defect

Type I orotic aciduria

Deficiency of both **orotate phosphoribosyl transferase and OMP decarboxylase (together called UMP Synthase)**

Type II orotic aciduria

Deficiency only of **orotidylate decarboxylase**

372 Self Assessment and Review of Biochemistry

Symptom box

Clinical Features of Orotic Acidurias
Manifest in the first year of life and is characterized by growth failure, developmental retardation, megaloblastic anemia, and increased urinary excretion of orotic acid.

Extra Edge

Drugs may Precipitate Orotic Aciduria
- Allopurinol.
- 6-Azauridine

Extra Edge

Antimetabolites used in Cancer Chemotherapy
Compounds with structural similarity to precursors of purines or pyrimidines (Synthetic Nucleotides), or compounds that interfere with purine or pyrimidine synthesis are called antimetabolites.
- Methotrexate: Inhibits dihydrofolate reductase competitively.
- Hydroxyurea inhibits ribonucleotide reductase
- 6-Mercaptopurine: Purine analog
- 6-Thioguanine: Nucleotide analog
- 6-Azaguanine: Nucleotide analog.

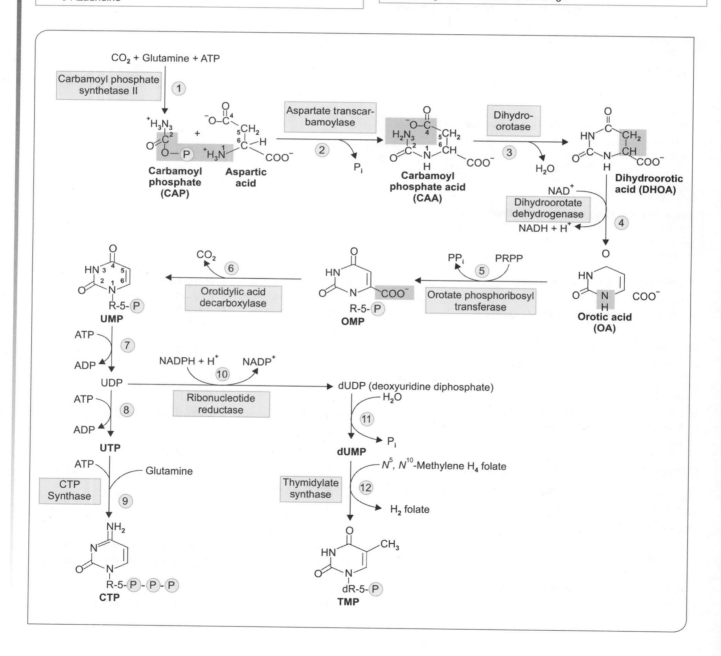

- Aza Serine: Glutamine Analog
- Diazenorleucine: Inhibit PRPP Glutamyl Amido Transferase
- Mycophenolic Acid^Q Potent Reversible uncompetitive inhibitor of IMP Dehydrogenase.
- 5-Fluorouracil (5FU)^Q Inhibits thymidylate synthetase
- 6 Aza-cytidine: Inhibits thymidylate synthetase
- 6 Aza –Uridine: Inhibits thymidylate synthetase
- **Cytosine Arabinoside**: Ribose is replaced by Arabinose

High Yield Points

- Synthetic nucleotide used in the treatment of Gout-Allopurinol
- Synthetic nucleotide used in the treatment of AIDS-Zidovudine
- Synthetic nucleotide used in the treatment of HerpeticKeratitis-5-iodo-deoxyuridine
- Synthetic nucleotide used to suppress immunologic rejection during organ transplantation-Azathioprine

Quick Review

- The amino acids that contribute to purine ring are Glycine, Aspartate and Glutamine.
- Rate limiting step of de novo purine synthesis is PRPP Glutamyl amidotransferase.
- Lesch Nyhan Syndrome is caused due to complete absence of HGPRtase enzyme.
- SCID is caused due to Adenosine Deaminase deficiency.
- The amino acid contributing to Pyrimidine ring is Glutamine and Aspartate.
- Both T cells and B cells are affected in SCID.
- Rate limiting step in pyrimidine biosynthesis is CPSII.
- Only mitochondrial step of pyrimidine synthesis is Dihydroorotate Dehydrogenase.
- Ribonucleotides are converted to deoxy ribonucleotide by Ribonucleotide reductase at Nucleotide diphosphate level.

Check List for Revision

1. An overall idea of this chapter is suffice.
2. Learn all high yield boxes and bold letters.

Review Questions MCQ

1. **Enzyme deficiency in Lesch Nyhan Syndrome?**
   (Recent Question Nov 2017)
   a. HGPRtase
   b. APRTase
   c. Adenosine deaminase
   d. Purine Phosphorylase

2. **Hyperuricemia is/are associated with:** *(PGI Nov 2016)*
   a. HGPRTase deficiency
   b. HGPRTase overactivity
   c. PRPP Synthetase deficiency
   d. G6PD Deficiency
   e. Glucose 6 Phosphatase deficiency

3. **A child presents with hyperuricemia and delayed developmental milestones. He also has the habit of biting fingers and nails. What is the most probable enzyme deficiency?** *(AIIMS Nov 2016)*
   a. HGPRtase deficiency
   b. Phenyl Alanine Hydroxylase
   c. Adenine Deaminase
   d. Hexosaminidase A

4. **End product of purine metabolism in non-primate mammals is:** *(AIIMS May 2008)*
   a. Uric acid
   b. Ammonia
   c. Urea
   d. Allantoin

5. **Deoxy ribonucleic acid is formed from:**
   a. Ribonuclease
   b. Ribonucleotide monophosphate
   c. Ribonucleotide diphosphate
   d. Ribonucleotide triphosphate

6. **What is involved in formation of d-TMP from d-UMP?** *(PGI June 07)*
   a. N^5, N^{10}-methylene tetrahydrofolate
   b. From imino folate
   c. N5 formyl folate
   d. Dihydro folate

7. **Inosinic acid is biological precursor:** *(Nimhans 97, JIPMER 04)*
   a. Uracil and thymine
   b. Purines and thymine
   c. Adenylic acid and guanylic acid
   d. Orotic acid and uridylic acid

8. **False regarding gout is:** *(AI 2001)*
   a. Due to increased metabolism of pyrimidines
   b. Due to increased metabolism of purines
   c. Uric acid levels may not be elevated
   d. Has a predilection for the great toe

9. **The enzyme deficient in Lesch-Nyhan syndrome is:** *(PGI June 99)*
   a. GTRT
   b. Glutaminase
   c. Transcarboxylase
   d. HGPRT

10. A 10-year-old child presents with history of rashes self mutilation family history positive. Which of the following investigations do you think may be suggestive of valuable for diagnosis? *(AI 2012)*
    a. Lead
    b. Alkaline Phosphatase
    c. LDH
    d. Uric acid

11. A ten-year-old child with aggressive behaviour and poor concentration is brought with presenting complaints of joint pain and reduced urinary output. Mother gives history of self mutilate his finger. Which of the following enzymes is likely to be deficient in this child? *(AI 2009)*
    a. HGPRTase
    b. Adenosine Deaminase
    c. APRTase
    d. Acid Maltase

12. A patient with increased Hypoxanthine and Xanthine in blood with hypouricemia which enzyme is deficient?
    a. HGPRtase
    b. Xanthine Oxidase
    c. Adenosine Deaminase
    d. APRtase

13. Choose the incorrect statement. Lesch-Nyhan Syndrome: *(Kerala 2015)*
    a. Affects young boys
    b. Presents with gouty arthritis
    c. The enzyme defect enhances the reutilization of purine bases
    d. Bizzare behavior of self mutilation

14. Hyperuricemia is not found in:
    a. Cancer
    b. Psoariasis
    c. Von Gierke's disese
    d. Xanthinuria

Answers to Review Questions

1. **a. HGPRTase** *(Ref: Harper 31/e page 320)*

2. **a, e. HGPRTase deficiency, Glucose 6 Phosphatase deficiency** *(Ref: Harper 31/e page 320)*

3. **a. HGPRTase deficiency** *(Ref: Harper 31/e page 320)*

    Lesch–Nyhan Syndrome^Q
    X-linked recessive disorder^Q
    Biochemical Defect
    Complete deficiency of HGPRTase
    Purine accumulates
    Purines degraded to uric acid
    Uric acid level increases
    Clinical Features
    - Hyperuricemia, Intellectual Disability
    - Dystonic Movement, Choreoathetosis, Dysarthric Speech
    - Compulsive self-mutilation^Q
    - Megaloblastic anemia also can occur in the Lesch–Nyhan syndrome, in which regeneration of purine nucleotides is blocked.

4. **d. Allantoin** *(Ref: Harper 31/e page 318)*
    - Humans convert adenosine and guanosine to uric acid
    - In mammals other than higher primates, uricase converts uric acid to the water-soluble product allantoin.
    - Humans lack uricase, the end product of purine catabolism in humans is uric acid.

5. **c. Ribonucleotide diphosphate** *(Ref: Harper 31/e page 312)*
    Reduction of the 2'-hydroxyl of purine and pyrimidine ribonucleotides, catalyzed by the ribonucleotide reductase complex, provides the deoxyribonucleoside diphosphates (dNDPs) needed for both the synthesis and repair of DNA

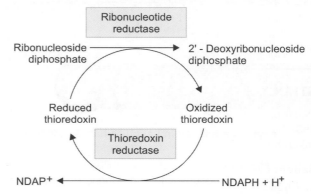

6. **a. and d. N^5N^{10} methylene tetrahydrofolate and Dihydrofolate** *(Ref: Harper 31/e page 314)*

    When TMP is formed from dUMP, N5 N10 Methylene THFA is converted to Dihydro folate

7. **c. Adenylic acid & Guanylic acid** *(Ref: Harper 31/e page 307)*
    - AMP and GMP is derived from IMP (Inosinic acid)

8. **a. Due to increased metabolism of pyrimidines**

    Remember Uric acid level may not be elevated always in gout.

9. **d. HGPRT** *(Ref: Harper 31/e page 320)*

10. **d. Uric Acid** *(Ref: Harper 31/e page 320)*

    Lesch Nyhan Syndrome
    Diagnosis
    - Hyperuricemia.
    - HGPRTase Enzyme activity in the Erythrocytes is deficient.

11. **a. HGPRtase** *(Ref: Harper 31/e page 320)*

12. **b. Xanthine Oxidase** (Ref: Harper 31/e page 321)

Hypouricemia

Hypouricemia and increased excretion of hypoxanthine and xanthine are associated with xanthine oxidase deficiency.

Lesch–Nyhan Syndrome

The Lesch–Nyhan syndrome, an overproduction hyperuricemia characterized by frequent episodes of uric acid lithiasis and a bizarre syndrome of self-mutilation, reflects a defect in **hypoxanthine-guanine phosphoribosyl transferase,** an enzyme of purine salvage.

Adenosine Deaminase Deficiency

Adenosine deaminase deficiency is associated with an immunodeficiency disease in which both thymus-derived lymphocytes (T cells) and bone marrow-derived lymphocytes (B cells) are sparse and dysfunctional. Patients suffer from severe immunodeficiency.

Purine Nucleoside Phosphorylase Deficiency

Purine nucleoside phosphorylase deficiency is associated with a severe deficiency of T cells but apparently normal B cell function.

13. **c. The enzyme defect enhances the reutilization of purine bases** (Ref: Harper 31/e page 320)

Lesch-Nyhan Syndrome is characterized by
- X-linked recessive inheritance
- Over production hyperuricemia
- Frequent episodes of uric acid lithiasis
- Bizarre syndrome of self mutilation. Defective reutilsation of Purine bases because of absence of HGPRtase enzyme.

14. **d. Xanthinuria** (Ref: Harper 31/e page 355)

10.3 ORGANIZATION AND STRUCTURE OF DNA

- Structure of DNA
- Supercoiling of DNA
- Noncanonical DNA Structures
- Topoisomerase
- Organization of DNA
- Central Dogma of Molecular Biology

STRUCTURE OF DNA

Watson, Crick, and Wilkins to propose a model of a DNA in 1953 based on the **X-ray diffraction** photographs of DNA taken by Rosalind Franklin.

Salient Features of Watson-Crick Model of DNA

1. Right handed double stranded DNA helix
2. **Base pairing rule**
   - Adenine always pairs with Thymine
   - Guanine pairs with Cytosine
3. Two strands are **antiparallel**
   The polarity of DNA is such that
   - One strand runs in 5' to 3' direction
   - Other strand runs in 3' to 5' direction
4. Hydrogen bonding
   Adenine pair with Thymine by 2 Hydrogen bonds. (A = T)
   Guanine pair with Cytosine by 3 Hydrogen bonds (G ≡ C)
5. Grooves of the DNA—Two types:
   1. Major groove
   2. Minor groove
      - Grooves often act as sites of DNA–Protein interaction needed for regulation of gene expression.
      - The DNA Protein interaction is via **hydrophobic interaction and ionic bond.**

Chargaff's Rule

In cellular DNA regardless of species

No. of Adenine = No. of Thymine

No. of Guanine = No. of Cytosine

From this sum of purine = Sum of pyrimidines

i.e. $A + G = C + T$

Different Types of DNA

There are 6 types of DNA
- A, B, C, D, E—are right handed
- **Z is left handed**

Characteristics	A DNA	B DNA	Z DNA
No. of base pairs per turns	11	10.5	12
Morphology	Broad and short	Longer and thinner	Elongated and thin
Base pair tilts The axis of helix	20° tilt	Base pair perpendicular to helix	9° tilt
Screw sense	Right handed	Right handed	Left handed

A-DNA

- X-ray diffraction studies on dehydrated DNA fibers revealed A form, called A-DNA.
- A-DNA is found in conditions of **low humidity and high salt concentration**.
- Right-handed double helix like B-DNA

Contd...

Contd...

B-DNA
- **Physiologically most common form**.
- Right handed double helix
- B-DNA is found in conditions of **high humidity and low salt concentration**.
- Highly flexible.

Z-DNA
- Phosphodiester backbone assume a zigzag form
- **Left-handed double helix**
- Seen in the **5' end of chromosomes**.
- Longer and thinner than B-DNA
- **12 bp per turn**
- Particularly seen in sequence of **alternating purine and pyrimidine**
- Particularly in $d(GC)_n$ sequence
- Methylation of Guanine and Cytosine residues stabilizes Z-form.
- Sequence that are not strictly alternating purine and pyrimidine also form Z-DNA on **methylation**.
- Z-DNA influences gene expression and regulation.

HIGH YIELD POINTS
- Physiologically most common is B-DNA.Q
- Under low salt and high degree of hydration B-DNA is usually found
- Under high salt concentration and low degree of hydration A DNA is usually found.
- The distance spanned by one turn of B-DNA is **3.4 nm (34A°)**
- The width of the double helix in B-DNA is **2 nm (20A°)**.

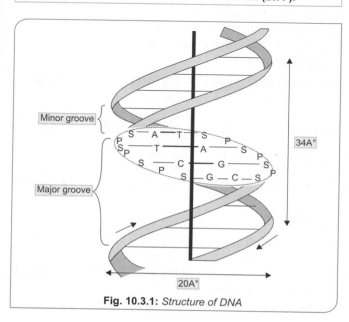

Fig. 10.3.1: *Structure of DNA*

NONCANONICAL DNA STRUCTURES

Triple-stranded DNAQ
- Triple-stranded DNA is generated by the hydrogen bonding of a third strand into the **major groove of B-DNA**
- The third strand forms hydrogen bonds with another surface of the double helix through so-called *Hoogsteen pairs.*Q
- **Poly (dA) and Poly (dT) strands** combine to form triple stranded DNA.

Four-Stranded DNA
- Four-Stranded Structure formed in DNA **high in guanine content.**
- Ends of eukaryotic chromosomes (telomeres) contain Guanine-rich sequences.
- A base pairing scheme for parallel four-stranded DNA, referred to as a **G-quartet DNA**Q
- **Hoogsteen pairs** are seen in four stranded DNA also.

HIGH YIELD POINTS
Methods to separate DNA
- DNA can be separated by HPLC, thin layer chromatography (TLC), paper chromatography and Gel electrophoresis.

Denaturation of DNA (Melting of DNA)
- The process by which two strands are separated into component strands.

Features of Denaturation
- Breaking of Hydrogen bonds
- **Phosphodiester bond is not broken**.
- Primary structure not altered, only secondary and tertiary structure altered.
- **Viscosity decreases**Q
- Increase in the optical absorbance of UV light at 260 nm by purine and pyrimidine bases, called **hyperchromicity.**Q

Melting Temperature (Tm)
The strands of a given molecule of DNA separate over a range of temperature. The midpoint is called melting temperature.

Factors influencing Tm
1. Base composition
   - More GC pairs more the Tm
2. Salt concentration
   - 10-fold increase of monovalent cation concentration increases the Tm by 16.6°C.
3. FormamideQ destabilize hydrogen bond, hence decreases Tm

4. **RNA duplex is more stable to denaturation than DNA duplex**

Application of Denaturation of DNA
- Measurement of increased optical absorbance at 260 nm is an indication of denaturation of DNA
- In recombinant DNA Technology
- Renaturation following denaturation obeying base pairing rule is applicable in various hybridization and blotting techniques.

ORGANIZATION OF DNA
- Genome in the prokaryotes^Q are loosely organized to structure called **nucleoid**
- In eukaryotes DNA is well-organized inside the nucleus.

Levels of Organization of DNA
I. DNA double helix
II. 10 nm chromatin fibril
III. 30 nm chromatin fibril
IV. Nuclear scaffold form (Interphase Chromosome)
- Non-condensed loop
- Condensed loop
V. Metaphase Chromosome

DNA Double Helix
- First level of organization of DNA
- The characteristics are same as that of Watson Crick model of DNA.
- Diameter is 2 nm

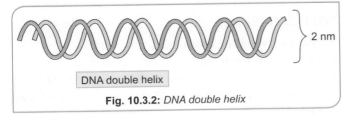

Fig. 10.3.2: *DNA double helix*

10 nm Chromatin Fibril
- Consist of nucleosomes separated by linker DNA
- Nucleosome is a nucleoprotein complex
- DNA double helix is wrapped nearly twice^Q (exactly **1.75 times**) over a histone octamer in **left-handed helix** to form a disc like structure.
- It is a **Solenoidal** supercoil
- Individual nucleosome are linked together by **30 bp segment** called linker.
- This gives a **Beads on a String appearance**^{Q} on electron microscopy.

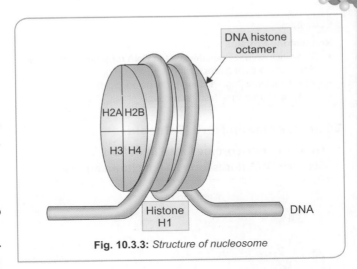

Fig. 10.3.3: *Structure of nucleosome*

HIGH YIELD POINTS

Nucleosome at a Glance
- No. of turns of DNA on Histone octamer—1.75 turns
- Direction of turn of the DNA over Histone—left-handed
- No. of base pairs in 1.75 turn of DNA—145-150 bp
- Diameter of nucleosome—10 nm
- No. of base pairs in the linker DNA—30 bp

Histones
- **Most abundant chromatin protein.**
- Small family of closely related basic proteins.
- The carboxyl terminal two-third is hydrophobic, while amino terminal one-third is rich in basic amino acids like **arginine and lysine**.
- Core histones are subject to at least six types of post-translational modifications.
- They are highly conserved among the species. Histones are divided into:
  1. Core Histones
  2. Linker Histones

Core Histones
- Core histones are **H2A, H2B, H3, and H4**.
- They form histone octamer.
- H3 and H4 forms tetramer, while H2A and H2B form dimers.
- (H3-H4)$_2$ tetramer associate with two (H2A-H2B) dimers to form histone octamer.

Linker Histones
- **H1 histone** which is seen in the linker region.
- This is loosely bound^Q to nucleosome.

EXTRA EDGE

Non-histone Proteins
- They include enzymes involved in DNA replication and repair, RNA synthesis and processing.
- Unlike histones, non-histone proteins are acidic.
- They are larger than histone proteins.

30 nm Chromatin Fibril (Solenoid)
- Groups of nucleosome form "DNA fibril"
- Six such DNA fibrils form 30 nm chromatin fibril.

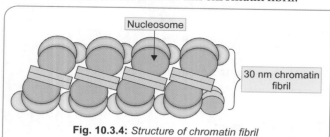

Fig. 10.3.4: Structure of chromatin fibril

Nuclear Scaffold associated Form or Interphase Chromosome
- 30,000 to 100,000 bp loops or domains anchored in a scaffolding or supporting matrix, called nuclear matrix.
- Loops can be a condensed loop or non-condensed loops.

Euchromatin[Q] and Heterochromatin[Q]

Euchromatin
- Chromatin is less **densely packed.**[Q]
- **Transcriptionally active**[Q]
- Chromatin stains **less densely**
- Also called **permissive chromatin**

Heterochromatin
- Chromatin is densely packed[Q]
- Transcriptionally in active[Q]
- Chromatin stains densely[Q]
- Also called **repressive chromatin**

Two types of heterochromatin

1. **Constitutive heterochromatin**
   - Always condensed
   - Essentially inactive
   - Seen in **Centromere** and chromosomal ends of the **telomere.**

2. **Facultative heterochromatin**
   - Is at times condensed, but at other times it is uncondensed and actively transcribed, e.g. one of the X chromosome in mammalian female.

The heterochromatic X chromosome decondenses during gametogenesis.

SUPERCOILING OF DNA

DNA can be in relaxed or Superhelical.
- Most cellular DNA is **unwound**
- Linear B DNA is relaxed
- This is thermodynamically most favored
- But biological activity of relaxed DNA, like replication, repair, etc. is reduced
- The biologically active form is superhelical
- The biologically active form is superhelical, but it is topologically strained isomer.

1. **Positive Supercoils**
2. **Negative Supercoils**

Positive Supercoils
Circular DNA twisted in the direction same as that original rotation, i.e. right-handed creates positive supercoils. Such DNA is said to be **over wound.**

Negative Supercoils
Circular DNA twisted in the **direction opposite from the clockwise** turns of the right-handed double helix, i.e. left handed creates negative supercoils.

Such DNA is said to be **underwound.**

Functions of Supercoils
- Supercoiling promotes packing of DNA into compact structures.
- Helps to generate regions with broken hydrogen bonds which facilitate DNA strand separation and facilitate replication, repair and recombination of the DNA.

TOPOISOMERASE

- **Nicking resealing enzyme.**
- Enzymes that can relax or insert supercoils.
- Enzymes that relieve torsional strains in the DNA.

Topoisomerases can be of two types: Type I and Type II.

Topoisomerase Type I

- Make transient **single stranded**[Q] break in a negatively supercoiled DNA double helix
- Phosphodiester bond is interrupted
- Topoisomerase remain bound to the phosphoryl group at incision site by a covalent bond, till the nick is resealed.
- ATP is not needed[Q]

There are two classes of Topoisomerases I, Type IA and Type IB Topoisomerase.

Topoisomerase Type II

- Dimeric enzyme that bind to double stranded DNA, make breaks in **both the strands**^{Q} of DNA.
- Can insert and remove supercoils.
- ATP is needed.^{Q}

High Yield Points

- Bacterial DNA Gyrases^Q is a subset of Topoisomerases type II
- All Topoisomerases type II relaxes supercoils in the DNA.
- Bacterial DNA Gyrases are the only subset of type II topoisomerases that can add negative supercoils.

Extra Edge

Some important prokaryotic topoisomerases

Prokaryotic topoisomerases	Type	Functions
E coli Topoisomerase I	IA	Relaxes negatively supercoiled DNA
E coli Topoisomerase III	IA	Relaxes negatively supercoiled DNA
E coli Topoisomerase IV	II	Relaxes negatively supercoiled DNA
E coli DNA Gyrase	II	Introduces negative supercoils. Relaxes either positive or negative supercoils

High Yield Points

- Almost all E coli Topoisomerases relaxes negative supercoils.
- Only E coli DNA Gyrase introduces negative supercoils.
- All Eukaryotic DNA Topoisomerases relax negative and positive supercoils.

Extra Edge

Bacterial Topoisomerases Inhibitors
1. Nalidixic Acid.
2. Fluoroquinolones
3. Novobiocin

Human Topoisomerase Inhibitors
Used in Cancer chemotherapy.

Human Topoisomerases that inhibit Type I Topoisomerases
- Irinotecan
- Topotecan

Contd...

Human Topoisomerases that inhibit Type II Topoisomerases
- Etoposide
- Adiramycin (Doxorubicin)
- Daunorubicin
- Idarubicin

CENTRAL DOGMA OF MOLECULAR BIOLOGY

Three process involved in Central Dogma of molecular biology are:

1. DNA replication (DNA to DNA)
2. Transcription (DNA to RNA)
3. Translation (RNA to protein)

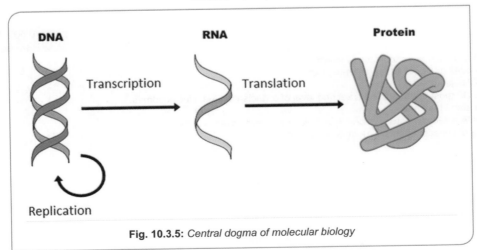

Fig. 10.3.5: *Central dogma of molecular biology*

Current Central Dogma of Molecular Biology

With the advances made in Human Genome Projects, the central dogma of molecular biology is changed.

Genome → Transcriptome → Proteome

- **Genome:** The complete set of genes of an organism.
- **Transcriptome:** The **complete set of RNA transcripts** produced by the genome of an organism.
- **Proteome:** The complete complement of proteins of an organism.

High Yield Points

Numerals in Molecular Genetics
- Total number of chromosome in humans—**46 (23pairs)**
- The number of base pairs in haploid set of chromosome is 3.0×10^9 bp (3 billion bp)
- Percentage of exons in human genome is approximately 1.14%
- The number of protein coding genes in human genome is **20,687.**
- The genes account for 10–15% of DNA.

Extra Edge

Repetitive Sequences in the Human DNA
Thirty percent of genome consist of repetitive sequence. More than half is unique or nonrepetitive sequences.
Repetitive sequences can be broadly classified into:

Highly Repetitive
5–500 bp repeated 1–10 million times in tandem per haploid genome. Seen clustered in **centromere and telomere**

Moderately Repetitive Sequences
In less than 1 million copies per haploid genome
Depending on the repeat size they are classified into:
i. Long interspersed repeat sequences (**LINEs**)—6–7 Kbp 20,000 to 50,000 copies pergenome
ii. Short Interspersed Repeat Sequence (**SINEs**)—70–300 bp 100,000 copies pergenome

Most LINEs and SINEs are transposable elements.
Alu family is an example of SINEs.

Quick Review

- Adenine always pair with Thymine by two hydrogen bonds.
- Guanine always pair with Thymine by three hydrogen bonds.
- Most stable DNA is B DNA.
- Type of DNA seen in high salt concentration and low degree of hydration is A DNA.
- During denaturation phosphodiester bond is not broken.
- Denatured DNA is hyperchromatic as it absorbs more UV light at 260 nm.
- The more GC pair the more is the melting temperature.
- Nucleosome is Histone octamer + 1.75 turns of DNA.
- Most abundant chromatin protein is histones.
- Euchromatin is transcriptionally active and less condensed.
- Heterochromatin is transcriptionally inactive and highly condensed
- Nicking resealing enzyme is Topoisomerase.
- Bacterial DNA Gyrase belongs to Topoisomerase II.
- Reverse transcription is not a process in central dogma of molecular biology.

Check List for Revision

- Structure of DNA is a must learn topic for all exams.
- Types of DNA—boxes and bold letters are must learn.
- Denaturation of DNA is important.
- Organisation of DNA most important is the second level of organisation, i.e. Nucleosomes.
- Learn everything about histones.
- Supercoiling of DNA a selective study of points in bold letters and boxes is suffice.
- An understanding of concept of central dogma is needed.

Review Questions MCQ

d. Mainly consists of left handed helix
e. 5'-3' Phosphodiester bonding is present

1. **What is the bond between the strands in the given diagram?** *(AIIMS Nov 2017)*

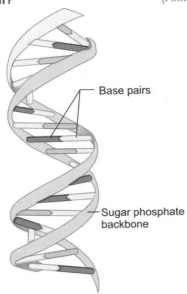

Base pairs
Sugar phosphate backbone

   a. Hydrogen bond
   b. Phosphodiester bond
   c. Covalent bond
   d. Glycosidic bond

2. **Two strands of the DNA are joined by:** *(PGI May 2016)*
   a. Glycosidic bond
   b. Hydrogen bond
   c. Covalent bond
   d. Ionic bond
   e. Van der Waal's force

3. **Highly repetitive DNA is found in:** *(PGI Nov 2015)*
   a. Cloning of DNA
   b. Microsatellite DNA
   c. Telomere
   d. Centromere
   e. DNA Transposons

4. **Which of the following is false?** *(PGI 2002)*
   a. Ratio of A:T & G:C is approximately equal to 1:1
   b. Ratio of A:G & T:C is approximately equal to 1:1
   c. A+T = G+C
   d. A+C = G+T
   e. A+G = C+T

5. **True about DNA structure:** *(PGI Nov 2010)*
   a. Purines are adenine, guanine and pyrimidines are uracil and cytosine
   b. Watson and Crick discovered structure in 1973
   c. Deoxyribose–phosphate backbone with bases stacked inside
   d. Mainly consists of left handed helix
   e. 5'-3' Phosphodiester bonding is present

6. **If a sample of DNA if adenine is 23% what will be the amount of guanine present?** *(PGI May 2013)*
   a. 23%
   b. 25%
   c. 46%
   d. 27%
   e. 54%

7. **True statements about DNA structure:** *(PGI June 06)*
   a. All nucleotides are involved in linkage
   b. Antiparallel
   c. Parallel
   d. Bases are perpendicular to DNA
   e. Attached by hydrogen I bond

8. **The two strands of DNA are held together by:** *(AIIMS Feb 97)*
   a. Van der Waal bond
   b. Hydrogen bond
   c. Covalent bond
   d. Ionic interaction

9. **Which form of DNA is predominantly seen in our body?** *(AI 1996)*
   a. A
   b. C
   c. B
   d. Z

10. **Chargaff rule state that:** *(Bihar 98, MP 04, UP 03)*
    a. A + G = T + C
    b. A/T = G/C
    c. A = U = T = G = C
    d. A + T = G + C

11. **A nucleic acid was analyzed and found to contain 32% adenine, 18% guanine, 17% cytosine and 33% thymine. The nucleic acid must be:** *(AIIMS May 06)*
    a. Single stranded RNA
    b. Single stranded DNA
    c. Double stranded RNA
    d. Double stranded DNA

12. **Triple bonds are found between which base pairs?** *(AI 2001)*
    a. A–T
    b. C–G
    c. A–G
    d. C–T

13. **At the physiological pH the DNA molecules are:** *(AIIMS Nov 02)*
    a. Positively charged
    b. Negatively charged
    c. Neutral
    d. Amphipathic

14. **Total number of genes in a human being is:** *(Kerala 2001, CMC 04, WB 98)*
    a. 800,000
    b. 50,000
    c. 100,000
    d. 30,000

15. **Triplex DNA is due to:** *(AIIMS May 2011)*
    a. Hoogsteen pairing
    b. Palindromic sequences
    c. Large no. of guanosine repeats
    d. Polypyrimidine tracts

16. About DNA which of the following is true:
    (JIPMER 2014)
    a. The nucleotide of one strand form bonds with nucleotide of opposite strand.
    b. Cytosine and Uracil differ by one ribose sugar
    c. The information from DNA is copied in the form of tRNA
    d. Each nucleotide pair includes two purines.

17. Which model of DNA was discovered by Watson and Crick? (Recent Question 2016)
    a. ADNA   b. BDNA
    c. CDNA   d. ZDNA

18. Total number of base pairs in human haploid set of chromosome: (Ker 2007)
    a. 3 million   b. 3 billion
    c. 33 billion d. 5 million

19. Proteins seen in chromosomes are called: (Ker 2006)
    a. Nucleotides   b. Histones
    c. Apoproteins   d. Glycoproteins

20. Euchromatin is the region of DNA that is relatively:
    (AI 2006)
    a. Uncondensed   b. Condensed
    c. Over condensed d. Partially condensed

21. The long and short arms of chromosomes are designated respectively as: (AI 2006)
    a. P and q arms   b. M and q arms
    c. q and p arms   d. l and s arms

22. True about Histone Proteins: (PGI May 2012)
    a. Ribonucleoprotein
    b. Present inside the nucleus
    c. Acidic
    d. Basic
    e. Glycoprotein

23. Y-chromosome is: (AIIMS May 2008)
    a. Metacentric
    b. Sub-metacentric
    c. Acrocentric
    d. Longer than the X-chromosome

24. Nucleosome consists of:
    (PGI May 2010, PGI May 2016)
    a. Histone   b. DNA
    c. RNA   d. DNA & RNA both
    e. Carbohydrate

25. Component of chromosome are: (PGI Dec 03)
    a. DNA   b. tRNA
    c. mRNA   d. rRNA
    e. Histones

26. The protein rich in basic amino acids, which functions in the packaging of DNA in chromosome, is: (AI 2003)
    a. Histones   b. Collagen
    c. Hyaluronic acid binding protein
    d. Fibrinogen

27. Random inactivation of X chromosome is:
    a. Lyonisation   b. Allelic exclusion
    c. Randomisation   d. Genomic imprinting

28. In the entire genome, the coding DNA constitutes how much? (AIIMS May 2014)
    a. 0.01   b. 0.02
    c. 0.25   d. 0.4

29. True about DNA hyperchromatism: (PGI Nov 2013)
    a. It is increase of absorbance
    b. Measured by absorbance at 260 nm (in a spectrophotometer)
    c. It occurs when the DNA duplex is denatured
    d. Double stranded DNA is more hyperchromic than ssDNA

- Between the DNA Hydrogen bond
- Between base and Sugar beta N Glycosidic bond, Nucleoside and Phosphate Ester bond
- Base stacking by van der Waal's forces

Answers to Review Questions

1. **a. Hydrogen bond**
   - The bond between the base pairs in the two strands of DNA is hydrogen bond.
   - Phosphodiester bond is between the nucleotides
   - Glycosidic bond is between Nitrogenous base and pentose sugar

2. **b. Hydrogen bond**
   - Bonds in the DNA
   - Between nucleotides-3' to 5' Phosphodiester bond

3. **c, d. Telomere, Centromere** (Ref: Harper 31/e page 340)

   DNA transposons are moderately repetitive sequences. Microsatellite are repetitive sequences 2-6 bp repeated up to 50 times. But they are not highly repetitive.

4. **b, c, d. Based on Chargaff's rule, Purines = Pyrimidines**

   Adenine pair with Thymine and Guanine with Cytosine, hence ratio of A:T & G:C is approximately equal to 1:1

   Option D is false because it is not as per Chargaff's rule as it says No of Purine = No of Pyrimidines

5. **c. Deoxyribose–phosphate backbone with bases stacked inside** *(Ref: Harper 31/e page 340)*

   Structure of DNA

   Elucidated by Watson and Crick based on the X-ray diffraction picture taken by Rosalind Franklin in 1953 and got Nobel prize in 1962.

   Salient Features of Watson—Crick Model of DNA
   Right handed.
   Double stranded DNA helix
   Purines are Adenine and Guanine and Pyrimidines are Cytosine and Thymine.
   Base Pairing Rule:
   - Adenine always pairs with Thymine.
   - Guanine pairs with Cytosine.

   Two strands are antiparallel
   - The polarity of DNA is such that.
     - One strand runs in 5' to 3' direction.
     - Other strand runs in 3' to 5' direction.

   Hydrogen Bonding:
   - Adenine pair with Thymine by 2 Hydrogen bonds (A =T).
   - Guanine pair with Cytosine by 3 Hydrogen bonds (G =C).

   Grooves of the DNA Two types:
   - Major groove
   - Minor groove

   Grooves acts as sites of DNA. Protein interaction needed for regulation of gene expression.

6. **d. 27%**

   Based on Chargaff' rule
   Purines = Pyrimidine
   No of Adenine = No of Thymine
   So Adenine + Thymine = 46% Guanine + Cytosine = 54%
   No of Guanine = No of Cytosine
   So the amount of Guanine = 54/2 = 27%

7. **a, b, d, e. All nucleotides are…, Antiparallel, Bases are perpendicular, attached by hydrogen 1 bond** *(Ref: Harper 31/e page 340)*

8. **b. Hydrogen bond** *(Ref: Harper 31/e page 340)*

9. **c. BDNA** *(Ref: Harper 31/e page 340)*
   - Physiologically most common is B-DNA.^{Q}
   - Under low salt and high degree of hydration B DNA is usually found
   - Under high salt concentration and low degree of hydration A-DNA is usually found.
   - The distance spanned by one turn of B-DNA is 3.4 nm (34A°)
   - The width of the double helix in B-DNA is 2 nm (20A°)

10. **a. A + G = T+ C** *(Ref: Harper 31/e page 340)*

    Chargaff's rule is Total number of Purines = Total number of Pyrimidines in a double stranded DNA.

11. **d. Double stranded DNA** *(Ref: Harper 31/e page 340)*

    The above nucleic acid is obeying Chargaff's rule as the number of Purines (32 + 18) = No of Pyrimidines (17 + 33) Double stranded RNA is not the answer as it does not contain Thymine.

12. **b. C–G**

    Three hydrogen bonds between C & G
    Two Hydrogen bonds between A & T.

13. **b. Negatively charged**

    DNA is negatively charged because of Phosphate group.

14. **d. 30,000**

    This is an old question, Now it is found that number of genes is approx 20,000.

15. **a. Hoogsteen pairing**
    (Ref: Textbook of Biochemistry with Clinical Correlation Thomas M Devlin 7/ep43)

    NONCANONICAL DNA STRUCTURES
    Triple-stranded DNA

    Triple-stranded DNA is generated by the hydrogen bonding of a third strand into the major groove of B-DNA Commonly seen in Polynucleotides, Poly (dA) and Poly (dT)

    The third strand forms hydrogen bonds with another surface of the double helix through so-called Hoogsteen pairs.

16. **a. The nucleotide of one strand form bonds with nucleotide of opposite strand**

    One starnd of DNA join with other strand of DNA by means of Hydrogen bond between the bases.
    Cytosine and uracil differ by amino group.
    The information in the DNA is copied in the form of mRNA
    Each nucleotide pair includes one purine and one pyrimidine.

17. **b. BDNA**

    Organization of DNA

18. **b. 3 billion** *(Ref: Harper 31/e page 377)*
    - Human haploid genome of each cell consist of 3×10^9 bp (3 billion bp)
    - Current estimates predict 20,687 protein coding genes.
    - Exome constitutes 1.14% of genome.
    - SNPs estimated is 10 million

19. **b. Histones** *(Ref: Harper 31/e page 371)*

    Histones are the most abundant histone proteins.

20. **a. Uncondensed** *(Ref: Harper 31/e page 354)*

    Euchromatin and Heterochromatin Euchromatin
    - Chromatin is less densely packed.
    - Transcriptionally active
    - Chromatin stains less densely Heterochromatin
    - Chromatin is densely packed.
    - Transcriptionally inactive.
    - Chromatin stains densely.

Two Types of Heterochromatin
1. Constitutive Heterochromatin:
   - Always condensed
   - Essentially inactive
   - Seen in Centromere and chromsomal ends of the telomere.
2. Facultative Heterochromatin:
   Is at times condensed, but at other times it is uncondensed and actively transcribed. e.g.: One of the X chromosome in mammalian female. The heterochromatic X chromosome decondenses during gametogenesis.

21. **c. q and p arms**
    (Ref: Emery's Elements of Medical Genetics 13/e p31)

22. **b, d. Present inside.., basic** (Ref: Harper 31/e page 351)

    Histones
    Most abundant Chromatin Protein.
    Small family of closely related basic Proteins.
    There are 5 classes of Histones
    - H1, H2A, H2B, H3, & H4

    Core Histones are
    - H3 + H4 + H2A + H2B
    - They form Histone Octamer.

    Linker Histones:
    - H1 histone which is seen in the linker region.
    - Loosely bound to nucleosome.

    Nonhistone proteins:
    - Most of which are acidic and larger than histones
    - The nonhistone proteins include enzymes involved in DNA replication and repair, and the proteins involved in RNA synthesis, processing, and transport to the cytoplasm.

23. **c. Acrocentric**
    (Ref: Emery's Elements of Medical Genetics p30,31)

    Karyotype
    Arrangement of chromosome in the decreasing order of length

Different Groups of Chromosome		
Groups	Chromosome	Description
A	1-3	Largest;1 and 3 Metacentric 2-Submetacentric
B	4,5	Large Submetacentric
C	6-12,X	Medium size, Submetacentric
D	13-15	Medium size, Acrocentric with satellite
E	16-18	Small; 16 is metacentric but 17 and 18 are Submetacentric
F	19-20	Small metacentric
G	21,22,Y	Small acrocentric

> **Points to Ponder**
> - X chromosome is Submetacentricc.
> - Y Chromosome is Small Acrocentric (G)
> - Most common is Submetacentric
> - Humans lack Telocentric Chromosome

24. **a, b. Histone, DNA** (Ref. Harper 31/e page 351)

    Nucleosome: Nucleoprotein complex
    DNA double helix is wrapped nearly twice over a histone octamer in left handed helix to form a disc like structure. Individual nucleosome are linked together by 30 bp segment called linker.
    This gives a **Beads on a String appearance** on electron microscopy.

25. **a, e. DNA, Histones** (Ref: Harper 31/e page 351)

26. **a. Histones** (Ref: Harper 31/e page 371)

27. **a. Lyonisation**

    Two factors that are peculiar to the sex chromosomes: (1) lyonization or inactivation of all but one X chromosome and (2) the modest amount of genetic material carried by the Y chromosome.

    In 1961, Lyon outlined the idea of X-inactivation, now commonly known as the Lyon hypothesis. It states that (1) only one of the X chromosomes is genetically active, (2) the other X of either maternal or paternal origin undergoes heteropyknosis and is rendered inactive, (3) inactivation of either the maternal or paternal X occurs at random among all the cells of the blastocyst on or about day 16 of embryonic life, and (4) inactivation of the same X chromosome persists in all the cells derived from each precursor cell.

    Theinactive X can be seen in the interphase nucleus as a darkly staining small mass in contact with the nuclear membrane known as the Barr body, or X chromatin. The molecular basis of X inactivation involves a unique gene called XIST, whose product is a noncoding RNA that is retained in the nucleus, where it "coats" the X chromosome that it is transcribed from and initiates a gene-silencing process by chromatin modification and DNA methylation. The XIST allele is switched off in the active X.

28. **a. 0.01**

    In the entire genome, the coding DNA constitutes 1.14% (~1%)

29. **a, b, c. It is increase of absorbance, Measured by absorbance at 260 nm (in a spectrophotometer). It occurs when the DNA duplex is denatured**

    During denaturation of DNA, there is increased in absorbance at 260 nm, measured by Spectrophotometry. This is called hyperchromicity.
    Ss DNA is more hyperchromic than dsDNA.

10.4 DNA REPLICATION AND REPAIR

☞ DNA Replication ☞ DNA Repair Mechanisms

DNA REPLICATION

Definition

The process by which copying of base sequence present in the parent strand to daughter strand, thereby passing the genetic information from parent to progeny is called **Replication**.

Salient Features of DNA Replication

1. Occurs in the **S Phase** of the cell cycle.
2. DNA strands separate and each acts as template strand on which complementary strand is synthesised.
3. Base pairing rule is obeyed.
4. **Semiconservative nature**—Proved by Meselson and Stahl Experiment.
   - Half of the parent strand is conserved in the daughter DNA.
5. New Strand is synthesised always in **5' to 3' direction**.
6. Overall DNA replication is **bidirectional**.
7. Synthesis of DNA in both strands are not similar.
   - Leading strand—The strand which DNA is continuously polymerised.
   - Lagging strand—The strand which DNA is discontinuously polymerised.
8. DNA replication is **semidiscontinuous**.

Enzymes Involved in the DNA Replication

1. **Topoisomerases**
   - Relieve torsional strain that results from helicase-induced unwinding of DNA.
   - Nicking resealing enzyme.
2. **Helicase**: **ATP driven** processive unwinding of DNA.
3. Single Strand Binding protein (SSB) prevent premature reannealing of dsDNA.
4. **DNA Primase:**
   - Initiates synthesis of RNA primers.
   - Special class of **DNA dependent RNA Polymerase**
5. **DNA Polymerase:** Catalyse the chemical reaction of DNA Polymerisation. Synthesise DNA only in 5' to 3' direction.
6. **DNA Ligase:** Seals the single strand nick between the nascent chain and Okazaki fragments **on lagging strand**.

Image-Based Information

Semiconservative Model of DNA Replication

- Semiconservative model of DNA Replication proved by Meselson and Stahl Experiment
- Half of the parent strand is conserved in the daughter strand.

DNA Polymerases[Q]

- Enzymes that catalyse Deoxyribonucleotide polymerisation.
- The initiation of DNA synthesis by DNA Polymerase always require priming by a short length of RNA, called Primer.

Three Important Properties of DNA Polymerase Complex

1. **Chain elongation**[Q]: Chain elongation accounts for the rate (in nucleotides per second; nucleotides/s) at which polymerization occurs
   Rate of chain elongation of DNA Pol III is 20–50 nucleotides/second.
2. **Processivity**[Q]: Processivity is an expression of the number of nucleotides added to the nascent chain before the polymerase disengages from the template. Processivity of DNA Pol III is 100 to >50,000 nucleotides
3. **Proofreading**[Q]: The proofreading function identifies copying errors and corrects them.

High Yield Points

- Proofreading function needs **3' to 5' exonuclease** activity
- Repair function needs **5' to 3' exonuclease activity**.

Prokaryotic DNA Polymerase

Three types of Prokaryotic DNA Polymerase:
1. Pol I
2. Pol II
3. Pol III

Prokaryotic DNA Polymerase	Function
Pol I	Removal of Primers and Gap filling on lagging strand DNA Proofreading DNA Repair Recombination
Pol II	DNA proofreading and repair
Pol III	Processive, leading strand synthesis Synthesis of Okazaki fragments DNA Proofreading

Role of DNA Sliding Clamp

Extra Edge

DNA Polymerase III associate with two identical β subunits of DNA **Sliding "clamp"** which increases the Pol III–DNA stability, processivity and rate of chain elongation.

High Yield Points

Bacterial DNA Polymerase—at a Glance
- Main replication DNA Polymerase is **DNA Polymerase III**
- DNA Polymerase with highest rate of chain elongation (**Most Processive**)^Q is **Pol III**
- DNA Polymerase with **proofreading activity**^{Q}—**Pol I, Pol II and Pol III**
- DNA Polymerase with repair activity—**Pol I and PolII**
- DNA Polymerase which fills the gap in the lagging strand **is PoI I**
- DNA Polymerase which polymerise Okazaki fragments **Pol III**
- DNA Polymerase which synthesize leading strand **Pol III**
- **Kornberg's Enzyme is DNAP I**, as it is Discovered by Arthur Kornberg
- Arthur Kornberg described the existence of DNA Polymerase I in E.coli
- **Klenow Fragment**^{Q}: DNA Polymerase in which 5' to 3' exonuclease activity is removed.

Eukaryotic DNA Polymerase

Mainly five types of Eukaryotic DNA Polymerase:
1. DNAPα
2. DNAPβ
3. DNAPγ
4. DNAPδ
5. DNAPε

Eukaryotic DNA Polymerase	Function
DNAP alpha	Primase
DNAP beta	**High fidelity** DNA repair
DNAP gamma	Mitochondrial DNA synthesis
DNAP delta	Lagging Strand Synthesis
DNAP epsilon	Leading Strand Synthesis

High Yield Points

- Eukaryotic DNA Polymerase involved in **DNA repair** are DNA Polymerase β (Main Repair enzyme), also **DNAP zeta**, DNAP eta and **DNAP kappa**
- Eukaryotic DNA polymerase **with proofreading** activity are **DNAP gamma, DNAP delta** and **DNAP epsilon**

Steps of DNA Replication

Identification of the origins of replication

Fixed points on the chromosome where replication begins are called Ori.

Single ori in bacteria

Multiple ori present in eukaryotes

Ori in different organisms
- In *E. coli*-ori C
- In Bacteriophage λ-ori λ
- In Yeast-Autonomous Replicating Sequence (ARS)
- In humans similar to Yeast.

There is an AT rich sequence adjacent to ori facilitating DNA unwinding.

Extra Edge

In eukaryotes ~80 bp AT rich sequence called **DNA Unwinding Element (DUE)**

- **Ori + dsDNA Binding Protein (dnaA)** opens the DNA Duplex
- **Unwinding** (denaturation) of dsDNA to provide an ssDNA template
- ori + ds binding protein causes local denaturation of DNA.
- This facilitates the further unwinding of DNA by **Helicase**.

Role of single strand binding protein (SSB)
- **Prevents the re-annealing** of the separated DNA strands.
- Human SSBs are called **Replication Protein A (RPA)**

Role of topoisomerase
- Strand separation create topological strain in the DNA, which is relieved by Topoisomerase.

Formation of the Replication Fork
- Unwinding of DNA forms replication bubble.
- A pair of replication fork is replication bubble.

Synthesis of RNA Primer
By the enzyme Primase
Synthesise 100–200 length ribonucleotides

- **DNA G** is the primase in case of Prokaryotes
- **DNAP α**^{Q} has primase activity in case of eukaryotes.

Initiation of DNA Synthesis and Elongation
- Two strands are synthesised in different manner^Q

Leading strand (continuous strand, forward strand) synthesis

The $3' \rightarrow 5'$ strand is called leading strand template
On this strand daughter strand (leading strand) is synthesised in **continuous**^{Q} manner in **5' to 3' direction**

Enzyme for leading strand synthesis
- In Prokaryotes by **DNA Polymerase III**
- In Eukaryotes by **DNA Polymerase ε**.

Lagging strand (discontinuous strand) (retrograde strand) synthesis
The $5' \rightarrow 3'$ **strand** is called **lagging strand template.**
On this strand daughter strand (Lagging strand) synthesised in discontinuous^Q manner

Small fragments of DNA are added in short spurts called Okazaki Fragment^Q synthesized in 5'–3' direction.

Okazaki fragment—facts
- By **DNAP III** in prokaryotes.
- By **DNAP δ** in eukaryotes.
- Length of Okazaki fragments in Prokaryotes—1000 to 2000 nucleotides
- Length of Okazaki fragments in Eukaryotes—100 to 250 nucleotides.

Removal of RNA Primers and gap filling of the lagging strand.
- In Prokaryotes
  - Removal of RNA Primer
  - Gap filling by **DNA polymerase-I**
- In eukaryotes
  - RNAse H removes the primer
  - DNAP δ fills the gap, where RNA Primer is removed.

Sealing the nick following gap filling:
- **DNA ligase:** Seals the single strand nick between the nascent chain and Okazaki fragments on lagging strand^Q

> **HIGH YIELD POINTS**
> - Time taken for replication in bacteria is 30 minutes.
> - Time taken for replication in entire human genome is 9 hours.

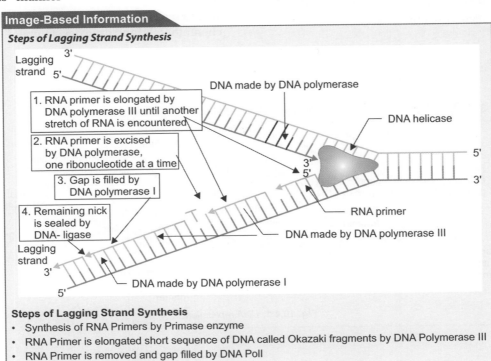

Steps of Lagging Strand Synthesis
- Synthesis of RNA Primers by Primase enzyme
- RNA Primer is elongated short sequence of DNA called Okazaki fragments by DNA Polymerase III
- RNA Primer is removed and gap filled by DNA PolI
- Nicks are sealed by DNA Ligase

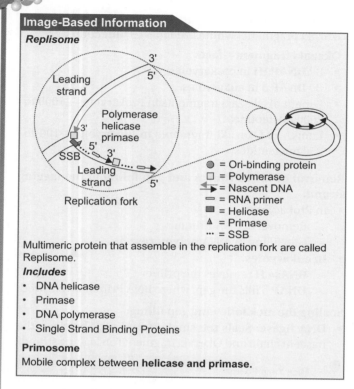

Image-Based Information

Replisome

Multimeric protein that assemble in the replication fork are called Replisome.

Includes
- DNA helicase
- Primase
- DNA polymerase
- Single Strand Binding Proteins

Primosome
Mobile complex between **helicase and primase**.

Telomere[Q] and Telomerase[Q]

- On the 5' end of the newly synthesized linear DNA, RNA primer is removed by RNAase H.
- This leaves a **gap at the 5' end of daughter strand**.
- In other words **3' end of the leading strand template is not replicated**
- This results in shortening of DNA with each cell division.
- This is prevented by presence of **Telomere[Q]** and **Telomerase.[Q]**

Telomeres

- The ends of chromosome contain structures called telomeres.
- Telomeres consist of short T-G repeats.
- Human telomeres have variable number of **tandem repeats of the sequence 5'TTAGGG-3'**.

Telomerase (Telomere Terminal Transferase)

- Enzyme which prevent shortening of DNA.
- Has an **intrinsic RNA primer[Q]**
- Has **Reverse Transcriptase[Q] (RNA Dependent DNA Polymerase)** activity.
- Present in **Germ line[Q], stem cells[Q], most cancer cells[Q]**.
- Absent from **most somatic cells[Q]**.

Clinical Significance of Telomerase

- Absence of Telomerase lead to **premature ageing**.
- In Cancer cells increased Telomerase activity.
- Telomere shortening is associated with **ageing, malignancy**.
- Telomerase has become an attractive target for cancer chemotherapy and drug development.

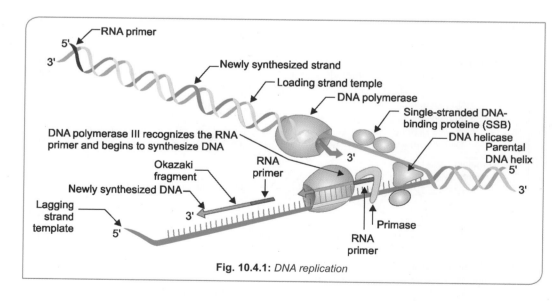

Fig. 10.4.1: *DNA replication*

Reverse Transcriptase
- Temin and Baltimore isolated this enzyme in 1970
- They are RNA Dependent DNA Polymerase
- Synthesize new DNA strand with RNA as template.
- Thus they reverse Central dogma of molecular genetics.
- These enzyme is important in RNA Viruses like Retroviruses.
- Telomerase have reverse transcriptase activity.

Image-Based Information
Eukaryotic and Prokaryotic Replication

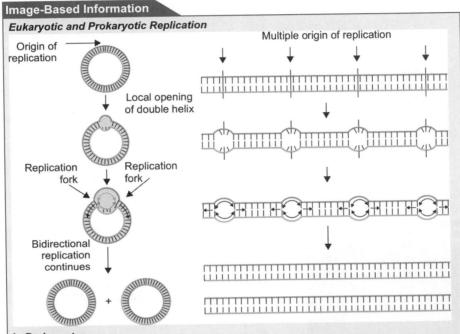

In Prokaryotes
- Circular DNA with single origin of replication
- Replication fork progressing bidirectionally

In Eukaryotes
- Very long linear DNA with multiple origin of replications
- Replication forks progressing bidirectionally

DNA REPAIR MECHANISMS

- DNA is subjected to a huge array of chemical, physical, and biological assaults on a daily basis.
- Repair of damaged DNA is critical for maintaining genomic integrity and thereby preventing the propagation of mutations.
- Eukaryotic cells contain five major DNA repair pathways.

The mechanisms of DNA repair include:
- Nucleotide Excision Repair, NER
- Mismatch Repair, MMR.
- Base Excision Repair, BER.
- Homologous Recombination, HR.
- Nonhomologous End-Joining, NHEJ repair.

EXTRA EDGE

Enzymes of DNA repair mechanism
Nucleotide Excision Repair
- ABC **Excinuclease**
- DNA PolI
- DNA Ligase

Base Excision Repair
- DNA **NGlycosylase**
- AP Endonuclease
- DNA PolI
- DNA Ligase

Mismatch Repair
- Dam Methylase
- **GATC Endonuclease**
- Mut H, Mut L, Mut S
- DNA Helicase
- SSB proteins
- DNA PolI
- DNA Ligase

DNA Damaging Agents	Defects in DNA	Repair Mechanism	Disorder Associated
1. Ionizing radiations 2. X rays[Q] 3. Antitumour drugs	1. Double Strand Breaks[Q] 2. Single Strand Breaks 3. Intrastrand cross links 4. Interstrand cross links	Nonhomologous End Joining (NHEJ)	Severe Combined Immunodeficiency (SCID)
		Homologous Recombination (HR)	Ataxia Telangectasia Like Disorder (ATLD) Nijimen Break Syndrome (NBS) Blooms Syndrome (BS) Werner Syndrome (WS) Rothmund Thomson Syndrome (RTS) Breast Cancer Susceptibility (BRCA1, BRCA2)
1. UV light[Q] 2. Chemicals	1. Bulky Adducts 2. Pyrimidine Dimers[Q]	**Nucleotide Excision Repair (NER)**[Q]	**Xeroderma Pigmentosa**[Q] **(XP)** Cockayne Syndrome (SS) Trichothiodystrophy (TTD)
1. Oxygen radicals 2. Alkylating agents	1. Abasic Sites 2. Single strand breaks 3. 8 oxoguanine lesions	Base Excision Repair (BER)	MUTYH–associated Polyposis (MAP)
1. Replication errors	1. Bases mismatch 2. Insertion 3. Deletion	**Mismatch repair (MMR)**	**Hereditary non-polyposis Colorectal Cancer (HNPCC)**[Q] (Lynch syndrome)

Image-Based Information

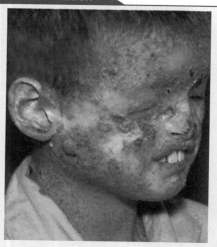

This is the picture of a boy who presented with blistering photosensitive lesions, diagnosed as a case of Xeroderma Pigmentosa. What is the molecular basis of this disorder?

- **Defect is in Nucleotide excision repair**

Nucleotide Excision repair in humans

To repair bulky lesions like **Pyrimidine dimers**, enzymes involved are:

- **Helicase** activity of transcription factor **(TfIIH)** which is a product of two XP genes (XP-B and XP-D)
- **Endonuclease** coded by XPG and XPF
- **Pol δ and Pol ε**
- **Ligase III**

Molecular defect in Xeroderma Pigmentosa

1. Mutation in any gene from XPA to XPG
2. Particularly **Helicase activity of TfIIH** a product of XPD is affected

Double Strand Break Repair Mechanisms (DSB)

- They are Homologous recombination (HR) and Non-Homologous End Joining Repair (NHEJ)

Homologous Recombination	Non-Homologous End Joining Repair
Major mechanism of DSB repair in yeast	Major mechanism of DSB repair mammals
Takes place between homologous chromosomes	Does not need a homologous Chromosome
Takes place before cell enter mitosis (S & G2/M phase)	Takes place before cell enter mitosis (G0/G1 phase)

Quick Review

- DNA Replication is bidirectional.
- The enzyme that unwinds the DNA is DNA Helicase.
- Kornberg's enzyme is DNA P I
- Klenow Polymerase is DNA Polymerase from which 5' to 3' exonuclease activity removed.
- SSB prevent reannealing of separated DNA.
- Multiple ori in humans.
- DNAP α has primase activity.
- Primer is synthesised by Primase
- Shortening of DNA is prevented by Telomerase.
- Telomerase is absent in somatic cells.
- SCID is caused due to a NHEJ, a double stranded break repair defect also.

Check List for Revision

- Enzymes of DNA Replication, particularly DNA Polymerase.
- Steps of DNA Replication an overall understanding is enough.
- Telomeres and Telomerase is recent trend question for exams.
- DNA Repair mechanism, the table given is must learn.

Review Questions MCQ

1. Which of the following is true about DNA Polymerase III? *(Recent Question Nov 2017)*
   a. It forms Okazaki fragments and it needs RNA primer
   b. It is needed for translation
   c. Bacteria can function without it
   d. Has DNA repair function

2. Telomerase is: *(PGI May 2017)*
   a. Reverse transcriptase
   b. Ribonucleoprotein
   c. Ribozyme
   d. Active in cancer cells and germ cells
   e. Carry its own RNA template

3. True about DNA Gyrase: *(PGI May 2017)*
   a. Prokaryotic Topoisomerase I
   b. Prokaryotic DNA Topoisomerase II
   c. Reverse transcriptase
   d. Restriction endonuclease
   e. RNA Polymerase

4. During DNA replication which bond breaks? *(PGI Nov 2016)*
   a. Phosphodiester bond
   b. Phosphate bond
   c. Hydrogen bond
   d. Glycosidic bond

5. False statements is/are: *(PGI May 2011)*
   a. In leading strands DNA is synthesized continuously
   b. Multiple origins of replication are possible for bacteria
   c. DNA replication proceeds in one direction
   d. Lagging strand stick by RNA primase
   e. DNA polymerase III–processive leading strand synthesis

6. True about Eukaryotic DNA replication compared to prokaryotic: *(PGI May 2013)*
   a. Conservative
   b. Semiconservative
   c. Unidirectional
   d. Bidirectional
   e. Semidiscontinuous

7. Incorrect statement is: *(PGI Nov 2010)*
   a. T4 DNA polymerase has 3'->5' exonuclease activity
   b. Klenow fragment of DNA polymerase I function is almost similar to T4 DNA polymerase
   c. Restriction endonuclease cut DNA chains at specific location
   d. Endonuclease cut DNA at 5'terminus
   e. Right handed helix of DNA is more common

8. Which DNA polymerase is involved in repair of mammalian DNA? *(PGI June 2009)*
   a. Alpha
   b. Beta
   c. Gamma
   d. Epsilon
   e. Delta

9. The gaps between segment of DNA on the lagging stand produced by restriction enzymes are joined/sealed by: *(AI 2009)*
   a. DNA Ligases
   b. DNA Helicase
   c. DNA topoisomerase
   d. DNA Phosphorylase

10. During replication of DNA, which one of the following enzymes polymerizes the Okazaki fragments? *(AI 2006)*
    a. DNA Polymerase I
    b. DNA Polymerase II
    c. DNA Polymerase III
    d. RNA Polymerase I

11. All of the following cell types contain the enzyme telomerase which protects the length of telomerase at the end of chromosomes, except: *(AI 2006)*
    a. Germinal
    b. Somatic
    c. Hemopoietic
    d. Tumor

12. DNA Polymerase with both replication and repair function is: *(Ker 2009)*
    a. I
    b. II
    c. III
    d. None of the above

13. Radiolabelled DNA was allowed to replicate twice in a non-radioactive environment. Which of the following is true? *(Ker 2008)*
    a. All the strands will have radioactivity
    b. Half of the DNA will have no radioactivity
    c. No strands will have radioactivity
    d. Three–fourth of the DNA replicated will have radioactivity

14. In which of the following phase, DNA doubling occurs: *(Ker 2006)*
    a. Gl phase
    b. S phase
    c. G2 phase
    d. M phase

15. Unwinding Enzyme in DNA synthesis:
    a. Helicase
    b. Primase
    c. DNA Polymerase
    d. Transcriptase

16. True about telomerase or telomere is/are: *(PGI Dec 03)*
    a. They are present at the ends of eukaryotic chromosome
    b. Increased telomerase activity favours cancer cells
    c. DNA dependent RNA polymerase
    d. DNA polymerase

17. **Action of Telomerase is:** *(CUPGEE 11)*
    a. DNA repair
    b. Longevity of cell
    c. Breakdown of telomere
    d. None

18. **Ends of chromosomes replicated by:** *(PGI Dec 06)*
    a. Telomerase
    b. Centromere
    c. Restriction endonuclease
    d. Exonuclease

19. **Highly repetitive DNA is seen in:** *(PGI June 03)*
    a. Cloning of DNA
    b. Microsatellite DNA
    c. Telomere
    d. Centromere

20. **Which enzymatic mutation is responsible for immortality of cancer cells:** *(AIIMS Nov 01)*
    a. DNA reverse transcriptase
    b. RNA polymerase
    c. Telomerase
    d. DNA polymerase

21. **Okazaki fragments are formed during the synthesis of:** *(AI 08)*
    a. dsDNA
    b. ssDNA
    c. mRNA
    d. tRNA

22. **Correct sequence of enzymes required for DNA formation is:** *(PGI June 01)*
    a. DNA polymerase → protein unwinding enzyme → DNA ligase → DNA Isomerase → Polymerase I
    b. Protein unwinding enzyme → polymerase I → DNA ligase → DNA isomerase → DNA polymerase
    c. RNA polymerase → DNA polymerase III → DNA polymerase I → DNA ligase
    d. RNA polymerase → DNA polymerase III → DNA ligase → exonuclease → DNA polymerase I

23. **True about DNA polymerase in eukaryotes:** *(PGI June 08)*
    a. Components are α, β, γ, δ, ε
    b. β associated with repair
    c. γ associated with repair
    d. δ associated with synthesis of mitochondria DNA
    e. α is abundant amount

24. **DNA polymerase have:** *(PGI June 03)*
    a. 3'-5' polymerase activity
    b. 5'-3' polymerase activity
    c. 3'-5' exonuclease activity
    d. 5'-3' exonuclease activity

25. **SCID is due to defect in:** *(AIIMS Nov 2017)*
    a. NHEJ
    b. Homologous Recombination
    c. Mismatch repair
    d. Nucleotide excision repair

26. **Xeroderma pigmentosa is due to:** *(Ker 2006)*
    a. Base excision defect
    b. Nucleotide excision repair
    c. SOS repair defect
    d. Cross linking defect

27. **UV light damage to the DNA leads to:** *(PGI Dec 05)*
    a. Formation of pyrimidine dimers
    b. No damage to DNA
    c. DNA hydrolysis
    d. Double stranded breaks

28. **Excessive ultraviolet (UV) radiation is harmful to life. The damage caused to the biological system by ultra-violet radiation I by:** *(AIIMS May04)*
    a. Inhibition of DNA synthesis
    b. Formation of thymidine dimers
    c. Ionization
    d. DNA fragmentation

29. **The primary defect in Xeroderma pigmentosa is:** *(AI 2000)*
    a. Formation of thymidine dimers
    b. Poly ADP ribose polymerase is defective
    c. Exonuclease is defective
    d. Formation of adenine dimers

30. **Which of the following is true regarding DNA double-strand breaks repair pathway:**
    a. Homologous recombination require a long homologous sequence to guide repair
    b. Non-homologous end-joining does not require a long homologous sequence to guide repair
    c. Homologous recombination repairs DNA before the cell enters mitosis
    d. Non-homologous end-joining repairs DNA before the cell enters mitosis

Answers to Review Questions

e. Non-homologous end-joining is prominent DSB repair mechanism in mammals.

1. **a. It forms Okazaki fragments and it needs RNA primer**
   (Ref: Harper 31/e page 363)

   Translation do not require DNA Polymerase. DNAP I and II are having DNA Repair activity.

2. **a, b, d, e. Reverse transcriptase, Ribonucleoprotein, Active in cancer cells and germ cells, Carry its own RNA template**
   (Ref: Harper 31/e page 363)

   Telomerase is not ribosome. Ribosome is the organelle where translation takes place.

3. **b. Prokaryotic DNA Topoisomerase II**
   (Ref: Harper 31/e page 364)

4. **a, c. Phosphodiester bond, Hydrogen bond**
   (Ref: Harper 31/e page 363)

   Phosphodiester bond breaks during removal of primer. Hydrogen bond breaks when two strands are separated.

5. **b. c. d., Multiple…, DNA…, Lagging…,**
   (Ref: Harper 31/e page 363)

   Multiple ori in eukaryotes
   DNA replication is bidirectional
   Lagging strand stick by DNA Ligase

   DNA Replication

   Salient Features of DNA Replication
   Occurs in the S Phase of the cell cycle.
   Each DNA Stand separate and each acts as template strand on which complementary strand is synthesized.
   Base pairing rule is obeyed.
   Semiconservative nature—Proved by Meselson and Stahl Experiment.
   New Strand is synthesized in the 5' to 3' direction.
   Synthesis of DNA in both strands are not similar.
   Leading strand-The strand which DNA is continuously polymerized.
   Lagging Strand—The strand which is DNA is discontinuously polymerized (Semi discontinuous)
   Replication proceeds from multiple origins in each chromosome in eukaryotes including humans (a total of as many as 100 in humans).
   Replication obeys polarity.
   Replication occurs in both directions along all of the chromosomes i.e. bidirectional in prokaryotes and eukaryotes.
   Both strands are replicated simultaneously. Replication process generates "**replication bubbles**"

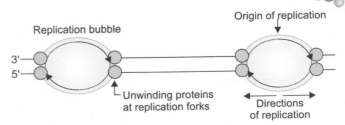

Fig. 10.4.2: Direction of replication

6. **b, d, e. Semiconservative, Bidirectional, Semi- discontinuous**
   (Ref: Harper 31/e page 363)

   This question means the common features between eukaryotic and prokaryotic DNA replication.

7. **d. Endonuclease cut DNA at 5' terminus**

   Endonuclease cut the DNA from within.
   Exonuclease cuts the DNA from ends.
   T4 DNA Polymerase similar to Klenow polymerase.

8. **b. beta** *(Ref: Harper 31/e page 364)*

9. **a. DNA Ligases** *(Ref: Harper 31/e page 364)*

10. **c. DNA Polymerase III** *(Ref: Harper 31/e page 363)*

    DNA Synthesis
    On 5'→3' Strand (Lagging Strand) (Discontinuous Strand) (Retrograde Strand)
    Synthesised in discontinuous manner
    Small fragments of DNA are in short spurts of 100–250 nucleotide (1000–2000 bp in prokaryotes.) called Okazaki
    Fragment synthesized in 5'—3' direction.
    - By DNAP III in prokaryotes.
    - By DNAP δ in eukaryotes.

11. **b. Somatic** *(Ref: Harper 31/e page 367)*

    Telomerase (Telomere Terminal Transferase)
    - Enzyme which prevent shortening of DNA.
    - Has an intrinsic RNA primer
    - Has Reverse Transcriptase (RNA Dependent DNA Polymerase) activity.
    - Present in Germ line, stem cells, most cancer cells
    - Absent from most somatic cells.

    Clinical Significance of Telomerase
    - Absence of Telomerase lead to premature ageing.
    - In Cancer cells increased Telomerase activity.
    - Telomerase has become an attractive target for cancer chemotherapy and drug development.

12. **a. I** *(Ref: Harper 31/e page 363)*

13. **b. Half of the DNA will have no radioactivity**
    (Ref: Harper 31/e page 363)

- Semiconservative Nature of DNA Replication proved by Meselson and Stahl states that half of the parent strand is conserved during replication in the daughter strand.
- After one replication all the DNA will have radioactivity.
- After two replication half of the DNA will have radioactivity.

14. **b. S Phase** *(Ref: Harper 31/e page 363)*

15. **a. Helicase** *(Ref: Harper 31/e page 364)*

Classes of proteins involved in DNA replication	
Protein	**Function**
DNA polymerases	Deoxynucleotide polymerization
Helicases	Processive unwinding of DNA
Topoisomerases	Relieve torsional strain that results from helicase-induced unwinding

Protein	**Function**
DNA primase	Initiates synthesis of RNA primers
Single-strand binding proteins	Prevent premature reannealing of dsDNA
DNA ligase	Seals the single strand nick between the Nascent chain and Okazaki fragments on lagging strand

16. **a, b. They are present at the ends of..., Increased telomerase...** *(Ref: Harper 31/e page 367)*

- Telomeres are present in the ends of eukaryotic chromosomes.
- Telomeres consists of TG repeats.
- Telomere shortening has been associated with malignant transformation and ageing
- Telomerase, multisubunit RNA template containing RNA Dependent DNA Polymerase (Reverse Transcriptases)

17. **b. Longevity of cell** *(Ref: Harper 31/e page 367)*

18. **a. Telomerase** *(Ref: Harper 31/e page 367)*

19. **c, d. Telomere and Centromere** *(Ref: Harper 31/e page 373)*

- In human DNA, at least 30% of genome consist of repetitive sequence.
- These sequences are clustered in the centromere and telomere,
- They are transcriptionally inactive.
- They are mostly having structural role in the chromosome.

20. **c. Telomerase** *(Ref: Robbins 9/e page 288)*

Telomerase is one of several factors that contribute to the endless replicative capacity (the immortalization) of cancer cells.

21. **a. dsDNA**

22. **c. RNA Polymerase → DNA polymerase III → DNA polymerase I → DNA ligase** *(Ref: Harper 31/e page 363)*

- The correct sequence of enzymes is Helicase, Primase, DNA Polymerase III, DNA Polymerase I on lagging strand.
- Helicase, Primase, DNA Polymerase III on leading strand.

23. **a, b. Components are ..., β associated with...** *(Ref: Harper 31/e page 363)*

Eukaryotic DNA Polymerase	Function
DNAP alpha	Primase
DNAP beta	DNA repair
DNAP gamma	Mitochondrial DNA synthesis
DNAP delta	Lagging Strand Synthesis
DNAP epsilon	Leading Strand Synthesis

24. **b, c, d. 5'-3' polymerase activity, 3'-5' exonuclease activity, 5'-3' exonuclease activity** *(Ref: Harper 31/e page 365)*

DNA Polymerase have 5' to 3' Polymerase activity, 3'-5' exonuclease (Proofreading) and 5'-3' exonuclease activity (repair) activity.

DNA Repair

25. **a. NHEJ** *(Ref: Harper 31/e page 370)*

26. **b. Nucleotide excision repair** *(Ref: Harper 31/e page 370)*

27. **a. Formation of pyrimidine dimers** *(Ref: Harper 31/e page 374)*

DNA lesions formed by UV light damage are Bulky adducts and Pyrimidine Dimers

28. **b. Formation of thymidine dimers** *(Ref: Harper 31/e page 374)*

29. **a. Formation of thymidine dimers** *(Ref: Harper 31/e page 374)*

- UV light radiation causes Bulky adducts and Pyrimidine dimers (Most common is Thymidine dimers)
- This is repaired by Nucleotide excision repair (NER)
- Defect in NER leads to Xeroderma Pigmentosa.

30. **a, b, c, d, e., Homologous..., Non-homologous..., Homologous..., Non-homologous..., Non-homologous end-joining...**

Double strand Break repair Mechanisms (DSB)

They are Homologous recombination (HR) and Non-Homologous End Joining Repair (NHEJ)

Homologous Recombination	Non-Homologous End Joining Repair
Major mechanism of DSB repair in yeast	Major mechanism of DSB repair mammals
Takes place between homologous chromosomes	Does not need a homologous Chromosome
Takes place before cell enter mitosis (S & G2/M phase)	Takes place before cell enter mitosis (G0/G1 phase)

10.5 TRANSCRIPTION

- Definition
- Transcription Cycle
- Salient Features of Transcription
- Comparison between Transcription and Replication
- Enzymes of Transcription
- Post-Transcriptional Modifications of mRNA or RNA Processing
- Promoters of Transcription
- Enhancers and Repressors

DEFINITION

The process by which RNA is synthesised from the DNA is called Transcription.

SALIENT FEATURES OF TRANSCRIPTION

- **Template Strand and Coding Strand**
  The strand that is transcribed or copied to the mRNA is referred to as **Template strand or nonsense strand.**
  The opposite strand is referred to as **Coding strand or Non-template strand or Sense Strand.**
- New RNA is synthesized in 5' to 3' direction
- **No primer is required**
- Only a part of the DNA strand is transcribed.

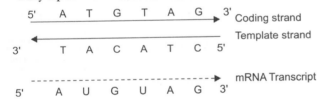

CONCEPT BOX

The non-template strand is called coding strand
- Primary transcript is complementary to the template strand.
- Hence the coding strand contains the same base sequence in the nascent mRNA except in the case of Thymine replaced by Uracil
- Hence non-template strand is called coding strand.

ENZYME OF TRANSCRIPTION

RNA Polymerase (RNAP)

They are DNA dependent RNA Polymerase

 HIGH YIELD POINTS

Differences between DNAP and RNAP
- No primer is needed in RNAP
- No proofreading activity in RNAP

Prokaryotic RNA Polymerase

- Only one type^Q of Prokaryotic RNA Polymerase
- Multisubunit Enzyme
- Core Enzyme + σ subunit = Holoenzyme (Eσ)
- Core Enzyme consists 2α and 1β and 1β' and ω subunit.
- **σ subunit^Q help RNA polymerase to bind to the promoter^Q site.**
- β subunit^Q is the catalytic subunit.
- β subunit$^\Theta$ binds the **Mg^{2+} ions.**

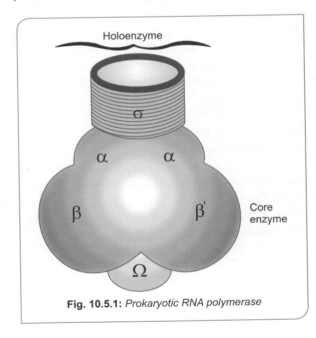

Fig. 10.5.1: *Prokaryotic RNA polymerase*

Eukaryotic RNA Polymerases

There are **three^Q** types of Eukaryotic RNA Polymerase. They are more complex than prokaryotic RNA polymerase with a number of subunits.

Eukaryotic RNA Polymerases and its Major Products

Form of RNA Polymerase	Sensitivity to α-Amanitin	Major Products of RNAP
RNA Polymerase I	Insensitive	• rRNA
RNA Polymerase II	High sensitivity	• mRNA • miRNA • SnRNA • lnc RNA
RNA Polymerase III	Intermediate sensitivity	• tRNA • 5S rRNA • certain sn RNA

PROMOTERS OF TRANSCRIPTION

Defined as the short conserved sequence in the coding strand of the DNA that specifies start site of the transcription. They are generally called as box or element.

Bacterial Promoters are

- **Pribnow Box**
  5'TATAAT 3' consensus sequence 10 bp upstream of the start site of transcription (–10 bp).
- **TGG Box**
  5'TGTTGACA3' sequence, 35 bp upstream of the start site of transcription (–35 bp).
- A third element AT rich element called Upstream promoter element called UP promoter element between –40 bp and –60 bp.

Eukaryotic Promoters are

- **Hogness Box[Q] or TATA box**
  5'TATAAAG 3' sequence 25 to 35 bp upstream of the start site of transcription (–25 to –35 bp)
- **CAAT Box**
  70 to 80 bp upstream of the start site of transcription
- GC-rich region (GCbox)

- Apart from this additional elements like In r (Initiator sequence) and DPE (Downstream promoter Element) can serve as promoters.

HIGH YIELD POINTS

- Usually promoters are located upstream of the start site of transcription.
- But promoters for RNA Polymerase III (transcribe tRNA) are located downstream within the gene.
- The promoter sequence (e.g. TATAAT) is on the coding strand of the DNA.
- Complementary sequence to the above promoter sequence is seen on the template strand.
- Σ subunit of RNA Polymerase bind to promoter on the template strand
- Almost all promoters are noncoding as they are upstream elements.
- Start site of Transcription is +1
- All nucleotides before start site are upstream elements. They are given numbers -1,-2,-3 etc
- All nucleotides after start site are downstream elements. They are given numbers, +2,+3 etc
- No zero is present in the DNA template

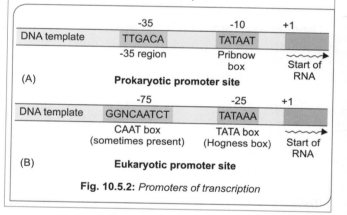

Fig. 10.5.2: Promoters of transcription

Transcription Unit

The region of DNA that includes signals of transcription initiation, elongation and termination.

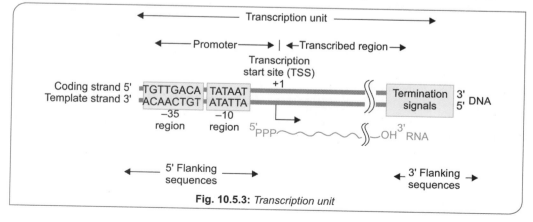

Fig. 10.5.3: Transcription unit

ENHANCERS AND REPRESSORS

Certain DNA elements facilitate or enhance initiation at the promoter and hence are termed enhancers and those which repress initiation at the promoter are called Repressors or silencers.

Enhancer elements, typically contain multiple binding sites for trans activator proteins.

Properties of Enhancers[Q]

- Can be located upstream or downstream of the transcription site.
- Work when located long distances from the promoter
- Work when upstream or downstream from the promoter
- Work when oriented in either direction
- Can work with homologous or heterologous promoters
- Work by binding one or more proteins
- Work by facilitating binding of the basal transcription complex to the cis-linked promoter.

Cis-acting Elements

- Promoters and enhancers that are located on the same DNA are called cis-acting elements.

Trans-acting Elements

- Trans-acting proteins are the products of a separate gene that interact with the cis-acting elements.

TRANSCRIPTION CYCLE

Steps of Transcription

Transcription can be Described in Six Steps:

1. **Template binding and closed RNA polymerase-promoter complex formation:**
   RNA polymerase (RNAP) binds to DNA and then locates a promoter (P) by means of sigma subunit.
2. **Open promoter complex formation:**
   Once bound to the promoter, RNAP melts the two DNA strands to form an open promoter complex.
   This complex is also referred to as the pre-initiation complex or PIC.
   Strand separation allows the polymerase to access the coding information in the template strand of DNA.

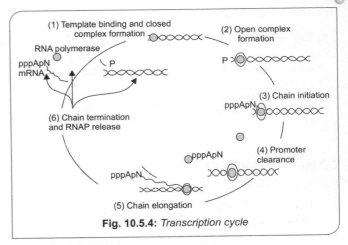

Fig. 10.5.4: *Transcription cycle*

3. **Chain initiation:**
   Using the coding information of the template RNAP catalyzes the coupling of the first base (often a purine) to the second, to form a dinucleotide.
4. **Promoter clearance:**
   After RNA chain length reaches ~10–20 nucleotides, the polymerase undergoes a conformational change.
   Then it moves away from the promoter, transcribing down the transcription unit.
5. **Chain elongation:**
   Successive residues are added to the 3'-OH terminus of the nascent RNA molecule until a transcription termination signal (**T**) is encountered.
6. **Chain termination and RNAP release:**
   By two methods:
   a. **ρ (Rho) factor dependent termination**[Q]
      Termination signal for transcription in the template strand is identified by ρ factor.
      (Rho) factor is an ATP dependent RNA-DNA helicase that disrupts the nascent RNA-DNA complex.
   b. **Intrinsic (Spontaneous) or ρ (Rho) factor independent termination.**
      RNA Polymerase identifies the termination signal on the template strand without the aid of ρ factor.

But for this termination the nascent RNA should have certain prerequisites.

1. GC rich region that forms a hairpin turn.
2. U rich region after the GC rich region. The binding of A-U is weak hence facilitate termination of transcription.

Transcription in Eukaryotes

> **EXTRA EDGE**
>
> Eukaryotic transcription require certain unique proteins apart from RNA Polymerases called **General transcription factors (GTFs)**.
> Pre Initiation complex formation require RNA Pol II and 6 GTFs (TFIIA, TFIIB, TFIID, TFIIE, TFIIF, TFIIH)
> **TFIID binds to TATA box** through its TBP (TATA binding Protein) subunit.

COMPARISON BETWEEN TRANSCRIPTION AND REPLICATION

- The general steps of initiation, elongation, and termination with 5'–3' polarity is present in both.
- Large, multicomponent initiation complexes are involved in both.
- Both obey Watson–Crick base-pairing rules.

Differences between Replication and Transcription

Replication	Transcription
Deoxyribonucleotides are added	Ribonucleotides are added
A is paired with T on the parent strand	U replaces T as the complementary base for A in RNA
Both the strands of DNA act as template	Only one strand^a of the DNA acts as the Template
Entire genome must be copied	Only portions of the genome are vigorously transcribed or copied into RNA
A primer is involved as DNA Polymerase cannot initiate DNA Synthesis de novo	**A primer is not involved** as RNA polymerases have the ability to initiate synthesis de novo
Highly active proofreading mechanism	No highly active proofreading mechanism
DNA dependent DNA Polymerase is the enzyme	DNA dependent RNA Polymerase is the enzyme

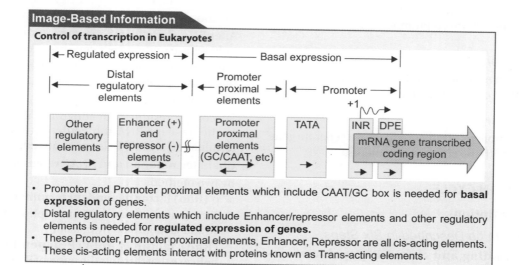

Image-Based Information
Control of transcription in Eukaryotes

- Promoter and Promoter proximal elements which include CAAT/GC box is needed for **basal expression** of genes.
- Distal regulatory elements which include Enhancer/repressor elements and other regulatory elements is needed for **regulated expression of genes.**
- These Promoter, Promoter proximal elements, Enhancer, Repressor are all cis-acting elements. These cis-acting elements interact with proteins known as Trans-acting elements.

POST-TRANSCRIPTIONAL MODIFICATIONS OF mRNA OR RNA PROCESSING

In prokaryotes

- mRNA are **not subjected** to post-transcriptional processing.
- Translation is started simultaneous with transcription.
- tRNA and rRNA of prokaryotes under go post-transcriptional modification.

In eukaryotes

- The RNA molecule synthesized by RNA polymerase is known as **Primary Transcript or Heteronuclear RNA (hnRNA)**
- h nRNA undergo extensive post-transcriptional modification.

Site: Nucleus

Post-transcriptional Processing of Primary Transcript of mRNA

- 7-methylguanosine capping at 5'end
- Addition of a poly-A tail at 3'end
- Removal of introns and joining of Exons called Splicing
- Methylations
- Alternative RNA Processing.

7-methylguanosine Capping at 5' End

This takes place in two steps:

1. Guanosine Triphosphate is attached to the 5' end of hnRNA
   - **By** an enzyme **Guanylyl transferase**
   - By an unusual **5'-5' triphosphate linkage**.
   - Takes place inside the **nucleus.**
2. Methylation of Guanosine Triphosphate
   - **By Guanine 7 methyl transferase.**
   - **S-adenosyl Methionine is the methyl donor.**
   - **Takes place inside the Cytosol. Functions of 5'capping**

Functions of 5' capping

Helps in

- 5' cap bind to Cap binding complex, participate in binding of mRNA to ribosome to initiate translation
- Helps to stabilize the mRNA.
- Prevents the attack of 5' to 3' exonuclease (Ribonuclease).

Cleavage and Addition of a Poly-A tail at 3' End

- 3'end of the hnRNA is cleaved Poly Atailis added to the cleaved site.
- Polyadenylate Polymerase is the enzyme.
- Takes place in the nucleus.
- Length of Poly A tail is up to 85–250 Adenine bases.

Functions of Poly A tail at the 3' end.

- Stabilize them RNA
- Prevents the attack of 3' to 5' exonuclease (Ribonuclease)
- Facilitate their exit from the nucleus.
- Poly A tail and its binding protein PAB-1 are required for efficient initiation of Protein Synthesis.

Removal of Introns and Joining of Exons Called Splicing

Intron: Intervening sequence that do not code for Amino Acid.

Exon^Q: Amino acid coding sequence.

Molecular machinery that carry out splicing is called Spliceosomes.

Spliceosomes

Consist of the primary transcript, five small nuclear RNAs (U1, U2, U4, U5, and U6) and more than 60 protein.

Spliceosome = snRNA + RNP + hnRNA (or mRNA precursor)

Small Nuclear RNA (Sn RNA)

- Uracil rich RNA which can act as enzymes, i.e. Ribozyme.
- U1, U2, U4, U5, U6 involved in mRNA processing.
- U6 is certainly essential, as yeast deficient in this Sn RNA is not viable.
- U7sn RNA is involved in production of correct 3'ends of histone mRNA that lacks Poly Atail.

SnRNP complex (Snurps) (Small nuclear ribonucleoprotein)

- SnRNP (Snurps) = SnRNA + Ribonucleo Protein (RNP)

Clinical Correlation

Systemic lupus erythematosus results from an autoimmune response in which the patient produces antibodies against host proteins, including snRNP (Snurps)

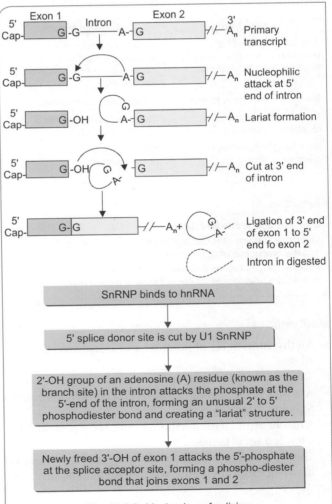

Fig. 10.5.5: *Mechanism of splicing*

High Yield Points

Mechanism of Splicing
- The introns have GU at 5' end and AG at 3'end
- The splicing start with a cut in the 5' splice donor site by U1SnRNP.
- An unusual 2'→5' phosphodiester bond is formed
- The excised intron is released as a lariat, which is degraded.
- RNA cleavage and ligation do not require ATP

Extra Edge

Self-Splicing Introns
- Certain introns itself has self splicing activity
- They are first discovered by **Thomas Cech** and his colleagues
- Because of Ribozyme activity
- They are Group I and Group II introns.

Comparison of Group I and Group II Introns

Group I Introns	Group II Introns
3' hydroxyl group of a **Guanosine** residue is used as nucleophile in the first step	2' hydroxyl group of an **adenosine** residue within the intron is used as nucleophile in the first step
No lariat structure is formed	**Lariat** structure is formed
Two transesterification reaction involved	Two **transesterification** reaction involved
No ATP is required as high energy cofactor	**No ATP** is required as high energy cofactor

Extra Edge

Effect of Splice site Mutations

Mutations at splice sites can lead to improper splicing (faulty splicing) and the production of aberrant proteins.

For example, mutations that cause the incorrect splicing of β-globin mRNA are responsible for some cases of **β-thalassemia**.

Methylations

- Methylation of N_7 of Adenine and 2' hydroxyl group of ribose
- Takes place in the cytoplasm.
- Alternative Processing of mRNA Precursor or Alternative Splicing
- A mechanism for producing a diverse set of proteins from a limited set of genes.
- The pre-mRNA molecules from same genes can be spliced in two or more alternative ways in different tissues.

Extra Edge

The mechanisms for Alternative processing of mRNA precursors
- Selective Splicing—selective inclusion or exclusion of exons
- Alternative 5' donor site—5' donor site of certain exons is changed.
- Alternative 3' acceptor site—3' acceptor site of certain exons is changed.
- Alternative Poly Adenylation site—Different site is used for Poly Adenylation.

Applications of Alternative mRNA Processing

- Generation of membrane bound or Secretory IgG by alternative polyadenylation sites.^{Q}
- Production of several tissue specific isoforms of tropomyosin from single mRNA transcript

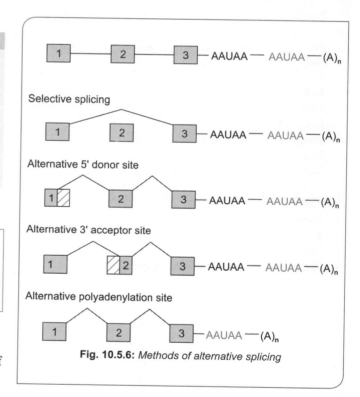

Fig. 10.5.6: *Methods of alternative splicing*

RNA Editing

- mRNA editing is an **exception to central dogma** of molecular genetics.
- Current estimate suggest that 0.01% of mRNA is edited in this fashion.

- mRNA editing is the process by which coding information is changed at the mRNA level by chemical modification of the nitrogenous bases present in the codons.
- Hence, the linear relationship between the coding sequence in DNA, the mRNA sequence, and the protein sequence is altered.

Example of RNA Editing
- Apolipoprotein B *(apoB)* gene and mRNA.

In the liver
- The single *apoB gene* is transcribed into an mRNA that directs the synthesis of a 512-kDa protein, apoB100 with 4536 amino acid residues.

In the intestine
- The same gene directs the synthesis of the primary transcript
- **A cytidine deaminase** converts a CAA codon in the mRNA to UAA at a single specific site
- Rather than encoding glutamine, this codon becomes a termination signal, and a truncated 242-kDa protein (apoB 48) with 2512 amino acid residue is the result.

Other examples of RNA editing
- Glutamate Receptor (Glutamine changed to Arginine)
- Trypanosome mitochondrial DNA.

Quick Review

- The sequence of bases in mRNA is same as that of coding strand except for T replaced by U.
- In prokaryotic RNA Polymerase polymerase, the catalytic subunit is β subunit.
- Sigma subunit of RNAP binds to promoter site to initiate transcription cycle.
- RNAP II has highest sensitivity towards α amanitin.
- Hogness box /TATA box is eukaryotic promoter located 25-35 base pairs upstream.
- Pribnow box is prokaryotic promoter located 10 bp upstream.
- RNA is synthesised fro 5' to 3' direction.
- Template strans is read in 3' to 5' direction.
- Enhancers and repressors are DNA elements that facilitate initiation at the promoter site.
- The site of post transcriptional modifications is nucleus.
- The cap at 5' end is 7 methyl guanosine.
- The poly A tails is at 3' end is added by Poly A polymerase.
- Spliceosome consist of Sn RNA + Proteins + heteronuclear RNA.
- The splicing activity is done by Spliceosome.
- Snurps = Sn RNA + Protein
- Generation of ApoB 48 and ApoB 100 is an example of RNA editing.

Check List for Revision
- Selective learning is possible in this chapter for NBE model exam.
- Salient features of transcription
- RNA Polymerases
- Names and location Promoters
- Properties of enhancers and repressors
- Difference between Transcription and Replication
- Post transcriptional processing bold letters
- Spliceosomes
- Bold letters of Group II introns
- Differential RNA processing.
- RNA editing

Review Questions MCQ

1. Which of the following do not require 5' capping? (AIIMS May 2018)
   a. mRNA for spectrin
   b. U6 sn RNA
   c. mRNA for p53
   d. tRNA for Alanine

2. Post-transcriptional modification includes: (PGI Nov 2018)
   a. Capping at 5'end
   b. Poly A tailing at 3' end
   c. All RNA undergo post transcriptional modification
   d. Intron excision by splicing
   e. Primarily occurs in cytoplasm

3. RNA Polymerase III synthesizes: (PGI May 2016)
   a. 28S rRNA
   b. 23S rRNA
   c. 5S rRNA
   d. tRNA
   e. mRNA

4. True about coding strand of DNA: (PGI Dec 03)
   a. Template strand
   b. Minus strand
   c. Runs at 5'-3' direction
   d. Runs at 3'-5' direction
   e. Plus strand

5. 5'TTACGTAC 3' after transcription what will be the RNA? (PGI June 08)
   a. 5'-TTACGTAC 3'
   b. 3'-TTACGTAC5'
   c. 5'-CATGCATT3'
   d. 3'-CATGCATT5'
   e. 5,-GUACGUAA 3'

6. Immunoglobulin molecule is synthesized by inmixed or separate due to: (AIIMS May 2012)
   a. Codominance
   b. Gene switching
   c. Allele exclusion
   d. Differential RNA processing

7. A four-year-old child is diagnosed with Duchenne muscular dystrophy, an X-linked recessive disorder, Genetic analysis shows that the patient's gene for the muscle protein dystrophin contains a mutation in its promoter region. What would be the most likely effect of this mutation? (Nov 2010)
   a. Tailing of dystrophin mRNA will be defective
   b. Capping of dystrophin mRNA will be defective
   c. Termination of dystrophin transcription will be deficient
   d. Initiation of dystrophin transcription will be deficient

8. Splicing activity is a function of (AIIMS Nov 2010)
   a. mRNA
   b. snRNA
   c. tRNA
   d. rRNA

9. Post-transcriptional modification include: (PGI Nov 2012)
   a. All RNA undergo post-transcriptional modification
   b. Capping of pre mRNA involves the addition of 7 methyl guanosine to the 5'end
   c. Poly A tail occur at 3'end
   d. Intron excision by spliceosome.
   e. Primarily occur in the cytoplasm

10. Not a product of transcription (PGI May 2011)
    a. tRNA
    b. mRNA
    c. rRNA
    d. cDNA
    e. New strand of DNA

11. Reverse transcriptase is: (PGI May 2011)
    a. DNA dependent RNA polymerase
    b. RNA dependent DNA polymerase
    c. DNA dependent DNA polymerase
    d. RNA dependent RNA polymerase
    e. RNA polymerase

12. Which type of RNA has the highest percent age of modified base? (AI 2006)
    a. mRNA
    b. tRNA
    c. rRNA
    d. snRNA

13. The sigma(s) submit of prokaryotic RNA polymerase: (AI 2006)
    a. Binds the antibiotic rifampicin
    b. Is inhibited by a-amanitin
    c. Specifically recognizes the promoter site
    d. Is part of the core enzyme

14. The base sequence of the strand of DNA used as a template has the sequence 5'GATCTAC 3'. What would be the base sequence of RNA product? (Ker 2012)
    a. 5'CTAGATG 3'
    b. 5' GAUCUAC3'
    c. 5'GTAGATC3'
    d. 5' GUAGAUC3'

15. DNA dependent RNA polymerase is seen in: (Ker 2008)
    a. Primase
    b. DNA polymerase I
    c. DNA polymerase III
    d. DNA gyrase

16. Strand of DNA from which mRNA is formed by transcription is called: (Ker 2006)
    a. Template
    b. Antitemplate
    c. Coding
    d. Transcript

17. On which of the following tRNA acts specifically?
    a. ATP
    b. Golgi body
    c. Specific amino acid
    d. Ribosome

18. In conversion of DNA to RNA, enzyme required: *(PGI June 08)*
    a. DNA-polymerase
    b. DNA Ligase
    c. DNA-polymerase III
    d. RNA polymerase
    e. Primase

19. RNA polymerase does not require: *(AI 2004)*
    a. Template (dsDNA)
    b. Activated precursors (ATP, GTP, UTP, CTP)
    c. Divalent metal ions (Mn^{2+}, Mg)
    d. Primer

20. The following is a generalized diagram of typical eukaryotic gene: *(AIIMS May 06)*

Promoter region	Polypeptide coding region

    Direction of transcription

    What is the most likely effect of a 2 bp insertion in the middle of the intron?
    a. Normal transcription, altered translation
    b. Defective termination of transcription, normal translation
    c. Normal transcription, defective mRNA splicing
    d. Normal transcription, Normal translation

21. In a DNA the coding region reads 5'-CGT-3'. This would code in the RNA as: *(AIIMS May 03)*
    a. 5'-CGU-3'
    b. 5'-GCA-3'
    c. 5'-ACG-3'
    d. 5'-UGC-3'

22. Cytoplasmic process during processing is: *(SGPGI 05, CMC 03)*
    a. 5'capping
    b. Poly A tailing
    c. Methylation of tRNA
    d. Attachment of CCA in tRNA

23. All are the processing reaction in tRNA, except: *(WB 03, Delhi 04, UP 05)*
    a. CCA tailing
    b. Methylation of bases
    c. Poly A tailing
    d. Trimming of 5'end

24. Introns are exised by: *(PGI Dec 05)*
    a. RNA splicing
    b. RNA editing
    c. Restriction endonuclease
    d. DNAase
    e. Helicase

25. A segment of eukaryotic gene that is not represented in the mature mRNA is known as: *(AI 2004)*
    a. Intron
    b. Exon
    c. Plasmid
    d. TATA box

26. An enzyme that makes a double stranded DNA copy from a single stranded RNA template molecule is known as: *(AI 2004)*
    a. DNA polymerase
    b. RNA polymerase
    c. Reverse transcriptase
    d. Phosphokinase

27. Function of Pseudouridine arm of tRNA: *(JIPMER Nov 2015)*
    a. Helps in initiation of translation
    b. Serves as the recognition site of amino acyl tRNA synthetase
    c. Recognises the triple nucleotide codon present in the mRNA.
    d. Helps in initiation of transcription

28. True about 3'exonuclease: *(PGI May 2014)*
    a. Cleave 3' end of DNA
    b. Cleave 5' end of DNA
    c. Cleave 3' end of RNA
    d. Cleave 5' end of RNA

29. Which is are verse transcriptase? *(JIPMER 2014)*
    a. Topoisomerase
    b. Telomerase
    c. RNA polymerase II
    d. DNA polymerase alpha

30. Which of the following is true regarding transcription except: *(PGI)*
    a. mRNA formed
    b. DNA polymerase enzyme is used
    c. RNA polymerase enzyme is used
    d. Eukaryotes possess 3 different types of RNA polymerase

31. ApoB 48 & ApoB 100 is synthesized from them RNA; the difference between them is due to: *(AIIMS May 2011)*
    a. RNA splicing
    b. Allelic exclusion
    c. Deamination of cytidine touridine
    d. Upstream repression

Answers to Review Questions

1. **d. tRNA for Alanine**

   mRNA has 5'capping, SnRNA has 5'capping is present but in the process of post transcriptional processing it is lost. But tRNA do not have 5' capping

2. **a, b, c, d Capping..., Poly A ..., all RNA undergo..., intron excision...** (Ref: Harper 31/e page 387,388)

   Capping at 5' end, Poly A tailing at 3' end intron excision are all post transcriptional modifications of mRNA. All eukaryotic RNAs undergo post transcriptional modifications.

   But the primary site of post transcriptional modifications is nucleus.

3. **c,d. 5SrRNA, tRNA** (Ref: Harper 31/e page 375)

Form of RNA Polymerase	Sensitivity to α-Amanitin	Major Products of transcription
RNA Polymerase I	Insensitive	rRNA
RNA Polymerase II	High sensitivity	mRNA, miRNA, SnRNA, lncRNA
RNA Polymerase III	Intermediate sensitivity	tRNA, 5s rRNA

4. **c, d, e. Runs in 5'- 3' direction, Runs at 3'- 5' direction, Plus strand** (Ref: Harper 31/e page 375)

   - Template strand is in 3' to 5' direction, or 5' to 3' direction, Minus strand, Antisense strand.
   - Coding strand is in 5' to 3' direction, or 3' to 5' direction Plus strand, Sense strand.
   - Whatever is the direction of template strand, for transcription, that strand is read always in 3' to 5' direction.

   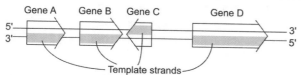

   This figure from Harper clearly shows template or coding strand can be either 5' to 3, or 3' to 5'. But the direction of transcription is 5' to 3'. The templates trand is read always in 3' to 5'direction.

5. **e. 5, GUACGUAA 3'**

   - Here question is to find the RNA, so all options with Thymine is not the answer. So answer is E.
   - The exact method is find the complementary sequence for the given DNA and at the places of T, replace by U
   - 5'TTACGTAC 3' is read in 3' to 5'direction
   - So RNA is 5' GUACGUAA3'

6. **d. Differential RNA processing** (Ref: Harper 31/e page 389)

   Alternative RNA Processing, which include Alternative Polyadenylation site in the μIg heavy chain primary transcript, results in two μ protein, μ_m and μ_s. One remain membrane bound to B lymphocyte (μ_m) and the other secreted (μ_s).

7. **d. Initiation of dystrophin transcription will be deficient** (Ref: Harper 31/e page 389)

   The fidelity and frequency of transcription is controlled by proteins bound to certain DNA sequences. These regions are termed promoters, and it is the association of RNAP with promoters that ensures accurate initiation of transcription. Promoters are responsible for initiation process of transcription.

8. **b. snRNA** (Ref: Harper 31/e page 387)

   Small Nuclear RNA (snRNA)

   Uracil rich RNA which can act as enzymes, i.e. Ribozyme. U1, U2, U4, U5, U6 involved in mRNA processing.

   U7 in processing histone mRNA

9. **a. b. c. d. All RNA undergo post ..., Capping of pre- mRNA ..., Poly A tail occur at 3' end, Intron excision by spliceosome**

   Post-transcriptional processing include:
   Primarily occur in the nucleus
   7-methylguanosine capping at 5' End
   This takes place in two steps Guanosine Triphosphate is attached to the 5' end of hnRNA By an enzyme Guanylyl transferase
   By an unusual 5'-5'triphosphate linkage
   Takes place inside the **nucleus**. Methylation of Guanosine Triphosphate By Guanine 7 methyl transferase.
   S-adenosyl Methionine is the methyl donor.
   Takes place inside the Cytosol.

   Addition of a Poly-A Tail at 3' End
   - Poly A tail is added to the 3'end of the hnRNA.
   - Polyadenylate Polymerase is the enzyme.
   - Takes place in the **nucleus.**
   - Length of Poly A tail is up to 200.

   Removal of introns and joining of exons called Splicing
   - Intron-intervening sequence that do not code for Amino Acid
   - Exon-Amino Acid Coding Sequence.
   - Molecular machinery that carry out splicing is called Spliceosome.

10. **d, e. DNA, New strand of DNA** (Ref: Harper 31/e page 387)

    cDNA is produced by reverse transcription New DNA strand produced by DNA replication

11. **b. RNA dependent DNA polymerase** (Ref: Harper 31/e page 380)

    Reverse Transcription
    - RNA converted to DNA is Reverse Transcription.
    - The enzyme is called Reverse Transcriptase
    - This is otherwise RNA dependent DNA Polymerase

Remember
DNA Polymerase is DNA dependent DNA Polymerase Reverse Transcriptase is RNA dependent DNA Polymerase Primase is DNA Dependent RNA Polymerase.

12. **b. tRNA** *(Ref: Harper 31/e page 346)*

Transfer RNA (tRNA)
- RNA which transfer amino acid from the cytoplasm to the ribosomal protein synthesising machinery
- Clover leaf shape in the secondary structure.
- L-shaped tertiary structure
- Single tRNA contains 74–95 nucleotides
- Cytoplasmic translation system possess 31 tRNA species
- Mitochondrial system possess 22 tRNAs

Contain significant proportion of nucleosides with unusual bases. They are
- Dihydrouridine (contain Dihydrouracil)
- Pseudouridine
- Inosine (contain Hypoxanthine.)
- Ribothymidine

13. **c. Specifically recognizes the promoter site** *(Ref: Harper 31/e page 380)*

Prokaryotic RNA Polymerase
Only one type of Prokaryotic RNA Polymerase Multi subunit Enzyme
- Core Enzyme + σ subunit = Holoenzyme (Eσ)
- Core Enzyme consists 2α and 1β and 1β' and ω subunit.
- σ subunit help RNA polymerase to bind to the promoter site.
- β subunit is the catalytic subunit.
- β subunit binds the Mg^{2+} ions.

14. **d. 5'GUAGAUC3'**

Read the strand in 3' to 5' direction. Write the complementary sequence in 5' to 3' direction obeying base pairing rule, except in the case of T replaced by U.

15. **a. Primase** *(Ref: Harper 30/e page 380)*

Primase enzyme synthesize RNA primer on the DNA strand, hence it is DNA dependent RNA Polymerase.

16. **a. Template strand** *(Ref: Harper 31/e page 375)*

17. **c. Specific amino acid** *(Ref: Harper 31/e page 346)*

At least one species of transfer RNA (tRNA) exists for each of the 20 amino acids
- tRNA molecules have extra ordinarily similar functions and three-dimensional structures. The adapter function of the tRNA molecules requires the charging of each specific tRNA with its specific amino acid.

18. **d. RNA polymerase** *(Ref: Harper 31/e page 375)*

Primase, DNA Polymerase and DNA Ligase for replication.

19. **d. Primer** *(Ref: Harper 31/e page 375)*
- RNA Polymerase does not require Primer.

20. **d. Normal transcription, normal translation**
- As the insertion is in the middle of the noncoding region, intron
- Transcription is normal,
- As the intron is spliced out and only exons join together, translation is also normal.

21. **a. 5'-CGU-3'** *(Ref: Harper 31/e page 913)*

As the question specifies that it is coding region, so the RNA product is same as coding region except in the case of T replaced by U.

22. **d. Attachment of CCA in tRNA** *(Ref: Harper 31/e page 376)*

Sites of Post-transcriptional processing tRNA
- Cleavage and attachment of CCA terminal at 3' end takes place in cytoplasm.
- Methylation of tRNA takes place in nucleus.

mRNA
- Poly A tail in Nucleus
- 5'MeGTP capping in nucleus and cytoplasm
- Methylation of some residues in cytoplasm.

rRNA
- Most common site is nucleolus.

23. **c. Poly A tailing**

Poly A tailing is a processing step of mRNA primary transcript.

24. **a. RNA splicing**

Removal of introns and joining of Exons called Splicing Intron: Intervening sequence that do not code for Amino Acid
Exon^Q: Amino acid coding sequence.
Molecular machinery that carry out splicing is called
Spliceosome.

Spliceosomes
Consist of the primary transcript, five small nuclear RNAs (U1, U2, U4, U5, and U6) and more than 60 protein.
Spliceosome = snRNA + RNP + hnRNA (or mRNA precursor).

25. **a. Intron** *(Ref: Harper 31/e page 389)*

26. **c. Reverse transcriptase**

27. **a. Helps in initiation of translation** *(Ref: Harper 31/e page 346)*

Arms of tRNA
1. **Acceptor Arm**
   Site of attachment of the Amino Acid
   3' end of the tRNA
   Has 3 unpaired nucleotide, CCA
   Carboxyl group of the amino acid is attached to the 3' hydroxyl group of the adenosyl moiety.

2. **Anticodon arm**

   Has the triplet nucleotide sequence complementary to the codon of the amino acid which the tRNA carries. The sequence is read from *3' to 5' direction.*

   Codon is read from 5' to 3' direction

   Codon of mRNA and anticodon in tRNA are antiparallel in their complementarity.

3. **DHU Arm**

   Contain Dihydrouracil residue. Acts as the recognition site for specific aminoacyl tRNA Synthetase.

4. **Pseudouridine Arm(TψCarm)** TψC stands for Ribothymidine, Pseudouridine, Cytidine Involved in the binding of aminoacyl tRNA to the Ribosomal surface. This helps in the formation of initiation complex.

28. **a, c. Cleaves 3' end of DNA, Cleaves 3' end of RNA**

    Exonuclease can cleave the nucleotide from the ends, 3' exonuclease can cleave 3' end of DNA and RNA. But m RNA are protected from cleavage by 3' Poly A tail. As the question is not specifying mRNA, a and c given as the answer.

29. **a, b, c, d., Topoisomerase, Telomerase, RNA..., DNA...**

    Reverse transcriptase is RNA dependent DNA Polymerase:

    Option A: Topoisomerase is a nicking resealing enzyme, not a reverse transcriptase

    Option B: Telomerase has reverse transcriptase activity.

    Option C: RNA polymerase II is an DNA dependent RNA Polymerase, not a reverse transcriptase

    Option D: DNAP alpha is a eukaryotic DNA dependent DNA Polymerase. But also has Primase activity. Primase is DNA dependent RNA Polymerase, not a reverse transcriptase.

30. **b. DNA polymerase enzyme is used***(Ref: Harper 31/e page 363)*

    - DNA Polymerase is the enzyme of replication.

31. **c. Deamination of cytidine to uridine**

    (Ref: Harper 31/e page 390)

    An example is the apolipoprotein B *(apoB)* gene and mRNA.
    - In the liver, the single *apo B* gene is transcribed into an mRNA that directs the synthesis of apo B100.
    - In the intestine, the same gene directs the synthesis of the primary transcript;
    - A cytidine deaminase converts a CAA codon in the m RNA to UAA at a single specific site
    - Rather than encoding glutamine, this codon becomes a termination signal apo B48 is the result.

10.6 DIFFERENT CLASSES OF RNA

- Classes of RNAs
- snRNA
- hnRNA
- mRNA
- miRNA and siRNA
- rRNA
- siRNA
- tRNA
- lncRNA

CLASSES OF RNAs

RNA exists in two major classes.

1. **Protein Coding**

2. **Nonprotein Coding**
   – Large noncoding RNA
   – Small noncoding RNA

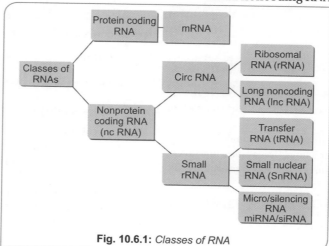

Fig. 10.6.1: *Classes of RNA*

Important RNAs

RNA	Types	Abundance	Stability
Protein Coding RNAs			
Messenger (mRNA)	~10⁵ Different species	2–5% of total	Unstable to very stable
Nonprotein coding RNA(nc RNA)			
Large nc RNAs			
Ribosomal (rRNA)	28S, 18S,	80% of total	Very stable
lnc RNA	~1000s	~1%–2%	Unstable to very stable
Small noncoding RNA (Small ncRNA)			
Small rRNA	5.8S, 5S	2%	Very stable
Transfer (tRNA)	~60 Different species	~15% of total	Very stable
Small nuclear (snRNA)	~30 Different species	1% of total	Very stable
Micro/ Silencing (mi/SiRNA)	100s–1000	<1% of total	Stable

MESSENGER RNA (mRNA)

- Most heterogeneous RNA
- Function as messenger conveying information to translation machinery.
- **2–5%** of total cellular RNA.
- 5' end capped by 7 Methyl Guanosine.
- 3' end non-genetically encoded by polymer of 85–250 adenylate residues.

RIBOSOMAL RNA (rRNA)

- Most abundant RNA is rRNA (80% of total RNA)
- Function—Forms protein synthesizing machinery called **Ribosome.**
- Ribosomal assembly is ribosomal RNAs associated with certain proteins (i.e., rRNA + Proteins).

Ribosomal Assembly in Prokaryotes^Q
70S Ribosome = 30S + 50S Subunits
In 30S Subunit = 16 S rRNA + Proteins
In 50S subunit = 23S rRNA + 5S rRNA + Proteins

Ribosomal Assembly in Eukaryotes^Q
80S Ribosome = 40S Subunit + 60S Subunit
60S Subunit = 28S rRNA + 5.8S rRNA + 5S rRNA + ~50 proteins
40S = 18S rRNA + ~30 proteins

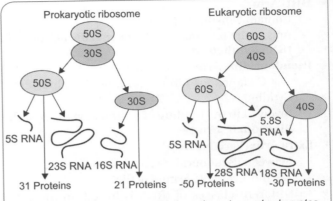

Fig. 10.6.2: *Ribosomal assembly in prokaryotes and eukaryotes*

TRANSFER RNA(tRNA)

- RNA which transfer amino acid from the cytoplasm to the ribosomal protein synthesising machinery.
- **Clover leaf shape** in the secondary structure.
- **L shaped** tertiary structure.
- Single tRNA contains 74–95 nucleotides.
- Cytoplasmic translation system possess **31 tRNA species**.
- Mitochondrial translation system possess **22 tRNA species**

> **EXTRA EDGE**
>
> tRNA contain significant proportion of nucleosides with unusual bases
> They are
> - Dihydrouridine (contain Dihydrouracil)
> - Pseudouridine
> - Inosine (contain Hypoxanthine)
> - Ribothymidine

Arms of tRNA

- **Acceptor Arm**
  - **Site of attachment of the Amino Acid**
  - **3' end** of the tRNA
  - Has 3 unpaired nucleotide, CCA
  - Carboxyl group of the amino acid is attached to the 3' hydroxyl group of the adenosyl moiety.
- **Anticodon Arm**
  - Anticodon arm has seven nucleotides, it recognizes 3 letter codon in mRNA.
  - *The sequence is read from 3' to 5' direction.*
  - *Codon is read from 5' to 3' direction*
  - *Codon of mRNA and anticodon in tRNA are anti parallel in their complementarity.*

- **DHU Arm**
  - Contain Dihydrouracil residue.
  - Acts as the recognition site for **specific amino acyl tRNA Synthetase**.
- **Pseudouridine Arm (TψC Arm)**
  - T ψ C stands for Ribothymidine, Pseudouridine, Cytidine
  - Involved in the **binding of amino acyl tRNA to the Ribosomal surface.**
- **Extra Arm (Variable Arm)**
  - Between Pseudouridine and Anticodon arm
  - The most variable feature of tRNA.
  - Different classes of tRNA is based on the extra arm.

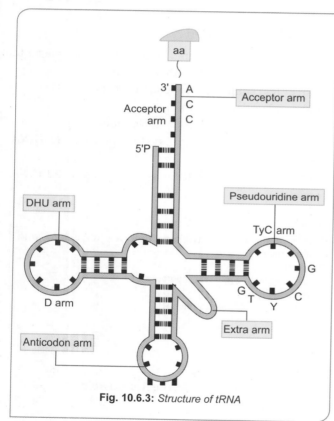

Fig. 10.6.3: *Structure of tRNA*

Post-transcriptional modification of tRNA precursor

- Standard bases (A,U,G,C) undergo methylation, reduction, deamination, rearrangement of glycosidic bond.
- Cleavage and attachment of CCA terminal at 3'end takes place in **cytoplasm.**
- Methylation of tRNA takes place in **nucleus.**

SMALL NUCLEAR RNA (snRNA)

- They belong to small RNAs of size 90 to 300 nucleotides
- 1% of total RNAs
- **They have ribozyme activity.**

Function

- mRNA Processing (U1, U2, U4, 5, 6 and U7 as a part of spliceosome)
- rRNA processing
- Gene regulation.

miRNA AND siRNA

- Small **noncoding** single stranded RNAs which are 21–22 nucleotide length.
- **Main function**[Q]: Post-transcriptional **regulation of gene expression** by **targeting mRNA by several distinct mechanism.**

We can discuss the following:

- Generation of miRNA and SiRNA
- Post-transcriptional modification of miRNA and siRNA
- Regulation of Gene expression by miRNA and siRNA
- Clinical correlations.

Micro RNA (miRNA)

Generation of miRNA

- Transcribed by RNA Polymerase II from miRNA encoding genes to PrimiRNA.

Post-transcriptional Modification of miRNA

- Pre-miRNA undergo extensive post-transcriptional processing as follows.
  - Pre-miRNA is subject to processing by **DROSHA-DGCR8 nucleases**, which trims 5'cap and 3' Poly A tail to generate Pre-miRNA.
  - The double stranded Pre-miRNA is transported to cytoplasm through nuclear pore, **Exportin-5.**
  - Pre-miRNA is further trimmed by Dicer nuclease (TRBP-Dicer) to form 21–22 nucleotide miRNA duplex.
  - One of the strand of duplex miRNA is selected.
  - The selected strand is loaded to RNA Induced Silencing complex (RISC).
  - Mature functional **21–22 nucleotide** miRNA is thus produced

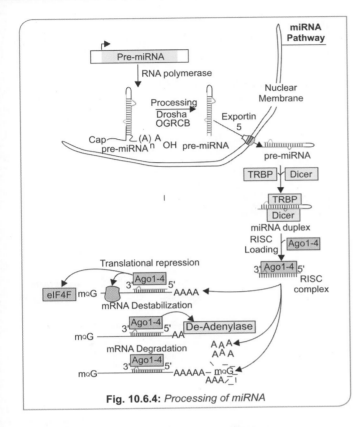

Fig. 10.6.4: *Processing of miRNA*

SILENCING RNA OR SMALL INTERFERING (siRNA)

Generation of siRNA
- Functional SiRNA is **generated endogenous or exogenous double stranded RNA.**
- Extracellular sources include RNA Viruses.

Post-transcriptional modification of siRNA
- Double stranded RNA are processed by Dicer nuclease.
- One strand is selected and loaded to RISC.

> **HIGH YIELD POINTS**
>
> RISC in miRNA is composed of Argonaute Proteins (Ago 1→4)
> RISC in siRNA is composed of Argonaute-2 Proteins

> **HIGH YIELD POINTS**
>
> **Specific Nucleases Involved in the Post-transcriptional Modification of miRNA and SiRNA**
> - DROSHA-DGCR8⁰ Nucleases
> - Dicer Nucleases

- **The regulation gene expression by miRNA and siRNA**
  - **By altering mRNA function. Regulation of gene expression by miRNA**
  - Binding of miRNA to mRNA by normal base pairing.
  - All mRNAs contain a *seed sequence* in their 3' untranslated region (UTR) that determines the specificity of miRNA binding and gene silencing.
  - If miRNA-mRNA base pairing **has one or more mismatches. Translation of cognate mRNA is inhibited**.
  - If miRNA-mRNA **base pairing is perfect, corresponding mRNA is degraded.**

miRNA modulate the function of target mRNA by three methods:
1. **Translation repression.**
2. **mRNA destabilization.**
3. **Promoting mRNA degradation directly.**

Regulation of gene expression by siRNA
Induces mRNA cleavage which inactivate target mRNA.

RNA Interference or RNAi
- Functional Consequence of Translation Arrest and mRNA degradation by miRNA/siRNA is **Silencing the gene expression or Gene Silencing.**
- This is otherwise called **RNA Interference.**
- **RNAi by miRNA/siRNA** is an example of gene knock down.

P Bodies
- Non-translating mRNA form ribonucleoprotein particles and they accumulate in cytoplasmic organelle called **P bodies.**
- Ribonucleoprotein or mRNP are mRNA, bound by specific packaging proteins.
- P bodies are sites of **translation repression and mRNA decay.**
- P bodies contain mRNA decapping enzymes, RNA helicases, RNA exonucleases, etc. for mRNA quality control.
- A portion of miRNA driven mRNA modulation takes place in **P bodies.**

Extra Edge

In 2006, Craig Mello and Andrew Fire were awarded Nobel Prize for silencing gene expression by miRNA.

miRNA and Cancer
- Can prevent cancer-by degrading mRNA of an oncogene. They are called Tumor Suppressivemi Rs
- Can cause cancer degrading mRNA of a tumour supressor gene called oncogenic mRNA (oncomir)

miRNA Associated with Cancer

Oncomirs
- miRNA that cause cancer are called Oncomirs
- Mir-21 is one of the widely studies oncogenic miRNA.
- miR-200 promote epithelial-mesenchymal transitions which is important in invasiveness and metastasis of tumour.
- miR-155 is over expressed in many human B cell lymphomas and indirectly up regulates genes like *MYC*.

Tumor suppressive miRs
- Deletions affecting certain tumor suppressive miRs, such as miR-15 and miR-16, are among the most frequent genetic lesions in chronic lymphocytic leukemia.
- Rare ovarian and testicular tumors, associated with germ line defects in *DICER*, a gene that encodes an endonuclease.

miRNA in DNA repair
- mir-34
- p53, the molecular policeman activate the expression of mir-34.

Extra Edge

Application of miRNA/siRNA

siRNA/miRNA and Transgenic mice
- Synthetic siRNA targeted against specific mRNA can be experimentally introduced into cell to study gene function by Gene knock down technology.
- SiRNA can be used as possible therapeutic agents to silence pathogenic genes, such as oncogenes involved in neoplastic transformation.

LONG NONCODING RNA (lncRNA)

- Transcribed by **RNA Polymerase II**
- They are >200 nucleotide length
- They modulate gene expression in many ways.

Regulation of gene expression by lncRNA

- Facilitate transcription factor binding and thus promote gene activation.
- Bind to transcription factors and thus prevent gene transcription, e.g. **Decoy lnc RNA**
  - The best known example of a repressive function involves XIST, which is transcribed from the X chromosome and plays an essential role in physiologic X chromosome inactivation.
- Facilitate Histone and DNA modification by directing methylases or acetylases.
- Act as scaffolding and stabilise secondary or tertiary structures and multisubunit complexes that influence chromatin structure.

Some recent facts about lncRNA

- lncRNAs may exceed coding mRNAs by 10- to 20-fold.

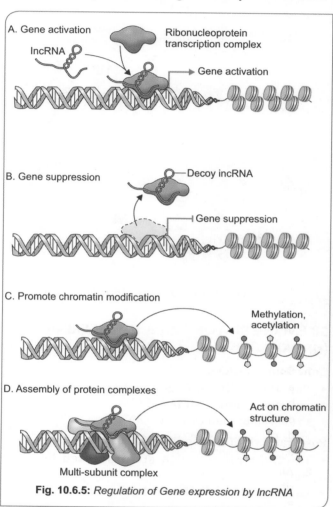

Fig. 10.6.5: *Regulation of Gene expression by lncRNA*

- It has recently been appreciated that many enhancers are sites of lncRNA synthesis, often increase transcription.
- Emerging studies are exploring the roles of lncRNA sin various human diseases, from atherosclerosis to cancer.

hnRNA (HETERONUCLEAR RNA)

- Primary transcript mRNA formed from DNA template.
- It undergoes post-transcriptional modification to form mature mRNA.

Newly Described Noncoding RNAs

Extra Edge

Small Nucleolar RNA (SnoRNA)
- RNA involved in eukaryotic **rRNA Processing** and assembly of ribosomes
- Present in the nucleolus.

Piwi-interacting RNAs (piRNAs)
- **The most common type of small noncoding RNA**
- Function: They have a role in post-transcriptional gene silencing (like miRNAs).

Long intervening noncoding RNAs (lincRNAs)
- Function: Regulate the factors that modify histones (epigenetic modifications) and thereby control gene expression.

Circular RNA (Circ RNA)
- From mRNA precursors and lncRNA precursors by RNA splicing.
- Belongs to large noncoding RNA
- Function–Regulation of gene expression

Quick Review

- All RNA except mRNA are noncoding RNAs
- Most abundant RNA is rRNA
- Pseudo uridine arm binds tRNA to ribosome surface.
- Main function of miRNA is post transcriptional regulation of gene expression.
- RNAi is RNA interference by miRNA and siRNA
- RNAi is gene knock down

Check List for Revision

- Recent trend topic for exams especially central institute exams.
- A selective reading is possible.
- All bold letters, boxes and tables are must learn.

Review Questions MCQ

1. **Components of RISC complex:** *(PGI Nov 2018)*
   a. Drosha
   b. rRNA
   c. Dicer
   d. miRNA
   e. Argonaute proteins

2. **All the following reactions are carried out by ribozyme except:** *(Recent Question Jan 2019)*
   a. Transesterification
   b. Ribonuclease P
   c. Peptidyl Transferase
   d. Poly A Polymerase

3. **Function of miRNA is/are:** *(PGI May 2014)*
   a. Gene silencing
   b. Gene activation
   c. Transcription inhibition
   d. Translation repression
   e. Cleavage of messenger RNA

4. **Normal role of MicroRNA is:** *(AI 2009)*
   a. Gene Regulation
   b. RNA splicing
   c. Initiation of Translation
   d. DNA conformation change

5. **Noncoding RNAs are:** *(PGI May 2012, May 2016)*
   a. siRNA
   b. miRNA
   c. tRNA
   d. mRNA
   e. rRNA

6. **Most common RNA is:** *(Ker 2011)*
   a. rRNA
   b. mRNA
   c. tRNA
   d. hnRNA

7. **Thymidylated RNA present in:** *(PGI June 01)*
   a. mRNA
   b. rRNA
   c. tRNA
   d. 16-s-RNA

8. **Met-tRNA would recognize:** *(PGI Nov 2009)*
   a. AUG
   b. GCA
   c. GUA
   d. UAC
   e. GAC

Answers to Review Questions

1. **c, e. Dicer, Argonaute protein** *(Ref: Harper 31e page 390 &391)*

   The components of RISC complex are TRBP-Dicer nuclease, miRNA/siRNA & Argonaute proteins. This complex facilitate cleavage of target mRNA.

2. **d. Poly A Polymerase**

   Ribozyme are RNA that act as catalyst
   Examples are
   - Ribonuclease P—post-transcriptional modification of tRNA.
   - Group II Introns—Transesterification reaction is involved in it
   - SnRNA in Spliceosome—Splicing of exons
   - Peptidyl Transferase—Peptide bond synthesis

3. **a, d, e. Gene silencing, Tranlation arrest, Cleavage of messenger RNA, miRNA and siRNA** *(Ref: Harper 31/e page 399)*

   Small noncoding single stranded RNAs which are 21-22 nucleotide length.

   Main function: Post-transcriptional regulation of gene expression by altering mRNA function.

   The regulation gene expression by miRNA
   By altering mRNA function.
   In 2006 Craig Mello and Andrew Fire were awarded Nobel Prize for silencing gene expression by miRNA
   Steps involved are:
   Binding of the miRNA to the target mRNA
   If the miRNA-mRNA base pairing has one or more mismatches, translation of the cognate "target mRNA" is inhibited **(TRANSLATIONAL ARREST)**
   - If the miRNA-mRNA base pairing is perfect over all 22 nucleotide.
   - The corresponding mRNA is degraded inside cytoplasmic organelle called **P Bodies. (mRNA DEGRADATION)**
   - Functional Consequence of Translation Arrest and mRNA degradation by miRNA is Silencing the gene expression or Gene Silencing.
   - This is otherwise called RNA Interference.

4. **a. Gene regulation** *(Ref: Harper31/e page 399)*

5. **a. a, b, c, d, e. siRNA, miRNA, t RNA, mRNA, rRNA** *(Ref: Harper 31/e page 390)*

   Noncoding RNAs (ncRNA)
   - Transfer RNA (tRNA)
   - Ribosomal RNA (rRNA)
   - Sno RNA
   - gRNA (Guide RNA)
   - miRNA
   - siRNA
   - lncRNA

6. **a. rRNA** *(Ref: Harper 31/e page 390)*

RNA	Types	Abundance	Stability
Ribosomal (rRNA)	28S, 18S, 5.8S, 5S	80% of total	Very stable
Messenger (mRNA)	~105 Different species	2–5% of total	Unstable to very stable
Transfer (tRNA)	~60 Different species	~15% of total	Very stable
Small RNAs			
Small nuclear (snRNA)	~30 Different species	1% of total	Very stable
Micro (miRNA)	100s–1000	<1% of total	Stable

   mRNA editing
   - The central dogma states that for a given gene and gene product there is a linear relationship between the coding sequence in DNA, the mRNA sequence, and the protein sequence.
   - Changes in the DNA sequence should be reflected in a change in the mRNA sequence and, depending on codon usage, in protein sequence.
   - mRNA editing is the process by which coding in formation is changed at the mRNA level by chemical modification of the nitrogenous bases present in the codons.
   - Hence the linear relationship between the coding sequence in DNA, the mRNA sequence, and the protein sequence is altered.
   - Thus it is an exception to Central Dogma of molecular genetics.

7. **c. tRNA** *(Ref: Harper 31/e page 346)*
   - **Pseudouridine Arm (T ψ C arm) of tRNA**
   - T ψ C stands for Ribothymidine, Pseudouridine, Cytidine

8. **a. AUG** *(Ref: Harper 31/e page 395)*
   - The first AUG sequence after the marker sequence is defined as the start codon.
   - AUG codon binds with met tRNA

10.7 TRANSLATION

- Definition
- Steps of Protein Synthesis
- Codon
- Inhibitors of Translation
- Genetic Code
- Post-translational Modifications

DEFINITION

- The process by which message in the genetic code in the mRNA is translated into sequence of amino acids in the proteins.

CODON

The triplet nucleotide sequence present in the mRNA representing specific amino acid.
- If 1 base represent 1 amino acid only 4 amino acid.
- If 2 base represent 1 amino acid 4^2 Amino acids, i.e. 16 amino acids
- If 3 base 4^3, i.e. 64 amino acids
- If 4 bases 4^4, i.e. 256 amino acids^Q

CONCEPT BOX

tRNA as an adapter molecule
- The language of nucleotide is translated to language of amino acid in translation
- The codon has no affinity towards the amino acid that it codes.
- tRNA acts as the adapter molecule between the codon and the specific amino acid.
- This is possible by two reasons
- Nucleotide sequence of codon is complementary to the anticodon in the tRNA.
- DHU arm recognize the specific Amino acyl tRNA synthetase, so that specific amino acid bind to the Acceptor arm of the tRNA.

For Example:
- UUU is the codon for Phenylalanine
- The tRNA that carries Phenylalanine has AAA in the anticodon arm.
- DHU arm recognizes the Phenylalanyl tRNA Synthetase.

GENETIC CODE

Definition

The relationship between a sequence of DNA and sequence of amino acid in the corresponding polypeptide is called Genetic Code.
- Cracking of Genetic Code was done by **Marshal Nirenberg and Har Gobind Khorana.**

Salient Features of Genetic Code

- **Triplet Codon:** Each amino acid is represented by triplet sequence.
- **Degenerate (Redundant)**^{Q}: More than 1 codon represent a single amino acid. Degeneracy of the codon lies in the **3rd Base.**^{Q}
- **Non overlapping:** Reading of genetic code do not involve overlapping sequence.

HIGH YIELD POINTS

Amino Acids with maximum number of codons (Six codons)
- Serine, Arginine, Leucine

Two amino acids with single codon
- AUG—Methionine
- UGG—Tryptophan

- **Unambiguous**
  – Any specific codon can represent only one amino acid.
- **Universal**
  – A specific codon represent a specific amino acid in all the species.
  – Exception to this rule—Codons of Mitochondrial DNA
- **Initiator codon**
  – In eukaryotes—AUG codes for Met
  – In Prokaryotes—AUG codes for **N-Formyl Methionine**
- **Terminator codons**
  – UAG—Amber
  – UGA—Opal
  – UAA—Ochre

Exceptions:
- UGA can be recoded to **Selenocysteine**.
- UAG can be recoded to **Pyrrolysine**.
- UGA codes for **Tryptophan in mitochondrial DNA.**

> **EXTRA EDGE**
>
> **Frequently asked doubts**
> If the anticodon arm has the anticodon with Inosine (the base in this Hypoxanthine), can it form hydrogen bond with any of the codons in the mRNA?
> Answer is Yes it can form Weak hydrogen bond with three bases
> They are Adenosine (A), Uridine (U) and C (Cytidine)
>
> ```
>                      3 2 1         3 2 1         3 2 1
> Anticodon      (3') G-C-I         G-C-I         G-C-I  (5')
>                     ≡ ≡ ≡         ≡ ≡ ≡         ≡ ≡ ≡
> Codon          (5') C-G-A         C-G-U         C-G-C  (3')
>                      1 2 3         1 2 3         1 2 3
> ```
>
> **Fig. 10.7.1:** *Codon-Anticodon relationship*
>
> The picture depicts three different codon pair relationship possible if anticodon contains Inosinate.

Wobbling Phenomenon

This was proposed by **Watson Crick**.

The base pairing at the 3rd nucleotide between the anticodon in the tRNA and Codon in the mRNA is not stringently regulated. This is called Wobbling phenomenon.

For example:

- Two codons for Arginine are AGA and AGG can bind with same tRNA having UCU in the anticodon arm.
- Base pairing at the third nucleotide is not always obeying **Watson Crick** base-pairing rule.
- Thus it is said degeneracy lies in the third base.
- This explains how 31 tRNA species can bind with 61 coding codons.

Cistron^Q

It is the smallest unit of genetic expression which code for a polypeptide chain.

Monocistronic—One cistron represents one Polypeptide, e.g. **Eukaryotic mRNA**

Polycistronic—One cistron represents more than one polypeptide, e.g. **Prokaryotic mRNA**

*Mnemonic: P for P (**P**rokayotic is **P**olycistronic)*

Polarity of Transcription

- The message in the mRNA is decoded from **5' end to 3' end**.
- The codon in the 5' end of mRNA corresponds to N
- Terminal amino acid of the polypeptide.

STEPS OF PROTEIN SYNTHESIS

- Charging of tRNA
- Initiation
- Elongation
- Termination

Charging of tRNA

- The process by which specific Amino acid is attached to the acceptor arm by specific aminoacyl tRNA Synthetase.
- **Specific aminoacyl tRNA Synthetase** enzyme is identified by **DHU arm.**
- The charging reaction has an error rate of **less than 10^{-4}**.
- Hence **aminoacyl tRNA Synthetase is considered as the proofreading mechanism of translation.**
- Two inorganic Phosphates are used in the charging of the tRNA.
- **Mg^{2+}** is the cofactor for aminoacyl tRNA Synthetase.

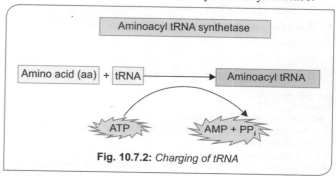

Fig. 10.7.2: *Charging of tRNA*

> **EXTRA EDGE**
>
> - In Prokaryotic translation initiator tRNA with UAC anticodon attaches Methionine or N Formyl Methionine?
>
> **Facts**
>
> - 5' AUG3' in bacterial mRNA recognises two tRNAs, i.e. tRNA met and tRNAf met.
> - If the AUG is initiator codon, N formyl methionine arrives at the ribosome.
> - This is formed by two successive reactions.
>   1. Methionine is attached to tRNAfMet
>   2. Transformylase transfer formyl group from N10 Formyl THFA to amino group of methionine residue.
> - So the answer is initiator tRNA with UAC anticodon, attaches Methionine to its acceptor arm. This Methionine is later formylated to N Formyl Methionine.

Initiation

Identification of Initiator Codon

By Marker Sequence—Consensus sequence that helps in the identification of initiator codon.

- **Prokaryotes: Shine Dalgarno sequence** (Identified by **John Shine and Lynn Dalgarno**)
- **Eukaryotes: Kozak sequence**

The first AUG sequence after the marker sequence is defined as the start codon.

AUG codon binds with met tRNA.

Initiation

- It is a multistep process.
- Facilitated by accessory proteins called **Initiation Factors.**
- In case of eukaryotes it is called eukaryotic Initiation Factors (eIF).

1. **Dissociation of the ribosome into its 40S and 60S** subunits
   - Two initiation factors delays the association of 40S Subunit and 60S Subunit
   - They are **eukaryotic Initiation Factor-3 (eIF-3) and eIF-1A.**

2. **Formation of 43S Preinitiation Complex.**
   Three steps
   1. First step involves binding of **GTP by eIF-2** to form **Binary Complex.**
   2. Second step involves binding of binary complex to **mettRNA^i** to form **Ternary Complex.**
   3. This ternary complex binds to 40S Subunit of Ribosome to form **43S Preinitiation complex.**

In Stressful conditions kinases are activated.

eIF2α is phosphorylated and protein synthesis is arrested during stressful conditions.

In Stressful conditions kinases are activated.

eIF2α is phosphorylated and protein synthesis is arrested during stressful conditions.

This explains how protein synthesis is decreased in glucose starvation, viral infection, etc.

3. **Formation of 48S Initiation Complex**
   - Binding of **43S Preinitiation complex to the mRNA** forms **48S Initiation Complex.**
   - Formation of 48S Initiation complex *require ATP Hydrolysis.*

> **EXTRA EDGE**
> **Factors that facilitate the binding of mRNA to 40S Subunit**
> - 5' methyl Guanosine cap and Cap binding Complex (Described below)
> - 3' Poly A tail and Poly A tail Binding Protein (PAB-1)
>
> **Cap Binding Complex eIF 4F Complex**
> - Consist of **eIF-4E and eIF4G-eIF4A**
> - This cap binding complex bind to 7-meG cap though **4E.**
> - This complex is very important in controlling the rate of translation.
> - 4E responsible for recognition of mRNA cap structure, is a rate limiting step of translation.
>
> Insulin and other mitogenic growth factors through AKT/PI3 Kinase pathway
>
> This phosphorylate 4E binding protein, this make 4E free. This facilitate 4F complex to bind mRNA cap through 4E. Thus Insulin increases the rate of initiation of translation.

4. Formation of 80S Initiation Complex

48S Initiation Complex + 60S Subunit of the ribosome form 80S Initiation Complex.

- Involves GTP Hydrolysis.
- Initiation factors are released
- So initiation complex consist of two subunits of ribosome, initiator tRNA and mRNA (No GTP or IFs)
- The initiator tRNA is in P site in the initiation complex.

3 sites are present in the 80S Ribosome:

1. **A site**-where the new Aminoacyl tRNA binds.
2. **P site**-where the growing peptidyl chain present
3. **E site**-where the deacylated tRNA is present

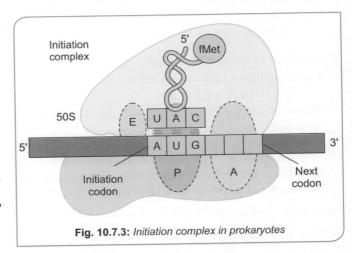

Fig. 10.7.3: *Initiation complex in prokaryotes*

Elongation of Polypeptide in Translation

Multi-step Process

Catalysed by proteins called **Elongation Factors.**

Involves 4 steps
1. Binding of Aminoacyl tRNA to the A site.
2. Peptide Bond Formation.
3. Translocation of ribosome on the mRNA.
4. Expulsion of deacylated tRNA from P and E site.

Binding of Aminoacyl tRNA to the A site
- tRNA carrying the specific amino acid binds to the **A site**.
- **Elongation factor EF-1** helps in the binding of tRNA^{aa}
- GTP is hydrolyzed to GDP.

Peptide Bond Formation
- The alpha amino group of the incoming amino acid in the A site forms peptide bond with the COOH group of the peptidyl tRNA in the P (Peptidyl or Polypeptide) site.
- As a result the growing peptide chain is attached to **A site**.
- Enzyme is **Peptidyl transferase,** a ribozyme which is a component of *28Sr RNA of 60S Subunit.*
- No energy is required for this peptide bond formation step
- The growing peptide chain is now in the A site.

Translocation of the ribosome on the mRNA
- **The ribosome move forwards**, then whole mRNA is shifted by a distance of one codon.
- **Peptidyl tRNA is translocated to P site** from A site.
- **A site is free** to receive the next incoming tRNA^{aa}.
- **Deacylated tRNA is on E site**.
- Translocation requires *EF-2 and GTP.*

Termination of Protein Synthesis
- **Releasing factor** helps in the termination.
- Stop codon is in the A site now.
- Releasing factor-1 (RF-1) recognises the stop codon in the A site.
- RF-1 is bound by RF-3 and **GTP**.
- This complex with **peptidyl transferase** promote hydrolysis bond between polypeptide chain and tRNA
- This involves **hydrolysis of GTP to GDP.**

Extra Edge - Eukaryotic Initiation Factors (eIF)

eIF-3 and eIF-1A	Bind to newly dissociated 40S ribosomal subunit. This allows translation initiation factor to associate with the 40S subunit.
eIF-2	Involved in the formation of binary complex, of 43S preinitiation complex.

Contd...

Contd...

eIF-4F	Cap binding complex Important in 48S Initiation complex
eIF-5	Hydrolysis of GTP bound to eIF-2 facilitate 60S association

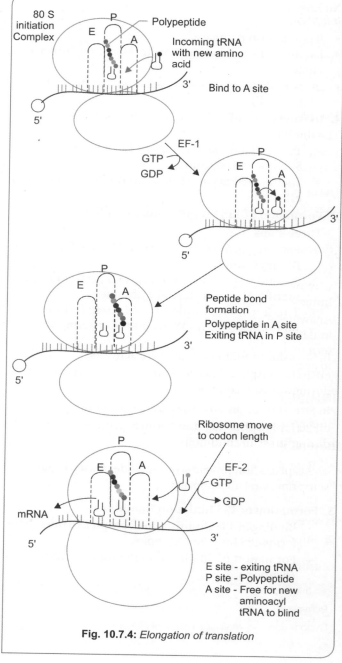

Fig. 10.7.4: Elongation of translation

Energetics of Peptide Bond Synthesis

- Charging of tRNA to tRNA^{aa}– 2 Inorganic Phosphates = 2 ATP equivalents.
- EF1—binding of tRNA^{aa} to A Site- 1GTP
- EF2—Translocation-1GTP

No Energy for actual peptide bond synthesis

Even though the actual peptide bond formation does not require energy, for the formation of one peptide bond hydrolysis of **4 inorganic Phosphates** is required. (2 ATP for charging of the amino acid 1GTP for EF-1 and 1GTP for EF-2 Translocation).

Regulation of Translation

The two control points of translation are
1. eIF 4E of 4F complex
2. eIF 2

Rate of Protein Synthesis Prokaryotes—18 Amino acids per second. Eukaryotes—6 Amino acids per second.

Frequently Asked Doubts

When energetics of 1 peptide bond synthesis calculated, why do not we take into account GTP used for the Initiation complex and Termination?

Answer is every time 1 peptide bond is synthesised initiation and termination is not repeated. Initiation and termination is a one time process. So we do not take in to account GTP required for initiation and termination.

INHIBITORS OF TRANSLATION

Bacterial Protein Synthesis Inhibitors

Reversible: Bacteriostatic

- Tetracyclins
- Chloramphenicol
- Erythromycin and Clindamycin

Irreversible Inhibitors—Bactericidal

- Streptomycin and other aminoglycosides

Mammalian Protein Synthesis Inhibitors

Inhibitor	Mechanism of action
Puromycin^Q	Structural analog of tyrosinyl tRNA.
Cycloheximide	Inhibit peptidyl transferase in the 60S ribosomal subunit.

Contd...

Inhibitor	Mechanism of action
Diphtheria toxin^Q	Exotoxin of Corynebacterium diphtheria Catalyzes ADP ribosylation of Elongation Factor-2 on the unique amino acid diphthamide in mammalian cell. This inactivate EF-2, specifically inhibits mammalian Protein synthesis.
Ricin^Q	From castor bean inactivates eukaryotic 28S ribosomal RNA.

POST-TRANSLATIONAL MODIFICATIONS

- Amino terminal and Carboxy terminal modification
- Loss of signal sequences
- Covalent Modifications of Aminoacyl residues
  - Gamma Carboxylation
  - Acetylation
  - Phosphorylation
  - Hydroxylation
  - Methylation
- Glycosylation of N-linked and O-linked Glycoprotein.
- Zymogen activation (Proteolytic processing)
- Addition of Isoprenyl group
- Formation of Disulfide crosslinks

High Yield Points

Polysome or Polyribosome

- Multiple ribosome on the same mRNA are called Polysome or Polyribosome.

Quick Review

- More than 1 codon can code for same amino acid is degeneracy.
- One specific codon for a specific amino acid cannot code for a different amino acid is unambiguous nature of genetic code.
- The stop codons that code for amino acids are UGA for SeCys and UAG for Pyrolys.
- The charging of tRNA need 2 inorganic phosphates.
- No energy is needed for peptidyl transferase activity.
- Peptidyl transferase is a ribozyme.
- For 1 peptide bond synthesis 4 inorganic phosphates are needed.

Check List for Revision

- For central institute exams like PGIMER a thorough knowledge is needed.
- For NBE model exams a selective study of bold letters is enough.
- Extra edge are all points which you can be selective.

Review Questions MCQ

1. **Components of 30s intiation complex:**
   (PGI Nov 2018)
   a. GTP
   b. mRNA
   c. IF 2
   d. EF 2
   e. ATP

2. **Which of the following are post-translational modification of polypeptides?** (PGI May 2017)
   a. Addition of base pair at the ends
   b. Deletion of base pairs at the end
   c. Phosphorylation of Serine
   d. Excision of a peptide
   e. Ubiquitination of Lysine

3. **Which RNA contain abnormal purine and Pyrimidine?**
   (PGI Nov 2016)
   a. tRNA
   b. 23S rRNA
   c. 16S rRNA
   d. m RNA
   e. 5S rRNA

4. **Components of 50S ribosomal subunit:**
   (PGI Nov 2016)
   a. 16S RNA
   b. 18S RNA
   c. 5.8S RNA
   d. 23S RNA
   e. 5S RNA

5. **A codon consists of:** (AIIMS 90, UP 99, WB 02)
   a. One molecule of aminoacyl-tRNA
   b. Two complementary base pairs
   c. Three consecutive nucleotide units
   d. Four individual nucleotides

6. **All are true of genetic code except:**
   (DNB 2001, Delhi 98, TN 95)
   a. Degenerate
   b. Universal
   c. Punctuation
   d. Nonoverlapping

7. **Wobble hypothesis–regarding the variation true is:**
   (PGI Dec 07)
   a. 3 – end of anticodon
   b. 5 – end of anticodon
   c. mRNA
   d. tRNA

8. **Components of 50S subunit is/are:** (PGI May 2011)
   a. 23S
   b. 28S
   c. 5S
   d. 5.8S
   e. 16S

9. **Ribosome 60S subunit contains:** (PGI Nov 2009)
   a. 5.8S subunit
   b. 23S subunit
   c. 28S subunit
   d. 16S subunit
   e. 18S subunit

10. **There are 20 amino acids with three codons in spite of the no of amino acids could be formed is 64 leading to that an amino acid is represented by more than one codon is called:** (AIIMS May 2012)
    a. Transcription
    b. Degeneracy
    c. Mutation
    d. Frameshift

11. **Genetic code has triplet of nucleotides each for one amino acid. When an amino acid is specified by more than one codon, it is called:** (AIIMS Nov 2012)
    a. Transcription
    b. Degeneracy
    c. Mutation
    d. Frameshift

12. **The polypeptide from poly(A) is:** (AIIMS Nov 2012)
    a. Polylysine
    b. Polyglycine
    c. Polyproline
    d. Polyalanine

13. **If constitutive sequence of 4 nucleotide codes for 1 amino acid, how many amino acid can be theoretically formed?** (Nov 2012)
    a. 4
    b. 64
    c. 16
    d. 256

14. **False about eukaryotic protein synthesis:**
    (AIIMS May 2009)
    a. N formyl^{Met} is the first tRNA to come into action
    b. mRNA read from 3' to 5'
    c. eIF-2 shifts between GDP TO GTP
    d. Capping helps in attachment of mRNA to 40S ribosome

15. **Termination process of protein synthesis is performed by all except:** (PGI May 2010)
    a. Releasing factor
    b. Stop codon
    c. Peptidyl transferase
    d. UAA codon
    e. AUG codon

16. **True about Ribozyme:** (AIIMS Nov 2012)
    a. Peptidyl Transferase activity
    b. Cuts DNA at specific site
    c. Participate in DNA Synthesis
    d. GTPase activity

17. **Part of eukaryotic DNA contributing to polypeptide synthesis:** (PGI May 2011)
    a. Exon
    b. Enhancer
    c. Leader sequence
    d. tRNA
    e. ncRNA

18. **Stop codons are:** (PGI Nov 2010)
    a. UAA
    b. UAG
    c. UGA
    d. UAC
    e. UCA

19. **Met-tRNA would recognize:** (PGI Nov 2009)
    a. AUG
    b. GCA
    c. GUA
    d. UAC
    e. GAC

20. **Which of the following statement is true?** *(Ker 2008)*
    a. N formyl methionine is the precursor of eukaryotic polypeptide synthesis
    b. Eukaryotic ribosomes are smaller than prokaryotic
    c. Identification of 5'cap of mRNA by IF4E is the rate limiting step
    d. Elongation factor 2 shuttles between ADP and ATP

21. **RNA polymerase differs from DNA polymerase:** *(Ker 2007)*
    a. It edits and synthesis
    b. Synthesize RNA primers
    c. Synthesis only in 5 to 3 direction
    d. Uses RNA templates

22. **The cellular component for protein synthesis is:** *(AI 97)*
    a. Smooth endoplasmic reticulum
    b. Rough endoplasmic reticulum
    c. Ribosomes
    d. Mitochondria

23. **Amber codon refers to:** *(AIIMS May 01)*
    a. Mutant codon
    b. Stop codon
    c. Initiating codon
    d. Codon for more than one amino acids

24. **Shine-Dalgarno sequence in bacterial mRNA is near:** *(AI 2004)*
    a. AUG codon
    b. UAA codon
    c. UAG codon
    d. UGA codon

25. **True regarding aminoacyl tRNA synthetase is A/E:** *(SGPGI 06)*
    a. Is accepting tRNA
    b. Implement genetic code
    c. Attachment of amino group to 5' end of tRNA
    d. Editing function

26. **In translation process, proofreading of mRNA is done by:** *(AIIMS Dec 97)*
    a. RNA polymerase
    b. Aminoacyl tRNA synthetase
    c. Leucine zipper
    d. DNA

27. **Which enzyme involved in translation is often referred to as 'Fidelity enzyme'?** *(AI 1998)*
    a. DNA polymerase
    b. RNA polymerase
    c. Aminoacyl tRNA synthetase
    d. Aminoacyl –reductase

28. **The hydrolytic step leading to release of polypeptide chain from ribosomes is catalysed by:** *(PGI June 02)*
    a. Stop codons
    b. Peptidyl transferase
    c. Releasing factors
    d. AUG codon
    e. Dissociation of ribosomes

29. **About peptidyl transferase true is:** *(JIPMER 2000)*
    a. Used in elongation and cause attachment of peptide chain to A- site of tRNA
    b. Used in elongation and cause attachment peptide chain to P site
    c. Used in initiation and cause 43S complex formation
    d. Used in initiation and cause 48S complex formation

30. **Termination is caused by all except:** *(AIIMS Dec 90)*
    a. RF-1
    b. UAA
    c. Peptidyl transferase
    d. 48S complex

31. **True about translation of proteinis:** *(PGI Dec 98)*
    a. It has 3 steps initiation, elongation, termination
    b. IF-2 prevent reassociation of ribosomal subunit
    c. IF-3 and 1A cause binding of initiation factors
    d. IF 2 has α and β units

32. **43S preinitiation complex include all except:** *(AIIMS Dec 93)*
    a. IF3
    b. IF1A
    c. IF2
    d. IF-4F

33. **IF 4F include all except:** *(PGI June 97, JIPMER 99)*
    a. 4A
    b. 4G
    c. 4E
    d. 4S

34. **For 1 peptide bond formation how many high energy phosphate bonds are required?** *(PGI Dec 07)*
    a. 0
    b. 1
    c. 2
    d. 3
    e. 4

35. **Vitamin required for post-translational modification of coagulants is:** *(AI 97)*
    a. Vitamin A
    b. Vitamin C
    c. Vitamin B_6
    d. Vitamin K

36. **Initiator tRNA is in which site of ribosome?** *(Recent Question 2016)*
    a. A site
    b. P site
    c. E site
    d. B site

Answers to Review Questions

1. **a, b, c. GTP, mRNA, IF2**

   In Prokaryotic protein synthesis during initiation process 30S initiation complex is formed which bind with 50S subunit to form 70S Initiation complex
   30S IC consist of IF 1, IF2, IF 3, GTP & mRNA

2. **c, d, e. Phsophorylation of Serine, Excision of a peptide, Ubiquitination of Lysine:** *(Ref: Harper 31/e page 394)*
   - Addition of base pair is a post-transcriptional modification

3. **a. tRNA** *(Ref: Harper 31/e page 395)*

4. **d, e. 23S RNA, 5S RNA** *(Ref: Harper 31/e page 399)*

5. **c. Three consecutive nucleotide units** *(Ref: Harper 31/e page 399)*
   - Each codon consists of a sequence of three nucleotides, i.e. it is a triplet code.

6. **c. Punctuation** *(Ref: Harper 31/e page 395)*

   Features of the Genetic Code
   - Degenerate
   - Unambiguous
   - Nonoverlapping
   - Notpunctuated
   - Universal

7. **b. 5 – end of anticodon** *(Ref: Harper 31/e page 395-397)*

   5' end of anticodon suggesting the binding of 3rd nucleotide of codon and anticodon is not strictly obeying the Watson Crick base pairing rule. This is called Wobble. It is between nucleotide in the 3' end of codon and 5' end of anticodon.

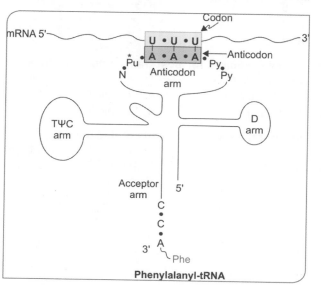

Phenylalanyl-tRNA

8. **a, c. 23S, 5S** *(Ref: Harper 31/e page 399)*

   Ribosomal Assembly in prokaryotes
   70S Ribosome = 30S + 50S Subunits
   In 30S Subunit 16S rRNA + Proteins
   In 50S rRNA 23S rRNA + 5S rRNA + Proteins
   Ribosomal Assembly in Eukaryotes
   80S Ribosome = 40S Subunit + 60S Subunit
   60S Subunit = 28S rRNA + 5.8S r RNA + 5S rRNA + ~50 proteins
   40S = 18S rRNA + ~30 proteins

9. **a, c. 5.8S subunit, 28S subunit** *(Ref: Harper 31/e p399)*

10. **b. Degeneracy** *(Ref: Harper 31/e p395)*

    Genetic Code—Salient features of Genetic Code
    - **Triplet Codon:** Each amino acid is represented by triplet sequence.
    - Degenerate (Redundant)—More than 1 codon represent a single amino acid. Degeneracy of the codon lies in the **3rd Base**.
    - Nonoverlapping— Reading of genetic code do not involve overlapping sequence.
    - Unambiguous—A codon can represent only one amino acid.
    - Universal—A specific codon represent a specific amino acid in all the species.

    Exception to this rule–Codons of Mitochondrial DNA
    - Initiator codon
      - In eukaryotes—AUG codes for Met
      - In Prokaryotes—AUG codes for N-Formyl Methionine
    - Terminator Codons
      - UAG—Amber
      - UGA—Opal
      - UAA—Ochre
    - Wobbling Phenomenon

    The base pairing at the 3rd nucleotide between the anticodon in the tRNA and Codon in the mRNA is not stringently regulated.

11. **b. Degeneracy** *(Ref: Harper 31/e page 395)*

12. **a. Polylysine** *(Ref: Harper 31/e page 395)*

    Poly A codes for Lysine Poly C codes for Proline Poly G codes for Glycine
    Poly U codes for Phenylalanine

13. **d. 256** *(Ref: Harper 31/e page 395)*

    Codon

    The triplet nucleotide sequence present in the mRNA representing specific amino acid.

    If 1 base represent 1 amino acid only 4 amino acid.

    If 2 base represent 1 amino acid 4^2 amino acids, i.e. 16 amino acids

    If 3 base 4^3, i.e. 64 amino acids

    If 4 bases 4^4, i.e. 256 amino acids

14. **a, b. N formyl^{Met} is the first tRNA to come into action, mRNA read from 3' to 5'** *(Ref: Harper 31/e page 395)*

Explanation

Option A
First tRNA to come into action is methionyl tRNA in eukaryotes and N Formyl tRNA in prokaryotes.

Option B
The translation of the mRNA commences near its 5' terminal with the formation of the corresponding amino terminal of the protein molecule.
The message is read from 5'–3' for translaton.

Option D
4E subunit of eIF–4F that is bound to the cap forms a circular structure that helps direct the 40S ribosomal subunit to the 5' end of the mRNA.
The poly (A) tail stimulates recruitment of the 40S ribosomal subunit to the mRNA.
Both Poly A tail and 5' cap is needed for the attachment of mRNA to the 40S subunit.

15. **e. AUG codon** *(Ref: Harper 31/e page 395)*

    Termination of Protein Synthesis
    Stop codons specifies the sites to stop translation. *Releasing factor-1* along with Releasing factor 3 helps in the termination
    Hydrolysis of GTP to GDP.
    Peptidyl Transferase also helps in the hydrolysis of the polypeptide chain.

16. **a. Peptidyl Transferase activity** *(Ref: Harper 31/e p396)*

    Ribozyme is RNA with catalytic activity.
    - E.g.: SnRNA in Spliceosome–Takes part in splicing of exons and removal of introns
    - Ribonuclease P—Cuts the RNA
    - **Peptidyl Tranferase:** Peptide Bond formation other options

    Cuts the DNA at specific site—Restriction Endonuclease
    DNA Synthesis—DNA Polymerase

17. **a. Exon**

18. **a, b, c. UAA, UAG, UGA** *(Ref: Harper 31/e page 395)*

    Terminator Codons/Stop Codons/Nonsense Codons UAG—Amber
    UGA—Opal UAA—Ochre

19. **a. AUG** *(Ref: Harper 31/e page 395)*

    The first AUG sequence after the marker sequence is defined as the start codon.
    AUG codon binds with met tRNA^I

20. **c. Identification of 5'cap of mRNA by IF4E is the rate limiting step** *(Ref: Harper 31/e page 396)*

    Rate limiting steps of Protein Translation are:
    eIF 2
    eIF 4F (specifically saying 4E of 4F complex)
    Option A: Methionine is the first amino acid in eukaryotic protein synthesis.
    Option B: Eukaryotic Ribosome is bigger than Prokaryotic ribosome.

    Option D: Elongation Factor 2 is associated with hydrolysis of GTP not ATP.

21. **b. Synthesize RNA primers** *(Ref: Harper 31/e page 396)*
    - Option aRNA Polymerase do not have editing function.
    - Optionc, both DNA Polymerase and RNA Polymerase synthesize in 5' to 3' direction.
    - Option d Both use DNA as template.
    - Option b, RNA primer is synthesized by RNA Polymerase.

22. **c >b. Ribosome > Rough endoplasmic reticulum**

    Rough endoplasmic reticulum is ER studded with Ribosome, there also in Ribosome protein synthesis takes place.

23. **b. Stop codon** *(Ref: Harper 31/e page 397)*
    - Amber is UAG stop codon.
    - Ochre is UAA stop codon.
    - Opal is UGA stop codon.

24. **a. AUG codon** *(Ref: Harper 31/e page 395)*

    Shine Dalgarno sequence is the marker sequence. The first AUG codon after Shine Dalgarno sequence is the start codon in bacteria. Similarly in eukaryotes there is Kozak sequence.

25. **c. Attachment of amino group to 5' end of tRNA** *(Ref: Devlin 7/e page 397)*

    The reaction of Aminoacyl tRNA synthetase
    Amino acid + ATP + Enzyme→Amino acid-AMP-Enzyme complex + PPi
    Amino acid-AMP-Enzyme complex + tRNA. Aminoacyl tRNA + AMP +Enzyme.
    - Aminoacyl tRNA synthetase first accept ATP and Amino acid
    - Then the enzyme accept the specific tRNA
    - Amino acid is attached to 3' hydroxyl adenosyl end of tRNA.
    - Two inorganic phosphates are used.
    - This enzyme is a part of editing mechanism of translation.

26. **b. Aminoacyl tRNA synthetase** *(Ref: Harper 31/e page 395)*
    - Aminoacyl tRNA synthetase which recognises the specific tRNA is a proofreading mechanism.

27. **c. Amino acyl tRNA synthetase** *(Ref: Harper 31/e page 395)*

28. **b, c. Peptidyl transferase, Releasing factors** *(Ref: Harper 31/e page 403)*
    - Releasing factor RF1 recognises that a stop codon resides in the A site.
    - RF1 bound by a complex consisting of releasing factor RF3 bound by GTP.
    - This complex, along with peptidyl transferase catalyses the hydrolysis of bond between peptide and tRNA occupying the P site.
    - Stop codon is not catalysing the hydrolytic cleavage, so it cannot be the answer.

29. **a. Used in elongation and cause attachment of peptide chain to A-site of tRNA** *(Ref: Harper 31/e page 403)*

- The α amino group of the new aminoacyl-tRNA in the A site carries out a nucleophilic attack on the esterified carboxyl group of the **peptidyl-tRNA** occupying the **P site** (**peptidyl or polypeptide site**). At initiation, this site is occupied by the initiator met-tRNAi.
- This reaction is catalyzed by a **peptidyl transferase**, a component of the 28S RNA of the 60S ribosomal subunit.
- This is another example of ribozyme activity.

30. **d. 48S complex** (Ref: Harper 31/e page 9401)
- Termination is helped by RF-1, RF-3, stop codon and Peptidyl transferase.

31. **a, c. It has 3 steps ..., IF-3 and 1A cause binding...** (Ref: Harper 31/e page 401)
- Like transcription, protein synthesis can be described in three phases, Initiation, Elongation and Termination.
- eIF3 and eIF 1 A bind to newly dissociated 40S ribosomal unit and prevent its reassociation with 60S subunit till translation initiation factors are all associated.
- eIF-2 is a part of binary complex.
- eIF-2 consist of α, β and γ-subunits.
- eIF is one of the control points of translation.

32. **d. IF-4F** (Ref: Harper 31/e page 399)
43 S Preinitiation complex consist of
- GTP-eIF2-tRNA^i + 40S subunit.
- This 43S Preinitiation complex is stabilised by eIF3 and eIF-1A.

33. **d. 4S** (Ref: Harper 31/e page 402)
- Cap binding complex consists of 4A, 4G, 4E

34. **a. 0** (Ref: Harper 31/e page 402)
- For peptide bond formation no energy is required as the aminoacyl tRNA is already activated.
- If the option doesn't have 0, then answer is 4.

35. **d. Vitamin K**
Gamma carboxylation of Clotting factors, require Vitamin K

36. **b. P site** (Ref: Harper 31/e page 404)
- Initiator tRNA is in the P site.
- New aminoacyl tRNA in the A site.

10.8 REGULATION OF GENE EXPRESSION

- Gene Expressions
- Regulation of Gene Expression of Different Levels
- Regulation of Gene Expression at Transcription Level
- Regulation of Gene Expression at DNA Level
- Epigenetics
- Gene Amplification
- Gene Silencing
- Gene Rearrangement
- Regulation of mRNA Stability
- Transposons
- Gene Switching

GENE EXPRESSION

- Organisms adapt to the environmental changes by altering gene expression.
- Most common regulation is modulation of transcription.

Housekeeping Gene (Constitutive Gene)

- Genes which are expressed at a constant rate in almost all the cells of the body.
- Required for the basal cellular function.
- For example: Enzymes for glycolysis.

Inducible Gene

- Genes which are expressed under special circumstances.
- Increase in response to an activator or inducer.
- Decrease in response to a repressor.

REGULATION OF GENE EXPRESSION AT DIFFERENT LEVELS

At the Level of Transcription

- Induction and repression (Explained by Operon Concept)
- Alternate mRNA processing (Described in Chapter Transcription)
- RNA editing (Described in Chapter Transcription)
- mRNA stability
- Noncoding RNA induced regulation (Described in Chapter Transcription)

At the level of DNA

- Gene amplification
- Gene rearrangement
- Transposition of DNA
- Epigenetic modifications.

GENE EXPRESSION AT TRANSCRIPTION LEVEL

Operon Concept

- Lactose Operon by **Francois Jacob and Jacques Monod** in 1961
- Operon is the linear array of genes involved in a metabolic pathway.
- For example: Lac operon concerned with lactose metabolism in prokaryotes.

Lac Operon Comprises

i. **Three structural genes coding for 3 proteins:**
   - Lac z—Beta galactosidase.
   - Lac y—Permease—a carrier protein that helps permeation of lactose to cell.
   - Laca—Thiogalactoside Transacetylase—function not known.

ii. **Regulatory genes comprises of:**
   - Lac promoter
   - Lac operator
   - Lac I encodes lac operon repressor protein.

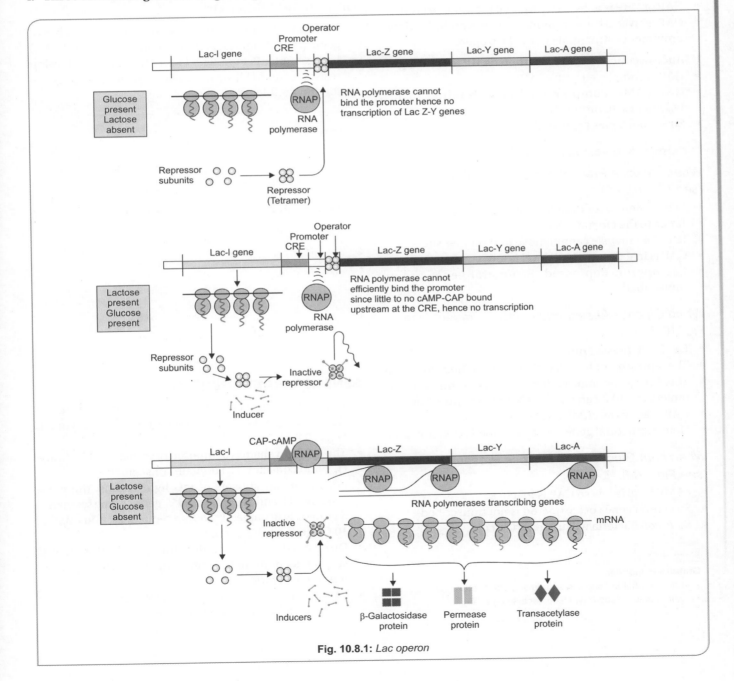

Fig. 10.8.1: *Lac operon*

Catabolite Repression in Lac Operon

CONCEPT BOX

Concept of Catabolite Repression in Lac Operon
- Glucose is a better energy source than lactose for *E. coli*. So when Glucose is present irrespective of presence of lactose, lac operon is switched off.

Mechanism of Action of CAP

- Catabolite repression is by catabolite repressor protein (CRP) or Catabolite activator protein (CAP).
- CAP is a positive regulator of lac operon.
- CAP or CRP is active only when it is bound to cAMP.
- CAP activates transcription of lac operon by binding to promoter of structural genes of lac operon.

If Glucose concentration low then cAMP level is high

- CAP is complexed with cAMP
- CAP-cAMP complex facilitates binding of RNA Polymerase to promoter
- Structural genes expressed.

Lac Operon in Various Conditions

When Glucose Present and Lactose Absent (see Fig. 10.8.1)

- Lac I → Repressor Protein active
- Binds to the Operator Site
- RNA Polymerase cannot move to Promoter site
- CAP is inactive as cAMP level is low
- Lac operon Repressed, Structural genes are not transcribed.

When Glucose Absent and Lactose Present (see Fig. 10.8.1)

- Lac I → Repressor protein.
- The repressor molecule, which has got affinity to lactose.
- This brings a conformational change in the repressor molecule and it can no more bind to the operator site.
- CAP is active as cAMP level is high.
- Hence structural genes are transcribed (Derepression).

When both Glucose and Lactose Present (see Fig. 10.8.1)

- cAMP level is low so CAP is inactive
- Structural genes of lac operon is not transcribed
- So *E. coli* bacteria will use only Glucose.

EXTRA EDGE

Gratuitous Inducer
- Lactose analogs capable of inducing Lac operon.
- For example: Isopropyl Thiogalactoside (IPTG)

CONCEPT BOX

Concept of Lac Operon
- Whenever Glucose is present irrespective of the presence or absence of lactose, the lac operon is switched off.
- Whenever Glucose is absent lac operon is on.

REGULATION OF GENE EXPRESSION AT DNA LEVEL

GENE AMPLIFICATION

- Gene amplification is the process by which number of gene available for transcription is increased (Fig.10.8.2)
- For example: Person who is on Methotrexate develop resistance to MTX by increasing the number of genes for dihydrofolate reductase.

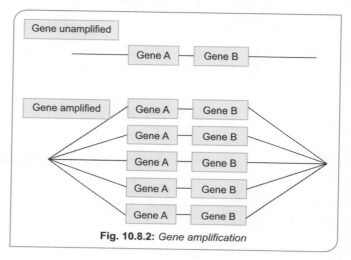

Fig. 10.8.2: *Gene amplification*

GENE REARRANGEMENT

- Different gene segments are brought together in different combination is called gene rearrangement. (Fig. 10.8.3).
- The IgG light chain is composed of variable (V_L), joining (J_L), and constant (C_L) domains or segments.
- For particular subsets of IgG light chains, there are roughly 300 tandemly repeated V_L gene coding segments, 5 tandemly arranged J_L coding sequences, and roughly 10 CL gene coding segments.
- During B-Lymphocyte development, different VJC segments combine to form *unique variable region in immunoglobulin.*
- This allows generation of 10^9-10^{11} different immunoglobulin from single gene.

Molecular Genetics

TRANSPOSONS (JUMPING GENE)

- Mobile DNA sequences that move to new positions within the genome
- Almost half of the genome are transposable elements
- Discovered by **Barbara McClintock**
- **Transposase** enzyme help in this process.

Transposons fall into two categories

Class I elements (Retro elements)

The elements which move to other location through an **RNA intermediate**. They are also known as **Retroposons**. 90% of transposable elements fall to this class.

Class II elements (DNA type elements)

Those DNA elements which can directly manipulate DNA so as to propagate to other location.

A simple transposon in bacteria is called **insertion sequence** contain:

1. Genes that codes for **enzyme transposase.**
2. Flanked by short inverted terminal repeats.

Transposons can

1. Activate or inactivate genes.
2. Cause insertion or deletion mutations.
3. Regulate gene expression.

GENE SWITCHING

The process by which one gene is switched off while a closely related gene take up its function.

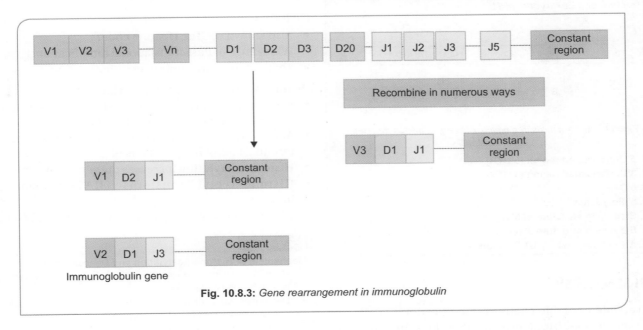

Fig. 10.8.3: *Gene rearrangement in immunoglobulin*

For example, during primary immune response genes for IgM is active but during secondary immune response genes for IgG is active.

GENE SILENCING

- The process by which a gene is switched off.
- By various **epigenetic mechanisms** (described later)
- By RNA interference (RNAi) by nc RNAs (described chapter Different classes of RNA).

REGULATION OF mRNA STABILITY

EXTRA EDGE

- mRNA exists in cytoplasm as **Ribonucleoprotein particle (RNP)**.
- Much of the **mRNA metabolism likely to occur in P bodies**.

RNP can help mRNA in two ways
- Proteins protect mRNA from Nuclease
- It can promote nuclease attack.

Contd...

Contd...

mRNA stability is determined by
1. **5'cap** that prevents 5' exonuclease attack
2. **3' Poly A tail** prevent 3' exonuclease attack
3. Other structures in coding region, 5' **Untranslated region**, 3' **Untranslated regions** of mRNA.
4. **Stem loop structure in 3' end region** prevents exonuclease attack in histone mRNA which lack Poly A tail.
5. Stem loop structure in 3' end is critical for regulation of mRNA encoding transferring receptor.
6. Presence **AU rich region** in the 3'UTR of certain mRNA shortens its half life of certain mRNA.
7. **Seed sequence in the 3' UTR** determine the specificity of binding of miRNA to mRNA.

Image-Based Information

Structure of a Eukaryotic mRNA

Fig. 10.8.4: *Structures in a typical eukaryotic mRNA that determines its stability*

A. 5'Untranslated region (5'UTR)
B. 3'Untranslated region (3'UTR)
C. 5'Cap
D. 3' Poly A tail
E. Stem loop structure in 5' region
F. Stem loop structure 3' region
G. AU rich region in 3' UTR region

EPIGENETICS^Q

Reversible heritable chemical modification^Q of DNA or histone or nonhistone proteins that does not alter DNA sequence itself.

- The term epigenetics means "above genetics" as the nucleotide sequence is unaltered.
- This is one of the recently discovered method of regulation of gene expression.

Epigenome—Constellation of covalent modification of DNA and histones that impact chromatin structure and modulation of gene expression.

Epigenetic Modifications
- DNA Methylation
- Post-translational modification of Histones.

DNA Methylation and Demethylations
- DNA methylation is usually restricted to **Cytosine residues in CpG dinucleotide of CpG islands.**
- Enzyme responsible for methylation is **Methyl transferases**.
- DNA methylation generally **decreases the gene expression** or Gene silencing.
- Acute **DNA demethylation, increases the rate of transcription.**

> **HIGH YIELD POINTS**
>
> **Gene Promoters and Regulation by Methylation—Demethylations**
> - Majority of gene promoters have high CG content.
> - CpG islands are the common sites of methylation-demethylations
> - Hence methylation-demethylations plays an important role in regulation of gene expression.
> - CpG islands of promoters are typically unmethylated, it favor transcription.

Histone Covalent Modifications
- Also known as **"The Histone code."**
- Wide range of Post-translational modifications.
- They are **dynamic and reversible.**

Histone Acetylation and Deacetylation
- Histone acetylation and deacetylation are the best understood modifications of histones.
- Acetylation occur on **lysine** residues in the amino terminal tails of histone molecules.
- **Histone acetylation** increases the gene transcription.
- **Histone Acetyl Transferase (HAT)** acetylate the histones.
- **Histone Deacetylases (HDAC)** deacetylate the histones.

> **CONCEPT BOX**
>
> **Mechanism of Histone Acetylation and Deacetylation**
> - Histone acetylation reduces the positive charge of amino terminal tails of histone molecules.
> - Decreases the binding affinity for the negatively charged DNA.
> - This increases euchromatin formation.
> - Transcription factors can bind to the promoters.
> - **This is Permissive Chromatin**
> - Histone deacetylation has the opposite effect, i.e. decreases transcription and increases heterochromatin formation.
> - This is called **Repressive Chromatin**

> **EXTRA EDGE**
>
> **Histone Methylation**
> - Addition of methyl group to specific lysine residues, but rarely to arginine residues.
> - Histone Methyl Transferases is the enzyme.
>
> **Functional consequence**—Alter chromatin configuration, which favor or decrease transcription.
>
> **Histone Phosphorylation**
> - **Serine** residues can be modified by phosphorylation
> - **Functional consequences**—Depending on the specific residue the DNA may be opened up for transcription or condensed to become inactive.

Other Histone Modification

Histone modification	Possible role of the modification
Acetylation of core histones[Q]	Increases transcription and increased expression of the gene
Phosphorylation of histone H1	Condensation of chromosomes during the replication cycle
ADP-ribosylation of histones	DNA repair
Methylation of histones	Activation and repression of gene transcription
Monoubiquitylation	Gene activation, repression, and heterochromatic gene silencing
Sumoylation of histones (SUMO; small ubiquitin-related modifier)	Transcription repression

> **EXTRA EDGE**
>
> **Biochemical Functions of Epigenetic Modification[Q]**
> 1. Regulation of Tissue specific gene expression.
> 2. X-chromosome inactivation (Facultative Heterochromatin) one of the two X-chromosomes in every cell of a female
> 3. Genomic Imprinting:
>   Gene inactivation on selected chromosomal regions of autosomes is called Genomic Imprinting. This is the cause of preferential expression of one of the parental allele.
>   *Maternal imprinting* refers to transcriptional silencing of the maternal allele.
>   Paternal imprinting implies that the paternal allele is inactivated. Imprinting occurs in the ovum or the sperm, before fertilization.
> 4. Ageing process.
> 5. Embryogenesis

Epigenetic Changes Causing Pathological Alteration

1. Fragile X syndrome
   - Promoter site Hypermethylation causing FMR-1 gene silencing
2. Cancer
   - *DNA methylation* and *histone modifications* dictate which genes are expressed, which in turn determines the lineage commitment and differentiation state of both normal and neoplastic cells.
   - **Local Promoter Hypermethylation** of Tumor Suppressor gene leads decreased expression of the tumor suppressor gene.
3. DNA methylation is considered as a defence mechanism that minimize the expression of retroviral incorporated sequences.
4. Genomic Imprinting.

Prader-Willi and Angelman Syndrome

The molecular basis of these two syndromes lies in the Genomic Imprinting.

Prader-Willi Syndrome

The facts
- A gene or set of genes on *maternal chromosome* 15q12 is imprinted (and hence silenced).
- The only functional allele(s) are provided by the paternal chromosome.

Causes of Prader-Willi Syndrome
- **Paternal deletion of Prader-Willi locus** located on Chr 15 (Most common)
- **Maternal Uniparental Disomy**
- Rarely due to imprinting defect in paternal chromosome.

Angelman Syndrome

The facts
- A distinct gene that also maps to the same region of chromosome 15 is imprinted on the *paternal chromosome*.
- Only the maternally derived allele of this gene is normally active.

Causes of Angelman Syndrome
- **Maternal deletion** of corresponding locus located on Chr 15 (Most common)
- **Paternal Uniparental Disomy**
- Rarely due to imprinting defect in maternal chromosome.

> **EXTRA EDGE**
>
> **Some examples of Tumor Suppressor Gene Silencing by Epigenetic Mechanism**
> - CDKN2A, a complex locus that encodes two tumor suppressors, p14/ARF and p16/INK4a p14/ARF is epigenetically silenced
> - Colon and Gastric cancers. p16/INK4a is silenced in a wide variety of cancers like, Bladder cancer, Head and Neck Cancer, ALL, Cholangiocarcinoma
> - BRCA-1 Silencing-Cancer Breast
> - VHL Silencing-Renal cell Cancer
> - MLH-1 Silencing-Colon Cancer

Molecular Methods to Detect Epigenetic Modifications in the Genome

Traditional Sanger sequencing alone cannot detect epigenetic modification.

1. **Methylation-specific PCR** can detect DNA methylations.
2. **DNA Chromatin Immune Precipitation (ChIP)** followed by Microarray hybridisation analysis or Direct Sequencing (**ChIP-Chip or ChIP-Seq**) detect histone modifications.
3. **Bisulphite sequencing.**
4. **Methylation sensitive restriction endonuclease digestion**.

Principle of Bisulphite Sequencing

- Sodium Bisulphite convert unmethylated Cytosine to Uracil, which function like thymine in base pairing.

- But methylated Cytosines are protected from modification, hence remain unchanged.

Procedure
- Treat genomic DNA with Sodium Bisulphite.
- Thus unmethylated (modified) DNA is discriminated from the methylated (unmodified) DNA on the basis of sequence analysis.

> **EXTRA EDGE**
>
> **Chromatin Immunoprecipitation (ChIP)**
> This method allows **precise localization of a particular protein** or modified protein on the DNA.
> For example: Acetylated, phosphorylated histones, etc. on a particular DNA sequence element in living cell.
>
> **RNA Immunoprecipitation (RIP)**
> An RNA immunoprecipitation method, performed like ChIP, which is used to score specific binding of a protein to a specific RNA in vivo.
> RIP uses formaldehyde cross linking to induce covalent attachment of proteins to RNA.
> CLIP: A method that uses UV cross-linking to induce covalent attachment of distinct proteins to specific RNAs in vivo.

- ChIP is chromatin Immunoprecipitation
- ChIP is hybridization on DNA ChiP or Microarray.

Therapeutic Application of Epigenetic Modification

Unlike DNA mutations, epigenetic changes are potentially reversible by drugs that inhibit DNA or histone-modifying factors.

> **EXTRA EDGE**
>
> 1. Drugs that inhibit DNA Methyl transferases (DNMT inhibitor)
>   - Azacytidine
>   - 5-aza-2'-deoxycytidine
>   - Decitabine.
> 2. Drugs that inhibit histone Deacetylases (HDAC Inhibitor)
>   - Vorinostat
>   - Valproic acid.

Quick Review
- Operon is a linear array of genes involved in a metabolic pathway.
- Whenever glucose is present irrespective of presence or absence of lactose, lac operon is switched off.
- CAP is active when bound to cAMP.
- CAP is positive regulator of lac operon.
- Histone acetylation causes euchromatin formation or Permissive chromatin.
- Seed sequence in 3'UTR of mRNA binds to mi RNA
- Epigenetics do no alter the sequence of nucleotide in the DNA.
- DNA methylation generally decrease the expression of gene.
- CpG islands in the promoters are the most common sites of DNA methylation.

Check List for Revision
- Operon concept and Epigenetics should be thoroughly read.
- Other topics mainly concentrate on bold letters.
- Extra edge topics are all for additional reading not must learn.

REVIEW QUESTIONS MCQ

1. **True about gene or function of gene:** *(PGI May 2017)*
   a. Not capable of independent expression
   b. Cistron is the smallest unit of gene expression
   c. Attachment to larger ribosome
   d. Attachment to tRNA
   e. RNA Polymerase attachment

2. **Which of the following are situated away from the coding region:** *(PGI June 06)*
   a. Promoter
   b. Enhancer
   c. Operator
   d. Structural gene

3. **Housekeeping genes are:** *(JIPMER 02, WB 03)*
   a. Inducible
   b. Required only when inducer is present
   c. Mutant
   d. Not regulated

4. **True among all is:** *(PGI 91)*
   a. Repressor is dimer and a positive regulator
   b. CRP is gratuitous inducer
   c. Lactose is positive regulator
   d. De-repression is due to presence of glucose
   e. Catabolite repression is mediated by CRP

5. **False statement is:** *(PGI 90, WB 03)*
   a. Repressor binds operator gene
   b. Regulator genes produce repressor subunits
   c. IPTG is inducer but not substrate
   d. Regulator gene is inducible

6. **Lac operon transcription is induced by:**
   (Delhi 03, TN 01)
   a. Glucose
   b. Glucose with inducer
   c. Inducer without glucose
   d. Both lactose and glucose

7. **All of the following statements about Lambda phage are true, except:** *(AI 2009)*
   a. In Lysogenic phase it fuses with host chromosome and remains dormant
   b. In Lytic phase it fuses with host chromosome and replicates
   c. Both Lytic and Lysogenic phase occur together
   d. In Lytic phase it causes cell lysis and releases virus particles

8. **True about transposons** *(PGI Nov 2013)*
   a. It has no effects on gene expression
   b. Also called jumping genes
   c. Mediated by enzyme transposase
   d. It is called retrotransposon when it involves an RNA intermediate

Epigenetics

9. **True about Chromatin Remodelling:** *(PGI Nov 2016)*
   a. Energy is required to displace histone octamer from DNA or translocate them on to neighbouring DNA segments
   b. Histone modification by specific enzyme
   c. Do not involve enzymes
   d. Aberration in chromatin remodeling proteins may be associated with cancers

10. **The following activity increases in DNA in a permissive chromatin:** *(AIIMS May 2018)*
    a. Methylation of CpG islands
    b. Phosphorylation
    c. Acetylation of histones
    d. Sumoylation

11. **CpG island in human genome is related to:**
    (AIIMS Nov 2016)
    a. tRNA synthesis
    b. DNA methylation
    c. DNA Acetylation
    d. Replication initiation

12. **Genes in CpG Island is in activated by:**
    (PGI Nov 2013)
    a. Methylation
    b. Metrylation
    c. Ubiquitisation
    d. Acetylation

13. **All are true DNA methylation except:** *(PGI Nov 2014)*
    a. It usually occurs in the cytosine
    b. Can alter the gene expression pattern in cells
    c. Role in genomic imprinting
    d. No role in carcinogenesis
    e. Essential for normal development

14. **All are true regarding epigenetics mechanisms except:** *(PGI May 2014)*
    a. Non inheritable
    b. Acetylation of histone
    c. Hereditary
    d. Methylation of DNA
    e. X chromosome inactivation

15. **Random inactivation of X chromosome is:**
    a. Lyonisation
    b. Allelic exclusion
    c. Randomisation
    d. Genomic imprinting

16. **True about 'X' chromosome inactivation:**
    (PGI Dec 06)
    a. XIST gene
    b. RNA interference
    c. Seen in male
    d. Seen in female

17. **Histone acetylation cause:** *(AIIMS May 2011)*
    a. Increased Heterochromatin formation
    b. Increased Euchromatin formation
    c. Methylation of cystine
    d. DNA replication

18. **Differential expression of same gene depending on parent of origin is referred to as:** *(AI 08)*
    a. Genomic imprinting
    b. Mosaicism
    c. Anticipation
    d. Nonpenetrance

19. **True about DNA methylation:** *(PGI Nov 2010)*
    a. Alteration in gene expression
    b. Genetic code remains intact
    c. Role in carcinogenesis
    d. Protective mechanism against cleaving by restriction endonuclease

20. **Epigenetics is a:** *(Recent Question)*
    a. Chemical modification of DNA
    b. Irreversible modification of DNA
    c. Change in nucleotide sequence
    d. Normal variation of nucleotides

21. **Methylation of Cytidine residues of DNA will cause:**
    (AIIMS May 2014)
    a. No Change
    b. Decrease in gene expression
    c. Mutation
    d. Increase in gene expression

Answers to Review Questions

1. **b, e. Cistron is the smallest unit of gene expression, RNA Polymerase attachment**
   - RNA Polymerase attaches to promoter region of gene.
   - Gene expression is possible.
   - Ribosome attaches to mRNA and tRNA not gene.

2. **b. Enhancer** *(Ref: Harper 31/e page 422)*

   Properties of Enhancers^Q
   - Can be located upstream or down stream of the transcription site.
   - Work when located long distances from the promoter
   - Work when upstream or downstream from the promoter
   - Work when oriented in either direction
   - Can work with homologous or heterologous promoters
   - Work by binding one or more proteins
   - Work by facilitating binding of the basal transcription complex to the cis-linked promoter

3. **d. Not Regulated** *(Ref: Harper 31/e page 410)*
   - The expression of some genes is constitutive, meaning that they are expressed at a reasonably constant rate and not known to be subject to regulation. These are often referred to as housekeeping genes.
   - An inducible gene is one whose expression increases in response to an inducer or activator, a specific positive regulatory signal

4. **c, e. Lactose is positive regulator, Catabolite repression is mediated by CRP** *(Ref: Harper 31/e page 410-412)*
   - Repressor is a tetramer.
   - Isopropyl Thiogalactoside (IPTG) is a gratuitous inducer.
   - Although Lactose is present in the cell, as long as glucose is present. E. coli does not activate lac operon. This is called Catabolite Repression. This is because of catabolite gene activator protein complexed with cAMP. This is complex and called cAMP Regulatory Protein (CRP).

5. **d. Regulator gene inducible** *(Ref: Harper 31/e page 412)*
   - Structural genes are inducible, not the regulator genes.
   - Lac I gene produces Repressor subunits.
   - IPTG is an inducer of lac-operon, but itself is not a substrate. This is called gratuitous inducer.

6. **c. Inducer without glucose** *(Ref: Harper 31/e page 412)*

   Lac operon is repressed when, the cell of E. coli contains
   - Glucose alone
   - Glucose and Inducer (lactose)

   Lac operon is induced if
   - Lactose alone is present.

7. **c. Both lytic and Lysogenic phase...** *(Ref: Harper 31/e page 412)*

   When lambda infects an organism of that species, it injects its 45,000-bp, double-stranded, linear DNA genome into the cell. Depending upon the nutritional state of the cell, the lambda DNA
   1. Will either **integrate** into the host genome (**lysogenic pathway**) and remain dormant until activated
   2. It will commence **replicating** until it has made about 100 copies of complete, protein-packaged virus, at which point it causes lysis of its host (**lytic pathway**).

   The newly generated virus particles can then infect other susceptible hosts.

   Poor growth conditions favor lysogeny while good growth conditions promote the lytic pathway of lambda growth.

8. **b, c, d. Also called jumping.., Mediated by enzyme..., It is called Retroposons...**

   Transposons (Jumping Gene)
   - DNA sequences that move to new positions within the genome
   - Discovered by Barbara McClintock.
   - Transposase enzyme help in this process.

   Retroposons
   - DNA Sequence move from one segment to another through an RNA Intermediate.
   - DNA segment is converted to RNA, RNA moves to another location, where it is reversely transcribed to a DNA.

Epigenetics

9. **b, d. Histone modification is by specific enzyme, Aberration in chromatin remodelling proteins may be associated with cancers**

   Chromatin remodelling can happen by histone modification by specific enzymes. This van alter gene expression hence it is important in carcinogenesis.

10. **c. Acetylation of histones**

    Effects of epigenetic modification
    - Methylation of CpG islands-Repressive chromatin
    - Phosphorylaion of histones –Chromatin condensation (Repressive chromatin) or opening of chromatin (Permissive chromatin)
    - Acetylation of histones–Permissive chromatin
    - Sumoylation–Repressive chromatin

11. **b. DNA methylation** *(Ref: Harrison 19/e page 431)*

    Gene promoters and regulation by methylation-demethylations

- Majority of gene promoters have high CG content.
- CpG islands are the common sites of methylation-demethylations
- Hence methylation-demethylations plays an important role in regulation of gene expression.
- CpG islands of promoters are typically unmethylated, it favour transcription.

12. a. Methylation *(Ref: Robbins 9/e p 3-5)*

Gene promoters and regulation by methylation-demethylations

a. Majority of gene promoters have high CG content.
b. CpG islands are the common sites of methylation-demethylations
c. Hence methylation-demethylations plays an important role in regulation of gene expression.
d. CpG islands of promoters are typically unmethylated, it favour transcription

13. d. No role in carcinogenesis *(Ref: Robbins 9/e page 3-5)*

Biochemical functions of Epigenetic Modification^Q

1. Regulation of Tissue specific gene expression
2. X chromosome inactivation (Facultative Hetero-chromatin) one of the two X-chromosomes in every cell of a female
3. Genomic Imprinting:
   Gene inactivation on selected chromosomal regions of autosomes is called Genomic Imprinting. This is the cause of preferential expression of one of the parental allele.
   Maternal imprinting refers to transcriptional silencing of the maternal allele.
   Paternal imprinting implies that the paternal allele is inactivated.
   Imprinting occurs in the ovum or the sperm, before fertilization.
4. Ageing Process

Epigenetic Changes Causing Pathological Alteration

1. Fragile X syndrome
   – Promoter site Hypermethylation causing FMR-1 gene silencing
2. Cancer
   – DNA methylation and histone modifications dictate which genes are expressed, which in turn determines the lineage commitment and differentiation state of both normal and neoplastic cells
   – Local Promoter Hypermethylation of Tumour Supressor gene leads decreased expression of the tumour suppressor gene.
3. DNA methylation is considered as a defence mechanism that minimise the expression of retroviral incorporated sequences.

14. a. Non inheritable *(Ref: Harper 31/e page 412)*

Reversible heritable chemical modification^Q of DNA or histone or nonhistone proteins that does not alter DNA sequence itself is called epigenetics.

15. a. Lyonisation

Two factors that are peculiar to the sex chromosomes: (1) lyonization or inactivation of all but one X chromosome and (2) the modest amount of genetic material carried by the Y chromosome.

In 1961, Lyon outlined the idea of X-inactivation, now commonly known as the Lyon hypothesis. It states that

(1) only one of the X chromosomes is genetically active,
(2) the other X of either maternal or paternal origin undergoes heteropyknosis and is rendered inactive, (3) inactivation of either the maternal or paternal X occurs at random among all the cells of the blastocyst on or about day 16 of embryonic life, and (4) inactivation of the same X chromosome persists in all the cells derived from each precursor cell.

The inactive X can be seen in the interphase nucleus as a darkly staining small mass in contact with the nuclear membrane known as the Barr body, or X chromatin. The molecular basis of X inactivation involves a unique gene called XIST, whose product is a noncoding RNA that is retained in the nucleus, where it "coats" the X chromosome that it is transcribed from and initiates a gene-silencing process by chromatin modification and DNA methylation. The XIST allele is switched off in the active X.

16. a, b, d. XIST gene, RNA interference, Seen in female

17. b. Increased euchromatin formation *(Ref: Harrison 19/e page 412)*

Euchromatin is transcriptionally active
Heterochromatin is transcriptionally inactive.

According to Harrison

Covalent post translational modifications of histones and other proteins play an important role in altering chromatin structure and, hence, transcription.

Histones can be reversibly modified in their amino-terminal tails, which protrude from the nucleosome core particle, by acetylation of lysine, phosphorylation of serine, or methylation of lysine and arginine residues.

Acetylation of histones by histone acetylases (HATs), for example, leads to unwinding of chromatin and accessibility to transcription factors.

Conversely, deacetylation by histone deacetylases (HDACs) results in a compact chromatin structure and silencing of transcription.

18. a. Genomic Imprinting *(Ref: Robbins 9/e page 172)*

Studies over the past two decades have provided definite evidence that, at least with respect to some genes, important functional differences exist between the paternal allele and the maternal allele. These differences result from an epigenetic process called *imprinting*.

19. a, b, c, d. Alteration in gene expression, Genetic code remains intact, Role in carcinogenesis, Protective mechanism
(Ref: Harper31/e page 412)

Epigenetics

Reversible heritable chemical modification of DNA or histone or nonhistone proteins that does not alter DNA sequence itself.

The term epigenetics means "above genetics" as the nucleotide sequence is unaltered.

This is one of the recently discovered method of regulation of gene expression.

This include:
a. DNA Methylation at Cytosine residues of CpG islands (Some consider this as a post-replicational modification)
b. Post-translational modification of Histones.

20. a. Chemical modification of DNA
(Ref: Harper 31/e page 412)

The term "epigenetics" means "above genetics" and refers to the fact that these regulatory mechanisms do not change the underlying, regulated DNA sequence, but rather simply the expression patterns of this DNA.

Epigenetics refers to reversible, heritable changes in gene expression that occur without mutation. Such changes involve post-translational modifications of histones and DNA methylation, both of which affect gene expression.

21. b. Decrease in gene expression
(Ref: Harper 31/e page 412))

- DNA Methylation generally cause decrease in gene expression.

10.9 MUTATIONS

- Definition
- Types of Mutations
- Mutation Detection Techniques

DEFINITION

Any permanent change in the primary nucleotide sequence regardless of its functional consequences.

Mutation rate ~10–10/bp per cell division.

TYPES OF MUTATIONS

I. Point Mutation

Single base changes in the nucleotide sequence in the gene.

Base Substitution

Replacement of a single nucleotide by another. These are the (most common type of mutation).

i. **Transition**^{Q}: A purine base replaced by a purine base or a pyrimidine replaced by another pyrimidine.
ii. **Transversion**^{Q}: A purine base replaced by another pyrimidine, or pyrimidine replaced by another purine.

Effects of Base Substitution

I. Silent mutation
   - If a mutation does not alter the polypeptide product of the gene.
   - Also called Synonymous Mutation.

II. Mis-sense mutation
   - The alteration in the nucleotide may result in the incorporation of a different amino acid.
   A mis-sense may be:
   i. Acceptable: No clinical symptoms, For example: Hb Hikari
   ii. Partially acceptable^Q, For example: Hb SQ
   iii. Unacceptable
      For example: Hb M

Another Classification of Mis-sense Mutation

1. Conservative mutation: A mis-sense mutation in which one amino acid is replaced by a similar amino acid.
2. Non-conservative mutation: A mis-sense mutation in which one amino acid is replaced by an amino acid with different characteristics.

For example: HbS.

II. *Nonsense Mutation*
   - A coding codon mutated to a nonsense codon result in premature termination of polypeptide chain.

III. *Splice Site Mutation*
   - Mutation at the Splice site results in Faulty Splicing.

IV. Promoter Site Mutation
- Results in altered gene expression.

V. Frame Shift Mutation
- Due to insertion or deletion of nucleotides that are not a multiple of three results in frame shift mutation.
- Reading frame is garbled.

VI. Null Mutation
- A mutation that lead to no functional gene product is called null mutation.

VIII. Constitutive Mutation
- Mutation in which a inducible gene mutated to house keeping gene or constitutive gene.

XIII. tRNA Suppressor Mutation
- The effect of mutation on mRNA can be suppressed by a mutant tRNA which has a mutant anticodon sequence.
- These mutant tRNA which can suppress the effect of mutation are called suppressor tRNAs.

> **EXTRA EDGE**
>
> **Classes of Mutation in Cystic Fibrosis**
> *Various mutations can be grouped into six "classes" based on their effect on the CFTR protein:*
> **Class I:** Defective protein synthesis.
> Complete lack of CFTR protein at the apical surface of epithelial cells.
> **Class II:** Abnormal protein folding, processing, and trafficking. The most common class II mutation is a deletion of three nucleotides coding for phenylalanine at amino acid position 508 (ΔF508).
> - Worldwide, this mutation can be found in approximately **70% of cystic fibrosis patients.**
>
> **Class III:** Defective regulation.
> - Mutations in this class prevent activation of CFTR by preventing ATP binding and hydrolysis.
>
> **Class IV:** Decreased conductance.
> - Mutations typically occur in the transmembrane domain of CFTR, which forms the ionic pore for chloride transport.
>
> **Class V:** Reduced abundance.
> - Mutations typically affect intronic splice sites or the CFTR promoter, such that there is a reduced amount of normal protein.
>
> **Class VI:** *Altered regulation of separate ion channels.*
> - Mutations in this class affect the regulatory role of CFTR.

New Mutation and Gonadal Mosaicism

In some **autosomal dominant disorders**, phenotypically normal parents have more than one affected children.
- This is a violation of Mendelian Inheritance
- This is an example of single gene disorder with non-Classic Inheritance.
- The sudden unexpected occurrence of a condition due to mutation of a gene is called new mutation.
- New mutation **occurs post zygotically during early (embryonic) development.**

If the mutation affects only cells destined to form the gonads, the gametes carry the mutation, but the somatic cells of the individual are completely normal. Such an individual is said to exhibit *germ line or gonadal mosaicism.*

Thus a phenotypically normal parent who has germ line mosaicism can transmit the disease-causing mutation to the offspring through the mutant gamete.

For example: Achondrodysplasia, Marfan's Syndrome, Neurofibromatosis.

MUTATION DETECTION TECHNIQUES

Test to Detect Mutations

Ame's Test
- Test to detect **mutagenicity**
- Special strains of **Salmonella typhimurium** have mutated histidine gene
- Hence they will grow only in medium containing Histidine gene.
- This is called reverse mutation.
- The number of colonies is proportional to the quantity of mutagens.

> **EXTRA EDGE**
>
> *Site Directed Mutagenesis*
> - Invented by Michael Smith in 1993

Principle

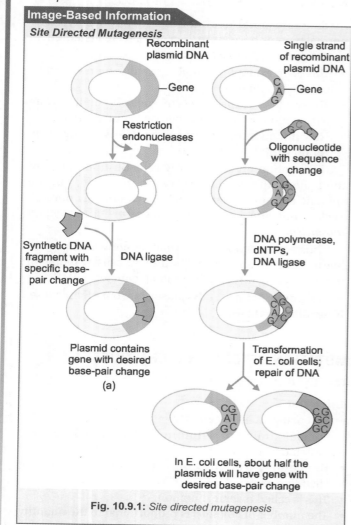

Fig. 10.9.1: *Site directed mutagenesis*

Two Approaches of Site Directed Mutagenesis

Approach-1
- Recombinant Plasmid is treated with Restriction endonuclease.
- Synthetic DNA fragment with specific base pairchange incorporated.
- Cloned in Expression vector.
- Study the effect of mutation in protein product.

Approach-2 (Oligonucleotide directed mutagenesis)
- Single strand of recombinant plasmid is selected.
- Oligonucleotide with sequence change is bound to plasmid.
- Oligonucleotide is extended with DNA Polymerase.

- Plasmid with desired base pair change is selected.
- Cloned in expression vector.

Techniques Used to Detect Mutations with DNA Sequence Alterations

I. First do a PCR amplification of the DNA, then different sequencing techniques can be used.

   a. *Sangers technique*—Still, 36 years after its Nobel-worthy invention by Frederick Sanger, Sanger sequencing is still considered the **"gold standard" for sequence determination**.

   b. *Pyrosequencing*

EXTRA EDGE

Pyrosequencing
Principle:
- When a nucleotide is incorporated into a growing DNA strand, there is release of pyrophosphate (PPi).
- In this technique, individual nucleotides (A,C,T, or G) one at a time into the reaction.
- If one or more nucleotides are incorporated into the growing strand of DNA, pyrophosphate is released.
- A secondary reaction involving luciferase that produces light, which is measured by a photodetector.

Advantage of Pyrosequencing
- More sensitive than Sanger sequencing
- Allowing for detection of as little as 5% mutated alleles in a background of normal alleles
- To analyze DNA obtained from cancer biopsies, in which tumor cells are often "contaminated" with large numbers of admixed stromal cells.

II. Restriction Fragment length Polymorphism:
- If a mutation affects a specific restriction site, this technique can be used.

Techniques Used to Detect Mutations that Affect Length of DNA

a. Amplicon length Analysis
b. Real-time PCR
c. Multiplex Ligation Dependent Probe Amplification.

Amplicon Length Analysis
- After doing PCR, the size of PCR products determined by gel electrophoresis.

Real-time PCR
- Detect and quantify the presence of particular nucleic acid sequences in "real time" (i.e., during the exponential phase of DNA amplification rather than post-PCR).

Molecular Genetics

Extra Edge

Multiplex Ligation Dependent Probe Amplification (MLPA)

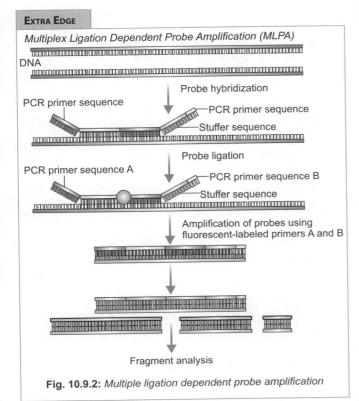

Fig. 10.9.2: *Multiple ligation dependent probe amplification*

Indication

High resolution technique to detect **mutations with large deletion and duplication**, which involves single exon, several exons or an entire gene.

For example, Duchenne Muscular dystrophy.

The procedure

- Three processes—**Hybridisation, Ligation, Amplification**
- Two fluorescently labeled oligonucleotide probe with stuffer sequence and PCR primer that hybridise adjacent to each other to a target sequence is added.
- After hybridization, the two oligonucleotides are joined by DNA **Ligase**
- **Amplified** using PCR
- The amplified products are separated by capillary electrophoresis and analysed.

Why the name?

MLPA blends DNA hybridization, DNA ligation, and PCR amplification
- Multiplex-because about 40 probes are added in a single reaction so multiple targets are amplified.
- Ligation-because adjacent probes are ligated.
- Amplification-because amplified.

Interpretation

- If certain exons are deleted then no hybridization, hence no amplified products of certain regions when compared to normal persons.
- If certain exons are duplicated if a quantitative PCR is done, then we see certain targets are more in number, when compared to control.

Uses

- To detect deletions and duplications of any size, including anomalies that are too large to be detected by PCR and too small to be identified by FISH.

Advantages of MLPA

- Can be performed on very small amounts of genomic DNA
- Each probe-set can be designed with identical primer sequences
- Many probe-sets can be applied and amplified in one reaction tube.

Other Mutation Detection Technique[Q]

Method	Type of mutation detected
Cytogenetic analysis	Numerical or structural abnormalities in chromosomes
Fluorescent in situ hybridization (FISH)	Numerical or structural abnormalities in chromosomes
Southern blot	Large deletion, insertion, rearrangement, expansions of triplet repeat, amplification
Polymerase chain reaction (PCR)	Expansion of triplet repeats, variable number of tandem repeats (VNTR), gene rearrangements, translocations; prepare DNA for other mutation methods
Reverse transcriptase PCR (RT-PCR)	Analyze expressed mRNA (cDNA) sequence; detect loss of expression
DNA sequencing *Gold standard mutation detection technique*	Point mutations, small deletions and insertions
Restriction fragment polymorphism (RFLP)	Point mutations, small deletions and insertions
Single-strand conformational polymorphism (SSCP)	Point mutations, small deletions and insertions
Denaturing gradient gel electrophoresis (DGGE)	Point mutations, small deletions and insertions
RNAse cleavage	Point mutations, small deletions and insertions
Oligonucleotide specific hybridization (OSH)	Point mutations, small deletions and insertions
Microarrays	Point mutations, small deletions and insertions, Genotyping of SNPs

Contd...

Self Assessment and Review of Biochemistry

Contd...

Method	Type of mutation detected
Protein truncation test (PTT)	Mutations leading to **premature truncations**
Pyrosequencing	Sequencing of whole genomes of microorganisms, resequencing of amplicons
Multiplex ligation-dependent probe amplification (MLPA)	Copy number variations

Quick Revision

- Most common mutation is point mutation.
- Most common point mutation is base substitution.
- A base substitution result is same amino acid is silent mutation.
- A base substitution results in a different amino acid is mis-sense mutation.
- A base substitution result a stop codon is nonsense mutations.
- Indels are insertions and deletion mutations.
- Indels are the deleterious mutations
- Insertion and deletion causes frame shift mutations
- Ame's test is a test to detect mutation.
- Premature truncation of proteins is detected by protein truncation test.
- Gold standard mutation detection technique is DNA sequencing.

Check List for Revision

- Types of mutation is must learn,
- Mutation detection technique mainly concentrate on bold letters and headings.
- Extra edge topics are optional for NBE model exams.

Review Questions MCQ

1. Which of the following is/are most severe/dangerous change in gene? *(PGI May 2014)*
   a. Deletion
   b. Insertion
   c. Mutation
   d. Translocation
   e. Duplication

2. No loss of genetic material occur in: *(AIIMS Nov 2012)*
   a. Deletion
   b. Insertion
   c. Substitution
   d. Inversion

3. True about Fragile-X syndrome: *(PGI June 2009)*
   a. Trinucleotide repeat sequence disease
   b. Chromosome–breakage
   c. X-chromosome defect
   d. Point-mutation
   e. Deletion

4. Base substitution mutations can have the following molecular consequence except: *(AI 2006)*
   a. Changes one codon for an amino acid into another codon for that same amino acid
   b. Codon for one amino acid is changed into a codon of another amino acid
   c. Reading frame changes downstream to the mutant site
   d. Codon for one amino acid is changed into a translation termination codon

5. Frame shift mutation is caused by: *(Ker 2007)*
   a. Deletion
   b. Point mutation
   c. Substitution
   d. Transversion

6. Cystic fibrosis mutation causing the reduced chloride conductance is: *(NBE Pattern Questions)*
   a. Class-1
   b. Class-2
   c. Class-3
   d. Class-4

7. X-ray causes DNA mutation by: *(PGI May 2014)*
   a. Double strand break
   b. Oxidation
   c. Pyrimidine dimer
   d. Intrastrand crosslinks

8. One of the following mutation is potentially lethal: *(Delhi 96)*
   a. Substitution of adenine for cytosine
   b. Substitution of methyl cytosine for cytosine
   c. Substitution of guanine for cytosine
   d. Insertion of one base

9. Sickle cell anemia is the clinical manifestation of homozygous genes for an abnormal hemoglobin molecule. The event responsible for the mutation in the B chain is: *(AIIMS 91, Kerala 90)*
   a. Insertion
   b. Deletion
   c. Nondisjunction
   d. Point mutation

10. Null mutation is: *(AI 2000)*
    a. Mutation occurring in non coding region
    b. Mutation that does not change the amino acid or end product
    c. Mutation that codes for a change in progeny without a chromosomal change
    d. Mutation that leads to no functional gene product

11. A mutation in the codon which causes a change in the coded amino acid, is known as: *(AIIMS May 02)*
    a. Mitogenesis
    b. Somatic mutation
    c. Mis-sense mutation
    d. Recombination

12. In a mutation if valine is replaced by which of the following would not result in any change in the function of protein? *(AIIMS May 02)*
    a. Proline
    b. Leucine
    c. Glycine
    d. Aspartic acid

13. Which of the following can be a homologous substitution for valine in the hemoglobin? *(AI 2004)*
    a. Isoleucine
    b. Glutamic acid
    c. Phenylalanine
    d. Lysine
14. Pyrimidine dimers are seen in:
    a. UV rays
    b. Xeroderma Pigmentosa
    c. Alkylating agents
    d. X-rays
15. Techniques used to detect Gene Mutation is/are:
    (PGI May 2012)
    a. RTPCR
    b. Denaturing gradient gel electrophoresis
    c. DNA sequencing
    d. Restriction fragment length polymorphism
    e. Single strand conformational polymorphism

Answers to Review Questions

1. **a, b. Deletion, Insertion**

   Deletion and insertion cause garbling of reading frame. Hence it is dangerous change in the polypeptide synthesized.

2. **d. Inversion** *(Ref: Robbins 9/e page 160)*

   Structural Anomalies in Chromosome
   1. Deletion
      Refers to loss of a portion of a chromosome. Most deletions are interstitial,
      Rarely terminal deletions may occur.
   2. Ring chromosome
      Is a special form of deletion.
      It is produced when a break occurs at both ends of a chromosome with fusion of the damaged ends.
      Ring chromosomes do not behave normally in meiosis or mitosis and usually result in serious consequences
   3. Inversion
      Refers to a rearrangement that involves two breaks within a single chromosome with reincorporation of the inverted, intervening segment.
      An inversion involving only one arm of the chromosome is known as *paracentric*.
      If the breaks are on opposite sides of the centromere, it is known as *pericentric*.
      Inversions are often fully compatible with normal development.
      No loss of genetic element.
   4. Isochromosome
      Break along the axis perpendicular to the axis of chromosome.
      One arm of a chromosome is lost
      The remaining arm is duplicated, resulting in a chromosome consisting of two short arms only or of two long arms.
      Loss of genetic element.
   5. Translocation
      A segment of one chromosome is transferred to another chromosome.
      Balanced reciprocal translocation
      ▫ There are single breaks in each of two chromosomes, with exchange of material.
      ▫ A balanced translocation carrier, however, is at increased risk for producing abnormal gametes.
      ▫ No loss of genetic element.
      ▫ Robertsonian translocation (or centric fusion)
      ▫ A translocation between two acrocentric chromosomes.
      ▫ Typically the breaks occur close to the centromeres of each chromosome.
      ▫ Transfer of the segments then leads to one very large chromosome and one extremely small one. Usually the small product is lost.

3. **a, c. Trinucleotide repeat sequence disease, X-chromosome defect**

4. **c. Reading frame changes downstream to the mutant site**
   (Ref: Harper 31/e page 398)

   Changes one codon for an amino acid into another codon for that same amino acid-Silent mutation
   Codon for one amino acid is changed into a codon of another amino acid—Mis-sense mutation
   Codon for one amino acid is changed into a translation termination codon—Nonsense mutation
   Reading frame changes downstream to the mutant site-Frameshift mutation

5. **a. Deletion** *(Ref: Harper 31/e page 398)*

   Frame shift Mutation
   Due to insertion or deletion of nucleotides that are not a multiple of three results in frameshift mutation.
   Reading frame is garbled

6. **d. Class-4**

   Class IV: *Decreased* conductance. These mutations typically occur in the transmembranedoma in of CFTR, which forms the ionic pore for chloride transport. There is a normal amount of CFTR at the apical membrane, but with reduced function. This class is usually associated with a milder phenotype.

7. **a, d. Double strand break repair, Intrastrand crosslinks**

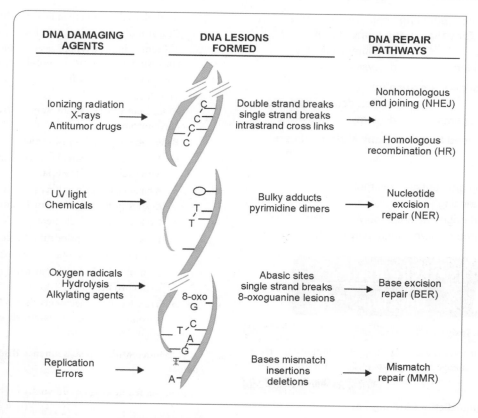

8. d. Insertion of base

Insertion or deletion of base can garble the reading frame, resulting in a frame shift mutation.

9. d. Point mutation

The mutation in HbS is an example of
- Point mutation
- Partially acceptable mis-sense mutation
- Transversion
- Base substitution.
- Nonconservative mutation

10. d. Mutation that leads to no functional gene product Null mutation

A mutation that lead to no functional gene product is called null-mutation.

11. c. Mis-sense mutation (Ref: Harper 31/e page 398)

I. Silent Mutation

If a mutation does not alter the polypeptide product of the gene.

Also called Synonymous Mutation.

II. Mis-sense mutation

The alteration in the nucleotide may result in the incorporation of a different amino acid.

12. b. Leucine

Valine and Leucine are branched chain nonpolar amino acids. So homologous substitution.

13. a. Isoleucine

Both are branched chain nonpolar amino acids.

14. b. Xeroderma Pigmentosa (Ref: Harper 31/e page 370)

- DNA damaging agent that results in pyrimidine dimer is U-V light chemicals.
- DNA repair mechanism that repair pyrimidine dimer is Nucleotide Excision repair.
- Defective Nucleotide excision repair mechanism causes Xeroderma pigmentosum, Cockayne Syndrome, Trichothiodystrophy.
- So pyrimidine dimers are seen in Disorders caused by defective nucleotide Excision repair.

15. a, b, c, d, e.

Test to Detect Mutations

1. Ame's Test

   Test to detect mutagenicity

   Special strains of Salmonella typhimurium have mutated histidine gene.

   Hence they will grow only in medium containing Histidine gene.

2. Site Directed Mutagenesis Michael Smith in 1993
   An oligodeoxyribonucleotide whose sequence is complementary to apart of known gene is synthesised.

Other Mutation Detection Technique

Method	Type of mutation detected
Cytogenetic analysis	Numerical or structural abnormalities in chromosomes
Fluorescent in situ hybridization (FISH)	Numerical or structural abnormalities in chromosomes
Southern blot	Large deletion, insertion, rearrangement, expansions of triplet repeat, amplification
Polymerase chain reaction (PCR)	Expansion of triplet repeats, variable number of tandem repeats (VNTR), gene rearrangements, translocations; prepare DNA for other mutation methods
Reverse transcriptase PCR (RT-PCR)	Analyze expressed mRNA (cDNA) sequence; detect loss of expression
DNA sequencing	Point mutations, small deletions and insertions

Contd…

Method	Type of mutation detected
Restriction fragment polymorphism (RFLP)	Point mutations, small deletions and insertions
Single-strand conformational polymorphism (SSCP)	Point mutations, small deletions and insertions
Denaturing gradient gel electrophoresis (DGGE)	Point mutations, small deletions and insertions
RNAse cleavage	Point mutations, small deletions and insertions
Oligonucleotide specific hybridization (OSH)	Point mutations, small deletions and insertions
Microarrays	Point mutations, small deletions and insertions Genotyping of SNPs
Protein truncation test (PTT)	Mutations leading to premature truncations
Pyrosequencing	Sequencing of whole genomes of microorganisms, resequencing of amplicons
Multiplex ligation-dependent probe amplification (MLPA)	Copy number variations

10.10 MITOCHONDRIAL DNA

- Structural Genes Coded by Mitochondrial DNA
- Unique Features
- Unique Genetic Code
- Clinical Correlations

INTRODUCTION

- 1% of cellular DNA is mitochondrial DNA.
- Mitochondria possess its own DNA and protein synthesizing machinery.
- Human mitochondria contains 2–10 copies of a small circular ~16 kbpds DNA molecule.
- Composed of Heavy (H) and Light (L) chain or strands.
- Contain **16,569 bp.**

Structural Genes Coded by Mitochondrial DNA

Mitochondrial DNA encodes 37 structural genes. They include

- **2 rRNAs**
- **22 mitochondrial tRNAs, large 16S rRNA and small 12S rRNA.**
- 13 protein subunits of respiratorychain.^{Q}
  1. Seven subunits of NADH Dehydrogenase (Complex I).
  2. Cytochrome b of Complex III.
  3. Three subunits of Cyt Oxidase (Complex IV).
  4. Two subunits of AT Psynthase.
  – *This constitutes almost 20% proteins of ETC complexes.*

Mitochondria has Unique Genetic Code

Codons^Q	Nuclear DNA code	Mitochondrial DNA code
AUA	Isoleucine	Methionine
UGA	Stop Codon	Tryptophan
AGA, AGG	Arginine	Stop codon

Unique Features of Mitochondrial DNA

1. **Mutation rate is very high because**
   – No Introns
   – No protective histones
   – No effective repair enzymes.
   – It is exposed to oxygen free radicals generated by oxidative phosphorylation.

2. **Non-Mendelian type of inheritance**
   - Cytoplasmic inheritance or Matrilineal inheritance. (Described in next chapter Patterns of inheritances)
   - Mitochondrial Disease with no maternal inheritance
   - Pearson Syndrome
   - Kearns-Sayre syndrome (KSS).
3. **Heteroplasmy**
   - Heteroplasmy is defined as the presence of normal and mutant DNA in different proportions in different cells.

Extra Edge

- Mitochondrial disease with homoplasy
  - Leber's hereditary optic neuropathy
  - Sensorineural deafness.

Quick Revision

- 1% of cellular DNA is mitochondrial DNA
- They are circular double stranded DNA.
- They code for <20% of proteins present in ETC.
- Mitochonrial DNA has high rate of mutations.
- They have maternal inheritance.

Check List for Revision

- This is a recent trend topic.
- The whole chapter is must learn.

Review Questions MCQ

1. **Choose the true statement about mit DNA:**
   (AIIMS May 2017)
   a. Few mutation compared to nuclear DNA
   b. It has 3×10^9 base pairs
   c. It receives 23 chromosomes from each parent
   d. It codes for less than 20% of the proteins involved in respiratory chain

2. **All are true about mitochondrial DNA except:**
   a. Contains 37 gene (PGI 2014)
   b. Transmit from mother to offsprings
   c. Transmit in classical Mendelian fashion
   d. Cause Leber hereditary optic neuropathy

3. **Mitochondrial DNA is:** (AI 2006)
   a. Closed circular   b. Nicked circular
   c. Linear   d. Open circular

Answers to Review Questions

1. **d. It codes for less than 20% of the proteins involved in respiratory chain**

   - In ETC there are around 67 subunits, out it 13 proteins are coded by Mit DNA, which comes around 19%.
   - Mitochondrial DNA has high mutation rate almost 5 to 10 times that of nuclear DNA.
   - It has 16,569 bp
   - Mitochondrial DNA has maternal inheritance

2. **c. Transmit in classical mendelian fashion**

   Mitochondrial DNA encodes Structural Genes
   - For 2 rRNAs
   - 22 mitochondrial tRNAs
   - 13 protein subunits of respiratory chain.
   - This includes 37 genes.

3. **a. Closed circular** (Ref: Harper31/e page 358-359)

   Mitochondrial DNA
   - Mitochondria possess its own DNA and protein synthesizing machinery.
   - Human mitochondria contains 2–10 copies of a small circular dsDNA molecule.
   - Composed of Heavy (H) and Light (L) chain or strands.
   - Contain 16,56 9 bp.
   - That makes upapproximately 1% of total circular DNA.

10.11 PATTERNS OF INHERITANCE

- Autosomal Dominant Inheritance
- X-linked Dominant Inheritance
- Autosomal Recessive Inheritance
- Y-linked Inheritance
- X-linked Recessive Inheritance
- Mitochondrial Inheritance

HIGH YIELD POINTS

Some terms used in Genetics
- **Genome:** The complete complement of genetic information in a living organism
- **Chromosome:** Physical Division of Genome.
- **Genes:** Functional Division of Genome.
- **Genetic Locus:** Location of a particular gene on the Chromosome.
- **Alleles**

In diploid organism there are two sets of chromosomes. Therefore there are 2 copies of each gene.
The different forms of the same gene that are found at the same locus are called alleles.
One allele is received from the father and the other allele from the mother.
They are responsible for alternate or contrast character.

- **Genotype** represents the set pattern of gene present in the cell.
- **Phenotype** is the observed character expressed by the gene.
- **Homozygous:** Both allele are defective.
- **Heterozygous:** One allele is normal and the other allele is defective.
- **Recessive mode of transmission:** Phenotypic expression of the disease only in the homozygous state.
- **Carrier State:** In recessive mode of inheritance if the person carries one abnormal gene it is not phenotypically expressed. Biochemically it is called Trait.
- **Dominant mode of Transmission:** Phenotypic expression even when one allele is abnormal or heterozygous state.
- **Autosomal** means defective gene is located in the autosomes. (Somatic)
- **Sex linked** means defective gene is located in the Sex chromosome.

Some Common Pedigree Symbols

Male	□
Female	○
Unknown sex	◇
Affected male	■
Affected female	●
Spontaneous abortion	△
Proband	□
Heterozygous male	◫
Heterozygous female	◐
Female carrier of X-linked trait	⊙
Consanginous union	□=○
Monozygotic twins	
Dizygotic twins	

AUTOSOMAL DOMINANT INHERITANCE

Disorder or trait which is manifested in the heterozygous state.

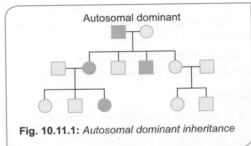

Fig. 10.11.1: *Autosomal dominant inheritance*

Characteristics of an Autosomal Dominant Inheritance

1. Males and females are affected in equal proportion.
2. Traced through many generation in the family tree. Hence called vertical transmission.
3. Usually one of the parents is affected (Exception is new mutation already explained).
4. Genetic risk is 50%, i.e. 50% of the progeny will be affected.

Skipping Generation

Incomplete penetrance in autosomal dominant inheritance is called skip ping generation.
That is, some individuals inherit the mutant gene but are pheno typically normal.

HIGH YIELD POINTS

Most common inheritance is **Autosomal Dominant 65%**
Followed by Autosomal Recessive 25% and X-linked Recessive 5%.

AUTOSOMAL RECESSIVE INHERITANCE

Disorder or trait which manifest in homozygous state.

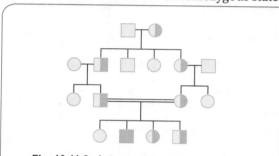

Fig. 10.11.2: *Autosomal recessive inheritance*

Characteristics of an Autosomal Recessive Inheritance

1. Males and females are affected in equal proportion.
2. Affected individuals are usually same generation (Hence called horizontal transmission).
3. Consanguineous marriage common.
4. You can find unaffected parents with affected progeny.

EXTRA EDGE

Pseudo Dominance in Autosomal Recessive Inheritance
- If an individual who is homozygous for an autosomal recessive disorder marry a heterozygous carrier, 50% chance of being affected.
- Resemble an autosomal dominant Pedigree.

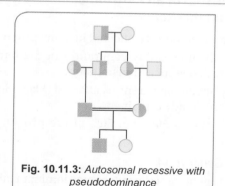

Fig. 10.11.3: *Autosomal recessive with pseudodominance*

X-LINKED RECESSIVE INHERITANCE

Mutant allele present in the X-chromosome.

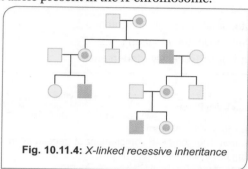

Fig. 10.11.4: *X-linked recessive inheritance*

Characteristics of X-linked Recessive Inheritance

1. Males are usually affected.
2. Females are usually carriers.
3. Affected males will have only carrier females
4. Carrier female will have affected males
5. Male to male transmission is never seen.

Knight Move Pattern of Inheritance

- In X-linked recessive inheritance affected male transmit the mutant allele to carrier female and never to a male.
- Carrier female transmit the mutant allele to the male who is affected.
- This pattern of transmission is called Knight move pattern or diagonal inheritance.

Difference between Hemizygous and Homozygous

- In X-linked or Y-linked inheritance, males with mutant allele does not have alternative allele in the homologous chromosome, as there is only one X and one Y-chromosome.
- Hence male with mutant allele on X or Y-Chromosome is called Hemizygous.
- Homozygous means mutant allele is present on both homologous chromosome.

X-LINKED DOMINANT INHERITANCE

Manifest in heterozygous female and hemizygous males.

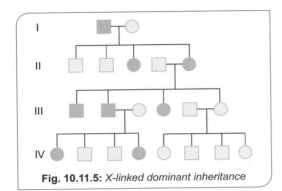

Fig. 10.11.5: *X-linked dominant inheritance*

Characteristics of X-Linked Dominant Inheritance

1. Resemble Autosomal dominant pedigree pattern as 50% chance of being affected.
2. Affected males transmit the disease to all females of next generation.
3. Male to male transmission is never seen.
4. This is because the son will receive only the Y chromosome from the father.

Y-LINKED INHERITANCE

- Mutant allele is present in the Y-chromosome.
- Other name for Y-linked inheritance is Holandric inheritance.

Characteristics of Y-linked Inheritance

1. Only males are affected
2. Only male to male transmission is seen.
3. The explanation is simple only Y-chromosome carries the mutant allele.

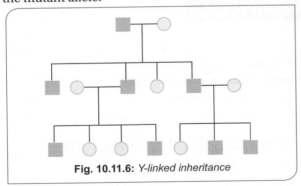

Fig. 10.11.6: *Y-linked inheritance*

Y-chromosome carries the genes:

1. *SRY* (Sex Reversal Y) codes Testis determining factors. Mutated cause Sex reversal.
2. *DAZ* (Deleted in Azoospermia).
3. *AZF* (Azoospermic factor gene) Mutated cause Azoospermia oroligospermia.

MITOCHONDRIAL INHERITANCE

Mutant allele present in the mitochondrial DNA.

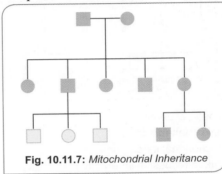

Fig. 10.11.7: *Mitochondrial Inheritance*

Characteristics of Mitochondrial Inheritance

1. Females transmit the disease to all her offsprings.
2. Males usually never transmit the disease.
3. This is called matrilineal inheritance.
4. Other name is cytoplasmic inheritance.

Quick Revision

- Males and females are equally affected in autosomal disorders.
- Vertical transmission in Autosomal dominant inheritance
- Horizontal transmission in Autosomal recessive inheritance.
- Genetic risk of 50% in Autosomal dominant inheritance
- Genetic risk of 25% in Autosomal recessive inheritance.
- Pseudodominance in autosomal recessive inheritance.
- Males are affected and females are carriers in X linked recessive.
- Affected males transmit the disease to all females in X linked dominant.
- Only males are affected in Y linked disorders.
- Females transmit the diasease to all the offsprings in mitochondrial inheritance.

Check List for Revision

- This is a frequently asked topic for image based question.
- Read the whole chapter thoroughly and understand the concept.

Review Questions MCQ

1. Identify the inheritance pattern:

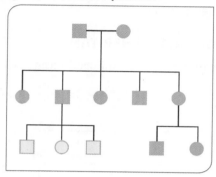

a. Autosomal dominant
b. Y linked inheritance
c. Mitochondrial inheritance
d. X linked recessive inheritance

2. Identify the inheritance pattern:

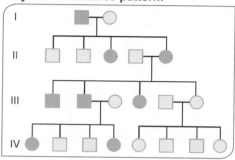

a. X linked dominant inheritance
b. Y linked inheritance
c. Mitochondrial inheritance
d. X linked recessive inheritance

Answers to Review Questions

1. **c. Mitochondrial Inheritance**

   Characteristics of Mitochondrial Inheritance
   - Females transmit the disease to all her offsprings
   - Males usually never transmit the disease

2. **a. x-linked dominant inheritance**

   Characteristics of x-linked Dominant Inheritance
   1. Resemble autosomal dominant pedigree pattern as 50% chance of being affected.
   2. Affected males transmit the disease to all females of next generation.
   3. Male to male transmission is never seen.
   4. This is because the son will receive only the Y chromosome from the father.

10.12 DNA POLYMORPHISM

- Introduction
- Repeat Length Polymorphism
- Different Types of Polymorphism
- Restriction Fragment Length Polymorphism
- Single Nucleotide Polymorphism
- Copy Number Variations

INTRODUCTION

- Normal variation in DNA sequence that have a frequency of at least 1% of population.
- Any two individuals share 99.9% of DNA sequence.
- DNA variation lies in this 0.1% (1 in 500–1000 bp)
- They are genetic markers that determines the uniqueness of an individual (i.e. like your fingerprints).

DIFFERENT TYPES OF POLYMORPHISM

1. Single Nucleotide Polymorphism.
2. Repeat length polymorphism or Short Tandem Repeats (Variable Number Tandem repeats).
   - Microsatellite repeats
   - Minisatellite repeats.
3. **Copy Number Variations**.
4. **Restricted Fragment Length Polymorphism**.

SINGLE NUCLEOTIDE POLYMORPHISM

- DNA variation in single base pair are called SNPs (or Snips).
- **Most common polymorphism**^{Q} (~90% of total polymorphisms).
- No. of SNPs in human genome is **10 million.**
- Occurrence is one nucleotide in every stretch of approximately 100–300 base pairs.
- SNPs that are in close proximity are inherited together (are linked) is called **haplotype**.
- Haplotype map information is referred to as **HapMap.**
- May occur anywhere in the genome within the exon, within the **intron** (most common).
- Used as genetic markers in Linkage and Association Studies.

SNP Genotyping Arrays

- Newer types of genomic arrays are designed to identify single nucleotide polymorphism (SNP) sites genome-wide.
- This technology is the mainstay of genome wide-association studies.

REPEAT LENGTH POLYMORPHISM OR SHORT TANDEM REPEATS OR VARIABLE NUMBER TANDEM REPEATS

- Short repetitive sequence in human DNA is called Repeat length Polymorphism.
- Most common one is **dinucleotide repeat involve AC to TG on other strand**.
  Depending on the repeat size it is divided into
  Microsatellite: Repeat size of **2–6 bp.**
    Total length the repeats extend is **usually <1 k bases.**
  Mini satellite: Repeat size of **15–70 bp**
    Total length the repeats extend **is usually 1 to 3 k bases**.

Application of Repeat Length Polymorphism

- Useful as genetic markers in Linkage and Association Studies.
- Familial diagnosis of disease like Polycystic Kidney Disease.
- Cancer genetics.
- Paternity testing.
- Forensic medicine.

Method to Detection of Repeat Length Polymorphism

- By allele specific PCR.

RESTRICTED FRAGMENT LENGTH POLYMORPHISM

- Special type of Polymorphism (Described later)
- DNA variations that create or abolish a restriction site.

COPY NUMBER VARIATION (CNV)

- Insertion or Deletion of a segment of genome (**involve 1 kb to several Mbs**).
- About **1500 CNVs** detected so far.
- Involve substantial **regions of the genome**, not single nucleotide.
- De novo CNVs observed among monozygotic twins.
- More recently detected.
- 50% occur in the **coding regions.**
- Responsible for human phenotypic diversity.

> **EXTRA EDGE**
>
> **Variant of Unknown Significance (VUS)**
> - Sequence alteration which are unclear whether it is a mutation or polymorphism are called VUS.

Quick Revision

- DNA polymorphism is normal variation in DNA sequence.
- Most common polymorphism is SNP.
- VNTR is Variable number tandem repeat.
- Short repeats of 2-6 bp size is microsatellite repeat.
- Short repeat of 15–70 bp size is minisatellite repeat.
- Repeat length polymorphism are assessed by allele specific PCR.

Check List for Revision

- Definition and types of polymorphism is must learn.
- All bold letters are the points to be highlighted for revision.

Review Questions MCQ

1. **The size of microsatellite repeat sequence is:** *(PGI Nov 2014)*
   a. Less than 1 kbp
   b. 2–6 kbp
   c. 1–3 kbp
   d. More than 3 kbp
   e. 5–20 bp

2. **Microsatellite sequence is:** *(AI 2006)*
   a. Small satellite
   b. Extra chromosomal DNA
   c. Short sequence (2–5) repeat DNA
   d. Looped-DNA

Answers to Review Questions

1. **a, b. Less than 1 kbp, 2–6 bp** *(Ref: Robbins 9/e page 179)*

   Repeat length Polymorphism or Short Tandem Repeats or Variable Number Tandem Repeats
   Short repetitive sequence in human DNA is called Repeat length Polymorphism.
   Most common one is dinucleotide repeat involve AC to TG on other strand.
   Depending on the repeat size it is divided in to Microsatellite—Repeat size of 2–6 bp, extend to <1 kbp Mini satellite–Repeat size of 15–70 bp.

2. **c. Short sequence (2–5) repeat DNA** *(Ref: Robbins 9/e page 179)*

   Repeat length Polymorphism or Short Tandem Repeats or Variable Number Tandem Repeats
   Depending on the repeat size it is divided into Microsatellite—Repeat size of 2–6 bp.
   Minisatellite – Repeat size of 15–70 bp.

CHAPTER 11

Molecular Biology Techniques

Chapter Outline

11.1 Recombinant DNA Technology
11.2 Amplification and Hybridization Techniques
11.3 Cytogenetic Techniques
11.4 DNA Sequencing Techniques, Transgenic Technique and Hybridoma
11.5 Other Molecular Biology Techniques and Recent Advances

CHAPTER 11

Molecular Biology Techniques

11.1 RECOMBINANT DNA TECHNOLOGY

- ☞ Definition
- ☞ Restriction Endonuclease
- ☞ Steps in Recombinant DNA Technology
- ☞ Probes
- ☞ Vectors
- ☞ Gene Library

DEFINITION

In vivo amplification technique used to get a clone of desired DNA fragment

To learn about Recombinant DNA Technology we should have knowledge about Vector, Restriction endonuclease, Chimeric DNA.

RESTRICTION ENDONUCLEASE

- **Werner Arber** discovered Restriction modification system in bacteria. It consist **of Restriction Endonuclease and a Site specific methylase.**
- Restriction endonuclease hydrolytically cleave polynucleotides internally at **specific pallindromic sites**.
- They belong to **Class III Hydrolases**.
- Restriction modification system **limit or restrict the expression by foreign viral (Bacteriophage) DNA through cleavage.**

Types of Restriction Endonucleases

- Type I Restriction endonuclease cleave at **random site.**
- Type II Restriction endonuclease cleave at **pallindromic sites.**

Palindrome

Palindrome is a sequence of duplex DNA that is the same when the two strands are read in opposite directions.
Examples:
GATCC & CCTAG   AATT & TTAA

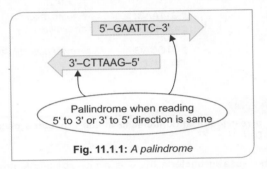

Fig. 11.1.1: *A palindrome*

Restriction Endonuclease as Molecular Biology Tools

- **Type II restriction endonuclease** discovered **by Hamilton Smith and Daniel Nathans** are used as biochemical tool of DNA manipulation.
- They are otherwise known as molecular scissors.
- Restriction endonucleases are named after the bacterium from which they are isolated.
- They cut the **dsDNA** at specific **palindromic** site to obtain
  1. **Sticky/staggered/Cohesive end**: These ends have an over hanging sequence.
  2. **Blunt ends**: These ends do not have overhanging sequence.

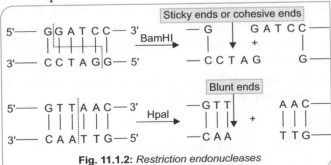

Fig. 11.1.2: *Restriction endonucleases*

Restriction Map

When a genomic DNA is treated with a specific Restriction endonuclease it cuts the DNA at specific sites to create a characteristic linear array of DNA. This is called **restriction map**.

Some examples of Restriction enzymes, its restriction site and Bacterial source

Eco R I	↓ GAATTC CTTAAC↑	*Escherichia coli*
EcOR II	↓ CCTGG GGACC↑	*Escherichia coli*
Hindi III	↓ AAGCTT TTCGAA↑	*Haemophilus influenzae*
Hpal	↓ GTTAAC CAATTC ↑	*Haemophilus parainfluenzae*
Pstl	↓ CTGCAG GACGTC ↑	*Providencia stuartii*

> **EXTRA EDGE**
>
> **Can a restriction endonuclease inside a bacteria cleave it own DNA?**
> Answer:
> No because in bacteria there is a site specific Methylase (described above) which methylate the all specific sites that it can hydrolytically cleave. So Restriction endonuclease cannot cut its own DNA.

VECTORS USED IN RECOMBINANT DNA TECHNOLOGY

Definition

A vector is a molecule of DNA to which the fragment of DNA to be cloned is joined.

Essential Properties of Vector

1. Capacity of **autonomous replication** within the host cell
2. Presence of **at least one restriction site** recognised by the Restrictionenzyme.
3. Presence of **at least one gene that confer antibiotic resistance** to select for the vector.

Plasmids

- **Circular, double stranded** DNA molecules seen in bacteria. (8–10 copies/cell).
- **Extra chromosomal**
- Each plasmid contains **an origin of replication** and can replicate in dependently
- They are **episomes**, i.e a genome above or outside the bacterium.
- Natural function is to **confer antibiotic resistance.**
- Carry **0.01-10 kbp** of DNA

Phages (Bacterial Viruses)

- Virus which infect the bacteria are called phages.
- They have a **linear DNA molecule**
- Carry DNA fragments up to **10–20 kbp**

Cosmids

- They are **plasmids which combine the features of Plasmid and Phages**
- They contain special genes called **cos site** (needed for packing lambda DNA into phage particles)
- Carry DNA fragments up to **30–50 kbp**.

BAC, YAC and PAC

Are artificially created chromosomes that can carry large DNA insert:

- **BAC**-Bacterial Artificial Chromosome
- **YAC**-Yeast Artificial Chromosome
- **PAC**-Artificial chromosome based on *E. coli* bacteriophage P1-based vectors

DNA Insert Size:

- BAC and PAC ____ 50–250 kbp
- YAC ____ 500–3000 kbp

Chimeric DNA or Recombinant DNA

DNA to be cloned + Vector DNA = Chimeric DNA

Passenger DNA

DNA to be cloned is called Passenger DNA or Foreign DNA.

Procedure to Prepare Chimeric DNA (See Fig. 11.1.2)

- Both the foreign DNA and vector DNA is treated with same Restriction endonuclease.
- This create sticky or blunt ends.
- This is religated using DNA Ligase.
- Thus chimeric DNA is produced.

Homopolymer Tailing

Technique used to overcome the problems inherent to sticky ends and blunt ends.

1. Problems with Sticky End
- Sticky ends of a vector may reconnect with themselves, with no net gain of DNA.
- Sticky ends of fragments also anneal so that heterogeneous tandem inserts form.
- Sticky-end sites may not be available or in a convenient position for the restriction endonuclease

2. Problems with Blunt Ends
- Blunt ends ligation is not directional.

> **Extra Edge**
> The enzyme that helps in blunt end ligation is Bacteriophage T4 Enzyme DNA Ligase.

Procedure of Homopolymer tailing
- To circumvent the problems of blunt ends, homopolymer tailing is used.
- New synthetic sticky ends are added using the enzyme **terminal transferase.**
- Poly d (G) is added to the 3' ends of the vector.
- Poly d (C) is added to the 3'ends of the foreign DNA using terminal transferase.
- Then the two molecules can only anneal to each other

This procedure is called homopolymer tailing.

Image-Based Information

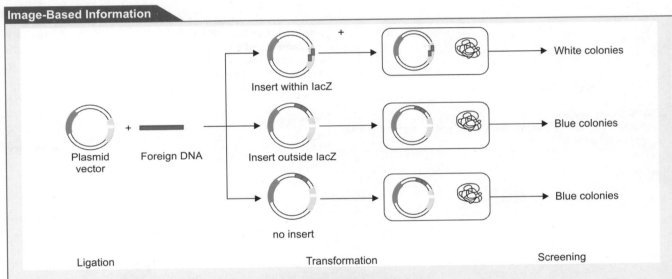

Fig. 11.1.3: *Blue white screening*

Blue-white Screening
- It is a quick and easy screening technique to detect successful ligation of DNA of interest to the vector (**Complementation**)
- Cells are grown in the presence of **X-Galactose**.
- If the ligation **was successful**, the bacterial colony will be **white**
- If not, the colony will be **blue**.

Recombinases: An Adjunct to Restriction Endonucleases

Catalyze specific incorporation of two DNA fragments that carry the appropriate recognition sequences and carryout homologous recombination, between relevant recognition site.

Examples of Recombinases and relevant recognition site

Recombinase	Recognition site
CRE Recombinase	Bacterial Lox P site
λ phage encoded INT protein	Bacteriophage λ att site
Yeast Flp Recombinases	Yeast FRT site

STEPS OF RECOMBINANT DNA TECHNOLOGY
- Isolation of specific DNA
- Selection of vector
- Synthesis of chimeric DNA
- Introduction of recombinant plasmid to bacteria
- Screening for Recombinant Vectors
- Selection of Specific DNA clones.

Uses of Recombinant DNA Technology in Clinical Medicine
- To understand molecular basis of diseases
- Preparation of Vaccines and Hormones

Self Assessment and Review of Biochemistry

- Diagnosis of infectious diseases
- Forensic medicine – reveal a criminal from specimens left on the scene of crime.

Crispr Cas 9 Genome Editing System

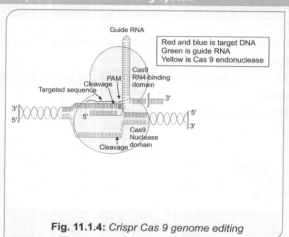

Fig. 11.1.4: *Crispr Cas 9 genome editing*

CRISPR Cas 9 System
- Clustered regularly interspersed short palindromic repeats associated gene 9.
- A prokaryotic immune system' conferring resistance to external genes from bacteriophage

Mechanism of action
- Guide RNA binds with target RNA and brings cas 9 endonuclease domain to target DNA.
- It cuts both the DNA strand.
- This results in a double strand break.
- This is corrected by any of the ds break repair mechanism (NHEJ or HR)

In Eukaryotes
- This system now emerged as Novel DNA/genome editing or gene regulatory system
- CRISPR use an RNA based targeting to bring Cas 9 nuclease to foreign DNA.
- This CRISPR- RNA Cas 9 complex then inactivates and degrade target DNA.
- This system can be used in eukaryotic cells, including humans.

Application of CRISPR Cas 9
- **Genomic editing**
- Targeted Mutagenesis—Gene knockout
- Modulation of gene expression.
- Gene deletion

EXTRA EDGE
*C2c2 is a new variant of CRISPR cas that **cleave RNA***

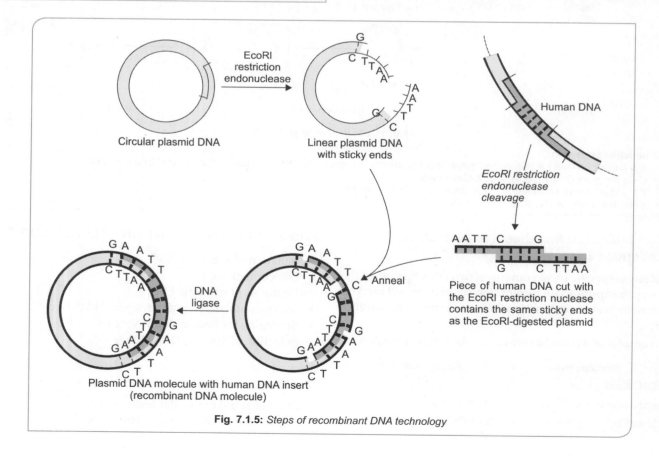

Fig. 7.1.5: *Steps of recombinant DNA technology*

cDNA^Q

DNA complementary to mRNA is called cDNA or Complementary DNA.

Procedure to Prepare cDNA

- Isolate them RNA.
- By the action of Reverse Transcriptase RNA-DNA hybrid is synthesized
- By RNAse H, RNA is digested
- By DNA Polymerase double stranded DNA is produced.
- Thus DNA complementary to mRNA is synthesized.

GENE LIBRARY/DNA LIBRARY

A collection of recombinant DNA clones generated from a specific source.

Types of Gene Library

Genomic DNA Library

- Prepared from total genomic DNA of an organism.
- By digestion of Genomic DNA by Restriction endonuclease.
- Then recombinant clones of such digested DNA is produced by recombinant DNA Technology.

cDNA Library

If a protein coding gene of interest is expressed at a high level in a particular tissue, mRNA transcribed from that gene is likely also present in high concentration in the cells of that tissue.

Reticulocyte mRNA is composed largely of mRNA that code for α-globin chain and β-globin chain.

cDNA obtained from these mRNA will be of globin chains. These mRNA can be used to create cDNA by using **reverse transcriptase**. Such cDNA from different organs can be cloned by Recombinant DNA technology to create cDNA libraries.

HIGH YIELD POINTS

Advantage of cDNA over Genomic DNA^Q
- Contains only **coding sequences**.
- Represent the mRNA in a tissue.
- Hence used to **study gene expression**

Disadvantage Extensive post-translational modifications is not possible in such proteins expressed by cDNAs

PROBES

Pieces of DNA or RNA labeled by various techniques to detect a complementary sequence.

Uses of Probes

- To detect DNA on Southern blot transfers
- To detect RNA on Northern blot transfers
- Can search libraries for a specific gene.

Labelling of DNA Probes

Two Types of Labels

Radioactive Labels

- Most commonly by Radioactive Phosphorus (32P)

Nonradioactive Labels

- Biotin label
- Fluorescent labels

Methods of Radio Labelling

1. End labelling at 5' end or 3' end of the probe.
2. Nick translation

Nick Translation

The technique used to produce radioactive labeled DNA probes.

Procedure

- Single stranded nick created and the dNucleotide is removed by DNAseI.
- Gap is filled by radio-labeled dNucleotide by DNA Polymerase.
- Thus radio labeled DNA probe is created.

Expression Vector

A vector in which the foreign DNA introduced by Recombinant DNA Technology synthesizes protein, i.e. the gene is expressed.

Uses of Expression Vectors

- To detect specific cDNA.
- To detect specific protein produced by a specific cDNA.
- To produce proteins like Insulin in large quantities.

Quick Revision

- Restriction endonucleases cut dsDNA at specific pallindromic sites.
- Plasmids are circular ds extrachromosomal DNA present in extrachromosomal location in bacteria.
- Blue white screening is for identifying successful ligation of foreign DNA to vector DNA called transformation.
- CRISPR cas 9 is a immune system in bacteria
- CRISPR cas 9 is a novel gene editing mechanism.
- cDNA library is more informative than genomic DNA library
- A collection of clones is called Gene library
- CRE recombinase bind to lox P site in bacteria.

Check List for Revision

- This chapter is a high yield chapter especially for PGI exam.
- Vectors, Restriction endonuclease, CRISPR cas 9 are must learn topics.

Review Questions MCQ

1. This image depicts *(AIIMS May 2018)*

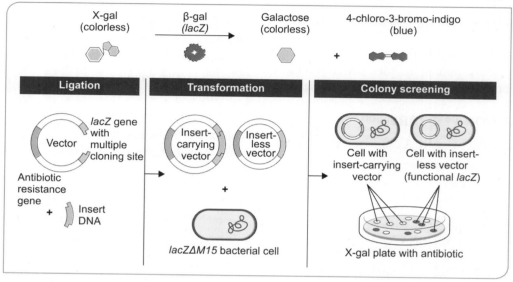

   a. Transformation
   b. Complementation
   c. Translation
   d. Hybridisation

2. cre-cis regulatory elements bind to what site? *(AIIMS May 2015)*

   a. RE site
   b. FTR site
   c. Lox P site
   d. INT site

3. cDNA used in gene amplification in bacteria of genomic DNA because: *(PGI May 2014)*

   a. Easy to replicate
   b. Human genome has many introns that cannot be removed by bacteria
   c. Promoter are not found
   d. Complete genome cannot be replicated

4. Restriction Endonuclease is used in: *(JIPMER 2013)*

   a. RFLP
   b. PCR
   c. FISH
   d. SDS-PAGE

5. Function of endonucleases: *(TN 97)*

   a. Cut DNA at specific DNA sequences
   b. To point out the coding regions
   c. Enhancers
   d. To find out antibiotic resistance

6. Enzymes used in DNA research programme are, except: *(PGI June 97)*

   a. Polymerase
   b. Exonuclease
   c. Nuclease
   d. Alkaline phosphatase
   e. None

7. In DNA transfer the vectors used from smallest to largest is: *(PGI Dec 07)*

   a. Cosmids, Plasmids, Bacteriophage
   b. Plasmids, Bacteriophage, Cosmids
   c. Bacteriophage, Cosmids, Plasmids
   d. Cosmids, Bacteriophage, Plasmids
   e. Plasmids, Cosmids, Bacteriophage

8. In gene cloning, largest fragment can be incorporated in: *(AIIMS Dec 95)*

   a. Plasmid
   b. Bacteriophage
   c. Cosmid
   d. Retrovirus

9. Function of restriction II enzyme: *(AI 2012)*

   a. Prevents protein folding
   b. Removing formed DNA
   c. Cleaves DNA at palindromic recognition site
   d. Negative supercoiling

10. **After digestion by restriction endonucleases DNA strands can be joined again by:** *(AIIMS May 2011) (Nov 2010)*
    a. DNA polymerase
    b. DNA ligase
    c. DNA topoisomerase
    d. DNA gyrase

11. **Starting material for production of insulin from bacteria is:** *(AIIMS May 2011)*
    a. Genomic DNA of lymphocytes
    b. m RNA of lymphocytes
    c. Genomic DNA of beta cell of pancreas
    d. mRNA of beta cells of pancreas

12. **True statement about Restriction Endonuclease:** *(PGI May 2012)*
    a. Palindromic sequence observed
    b. Protects bacteria from infection by virus
    c. Present only in Eukaryotes
    d. Restrict replication of DNA

13. **True about Gene Library:** *(PGI Nov 2009)*
    a. Library of Gene books
    b. Plasmid with copies of different genes
    c. Computer database with all gene knowledge
    d. Collection of gene copies of one organism as completely as possible in bits and pieces.
    e. DNA fragments.

14. **Restriction enzymes:** *(PGI Nov 2009)*
    a. Prevent elongation of DNA
    b. Break DNA to create sticky end
    c. Cuts at palindromic sites
    d. Restriction sites are not specific
    e. Breaks at sugar-phosphate bond

15. **Correct statements regarding restriction endonuclease is/are:** *(PGI Dec 03)*
    a. Restriction endonuclease recognizes specific sites of DNA sequence
    b. Restriction endonuclease recognizes short sequence of DNA
    c. It acts at 5' – 3' direction
    d. It acts at 3' – 5' direction

16. **True about the function of Restriction endonuclease:** *(PGI Dec 06)*
    a. Cut both the strands of dsDNA
    b. The cut ends produced are sticky
    c. The cut ends produced are blunt
    d. Cuts single strand of DNA

Answers to Review Questions

1. **b. Complementation**

   This is blue white screening used in the screening of recombinant vectors for inserted DNA fragments. It is done by making use of phenomenon called as complementation.

   In this case alpha complementation of beta galactosidase gene. The foreign DNA is inserted to lac Z gene so no functional galctosidase enzyme produced.

   Recombinant vector with proper insert at Lac Z gene will be white colonies.

   Recombinant vector with no insert or improper insert will be blue colonies.

   Transformation is introduction of recombinant vector to host cell.

2. **c. Lox P site** *(Ref: Harper 31/e page435)*

Recombinases	Recognition site
CRE Recombinase	Bacterial Lox P site
λ phage encoded INT protein	Bacteriphage λ att site
Yeast Flp Recombinases	Yeast FRT site

3. **a, b, d. Easy to replicate, Human genome has many introns that cannot be removed by bacteria, Complete genome cannot be replicated** *(Ref: Harper 31/e page438)*

cDNA

DNA complementary to mRNA is called cDNA or Copy DNA or Complementary DNA.

Procedure to prepare cDNA:
- Isolate them RNa
- By the action of Reverse Transcriptase RNA-DNA hybrid is synthesized
- By RNAse H RNA is digested
- By DNA Polymerase double stranded DNA is produced.

Advantage of cDNA over genomic DNA
- Contains only coding sequences.
- Represent the mRNA in a tissue.
- Hence used to study gene expression.

4. **a. RFLP**

   In RFLP, DNA is cut using a Restriction endonuclease. They cut DNA at specific pallindromic sites. So it produces a characteristic restriction map.

5. **a. Cut DNA at specific DNA sequence** *(Ref: Harper 31/e page433)*

   - Restriction endonuclease cut DNA at specific palindromic sites.
   - These enzymes are isolated from bacteria.
   - They restrict the entry of phages into the bacteria.

6. **None** *(Ref: Harper 31/e page434)*

Enzyme	Reaction	Uses
Alkaline phosphatase	Dephosphorylates 5' ends of RNA and DNA	Removal of 5'-PO_4 groups prior to kinase labeling; also used to prevent self-ligation
DNA ligase	Catalyzes bonds between DNA molecules	Joining of DNA molecules
DNA polymerase I	Synthesizes double-stranded DNA from single-stranded DNA	Synthesis of double-stranded cDNA; Nick translation; Generation of blunt ends from sticky ends
Thermostable DNA polymerases (Taq Polymerase)	Synthesize DNA at elevated temperatures (60–80°C)	Polymerase chain reaction (DNA synthesis)
DNase I	Under appropriate conditions, produces single-stranded nicks in DNA	Nick translation Mapping of hypersensitive sites Mapping protein—DNA interactions
Exonuclease III	Removes nucleotides from 3' ends of DNA	DNA sequencing Mapping of DNA—protein interactions
λ Exonuclease	Removes nucleotides from 5' ends of DNA	DNA sequencing
Polynucleotide kinase	Transfers terminal phosphate (λ position) from ATP to 5'-OH groups of DNA or RNA	^{32}P end-labeling of DNA or RNA
Reverse transcriptase	Synthesizes DNA from RNA template	Synthesis of cDNA from mRNA; RNA (5' end) mapping studies
S1 nuclease	Degrades single-stranded DNA	Removal of "hairpin" in synthesis of cDNA; RNA mapping studies (both 5' and 3' ends)
Terminal transferase	Adds nucleotides to the 3' ends of DNA	Homopolymer tailing
CRISPR-Cas9	RNA targeted DNA directed Nuclease	Genome editing and modulation of gene expression

7. **b. Plasmids, Bacteriophage, Cosmids**
(Ref: Harper 31/e page 433)

The DNA insert size in ascending order is Plasmid< Phage<Cosmids<BAC/PAC<YAC

Cloning capacity of common cloning vectors

Vector	DNA insert size (Kbp)
Plasmid	0.01–10
Lambda Phage	10–20
Cosmid	35–50
BAC, PAC	50–250
YAC	500–3000

8. **c. Cosmid** (Ref: Harper 31/e page434)

9. **c. Cleaves at palindromic recognition site.**
(Ref: Harper 31/e page434)

10. **b. DNA Ligase** (Ref: Harper 31/e page434)

Enzymes involved in the DNA Replication

1. **Topoisomerases**
   - Relieve torsional strain that results from helicase-induced unwinding of DNA.
   - Nicking Resealing Enzyme

2. **Helicase:** ATP driven processive unwinding of DNA
3. **Single Strand Binding Protein (SSB)** Prevent premature reannealing of dsDNA
4. **DNA Primase:**
   - Initiates synthesis of RNA primers.
   - Special class of DNA dependent RNA Polymerase
5. **DNA Polymerase:** Catalyse the chemical reaction of DNA Polymerisation. Synthesise DNA only in 5' to 3' direction.
6. **DNA Ligase:** Seals the single strand nick between the nascent chain and Okazaki fragments on lagging strand.

11. **d. mRNA of Beta cells of Pancreas**
(Ref: Vasudevan and Sreekumari 7/e page626)

12. **a, b. Palindromic Sequence Observed, Protects bacteria from infection by virus** (Ref: Harper31/e page434)

Restriction Enzymes

- Recognizes and cleaves a specific palindromic double-stranded DNA sequence that is typically 4–7 bp long.
- These DNA cuts result in blunt ends (e.g. Hpa I) overlapping (sticky or cohesive) ends (e.g. Bam HI)
- Are a key tool in recombinant DNA research.
- These enzymes were called restriction enzymes because their presence in a given bacterium restricted the growth of certain bacterial viruses called bacteriophages.

- Each enzyme recognizes and cleaves a specific double-stranded DNA sequence that is typically 4–7 bp long.
- Restriction endonucleases are present only in cells that also have a companion enzyme, **site-specific DNA methylases**, that site-specifically methylates the host DNA, rendering it an unsuitable substrate for digestion by that particular restriction enzyme. Thus prevent the digestion of host DNA.

Restriction enzymes are named after the bacterium from which they are isolated

- For example, Eco R I is from Escherichia coli, and BamHI is from *Bacillus amyloliquefaciens*
- The first three letter sin the restriction enzyme name consist of the first letter of the genus(E) the first two letters of the species (co).
- These may be followed by a strain designation (R)
- **A roman numeral (I) to indicate the order of discovery (e.g. *Eco R I* and *Eco R II*).**

13. **d, e. Collection of gene copies of one organism as completely as possible in bits and pieces, DNA fragments.**
    (Ref: Harper 31/e page 434)

Gene Library
A collection of recombinant DNA clones generated from a specific source.

Two types of Gene Library:
1. Genomic DNA Library:
   - Prepared from total genomic DNA of an organism.
   - By digestion of Genomic DNA by Restriction Endonuclease.
   - Then recombinant clones of such digested DNA is produced by recombinant DNA Technology.
2. c DNA library
   - cDNA is prepared from mRNA by the action of Reverse Transcriptase
   - The recombinant clones for cDNA are produced by Recombinant DNA Technology.

Advantage of cDNA over Genomic DNA
- Contains only coding sequences.
- Represent the mRNA in a tissue.
- Hence used to study gene expression.

14. **a, b, c, e. Prevent elongation of DNA, Break DNA to create sticky end, Cuts at palindromic sites, Breaks at sugar-phosphate bond.**
    (Ref: Harper 31/e page 434)

15. **a. Restriction endonuclease recognizes specific sites of DNA sequence**
    (Ref: Harper 30/e page457)

16. **a, b, c. Cut both the strands..., The cut ends produced are sticky, The cut ends produced are blunt** *(Ref Harper 31/e page434)*

Restriction endonuclease cut both strands of DNA. They produce sticky ends or blunt ends depending on which restriction endonuclease act on the DNA.

11.2 AMPLIFICATION AND HYBRIDIZATION TECHNIQUES

- Amplification Techniques
- Hybridisation Techniques

AMPLIFICATION TECHNIQUES

Classification of Amplification Techniques

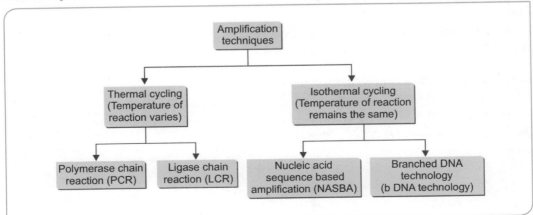

Another Classification

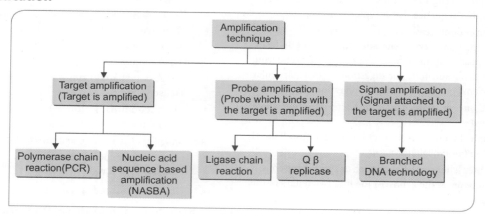

Polymerase Chain Reaction

- Revolutionary technique invented by **Karry B Mullis**[Q].
- He got Nobel Prize for this in **1993**.
- The polymerase chain reaction (PCR) is a **test tube method for amplifying a selected DNA sequence.**
- Target amplification by thermal cycling.
- **Exponential**[Q] amplification of the sample.
- The number of samples after n number of cycles is 2^n.
- One cycle of PCR require **<5 to 10 minutes**.
- 20 cycles result in million-fold amplification of the target DNA.
- Product obtained by amplification is called **Amplicon**.

The instrument that takes samples through the multiple steps of changing temperature in PCR Cycle is called **Thermocycler**.

Steps of PCR Cycle[Q]

1. Initial denaturation – where the entire DNA is denatured. This occur only one time.
2. **Denature the DNA:** The target DNA to be amplified is heated to separate the double-stranded target DNA into single strands.
3. **Annealing of primers to ssDNA:** The separated strands are cooled and allowed to anneal to the two primers (one for each strand).
4. **Extension of the Primer:** Synthesize new chains complementary to the original DNA chains.

- Steps 2, 3, & 4 is repeated several times

Steps	Temperature (°C)	Time (s)
Initial denaturation	95	3 min
Denaturation	90–96	20–60
Annealing	50–70	20–90
Extension	68–75	10–60

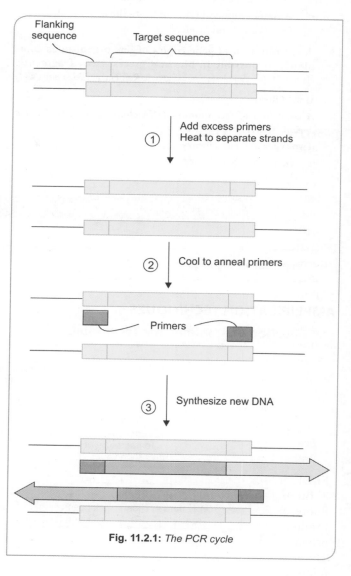

Fig. 11.2.1: *The PCR cycle*

Pre-requisites of PCR
- Sample DNA to be amplified
- Deoxynucleotides.
- Thermostable Polymerase: **Taq Polymerase**[Q] obtained from Thermus Aquaticus found in hot springs.
- Primer.
- $MgCl_2$ and KCl
- TaqPolymerase **lack proofreading** activity

Variants of PCR

Reverse Transcriptase PCR (RT PCR)
- It is the PCR amplification of a reverse transcription product
- RT PCR amplifies very small amounts of any kinds of RNA (**mRNA, rRNA, tRNAetc.**).
- cDNA copies of mRNA generated by a retroviral reverse transcriptase.
- This cDNA are amplified as in usual PCR.
- Tth Polymerase (from Thermus thermophilus) is used.
- Method used to obtain relative expression of gene in a cell.

Real Time (Homogenous or Kinetic) PCR
- This is a type of **Quantitative PCR**.
- Methods used to amplify and at the same time to quantitate the amount of target sequence during the **exponential phase** of PCR cycle (called Realtime). It is not after the amplification is over.

Methods used to quantitate the number of copies DNA present within the genome.

Intercalating Dyes:
- **Ethidium Bromide** that intercalate to dsDNA and fluoresce
- **SYBR Green**[Q] has robust fluorescence, less toxicity and more specificity when compared to Ethidium Bromide.

Sequence Specific Probes:
Generate fluorescence when they hybridize to target sequence. So they have more sensitivity than intercalating dyes.
- TaqMan,
- Molecular beacon
- **Fret Probes:** Fluorescence Resonance Energy Transfer Probes

FRET probes utilises two specific probes, one with 3' fluorophore (donor) and other with 5' catalyst for fluorescence (acceptor). When donor and acceptor are brought within 1 to 10 nm, by specific DNA binding, excitation energy is transferred from donor to acceptor. The acceptor will then loses energy in the form of heat or fluorescence called sensitised emission.

Nested PCR
When the DNA to be amplified is present in low concentration compared to total DNA in the sample, this technique is used. Nested PCR uses two sets of amplification primers to amplify single target in two PCR runs.
- First round of PCR with outer primers
- Second round of PCR with inner primers that bind to sequence inside of binding site of outer primers.
- This increases the sensitivity and specificity of PCR

Multiplex PCR
- Simultaneous amplification of many targets in one reaction
- Uses more than one pair of primers.

Arbitrary PCR (Arbitrarily Primed Polymerase Chain Reaction) (AP PCR)
Also known as Randomly amplified Polymorphic DNA or Random amplification of Polymorphic DNA (RAPD).

Main Features:
- Uses short random primers (Usually 10–15 bases)
- Amplifies anonymous stretches of DNA under low stringency
- Reproducibility is low.

Applications of PCR

1. Used in Forensic Medicine
2. Viruses that have long latency period especially HIV are difficult to detect at early stage but PCR offers rapid and sensitive method for detecting viral RNA/DNA sequences.
3. Real time PCR allows to determine viral load.
4. To make **prenatal diagnosis of disease like Cystic fibrosis**. The mutant allele that three base deletion when compared to normal allele. So by detecting the size of amplified products using DNA electrophoresis, normal, carrier and affected individuals can be detected.
5. To detect length polymorphisms like VNTR.
6. PCR allows synthesis of mutant DNA in sufficient amount which can be later sequenced to detect mutation.
7. To establish precise tissue types for transplants
8. To study evolution, using DNA from archeological samples.
9. For quantitative RNA analyses after RNA copying and mRNA quantitation by the so-called Reverse Transcriptase PCR method.
10. **To score in vivo protein:** DNA occupancy using chromatin immuno precipitation assays (ChIP)
11. To facilitate New Generation Sequencing, by amplifying the DNA.

EXTRA EDGE: OTHER AMPLIFICATION TECHNIQUES

Nucleic Acid Sequence-based Amplification (NASBA)

This is a transcription-based amplification technique.
Isothermal Reaction

Steps of NASBA
1. RNA is the target.
2. ARNA–cDNA hybrid is synthesized from RNA target by Reverse transcriptase.
3. RNase-H removes the RNA.
4. ssDNA results
5. Then ds DNA is produced
6. DNA is transcribed by RNA Polymerase to produce millions of copies of RNA.

Enzymes needed are
- Reverse Transcriptase
- T7 RNA Polymerase
- RNaseH.

Advantage

Detection of RNA and Quantification of RNA is possible. Especially in microbiology (M. tuberculosis, Chlamydia Trachomatis, HIV, Cytomegalovirus).

Ligase Chain Reaction (LCR)

Probe/Primer amplification technique. This technique needs a thermal cycler

Steps of LCR
1. Primers are attached to the target, immediately adjacent to each other.
2. DNA ligase attach the adjacent primers.
3. Ligated primers can serve as a template for annealing.
4. Next set of adjacent primers are annealed, then ligated.
5. So product of LCR is ligated primer.

Uses
- Even 1 base pair mismatch will prevent ligation of primers.
- So LCR can detect point mutations.

Q β Replicase

- **Probe amplification** technique
- Amplify probes that have specificity to target.
- The method is so named because of the enzyme used to amplify the probe, **Q β Replicase**.
- It is a RNA dependent RNA Polymerase from bacteriophage Qβ.
- Target can be DNA or RNA.
- Used to amplify nucleic acid associated with infectious-organism (Chlamydia, CMV, HIV and Mycobacterium).

Branched DNA Technology

Signal Amplification technique Target can be DNA or RNA.

Steps
1. Target nucleic acid bind to capture probe that are fixed to solid support.
2. Then Extender probes bind to target as well as to amplifier probes.
3. Amplifier probe has multiple alkaline Phosphatase labelled nucleotides.
4. Chemiluminescence is measured.

Uses

Qualitative and Quantitative detection of viruses (Hepatitis B, Hepatitis C, HIV).

HYBRIDISATION AND BLOT TECHNIQUES

Southern Blot

- Devised by **Edward Southern in 1975**.
- Technique to detect **specific DNA Segment**

Principle
- Based on specific base pairing rule of complementary nucleic acid strands.
- It is a **DNA-DNA Hybridisation**.

Steps of Southern Blot
- Duplex DNA isolated.
- Treated with Restriction Endonuclease.
- DNA is fragmented.
- Fragmented DNA separated by Agar gel electrophoresis.
- Treated with NaOH to denature the DNA.
- Denatured DNA fragments transferred to Nitrocellulose membrane or nylon paper (Blot)
- Add Radio labelled or Fluorescent cDNA probes
- Detection of Hybridization by imaging (DNA-DNA Hybridisation).

Uses of Southern Blot
- Identification of specific viral or bacterial DNA in the infected sample.
- Screening test to detect inborn errors.
- Detect DNA mutations such as Large insertion, Deletion, Trinucleotide repeat expansion, point mutation and rearrangements of nucleotide
- Alterations in gene (deletion, in forensic medicine, to analyze DNAs from specimens at the scene of crime-blood, semen, saliva, etc.

Northern Blot

- Technique used to detect specific RNA
- Principle: RNA- DNA hybridization technique.
- Radioactive/fluorescent labelled cDNA Probes used.

Uses
- Used to detect specific gene expression in specific tissues.

Western Blot (Immunoblot) Analysis for Proteins
Technique to detect specific Protein in a sample.

Principle:
- Antigen antibody Interaction.
- Radioactive labeled Antibody used.

Uses:
- Identification of a specific protein in sample.
- Detection of Viral pathogens by identifying viral proteins- HIV virus or Hepatitis B.

South Western Blotting

To examine Protein–DNA Interaction

Dot Blot Technique

- The step, blotting to nitrocellulose membrane is avoided.
- The sample is directly applied to slots on a specific blotting apparatus containing nylon membrane. This is also called *slotblot*.

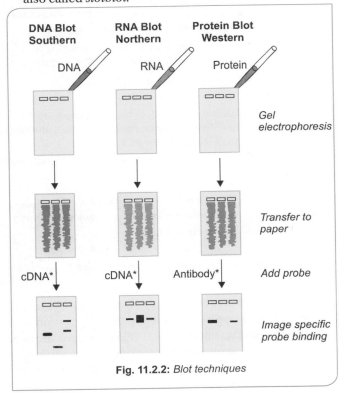

Fig. 11.2.2: *Blot techniques*

Quick Revision

- PCR and LCR are Thermal recycling amplification technique.
- NASBA and B DNA Technology are isothermal cycling.
- Target is amplified in PCR, PCR and NASBA,
- RT PCR is a technique to detect and quantify any kinds of RNA.
- Mg^{2+} & K^+ are the cations of PCR.
- RealTime PCR is the quantitative PCR. Southern blot detect a specific DNA
- Northern blot detect a specific RNA hence study gene expression as well.
- Western blot detect a specific protein.
- South western blot detect DNA Protein interaction.

Check List for Revision

- This is a high yield topic
- PCR and Variant of PCR, Blot techniques are must learn.
- Extra edge topics are optional

Review Questions

PCR

1. **Which of the following is used as proof reading in PCR?** *(PGI Nov 2018)*
   a. Pfu
   b. Taq polymerase
   c. Topoisomerase
   d. Helicase

2. **Real time PCR is used for:** *(AIIMS May 2013)*
   a. Multiplication of RNA
   b. Multiplication of specific segment of DNA
   c. Multiplication of Protein
   d. To know how much amplification has occurred

3. **Quantitative DNA analysis/estimation is done by:** *(AIIMS May 2012)*
   a. pH meter
   b. Sphymometer
   c. Spirometer
   d. Spectrometer

4. **All are added to PCR, except:** *(AIIMS Nov 2012)*
   a. Deoxynucleotide
   b. Dideoxynucleotide
   c. Thermostat DNAP
   d. Template DNA

5. **For PCR which of the following is not required?** *(AIIMS May 2007)*
   a. Taq polymerase
   b. d-NTP
   c. Primer
   d. Radiolabeled DNA probe

6. **SYBR Green Dye is used for:** *(AIIMS May 2008)*
   a. HPLC
   b. Immunofluorescence
   c. PCR
   d. ELISA

7. **True about DNA polymerase used in PCR:**
   (PGI May 2013)
   a. Obtained from virus
   b. Obtained from bacteria
   c. Used for joining two strands
   d. It is heat stable
   e. Add nucleotide

8. **Enzyme(s) used in polymerase chain reaction is/are:**
   (PGI May 2011)
   a. Restriction endonuclease
   b. DNA polymerase
   c. Alkaline phosphate
   d. RNA polymerase
   e. Reverse transcriptase

9. **True about PCR all except:** *(PGI June 2009)*
   a. Carried out by thermostable DNA-polymerase
   b. Exponential amplification
   c. Additive amplification
   d. Specific amplification
   e. Single-stranded DNA required

10. **PCR is used in:** *(Ker 2006)*
    a. Medicolegal cases
    b. Amplification of gene
    c. Identification of organism
    d. All of the above

11. **Which of the following is used in PCR?** *(AIIMS Nov 07)*
    a. Ca^{++}
    b. Mg^{++}
    c. Li^+
    d. Na^+

12. **In PCR Acquaticus thermophilus is preferred over E. coli, because:** *(PGI Dec 07)*
    a. Thermostable at temperature at which DNA liquefies
    b. Proof reading done
    c. Done in more precisely
    d. Does not require primer
    e. Better DNA replication

13. **Which of the following is not true about PCR?**
    (PGI 2003)
    a. Thermostable enzyme is used
    b. Annealing is done after DNA denaturation
    c. Specific primers are required
    d. Required at least 1st week time for synthesis
    e. DNA polymerase has to be added to each cycle

14. **In PCR true is/are:** *(PGI June 08)*
    a. Thermostable enzyme is needed
    b. 2 n copies formed after 'n' numbers of multiple
    c. Nonspecific
    d. Thermolabile enzyme
    e. Primer is needed

Blot Techniques

15. **Northern blot is for:** *(AIIMS Nov 2018)*
    a. DNA
    b. RNA
    c. Protein
    d. DNA Protein interaction

16. **Western blot detects:** *(AIIMS Nov 2009)*
    a. DNA
    b. RNA
    c. Protein
    d. mRNA

17. **Confirmatory test for proteins are:** *(PGI May 2014)*
    a. Western blot
    b. ELISA
    c. Chip assay
    d. Dot blot

18. **Which is the test used to identify mRNA?**
    (JIPMER Nov 2015)
    a. Southern blot
    b. Northern blot
    c. Western blot
    d. South Western blot

Answers to Review Questions

PCR

1. **a. Pfu**

   Pfu DNA polymerase is a thermostable enzyme that is isolated from Pyrococcus furiosus. When compared to Taq polymerase apart from 5' to 3' DNA polymerase activity Pfu has 3. to 5' exonuclease is proof reading activity also. So Pfu is used in high fidelity DNA amplification.

2. **d. To know how much amplification has occurred**
   (Ref: Tietz Textbook of Clinical Chemistry 4/e page1446)

   Real-Time (Homogeneous, Kinetic) Polymerase Chain Reaction

   Real-time PCR describes methods by which the target amplification and detection steps occur simultaneously in the same tube. Thus it is a method of quantitative PCR or qPCR.

   Quantitation is done in the exponential phase of amplification.

3. **d. Spectrometer**

   Absorbance of UV light at 260 nm can be used to estimate DNA Can be done using spectrophotometer or simply spectrometer. Other options are self explanatory.

4. **b. Dideoxynucleotides**
   (Ref: Tietz Textbook of Clinical Chemistry 4/e page1446)

   Pre requisites of PCR
   - Sample DNA to be amplified
   - Deoxynucleotides.
   - Thermostable Polymerase—Taq Polymerase obtained from Thermus Aquaticus found in hot springs.
   - Primer.
   - $MgCl_2$, KCl

5. **d. Radiolabeled DNA Probe**
   (Ref: Tietz Textbook of Clinical Chemistry 4/e page1446)

6. **c. PCR** (Ref: Tietz Textbook of Clinical Chemistry 4/e page 1446)

Real Time PCR

This is a type of Quantitative PCR.

Methods used to quantitate PCR Products in Real time PCR

Intercalating Dyes—Ethidium Bromide, SYBR Green. Sequence Specific Probes—TaqMan, Molecular beacon Fret Probes—Fluorescence Resonance Energy Transfer Probes

7. **b, d, e. Obtained from bacteria, It is heat stable, Add nucleotide**
   (Ref: Tietz Textbook of Clinical Chemistry 4/e page1446)

DNA Polymerase used in PCR is TaqPolymerase. This is thermostable polymerase obtained from bacteria Thermophilus Aquaticus. TaqPolymerase add nucleotide during extention of PCR Cycle.

8. **b. DNA Polymerase**
   (Ref: Tietz Textbook of Clinical Chemistry 4/e page1446)

Pre requisites of PCR:
Sample DNA to be amplified
Deoxy nucleotides.
Thermostable Polymerase—Taq Polymerase obtained from Thermus Aquaticus found in hot springs.
Primer.
$MgCl_2$

9. **c. Additive amplification.**

Polymerase Chain Reaction
Revolutionary technique invented by Karry B Mullis in 1989. He got nobel prize for this in 1993. The polymerase chain reaction (PCR) is a test tube method for amplifying a selected DNA sequence.

Exponential amplification of the sample.
- The number of samples after n number of cycles is 2n.
- One cycle of PCR require 20-30 seconds.
- 20 cycles result in million-fold amplification of the target DNA.
- Product obtained by amplification is called Amplicon.
- The instrument that takes samples through the multiple steps of changing temperature in PCR Cycle is called Thermocycler.

10. **d. All of the above**
    (Ref: Tietz Textbook of Clinical Chemistry 4/e page1446)

11. **b. Mg^{++}**
    (Ref: Tietz Textbook of Clinical Chemistry 4/e page1446)

12. **a. Thermostable at temperature at which DNA liquefies**
    (Ref: Tietz Textbook of Clinical Chemistry 4/e page1446)

13. **d, e. Required at least 1st week time for synthesis, DNA polymerase has to be added to each cycle**
    (Ref: Tietz Textbook of Clinical Chemistry 4/e page1446)

14. **a, b, e. Thermostable enzyme is needed, 2n copies formed after 'n' numbers of multiple, Primer is needed**

Blot Techniques

15. **b. RNA** (Ref: Harper 31/e page439)

Blot Techniques
Southern Blot
- Devised by Edward Southern in1975.
- Technique to detect specific DNA Segment
- Principle: Based on specific base pairing rule of complementary nucleic acid strands.
- It is a DNA-DNA Hybridisation.
- Northern Blot:
- Technique used to detect specific RNA
- Principle: RNA- DNA hybridization technique.
- Radioactive labeled cDNA Probes used.

Western Blot (Immunoblot) analysis for Proteins
- Technique to detect specific Protein in a sample.
- Antigen antibody Interaction.
- Radioactive labeled Antibody used.

South Western Blotting
To examine Protein –DNA Interaction Dot blot Technique
The step, blotting to nitrocellulose membrane is avoided. The sample is directly applied to slots on a specific blotting apparatus containing nylon membrane. This is also called slotblot.

16. **c. Protein** (Ref: Harper 31/e page439)

17. **a, b, c, d. Western blot, ELISA, Chip assay, dot blot**
    (Ref: Harper 31/e page439)

Western blot, ELISA, Chip assay and dot blot is based on Antigen antibody interaction. Hence they are confirmatory test for proteins. Chip is the other name for Microarray. Just like DNA Chip, where DNA–DNA Hybridisation is done, there Protein Microarray or Protein Chip where Antigen antibody interaction is done.

Dot blot technique–In blotting technique, the blotting to nitrocellulose membrane is avoided, instead they are arranged in different slots. This is called dotblot technique. This is applied to detect protein.

18. **b. Northern blot** (Ref: Harper 31/e page439)
- Southern blot detect DNA by DNA-DNA hybridisation
- Northern blot detect RNA by RNA-cDNA hybridization
- Western blot detect Protein by Antigen antibody Interaction.

11.3 CYTOGENETIC TECHNIQUES

- Classification
- Banding Technique
- FISH
- Microarray
- Array CGH

The word chromosome is a combination of two Greek words, "**Chroma**"-means '**colour**' and "**Somes**"-means '**body**'. So chromosomes are coloured bodies.

Arms of Chromosome
- p arm is the short arm, p stands for petite
- q arm is the long arm, q is the next letter after p
- q arm is also called g arm, g stands for grand e

CLASSIFICATION OF CYTOGENETIC TECHNIQUES

Can be divided into
- Conventional Cytogenetic techniques—Chromosome Banding
- Molecular Cytogenetic Techniques—FISH
- Array-based Techniques (Cytogenomic techniques)
  i. Comparative Genomic Hybridisation (CGH) array
  ii. Single Nucleotide Polymorphism (SNP) array

Samples for Chromosome Analysis
- Prenatal (fetal) chromosome
- Amniocytes by Amniocentesis
- Chorionic villi by Chorionic villus Sampling—Allows early detection of anomalies <10 wks
- Fetal Blood by Percutaneous Umbilical cord sampling (PUBS) in late second trimester

For Preimplantation Detection of Anomalies
- Analysis of **Blastomere**
- **Other Samples in adults are**
  1. Cultured skin fibroblast
  2. Bone marrow
  3. Peripheral Lymphocyte

Timing for Chromosome Analysis Artificially arrested in mitosis during Metaphase (or prometaphase)
- Metaphase arrest by-**N deacetyl N-methyl Colchicine (Colcemid)**

CONVENTIONAL CYTOGENETICS: BANDING TECHNIQUES

- Cytogenetic analysis is most commonly carried out on cells in mitosis, requiring dividing cells.
- Halting mitosis in metaphase is essential, because chromosomes are at their most condensed state during this stage of mitosis.
- The banding pattern of ametaphase chromosome is easily recognizable and is ideal for karyotyping.

Important Banding Techniques

G Banding
- Other name **G band Trypsin Giemsa (GTG) Banding**
- The chromosomes are treated with Trypsin
- Stained by Giemsa
- Produce Dark (G+) and Light (G-)
- Visualised under light Microscope
- Most commonly used.

R (Reverse) Banding
- Stained by **Giemsa, Acridine Orange**
- Yield light and dark band which are reverse of G Banding
- Visualized under light microscope
- Used for detecting rearrangements at the terminal end of chromosome.

C Banding
- Centromeric Heterochromatin Banding
- Pretreated with acid followed by alkali
- Stained by Giemsa
- Visualized under light microscope
- Centromeric and Heterochromatic regions are preferentially stained
- To study Chromosomal translocation in the centromere.

Q Banding
- Stained by Quinacrine Mustard
- Visualised under Fluorescence Microscope
- Bands similar to G banding
- Fluorescent Bands are obtained.

Disadvantages of conventional banding techniques
- Deletions smaller than several million base pairs are not routinely detectable by standard G-banding techniques
- Chromosomal abnormalities with indistinct or novel banding patterns can be difficult or impossible to interpret
- To carry out cytogenetic analysis, cells must be dividing, which is not always possible to obtain (e.g. in autopsy or tumour material that has already been fixed).

MOLECULAR CYTOGENETIC TECHNIQUE: FLUORESCENT IN SITU HYBRIDISATION (FISH)

Simple detection of specific genetic information in **a morphologically intact tissue**, cell or chromosome using fluorescent probes.

To metaphase spread of chromosomes on a glass slide fluorescent probe is added.

Advantages of FISH
- FISH permits determination of the number and location of specific DNA sequences in human cells.
- FISH can be performed on metaphase chromosomes, as with G-banding, but can also be performed on cells not actively progressing through mitosis.
- FISH performed on **nondividing cells** is referred to as **interphase or nuclear FISH**.

Disadvantages of FISH
- FISH requires a preselection of an informative molecular probe prior to analysis.
- So a **prior knowledge of the anomaly is needed**.

Uses of FISH[Q]
- Detection of **numeric abnormalities** of chromosomes (aneuploidy)
- The demonstration of **subtle microdeletions**.
- Detection of **complex translocations** not detectable by routine karyotyping
- For analysis of **gene amplification**, e.g. *HER2/NEU* in breast cancer or *N-MYC* amplification in neuroblastomas
- For mapping newly isolated genes[Q] of interest to their chromosomal loci.

Advantages of Interphase FISH
- Growing cells in culture is not needed.
- **More rapid** method than metaphase FISH.
- **Sensitivity is higher** than metaphase FISH.
- Commonly used in prenatal samples, tumours, haematological malignancies, etc.

Chromosome Painting
- It is an **extension of FISH.**
- Probes are prepared for **entire chromosomes**.
- Different chromosomes are identified by different fluorescently labelled probes.
- The number of chromosomes that can be *detected simultaneously* by chromosome painting is limited by the availability of fluorescent dyes.

Multicolor FISH or Spectral Karyotyping (SKY)
- Similar to Chromosome painting in which probe is prepared for entire Chromosome.
- But unlike chromosome painting, a combination of five fluorochromes are used.
- 23 distinct mixtures of **5 fluorophores** to create a unique "colour" for each chromosome.
- By appropriate computer-generated signals, the entire human genome can be visualized.

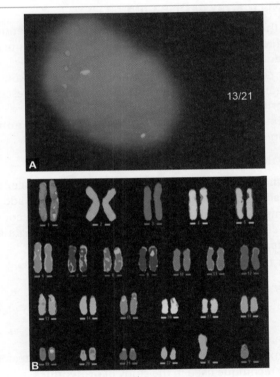

Figs. 11.3.1A and B: *Spectral karyotyping*

ARRAY-BASED METHODOLOGIES (CYTOGENOMICS)

Array-based methods were introduced into the clinical lab beginning in 2003 and quickly revolutionized the field of cytogenetics.

They are
1. CGH array
2. Microarray
3. SNP array

Advantages of Array-based Techniques

- Permit analysis of many regions of the genome in a single analysis
- **High resolution** over standard cytogenetics.

MICROARRAY TECHNIQUE

- Other names are **DNA Chip**
- DNA microarrays contain thousands of known immobilized DNA sequences organized in an area not larger than a microscope slide.
- The fluorescently tagged sample to be sequenced is added.

DNA Microarray or DNA Chip

- To the DNA Chip containing **known oligonucleotide sequence**, fluorescently labelled unknown oligonucleotide is added.
- The computerised algorithms can decode the unknown oligonucleotide, by detecting the location of fluorescent hybridization pattern on the chip.
- This technique is used for genotyping orgenome sequencing.

RNA Microarray

- To **the known array of oligonucleotide, fluorescently labelled cDNA** prepared from unknown mRNA is added
- Computerized algorithms can then rapidly "decode" the cDNA sequence based on fluorescent hybridization pattern on the chip.
- This technique is used for gene expression studies.

Protein Microarray

- Immobilised **known antibodies** placed on the glass slide.
- Fluorescently tagged target protein added.
- By antigen-antibody interaction target protein detected.
- This technique used in the study of **proteomics**.

Uses of Microarrays[Q]

- To analyze a DNA sample for the presence of **gene variations or mutations (genotyping)**
- To determine the patterns of mRNA production (**gene expression analysis**)
- To study Proteins (**Protein Microarray**)
- To **SNP geno typing**.

ARRAY-BASED COMPARATIVE GENOMIC HYBRIDIZATION (ARRAY CGH)

- This is also a hybridisation technique done on a Microarray or DNA Chip, hence is the name array- based.
- Here two Genomes are compared, hence the name Comparative Genomic Hybridisation.

Procedure

In array CGH the test DNA and a reference (normal) DNA are labelled with two different fluorescent dyes (most commonly Cy5 and Cy3, which fluoresce red and green, respectively).

The differentially labelled samples are added to a DNA Chip spotted with entire human genome (usually cover all 22 autosomes and the X chromosome) at regularly spaced intervals.

- If the contributions of both samples are equal for a given chromosomal region
  - Then all spots on the array will fluoresce yellow
  - The result of an equal admixture of green and red dyes.
- If the test sample shows an excess of DNA at any given chromosomal region (such as resulting from an amplification).
  - There will be a corresponding excess of signal from the dye with which this sample was labeled.
- If the test sample show deletion
  - There will be an excess of the signal used for labelling the reference sample (green).

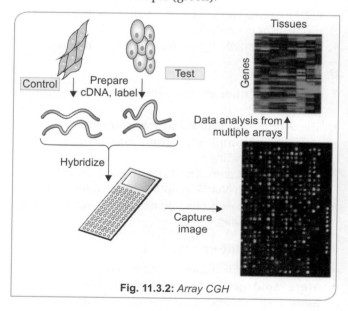

Fig. 11.3.2: *Array CGH*

Uses of Array CGH

1. Detect Gene Amplification

Molecular Biology Techniques

2. Detect Gene Deletion.
3. Detect Copy number variations.

Hence used diseases of unknown etiology like Cancer, Autism, Mental Retardation, Child with dysmorphic features, etc.

Remember: Array CGH cannot detect Balanced Translocations.

SNP Array

SNP platforms use arrays to find out SNPs that a redistributed across the genome. (Please *see* DNA Polymorphisms in Chapter Regulation of Gene expression, to know about SNPs)

Uses

- Used in genome-wide association studies to identify disease susceptibility genes.
- To identify genomic deletions and duplications
- To detect regions of the genome that have an excess of homozygous genotypes and absence of heterozygous genotypes (e.g. CC and TT genotypes only, with no CT genotypes).

Comparison of Different Cytogenetic Techniques

Method	Requires growing cells	Detects deletion and duplication	Detects balanced structural rearrangements	Detects uniparental disomy	Lower limits of detection
G Banding	Yes	Yes	Yes	No	5–10 mb
Metaphase FISH	Yes	Yes	Yes	No	40–250 thousand mb
Interphase FISH	No	Yes	Some	No	40–250 thousand mb
CGH array	No	Yes	No	No	Single Exon or Single gene
SNP array	No	Yes	No	Some	Single Exon or Single gene

Extra Edge

Methods to detect aneuploidy

- Conventional Karyotyping
- FISH
- Microarray based CGH
- Multiple Ligation dependent Probe Amplification (MLPA)
- Real time PCR
- Quantitative Fluorescent PCR (QF-PCR)
- Multiplex amplifiable probe hybridization (MAPH)

Quick Review

- FISH can detect structural abnormalities of chromosome.
- Identification of newly isolated gene to its correct chromosomal location is possible by multicolor FISH.
- Interphase FISH is more rapid.
- DNA microarray or DNA chip technique can detect an unknown oligonucleotide.
- Array CGH can compare two genomes.
- Array CGH can be used to find out genomic abnormalities in diseases wit unknown etiology.
- Array CGH cannot detect balanced translocation

Check List for Revision

- FISH Technique, Microarray and array CGH are must learn topic for central institute exams.

Review Questions MCQ

1. An eight year old child showing small blue round cells consistent with Ewings sarcoma. The best method to confirm translocation on t (11.22) in this case is:
   (AIIMS May 2018)
   a. FISH
   b. Conventional karyotyping
   c. Next generation sequencing
   d. PCR

2. Which method is used to locate a known gene locus?
   (AIIMS May 2013)
   a. FISH
   b. CGH
   c. Chromosome painting
   d. RT-PCR

3. Light microscopy resolution to visualise chromosomes:
   (AIIMS May 2013)
   a. 500 kb
   b. 5 mb
   c. 50 mb
   d. 5 kb

4. Test to differentiate in the chromosome of normal and cancer cell:
   (AIIMS Nov 2012)

a. PCR
b. Comparative genomic hybridisation
c. Western blotting
d. Southern blotting

5. Karyotyping under light microscopy is done by: *(AIIMS Nov 2009)*
   a. R banding
   b. Q banding
   c. G banding
   d. C banding

6. Rapid method of chromosome identification in intersex is: *(AIIMS May 2008)*
   a. FISH
   b. PCR
   c. SSCP
   d. Karyotyping

7. Which of following techniques is used for detection of variation in DNA sequence and Gene expression?
   a. Northern blot *(AI 2010)*
   b. Southern blot
   c. Western blot
   d. Microarray

8. Which of the following tests in not used for detection of specific aneuploidy? *(AI 2010)*
   a. FISH
   b. RT-PCR
   c. QF-PCR
   d. Microarray

9. For isolating a gene of long DNA molecules (50-100 KB) following is used: *(AIIMS Dec 98)*
   a. Chromosome walking
   b. Sanger's sequencing
   c. RFLP
   d. SSLP

Answers to Review Questions

1. **a. FISH** *(Ref: Robbins 9/e page177, 178)*

   FISH is used to study structural abnormalities of the chromosomes

2. **a. FISH** *(Ref: Robbins 9/e page177)*

   Robbins 9th edition gives the following description
   FISH uses DNA probes that recognize sequences specific to particular chromosomal regions.

Method	Requires growing cells	Detects deletion and duplication	Detects balanced structural rearrangements	Detects uniparental disomy	Lower limits of detection
G banding	Yes	Yes	Yes	No	5–10 mb
Metaphase FISH	Yes	Yes	Yes	No	40–250 thousand mb
Interphase FISH	No	Yes	Some	No	40–250 thousand mb
CGH array	No	Yes	No	No	Single Exon or Single gene
SNP array	No	Yes	No	Some	Single Exon or Single gene

3. **b. 5 mb** *(Ref: Emery's Elements of Medical Genetics 13/e page33)*

   G Banding generally provides high quality Chromosome analysis with approximately 400-500 bands per haploid set. Each of these band corresponds on an average to approximately 6000-8000 kilo bases (i.e. 6-8 mb) the rapid analysis of prenatal samples from cells obtained through amniocentesis. Such probes are available for chromosomes 13, 18, and 21 and for the sex pair X and Y.
   Interphase FISH^Q

   Allows rapid diagnosis within 24-48 hours. Especially useful for amniocytes.

4. **b. Comparative genomic hybridisation** *(Ref: Robbins and Cotran Pathologic basis of disease 9/e page178,179)*

   Array-Based Comparative Genomic Hybridization (Array CGH)

   This is also a hybridization technique done on a Microarray or DNA Chip, hence is the name array-based.

   Here two Genomes are compared, hence the name Comparative Genomic Hybridization.

   Uses of array CGH
   - Detect Gene Amplification (e.g. Microduplication)
   - Detect Gene Deletion. (e.g. Subtelomeric deletion, Micro deletion)
   - Detect Copy number variations
   - Detect Aneuploidy
   - Hence used diseases of unknow netiology like Cancer, Autism, Mental Retardation, Child with dysmorphic features, etc.

5. **c. G Banding** *(Ref: Emery's Elements of Medical Genetics 13/e page30,31)*

R-Banding-G -Banding & C Banding are done under light microscopy.

As G Banding is the most common method, it is the correct answer.

6. **a. FISH** *(Ref: Robbins 9/e page177, 178)*

Certain probes used in FISH hybridize to repetitive sequences located to the pericentromeric regions. These probes are useful for the rapid identification of certain trisomies in interphase cells of blood smears, or even in FISH requires prior knowledge of the one or few specific chromosomal regions suspected of being altered.

Genomic abnormalities canal so be detected without prior knowledge of what these aberrations may be, using a global strategy such as array CGH.

From the above description it is concluded that

To locate a known gene locus FISH can be used.

To locate an unknown gene locus, a global strategy like CGH is used.

7. **d. Microarray** *(Ref: Robbins 9/e page177,178)*
   - DNA microarray can be used to detect DNA sequence variations
   - RNA Microarray can detect Gene expression.

8. **b. RT-PCR** *(Ref: Henrys Clinical Diagnosis and Management by Laboratory Methods 22/e 1294,1297)*

Use of a DNA method, QF-PCR, in the prenatal diagnosis of fetal aneuploidies. J ObstetGynaecol Can. 2011 Sep; 33(9):955-60.

- Quantitative Fluorescent PCR is a Real time PCR in which fluorescent dyes are used for quantitation.
- RT-PCR is used to amplify an RNA target. Not a method to detect aneuploidy.

Methods of aneuploidy detection
- Conventional Karyotyping
- FISH
- Microarray based CGH
- Multiple Ligation dependent Probe Amplification (MLPA)
- Real time PCR
- Quantitative Fluorescent PCR (QF-PCR)
- Multiplex amplifiable probe hybridization (MAPH)

9. **a. Chromosome walking** *(Ref: Harper 30/e page463)*

Chromosome Walking

Method to isolate and clone target DNA from a long segment of DNA

In Chromosome walking, a fragment representing one end of along piece of DNA is used to isolate another fragment, that overlaps the first but extends the first. This process continued till the target DNA is isolated.

Cystic fibrosis (CF) gene was the first to be isolated solely by chromosome walking.

11.4 DNA SEQUENCING TECHNIQUES, TRANSGENIC TECHNIQUE AND HYBRIDOMA

- DNA Sequencing Techniques
- Transgenic Technique
- Hybridoma Technique

DNA SEQUENCING TECHNIQUES

1. **Sanger's Technique (Controlled Chain termination Method)**
   - Chain terminators used **2'3'Dideoxynucleotides**
   - DNA Polymerase used is **Klenow Polymerase**
   - This can be automated

Principle:
   - Introduction of 2'3' dideoxynucleotides will terminate the DNA synthesis because no free 3'OH group for the formation of next phosphodiester bond.

2. **Maxam and Gilbert Chemical Cleavage Method**
3. **Pyrosequencing (Discussed along with Mutation detection techniques)**
4. **Next Generation Sequencing (NGS).**

Term used to describe several newer DNA sequencing technologies that are capable of producing large amounts of sequence data in a **massively parallel manner**.

Comparison of Sanger's Sequencing and Next Generation Sequencing

Sanger's Sequencing	New Generation Sequencing
High cost	Low cost
Single, simple homogeneous template DNA	No such requirement, sample from almost any source can be used
Provides an "average" result for a DNA sample	Well suited for heterogeneous DNA Sample
Uninterpretable results if sample of DNA is heterogeneous	

Three Basic Processes of NGS

Spatial separation
At the beginning of the procedure, individual input DNA molecules are physically isolated from each other in space. The specifics of this process are platform-dependent.

Local amplification
After separation, the individual DNA molecules are amplified in situ using a limited number of PCR cycles.

Parallel sequencing
The amplified DNA molecules are simultaneously sequenced by the addition of polymerases and other reagents, with each spatially separated and amplified original molecule yielding a "read" corresponding to its sequence.

> **Extra Edge**
>
> **Some new sequencing techniques CAGE (Cap analysis of gene expression)**
> A method that allows the selective capture, amplification, cloning, and sequencing of mRNAs via the 5'Cap structure RNA-Seq
> mRNAs are converted to cDNAs using reverse transcription, and these cDNAs are amplified and directly sequenced. This method is termed **RNA-Seq**.
> **GRO-Seq (Global Run-on Sequencing)**
> A method where nascent transcripts are specifically captured and sequenced using NGS sequencing.
> This helps to map location of active transcription complexes.
> **NET-seq (native elongating transcript sequencing)**
> This technique allows for sequencing of RNA within elongating RNA polymerase-DNA-RNA ternary complexes. This helps in genome-wide analysis of transcription in living cells.
> **WES (Whole Exome Sequencing)**
> Since exons comprise only about 1% of the human genome, the exome represents a much smaller and more tractable target than the complete genome.
> Whole Exome sequencing has emerged as an alternative to whole genome sequencing as a means for diagnosing rare or cryptic genetic diseases.

TRANSGENIC ANIMALS

Foreign genes can be introduced into fertilized egg. Animals that develop from such fertilized egg is called transgenic animal.

- The gene of interest is a cloned recombinant DNA with its own promoter and a different promoter which can be selectively regulated.

Transgenic Models of Animals

Several organisms have been studied extensively as genetic models.

Musculus (mouse), Drosophila melanogaster (fruit fly), Caenorhabditis elegans (nematode), Saccharomyces cerevisiae (Baker's yeast), and **Escherichia coli** (colonic bacterium).

Different Strategies of Genetic Modification

1. Injection of the transgene into the male pronucleus of fertilised ovum.
2. By homologous recombination in embryonic stem cell. This is called targeted mutagenesis.
3. Forward genetics
4. Animal cloning.

Direct Injection of Transgene

Features

- Pronuclear injection of transgene
- This technique is commonly used
- **Random integration** of transgene.
- Genomic DNA or cDNA constructs can be injected.
- Variable copy numbers of transgene.
- Variable expression in each individual founder.
- Gain-of-function models produced due to over expression using tissue-specific promoters
- Loss-of-function models produced using antisense and dominant negative transgenes.
- **Inducible expression** possible.
- Applicable to **several species**.

Targeted Mutagenesis

Basic Principle of Targeted Mutagenesis

The normal gene in a very low percentage of ES cells may be replaced by the neo-disrupted recombinant gene (transgene) by *homologous recombination.*

Targeted Mutagenesis can be of different types

Gene Knock out: Endogenous gene is replaced by mutated transgene by homologous recombination in embryonic stem cell.

Gene Knock in: Mutated endogenous gene is replaced by normal transgene by homologous recombination in embryonic stem cell.

Gene Knock down: siRNA or miRNA induced gene silencing called RNA interference or RNAi.

Gene Knock up: Using transcription factors, transcription of gene is increased.

Steps for Generation of a Knock out Mouse

1. Inactivating a recombinant purified gene.
2. Culture embryonic stem (ES) cell from mice.
3. Transfection of ES with cloned mutated non-functional gene.
4. A few cultured embryonic cells contain the non-functional gene through homologous recombination.

5. Isolate ES cells with altered gene.
6. Microinjection of altered ES cells into mouse blastocyst.
7. Implantation of blastocyst into foster mother.
8. **Breed to many generation** till offspring with altered gene in its germ cell.

Features of targeted mutagenesis by homologous recombination
- Site specific integration of transgene.
- Predominantly used in mice.
- In gene knock out, tissue-specific knock-out possible.
- But gene knock in can accurately model human disease.

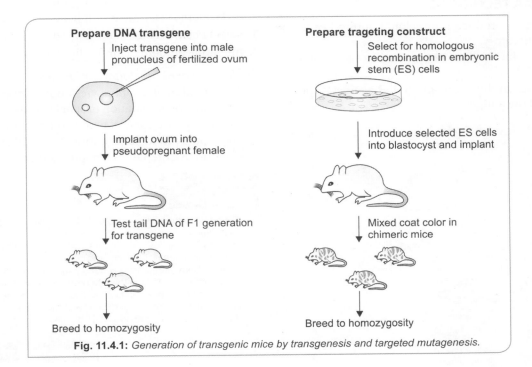

Fig. 11.4.1: *Generation of transgenic mice by transgenesis and targeted mutagenesis.*

> **EXTRA EDGE**
>
> Latest technique of gene modulation in targeted mutagenesis are
> - miRNA/SiRNA mediated Gene Silencing called as Gene Knock down.
> - CRISPR Cas 9 mediated gene editing (loss of function of gene otherwise knock down or gain of function of genes)

Forward Genetics
- Mutations created randomly by ENU (*N*-ethyl-*N*-nitrourea)
- A phenotype selected.
- Then its genetic characterisation done.
- Useful for identifying novel genes.

Animal Cloning

Nucleus of an oocyte is removed. This enucleated oocyte is fused nucleus of a somatic cell. The fused cell is implanted into the uterus of a surrogate mother. This also called Nuclear transfer.

Characteristics of Animal Cloning
- Successful in several mammalian species including sheep, mice, cows, monkeys.
- Cloning of genetically identical individuals is possible.
- May affect lifespan.
- Lots of Ethical concerns.

Successful Stories of Animal Cloning
- Ian Wilmut and Keith Campbell cloned sheep named Dolly in 1996.
- First lamb born to Dolly is Bonnie.

Uses of Transgenic Animals
- Study of DNA regulatory elements of a gene.
- Study of functions of gene (role of oncogene in induction of tumorigenesis).
- Study of disorders susceptible for Gene dosage (over expression or under expression).

- Used as a precursor to gene therapy.
- For studying the physiologic effects of insertion or deletion of a particular gene.
- Providing unique genetic models for doing animal experiments in pathology and Pharmacology.

HYBRIDOMA TECHNIQUE

- Technique to produce monoclonal antibodies in the clinical laboratory.
- Introduced by Georges Kohler and Cesar Milstein in 1975.
- Monoclonal antibody are antibodies against a specific epitope of the antigen.

Steps of Hybridoma Technique

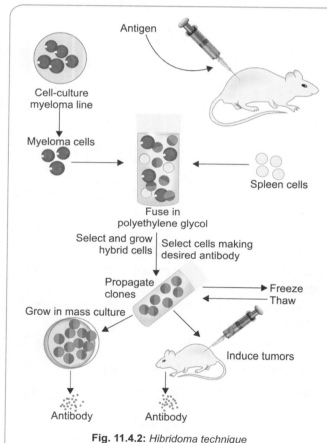

Fig. 11.4.2: *Hibridoma technique*

1. Spleen cells from immunised animal fused with mice myeloma cell to produce hybrid cell. (Poly Ethylene Glycol^Q, PEG -1500 is the fusing agent).
2. Grown in HAT (Hypoxanthine, Aminopterineand Thymidine) medium.
    a. Unfused normal cells die as they lack multiplication potential
    b. Unfused myeloma cells die as Aminopterine, a folate antagonist inhibit de novo purine synthesis and hence DNA synthesis as they lack HGPRTase enzyme
    c. Fused normal cell and myeloma cell survive as normal cell provide HGPRTase enzyme
    d. Myeloma cell have multiplication potential.
3. Select Positive clones and expand.
4. Indefinite amount of antibodies harvested.

Application of Monoclonal Antibodies

- Enumeration of Lymphocyte subpopulation
- Quantitative preparation of specific cells.
- Nephelometric assay of blood components.
- To prepare Monoclonal antibodies for ELISA.
- Quantitative preparation of pure antigens.

Quick Review

- Sanger's sequencing is also called controlled chain termination technique.
- Dideoxynucleotides are used for chain termination,
- NGS is next generation sequencing
- RNAi or RNA interference is gene knock down technique.
- Hybridoma technique is used to produce monoclonal antibody
- HAT medium is Hypoxanthine, Aminopterine, Thymidine medium.
- In HAT medium fused normal cell and myeloma cell survive as the normal cell can provide HGPRtase enzyme.

Check List for Revision

- This is a chapter important for central institute exams.
- For NBE model exams an overall idea is enough

REVIEW QUESTIONS MCQ

1. **Klenow fragment lack the activity of:**
   (AIIMS May 2018)
   a. 3'-5' exonuclease
   b. 5'-3' exonuclease
   c. DNA polymerase 5'-3'
   d. All the above

2. **Correct statement about Restriction fragment gene:**
   (PGI May 2011)
   a. Detected by Southern blot
   b. Detected by Northern blot
   c. Used for identification of gene for genomic mapping
   d. RFLP is DNA variation sequence

3. **DNA finger printing was founded by:** (PGI June 06)
   a. Watson
   b. Galton
   c. Jeffrey
   d. Sanger
   e. Wilkins

4. The following methods can be used to detect the point mutation in the beta globulin gene that causes sickle cell anemia, except: *(AI 06)*
   a. Polymerase chain reaction with allele-specific oligonucleotide hybridization
   b. Southern blot analysis
   c. DNA sequencing
   d. Northern blot analysis

5. DNA fingerprinting is based on possessing in DNA of: *(PGI Dec 08)*
   a. Constant tandem repeat
   b. Variable number tandem repeat
   c. Non-repetitive sequence
   d. Exon
   e. Intron in eukaryotes

6. RFLP, true is/are: *(PGI Dec 06)*
   a. Detects mutation
   b. Recognizes trinucleotide repeats
   c. Detects deletion
   d. Blunt ends are produced
   e. Always short ends are produced

7. Silver staining is used for: *(PGI May 2016)*
   a. DNA
   b. RNA
   c. Protein
   d. Karyotype analysis
   e. Collagen

Transgenic Animals

8. RNAi in gene expression denotes: *(AIIMS May 2015)*
   a. Knock down
   b. Knock up
   c. Knock in
   d. Knock out

9. True statement about transgenic mice: *(PGI Nov 2010)*
   a. Developed from DNA insertion into fertilized egg
   b. Have same genome as parents except one or more genes
   c. Identical genome to parent mice
   d. Produced by breeding over several generations.
   e. Homozygous are selected

10. The function of a gene is determined by: *(AI 2012)*
    a. Southern blot
    b. Western blot
    c. Inserting in transgenic mice
    d. Inserting as a knock out gene

Answers to Review Questions

1. **b. 5' -3' exonuclease activity**
   - In DNA Sequencing Klenow polymerase used. It is DNAP I from which 5' -3' exonuclease activity is removed.
   - Kornberg's enzyme is DNA Polymerase I which is first discovered by Arthur Kornberg.

2. **a, c, d. Detected by Southernblot, Used for identification of gene for genomic mapping, RFLP is DNA variation sequence**
   (Ref: Lippincott Illustrated Biochemistry 6/e page475)

   The procedure of RFLP
   - DNA is extracted from the cell
   - DNA is cleaved by a specific Restriction endonuclease
   - DNA fragments obtained are separated by Agarose Gel Electrophoresis
   - DNA fragments are denatured
   - Transferred to Nitrocellulose membrane (Southern blot)
   - Treated with Radiolabelled probe.
   - DNA variation detected by looking for hybridisation by autoradiography.

3. **c. Jeffrey**

4. **d. Northern blot analysis**

Methods to detect point mutation

Method	Type of Mutation Detected
Southern blot	Large deletion, insertion, rearrangement, expansions of triplet repeat, amplification
DNA sequencing	Point mutations, small deletions and insertions
Restriction fragment polymorphism (RFLP)	Point mutations, small deletions and insertions
Single-strand conformational polymorphism (SSCP)	Point mutations, small deletions and insertions
Denaturing gradient gel electrophoresis (DGGE)	Point mutations, small deletions and insertions
RNAse cleavage	Point mutations, small deletions and insertions
Oligonucleotide specific hybridization (OSH)	Point mutations, small deletions and insertions
Microarrays	Point mutations, small deletions and insertions Genotyping of SNPs

5. **b. Variable number tandem repeat**

   DNA Finger printing
   The use of normal genetic variation in the DNA (SNP or VNTR or RFLP) to establish a unique pattern of DNA fragments for an individual.
   This is also called DNA Profiling.

Most commonly used is VNTR or repeat length polymorphism. The process of DNA finger printing was invented by Alec Jeffreys in 1985.

Primer is needed

6. **a, c, d. Detects mutation, Detects deletion, Blunt ends are produced** *(Ref: Harper 30/e page463)*

   Two type of DNA Variations that result in RFLP are
   1. Single Nucleotide Polymorphism (SNP)
   2. Variable Number Tandem Repeat (VNTR) Uses of RFLP
   - Tracing chromosomes from parent to offspring.
   - Prenatal diagnosis of diseases.
     - Direct diagnosis of sickle cell disease using RFLP.
     - Indirect, prenatal diagnosis of phenyl ketonuria
   - Medicolegal uses.
   - Detect mutations, point mutation, insertion, deletion.

7. **a, b, c, d, e. DNA, RNA, Protein, Karyotypeanalysis, Collagen**

   Silver staining is originally used for Protein, but can be used for nucleic acid, Karyotype analysis and Collagen.

Transgenic Animals

8. **a. Knock down** *(Ref: Harper 30/e page447)*

   Gene Knock out: Endogenous gene is replaced by mutated transgene by homologous recombination in embryonic stem cell.
   Gene Knock in: Mutated endogenous gene is replaced by normal transgene by homologous recombination in embryonic stem cell.
   Gene Knock down: siRNA or miRNA induced gene silencing called RNA interference or RNAi

   Gene Knock up: Using transcription factors, transcription of gene is increased.

9. **a, b, d, e. Developed from DNA insertion into fertilized egg, Have same genome as parents except one or more genes, Produced by breeding over several generations, Homozygous are selected**
   (Ref: Text book of biochemistry Thomas M Devlin 7/e page292, 293)

   Different Strategies of Genetic Modification
   Injection of the transgene into the **male pronucleus of fertilised ovum**.

   Steps for generation of a Knock out Mouse
   - Inactivating a recombinant purified gene.
   - Culture embryonic stem (ES) cell from mice.
   - Transfection of ES with cloned mutated non-functional gene.
   - A few cultured embryonic cells contain the non-functional gene through homologous recombination.
   - Isolate ES cells with altered gene.
   - Microinjection of altered ES cellsinto mouse blast ocyst.
   - Implantation of blastocyst into foster mother.
   - **Breed to many generation till offspring with altered gene in its germ cell.**
   - Homozygous are selected.

10. **d > c. Inserting as a knockout gene > Inserting in transgenic mice.**
    (Ref: Text book of biochemistry Thomas M Devlin 7/e page292, 293)

    Inserting in transgenic mice site specific integration is not possible.
    Knockout transgenesis is by homologous recombination.
    So site specific integration possible. So d > c.

11.5 OTHER MOLECULAR BIOLOGY TECHNIQUES AND RECENT ADVANCES

- Antisense Oligonucleotide Technique
- RFLP
- DNA Fingerprinting
- DNA Footprinting
- Chromosome Walking
- DNA Electrophoresis
- Gene Expression Analysis
- Stem Cell Biology
- Gene Therapy
- Human Genome Project
- Linkage and Association Studies

ANTISENSE NUCLEIC ACID TECHNIQUE

Antisense nucleic acid binds with target mRNA, it is selectively destructed or inhibited.

Antisense Nucleic Acids

- Antisense nucleic acid can be RNA or DNA
- It can be natural or synthetic.
- It is complementary to mRNA.
- SiRNA can be used.

Uses

- To study function of a gene.
- Treat viral infection like HIV.
- Antisense oligonucleotide therapies are used in treatment of Cancers, Hyperlipoproteinemias, etc.

RESTRICTION FRAGMENT LENGTH POLYMORPHISM (RFLP)

- Commonly called as 'riflip'

Definition

An inherited difference in the pattern of restriction map produced by the digestion of a specific restriction endonuclease is called Restriction fragment length polymorphism.

This is because of certain DNA variations present in the genome that create a new restriction site (the site where the restriction endonuclease cleave) or abolish a restriction site.

Two Type of DNA Variations that Result in RFLP are

1. Single Nucleotide Polymorphism (SNP)
2. Variable Number Tandem Repeat (VNTR)

Certain Mutation can also results in RFLP if it alter the Restriction Site

Sickle cell anemia is an example of a genetic disease caused by a point mutation. The sequence altered by the mutation abolishes the recognition site of the restriction endonuclease MstII.

Procedure of RFLP

- DNA is extracted from the cell.
- DNA is cleaved by a specific Restriction endonuclease.
- DNA fragments obtained are separated by Agarose Gel Electrophoresis
- DNA fragments are denatured
- Transferred to Nitrocellulose membrane (Southern blot)
- Treated with Radiolabelled probe.
- DNA variation detected by looking for hybridisation by autoradiography.

Uses of RFLP

Tracing chromosomes from parent to offspring.
Prenatal diagnosis of diseases.
- Direct diagnosis of sickle cell disease using RFLP.
- Indirect, prenatal diagnosis of phenylketonuria Medicolegal uses.

Detect mutations, point mutation, insertion, deletion.

DNA FINGERPRINTING

- The use of normal genetic variation in the DNA (SNP or VNTR or RFLP) to establish a unique pattern of DNA fragments for an individual.
- This is also called **DNA Profiling**.
- Most commonly used is **VNTR or repeat length polymorphism**.
- The process of DNA finger printing was invented **by Alec Jeffreys in 1985**.

DNA FOOTPRINTING

- DNA with **protein bound is resistant to digestion by DNase enzymes**.
- When a sequencing reaction is performed using such DNA, a protected area, representing the "footprint" of the bound protein, will be detected.
- Because nucleases are unable to cleave the DNA directly bound by the protein.

CHROMOSOME WALKING

- Method to isolate and clone target DNA from a long segment of DNA
- In Chromosome walking, a fragment representing one end of a long piece of DNA is used to isolate another fragment, that overlaps the first but extends the first.
- This process continued till the target DNA is isolated.
- Cystic fibrosis (CF) gene was the first to be isolated solely by chromosome walking.

Technique of Chromosome Walking

- Gene X to be isolated from a long of DNA
- The exact location of Gene X is also not known.
- A probe directed against a fragment of DNA in the 5' end is available.
- The initial probe hybridise fragment 1.
- This is used to detect a probe to detect fragment 2.
- This process repeated until the fragment that contain Gene X is reached.

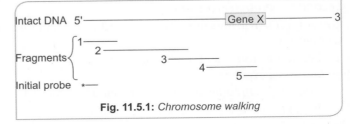

Fig. 11.5.1: *Chromosome walking*

Chromosome Jumping

- By pass regions difficult to clone, such as those containing repetitive DNA, that cannot be easily mapped by chromosome walking.

DNA ELECTROPHORESIS

In an electric field DNA migrate towards positive electrode. Rate of migration depends on the size of the DNA. Because each nucleotide has one negative charge, the charge to mass ratio of molecules with different sizes remains constant. So DNA fragments migrate at speeds inversely related to their size.

Gel Electrophoresis

1. Agarose gel electrophoresis
2. Pulsed field Gel electrophoresis: Very large (mega base pair) DNA molecules, pulses of current applied to the gel enhance migration.
3. Polyacrylamide Gel Electrophoresis (PAGE)-Very small fragments and single stranded DNA are best resolved in PAGE. Main advantage of PAGE is high resolution for small fragment.

METHODS OF GENE EXPRESSION ANALYSIS

Analysing the mRNA and Protein products is Gene expression analysis.

Methods to Analyse mRNA
- Northern blot
- cDNA Microarray

Methods to Analyse Proteins
- ELISA
- Western blot
- Study of Proteomics.

Techniques to Study Proteomes

First Generation Proteomics
- To purify proteins sample-SDS PAGE or 2D Electrophoresis
- To determine the amino acid sequence: End group analysis like Sanger's reagent and Edman's reagent.

Second Generation Proteomics

To purify protein in the sample Nanoscale Chromatographic techniques

To determine amino acid sequence:
- Mass Spectrometry
- MudPIT (Multidimensional Protein identification Technology).

> **Extra Edge**
> **MudPIT**
> Successive rounds of chromatography to resolve the peptides produced from complex biologic sample into simpler fractions, that can be analysed separately by Mass Spectrometry.

EXTRA EDGE: RECENT ADVANCES STEM CELL BIOLOGY

Stem Cell

Definition

The most widely accepted stem cell definition is a cell with a unique capacity to produce unaltered daughter cells (Self Renewal) and to generate specialized cell types (potency).

Totipotent cells
- Can form an entire organism autonomously.
- Only a fertilized egg (zygote) possesses this feature.

Pluripotent cells
- Can form almost all of the body's cell lineages (endoderm, mesoderm, and ectoderm), including germ cells, e.g. Embryonic Stem (ES) cells.

Multipotent cells
- Can form multiple cell line a gesbutcannot form all of the body's cell lineages, e.g. Hematopoietic Stem (HS) cells.

Oligopotent cells
- Can form more than one cell lineage but are more restricted than multipotent cells. Oligopotent cells are sometimes called progenitor cells or precurs or cells, e.g. Neural Stem cells.

Unipotent cells or monopotent cells
- Can form a single differentiated cell lineage, e.g. spermatogonial stem (SS) cells.

Nuclear reprogramming
- The reversal of the terminally differentiated cells to totipotent or pluripotent cells.

Trans differentiation
- Lineage-committed multipotent cells, possessing the capacity to differentiate into cell types outside their lineage restrictions.

Some Important Stem Cells

Unrestricted somatic stem cells (USSC)^a	Mononuclear fraction of **cord blood**	USSCs can differentiate into a variety of cell types in vitro and can contribute a variety of cells types in in vivo transplantation experiments
Induced pluripotent stem cells (iPS, iPSC)	Variety of terminally differentiated cells and tissue stem cells.	A number of somatic cell types can be converted into iPS cells using different combinations of transcription factors and treatment with small molecule.

GENE THERAPY

Novel area of therapeutics

Intracellular delivery of genes to generate a therapeutic effect by correcting an existing abnormality

Divided into

1. Somatic cell gene therapy-Gene is introduced into somatic cells.
2. Germ cell gene therapy (Transgenic animal)

Methods of Gene Delivery^Q

Chemical methods of gene delivery.

DEAE-dextran

- Diethylaminoethyl-dextran (DEAE Dextran) is a poly cationic derivative of the carbohydrate polymer, dextran.
- Because of its positive charge, DEAE-dextran is able to bind to the anionic phosphodiester back bone of DNA.
- The resultant complex maintains an overall cationic charge and is able to bind to negatively charged cell membrane surfaces.
- Subsequently, the complex is internalized, presumably by endocytosis.

Calcium Phosphate[Q]

- Mixing DNA with calcium chloride, and then carefully adding this mixture to a phosphate buffered saline solution followed by incubation at room temperature.
- This generates DNA-containing precipitate, which is then dispersed onto cultured cells
- The precipitate is then taken into the cells via endocytosis or phagocytosis.

Cationic Lipids (Lipofection)

When lipids mixed with DNA in water, the lipids formed hollow spheres, called liposomes, with DNA entrapped in the aqueous center.

When these liposomes were added to cells growing in vitro, some of the liposomes would fuse with cellular plasma membranes and β-catenin to the cells via endocytosis.

Physical Methods of Gene Delivery

Microinjection

It entails the direct injection of DNA into the nuclei of target cells using fine glass needles under microscopy.

Electroporation[Q]

Electroporation is a method of introducing nucleic acids into cells by exposing the cells to a rapid pulse of high-voltage current, causing pores in the cell membrane to open temporarily.

Gene Gun[Q]

Plasmid DNA is coated onto metal microparticles and then blasted into cells using either electrostatic force or gas pressure.

HUMAN GENOME PROJECT

In 1990, the United States launched a multibillion Dollar effort, the **Human Genome Project**, for the express purpose of developing the automated **high-throughput** techniques, instrumentation, and data mining software necessary to determine the entire DNA sequence of the *Homo sapiens* genome.

Bioinformatics

The discipline concerned with the collection, storage, and analysis of biologic data, mainly DNA and protein sequences.

> **EXTRA EDGE**
>
> **"OMICS of Molecolar Biology**
> **Genome:** The complete set of genes of an organism
> **Genomics:** In depth study of the structures and functions of genomes.
> **Transcriptome:** The complete set of RNA transcripts produced by the genome during a fixed period of time.
> **Transcriptomics:** The comprehensive study of transcriptome.
> **Proteome:** The complete compliment of proteins of an organism.
> **Proteomics:** The systematic study of structures and functions of proteomes and their variation in health and disease.
> **Glycome and Glycomics**
> The glycome is the total complement of simple and complex carbohydrates in an organism. Glycomics is the systematic study of the structures and functions of glycomes such as the human glycome.
> **Lipidome and Lipdomics**
> The lipidome is the complete complement of lipids found in an organism. Lipidomics is the in-depth study of the structures and functions of all members of the lipidome and of their interactions, in both health and disease.
> **Metabolome and Metabolomics**
> The metabolome is the complete complement of metabolites (small molecules involved in metabolism) present in an organism. Metabolomics is the in-depth study of their structures, functions, and changes in various metabolic states.
> **Nutrigenomics**
> The systematic study of the effects of nutrients on genetic expression and of the effects of genetic variations on the metabolism of nutrients.
> **Pharmacogenomics**
> The use of genomic information and technologies to optimize the discovery and development of new drugs and drug targets.
> **Exome:** The nucleotide sequence of the entire complement of mRNA exons expressed in a particular cell, tissue, organ or organism.

Bioinformatic and Genomic Resources

The large collection of databases that helps in contemporary molecular, biochemical, epidemiological, and clinical research.

UniProt KB

The UniProtKnowleldgebase, UniProtKB, is jointly sponsored by the Swiss Institute of Bioinformatics and the European Bioinformatics Institute.

UniProtKB's stated objective is "to provide the scientific community with a comprehensive, high-quality and freely accessible resource of protein sequence and structural information".

It is organized into two sections.
- Swiss-Prot contains entries whose assigned functions, domain structure, post-translational modifications, etc. have been verified by manual curation
- TrEMBL, on the other hand, contains empirically determined and genome derived protein sequences whose potential functions have been assigned, or annotated, automatically—solely on the basis of computer algorithms.

Thus, while TrEMBL currently includes more than 80 million entries, Swiss-Prot contains slightly more than 500,000.

GenBank
- The genetic sequence database
- The goal of GenBank, of the National Institutes of Health (NIH), is to collect and store all known biological nucleotide sequences and their translations in a searchable form.

PDB
- The RCSB Protein Data Base (PDB) is a repository of the three dimensional structures of proteins, polynucleotides, and other biological macromolecules.
- The PDB presently contains over 95,000 three-dimensional structures for proteins.

Tagged SNPs
- When sets of SNPs localized to the same chromosome are inherited together in blocks, the pattern of SNPs in each block is termed a haplotype.
- TagSNPs, is a subset of the SNPs in a given block sufficient to provide a unique marker for a given haplotype.
- Selected regions are then subject to more detailed study to identify the specific genetic variations that contribute to a specific disease or physiologic response.

HapMap
- **Haplotype Map (HapMap) Project**, a comprehensive effort to identify SNPs associated with common human diseases and differential responses to pharmaceuticals.
- The long-term goal of the project is to provide earlier and more accurate diagnosis of potential genetic risk factors that leads to improved prevention and more effective patient management.

ENCODE
- National Human Genome Research Institute (NHGRI) initiated the **ENCODE (Encyclopedia of DNA Elements) Project.**
- ENCODE is a collaborative effort that combines laboratory and computational approaches to identify every functional element in the human genome.
- These include mapping sites of DNA methylation, assessing local histone methylation, etc.

Entrez Gene
Entrez Gene, a data base maintained by the National Center for Biotechnology Information (NCBI)
- This provides a variety of information about individual human genes.
- The information includes the sequence of the genome in and around the gene, exon-intron boundaries, the sequence of the mRNA (s) produced from the gene, and any known phenotypes associated with a given mutation of the gene in question.

> **EXTRA EDGE**
> **dbGAP**
> dbGAP, the Database of Genotype and Phenotype, is an NCBI database that complements Entrez Gene.
> dbGAP compiles the results of research into the links between specific genotypes and phenotypes.

LINKAGE AND ASSOCIATION STUDIES

Principle of Linkage Analysis and Association Studies

Based on the concept of linkage
- When two genes are close together on a chromosome, they are transmitted together unless a recombination event separates them.
- Genes which are close together are less likely to be separated by a recombination event.

Primary Strategies for Mapping Genes that Cause or Increase Susceptibility To Human Disease

1. Classic linkage analysis can be performed based on a known genetic model or, when the model is unknown, by studying pairs of affected relatives
2. Disease genes can be mapped using allelic association studies (Genome Wide Association Studies, GWAS)

Linkage Analysis
- The genetic marker loci very close to the disease allele, that are transmitted through pedigrees called linkage disequilibrium.
- Assess the genetic marker loci in family members having the disease or trait of interest
- With time it becomes possible to define a panel of marker loci, all of which co-segregate with the putative diseases

- Hence Linkage analysis facilitates localization and cloning of the disease allele.
- The genetic markers used are, SNPs and repeat-length polymorphisms (minisatellite and microsatellite repeats).

GWAS

- In GWAS large cohorts of patients with and without a disease (rather than families) are examined across the entire genome for genetic variants or polymorphisms that are over-represented in patients with the disease
- This identifies regions of the genome that contain a variant gene or genes that confer disease susceptibility.
- The causal variant within the region is then provisionally identified using a "candidate gene" approach.
- By this approach, in the provisionally identified region, genes are selected based on how tightly they are associated with the disease and whether their biologic function seems likely to be involved in the disease under study.
- Thus localize the causal gene or in some instances, functional polymorphism associated with it.

Comparison of Linkage and Association Studies

Linkage study	Association study
Study conducted in the same family or same sibships	Compare a population of affected individuals with a control population
Analysis of monogenic traits	Suitable for identification of susceptibility genes in polygenic and multifactorial disorders
Difficult to obtain sufficient statistical power for complex traits	More Statistical Power for Complex multigenic disorder

Check List for Revision

- This chapter is an extra edge chapter.
- All topics given as extra edge are optional.
- For central institute exams like PGI, these topics are important.

Review Questions

1. **Gene editing can be done by various methods like hypermethylation and amplification. Which of the following will not change the genetic code?**
   (AIIMS May 2017)
   a. Epigenetics b. CRISPR
   c. GenXpert   d. TALEN

2. **RFLP used in surgical ICU to identify Staph aureus. The restriction site of the restriction endonuclease HIND III can be:** *(AIIMS May 2017)*
   a. AAGAAG
      TTAGGT
   b. AAGAGA
      GAAGCA
   c. AAGCTT
      TTCGAA
   d. AAGGAA
      CCTTGA

3. **CRISPR is:** *(Central institute exam May 2017)*
   a. It is a type of bacterial defense mechanism in bacteria against phages/viruses
   b. It is a type of bacterial defense mechanism in virus against bacteria
   c. It is an anticaspase used against bacteriophages in humans

4. **Best assessment of a protein binding regions on a DNA molecule can be done by:** *(PGI Nov 2016)*
   a. DNA footprinting   b. RTPCR
   c. Microarray   d. Western blotting
   e. Northern blotting

Gene Therapy

5. **Natural methods of horizontal gene transfer in bacteria:** *(PGI Nov 2009)*
   a. Transformation   b. Transduction
   c. Conjugation   d. Electroporation
   e. Mutation

6. **Methods of introducing gene in target cells are all except:** *(AIIMS Nov 2010)*
   a. Electroporation   b. Transfection
   c. Site directed recombination
   d. FISH

7. **The first gene therapy (somatic enzyme) was successfully done in:** *(PGI Dec 07)*
   a. SCID
   b. Phenylketonuria   c. Thalassemia
   d. Cystic fibrosis   e. Alkaptonuria

8. **Gene therapy methods are:** *(PGI June 03)*
   a. Electroporation   b. Intranuclear injection
   c. Site directed mutagenesis
   d. Retrovirus

9. **Purpose of gene therapy:** *(PGI June 03)*
   a. Replacement of abnormal gene by normal gene
   b. Replacement of normal gene by abnormal gene
   c. Knock out of abnormal gene
   d. Introduction of viral gene

Bioinformatics

10. The following are used to study pathological genome except:
    a. GenBank
    b. Entrez gene
    c. HapMap
    d. BLAST

11. Study of structure and products of gene is: *(PGI Dec 05)*
    a. Genomics
    b. Proteomics
    c. Bioinformatics
    d. Cytogenetics
    e. Pharmacogenomics

12. Study of multiplication of proteins in disease process is called: *(AI 07)*
    a. Proteomics
    b. Genomics
    c. Glycomics
    d. Nucleomics

13. What biologist uses to diagnose and treat diseases with disorders with multigenic inheritance?
    a. Gene card
    b. Tag SNPs
    c. Flipped card
    d. Virtual Cell

14. Which of the following statement is true about Linkage analysis? *(AIIMS Nov 07)*
    a. Detection of characteristic DNA polymorphism in a family associated with disorders
    b. Useful to make pedigree chart to show affected and non-affected family members
    c. Used to make a pedigree chart to show non-paternity
    d. Non gene mapping method of genetic study

Answers to Review Questions

1. **a. Epigenetics**

   Gene addition therapy
   Definition: Normal gene transferred to a target tissue to drive expression of a gene product with therapeutic effects.

   Genome editing
   Definition: Mutation is corrected in situ, generating a wild type copy under the control of the endogenous regulatory signals.
   - CRISPR, TALEN, Zinc finger nucleases are gene editing techniques hence alter genetic code as it edit site of mutation.

   Gene knock down
   siRNA or short hairpin RNA used to knock down the expression of deleterious gene.
   Eg: Huntingtin gene in Huntington's disease, Hepatitis C genome

   Other options
   - GeneXpert is Nucleic Acid Amplification technique used to detect bacteria like *Mycobacterium tuberculosis*.
   - Epigenetics only chemically modify DNA, does not alter nucleotide sequence.

2. **c. AAGCTT TTCGAA**

   Among the given options only c is the pallindromic sequence.

Endonuclease	Sequence recognized cleavage sites shown	Bacterial source
BamHI	↓ GGATCC CCTACC ↑	*Bacillus amyloliquefaciens* H
BgIII	↓ AGATCT TCTAGA ↑	*Bacillus globigii*
EcoRI	↓ GAATTC CTTAAC ↑	*Escherichia coli* RY13
EcoRII	↓ CCTGG GGACC ↑	*Escherichia coli* R245
HindIII	↓ AAGCTT TTCGAA ↑	*Haemophilus influenzae* Rd

3. **a.** It is type of bacterial defense mechanism in bacteria against phages/viruses

4. **a, c. DNA footprinting, Microarray**
   a. Protein bound regions are assessed by DNA footprinting and Chromatin immunoprecipitation (ChIP)
   b. The protein bound DNA sequence identified are then sequenced by two methods
      1. Direct sequencing called ChIP in Seq
      2. Microarray called Chip on ChIP

Gene Therapy

5. **a, b, c. Transformation, Transduction, Conjugation**
   - Electroporation is an artificial method of gene delivery
   - Trinucleotide..., X chromosome...

6. **d. FISH** *(Ref: lib.store.yahoo.net/.../How-to-Choose-the-Optimal-Gene-Delivery-Method.pdf)*

   Electroporation
   Electroporation is a method of introducing nucleic acids into cells by exposing the cells to a rapid pulse of high-voltage current, causing poresin the cell membrane to open temporarily.

   This allows exogenous DNA to pass through the pores and into the cytoplasm of the cells.

   Typically, the gene transfer efficiency is relatively low, and electroporation frequently results in a high incidence of cell death.

Transfection: Introduction of gene through nonviral vectors.
Transduction: Introduction of gene through viral vectors. Site directed recombination: The method by which endogenous gene is replaced by passenger gene by homologous recombination.
Fluorescent In Situ Hybridisation (FISH) –Not a method of gene delivery.

7. **a. SCID**

Dr French Anderson did the first gene therapy to treat SCID.

8. **a, b, c, d. Electroporation, Intranuclear injection, Site directed mutagenesis, Retrovirus**

9. **a. Replacement of abnormal gene by normal gene**

Novel area of therapeutics
Intracellular delivery of genes to generate a therapeutic effect by correcting an existing abnormality
Divided into
1. Somatic cell gene therapy: Gene is introduced into somatic cells.
2. Germ cell gene therapy (Transgenic animal)

Bioinformatics

10. **d. BLAST** (Ref: Harper 30/e page 98,99)

Bioinformatic and Genomic Resources
The large collection of databases that have been developed for the assembly, annotation, analysis and distribution of biological and biomedical data reflects the breadth and variety of contemporary molecular, biochemical, epidemiological, and clinical research. The prominent bioinformatics resources: UniProt, GenBank, and the Protein Database (PDB) represent three of the oldest and most widely used bioinformatics data bases

Uniprot: The world's most comprehensive resource on protein structure and function
Genbank: The store all known biological nucleotide sequences and their translations in a searchable form
PDB (Protein Data Base) a repository of the three- dimensional structures of proteins, polynucleotides, and other biological macromolecules
HapMap is a comprehensive effort to identify SNPs associated with common human diseases and differential responses to pharmaceuticals
ENCODE (Encyclopedia of DNA Elements) Project- Identification of all the functional elements of the genome will vastly expand our understanding of the molecular events that underlie human development, health, and disease.
Entrez Gene provides a variety of information about sequence of the genome in and around the gene, exon- intronboundaries, the sequence of them RNA (s) produced from the gene, and any known phenotypes associated with a given mutation of the gene inquestion.
dbGAP (Database of Genotype and Phenotype), compiles the results of research into the links between specific genotypes and phenotypes BLAST (Basic Local Alignment Search Tool) is a method to identify protein by homology. This is not a bioinformatic data base.

11. **a. Genomics** (Ref: Harper 30/e page 98,99)
    - Bioinformatics: The discipline concerned with the collection, storage, and analysis of biologic data, mainly DNA and protein sequences
    - Genome: The complete set of genes of an organism
    - Genomics: In depth study of the structures and functions of genomes.
    - Transcriptome: The complete set of RNA transcripts produced by the genome during a fixed period of time.
    - Transcriptomics: The comprehensive study of transcriptome.
    - Proteome: The complete compliment of proteins of an organism.
    - Proteomics: The systematic study of structures and functions of proteomes and their variation in health and disease.
    - Exome: The nucleotide sequence of the entire complement of mRNA exons expressed in a particular cell, tissue, organ or organism.

12. **a. Proteomics** (Ref: Harper 30/e page98,99)

13. **b. Tag SNP** (Ref: Robbins 9/e page179)

Linkage and Association Studies
There are two primary strategies for mapping genes that cause or increase susceptibility to human disease:
1. Classic linkage can be performed based on a known genetic model or, when the model is unknown, by studying pairs of affected relatives
2. Disease genes can be mapped using allelic association studies (Genome Wide Association Studies, GWAS)

Genome-wide association studies (GWAS) have elucidated numerous disease-associated loci and are providing novel insights into the allelic architecture of complex traits.
These studies have been facilitated by the availability of comprehensive catalogues of human single-nucleotide polymorphism (SNP) haplotypes generated through the HapMap Project.
The data generated by the HapMap Project are greatly facilitating GWAS for the characterization of complex disorders. Adjacent SNPs are inherited together as blocks, and these blocks can be identified by genotyping selected marker SNPs, so-called Tag SNPs, thereby reducing cost and workload.

14. **a. Detection of characteristic DNA polymorphism in a family associated with disorders** (Ref Robbins 9/e page 177–179)

Linkage analysis
- The genetic marker loci very close to the disease allele, that are transmitted through pedigrees called linkage disequilibrium.
- Assess the genetic marker loci in family members having the disease or trait of interest
- With time it becomes possible to define a panel of markerloci, all of which co-segregate with the putative diseases
- Hence Linkage analysis facilitates localization and cloning of the disease allele.

The genetic markers used are, SNPs and repeat-length polymorphisms (minisatellite and microsatellite).